Chemistry at Extreme Conditions

Chemistry at Extreme Conditions

M. Riad Manaa

Lawrence Livermore National Laboratory,
Livermore, CA, USA

2005

ELSEVIER

Amsterdam · Boston · Heidelberg · London · New York · Oxford · Paris
San Diego · San Francisco · Singapore · Sydney · Tokyo

ELSEVIER B.V.
Radarweg 29
P.O. Box 211, 1000 AE Amsterdam
The Netherlands

ELSEVIER Inc.
525 B Street, Suite 1900
San Diego, CA 92101-4495
USA

ELSEVIER Ltd.
The Boulevard, Langford Lane
Kidlington, Oxford OX5 1GB
UK

ELSEVIER Ltd.
84 Theobalds Road
London WC1X 8RR
UK

1st edition 2004

Library of Congress Cataloging in Publication Data
A catalog record is available from the Library of Congress.

British Library Cataloguing in Publication Data
A catalogue record is available from the British Library.

ISBN: 0-444-51766-9

∞ The paper used in this publication meets the requirements of ANSI/NISO Z39.48-1992 (Permanence of Paper).
Printed in The Netherlands.

PREFACE

Although human life is confined to a narrow range of pressure of a few atmospheres and scores of degrees of temperature, violent events and processes that occur at multiple orders of magnitude of these two variables are taking place continuously within our planet and the vast universe. From a molecular perspective, chemical events occur as bonds are broken and others are formed, with the nuclei circumventing energetic barriers. The effect of temperature is to accelerate the motion in crossing these barriers. The effect of pressure, however, is diabolic: it changes the structure of the barrier height, so while it is decreased in some cases, energetic barriers are increased in others. The combination of both high-temperature and pressure on a system, such as in hot and dense fluids, alters the chemical transformation in a manner that is markedly different from that which we encounter in gas-phase chemistry. Chemical processes that occur in the pressure regime of 0.5 – 200 GPa and temperature range of 500 – 5000 K include such varied phenomena as comet collisions, synthesis of super-hard materials, detonation and combustion of energetic materials, and organic conversions in the interior of planets.

Recent experimental advances in high-pressure technology, shock physics, ultrafast laser spectroscopy, advanced light sources, and laser heating techniques have initiated exciting research as to the nature of the chemical bond in transient processes at extreme conditions. High-pressure studies of up to 350 GPa can now be conducted in a diamond anvil cell, and fast laser heating can elevate the temperature of the compressed materials to 4000 K. Similarly, shock experiments can simultaneously achieve a range of high-pressure and temperature, and provide results on the hydrodynamic behavior of materials. Diagnostics of ultrafast laser spectroscopy of Raman, Infrared, UV, and x-ray diffraction, an increasing number of which have been developed and make use of advanced synchrotron radiation sources at national and international laboratories, are now available to map out the chemical transformations that result under extreme conditions.

On the theoretical front, the use of molecular dynamics simulations in combination with first principle, semi empirical, or classical fields have also emerged as viable tools to access the short time scale for chemical events of dense fluids at high-temperature. These methods not only complement experimental work, but also predict the early chemical transformations and decomposition products of both simple and complex organic systems. A multitude of these methods allow for tailoring the size and duration of the process: from a hundred or so atoms and a few picoseconds that can be treated with ab initio density functional methodology, to simulating several thousands atoms for a few tens of picoseconds with the use of reactive force fields and tight-binding methods. Furthermore, multi-scale techniques to describe shock compression processes for extended time that combine atomistic molecular dynamics calculations with macroscopic description of condensed matter are continuously being developed.

This book contains both experimental and computational contributions to the study of chemistry and materials at elevated conditions of temperature and pressure, along with applications in a variety of disciplines. Chapters have been organized in a not altogether capricious fashion under four broad themes of applications. Topics of biological and bioinorganic systems are dealt with in the first four chapters. Experimental works on the transformations in small molecular systems such as CO_2, N_2O, H_2O and N_2 are presented in chapters 5–8. Theoretical methods and computational modeling of shock-compressed materials are the subject of chapters 9 through 12. Finally, chapters 13 through 17 present experimental and computational approaches in energetic materials research. It is hoped that this assortment of topics will provide an insight into the active and exciting field of research of chemistry at extreme conditions.

M. Riad Manaa
Lawrence Livermore National Laboratory
Livermore, California
2004

TABLE OF CONTENTS

Chemistry at Extreme Conditions
M. Riad Manaa (Editor)

Chapter 1

Pressure - Temperature Effects on Protein Conformational States

Karel Heremans

Department of Chemistry, Katholieke Universiteit Leuven, B-3001 Leuven, Belgium

1. INTRODUCTION

Proteins are unique among the biological macromolecules and have attracted active interest from various disciplines. For the physicist it is the structure that has the characteristics of order as well as of disorder. The chemist is attracted by the unique properties that show up in the catalytic activity of enzymes and in the conversion of chemical into mechanical energy in muscle. Biologists tend to put the emphasis on the functional role of proteins.

The pioneering work in high pressure protein research is that of Bridgman who observed that a pressure of several 100 MPa will give egg white an outlook similar, but not identical, to that of a cooked egg [1]. Since Bridgman worked on various pure substances, he may not have realized that the pressure effect on egg white has basically to do with an aqueous solution of a protein where water plays a vital role. In addition, he made the unexpected observation that the ease of the pressure-induced coagulation increases at low temperatures. In other words he observed a negative activation energy for a chemical process. Also when the egg white was taken to 1.2 GPa into the ice VI phase, the coagulum did not seem to be affected by the freezing.

The modern era starts with the seminal paper by Suzuki [2] with the systematic observations on the stability of ovalbumin, the main protein component of egg white, and hemoglobin. These proteins show an elliptic stability phase diagram in the temperature-pressure plane. Nowadays it is well known that proteins in solution are marginally stable under conditions of high temperature and pressure. By contrast there is the observation that certain bacteria live under extreme temperature conditions and it is well known that bacteria can survive in the deepest parts of the ocean.

Hayashi et al. [3] analyzed the effect of pressure on egg yolk. It is well known from the preparation of a hard boiled egg, that the yolk becomes solid at a slightly higher temperature than the white. The reverse is true for the effect of pressure: the yolk becomes solid at a lower pressure than the white. It is now clear that these observations are the consequence of the unique behavior of proteins.

If the conditions for equilibrium or isokineticity are plotted versus temperature and pressure, a phase or stability diagram is obtained with an elliptical shape [4,5]. One of the practical consequences is the possible stabilization against heat unfolding by low pressures. This has been observed in several proteins and enzymes [6,7] but, as we shall discuss later, there are some notable exceptions. Interestingly, this also applies to the effect of pressure on the heat gelation of starch [8,9]. Of special interest is the observation that the inactivation kinetics of microorganisms shows diagrams similar to those of proteins [10]. This suggests that proteins are the primary targets in the pressure and temperature inactivation of organisms. This is shown schematically for a bacterium and a yeast in Fig. 1.

The physico-chemical viewpoint of using extreme conditions is to explore the effect of temperature and pressure on the conformation, the dynamics and the reactions of biomolecules. The unique properties of biomolecules are determined by the delicate balance between internal interactions which compete with interactions with the solvent. The primary source of the dynamical behavior of biomolecules is the free volume of the system and this may be expected to decrease with increasing pressure. As temperature effects act via an increased kinetic energy as well as free volume, it follows that the study of the combined effect of temperature and pressure is a prerequisite for a full understanding of the dynamic behavior of biomolecules. By intuition, pressure effects should be easier to interpret than temperature effects.

2. LIFE AT EXTREME CONDITIONS

Life as we know it is connected with the presence of liquid water, the energy input from the sun and the chemical control of the energy flow. The chemistry of living systems is characterized by redox reactions of a limited number of metals and the organic chemistry is restricted to specific biomolecules. Moreover, biological systems show a very great internal physical and chemical heterogeneity, and a dynamic exchange between processes internal to the objects and the world outside of them [11].

Extreme conditions are defined in terms hostile to human beings. Extremes of a chemical nature are, for example, low water activity, salinity, acidity, gases, high concentrations of metals or organic solvents, radiation, etc. In general organisms cope with these external conditions by maintaining non-extreme internal conditions or by evolving very effective repair mechanisms [12].

Temperature and pressure extremes require different strategies. Cellular lipids, proteins and nucleic acids are sensitive to high temperatures. Hyperthermophile bacteria have ether lipids instead of the more hydrolysis sensitive ester lipids in mesophiles [13]. Enzymes from hyperthermophiles show an unusual thermostability in the laboratory, and an important aspect of protein chemistry research is to find out the stabilizing principles. Crude cell extracts of hyperthermophiles show the presence of heat inducible proteins, called chaperones, which assist in the folding of proteins during cellular synthesis. Molecular details for cold adaptation of enzymes have been reported but are less extensively studied [14].

The maximum pressure that microbial cells can cope with is of the order of 100 MPa [15,16]. But pressure sensitive processes such as motility, transport, cell division, cell growth, DNA replication, translation and transcription are affected at much lower pressures. Thus far it is clear that many deep-sea bacteria have genes for the production of polyunsaturated fatty acids. These lipids remain fluid up to higher pressure than the more saturated ones. In the future one may expect more details on the effect of pressure in the fields of proteomics, genomics and metabolomics.

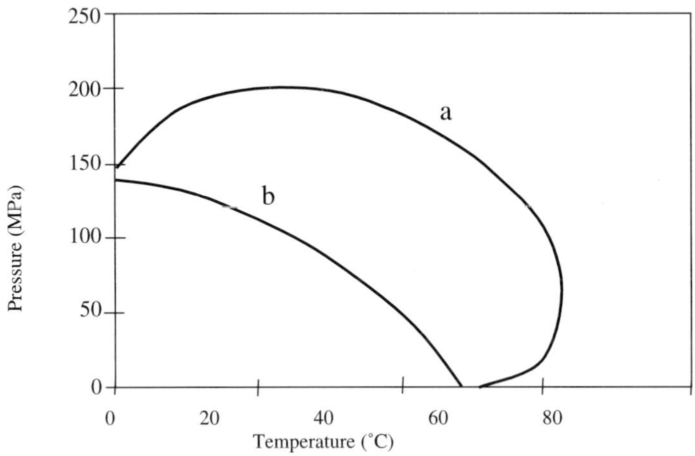

Figure 1. A schematic representation of the isokinetics of the survival of a bacterium and yeast as a function of pressure and temperature. (a): *Escherichia coli*. Redrawn after [24], (b): *Zygosaccharomyces bailii*. Redrawn after [25].

A recent report with the observation of microbial activity at GPa pressure came as a surprise [17]. A critical comment with reply came shortly afterwards [18]. One possible explanation for most of the described effects may come from the limited availability of water in the highly concentrated suspension of bacterial cells that were observed in the small volume of the diamond anvil cell.

One fascinating example of the crucial role of water in the behavior of organisms under extreme conditions is given by small organisms called Tardigrades. These animals, composed of about 40,000 cells, become immobile and shrink into a special state when the humidity of the surroundings decreases. In such a state they can survive temperatures from –253°C up to 151°C and pressures up to 600 MPa [19]. In the normal state they are killed at 200 MPa. This behavior is quite similar to that of bacterial spores and dry proteins where pressures of more than 1 GPa are not able to provoke any changes [20,21]. The fact that the Tardigrades can undergo a transformation to an extreme dry state may be much more exceptional than the fact that they are resistant to extremes of temperature and pressure in the dry state.

Small multicellular organisms are sensitive to very low pressures. The swimming activity of larvae of tadpoles can be reduced by 2.5 % ethanol in the medium. The activity can be restored by pressures up to 28 MPa [22]. Macdonald and Fraser [23] reported effects by pressures of 20 kPa or less on aquatic animals at the level of growth and or metabolism. The authors concluded that cells are able to respond to micropressures also through mechanical processes.

3. THE PROTEIN VOLUME

From a macroscopic point of view, the volume is a measure of that part of space that is inside a given surface. On the molecular and atomic level there is no well defined surface and it follows that a definition of the volume can use different approaches. The first one, the partial molar volume, is the phenomenological one and this is used in thermodynamics and in experimental work. The second one defines a surface such as the van der Waals or any other calculable surface, from which the volume is obtained. This approach is the one that is used in molecular dynamics simulations or other computer calculations. The partial molar volume of a solute molecule or ion is defined as the change in volume of the solution by the addition of a small amount of the solute over the number of moles of added solute keeping the amount of the other components constant. It is not equal to the volume of the molecule or the ions since it includes also the interaction with the solvent. This may be seen from the fact that the partial molar volume for a salt such as $MgSO_4$ is negative at high dilution because of the strong electrostriction of the solvent around the ions. For the same reason the volume of the uncharged glycolamide is larger (56.2 mL/mol) than that of the amino acid glycine (43.5 mL/mol). Following Kauzmann [26], the partial molar volume of a protein in solution may be defined as:

$$V_{protein} = V_{atom} + V_{cavities} + V_{hydration} \tag{1}$$

In this expression V_{atom} and $V_{cavities}$ are the volumes of the atoms and the cavities respectively and $\Delta V_{hydration}$ is the volume change of the solution resulting from the interactions of the protein molecule with the solvent. More defined models for the partial molar volumes of proteins are discussed by Chalikian [27,28]. Care should be taken if quantities derived from the volume (such as compressibility and thermal expansion) are interpreted on the molecular level. The experimental results may depend on the sensitivity range of the method used. Global measurements such as ultrasonics detect the whole molar volume, while some local probes may feel only the change of the protein interior volume.

3.1. Cavities and hydration

As the volumes of the atoms may be considered, as a first approximation, to be temperature and pressure independent, it follows that both the thermal expansion and the compression are composed of two main terms, the cavity and the hydration. For the thermal expansion this gives:

$$\delta V / \delta T = \delta V_{cavities} / \delta T + \delta \Delta V_{hydration} / \delta T \tag{2}$$

For the compression we obtain:

$$\delta V / \delta p = \delta V_{cavities} / \delta p + \delta \Delta V_{hydration} / \delta p \qquad (3)$$

An estimate of the contribution of each factor is not easy to evaluate and relies on assumptions that are not easy to check experimentally. However, the compressibility of amino acids is negative because of changes in the solute solvent interaction (i.e. the amino acid solution is less compressible than the pure solvent). It follows then that the contribution of cavities compensates this effect so that the compressibility of the protein in solution becomes positive. A more quantitative estimate is possible when one makes assumptions about the compressibility of the hydrational water [29]. Most compressibility data are obtained from ultrasound velocimetry on dilute solutions of proteins and the interpretation concentrates on the hydration effects.

3.1.1. Cavities

On the basis of the low compressibilities and the average high packing densities, the protein interior is often considered as a solidlike material with little free volume. However when one considers the free volume distribution, proteins look more like liquids and glasses [30]. The free volume is often called cavities, voids or pockets. Their role in the volume changes of protein reactions or interactions was suggested by Silva and Weber [31]. A recent review paper emphasizes the similarities between the role of hydration and cavities in protein-protein interactions and protein unfolding [32].

As for the volumes of the atoms, the thermal expansion and compressibility is composed of two main terms, the cavity and the hydration. An estimate of the contribution of each factor relies on assumptions that are not easy to check. An estimate of the expansion or compression of the cavities should be possible with positron annihilation lifetime spectroscopy. This approach has proven to be a useful tool for determining the size of cavities and pores in polymers and materials. The lifetime is sensitive to the size of the cavity in which it is localized. A number of empirical relations correlate the distribution of the lifetime and the free volume [33]. Data on the pressure effect on the lifetime are only available for polymers. The results suggest that there may be a considerable contribution of the reduction in cavity size to the compressibility of a protein.

The compressibility of a protein may also be obtained from fluorescence line-narrowing spectroscopy at 10 K low temperatures. Under these conditions one does expect the hydrational changes not to play a very prominent role. Nevertheless the compressibilities that are obtained under such conditions are of the same order of magnitude as those obtained at ambient conditions [34]. This points to important contributions from the cavities to the compressibility and the thermal expansion. The observed pressure-induced amorphization in inorganic substances [35], liquid crystals [36], synthetic polymers [37,38] and starch [9] also support this hypothesis.

In collaboration with K. Süvegh, T. Marek (Lorand Eötvös University) and L. Smeller (Semmelweis University) at Budapest, we have started to measure the changes in cavities in lysozyme with positron annihilation lifetime spectroscopy [39] as a function of temperature and pressure. The pressure results suggest a correlation with recent high pressure NMR data [40,41]. The temperature data suggest a correlation with the compressibility data from ultrasound [29].

Recent progress in X-ray diffraction of protein crystals in the diamond anvil cell will also make it possible to obtain quantitative information on the cavities [42, 43]. Optical spectroscopy [44] and neutron scattering [45] should also be valuable tools to probe the role of cavities. High-pressure molecular dynamics simulations should also allow estimating the contributions of the hydration and the cavities. High-pressure simulations on the small protein, bovine pancreatic trypsin inhibitor, indicate an increased insertion of water into the protein interior before unfolding starts to occur [46,47].

3.1.2. Hydration

The role of water in the conformation, the activity and the stability of proteins has been investigated with many experimental and theoretical approaches. Because of its importance it has been coined as the "21st amino acid". There is now sufficient experimental evidence for the fact that dry proteins do not unfold by increased temperature or pressure [21]. Low levels of hydration give rise to a glassy state and the temperature of the glass transition depends on the amount of water as observed for synthetic polymers. Water can therefore be considered as a plasticizer of the protein conformation. Whereas hydrophobic interactions have dominated the interpretation of the data, hydrogen bond networks of water may also play a predominant role in water-mediated interactions [48,49].

The influence of various cosolvents on protein stability has been discussed by Timasheff [50]. There has been a considerable debate in the literature on the number of water molecules that are taking part in protein-protein or protein-DNA interactions as estimated by various methods. A recent theoretical analysis suggests that the osmotic stress method may overestimate the number of waters involved [51]. These models assume that the cavities that are formed at the interface between macromolecules do not contribute to the measured volume changes as suggested by Silva and Weber [31].

Although pressure studies are limited, it seems that the stabilizing effect of organic cosolvents against temperature unfolding are also found against pressure unfolding [52]. Kinetic studies under pressure of the folding of staphylococcal nuclease in the presence of xylose, show that the sugar effect is primarily on the folding step suggesting that the transition state, a dry molten globule state, is close to the folded state [53].

3.2. Compressibility

The partial molar isothermal compressibility, β_T, is defined as the relative change of the partial molar volume, V, with pressure:

$$\beta_T = -\left(\delta \ln V / \delta p\right)_T \quad (4)$$

The compressibility is a thermodynamic quantity of interest not only from a static but also from a dynamic point of view. Its relevance to the biological function of a protein can be understood through the statistical mechanical relation between the isothermal compressibility β_T and volume fluctuations:

$$\langle \delta V^2 \rangle = k_B T V \beta_T \tag{5}$$

Table 1. Thermal expansion, compressibility and heat capacity for some liquids compared with proteins. Data for proteins are in dilute aqueous solutions.

	Thermal expansion 10^{-7}/K	Compressibility 1/Mbar	Heat capacity (Cp) kJ/kg K
Water	210	46	4.2
Hexane	1380	166	2.3
Benzene	1220	96	1.7
Proteins	40-110	2-15	0.32-0.36

Because of the small size of the protein, the volume fluctuations are relatively large. It seems that the expansion and contraction of the cavities is the only way to generate these volume fluctuations. The biological relevance of the volume, as well as the energy and volume-energy fluctuations that will be considered in the following sections, can be illustrated by referring to a number of processes that are related to the dynamical properties of proteins. These include the opening and closing of binding pockets in enzymes, the allosteric effects, the conversion of chemical in conformational energy in muscle contraction, the biological synthesis of proteins and nucleic acids, and the transport of molecules trough membranes. Data for the compressibility of proteins are given in Table 1 and compared with data for some liquids.

The most frequently used method to obtain compressibilities of proteins in dilute solution is ultrasound [27,28]. Unfortunately this method is not easily applicable to high pressure studies [54]. The interpretation of the ultrasound data is based on the assumption of the additivity of the compressibilities of the solvent and the solute. The assumption however that the sound propagation of the components are additive, gives rise to the conclusion that the contribution of the hydration is usually overestimated [55,56]. Computer simulations support this conclusion in that they indicate that the experimental compressibilities obtained from ultrasound can be largely accounted for by the intrinsic compressibility [57].

Protein compressibilities that are obtained at liquid helium temperatures from fluorescence line-narrowing spectroscopy are of the same order of magnitude as those obtained at ambient conditions [34]. This also points to the intrinsic compressibility as the most important contribution. High-pressure NMR studies on many proteins suggest that the fluctuation of the structure is cavity-based [58]. Similarly, it should be possible to obtain dynamical information from X-ray crystallography at variable pressure and thus probe the role of the cavities to the compressibility [42]. Gekko and coworkers [59] have determined the compressibility of several proteins that contain disulfide bridges. Upon reduction of the disulfide bridges, the compressibility decreases and a further reduction is observed upon lowering the pH from 7 to 2.

Optical spectroscopy can also be used to determine the compressibility of the environment of specific absorption centers. Jung et al. [60] have determined the compressibility of the heme pocket of cytochrome P450cam from the pressure-induced frequency shifts of the heme bound ligand CO. Small angle neutron scattering also provide compressibility data [61].

Normal mode analysis of the mechanical properties of a triosephosphate isomerase-barrel protein suggests that the region between the secondary structures plays an important role in the dynamics of the protein. The beta-barrel region at the core of the protein is found to be soft in contrast to the helical, strand and loop regions [62]. A detailed discussion of other properties of proteins as mechanically highly non-linear systems is given by Kharakoz [63].

An adhesive-cohesive model for protein compressibility has been proposed by Dadarlat and Post [57]. This model assumes that the compressibility is a competition between adhesive protein-water interactions and cohesive protein-protein interactions. Computer simulations suggest that the intrinsic compressibility largely accounts for the experimental compressibilities indicating that the contribution of hydration water is small. The model also accounts for the correlation between the compressibility of the native state and the change in heat capacity upon unfolding for nine single chain proteins.

3.3. Thermal expansion

The partial molar expansion, α, is defined as the relative change of the partial molar volume with temperature:

$$\alpha = \left(\delta \ln V / \delta T \right)_p \tag{6}$$

As for the compressibility, the thermal expansion is considered to be composed of two main terms, the cavity and the hydration terms. A quantitative analysis is possible when one makes assumptions about the thermal expansion of the hydrational water [64].

The thermal expansion can be related to the fluctuations of the system. However this relation is not as widely known as the others mentioned above. The thermal expansion is proportional to the cross-correlation of the volume and energy fluctuations:

$$\langle \delta H \delta V \rangle = k_B T^2 V \alpha \tag{7}$$

This agrees with the intuitive picture, that the thermal expansivity characterizes some kind of coupling between the thermal (H) and the mechanical parameters (V).

Data for the thermal expansion of proteins are given in Table 1 and compared with data for some liquids.

In contrast to the compressibility, the thermal expansion of proteins has received much less attention. Frauenfelder et al. [65] have estimated the thermal expansion of myoglobin from the refined X-ray structure at 80 and 255-300K. They conclude that the expansion comes mainly from the subatomic free volumes between the atoms. The expansion obtained from the temperature dependence of the vibrational frequency shifts of the hydrogen bonds support these data [66]. On the other hand, 3D NMR solution spectra of lysozyme suggest that the internal cavity shrinks between 35 and 4°C [67].

The thermal expansion of lysozyme in solution between 10 and 60°C was obtained from the partial molar volume. The intrinsic expansion of the protein was called anomalous since it did not behave like an ordinary solid or like a typical liquid [64]. Pressure perturbation calorimetry (PPC) is a relatively new experimental approach that allows one to obtain the thermal expansion (α) from the heat released or absorbed (Q) after short positive or negative pressure pulses (ca 5 bar) on proteins solutions from the relation [68-70]

$$\Delta Q = -T\Delta P g_s v_s(\alpha_s - \alpha) \tag{8}$$

where the index s is for the solute. With the assumption that the product $g_s v_s$ is temperature independent, it is found that the thermal expansion decreases with increasing temperature as observed from the temperature dependence of the partial molar volume [64]. Other literature reports however, show no temperature dependence [71].

Gekko et al. [59] have observed from the temperature dependence of the molar volume that the reduction of disulfide bonds in proteins invariably leads to an increase of the thermal expansion. This is interpreted as a decrease in the internal cavity and an increase in the surface hydration.

3.4. Heat capacity

The partial molar heat capacity, C_p, is defined as:

$$C_p = \left(\delta H/\delta T\right)_p \tag{9}$$

Note that in contrast to the partial molar volume, this quantity is not a relative one. This follows from the fact that the absolute value of the partial molar enthalpy cannot be determined. In a thermodynamic system with constant T and P, the isobaric heat capacity can be regarded as the measure of the enthalpy fluctuations of the system:

$$\langle \delta H^2 \rangle = k_B T^2 C_p \tag{10}$$

The partial molar heat capacity can be considered to be composed of intrinsic and hydration contributions. The intrinsic component contains contributions from covalent and non-covalent interactions. It has been shown that about 85% of the total heat capacity of the native state of a protein in solution is due to the covalent structure [72]. Changes in the heat capacity upon unfolding are therefore primarily interpreted as due to changes in the hydration. A physical picture of energy fluctuations means changing the conformation between ordered and less ordered structures. This can be achieved by hindered internal rotations, low frequency conformational fluctuations or high frequency bond stretching and bending modes. Data for the heat capacity of proteins are given in Table 1 and compared with data for some liquids.

Dadarlat and Post [57] have found an interesting correlation between the heat capacity changes of unfolding and the compressibility of the native state of proteins. This can be rationalized from a similar dependence of these quantities on the distribution of the atom types, polar/charged versus nonpolar.

Table 2. Volume changes (mL/mol) for the formation of molecular complexes obtained from the pressure dependence of the equilibrium constants [75].

Reagents*	Solvent	ΔV
FMN/AMP	water	- 4.0
FMN/I$^-$	water	- 1.7
FAD	water	- 4.6
FTME (n=3)	water	- 1.8
FTME (n=5)	water	- 4.8

* FMN: flavin mononucleotide; AMP: adenylic acid; FAD: flavin adenin dinucleotide; FTME: N-flavinyl tryptophan methylester.

3.5. The Grüneisen parameter

Assuming that the compressibility, the thermal expansion and the heat capacity reflect the same intrinsic properties of proteins, one may expect a close relation between them. For solids and synthetic polymers, the relation between the volume, V, the thermal expansion, α, the compressibility, β_T, and the heat capacity at constant volume, C_v, is given by the Grüneisen parameter γ:

$$\gamma = \frac{V\alpha}{C_v \beta_T} \tag{11}$$

This volume-independent parameter can be obtained from pressure-induced vibrational wavenumber shifts in solids or polymers [73].

$$\gamma = -\frac{\delta \ln \nu}{\delta \ln V} \tag{12}$$

If this parameter is assumed to be the same for all vibrations, one can obtain a bulk thermodynamic definition for γ. The bulk Grüneisen parameter is found to be about 4 for polymers from the effect of pressure on the velocity of sound. The data suggest that for the heat capacity only the interchain contribution should be taken into account. With this assumption, an order of magnitude calculation shows that the bulk Grüneisen parameter for proteins is of the same order of magnitude as that of polymers. This suggests that the thermal expansion and the compressibility of proteins reflect primarily the movement between the secondary structures. These movements are reflected in the low frequency part of the vibrational spectrum. Unfortunately, no experimental data are available on the effect of pressure on these vibrations, except for small hydrogen bonded molecules [74]. However, the assumption that the Grüneisen parameter reflects primarily the intrinsic contributions to the compressibility, the thermal expansion and the heat capacity, indicates that changes in the hydration play a minor role.

3.6. Intermolecular interactions in water

The molecular interpretation of thermodynamic data of temperature and pressure effects on proteins and their reactions is based on the data obtained from small molar mass model compounds in water. Weber and Drickamer [75] have pointed out the role of mechanical effects on the volume of association of molecular complexes by introducing molecular spacers that prevent molecules to get in close contact. As can be seen from Table 2, these mechanical effects can show up considerably in the volume changes. It is clear that such effects should also influence hydrophobic interactions in proteins.

Hydrophobic interactions have become very popular and, although their molecular origin is still debated, their role is considered central in chemistry and biology [76,77]. A crucial point is that these interactions are isotropic, i.e. there is no orientational preference for this interaction. In contrast, dipolar interactions, and in particular hydrogen bonds are highly directional attractive interactions. Hydrogen bonding networks are also very characteristic for proteins and their interactions with water. Rather than in interpreting the thermodynamic data in terms of hydrophobic interactions, the alternative view, hydrogen bonded networks, should also be considered [48].

Given the directional character of hydrogen bonds, it would not be very surprising if changes in hydration contribute considerably to the observed volume changes of many biological reactions. Estimates of the changes in hydration from high pressure measurements, from osmotic stress analysis and from preferential hydration experiments have shown considerable discrepancies and are heavily debated in the literature. Shimizu [51] has recently analysed some data with the Kirkwood-Buff theory and concluded that the osmotic stress method may overestimate hydration number changes.

4. PROTEINS: STABILITY CONDITIONS

The stability of proteins can be viewed from kinetic as well as from thermodynamic considerations. Here we give the thermodynamic description and note that the kinetic description would be equivalent in view of the thermodynamic basis of the classical transition state theory. An example of the treatment of kinetic data of the stability of enzymes is given by Weemaes et al. [78].

4.1. The phase diagram

The first systematic experimental evidence for the unusual behavior of proteins with respect to temperature and pressure came from the kinetic studies of Suzuki [2] and thermodynamic studies by Hawley [4]. The mathematical implications and assumptions that are usually made in the analysis of the data have recently been discussed [79]. Of particular interest is the cold unfolding of proteins that was predicted on the basis of the model describing the temperature dependence of the reaction enthalpy accounting for the large increase in heat capacity upon unfolding. As we shall see, the effects of heat, cold and pressure on proteins are interconnected.

The elliptic phase diagram on the P-T plane is characteristic for proteins. Mathematically this shape originates from the fact that in a Taylor expansion of the free energy difference, ΔG, as a function of temperature and pressure, the second order terms give a significant contribution. Physically these second order terms are proportional to $\Delta\beta$, ΔC_p, $\Delta\alpha$, the changes in the compressibility, the heat capacity and the thermal expansion, respectively, between the unfolded (U) and the native (N) state of the protein:

N (Native) $\leftrightarrow$ U (Unfolded)

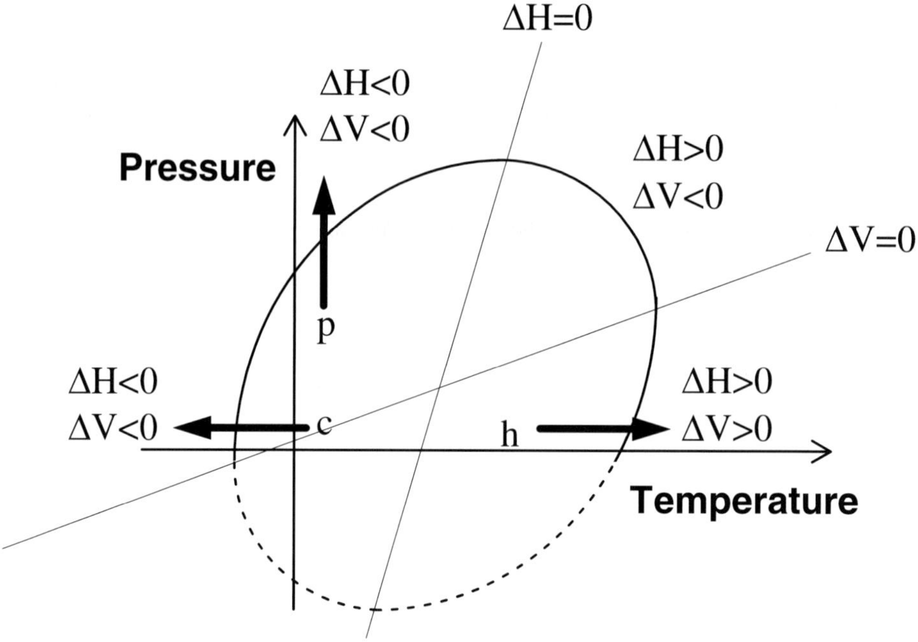

Figure 2. The elliptical temperature-pressure stability phase diagram characteristic for proteins. After Suzuki [2] and Hawley [4]. Note the thermodynamic similarities between the cold, c, and pressure, p, unfolding and the contrast with heat, h, unfolding.

The pressure and temperature dependence of ΔG ($\Delta G = G_{Unfolded} - G_{Native}$), the difference in free energy between U and N is given by

$$d(\Delta G) = -\Delta S dT + \Delta V dp \tag{13}$$

Integration at constant pressure, gives the expression for the temperature dependence of the free energy assuming that ΔC_p is temperature independent:

$$\Delta G(p_0,T) = \Delta G_0 - \Delta S_0(T - T_0) + \Delta C_p\left((T - T_0) - T\ln {T}/{T_0}\right) \tag{14}$$

ΔG_0 refers to the conditions p_0 and T_0. The pressure dependence can be taken into account starting from the volume change that is not only pressure but also temperature dependent:

$$\Delta V(p,T) = \Delta V + \Delta\alpha^*(T - T_0) - \Delta\beta^*(p - p_0) \tag{15}$$

ΔV refers here to p_0 and T conditions. It follows that the pressure and temperature dependence of the free energy change contains a cross term in temperature and pressure:

$$\begin{aligned}\Delta G(p_0,T) = {} & \Delta G_0 - \Delta S_0(T - T_0) - \Delta C_p\left((T - T_0) - T\ln {T}/{T_0}\right) \\ & + \Delta V_0(p - p_0) - (\Delta\beta^*/2)(p - p_0)^2 + \Delta\alpha^*(T - T_0)(p - p_0)\end{aligned} \tag{16}$$

In this equation $\Delta\beta^*$ is the compressibility factor difference ($\beta^*=\beta V$) and $\Delta\alpha^*$ the difference of the thermal expansion factor ($\alpha^*=\alpha V$) of the denatured and native states of proteins. An important assumption in the derivation of this equation is the temperature and pressure independence of $\Delta\alpha^*$, $\Delta\beta^*$ and ΔC_p. The $\Delta G=0$ curve is an ellipse on the P-T plane and it describes the equilibrium border between the native and denatured state of the protein. This curve is known as the phase or stability diagram. This is visualized in Fig. 2. The diagram illustrates the interconnection between the cold, heat and pressure unfolding of proteins.

An equation similar to the Clausius-Clapeyron equation can be obtained for the slope of the phase boundary

$$\delta T / \delta p = \Delta V_0 + \Delta\beta^*(p - p_0) + \Delta\alpha^*(T-T_0) \;/\; \Delta S_0 - \Delta\alpha^*(p - p_0) + \Delta C_p(T - T_0)/T_0 \tag{17}$$

This expression reduces to the classical Clausius-Clapeyron equation when the difference in compressibility, thermal expansion and heat capacity vanish as is observed for most phase transitions in lipids [80].

The shape of the phase diagram for proteins is of considerable interest since, as pointed out previously, it not only contains thermodynamic information on protein unfolding but also information on the difference in the dynamics between the native and unfolded state of the volume and energy fluctuations and on the coupling between volume and energy fluctuations. These data may be obtained from fitting the free energy equation to the experimental data [4] or directly from the temperature and pressure dependence of the partial molar volume of the protein [81]. Changes in heat capacity can be obtained from calorimetry and the thermal expansion can be obtained with high precision from pressure perturbation calorimetry. These techniques considerably expand our knowledge on the phase diagram of proteins [82].

As can be seen from the schematic representation in Fig. 3, two types of phase diagrams can be observed. In one case pressure stabilizes the protein against thermal unfolding, in the other case there is a destabilization. The detailed mechanism of the difference between the two possibilities is at present not clear. For cytochrome c, Chalikian et al. [83] have recently shown that for the native-to-unfolded transition the native conformation is stabilized by pressure whereas for the molten globule-to-unfolded transition the molten globule conformation, at pH 2, is destabilized. Whether this applies to other proteins remains to be determined but the observation suggests that the assumption of the two state unfolding, which is at the basis of the phase diagram, might not be generally valid. Spectroscopic studies also suggest that the conformation of the temperature and pressure unfolded protein are not similar. NMR experiments [84], small-angle-X-ray scattering [82], fluorescence [85] and Fourier transform infrared spectroscopy [86] suggest that the pressure and temperature denaturation lead to a different state. Fluorescence spectroscopy suggests that the temperature induced unfolding can be presented as an exposure of the protein core to water, whereas the pressure induced unfolding results from the pressure induced penetration of water into the protein matrix followed by a disruption of the chain-chain interactions [85].

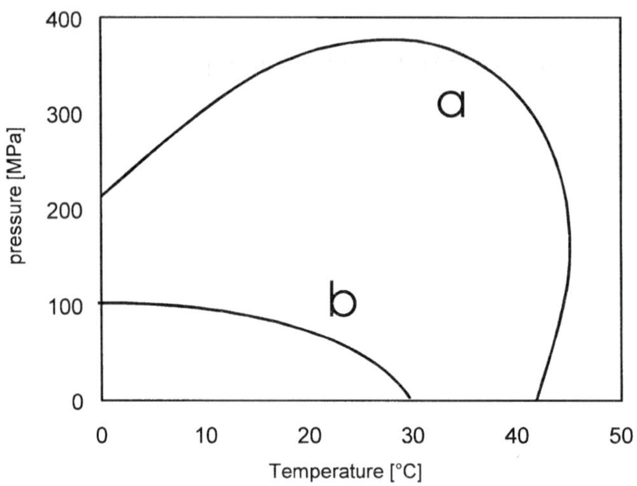

Figure 3. Schematic representation of the two possible stability diagrams observed for proteins. (a): Chymotrypsinogen. After Hawley [4], (b): Ribonuclease A. After Brandts et al. [87]. Note the similarity with the diagram for the survival of microorganisms shown in Fig. 1.

The re-entrant or Tammann loop-shape phase diagram as observed in proteins is also found in other systems and has been connected to exothermic disordering [88]. In this particular case, nematic – smectic A transitions in liquid crystals and the phase behaviour of a crystalline polymer, poly(4-methyl-pentene-1), the phase behaviour can be understood by considering the vibrational and configurational contributions to the volume and the entropy of the material in the various phases.

Inverse melting, another expression for the same phenomena, has now been computer modelled with a statistical mechanical model [89]. However, the challenge remains to simulate the crucial role of water in the phase diagram of proteins. A possible mechanism has been proposed for the cold denaturation of proteins at high pressure [90].

4.2. Pressure unfolding and chemical unfolding

It is of particular interest to note that the temperature dependence of the pressure-induced unfolding and the urea-induced unfolding of proteins at ambient pressure give rise to negative activation energies, i.e. these processes go faster at lower temperatures. Bridgman already noted this effect and found it quite unexpected [1].

Formally the effect of a cosolvent, e.g. urea, can be analyzed by the equations used for the phase diagram with the inclusion of a term that accounts for the concentration dependence of the change in free energy, $m = \delta\Delta G/\delta[cosolvent]$

$$d(\Delta G) = -\Delta S dT + \Delta V dp + m\,[cosolvent] \tag{18}$$

A detailed discussion of the use of this equation has been given elsewhere [52,91]. If we compare the mechanisms that have been proposed, then it is evident that the first step, the exothermic binding of urea or water, must be the source of the negative activation energies. In the pressure induced unfolding the binding of water to the protein is exothermic giving rise to the observed negative activation energies under pressure

$$N + H_2O \leftrightarrow N{*}H_2O \rightarrow U$$

In the urea-induced unfolding the binding of urea to the protein is exothermic at ambient pressure

$$N + urea \leftrightarrow N{*}urea \rightarrow U$$

The large positive activation energies observed for temperature induced unfolding of proteins is attributed to the unfolding step

$$N \leftrightarrow U + H_2O \rightarrow U{*}H_2O$$

These mechanisms are consistent with the differences that have been observed between the temperature and pressure induced unfolding of proteins with various experimental approaches.

5. WATER SOLUBLE POLYMERS AS MODEL SYSTEM

The re-entrant phase diagram experimentally observed for proteins has also been observed for a number of water-soluble polymers. As expected, the sidechains of the polymers are able to form hydrogen bond interactions with water. Poly(ethylene oxide) [92] and poly(vinyl methyl ether) [93] form weak hydrogen bonds via their ether groups. Poly(N-vinylpyrrolidone), poly(N-isopropylacrylamide) [94], poly(N-vinylisobutyramide) [94], poly(N-vinylcaprolactam) and poly(2-acrylamido-2-methyl-1-propanamide) [95] contain amide groups in their side chains. It is well known that hydrogen bonds to ethers are much weaker than those formed with amides. It is clear that these polymers also show hydrophobic interactions.

The pressure-temperature diagram for poly(ethylene oxide) and poly(N-vinylpyrrolidone) in water shows similar re-entrant phase behaviour with exothermic melting as observed in proteins [96]. At low temperature and pressure the polymer is in the random coil state (one phase region) and collapses at high temperature and/or pressure (two phase region). It is of interest to note that the polymers show considerable aggregation in the two phase region. This aggregation can be reduced by the addition of the detergent sodium dodecyl sulphate without noticeable effect on the temperature or the pressure at which the transition takes place.

The introduction of polymer-water and water-water interactions are important in the modelling of the temperature-concentration behaviour of poly(ethylene oxide) [92]. This suggests that the strong and highly directional hydrogen bond formation in these systems seems to be more important than the hydrophobic interactions. It would of particular interest to know whether high pressure simulations support this hypothesis.

6. PROTEIN-PROTEIN INTERACTIONS

In this section we concentrate on the effect of pressure on the possible interactions that a protein can undergo under given chemical and physical conditions. The possible conformational states and interactions that a protein may undergo are shown in Fig. 4. One type of aggregation, amyloid formation, has attracted considerable attention in recent years because of its possible relevance for a number of molecular diseases.

One important aspect is solution condition. In most laboratory experiments dilute protein solutions are used in dilute salt or cosolvent conditions. These conditions are widely different from those found in the living cell. In the cell molecular crowding from the presence of macromolecules may have a profound influence on protein-protein interactions [99]. In addition low molar mass compounds may accumulate under certain conditions in the cells of certain plants, animals and microorganisms to cope with environmental stresses. The presence of these molecules may have important contributions to the dynamics, the stability and the interactions of proteins [100]. While the effect of these molecules has been studied widely on the temperature stability of proteins, their detailed study on the effect of the pressure stability is just at a beginning [101].

6.1. Crystallization

The observation that pressure (200 MPa) accelerates dramatically the crystallization of glucose isomerase [102] has stimulated a number of studies into the application of this approach. On the other hand the crystallization of subtilisin is retarded by the same pressure and was attributed to the increase of the solubility of the protein with pressure [103]. A recent study by Kadri et al. [104] on lysozyme points to the pH as an important parameter to influence the solubility and the crystallization process under pressure. The tetragonal crystal form transforms into an orthorhombic one at high pressure.

Technically it is now possible to take proteins crystals up to a pressure of 1 GPa [105]. This allows one to determine the compressibility of the crystals. One point of interest is that the stability of lysozyme in the crystalline state is higher (1 GPa) than in solution (0.5 GPa) [106]. Fourme et al. [42] noticed that the order in crystals of proteins and protein complexes, such as cubic cowpea mosaic virus, increases at high pressure thus increasing the resolution. These studies open new ways to investigate the behavior of proteins under pressure. The hydration is one of the topics that is certainly of interest to study. Computer simulations have already provided considerable help in interpreting the experimental data [107].

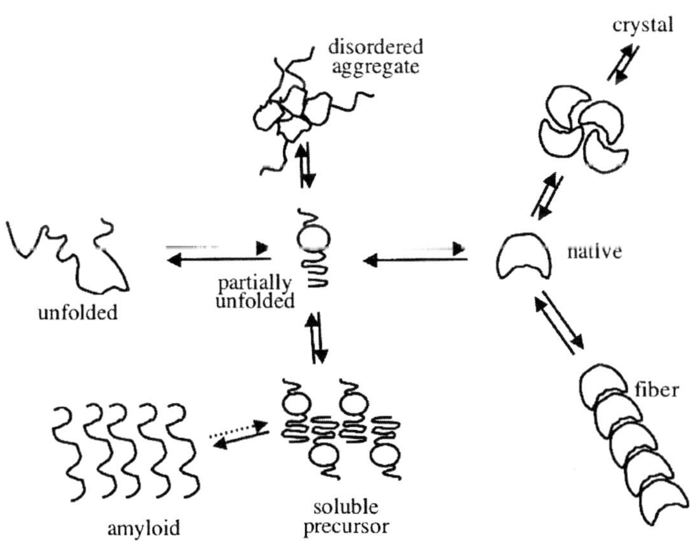

Figure 4. A selection of the possible conformational states that a protein may show and the interactions that these may undergo depending on chemical (pH, salt, cosolvent, etc.) and physical (pressure, temperature) conditions. Redrawn after Dobson [97,98].

6.2. Oligomerization

It has been known for a long time that protein-protein interactions are, in general, accompanied by positive volume changes. Silva and Weber [31] and Silva et al. [32] have suggested an analogy between the positive volume changes for the association of proteins and the positive volume changes for the folding of proteins. In both cases they suggest a combined effect of the formation of solvent-excluding cavities and the release of bound water molecules. Organic colsovents, such as glycerol, also stabilize protein-protein interactions against pressure dissociation [32].

Many molecular assemblies that play a fundamental role in biomolecular processes are sensitive to pressure. Thus pressure has a very strong effect on the association of the proteins in ribosome's, actin, microtubuli, bacterial flagellin and other macromolecular assemblies. Temperature may have a strong influence on the pressure effect as demonstrated in the case of microtubules [108]. In some cases, very small pressures have an effect, visible with the optical microscope, on the shape and the behavior of living cells [109]. When the pressures are mild, the effects may be reversible. Studies on small living systems should teach us a lot on the effect of pressure on living matter.

Weber et al. [110] have presented a model that accounts for the concentration independence of the pressure dissociation of virus particles and the partial restoration of the concentration dependence in the presence of urea concentrations that are below the concentration that denatures the protein. Under certain conditions a transition may be observed from the deterministic assembly of the virus particle from the subunits towards the normal stochastic assembly process. A combined effect of specific ligands, pressure and temperature may therefore help in designing new strategies for the design of vaccination procedures.

6.3. Aggregation and fibril formation

The connection between protein folding and the misfolding that occurs under certain conditions and that gives rise to specific fibril formation is now one of the most fascinating topics of protein research [98]. The aggregation of misfolded proteins seems now to be a common feature of many proteins; some of them may lead to what is now commonly called conformational diseases. Highly soluble proteins are converted to insoluble filamentous fibers that are rich in β-pleated structures. The kinetics of the fibril formation is that of a nucleation dependent polymerization, similar to the process of crystallization. Atomic resolution of the structure of an amyloid-forming peptide that has a dehydrated β-sheet structure has not been obtained up to now [111]. One possibility is that the structure may be of the type of β-helices that has been identified as a new structural component in enzymes such as pectinesterases [112]. It is of considerable importance that many proteins that have now been observed to form fibril aggregates, under the right conditions, are found to avoid the formation of disordered aggregates. Several studies have also indicated that pressure is a very valuable tool for the study of the mechanism of aggregation [113-117].

The stabilizing (destabilizing) effect of pressure on the thermal denaturation of proteins has been associated with the presence (absence) of aggregation of the unfolded protein [79]. Intermolecular aggregation is indeed one of the most commonly observed effects of thermal denaturation. The temperature of unfolding may however be lowered considerably by a previous high pressure treatment at ambient temperatures [118]. Since the pressure-induced degree of unfolding is in general less than that induced by temperature, these experiments suggest that the state that is prone to unfolding is not the completely unfolded state but intermediate between the native and the random coil state. In the case of interferon-γ we have observed an extensive aggregation induced by pressure *after* pressure release [119]. Making the native structure more flexible by reducing the disulfide bridges makes the molecule more prone to aggregation. This suggests that the flexibility of the intermediate state is of particular importance for the aggregation process [120].

The fibers that are formed starting from several proteins show a characteristic infrared spectrum in the amide I region that is typical for a β-sheet structure. Structural details at the atomic level have been obtained for an amyloid forming peptide from yeast prion Sup35 [121]. The structure shows a very high density of hydrogen bonding and a lack of water of solvation. This suggests highly compact structures for the fibrils.

In the course of our studies on the fibril formation in insulin, we have discovered several interesting possibilities to probe the unusual properties of the fibers. In the first place it is possible to follow the formation of the fibers with ultrasound velocimetry. As can be seen from Fig. 5 the decrease in adiabatic compressibility takes place at the same temperature where the β-structures start to form. The dimension of the fibers was verified by atomic force microscopy.

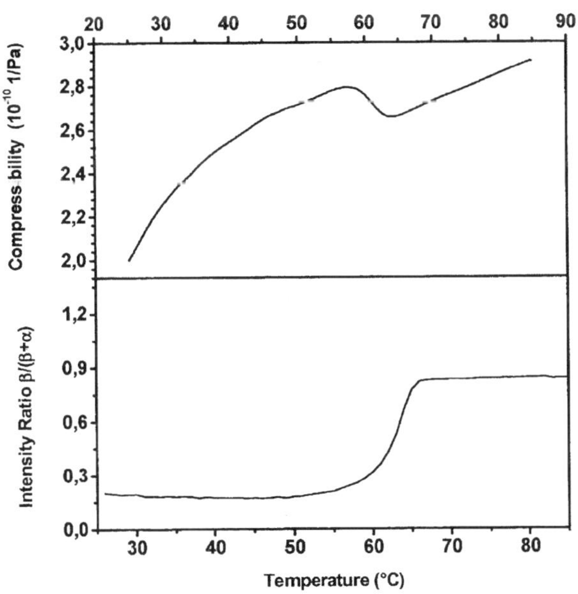

Figure 5. Correlation between the change in secondary structure of insulin as obtained from FTIR spectroscopy and the change in compressibility as obtained from ultrasound. Data were not corrected for the change in density with temperature [122].

Dzwolak et al. [123] have followed the aggregation of insulin under similar solution conditions and found an increase in the heat capacity and a decrease in thermal expansion from pressure perturbation calorimetry.

Secondly, when the fibers are formed in H_2O and then transferred to D_2O, a subsequent compression up to more than 1 GPa does not induce an extensive exchange of H for D. This indicates that the core of the fibrils is extremely resistant to pressure, a phenomenon that is only observed in very limited number of proteins [124,125]. On the basis of the available structural information one would expect these properties to be general for fibrils formed from a wide variety of proteins.

7. OTHER BIOPOLYMERS

While this chapter concentrates on proteins it is useful to have a short look at other biopolymers and try to find a rational basis for the similarities and differences with the properties observed for proteins. We focus on starch, nucleic acids and phospholipids. We note from the outset the similarities in the pressure temperature diagram between starch and proteins.

7.1. Starch

Starch granules are semicrystalline particles mainly composed of amylose and amylopectin. Heat induced changes in starches are well characterized and reveals that the amorphous growth rings undergo hydration during gelation. The application of 200 MPa induces an upward shift of 5°C for the gelatinization temperature [8]. Even when high pressures and high temperatures may induce gelatinization, their effect on granule structure is not similar.

Starch shows absorption bands in the infrared that are sensitive to polymer conformation and hydration. Infrared spectroscopy and the high pressure diamond anvil cell can be used to follow *in-situ* the gelatinization of starch granules as a function of temperature and pressure [9]. The shape of the gelatinization diagram of starch is very similar to the stability diagram for the denaturation of proteins as given in Fig. 2, and the cloud point of water soluble synthetic polymers [94]. This is quite remarkable given the major differences in composition between these polymers.

For the water soluble synthetic polymers a pressure and temperature induced phase separation is observed resulting in pressure-temperature stability diagrams with a shape similar as that observed for starch. This is interpreted as arising from pressure and temperature effects on hydrogen bonding and hydrophobic interactions. The situation for starch and proteins is exactly the opposite. Here a change from the helical to the coil form is proposed induced by high temperature or high pressure. It is assumed that the solvation is the first step in the unfolding. For proteins the role of hydrogen bonding and hydrophobic interactions has been assumed to be the primary source of the reentrant behavior. For starch it seems that the imperfect packing of amylose and amylopectin chains in the starch granule plays an important role in the observed changes. We also note the important role of water in this process. In ethanol no pressure induced swelling of starch is observed.

Other polysaccharides show quite different pressure effects on the gel to sol temperature induced transitions with interesting differences between carrageenans and agarose. For kappa-carrageenan and iota-carrageenan, the melting temperature decreases with increasing pressure up to 300 MPa. The reverse is observed for agarose up to 200 MPa.

7.2. Nucleic acids

Earlier studies on the effect of pressure on the melting temperature of the double helix into single strands showed very small effect. The dTm/dP values are of the order of 3°C/100 Mpa [126]. This implies that the volume change for the melting process is positive. The volume changes depend on the salt concentration and the temperature. Whereas the enthalpy change is always positive, the volume change is zero at about 50°C. It is positive above that temperature and negative below it. The dTm/dP values depend therefore on the melting temperature.

Chalikian and coworkers [127] have proposed a phase diagram for nucleic acids on the basis of thermodynamic data. The diagram is amenable to experimental testing.

Protein-nucleic acid interactions are the basis for the regulation of transcription, translation, replication, virus assembly and recombination. The interactions are van der Waals, electrostatic and hydrogen bonds. In many cases the sequence-specific DNA-binding proteins are dimers and the effects of pressure on the subunit interactions suggests that a large surface area is buried upon association and binding to the specific DNA [128]. Because of the large binding constants, it is difficult to dissociate the protein-DNA complex under pressure conditions where the protein remains intact. The structure of the DNA remains stable in the 1 GPa pressure range.

7.3. Phospholipids

The effect of pressure on the melting temperature (T_m) of lipids and biomembranes is largely determined by the presence of hydrocarbons and therefore given by the Clausius-Clapeyron equation

$$dT_m/dP = T_m\Delta V/\Delta H \tag{19}$$

Since the volume (ΔV) and enthalpy change (ΔH) on melting are in general positive one can expect an increase in melting temperature with increasing pressure. For many organic compounds dT_m/dP is ca. 15 K/100 MPa. As a first approximation, the effect of a pressure increase by 100 MPa would then correspond to the physical effect of a decrease in temperature by 15 K. The volume (ΔV) and enthalpy change (ΔH) for lipids are smaller than those observed for aliphatic hydrocarbons. An illustration of the Clausius-Clapeyron equation may be found in the effect of pressure on the temperatures of the transitions in phospholipids as given by Winter [80].

Whereas the transition temperature of the lipids depends on the length of the hydrocarbon chain, the rate at which the temperature changes with pressure is almost independent on the length (dT/dP = 20 K/100 MPa). It is of interest to note that a higher degree of unsaturation of the hydrocarbon chain lowers the dT/dP ratio to 14 K/100 MPa. The dT/dP ratio is very little dependent on the pressure, except for the formation of lipid interdigitated phases.

When proteins are incorporated in membranes the situation is more complex. In biomembranes it has been found that the physical state of the lipids that surround the membrane proteins play a crucial role in the activity of membrane bound enzymes. In addition, the integrity of the membrane of living organisms, such as bacteria, is very sensitive to pressure. This may explain the sterilization by pressure. It has been observed that the temperature and pressure dependence of the activity of membrane bound enzymes such as the Na^{+}-K^{+}-ATPase and the Ca^{++}-ATPases are non-linear [129].

Many studies have demonstrated that the membranes of piezophilic microorganisms are significantly more fluid as the ratio of unsaturated/saturated lipids increases [16]. This provides an advantage at higher pressures against the decrease in fluidity from the restricted molecular motion of the hydrocarbons. The increased presence of unsaturated lipids may play a role in maintaining the proper membrane fluidity under high pressure.

8. CELLULAR PROCESSES AND ORGANISMS

The sensitivity of supramolecular assemblies such as microtubuli and flagellin suggests that we may expect that cells are quite sensitive to pressure. Without much hesitation we may consider the cell as a molecular factory. This is supported by the observations on viruses. Very small pressures induce changes that make viruses biologically inactive although the structure may be little affected. The molecular adaptation mechanisms that play a role in the composition of bacterial membranes may be interpreted along the same lines.

The observed role of water in the stability of macromolecules is also found in cells. It is well known that bacterial spores are very difficult to inactivate by high pressures. On the other hand the germination of spores takes place at pressures of the order of 100 Mpa [20]. We have already mentioned the example of the role in water in the effect of pressure on the activity of Tardigrades [19].

Pressure may have delicate effects on the metabolism and the gene expression of cells [130-132]. The description of these effects assumes a complete knowledge of the metabolic pathways and the genetic information of the cell. Such studies are now under way for yeast cells. Alternating cycles of exposure to high pressure and outgrowth of surviving populations can be used to select for highly pressure-resistant mutants of some bacteria [133]. In some cases a single pressure pulse on the spores has given rise to a transmutation of bacteria [134]. The interesting point is that the bacteria themselves are not sensitive to such treatment.

All these observations point to the possibility of an understanding of pressure effects of the biology of the cell from an understanding of the effects of pressure on the interactions between macromolecules and their reactions. But it is also obvious that cell biology may need more theory to make sense of the complex networks of metabolism. Rate and equilibrium data are needed on reactions and interactions between molecules. Computational modeling of cell biology may stimulate new experiments and help in the understanding of the present results.

The final step in our understanding should be the explanation of the sensitivity of different microorganisms for a pressure treatment. This may be a problem of understanding not only the metabolism, the gene expression but also more complex phenomena such as unfolding of critical enzymes followed by phase separation that the cell cannot cope with anymore by the available repair mechanisms. It would be of interest to look into the differences between eukaryotes and the prokaryotes. It is to be noted that many extremophiles are of the latter type.

9. CONCLUSION: SPECIFICITY OF PRESSURE EFFECTS?

As shown in this chapter, pressure effects on proteins are on intramolecular dynamics as well as on intramolecular and intermolecular interactions. Because the effects are less severe than temperature effects, there are good reasons to study pressure effects on biological macromolecules and compare them with temperature and chemical effects.

There is first the thermodynamic or energetic point of view. In contrast to increasing temperature, which focuses on energy and volume effects due to thermal expansivity, pressure effects are mainly on the volumetric aspects via the compressibility of the system. Apart from this equilibrium aspect, information is also obtained on the dynamics of the system in view of the relation between the compressibility and the volume fluctuations. Secondly, from a mechanistic point of view these volume fluctuations, which are of the order of 1 % or less of the total volume of a protein, arise from contributions of the internal cavities and the interactions with water. Recent computer simulations suggest that the contribution of the cavities to the compressibility may be much larger than assumed up to now. Pressure-induced effects in proteins may then be largely attributed to the penetration of water into the protein structure and this results in unfolding. In the case of heat-induced denaturation, unfolding may take place first followed by water penetration. The fact that unfolding is the first step in the heat unfolding may explain the observation that aggregation results more frequently from heat than from pressure treatment.

The reduction of living systems to molecules has proved to be a fruitful approach provided one keeps in mind the simplifications that are introduced. Do we then need new strategies for the study of the complexity of life? As expected, opinions are divided. But the observation that stability diagrams for bacteria, viruses and other complex systems that are similar to those obtained for proteins suggests that proteins are the primary targets in the pressure and temperature behavior of organisms.

A better understanding of the behavior of proteins and other biomolecules under extreme conditions may be expected from a concerted application of a number of existing and new experimental approaches. New developments in science usually imply fresh views at existing experimental data. These new views stimulate new experiments. And the best experiments in science are those that lead to the least expected results. The simple experiments that Bridgman performed on egg white illustrate this in a unique way.

REFERENCES

[1] P. W. Bridgman, J. Biol. Chem., 19 (1914) 511-512.
[2] K. Suzuki, Rev. Phys. Chem. Japan, 29 (1960) 91-98.
[3] R. Hayashi, Y. Kawamura, T. Nakasa, O. Okinaka, Agr. Biol. Chem., 53 (1989) 2935-2939.
[4] S. A. Hawley, Biochemistry, 10 (1971) 2436-2442.

[5] A. Zipp and W. Kauzmann, Biochemistry, 12 (1973) 4217-4228.
[6] Y. Taniguchi and K. Suzuki, J. Phys. Chem., 87 (1983) 5185-5193.
[7] A. Weingand-Ziadé, F. Renault and P. Masson, Biochim. Biophys. Acta, 1340 (1997) 245-252.
[8] J. Thevelein, J. A. Van Assche, K. Heremans and S. Y. Gerlsma, Carbohydrate Res., 93 (1981) 304-307.
[9] P. Rubens and K. Heremans, Biopolymers, 54 (2000) 524-530.
[10] C. Hashizume, K. Kimura and R. Hayashi, Biosci. Biotech. Comm., 59 (1995) 1455-1458.
[11] R. Lewontin, The Triple Helix (Gene, Organism & Environment), Harvard University Press, 2000.
[12] L.J.Rothschild and R.L. Mancinelli, Nature, 409 (2001) 1092-1101.
[13] K.O. Stetter, FEBS Lett., 452 (1999) 22-25.
[14] J.W. Deming, Curr. Opinion Microbiol., 5 (2002) 301-309.
[15] A.A. Yayanos, Proc. Natl. Acad. Sci. USA, 83 (1986) 9542-9546.
[16] D.H. Bartlett, Biochem. Biophys. Acta, 1595 (2002) 367-381.
[17] A. Sharma, J.H. Scott, G.D. Cody, M.L. Fogel, R.M. Hazen, R.J. Hemley and W.T. Huntress, Science, 295 (2002) 1514-1516.
[18] A.Yayanos, Science, 297 (2002) 295.
[19] K. Seki and M. Toyoshima, Nature, 395 (1998) 853-854.
[20] B. Sojka and H. Ludwig, Pharm. Ind., 59 (1997) 436-438.
[21] K. Goossens, L. Smeller, J. Frank and K. Heremans, Eur. J. Biochem., 236 (1996) 254-262
[22] F.H. Johnson and E.A. Flagler, Science, 112 (1951) 91-92.
[23] A.G. Macdonald and P.J. Fraser, Comp. Biochem. & Physiol., A 122 (1999) 13-36.
[24] H. Ludwig, W. Scigalla & B. Sojka. In J. L. Markley, C. Royer & D. Northrup (Eds.) High Pressure Effects in Molecular Biophysics and Enzymology. Oxford University Press (1996) pp. 346-363.
[25] K.M.F.A. Reyns, C.C.F. Soontjens, K. Cornelis, C.A. Weemaes, M.E. Hendrickx and C.W. Michiels, Int. J. Food Microb., 56 (2000) 199-210.
[26] W. Kauzmann, Adv. Prot. Chem., 14 (1959) 1-63.
[27] N. Taulier and T.V. Chalikian, Biochim. Biophys. Acta, 1595 (2002) 48-70.
[28] T.V. Chalikian, Annu. Rev. Biophys. Biomol. Struct., 32 (2003) 207-235.
[29] K. Gekko and Y. Hasegawa, J. Phys. Chem., 93 (1989) 426-429.
[30] J. Liang and K.A. Dill, Biophys. J., 81 (2001) 751-766.
[31] J.L. Silva and G. Weber, Ann. Rev. Phys. Chem., 44 (1993) 89-113.
[32] J.L. Silva, D. Foguel and C.A. Royer, Trends Biochem. Sci., 26 (2001) 612-618.
[33] R. B. Gregory. In Protein-Solvent Interactions, R.B. Gregory (Ed.) Marcel Dekker, N.Y. (1995) pp.191-264.
[34] L. Smeller and J. Fidy, Biopys. J., 82 (2002) 426-436.
[35] Y. Katayama, T. Mizutani, W. Utsumi, O. Shimomura, M. Yamakata and K. Funakoshi, Nature, 403 (2000) 170-173.
[36] P.E. Cladis, R.K. Bogardus, W.B. Daniels and G.N. Taylor, Phys. Rev. Lett., 39 (1977) 720-723.
[37] S. Kunugi, K. Takano, N. Tanaka, K. Suwa & M. Akashi, Macromolecules, 30 (1997) 4499-4501.
[38] S. Rastogi, G.W.H. Höhne and A. Keller, Macromolecules, 32 (1999) 8897-8909.
[39] K. Süvegh, A. Vértes and T. Hyodo, Adv. Molec. Struct. Res., 5 (1999) 313-357.
[40] Y.O. Kamatari, H. Yamada, K. Akasaka, J.A. Jones, C.M. Dobson and L.J. Smith, Eur. J. Biochem., 268 (2001) 1782-1793.
[41] M. Refaee, T. Tezuka, K. Akasaka and M.P. Williamson, J. Mol. Biol., 327 (2003) 857-865.
[42] R. Fourme, I. Ascone, R. Kahn, M. Mezouar, P. Bouvier, E. Girard, T.W. Lin and J.E. Johnson, Structure,10 (2002) 1409-1414.
[43] P. Urayama, G.N. Phillips and S.M. Gruner, Structure, 10 (2002) 51-60.
[44] A. Gall, A. Ellervee, B. Robert and A. Freiberg, FEBS Lett., 560 (2004) 221-225.
[45] W. Doster and R. Gebhardt, Chem. Phys., 292 (2003) 383-387.

[46] B. Wroblowski, J.F. Diaz, K. Heremans and Y. Engelborghs, Proteins, 25 (1996) 446-455.
[47] E. Paci, Biochem. Biophys. Acta, 1595 (2002) 185-200.
[48] A. Cooper, Biophys. Chem., 85 (2000) 25-39.
[49] G.A. Papoin, J. Ulander, M.P Eastwood, Z. Luthey-Schulten and P.G. Wolynes, Proc. Natl. Acad. Sci.USA, 101 (2004) 3352-3357.
[50] N. S. Timasheff, Biochemistry, 41 (2002) 13473-13482.
[51] S. Shimizu, Proc. Natl. Acad. Sci. USA, 101 (2004) 1195-1199.
[52] G. Pappenberger, C. Saudan, M. Becker, A.E. Merbach and T. Kiefhaber, Proc. Natl. Acad. Sci. USA, 97 (2000) 17-22.
[53] K. J. Frye and C. A. Royer, Protein Sci., 6 (1997) 789-793.
[54] V. Chalikian, A.P. Sarvazyan, Th. Funck, Ch.A. Cain and K.J. Breslauer, J. Phys. Chem., 98 (1994) 321-328.
[55] H. Pfeiffer and K. Heremans, Chem. Phys. Lett., 361 (2002) 226-230.
[56] H. Pfeiffer and K. Heremans. In R. Winter (Ed.) Advances in High Pressure Bioscience and Biotechnology II, (Springer Verlag. 2003) pp. 481-484.
[57] V. M. Dadarlat and C. B. Post, Proc. Natl. Acad. Sci. USA, 100 (2003) 14778-14783.
[58] K. Akasaka, Biochemistry, 42 (2003) 10875-10885.
[59] K. Gekko, A. Kimoto and T. Kamiyama, Biochemistry, 42 (2003) 13746-13753.
[60] C. Jung, S.A. Kozin, B. Canny, J-C. Chervin and G. Hui Bon Hoa, Biochem. Biophys. Res. Comm., 312 (2003) 197-203.
[61] C. Loupiac, M. Bonetti, S. Pin and P. Calmettes, Eur. J. Biochem., 269 (2002) 4731-4737.
[62] N. Kobayashi, T. Tamato and N. Go, Proteins, 28 (1997) 109-116.
[63] D.P. Kharakoz, Biophys. J., 79 (2000) 511-525.
[64] M. Hiebl and R. Maksymiw, Biopolymers, 31 (1991) 161-167.
[65] H. Frauenfelder, H. Hartmann, M. Karplus, I.D. Kuntz, Jr., J. Kuriyan, F. Parak, G.A. Petsko, D. Ringe, R.F. Tilton, Jr., M.L. Connelly and N. Max, Biochemistry, 26 (1987) 254-261.
[66] F. Demmel, W. Doster, W. Petry and A. Schulte, Eur. Biophys. J., 26 (1997) 327-335.
[67] H. Kumeta, A. Miura, Y. Kobashigawa, K. Miura, C. Oka, N. Nemoto, K. Nitta and S. Tsuda, Biochemistry, 42 (2003) 1209-1216.
[68] L-N. Lin, J.F. Brandts, J.M. Brandts and V. Plotnikov, Anal. Biochem., 302 (2002) 144-160.
[69] R. Ravindra and R. Winter, Z. Phys. Chem., 217 (2203) 1-23.
[70] R. Ravindra, C. Royer and R. Winter, Phys. Chem. Chem. Phys., 6 (2004) in press.
[71] J. Rösgen and H-J. Hinz, Biophys. Chem., 83 (2000) 61-71.
[72] J. Gomez, V. J. Hilser, D. Xie and E. Freire, Proteins, 22 (1995) 404-412.
[73] J. J. Flores and E. L. Chronister, J. Raman Spectrosc., 27 (1996) 149-153.
[74] K. Goossens, L. Smeller and K. Heremans, J. Chem. Phys., 99 (1993) 5736-5742
[75] G. Weber and H.G. Drickamer, Q. Rev. Biophys., 16 (1983) 89-112.
[76] G. Hummer, S. Garde, A.E. Garcia, M.E. Paulaitis and L. R. Pratt, Proc. Natl. Acad. Sci. USA, 95 (1998) 1552-1555.
[77] N. T. Southall, K. A. Dill and A.D.J. Haymet, J. Phys. Chem., B 106 (2002) 521-533.
[78] C. Weemaes, S. De Cordt, K. Goossens, L. Ludikhuyze, M. Hendrickx, K. Heremans & P. Tobback, Biotechnol. Bioeng., 50 (1996) 49-56.
[79] L. Smeller, Biochim. Biophys. Acta, 1595 (2002) 11-29.
[80] R. Winter, Biochem. Biophys. Acta, 1595 (2002) 160-184.
[81] H. Seemann, R. Winter and C.A. Royer, J. Mol. Biol., 307 (2001) 1091-1102.
[82] R. Ravindra and R. Winter, ChemPhysChem, 4 (2003) 359-365.
[83] D.N. Dubins, R. Filifil., R.B. Macgregor Jr. and T.V. Chalikian, Biochemistry, 42 (2003) 8671-8678.
[84] J. Zhang, X. Peng, A. Jonas and J. Jonas, Biochemistry, 34 (1995) 8361-8641.
[85] H. Lesch, H. Stadlbauer, J. Friedrich and J.M. Vanderkooi, Biophys. J., 82 (2002) 1644-1653.
[86] F. Meersman, L. Smeller and K. Heremans, Biophys. J., 82 (2002) 2635-2644.
[87] J. F. Brandts, R. J. Oliveira and C. Westort, Biochemistry, 9 (1970) 1038-1047.

[88] G.P. Johari, Phys. Chem. Chem. Phys., 3 (2001) 2483-2487.
[89] M. R. Feeney, P.G. Debenedetti and F.H. Stillinger, J. Chem. Phys., 119 (2003) 4582-4591.
[90] M.I. Marques, J.M. Borreguero, H.E. Stanley and N. V. Dokholyan, Phys. Rev. Lett., 91 (2003) 131803.
[91] J. A. Kornblatt and M.J. Kornblatt, Biochem. Biophys. Acta, 1595 (2002) 30-47.
[92] E.E. Dormidontova, Macromolecules, 35 (2002) 987-1001.
[93] J. Zhang, B. Berge, F. Meeussen, E. Nies, H. Berghmans and D. Shen, Macromolecules, 36 (2003) 9145-9153.
[94] S. Kunugi, K. Takano, N. Tanaka, K. Suwa and M. Akashi, Macromolecules, 30 (1997) 4499-4501.
[95] J. Wang, F. Meersman, R. Esnouf, M. Froeyen, R. Busson, K. Heremans and P. Herdewijn, Helv. Chim. Acta, 84 (2001) 2398-2408.
[96] T. Sun and H.E. King, Jr., Macromolecules, 31 (1998) 6383-6386.
[97] C. M. Dobson, Phil. Trans. Roy. Soc. Lond., 356 (2001) 133-145.
[98] C. M. Dobson, Nature, 426 (2003) 884-890.
[99] R.J. Ellis, Curr. Opin. Struct. Biol., 11 (2001) 114-119.
[100] Y. Qu and D. W. Bolen, Biochemistry, 42 (2003) 5837-5849.
[101] H. Herberhold, C.A. Royer and R. Winter, Biochemistry, 43 (2004) 3336-3345.
[102] K. Visuri, E. Kaipainen, J. Kivimaki, H. Niemi, M. Leisola and S. Palosaari, Bio/Technol. (1990) 547-549.
[103] J.N. Webb, R.Y. Waghmare, J.F. Carpenter, C.E. Glatz and T.W. Randolph, J. Cryst. Growth, 205 (199) 563-574.
[104] A. Kadri, M. Damak, B. Lorber, R. Geige and G. Jenner, High Pressure Research, 23 (2003) 485-491.
[105] A. Katrusiak and Z. Dauter, Acta Cryst., D52 (1996) 607-608.
[106] K. Heremans and P. T. T. Wong, Chem Phys. Lett., 118 (1985) 101-104.
[107] F. Merzel and J.C. Smith, Proc. Natl. Acad. Sci. USA, 99 (2002) 5378-5383.
[108] Y. Engelborghs, K.A.H. Heremans, L.C.M. De Maeyer and J. Hoebeke, Nature, 259 (1976) 686-689.
[109] M. Kirschner, J. Gerhart and T. Mitchison, Cell, 100 (2000) 79-88.
[110] G. Weber, A.T. Da Poian and J.L. Silva, Biophys. J., 70 (1996) 167-173.
[111] R. Diaz-Avalos, C. Long, E. Fontano, M. Balbirnie, R . Grothe, D. Eisenberg and D.L.D. Caspar, J. Mol.Biol., 330 (2003) 1165-1175.
[112] R. Wetzel, Structure, 10 (2002) 1031-1036.
[113] A.D. Ferrão-Gonzales, S.O. Souto, J.L. Silva and D. Foguel , Proc. Natl. Acad. Sci. USA, 97 (2000) 6445-6450.
[114] Y-S. Kim, T.W. Randolph, F.J. Stevens and J.F. Carpenter, J. Biol. Chem., 277 (2002) 27240-27246.
[115] J. Torrent, M.T. Alvarez-Martinez, F. Heitz, J.P. Liautard, C. Balny and R. Lange, Biochemistry, 42 (2003) 1318-1325.
[116] D. Foguel, M.C. Suarez, A.D. Ferrao-Gonzales, T.C.R. Porto, L. Palmieri, C.M. Einsiedler, L.R. Andrade, H.A. Lashuel, P.T. Lansbury, J.W. Kelly and J.L. Silva, Proc. Natl. Acad. Sci. USA, 100 (2003) 9831-9836.
[117] T.N. Niraula, T. Konno, H. Li, H. Yamada, K; Akasaka and H. Tachibana, Proc. Natl. Acad. Sci. USA,101 (2004) 4089-4093.
[118] L. Smeller, P. Rubens and K. Heremans, Biochemistry, 38 (1999) 3816-3920.
[119] K. Goossens, J. Haelewyn, F. Meersman, M. De Ley and K. Heremans, Biochem J., 370 (2003) 529-535.
[120] F. Meersman and K. Heremans, Biophys. Chem., 104 (2003) 297-304.
[121] M. Balbirnie, R. Grothe and D.S. Eisenberg, Proc. Natl. Acad. Sci. USA, 98 (2001) 2375-2380.
[122] C. Dirix, F. Meersman, H. Pfeiffer and K. Heremans. Unpublished observations.
[123] W. Dzwolak, R. Ravindra, J. Lendermann and R. Winter, Biochemistry, 42 (2003) 11347-11355.

[124] P. Fusi, K. Goossens, R. Consonni, M. Grisa, P. Puricelli, G. Vecchio, M. Vanoni, L. Zetta, K. Heremans and P. Tortora, Proteins, 29 (1997) 381-390.
[125] L. Smeller, F. Meersman, J. Fidy and K. Heremans, Biochemistry, 42 (2003) 553-561.
[126] S.A. Hawley and R.M. Macleod, Biopolymers, 13 (1974) 1417-1426.
[127] D.N. Dubins, A. Lee, R.B. Macgregor and T.V. Chalikian, J. Am. Chem. Soc., 123 (2001) 9254-9259.
[128] J.L. Silva, A.C. Oliveira, A.M. Gomes, L.M.T.R. Lima, R. Mohana-Borges, A.B.F. Pacheco and D.Foguel, Biochem. Biophys. Acta, 1595 (2002) 250-265.
[129] K. Heremans and F. Wuytack, FEBS Lett., 117 (1980) 161-163.
[130] H. Iwahashi, H. Shimizu, M. Odani and Y. Komatsu, Extremophiles, 7 (2003) 291-298.
[131] A. Molina-Hoppner, T. Sato, C. Kato, M.G. Ganzle and R.F. Vogel, Extremophiles, 7 (2003) 511-516.
[132] P.M.B. Fernandes, T. Domitrovic, C.M. Cao and E. Kurtenbach, FEBS Letters, 556 (2004) 153-160.
[133] K.J.A. Hauben, D.H. Bartlett, C.C.F. Soontjens, K. Cornelis, E.Y. Wuytack and C.W. Michiels, Appl.Environ. Microbiol., 63 (1997) 945-950.
[134] I. Sorokine-Durm, K. Werner, M. Schauer and H. Ludwig. In R. Winter (Ed.) Advances in High Pressure Bioscience and Biotechnology II. Springer Verlag. 2003, pp. 286-291.

Chemistry at Extreme Conditions
M. Riad Manaa (Editor)

Chapter 2

High Pressure Effects in Molecular Bioscience

Roland Winter

University of Dortmund, Physical Chemistry I, Otto-Hahn-Straße 6, D-44227 Dortmund

1. INTRODUCTION

Biologists often do not take into account that the greatest portion of our biosphere is in the realm of environmental extremes. For several decades now, the limits of the existence of life have been pushed to unexpected extremes of pressure, temperature, pH, salinity, etc. [1-7]. Hydrostatic pressure significantly influences the structural properties and thus functional characteristics of cells; yet this has not prevented the invasion of cold and high pressure habitats by deep-sea organisms. 70 % of the surface of the earth is covered by oceans, and the average pressure on the ocean floor is about 380 bar. In 1884 Certes was the first to discover the existence of microorganisms in deep-sea sediments. Deep-sea sediments and hydrothermal vents are densely crowded with barophilic-thermophilic species (i.e., pressure- and heat-adapted species). Psychrophilic-barophilic species (cold-adapted), which live at ~2 °C, are found on the deepest ocean floor (~11000 m) in the Mariana Trench and in deep-sea sediments. Close to hydrothermal vents, generally several thousand meter under the sea surface, organisms far more complex than bacteria can be found in conditions of high hydrostatic pressure (HHP) (~300 bar), high temperature (up to 120 °C), high concentrations of hydrogen sulfide, and the absence of light. These organisms (crustaceans, tube worms, sponges, etc.) live in symbiosis with chemoautotrophic bacteria, which produce energy by oxidizing sulfur compounds. Those conditions that are commonly considered extreme, actually are bursting with forms of life that, quite amazingly, are not so different from those found on the surface. Deep sea organisms have evolved several characteristics: They have more fluid membranes, more stable proteins, and biomacromolecular complexes leading to optimal biological activity under HHP. Most bacteria living below 2000 m (200 bar) are strictly barophilic, whereas those that live above 2000 m are pressure-tolerant. For example, a bacterial strain (MT41) that was collected at 10476 m (~1 kbar) depth cannot live at pressure lower than 38 MPa (380 bar) [8]. It is worth noting that surface species (mesophilic species, 1 bar adopted) may also show unexpected capacities of adaptation to high pressure.

Interest in pressure as a thermodynamic and kinetic variable has been growing also in physico-chemical studies of biological materials in recent years [9-18]. The fundamental reasons why it can be desirable to carry out high pressure experiments on these systems are: i) Changing temperature of a biochemical system at atmospheric pressure produces a simultaneous change in thermal energy and volume; therefore, to separate thermal and volume effects, one must carry out high pressure experiments. ii) Because noncovalent interactions play a primary role in the stabilization of biochemical systems, the use of pressure allows one to change, in a controlled way, the intermolecular interactions without the major perturbations produced by changes in temperature or cosolvent concentration. iii) Pressure affects chemical equilibria and reaction rates; the following standard equations define the reaction volume, ΔV°, and the activation volume, $\Delta V^{\neq}$, respectively

$$\Delta V^{\circ} = -\left(\frac{RT\partial \ln K}{\partial p}\right)_T, \Delta V^{\neq} = -\left(\frac{RT\partial \ln k}{\partial p}\right)_T \tag{1}$$

where K is the equilibrium constant, and k is the reaction rate. The behavior of all systems under high pressure is governed by Le Châtelier's principle, which predicts that the application of pressure shifts an equilibrium towards the state that occupies a smaller volume, and accelerates processes for which the transition state has a smaller volume than the ground state. For example, if a reaction is accompanied by a $\Delta V^{\neq}$ value of -100 mL mol^{-1}, it is enhanced more than 3000-fold by applying a pressure of 2 kbar. With the knowledge of ΔV° and $\Delta V^{\neq}$ values, one can draw valuable conclusions about the nature of the reaction and its mechanism. iv) The viscosity of the solvent can be changed continuously by pressure. v) Pressure-dependent studies often led to the discovery of new phases and processes. vi) One can extend the range of conditions and carry out experiments at subzero temperatures and in the supercritical state. According to the high pressure phase diagram of water, even at –15 °C water is still a liquid. Therefore, protein solutions can be measured at subzero temperatures to investigate their cold-denaturation behavior.

Pressures used to investigate biochemical systems range from 0.1 MPa to about 1 GPa (0.1 MPa = 1 bar, 1 GPa = 10 kbar). Such pressures only change intermolecular distances and affect conformations, but do not change covalent bond distances or bond angles. In fact, pressures in excess of 30 kbar are required to change the electronic structure of a molecule. The covalent structure of low molecular mass biomolecules (peptides, lipids, saccharides), as well as the primary structure of macromolecules (proteins, nucleic acids and polysaccharides), is not perturbed by pressures up to about 20 kbar. Pressure acts predominantly on the conformation and supramolecular structures of biomolecular systems.

Besides the general physico-chemical interest in using high pressure as a tool for understanding the phase behavior, structure and energetics of biomolecules, high pressure is also of biotechnological (e.g., high pressure food processing) and interest [9-14, 16].

In this review, we first briefly summarize pressure effects in molecular biology and applications in biotechnology. We then discuss some experimental techniques and finally we discuss in more detail results of studies on the high pressure structure and phase behavior of biomolecular systems, such as nucleic acids, lyotropic lipid mesophases, model biomembrane systems and proteins.

2. EXPLOITATION OF PRESSURE EFFECTS IN MOLECULAR BIOLOGY AND BIOTECHNOLOGY

Without the help of sophisticated apparatuses, humans cannot withstand pressures exceeding about 5 bar, i.e., the pressure at 50 m under the sea surface. Surprisingly, the pressure resistance of isolated cells is considerably higher than that of whole organisms and is only slightly inferior to that of isolated biomacromolecular assemblages. Cells may remain viable up to about 1 kbar. Some species can even reversibly support much higher pressures, such as *Sacharomyces cerevisiae* [3, 14]. The accidents that occur to divers upon decompression are primarily due to the nucleation of gas bubbles in the direct effects of HHP on the cells. In cells, the absence of fluids containing dissolved gas can explain their resistance to pressure. In the following, we essentially focus on bacterial cells.

Pressure stress affects all levels of **cellular physiology** including bacterial metabolism, membrane physiology, transport, transcription and translation [3, 7]. Molecular assemblages generally are dissociated by hydrostatic pressures lower than those necessary to denature proteins. The pressures sufficient to dissociate some assemblages often are less than 1 kbar, whereas many monomeric proteins generally resist pressures up to ~4 kbar. Some oligomeric proteins are rather fragile and dissociate at pressures even lower than 1 kbar. There are many examples of pressure-induced oligomer dissociation, such as repressors and transcription factors. The cytoplasmic membrane is also a complex heterogeneous aggregate structure containing various types of macromolecules which can be disturbed by high pressure in their structure and function. The integrity and functionality is vital for the cell, e. g. in energy generation, selective transport, maintenance of osmotic pressure and intracellular pH. Although some ion transporters are unaffected or even activated by pressure, certain other channels and pumps are inactivated at moderate pressures [3, 19, 20]. For example, the sodium pump (Na^+/K^+-ATPase), an α-β dimer, is reversibly inactivated between 1 and 2.5 kbar (Fig. 1) [20].

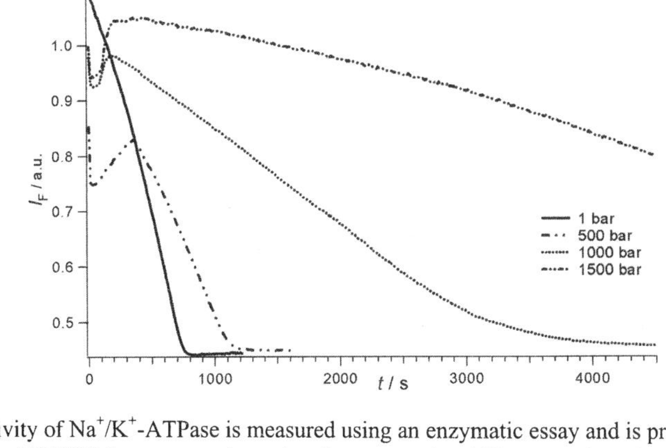

Fig. 1. The activity of Na^+/K^+-ATPase is measured using an enzymatic essay and is proportional here to the decrease of NADH fluorescence intensity, I_F, measured at 460 nm at selected pressures and 37 °C [19].

Figure 1 presents the Na^+/K^+-ATPase activity (k) data at 37 °C as a function of pressure. The data show that the activity of Na^+/K^+-ATPase is inhibited reversibly by pressures below 2 kbar. At higher pressures, the enzyme is inactivated irreversibly. The plot of ln$k$ *vs.* pressure is essentially linear at longer timescale, thus allowing for the calculating of an apparent activation volume of the pressure-induced inhibition reaction, which amounts to $\Delta V^{\neq} = 47$ mL mol^{-1}. The inhibition might be due to a pressure-induced ordering of the acyl-chains or a fluid to gel phase transition of the lipid matrix, but might also be caused by subunit dissociation. This would be in accord with the findings of Chong et al. [21] and with the data obtained by Cornelius [22], who reconstituted the ATPase into liposomes of different acyl chain lengths and degree of unsaturation, and found a significant influence of the lipid matrix on the activity of the Na^+/K^+-ATPase. Pump inactivation appears to depend directly on the membrane fluidity. A decrease in fluidity due to HHP hinders conformational transitions of the enzyme. In fact, changes in membrane fluidity seem to play a major role in the adaptation of deep-sea species to high pressure.

The cytoskeleton appears to be the first affected organelle by pressure, whereas the nucleus seems to be one of the most resistant organelles [3]. Relatively moderate hydrostatic pressures (200-400 bar) disorganize the cytoskeleton, which leads to changes in cell shape and loss of motility. The disruption of the cytoskeletal apparatus by HHP makes the plasma membrane leaky and often leads to changes ("rounding up") of the cell shape. It has also been shown that elevated pressures dissociate and inactivate ribosomes, which has been considered to be a major reason for the inhibition of protein synthesis by pressure. In fact, protein synthesis seems to be one of the most barosensitive cellular functions, which is blocked by about 700 bar [3].

Gene expression has been shown to be affected by HHP at the levels of both transcription and translation. In most cases, gene expression is inhibited by HHP. However, some specific pressure-shock proteins are synthesized. *In vitro* studies show that the posttranslational ribosomal complexes are affected above 700 bar. The production of more than fifty pressure-induced proteins after an abrupt pressure shift, from 1 to 550 bar, has been reported [23]. Interestingly, some of these proteins are well-known heat-shock proteins (HSp) or cold-shock proteins, such as Hsp70, GroEL, GroES, Hsc70, DnaK. For that reason, the response to these different shocks often is termed the "universal stress response". *Sacharomyces cerevisiae* induces the production not only of a heat-shock protein, but also of sugar (trehalose), which appears to play a role in the protection against pressure.

Although the occurrence of bacteria in the deep sea has been known for more than a century, the first barophilic bacterium (defined as an organism that grows well at pressures above 400 bar) was isolated only in 1979 by Yayanos et al. [24]. In recent years, proteins and genes present in various deep-sea-adapted bacteria have been studied, and pressure-tolerant and pressure-regulated genes have been identified [7,12,25,26]. They encode proteins that are implicated in the adaptation to living over a wide range of depths. Resistance to cold and pressure probably involves similar changes to organisms. Hence, many species are both barophilic and thermophilic. Moreover, their thermotolerance increases with pressure. For example, the temperature of the maximal growth rate of the archaeon *Thermococcus peptonophilus* is 85 °C under 300 bar, but it rises to 90-95 °C under 450 bar [27]. Deep-sea adapted microorganisms may be very useful in new applications of biotechnology. The genes and proteins from deep-sea barophilic bacteria are adapted to high-pressure conditions, so they could be used for the development of high-pressure bioreactors, for example.

To conclude this part, the most pressure sensitive cellular targets are affected instantly during the ramp above ~500 bar and appear to be the bacterial membrane and macromolecules requiring associations to function in cellular organization. Changes in the proteome observed indicate that general and energy metabolism is turned on to compensate for impairment of ion gradients and transport across the membrane. During pressure holding times, intracellular changes other than high pressure, e. g. loss of ions or drop in pH, are synergic stress factors promoting irreversible and denaturing effects followed by cell death.

High hydrostatic pressures have already been exploited in diverse areas of **biotechnology**, including biomolecular product extraction, modifying the catalytic behavior of enzymes, altering the metabolism and gene expression of microorganisms, and in food processing [9-13,16] to preserve nutritional and taste qualities of foodstuffs and to prevent microorganisms' proliferation. In 1992, the Meidiya Food Company in Japan released high pressure-processed jams on the food market, followed by high pressure-processed fruit jellies and sauces. The rationale behind the use of high pressure instead of high temperature in food processing is the improved preservation of food taste, flavor and color. This is based on the stability of the covalent structures of proteins, saccharides, vitamins, lipids and pigments to high pressures, in contrast with their relative instability towards increased temperatures. One of the most important roles played by pressure here is sterilization. Studies of the effects of high hydrostatic pressure on microorganisms of food showed that short-term treatment with pressures of several kbar reduces the bacterial content in foods by several orders of magnitude. Microorganisms differ significantly in their ability to withstand pressure: bacterial spores and some viruses are among the most resistant and can survive pressures higher than 10 kbar. By analogy to pasteurization, these sterilization procedures are also termed "pascalization". Improved high-pressure food processing methods may also result from the possibility of operating at temperatures below 0 °C without freezing. Meanwhile it has been learnt that the combined use of high pressure and temperature can be even more effective in developing high quality foods.

Reasons for pressure-induced changes in the rate of **enzyme-catalysed reactions** may be classified into three main groups: 1) changes in the structure of the enzyme, 2) changes in the reaction mechanism, and 3) the effect of a particular rate-determining step on the overall rate. For example, the hydrolysis of an anilide in reversed micelles, catalysed by α-chymotrypsin, shows a negative value of $\Delta V^{\neq}$. Consequently, an increase in pressure from atmospheric pressure to 2 kbar results in a sevenfold increase in the reaction rate. By contrast, hydrolysis of an ester by α-chymotrypsin has a positive value of $\Delta V^{\neq}$, and an increase in pressure thus leads to a more than tenfold deceleration in the rate of reaction ([13], N. Klyachko in [10,11], [28,29]). Further opportunities for manipulating reaction rates by altering pressure arise from the use of organic solvents as media for enzyme reactions. Catalysis in organic media is very sensitive to the hydration of enzymes which, among other factors, can be influenced by pressure. An additional exciting area of biotechnological research into the use of elevated pressure is biocatalysis in supercritical fluids. Carbon dioxide, the most widely used supercritical fluid, dissolves many enzyme substrates that are poorly soluble in water. High pressure is necessary to generate the supercritical state, and can also be used in the control of properties such as specificity and enantioselectivity [13].

In response to the fast growing applications, basic bioscience at artificially generated high pressure and its effect on biological materials increases its importance and needs further exploration by scientists of different disciplines including biochemistry, cell biology, molecular biology, biophysics, engineering, food science, medicine and pharmacology.

3. EXPERIMENTAL TECHNIQUES FOR HIGH PRESSURE RESEARCH

There are several reviews discussing the wide spectrum of biochemical problems that can be investigated by high pressure techniques [9-11,17, 30-54]. The techniques that have been used most extensively in the investigation of biological systems at high pressure include fluorescence methods, vibrational, optical absorption and nuclear magnetic resonance (NMR) spectroscopy, light scattering, X-ray and neutron diffraction. As only few commercial instruments are available, several laboratories have undertaken the design and construction of their own instrument. We will describe some of those necessary for understanding the examples presented in the following chapters.

3.1. X-ray- and neutron scattering-techniques and theoretical background

The development of synchrotron radiation sources from multi-GeV electron and positron storage rings increased the flux on the sample by factors of more than 10^5 when compared to conventional laboratory X-ray sources. Use of these sources coupled with efficient electronic detectors made it possible to start collecting small-angle X-ray scattering (SAXS) and wide-angle X-ray scattering (WAXS) data on rapid time scales, which can now go down to tens of ms or even below, thus allowing to perform also kinetic structural investigations.

For the high pressure X-ray studies, flat diamond cells are generally used. The X-ray pressure cell (Fig. 2) is home-built and made from stainless steel or a Ni-Cr-Co alloy (NIMONIC 90) of high tensile strength [53-55]. It has a high-pressure connection to the pressurizing system and a bore for a thermocouple. Temperature control is achieved by circulating water from a thermostat through the outside jacket of the vessel. For pressures up to 6-8 kbar, flat diamond windows of 0.5-1 mm thickness are used. The window holders are sealed with Viton-O-rings and are tightened by closure nuts. The sample of 40 μL volume is held in a PTFE-ring that is closed with two mylar foils glued on both sides of the ring to separate the sample from the pressurizing medium (distilled water). The pressurizing system consists of a Heise Bourdon gauge (or an electronic pressure gauge) and a hand-operated pressure generator. Pressure jumps are performed by a computer-controlled opening of an air-operated valve between the high pressure cell and a liquid reservoir. With this pressure jump apparatus, fast jumps (< 5 ms) are possible with variable amplitudes. To minimize adiabatic temperature changes in the course of a kinetic experiment (ca. 2 $mK \cdot bar^{-1}$ under pure adiabatic conditions), the high pressure sample cell was constructed to hold only a very small volume of the pressurizing medium. The pressure jump technique has been shown to offer several advantages over the temperature jump approach: 1) Pressure propagates rapidly so that sample inhomogeneity is a minor problem. 2) Pressure jumps can be performed bidirectional, i.e. with increasing or decreasing pressure. 3) In the case of fully reversible structural changes of the sample, pressure jumps can be repeated with identical amplitudes to allow for an averaging of the diffraction data over several jumps and an improvement of the counting statistics.

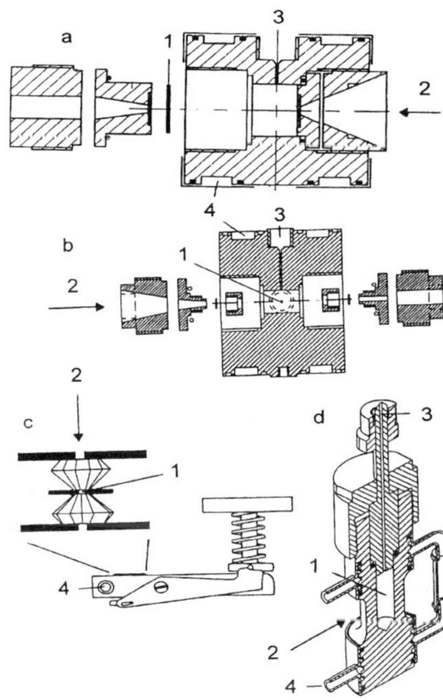

Fig. 2. High pressure sample cells: a) cell made from stainless steel with Be-windows for X-ray scattering studies up to 2 kbar, b) cell made from NIMONIC 90 alloy with diamond windows for X-ray scattering studies up to 8 kbar, c) diamond anvil cell (DAC) for X-ray scattering studies up to 20 kbar, and d) cell made from an Al alloy for neutron scattering studies up to 2.5 kbar (1: sample, 2: X-ray or neutron beam, 3: high pressure connection, 4: thermostating water circuit).

In the following paragraph, the basic theoretical concepts for analysing scattering patterns from partially ordered systems (membranes, lipid mesophases) and particles (proteins) in solution are discussed.

3.1.1. Partially ordered systems: membrane and lipid mesophase diffraction patterns

In biophysical studies, lipid bilayer model systems consisting of a few components are generally studied. For investigation of the model biomembrane systems, X-ray (and neutron) diffraction is the most powerful tool for characterizing the topology and packing of the membrane and for following up its changes in environmental conditions, such as ionic strength, temperature and pressure. Several sample preparation techniques may be used for structural investigations [55, 56]. Lipid multi-bilayer films can be prepared and deposited on solid surfaces, such as glass slides, where they adopt a predominant orientation parallel to the bilayer surface. The slide, the level of hydration (relative humidity) of which has to be carefully controlled, is then aligned in an X-ray beam that hits the film tangentially. In the diffraction pattern, equidistant intensity spots on the meridian of the X-ray film exhibit that we have a one-dimensional periodicity along the direction perpendicular to the multilayer film. Off-axis to the equator, reflection spots appear, which originate from the packing of the lipid acyl chains. Instead of lipid films, also multilamellar vesicles in solution can be studied. The lipids form onion-like shaped spherical structures of stacked bilayers with an interlamellar water layer in between. The diffraction pattern then consists of concentric equidistant rings centered around the origin of the diffraction pattern. In the following, we focus on the solution scattering of multilamellar lipid systems. Regarding the diffraction of X-rays or neutrons, the lipid-water dispersions are then equivalent to powder samples that are composed of many randomly oriented microcrystals. Thus, Bragg's condition is automatically fulfilled, and all possible diffraction peaks are simultaneously recorded (Table 1). While the positions of the diffraction peaks are related to periodic distances within the lyotropic lipid mesophase, their sharpness or width reflects the extent of this periodicity over large distances. The measured reciprocal spacings are given by

$$s = \frac{2}{\lambda}\sin\theta \tag{2}$$

(2θ scattering angle, λ wavelength of radiation). If a lipid-water phase is lacking any periodic structure, diffuse small-angle scattering is observed only.

Lamellar lipid-water mesophases (denoted as L or P) form alternating layers of lipid and water molecules. This quasi one-dimensional periodic structure exhibits diffraction patterns in the small-angle regime that are described by the equation

$$s_n = n\frac{1}{d} \tag{3}$$

where $n = 1, 2, 3, \ldots$ and d is the lamellar repeat distance of this one-dimensional lattice, which is the thickness of the lipid bilayer plus that of the adjacent water layer.

Non-lamellar lipid mesophases (Fig. 4) may also be identified by their characteristic small-angle diffraction pattern. The structure of the inverse hexagonal lipid-water mesophase (denoted as H_{II}) is based on cylindrical water rods, which are surrounded by lipid monolayers. The rods are packed in a two-dimensional hexagonal lattice with Bragg peaks positioned at

$$s = \frac{2}{\sqrt{3}a}\sqrt{h^2 + k^2 + hk}\,. \tag{4}$$

The lattice constant a is here the distance between the centers of two neighboring rods and h, k are the Miller indices. The hexagonal lipid phases are easily distinguished from lamellar phases by their ratio of Bragg peak positions, which is $1:\sqrt{3}:2:\ldots$.

Bragg peaks of cubic lipid structures may be observed at

$$s = \frac{2}{a}\sqrt{h^2 + k^2 + l^2} \tag{5}$$

where a is the cubic lattice constant (Table 1). The Miller indices h, k, l depend on the lattice type (primitive, body-centered, face-centered) and the symmetry elements of the cubic structure. In Table 1 the relative Bragg peak positions of a variety of lipid mesophases including cubic phases are summarized.

Table 1. Miller indices *(hkl)* und ratio of Bragg peak positions of cubic (Q) and lamellar (L) lipid structures.

$P4_332$ (Q^{212})	Pm3n (Q^{223})	Pn3m (Q^{224})	Fm3m (Q^{225})	Fd3m (Q^{227})	Im3m (Q^{229})	Ia3d (Q^{230})	L	ratio
							100	1
110	110	110			110			$\sqrt{2}$
111		111	111	111				$\sqrt{3}$
	200	200	200		200		200	2
210	210							$\sqrt{5}$
211	211	211			211	211		$\sqrt{6}$
220	220	220	220	220	220	220		$\sqrt{8}$
221		221					300	3
310	310	310			310			$\sqrt{10}$
311		311	311	311				$\sqrt{11}$
222	222	222	222	222	222			$\sqrt{12}$
320	320							$\sqrt{13}$
321	321	321			321	321		$\sqrt{14}$
400	400	400	400	400	400	400	400	4
410	410	410						$\sqrt{17}$
411	411	411			411			$\sqrt{18}$
331		331	331	331				$\sqrt{19}$
420	420	420	420		420	420		$\sqrt{20}$
421	421	421						$\sqrt{21}$
332		332			332	332		$\sqrt{22}$

The Bragg peaks appearing in addition in the wide-angle scattering region are related to the packing of the lipid acyl chains in a monolayer. Generally, this packing can be described as a centered rectangular lattice with lattice constants a_{rec} and b_{rec}, which are calculated from the observed d-spacings using

$$a_{rec} = 2d_{20}, \; b_{rec} = \frac{d_{11}}{\sqrt{1-[d_{11}/(2d_{20})]^2}}. \tag{6}$$

When the lipid chains are rigidly packed in a hexagonal lattice, only a single wide-angle Bragg peak is usually observed. Then, the lattice constant (chain-chain distance) may be obtained from Eq. 4. In the case of fluid-like disordered lipid layers, only a very broad wide-angle peak is observed around $s \approx 0.25$ Å^{-1}.

3.1.2. SAXS studies by solutions of biological macromolecules (proteins)

Details of the experimental techniques for studying dilute protein solutions are discussed elsewhere [47-49,57]. Briefly, the SAXS experiments were performed at synchrotron X-ray beamlines, such as ESRF or APS. The scattering data are presented as a function of momentum transfer $Q = (4\pi/\lambda)\sin\theta$ (2θ scattering angle, λ wavelength of radiation). The overall conformation of the protein can be obtained by measuring its radius of gyration R_g and pair distance distribution function $p(r)$ [57-60]. The pair distance distribution function $p(r)$, which depends on the molecular particle shape and on the intra-particle scattering distribution, is given by the indirect Fourier transform of the measured scattered intensity. For a particle of uniform electron density, it is given by

$$p(r) = \frac{1}{2\pi^2}\int_0^{\infty} I(Q)Qr\sin(Qr)\mathrm{d}Q\,. \tag{7}$$

The function $p(r)$ represents the frequency of vector length r connecting small volume elements within the volume of the scattering particle, that is, the protein molecule, with maximum dimension D_{max}. As the use of the Guinier approximation for the determination of the radius of gyration, R_g, might lead to errors caused by concentration or aggregation effects, we used the normalized second moment of the pair distance distribution function for calculation of R_g:

$$R_g^2 = \frac{\int_0^{D_{max}} p(r)r^2\mathrm{d}r}{2\int_0^{D_{max}} p(r)\mathrm{d}r} \tag{8}$$

3.2. High pressure FT-IR spectroscopy of proteins

High pressure FT-IR spectra were recorded with a Nicolet Magna 550 spectrometer equipped with a liquid nitrogen cooled HgCdTe-detector [48-50]. Powdered α-quartz was placed in the hole of the steel gasket of a diamond anvil cell (Fig. 2c) and changes in pressure were quantified by the shift of the quartz phonon band at 695 cm^{-1}. The protein was dissolved in D_2O containing pressure-insensitive bis-Tris buffer. To ensure complete H/D exchange, the protein solution was heated up to a temperature close to its denaturation temperature for one hour and then stored overnight at room temperature. Fourier self-deconvolution was performed with a resolution enhancement factor of 1.8 and a bandwidth of 15 cm^{-1}. The fractional intensities of the secondary structure elements were calculated from a bandfitting procedure assuming Gaussian-Lorentzian line shape functions.

3.3. High-pressure NMR spectroscopy

Advances in superconducting magnet technology in the last decades have resulted in the development of superconducting magnets capable of attaining a high homogeneity of the magnetic field over the sample volume, so that even without sample spinning, high resolution can still be achieved. The ability to record high-resolution NMR spectra on dilute spin systems opened a new field of high-pressure NMR spectroscopy which deals with pressure effects also on biochemical systems. The literature on the techniques and the equipment available for experiments at high pressure has significantly increased in recent years. Comprehensive reviews can be found in refs.[10, 17, 61-63]. Generally, the pressure is generated outside the NMR probe. Pressure is transferred from the pressure-generating equipment via a capillary tube to the sample inside the probe by a pressurizing medium. The main experimental obstacle for performing NMR experiments at high pressure is the construction of reliable high-pressure probes. Essentially two approaches have emerged for the measurement of NMR spectra at high hydrostatic pressure [10, 17, 61-63]. One approach is a pressure sample tube (strengthened glass cells, made from borosilicate, quartz capillaries or sapphire) which can be placed directly in a standard spectrometer probe. Such systems give good resolution but often suffer from a good temperature control and a limited pressure range of about 2-3 kbar. The alternative approach is the manufacture of a dedicated high-pressure probe head which replaces the ambient pressure commercial probe head. Both, radio frequency (rf) coil and sample are immersed in the high-pressure fluid and the coil is wound on a support placed closely around the sample tube. Here, the sample cannot be made to spin, so in order to obtain good resolution from the static sample, the tube diameter has to be kept small. The system we have been using is of the latter type and described in the following. The material for any high-pressure NMR vessel must be nonmagnetic and of high mechanical strength. The most commonly used materials for this purpose are beryllium copper (Berylco) and titanium alloys which allow building pressure vessels up to pressures of about 8-10 kbar. In such an autoclave-style NMR vessel, rf-pulses enter the pressure vessel through electrical feedthroughs. Since the tuning–matching capacitor network is typically located outside the pressure vessel, the feedthroughs become part of the rf-circuit. The type, as well as the shape of the sample cell, i.e., a piston-based design or a bellow system, is related to the type of the experiment performed. Figure 3 shows a schematic drawing of one type of vessel we have been using in our laboratory for ^{1}H and ^{2}H NMR measurements. The high-pressure probe was constructed to fit into the room temperature bore of a wide-bore superconducting magnet, $\varnothing$ = 72 mm, operating at a magnetic field of 9.4 T. The pressure vessel is made of a titanium alloy, Ti-6Al-4V, and has an external diameter of 64 mm, an inner diameter of 18 mm and an inner length of 140 mm. A recess is cut in the outside of the vessel for the circulation of the thermostating liquid (an ethylene glycol/water mixture). It is jacketed with a brass cylinder with an outer diameter of 70 mm. On top of the pressure vessel are three threaded connections, one for pressurization, the others for inlet and outlet of the thermostating liquid. The high-pressure plug at the bottom of the vessel is sealed with an O-ring made of Viton. For pressures above ~3 kbar, Bridgman seals or c-seals may be used to contain the pressure. The plug has two threaded connections for the electrical feedthroughs and a drilling for a thermocouple which measures the temperature close to the sample. At the bottom plug driver, the rf-box is attached, containing the tuning and matching circuit, e.g. for a ^{1}H (400 MHz) observation frequency and a ^{2}H (61.4 MHz) field lock. Both resonance frequencies can be adjusted using screwdrivers from outside the magnet. The high-pressure electrical feedthroughs consist of a low capacitance signal transmission cable which is sealed with a steel cone soldered to the metal surface or of Berylco 25 which is electrically isolated by a polymer material. The internal sample holder contains a two turned saddle-shaped rf-coil, home-made from silver-coated copper wire, and the sample tube, made of a shortened commercial NMR glass tube of outer diameter 5 mm, which is fused to a precision glass tube (inner diameter 8.00 mm, wall thickness 1.6 mm) which separates the sample from the pressurizing liquid by a movable Tekapeek piston, sealed with two Viton O-rings. The pressure is generated by a standard system consisting of a hand pump and valves and a Heise-Bourdon gauge using liquid CS_2, tetrachloroethylene or methylcylcohexane as pressurizing medium for proton or deuteron studies, respectively. The current apparatus is capable of temperature control in a range from −20 up to 80 °C with a maximum working pressure of 3000 bar (6000 bar can be reached after special treatment of the Ti-alloy).

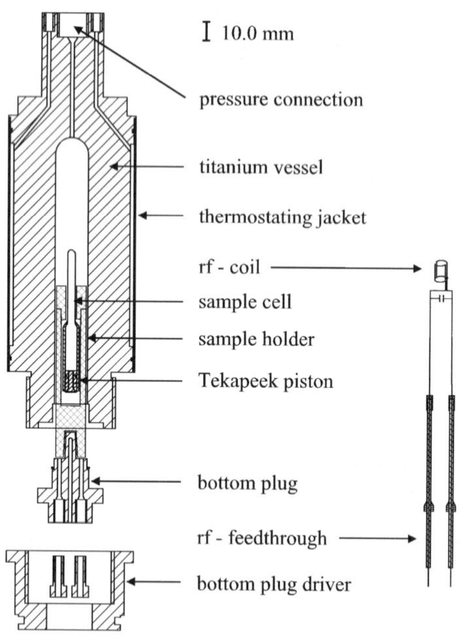

Fig. 3. Schematic drawing of the high pressure NMR autoclave. [62, 63]

The ^{2}H-NMR experiments were performed using the quadrupolar echo pulse sequence $\pi/2_x$-τ_1-$\pi/2_y$-τ_2-acquisition with phase-cycling and quadrature detection. A Bruker MSL 400 spectrometer was used for the high pressure studies operating at a resonance frequency of 61.4 MHz. In the liquid-crystalline phase, perdeuterated lipids display ^{2}H-NMR spectra, which are superpositions of axially symmetric quadrupolar powder patterns of all C–D bonds. From the sharp edges, the quadrupolar splittings

$$\Delta\omega_Q = \frac{3}{4}\frac{e^2qQ}{h} S_{CD} \tag{9}$$

can be obtained, where e^2qQ/h (the static quadrupole coupling constant) for ^{2}H(D) in C–D bonds is about 167 kHz, and the C–D bond order parameter S_{CD} for distinguishable deuterons can be calculated, which is defined by

$$S_{CD} = \frac{1}{2}\langle 3\cos^2\theta - 1\rangle . \tag{10}$$

Here θ is the instantaneous angle between a given C–D bond vector and the axis of rotational symmetry of the molecules, i.e., the bilayer normal. The brackets denote an average over the time scale of the experiment (~10^{-5} s) so that S_{CD} is the time-averaged orientation of the particular C–D bond with respect to the bilayer normal. Furthermore, in all phases studied the first spectral moment M_1 of the ^{2}H NMR spectra can be calculated and the weighted mean splitting of the ^{2}H NMR spectrum can be obtained, which is proportional to the average chain orientational order parameter of the lipid, using:

$$M_1 = \frac{\int_0^{\infty} \omega \cdot f(\omega)\mathrm{d}\omega}{\int_0^{\infty} f(\omega)\mathrm{d}\omega} = \frac{4\pi}{3\sqrt{3}}\langle\Delta\omega_Q\rangle \propto \langle S_{CD}\rangle \tag{11}$$

where $f(\omega)$ is the deuteron spectrum.

4. PRESSURE EFFECTS ON BIOMOLECULAR SYSTEMS

4.1. Pressure effects on nucleic acids

Due to the stabilizing effect of HHP on DNA hydrogen bonds, the duplex to single strand transition temperature (melting temperature, T_M) increases under pressure [3]. In other words, HHP increases the resistance of double-stranded and triple-stranded DNA to denaturation into single strands [64, 65]. The double-helix structure of DNA generally does not unwind until the applied pressure exceeds 10 kbar. Stacking interactions, which have been shown to produce a negative volume change, are also a cause of stabilization of the double helix by HHP and, consequently, of the increase in T_M. In a recent paper by Dubins et al. [66], the observation was made that the effect of pressure on the stability of nucleic acid complexes strongly depends on its melting temperature T_M and must always be defined in the context of the solution ionic strength and a specific pressure-temperature domain. Nucleic acid duplexes within T_M values below ~50 °C are destabilized by pressure, while those – which is generally the case – with higher T_M values are stabilized by pressure.

High hydrostatic pressure has also been shown to induce the B- to Z-form transition of DNA double helices. The B-form has a larger molecular volume than in Z-form because it contains more water, and because this water is also structured. Only little information is available about RNA, which seem to be rather pressure stable as well. [67]

4.2. Lipid mesophases and model biomembrane systems upon pressurization

Lyotropic lipid mesophases are organized biomolecular systems formed by amphiphilic molecules, such as phospholipids, in the presence of water. They exhibit a rich structural polymorphism, depending on their molecular structure and environmental conditions, such as water content, pH, ionic strength, temperature and pressure [10, 11, 68-70, 30-46]. The basic structural element of biological membranes consists of a lamellar phospholipid bilayer matrix (see Fig. 4). In the lamellar structure, the interfaces are flat and are periodically stacked in multilamellar vesicles. Due to the large hydrophobic effect, most phospholipids associate in water even at very low concentrations ($< 10^{-12}$ mol L^{-1}). Not only the entire biomembrane is very complex, containing a variety of different lipid molecules and a host of proteins performing versatile biochemical functions, but also the simplest lipid bilayer consisting of only one or two kinds of lipid molecules is already very complex by the standards of what is understood about the statistical mechanics of phases and phase transitions.

Saturated phospholipids often exhibit two thermotropic lamellar phase transitions, a gel-to-gel ($L_{\beta'}/P_{\beta'}$) pretransition and a gel-to-liquid-crystalline ($P_{\beta'}/L_\alpha$) main transition at a higher temperature T_m (for the structures see Fig. 4). In the fluid-like L_α-phase, the acyl chains of the lipid bilayers are conformationally disordered, whereas in the gel phases, the chains are more extended and ordered. Because the average end-to-end distance of disordered hydrocarbon chains in the L_α-phase is smaller than that of ordered (all-trans) chains, the bilayer becomes thinner during melting at the $P_{\beta'}/L_\alpha$-transition, even though the lipid volume increases. In addition to these thermotropic phase transitions, pressure-induced phase transformations have also been observed [10,11,30-46]. Upon compression, the lipids adopt to volume restriction by changing their conformation and packing. Lipid systems are these biological systems which are among the most pressure sensitive and in general they easily undergo phase transformations under changes of environmental conditions.

It is now well known that many biological lipid molecules also form nonlamellar liquid-crystalline phases (see Fig. 4) [70-73]. Lipids, which can adopt a hexagonal phase, are present at substantial levels in biological membranes, usually with at least 30 mol% of the total lipids. Some lipid extracts, such as those from archaebacteria (*S. solfataricus*), also exhibit cubic liquid-crystalline phases [73]. It is generally assumed that the nonlamellar lipid structures, such as the inverse hexagonal (H_{II}) and cubic (Q_{II}) lipid phases, are also of signifiant biological relevance. Fundamental cell processes, such as endo- and exocytosis, fat digestion, membrane budding and fusion, involve a rearrangement of biological membranes where nonlamellar lipid structures are probably involved. Also static cubic structures (cubic membranes) might occur in biological cells [73-75].

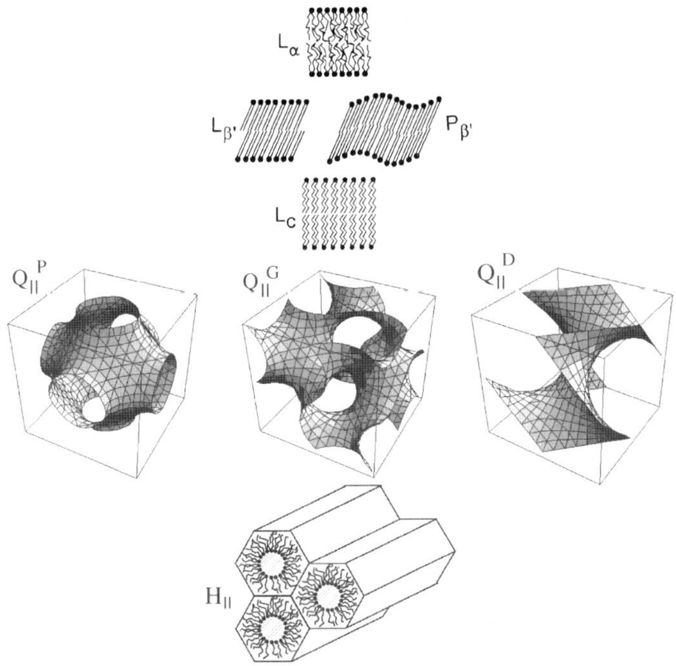

Fig. 4. Schematic drawing of lipid-water mesophases (L_c, lamellar crystalline; $L_{\beta'}$, $P_{\beta'}$, lamellar gel; L_α, lamellar liquid-crystalline; Q_{II}^G, Q_{II}^D, Q_{II}^P, inverse bicontinuous cubics; H_{II}, inverse hexagonal). The cubic phases are represented by the G, D, and P minimal surfaces, which locate the midplanes of fluid lipid bilayers.

At a macroscopic level, lipid cubic phases are very viscous and optically isotropic. They can be sorted in two main classes: bicontinuous and micellar. In the first class the amphiphiles build bicontinuous cubic phases of type II (Q_{II}), which can be visualized in terms of highly convoluted lipid bilayers, which subdivide three-dimensional space into two disjointed polar labyrinths separated by an apolar septum. The cubic symmetries most commonly observed are Ia3d, Pn3m and Im3m. The structures of these phases, Q_{II}^G, Q_{II}^D and Q_{II}^P, are closely related to the Schoen Gyroid (G), the Schwarz D and the Schwarz P infinite periodic miminal surfaces (IPMS). An IPMS is an intersection-free surface periodic in three dimensions with a mean curvature that is everywhere zero. The surface, that sits at the lipid bilayer midplane, thus separates the two interpenetrating but not connected water networks. In the second class of cubic structures, the amphiphiles build an infinite number of finite aggregates, lipid globules or micelles, organized in regular packings of various symmetries [76].

Generally, the molecular organization within a lipid aggregate can be understood in terms of a balance of attractive and repulsive forces acting at the level of the lipid polar headgroups and non-polar acyl chains (Fig. 5). Within the headgroup (lipid/water) region there is an effectively attractive force $F_{\text{lipid/water}}$, which arises from the unfavorable contact of the hydrocarbon chains with water (the hydrophobic effect); attractive contributions from hydrogen bonding, as in the case of phosphatidylethanolamines, may also be present. The repulsive headgroup pressure F_{head} is the result of hydrational, steric, and – for charged headgroups – electrostatic contributions. For the acyl chain region, the attractive van der Waals interactions among the CH_2-groups are compensated by the repulsive lateral pressure, F_{chain}, owing to entropy-driven thermally activated dihedral angle isomerizations. In a membrane bilayer at equilibrium, the various lateral pressures are balanced. An imbalance of attractive and repulsive forces at the level of the headgroup and acyl chains within a given lipid monolayer yields a spontaneous curvature. Clearly, the two monolayers cannot be simultaneously at a free energy minimum with regards to their intrinsic or spontaneous curvature. It follows that a symmetrical bilayer composed of nonlamellar forming lipids is under a condition of curvature elastic stress. This curvature free energy associated with the membrane lipid-water interface then leads to the formation of nonlamellar (cubic and/or H_{II}) phases. The spontaneous curvature increases, for instance, if the lateral chain pressure increases due to an increase of trans/gauche isomerisations of the acyl chains at high temperatures, or if the level of headgroup hydration decreases (e.g., due to Ca^{2+} adsorption at the polar/apolar interface).

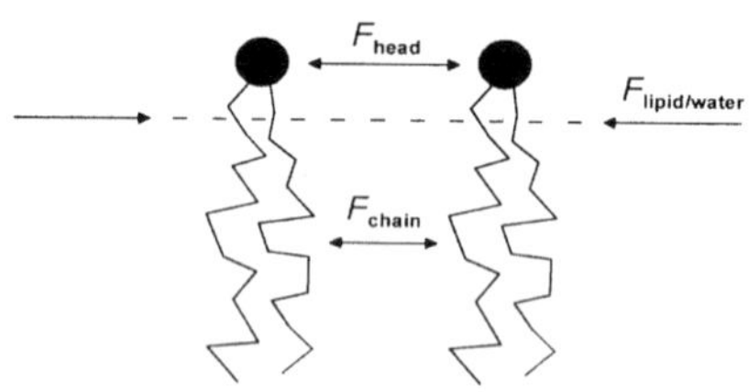

Fig. 5. Illustration of the balance of lateral forces across a lipid monolayer.

Until now, no full theoretical description of the lyotropic lipid phase behavior exists, though some progress has been made in recent years [77-84]. Often a concept is used that can be explained in terms of a small set of parameters, irrespective of the precise chemical nature of the lipid molecule. Following Helfrich (1978), the surface curvature energy contribution associated with amphiphile films is described in terms of three curvature elastic parameters: the spontaneous mean curvature H_s, the mean curvature modulus κ_m, and κ_G, the Gaussian curvature modulus. Applying differential geometry, the surface energy per unit area for small curvatures is given by

$$g_{\text{bend}} = 2\kappa_{\text{m}}\left\langle (H - H_{\text{s}})^2 \right\rangle + \kappa_{\text{G}}\left\langle K \right\rangle. \tag{12}$$

$H = (C_1 + C_2)/2$ is the mean interfacial curvature, which is equal to half the sum of the principal curvatures $C_1 = 1/R_1$ and $C_2 = 1/R_2$ at the interface, and $K = C_1C_2$ is the Gaussian curvature at the interface, given by the product of the principal curvatures C_1 and C_2 of the interface (e.g., $C_1 = C_2 = 0$ for a planar bilayer; $C_1 < 0$, $C_2 = -C_1 > 0$, i.e., $H = 0$, $K < 0$, for a hyperbolic saddle surface). The spontaneous mean curvature H_s is the mean curvature the lipid aggregate would wish to adopt in the absence of external constraints, and κ_m tells us what energetic cost there would be for deviations away from this. Besides the curvature energetic contribution, there will be other energetic contributions. Due to the desire to fill all the hydrophobic volume by the amphiphile chains (due to the hydrophobic effect), there will be a contribution quantifying an eventual packing frustration. A further, but minor, contribution is due to interlamellar interactions. The curvature elastic energy is believed to be the crucial term governing the stability of nonlamellar phases and the ability of lipid membranes to bend, in particular at high levels of hydration. To probe the concept of any energetic description and the resultant set of parameters necessary to provide a general explanation of a universal lyotropic phase behavior, one needs to scan the appropriate parameter space experimentally. To this end, pressure dependent studies have proven to be a very valuable tool.

4.2.1. Single-component lipid systems

Lamellar phase transitions. The increase in entropy with lipid chain rotational disorder, the increase in intermolecular entropy, and the increase in lipid headgroup hydration are the driving forces for the gel-fluid transition of lipid membranes (denoted as L_β-, $L_{\beta'}$- or $P_{\beta'}$-L_α transition, see Fig. 4). In a simple way, this transition is often interpreted as the melting of the lipid hydrocarbon chains. Opposing this chain melting is the increase of the internal energy due to increasing rotational isomerism, the decrease of the van der Waals-attraction between the hydrocarbon lipid chains, and the increase of the polar-apolar interface resulting from the lateral bilayer expansion due to increasing chain isomerism. The balance of all these contributions to the system free energy, which depends on the lipid molecular structure, determines the main (melting) transition temperature, T_m, of a lipid bilayer. Generally, the lamellar gel phases L_β, $L_{\beta'}$, and $P_{\beta'}$ prevail at high pressures and low temperatures, whereas the lamellar fluid (liquid-crystalline) phase L_α is found as the pressure is lowered and/or the temperature is raised (Fig. 6). The compressibility of the $P_{\beta'}$ gel phase is substantially lower than that of the liquid-crystalline phase (for $T = 30$ °C: $\kappa_T(P_{\beta'}) \approx 5\cdot10^{-5}\cdot\text{bar}^{-1}$ and $\kappa_T(L_\alpha) \approx 13\cdot10^{-5}\ \text{bar}^{-1}$). The main transition is accompanied by a well pronounced 3 % change in volume, which is mainly due to changes of the chain cross-sectional area, because chain disorder increases drastically at the transition. The volume change ΔV_m at the main transition decreases slightly with increasing temperature and pressure along the main transition line.

A common slope of ~22 °C/kbar has been observed for the gel-fluid phase boundary of the saturated phosphatidylcholines DMPC, DPPC, DSPC as shown in Fig. 6. Using the Clapeyron relation, $dT_m/dp = T_m\Delta V_m/\Delta H_m$, the positive slope can be explained by an endothermic enthalpy change, ΔH_m, and a volume increase, ΔV_m, for the gel-fluid transition, which have indeed been found in direct measurements of these thermodynamic properties. For example, for DMPC an enthalpy change of $\Delta H_m = 26$ kJ mol^{-1} and an entropy change of $\Delta S_m = 86$ J mol^{-1} K^{-1} has been determined. High pressure DTA experiments on this system revealed, that ΔH_m does not change significantly up to about 1.5 kbar. Similar transition slopes have been found for the mono-*cis*-unsaturated lipid POPC, the phosphatidylserine DMPS, and the phosphatidylethanolamine DPPE. Only the slopes of the di-*cis*-unsaturated lipids DOPC and DOPE have been found to be markedly smaller. While the lengths of the lipid hydrocarbon chains and the type (chemical structure) of the lipid headgroup do not affect the slope of the main transition significantly, these parameters determine the transition temperature. The two *cis*-double bonds of DOPC and DOPE lead to very low transition temperatures and slopes, presumably as they impose kinks in the linear conformations of the lipid acyl chains. Thus, significant free volume is created in the bilayer and the ordering effect of high pressure is reduced.

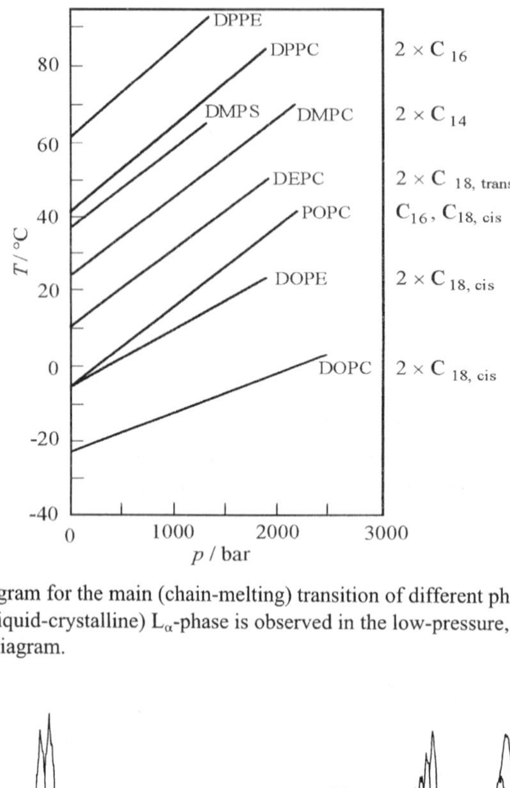

Fig. 6. *T,p*-phase diagram for the main (chain-melting) transition of different phospholipid bilayer systems. The fluid (liquid-crystalline) L_α-phase is observed in the low-pressure, high-temperature region of the phase diagram.

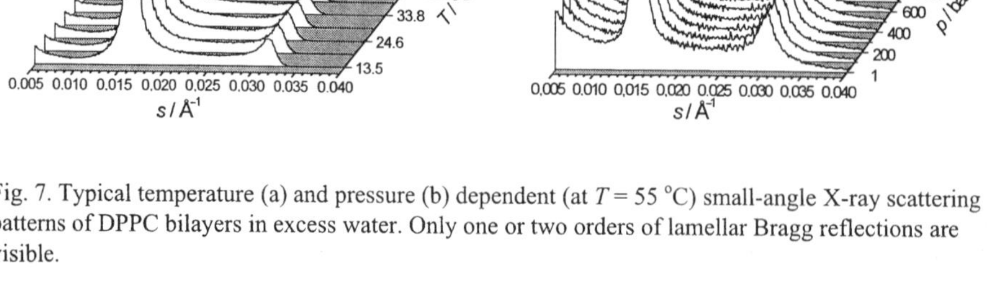

Fig. 7. Typical temperature (a) and pressure (b) dependent (at $T = 55$ °C) small-angle X-ray scattering patterns of DPPC bilayers in excess water. Only one or two orders of lamellar Bragg reflections are visible.

As an example, Fig. 7a shows small-angle diffraction data of a DPPC bilayer in excess water as a function of temperature. Clearly the pretransition as well as the main lipid phase transition as a relatively sharp shift of the Bragg peak positions are observed at about 35 and 42 °C, respectively. The lamellar lattice constant increases from ~63 Å in the $L_{\beta'}$-phase to ~72 Å in the ripple gel phase $P_{\beta'}$. Because of the highly disordered chains in the fluid L_{α}-phase, the bilayer thickness decreases to a lattice constant of about 66 Å. Figure 7b shows some pressure dependent data. In DPPC dispersions at 55 °C, a shift to lower scattering vectors together with a change in the lineshape is observed at 800 bar which is due to the pressure-induced L_{α} to $P_{\beta'}$ phase transition; the corresponding lattice constant increases from 68 to 71 Å. Further increase in pressure leads to a pressure-induced interdigitated gel phase, $L_{\beta i}$ around 1400 bar, where the lipid acyl chains from opposing monolayers partially interpenetrate, which leads to a decrease of the lamellar repeat period to about 50 Å. At around 2.8 kbar the transition to the Gel III phase occurs at this temperature with a lattice constant which is about 10 Å larger [44].

It is noted that applying high pressure can lead to the formation of additional gel phases, which are not observed under ambient pressure conditions, such as the interdigitated high pressure gel phase $L_{\beta i}$ found for phospholipid bilayers with acyl chain lengths $\geq C_{16}$ [31, 35, 44]. To illustrate this phase variety, the results of a detailed X-ray diffraction and FT-IR spectroscopy study of the *p*,*T*-phase diagram of DPPC in excess water is shown in Fig. 6. The structures of the L_{α}, $P_{\beta'}$ (Gel 1), $L_{\beta'}$ (Gel 2), Gel 3, Gel 4 and Gel 5 phase are illustrated schematically in Figs. 4, 8. In the Gel 5 phase, the multiamellar vesicle has lost essentially all the interlamellar hydrating water, which now coexists as bulk frozen water (ice VI).

Very little is known about the motions of lipid bilayers at elevated pressures. Of particular interest would be the effect of pressure on lateral diffusion, which is related to biological functions such as electron transport and some hormone-receptor interactions. Pressure effects on lateral diffusion of pure lipid molecules and of other membrane components have yet to be carefully studied, however. Figure 9 shows the pressure effects on the lateral self diffusion coefficient of sonicated DPPC and POPC vesicles [86]. The lateral diffusion coefficient of DPPC in the liquid-crystalline (LC) phase decreases, almost exponentially, with increasing pressure from 1 to 300 bar at 50 °C. A sharp decrease in the *D*-value occurs at the LC to GI phase transition pressure. From 500 bar to 800 bar in the GI phase, the values of the lateral diffusion coefficient ($\sim 1 \cdot 10^{-9}$ cm^2 s^{-1}) are approximately constant. There is another sharp decrease in the value of the lateral diffusion coefficient at the GI-Gi phase transition pressure. In the Gi phase, the values of the lateral diffusion coefficient ($\sim 1 \cdot 10^{-10}$ cm^2 s^{-1}) are again approximately constant.

The data presented in this chapter demonstrate that biological organisms could modulate the physical state of their membranes in response to changes in the external environment by regulating the fractions of the various lipids in a cell membrane differing in chain length, chain unsaturation or headgroup structure ("homeoviscous adaption"). However, nature has further means to regulate membrane fluidity, such as by changing the membrane concentrations of cholesterol (if present) or various proteins. In fact, many studies have demonstrated that membranes are significantly more fluid in barophilic and/or psychrophilic species. This is principally a consequence of an increase in the unsaturated/saturated lipid ratio. Interestingly, in both atmospheric and high-pressure adapted species, the plasma membrane appears to play a key role in the defense against pressure shocks. Transition between different local membrane structures, induced by physical changes in the cell environment, could act as signals to trigger transduction cascades ([3] and refs. therein).

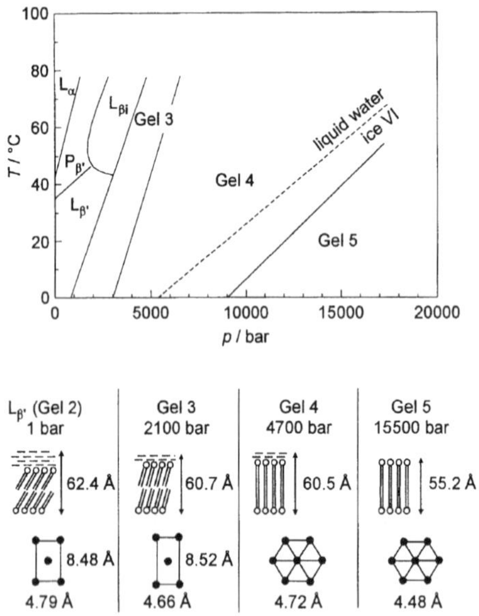

Fig. 8. *T,p*-phase diagram of DPPC bilayers in excess water and schematic drawing of the lamellar lattice constant and lipid packing in the bilayer plane of DPPC gel phases at 23 °C [44,85]. It is noteworthy that an additional crystalline gel phase (L_c) can be induced in the low-temperatue regime after prolonged cooling.

Nonlamellar phase transitions. For a series of lipid molecules, also nonlamellar lyotropic phases are observed as thermodynamically stable phases or as long-lived metastable phases after special sample treatment. We will discuss lipid-water systems with the lipids taken from different groups of amphiphilic molecules. Contrary to DOPC which shows a lamellar L_β-L_α transition only (Fig. 6), the corresponding lipid DOPE with ethanolamine as headgroup exhibits an additional phase transition from the lamellar L_α to the nonlamellar, inverse hexagonal H_{II} phase, when it is dispersed in water. As pressure forces a closer packing of the lipid chains, which results in a decreased number of *gauche* bonds and kinks in the chains, both transition temperatures of the L_β-L_α and the L_α-H_{II} transitions increase with increasing pressure. In Fig. 10a, the *T,p*-phase diagram of DOPE in excess water is displayed, which has been obtained by SAXS and WAXS experiments using a diamond anvil cell. The slope of the L_α-H_{II} transition is almost three times larger than that of the L_β-L_α transition; values of about 40 and 14 °C/kbar have been found, respectively. The L_α-H_{II} transition observed in DOPE/water and also in egg-PE/water (egg-PE is a mixture of different phosphatidylethanolamines) is the most pressure-sensitive lyotropic lipid phase transition found to date. The reason why this transition in DOPE has such a strong pressure dependence could be conjectured to be the strong pressure dependence of the chain length and volume of its unsaturated chains.

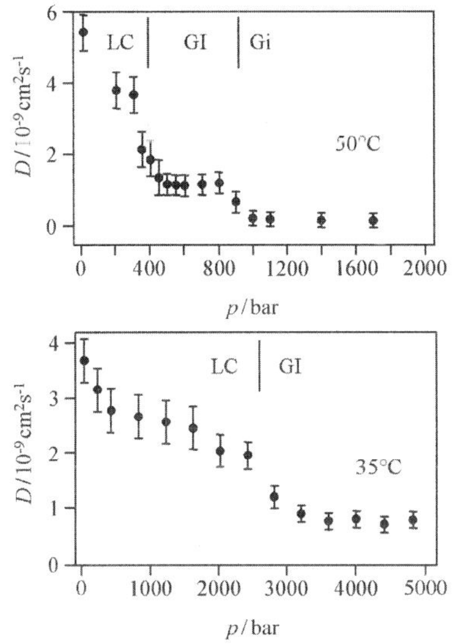

Fig. 9. Lateral self-diffusion constant D of DPPC (top) and POPC (bottom) in sonicated vesicles as a function of pressure at 50 °C and 35 °C, respectively (after ref. [86]).

Interestingly, in the DOPE/water system inverse cubic phases, Q_{II}^{D} and Q_{II}^{P}, can be induced in the region of the L_α-H_{II} transition by subjecting the sample to extensive temperature or pressure cycles at temperature and pressure conditions close to the phase transition [87, 88]. It has been shown that for conditions which favor an intermediate spontaneous curvature of a lipid monolayer, the topology of an inverse bicontinuous cubic phase can have a similar or even lower free energy than the lamellar or inverse hexagonal phase, as the cubic phases are characterized by a low curvature free energy and do not suffer the extreme chain packing stress predominant in the H_{II}-phase. Metastable cubic phases might be formed via defect structures, which occur in passing the L_α-H_{II} transition, such as interlamellar micellar intermediates (IMI) or stalks [89, 90]. Membrane fusion as a part of specific biological reactions probably also involves the formation of intermediates, such as stalks. The energy of these intermediates and consequently the rate and extent of the fusion depend on the propensity of the membrane monolayers to bend. To be best suited for fusion, lipid bilayer membranes should be asymmetrical with the contracting monolayer composed of H_{II}-phase-promoting lipids (e.g., cholesterol, PE's with a negative spontaneous curvature towards the water) and the expanding monolayer composed of micelle-forming lipids (e.g., lysolipids). The lipid composition of biological membranes is highly regulated and may be altered by enzymatic attacks. One of the possible functional roles of the transbilayer asymmetry will thus be the regulation of membrane fusion induced by a spontaneous monolayer curvature. However, fusion proteins may affect the energy of these proposed structures or perhaps give rise to other intermediate structures in the biological process of membrane fusion.

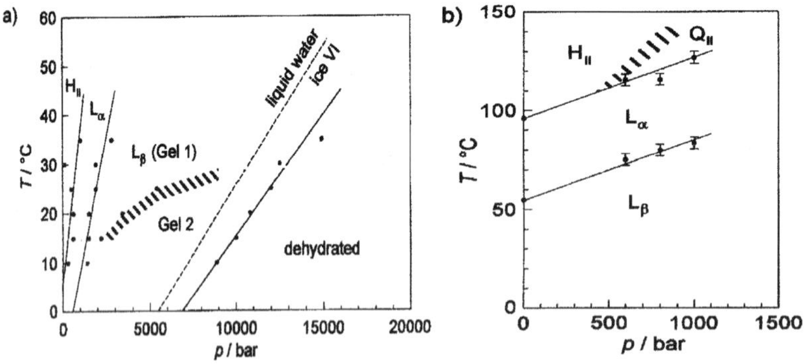

Fig. 10. *T,p*-phase diagram of a) DOPE and b) DTPE in excess water (the phase boundaries of cubic phases Q_{II} are somewhat tentative).

As found for DOPE/water, the thermotropic phase order of DTPE in excess water (Fig. 10b) is L_β, L_α, H_{II} at atmospheric pressure. However, at pressures higher than about 500 bar an additional, intermediate region of inverse cubic phases, Q_{II}^D and Q_{II}^P, is observed [46]. Although the transition region from the cubic phases to the inverse hexagonal phase cannot clearly be resolved, this finding indicates a triple point in the *T,p*-phase diagram and again illustrates the nonequivalence of the effects of temperature and pressure on lyotropic liquid-crystalline phases.

Two further examples of single-component lyotropic lipid systems exhibiting nonlamellar phases are discussed: ME and MO dispersed in excess water. ME and MO are intermediates in the fat digestion and differ only in the configuration of the double bond in their single acyl chain, which is *trans* in ME and *cis* in MO. In contrast to the preceding examples, ME and MO form spontaneously cubic phases over wide ranges of temperature and hydration [38,91,92]. The *T,p*-phase diagrams of ME and MO in excess water are presented in Fig. 11. As can clearly be seen, the single change in the acyl chain double bond configuration, from *trans* (ME) to *cis* (MO), causes a dramatic change in the observed phase behavior. In the system MO-water, the cubic Q_{II}^D-phase is stable over wide ranges of temperature and pressure. The *cis* configuration of MO leads to a more wedge-like molecular shape and a strong tendency for a MO monolayer to curve toward the water. Hence, the formation of lamellar phases, which requires a cylindrical molecular shape, is disfavored. Analysis of the infrared CH_2 stretching and wagging modes for evaluation of conformational states in the various disordered (L_α, Q_{II}^P, Q_{II}^D) phases of ME reveals different populations of *gauche* conformers and kinks in these fluid-like phases [92]. From the analysis of the carbonyl stretching mode vibration also small differences in the level of hydration of different bicontinuous cubic phases have been detected. Compared to the Q_{II}^D phase of ME, the lipid chains of the body centered cubic phase Q_{II}^{P} seem to contain a slightly higher population of *gauche* sequences, and a slightly lower level of hydration of the carbonyl group.

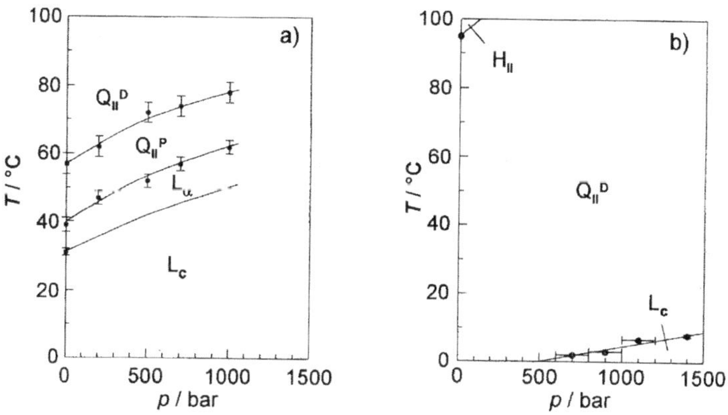

Fig. 11. *T,p*-phase diagram of a) ME and b) MO in excess water.

4.2.2. More-component lipid mixtures

Fatty acids/phosphatidylcholines. The addition of fatty acids to aqueous phosphatidylcholine dispersions changes drastically the observed *T,p*-phase behavior. Dispersions of pure phosphatidylcholines merely exhibit lamellar phases. Nonlamellar, inverse hexagonal and inverse cubic phases can be induced by adding fatty acids, such as lauric acid (LA), myristic acid (MA), palmitic acid (PA), or stearic acid (SA). Fatty acids influence also the fusogenicity of biological membranes, because they relieve the formation of nonlamellar intermediate structures, which have to occur in the process of membrane fusion. The gel to fluid-phase transition temperature of a DMPC/MA(1:2) mixture is 25 °C above the main transition temperature, T_m, of pure DMPC dispersions, which is 23.9 °C. The reason for this observation is that the fatty acid molecules act as spacers between the lecithin molecules and reduce the steric repulsion between the relatively bulky lecithin headgroups. The resulting change in the lateral pressure profile across a monolayer results in fluid nonlamellar phases being energetically favored over the fluid lamellar L_α-phase at temperatures above the chain melting temperature. However, the lamellar L_α-phase has been found in the system DLPC/LA and under nonequilibrium conditions after pressure jumps from the gel (L_β) to the fluid (Q_{II}, H_{II}) phase region. In the liquid-crystalline phase region of DMPC/MA(1:2), DPPC/PA(1:2), and DSPC/SA(1:2) mixtures, no liquid-crystalline lamellar phase is observed under equilibrium conditions, and the lamellar gel phase directly transforms to nonlamellar liquid-crystalline phases. In Fig. 12, the *T,p*-phase diagrams of the 1:2-mixtures DLPC/LA, DMPC/MA, DPPC/PA, and DSPC/SA in excess water are presented [93]. In the latter three systems, a phase separation occurs at low temperatures, leading to a lamellar crystalline phase L_c composed of the pure fatty acid and a lamellar crystalline phase L_c^{com} being a mixture of the fatty acid and the phosphatidylcholine lipid. At temperatures above the chain melting, the inverse hexagonal phase H_{II} is found to be the stable liquid-crystalline phase in DSPC/SA(1:2) and DPPC/PA(1:2) dispersions, whereas this phase coexists with cubic phases in the case of DMPC/MA(1:2) and is even replaced by the pure cubic Q_{II}^{P}-phase over a limited temperature interval in the DLPC/LA(1:2) system. This observed trend in the phase behavior is largely a result of the decreasing lipid chain length from DSPC/SA to DLPC/LA. The large splay of the fluid acyl chains in the DSPC/SA and DPPC/PA systems leads to a large spontaneous negative curvature of the monolayers that can only be adopted by the structure of the H_{II}-phase. On the contrary, the shorter lipid chains in the DMPC/MA and DLPC/LA systems increase the chain packing stress of the inverse hexagonal phase so that the formation of cubic phases is favored.

It has been shown that the tendency for interfacial curvature can be reduced dramatically by a decreased fatty acid fraction in the lecithin/fatty acid mixtures rather than by an increase in pressure [93]. The marked differences between the effects of pressure and monolayer composition on the phase behavior of lecithin/fatty acid mixtures reflect the fact that compositional variations cause large changes in the lateral pressure between amphiphiles, whereas hydrostatic pressure does not. Hence, pressure provides an extremely fine resolution parameter for probing the stability and geometry of lyotropic lipid mesophases.

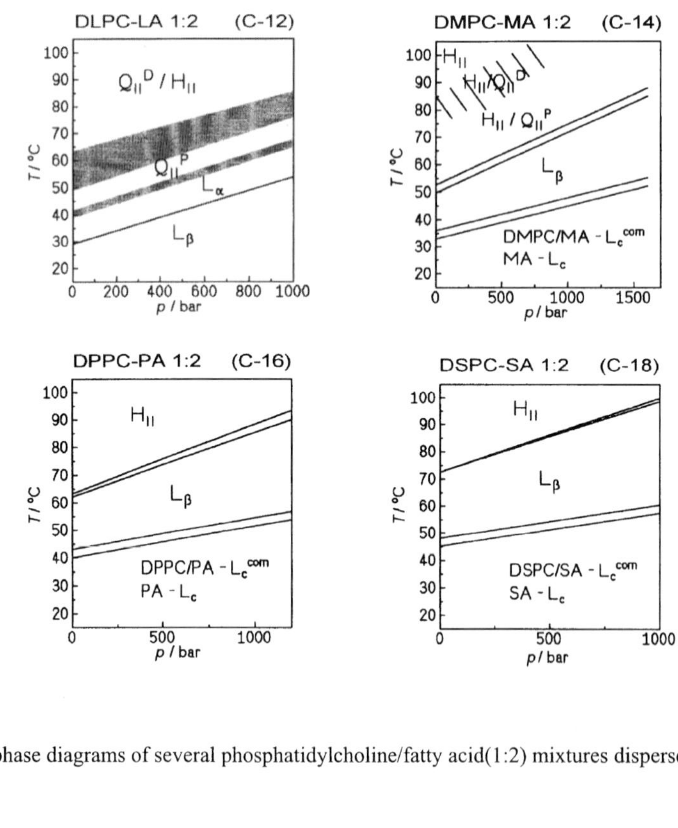

Fig. 12. *T,p*-phase diagrams of several phosphatidylcholine/fatty acid(1:2) mixtures dispersed in excess water.

Phase-separated phospholipid mixtures. Phase diagrams of binary mixtures of saturated phosphatidylcholine lipids are typically characterized by a lamellar gel phase at low temperatures, a lamellar fluid phase at high temperatures, and an intermediate fluid-gel coexistence region, i.e., a phase separation region. Small-angle neutron scattering (SANS) experiments using multilamellar vesicles have shown the presence of large-scale fractal structures of coexisting gel and fluid regions in binary lipid membranes [94]. The mixtures investigated in these SANS studies were DMPC/DPPC(1:1), DMPC/DSPC(1:1), and DLPC/DSPC(1:1) in excess water. The mismatch, *m*, of the acyl chain lengths of the two components, defined as the ratio (difference in C-atom number)/(C-atom number of shorter chain), is increasing from 2/14, to 4/14, to 7/14, respectively, and is the origin for the appearance of the corresponding temperature-composition phase diagrams, as shown in Fig. 13. The narrow fluid-gel coexistence region in the DMPC/DPPC system indicates a nearly ideal mixing behavior of the two components (isomorphous system). In comparison, the coexistence region in the DMPC/DSPC system is broader and manifests pronounced deviations from ideality. In both systems, the lipids are completely miscible in the all-gel and all-fluid phase. The strongest deviation from ideality occurs in mixing DLPC and DSPC, where the acyl chain mismatch is so large that the components are essentially immiscible in the gel phase at low temperatures (peritectic system). As the temperature is raised and the three-phase line is crossed, e.g., at mole fraction $x = 0.5$, the gel$_1$ phase (consisting mainly of DLPC) melts and lipid bilayers with fluid and gel$_2$ regions form.

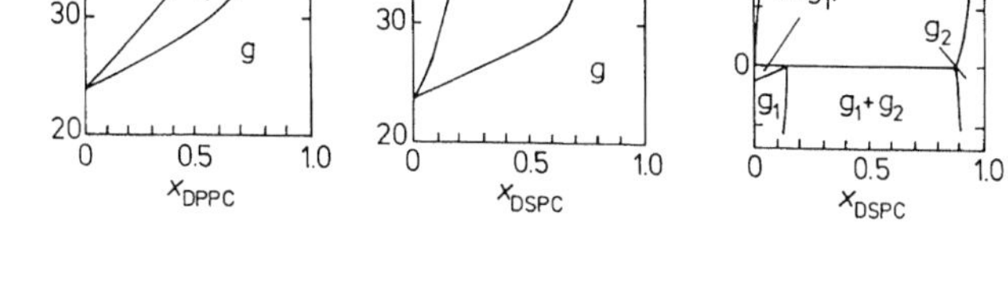

Fig. 13. Temperature-composition phase diagrams of the binary mixtures a) DMPC/DPPC (di-C_{14}/di-C_{16}), b) DMPC/DSPC (di-C_{14}/di-C_{18}), and c) DLPC/DSPC (di-C_{12}/di-C_{18}) dispersed in excess water (g, gel phase; f, fluid phase; *x*, mole fraction).

In order to observe the concentration fluctuations caused by the gel-fluid phase coexistence in the above-mentioned binary phospholipid mixtures, SANS studies in combination with the H/D substitution technique were performed. Under these so-called matching conditions, no SANS signal is obtained for homogeneously mixed lipids in the all-gel or all-fluid phases, since then the scattering length density is constant over the whole sample. However, in the case of gel-fluid phase heterogeneities, SANS occurs due to the different compositions and scattering length densities of the two phases. As an illustrative example, several SANS curves of an equimolar DMPC-d_{54}/DSPC mixture in excess water are plotted as a function of temperature in Fig. 14. By comparison with the corresponding phase diagram, one clearly sees that significant SANS occurs within the gel-fluid coexistence region only. The SANS intensity within the coexistence region may be analyzed by plotting $\ln(d\Sigma/d\Omega)$ *vs.* $\ln Q$. The obtained straight lines over the whole Q-range studied are indicative of a fractal-like structure of the sample. In the case of surface fractals, which are scatterers having a fractal surface only, the scattering law is given by $(d\Sigma/d\Omega) \sim Q^{D_s-6}$, with D_s being the surface-fractal dimension ($2 < D_s < 3$). If the scatterers in the sample are mass fractals, the SANS intensity is described by $(d\Sigma/d\Omega) \sim Q^{D_m}$, with D_m being the mass-fractal dimension ($0 < D_m < 3$). From the double-logarithmic plots of Fig. 14 at 1 bar a slope of -3.3 ± 0.1 is obtained which yields a surface-fractal dimension of $D_s = 2.7 \pm 0.1$. Two-photon excited fluorescence microscopy studies showed that these heterogeneous structures extend even up to the μm length scale [95].

The results obtained imply that the real membrane structure in the gel-fluid coexistence region of binary phosphatidylcholine mixtures deviates strongly from the simple structure of large gel and fluid domains separated by smooth boundaries, which is expected from the equilibrium phase diagrams of the lipid mixtures and which is generally observed for macroscopically large binary fluid mixtures. The heterogeneous membrane structures observed in the two-phase coexistence regions might be due to interfacial wetting effects, i.e., the interface between coexisting gel and fluid phase domains is found to be enriched by one of the lipid species leading to a decrease of the interfacial tension and hence to a stabilization of nonequilibrium lipid domains. These and further results for similar binary lipid systems suggest that such heterogeneous and fractal-like domain morphologies might be a rather common phenomenon. Depending on the acyl chain mismatch of the lipid components, the clusters scatter like surface or mass fractals implying that gel and fluid domains are correlated across many bilayers in a vesicle, and that segregation into a minority and a majority phase occurs. In a phenomenological way, it is interesting to point out that with increasing nonideality in the mixing behavior of the two lipid components the fractal dimension is decreasing and switching from the surface to the mass type.

With increasing pressure, the gel-fluid coexistence region of a DMPC/DSPC mixture is shifted toward higher temperatures (Fig. 15) [96]. A shift of about 22 °C/kbar is observed, similar to the slope of the gel-fluid transition line of the pure lipid components (Fig. 6). We noted that the mixing behavior of DMPC and DSPC deviates even more from ideality at 1000 bar than at ambient pressure, as can be inferred from the significant small-angle scattering intensity observed at 1000 bar and 21 °C, i.e., below the gel-fluid coexistence region.

Certainly, more complex lipid systems, such as the three-component "raft"-mixtures may represent more realistic models for biomembrane systems. Their pressure dependent lateral organization and phase behavior has not been studied yet, however. Some data are available on pressure effects on lipid extracts from natural membranes, such as bipolar tetraether liposomes composed of the polar lipid fraction E (PLFE) isolated from the thermoacidophilic archaeon *Sulfolobus acidocaldarius*. The SAXS data on PLFE multilamellar vesicles also exhibit several temperature dependent lamellar phases, and, in addition, the existence of cubic structures at high temperature. Also a variety of new gel-like phases is observed at elevated pressures, thus showing a rich polymorphism also in PLFE liposomes [97].

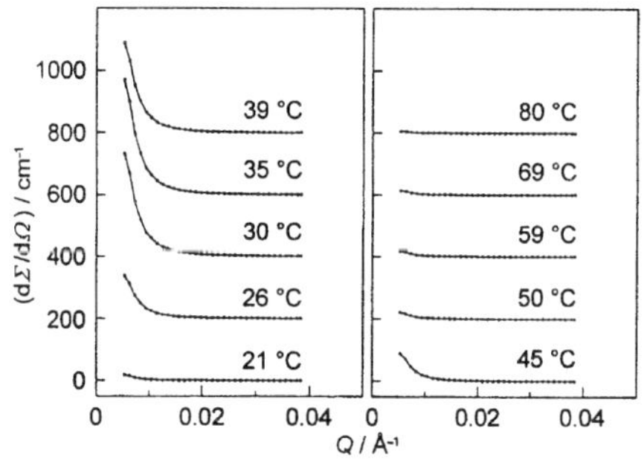

Fig. 14. SANS curves of a contrast-matched DMPC-d_{54}/DSPC(1:1) mixture dispersed in excess water at 1 bar and selected temperatures.

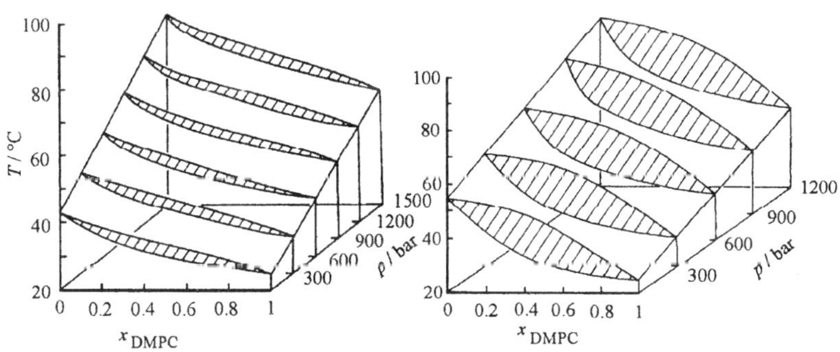

Fig. 15. T,x-phase diagram of the equimolar binary lipid mixtures DMPC/DPPC and DMPC/DSPC in excess water as a function of pressure.

4.2.3. Effect of additives

Cholesterol effect. Natural biological membranes consist of lipid bilayers, which typically comprise a complex mixture of phospholipids and sterol, along with embedded or surface associated proteins. The sterol cholesterol is an important component of animal cell membranes, which may consist of up to 50 mol% cholesterol. Cholesterol thickens a liquid-crystalline bilayer and increases the packing density of lipid acyl chains in the plane of the bilayer in a way that has been referred to as a "condensing effect". Increasing cholesterol concentration leads to a drastic reduction of the main transition enthalpy, ΔH_m, until at cholesterol contents higher than 30 - 50 mol% the main transition vanishes.

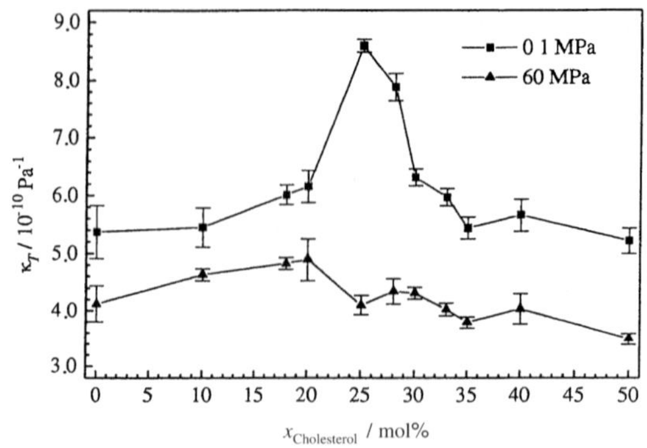

Fig. 16. Isothermal compressibility data of DPPC–cholesterol mixtures as a function of cholesterol concentration and pressure at $T = 50$ °C [98].

Figure 16 depicts the isothermal compressibility (κ_T) of DPPC in the fluid phase (at 1 bar) and in the pressure-induced gel phase at 600 bar as a function of cholesterol concentration. κ_T first increases with increasing cholesterol concentration up to 25 mol% cholesterol, where κ_T has increased by ~60 %. Upon further incorporation of cholesterol, κ_T decreases again, and at 50 mol% cholesterol, the κ_T value corresponds to that of the pure lipid bilayer. The initial increase of κ_T up to 25 mol% cholesterol might be due to an increase of free volume in the middle of the lipid bilayer region which is due to the differences in hydrophobic lengths of DPPC and cholesterol. At high cholesterol concentrations, above 25 mol% cholesterol, the drastic increase in order parameter overcompensates the free volume effect. In the pressure-induced gel-phase at 50 °C and 600 bar, κ_T increases by ~18 % up to a concentration of 20 mol% cholesterol. Further increase of cholesterol incorporation leads to a decrease of κ_T again, reaching values which are significantly lower than that of the pure DPPC gel-phase lipid bilayer. The initial increase of κ_T is probably connected with an increase of a small population of gauche conformers upon incorporation of the sterol leading to an increase in volume fluctuations. At 50 °C, probably a transition from a L_α+$L_{o\alpha}$ to a $L_{o\alpha}$+$L_{o\beta}$ coexistence region occurs around 21 mol% cholesterol (L_α fluid phase of pure DPPC, $L_{o\alpha}$ liquid ordered DPPC–cholesterol phase, $L_{o\beta}$ gel-like ordered DPPC–cholesterol phase). We note that the increase in κ_T at 50 °C is found for cholesterol concentrations that correspond to those in the transition region between these two coexistence regions. Interestingly, a marked change of κ_T occurs in the transition region only. At 50 °C and 600 bar, pure DPPC is in the gel phase. Under these conditions, $\kappa_T = 4.1\cdot10^{-10}$ Pa^{-1}. Up to 20 mol% cholesterol, κ_T slightly increases to $4.9\cdot10^{-10}$ Pa^{-1}. Under these conditions, the DPPC–cholesterol system might be in a pressure-induced L_β+$L_{o\beta}$ two-phase region. At higher cholesterol concentrations (> 22 mol%), the system is in the $L_{o\alpha}$+$L_{o\beta}$ state, probably even up to pressures of 700 bar. For 50 mol% cholesterol, no phase transitions are observed anymore in corresponding SAXS data at pressures up to ~2 kbar.

Measurements of the acyl chain orientational order of the lipid bilayer system by measuring the steady-state fluorescence anisotropy r_{ss} of the fluorophore 1-(4-trimethylammonium-phenyl)-6-phenyl-1,3,5-hexatriene (TMA-DPH) clearly demonstrate the ability of sterols to efficiently regulate the structure, motional freedom and hydrophobicity of biomembranes [99]. The pressure dependencies of r_{ss} of TMA-DPH labeled DPPC and DPPC/cholesterol are shown in Figs. 17. r_{ss} of pure DPPC at $T = 50$ °C increases slightly up to about 400 bar, where the pressure-induced liquid-crystalline to gel phase transition starts to take place. Since r_{ss} reflects the order parameter of the upper acyl chain region for the fluorophor, these results indicate that increased pressures cause this region to be ordered in a manner similar to that which occurs on decreasing the temperature. Addition of increasing amounts of cholesterol leads to a drastic increase of r_{ss} values in the lower pressure region, whereas the corresponding data at higher pressures in the gel-like state of DPPC are slightly reduced. For concentrations above about 30 mol% cholesterol, the phase transition can hardly be detected any more. At a concentration of 50 mol% cholesterol, r_{ss} values are found to be almost independent of pressure up to 1 kbar.

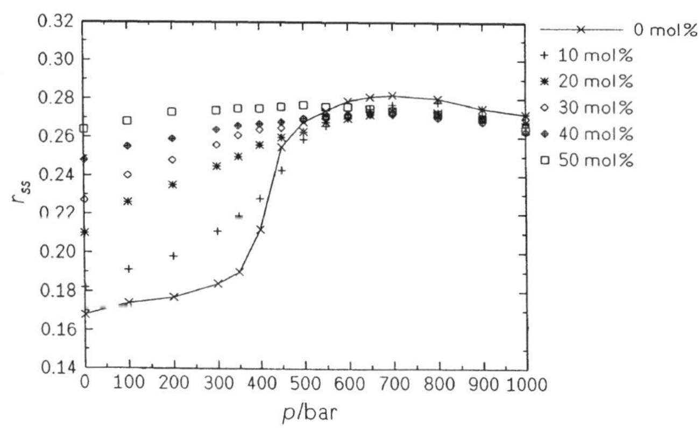

Fig. 17. Pressure dependence of the steady-state fluorescence anisotropy r_{SS} of TMA-DPH in DPPC/cholesterol unilamellar vesicles at different sterol concentrations ($T = 50$ °C).

The excited-state lifetime τ_F of TMA-DPH is a function of the dielectric permittivity of the solvent cage. As the fluorophore resides in the interfacial region of the membrane, it experiences quenching by probe-water interactions. Information on the hydration level at the location of the fluorophore position in the bilayer can thus be obtained by measurements of τ_F. Increasing pressure results in longer fluorescence lifetimes of TMA-DPH in DPPC vesicles in their fluid-like state. At 55 °C, τ_F increases slightly from 2.5 ns at 1 bar to 3.5 ns at 700 bar. Addition of for example 30 mol% cholesterol causes a 2.5-fold increase of τ_F at atmospheric pressure, and hydrostatic pressure increase has only a small effect on fluorescence lifetime in the mixture. Incorporation of the sterol thus reduces the probability of water penetration into the bilayer, thus leading to longer fluorescence lifetimes. These results indicate that the incorporation of cholesterol into the DPPC bilayer leads to a significant increase in hydrophobicity of the membrane. An increase in pressure up to the 1 kbar range is much less effective in suppressing water permeability than cholesterol embedded in fluid DPPC bilayers at these levels of concentration.

Due to both its amphiphilic character and its size (intermediate between that of the shortest and longest acyl chains, C_{16} and C_{24}), cholesterol is inserted in phospholipid membranes and can potentially regulate the effects of the external shocks on the physical state of these membranes. The results shown above and further FT-IR pressure studies [37] thus clearly demonstrate the ability of sterols to efficiently regulate the structure, motional freedom and hydrophobicity of biomembranes, so that they can withstand even drastic changes in environmental conditions, such as in temperature and external pressure. Bilayers thus could regulate their fluidity by an adjustment of their cholesterol composition as well as by a lateral redistribution of their various lipid components (bilayer and non-bilayer favoring ones), and saturated (ordered) and unsaturated (fluid) domains. An increase in the cholesterol level in a membrane reduces the effect of variations in pressure, probably because of the reduction in motional freedom of the head-group region.

Salt effect. It is well known that the effect of inorganic ions on the main transition temperature of lipid bilayers depends on the nature of the ions. The effect is especially dramatic when Ca^{2+}-ions are adsorbed on negatively charged membranes, because of the formation of more or less stable complexes between the divalent ion and the phosphate group. DSC measurements on DMPC/Ca^{2+} dispersions revealed, that increasing Ca^{2+} concentration leads to an increase in main transition temperatures with little change in transition enthalpy, and also to an increase of the $L_{\beta'}$-$P_{\beta'}$ gel to gel transition temperature, until both transitions finally merge at high Ca^{2+} concentration. The main transition is slightly shifted towards smaller pressures with increasing temperature in comparison to that of pure DMPC dispersions. Otherwise, the transition slope dT_m/dp is parallel to that of pure DMPC up to pressures of about 2 kbar, and the volume change ΔV_m at the main transition is of similar magnitude. A similar behavior has been observed for the negatively charged lipid DMPS with addition of Ca^{2+} [100].

Local anesthetics. Anesthetics interact with membranes and increase the gel to liquid-crystalline transition of fully hydrated bilayers. They induce a volume expansion which has the opposite effect of HHP and so they antagonize the effect of HHP on membranes' fluidity and volume, making membranes more fluid and expanded. The application of HHP to membrane-anesthetic systems may even result in the expulsion to the aqueous environment. The local anesthetic tetracaine (TTC) can be viewed as a model system for a large group of amphiphilic molecules. From volumetric measurements on a sample containing e.g. 3 mol% TTC, it has been found that the main tansition at ambient pressure shifts to a lower temperature. The expansion coefficient α drastically increases relative to that of the pure lipid system in the gel phase, and the incorporation of the anesthetic into the DMPC bilayer causes an about 15 % decrease of ΔV_m relative to that of the pure lipid system. The addition of 3 mol% TTC shifts the pressure-induced liquid-crystalline to gel phase transition towards somewhat higher pressures. Larger values for the compressibilities are found for both lipid phases by addition of 3 mol% TTC, and there is no apparent difference in the coefficient of compressibility between the gel and liquid-crystalline phases. Comparison of the IR spectra of DMPC and DMPC/TTC mixtures at pH 5.5 as a function of pressure shows an abrupt appearance of a band at around 1685 cm^{-1} for $p \geq 2.5$ kbar, when TTC is incorporated. This high pressure induces partial expulsion of the TTC from the membrane [100]. These kind of results might be of interest in understanding barotropic phenomena in cell membranes such as the antagonistic effect of hydrostatic pressure against anesthesia *in vivo*.

Incorporation of polypeptides. We also investigated the effect of the incorporation of the model channel gramicidin D (GD) on the structure and phase behavior of phospholipid bilayers of different chain-lengths [101]. Gramicidin is a linear hydrophobic polypeptide antibiotic that forms specific channels in lipid membranes for the transport of monovalent cations. Bilayers containing this channel-forming polypeptide are often used as a model for protein-lipid interaction studies. Pressure has been applied so as to be able to finely tune the lipid chain-lengths and conformation. Infrared spectral parameters obtained by FT-IR spectroscopy and data obtained from X-ray diffraction and ^{2}H-NMR experiments were used to detect structural changes upon incorporation of GD into DMPC, DPPC and DSPC lipid bilayers. Analysis of the infrared amide I band frequencies allowed us to determine the corresponding peptide structure adopted in the lipid environment. Gramicidin is highly polymorphic, being able to adopt a wide range of structures with different topologies. Common forms are the dimeric single-stranded right-handed $\beta^{6.3}$-helix with a length of 24 Å, and the antiparallel double-stranded $\beta^{5.6}$-helix, being approximately 32 Å long. For comparison, the hydrophobic fluid bilayer thickness is about 30 Å for DPPC bilayers, and the hydrophobic thicknesses of the gel phases are 4-5 Å larger. Depending on the GD concentration, significant changes of the lipid bilayer structure and phase behavior were found, such as the disappearance of some of the gel phases formed by the pure lipid bilayer systems, and the formation of broad two-phase coexistence regions at higher GD concentrations. In the liquid-crystalline phases of the phospholipid bilayers, generally more orientational order is induced in the lipid molecules by the incorporation of GD. Vice versa, also the peptide conformation is influenced by the lipid environment. Depending on the phase state and lipid acyl chain-length, GD adopts at least two different types of quaternary structures in the bilayer environment, a double helical "pore" and a helical dimer "channel". With regard to the changes of the bilayer thickness at the gel/fluid main phase transition of DPPC and DSPC, we showed that the conformational equilibrium of the peptide is changed. In gel-like DPPC and DSPC bilayers, the equilibrium of the GD species in the lipid bilayer is shifted in favor of the double helical configuration [101]. These changes may be attributed to the ability of the double helical conformation to tolerate more hydrophobic mismatch than the helical dimer, perhaps due to increased numbers of stabilizing intermolecular hydrogen bonds. We were able to construct a tentative p,T-phase diagram for the DPPC-GD (4.7 mol%) mixture up to pressures of 4000 bar, which is shown in Fig. 18. Gramicidin insertion clearly has a significant influence on the lipid topology and phase behavior. To avoid large hydrophobic mismatch, the lipid topology and dynamics is altered and broad fluid-gel coexistence regions are formed. In these phase-separated regions, interfacial adsorption, wetting layer formation and condensation phenomena may be operative. This example clearly demonstrates not only that the lipid bilayer structure and phase behavior drastically depends on the polypeptide concentration, but also that the peptide conformation can significantly be influenced by the lipid environment. No pressure-induced unfolding of the polypeptide is observed up to 10 kbar.

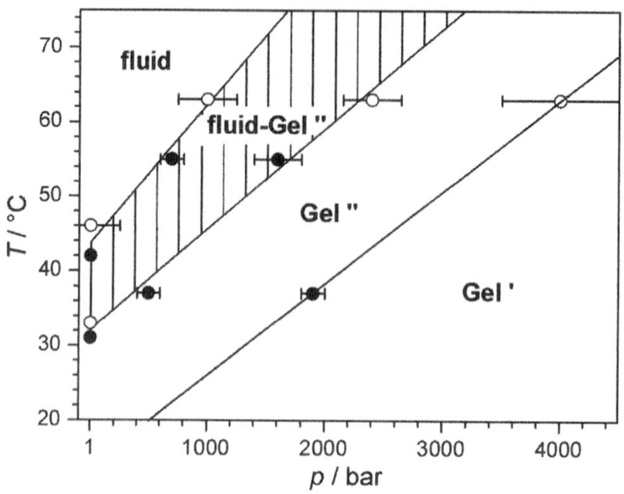

Fig. 18. Phase diagram of DPPC-GD (4.7 mol%) in excess water as obtained from SAXS data and FT-IR studies.

^{2}H-NMR spectral parameters were used to detect more detailed structural and dynamic changes upon incorporation of the polypeptide into the lipid bilayers. Figure 19 exhibits pressure dependent spectra a d_{62}-DPPC and the d_{62}-DPPC–GD mixture at T = 55 °C. The pressure-induced main phase transition of pure d_{62}-DPPC is identified at a pressure of 650 bar as an increase of the first spectral moment M_1 and a drastic change in the lineshape from a motionally averaged spectrum in the liquid-crystalline phase to a rigid lattice type ^{2}H-NMR spectrum. At ~1500 bar, the interdigitated gel phase is formed. Compared to the pure lipid system, the ^{2}H-NMR lineshapes of the d_{62}-DPPC–GD (4.7 mol%) mixture are markedly different, showing features of both, motionally averaged and rigid lattice type spectra over a wide pressure range. At high pressures, the lineshapes of the mixture are no longer comparable to the ones obtained by the pure lipid dispersion and show no indication of the Gi ($L_{\beta i}$) phase any more. The pressure dependence of the segmental C–^{2}H order parameters is depicted in Fig. 20. The chain order parameter values, S_{CD}, increase with increasing pressure, in particular at the terminal methylene segments and in the fluid to gel phase transition region. For example, at p = 600 bar the order parameter has increased by ~20% in the plateau region (in the upper part of the acyl chains) and by 40-50 % for the methylene groups at the end of the chains, i.e. in the inner part of the lipid bilayer. The hydrophobic thickness D_C of the pure DPPC monolayer increases from 13.7 Å at 1 bar to 14.7 Å at 600 bar; for the d_{62}-DPPC–4.7 mol% GD mixture, D_C increases from 14.2 to 15.6 Å. The average chain cross sectional area $\langle A \rangle$ decreases concomitantly from ~66 Å^2 in pure DPPC at 1 bar to 60 Å^2 at 600 bar. Adding 4.7 mol% GD to the DPPC bilayer at 55 °C yields values of 63.5 Å^2 at ambient pressure and 55.5 Å^2 at 600 bar, respectively. The data thus clearly shows that, depending on the peptide concentration, the conformation of the temperature- and pressure-dependent lipid bilayer is significantly altered by the insertion of the polypeptide [63].

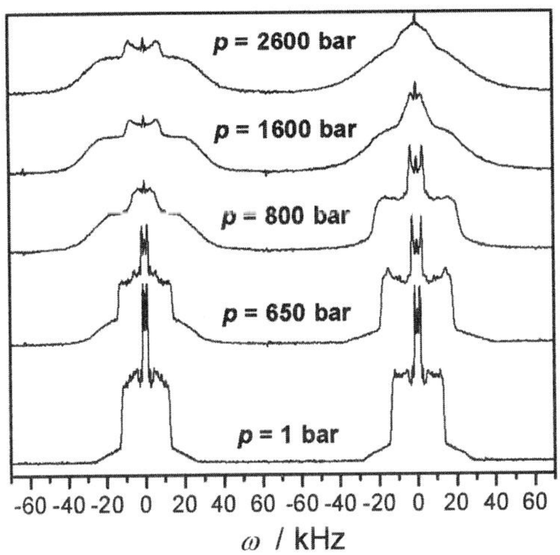

Fig. 19. ^{2}H NMR spectra of pure d_{62}-DPPC (left) and the d_{62}-DPPC–GD (4.7 mol%) mixture (right) at $T = 55$ °C and selected pressures.

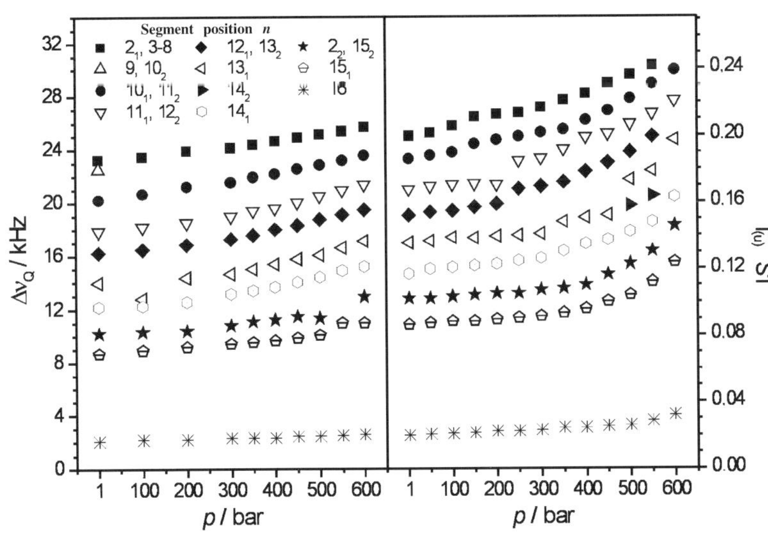

Fig. 20. Measured segmental order parameter profiles of d_{62}-DPPC (left) and the d_{62}-DPPC–GD (4.7 mol%) mixture (right) as a function of pressure at 55 °C.

For large integral and peripheral proteins, however, pressure-induced changes in the physical state of the membrane may lead to a weakening of protein-lipid interactions and they may even be released from the membranes. This phenomenon may form the basis of a new method for extracting proteins from membranes. Disaggregation of lipid-protein assemblies by high pressure would have the advantage of avoiding the addition of surfactants, and thus favors the preservation of the native-like conformation of the isolated proteins. By using high pressure extraction, it was possible, for example, to isolate protein kinase C and other lipid-interacting proteins in complexes with essential lipid molecules [102]. High pressure may also be used to study structural and kinetics aspects of membrane proteins, such as ATP-synthase (see Mignaco et al. in ref. [11]).

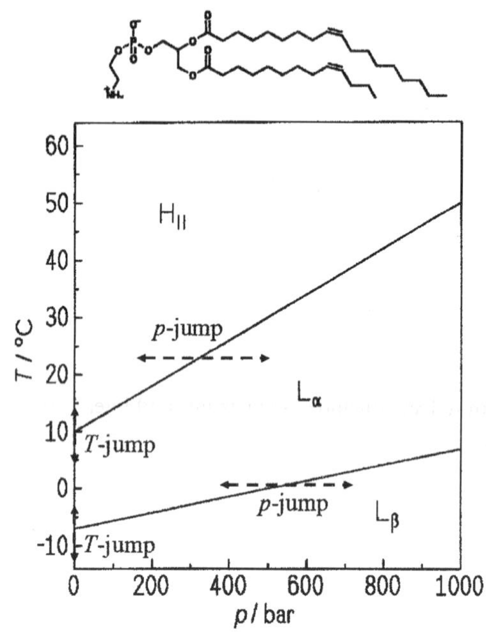

Fig. 21. *T,p*-phase diagram of DOPE (molecular structure on top) in excess water. The dashed arrows indicate how by using fast *T*- or *p*-jumps the kinetics of the different phase transformations can be studied.

4.2.4. Kinetics of phase transformations in lipid systems

Phase transitions between lipid mesophases must be associated with deformations of the interfaces which, very often, imply also their fragmentation and fusion so that not only the symmetry changes but also the topology of the lipid organization. Depending on the topology of the structures involved, transition phenomena of different complexity are observed. In addition, the transition rates and mechanisms depend on the level of hydration of the structures involved and on the forces driving the transition. We used the synchrotron X-ray diffraction technique to record the temporal evolution of the structural changes after induction of the phase transition by a pressure jump across the phase boundary (Fig. 23) [103, 104]. We discuss two representative examples, only.

First, we present pressure jump experiments carried out in DEPC-water dispersions to study the L_β-L_α gel-to-fluid main transition, which occurs at T_m = 12 °C at ambient pressure and which has a pressure dependence of dT_m/dp = 20 °C $kbar^{-1}$. Selected SAXS diffraction patterns at 18 °C after a pressure jump from 200 bar (L_α phase) to 370 bar (L_β phase) are depicted in Fig. 22. An intermediate structure is clearly observable here. The first-order (001) Bragg reflection of the initial L_α-phase (a = 66 Å) vanishes in the course of the pressure jump (< 5 ms). The first diffraction pattern collected after the pressure jump exhibits a Bragg reflection of a new lamellar structure L_x with a slightly larger d-value, which increases with time. The lattice constant of the L_β phase formed is 78 Å. The transition is complete after about 15 s. In equilibrium measurements, no such intermediate lamellar structure is detectable.

Experiments for investigating the lamellar-H_{II} transition kinetics have been performed, for example, on DOPE dispersions. The T,p-phase diagram of DOPE in excess water is depicted in Fig. 21. Figures 22, 23 show the diffraction patterns and lattice parameters at 20 °C after a pressure jump from 300 to 110 bar. Clearly, the (001) reflection of the L_α phase and the (10) reflection of the developing H_{II} phase can be identified. In this case, a two-state mechanism is observed. Interestingly, we find that successive pressure jumps lead to an acceleration of the phase transition kinetics. The half transit time decays from 8.5 s for the first pressure jump to 5.3 s after the fourth jump. After the pressure jump, an induction period of several seconds is observed before the first Bragg reflections of the newly formed H_{II} phase appear. Upon successive pressure cycles, this induction period decreases. An explanation for this phenomenon might be the formation of defect structures, such as inverted intermicellar intermediates, which are formed during the pressure cycles and which have not healed between successive pressure cycles. This observation also shows that the history of sample preparation plays an essential role in these kinds of studies. It has been found that with increasing pressure jump amplitude, the induction period decreases and the rate of phase transformation increases.

Generally, as has also been found in studies of pressure and temperature jump induced phase transitions of other systems [103-107], these results show that the relaxation behavior and the kinetics of pressure-induced lipid phase transformations depend drastically on the topology of the lipid mesophase, and also on the temperature and the driving force, i.e., the applied pressure jump amplitude Δp. Often multicomponent kinetic behavior is observed, with short relaxation times (probably on the nanosecond to microsecond time-scale) in a burst phase referring to the relaxation of the lipid acyl chain conformation in response to the pressure change, which leads to the small changes in the observed lattice constants right after the pressure jump. The longer relaxation times measured here are due to the kinetic trapping of the system. In most cases the rate of the transition is limited by the transport and redistribution of water into and in the new lipid phase, rather than being controlled by the time required for a rearrangement of the lipid molecules. This can be inferred from the lattice relaxation experiments performed in the lipid one-phase regions and by modelling the data using simple hydrodynamics. The obstruction factor of the different structures, especially in cases where nonlamellar (hexagonal and cubic) phases are involved, controls the different kinetic components. In addition, nucleation phenomena and domain size growth of the structures evolving play a role. In many cases (e.g., DEPC, DMPC-MA), a digression of the mechanism of phase transformation observed under slow-scanning equilibrium conditions appears under high free energy gradients (here large pressure jump amplitudes), and the high driving force drives the system through a correlated ordered intermediate state.

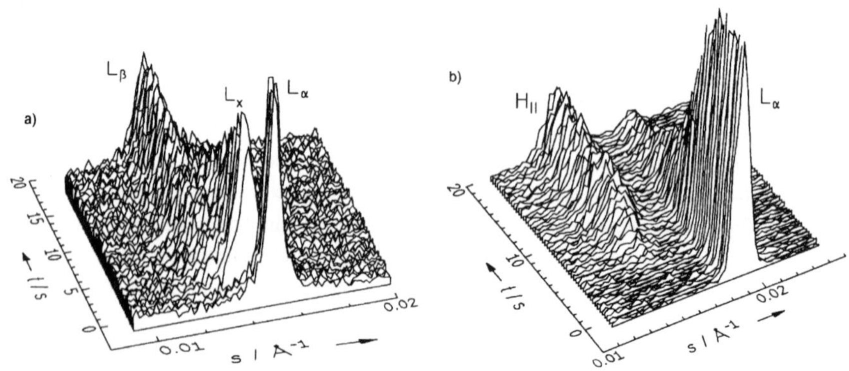

Fig. 22. Diffraction patterns of a) DEPC in excess water after a pressure jump from 200 to 370 bar at 18 °C, and b) of DOPE in excess water after a pressure jump from 300 to 110 bar at 20 °C.

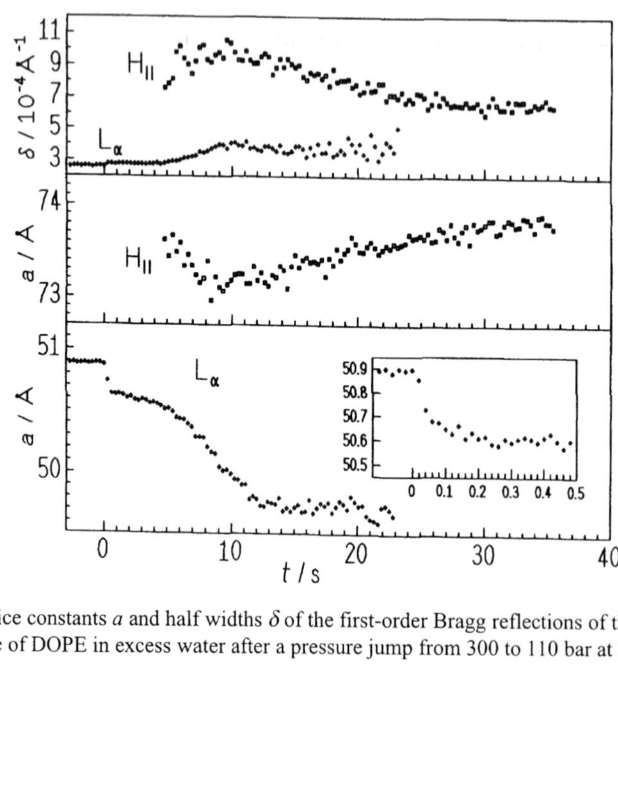

Fig. 23. Lattice constants a and half widths δ of the first-order Bragg reflections of the L_α-phase and the H_{II}-phase of DOPE in excess water after a pressure jump from 300 to 110 bar at 20 °C.

In cases, where the transition occurs without change in water content within the mesophase, such as in the inter-cubic $Q_{II}^{G} \rightarrow Q_{II}^{D}$ transition of the system DLPC-LA (1:2) at a fixed water composition, the kinetics may be much faster. As mechanism for this cubic-cubic transformation, a stretching mechanism has been proposed whereby each 4-fold junction in the Q_{II}^{D} phase is formed by bringing together two 3-fold junctions in the Q_{II}^{G} phase (Fig. 24a) [107]. Recently, it has been suggested that such continuous cubic transitions could also involve non-cubic (tetragonal, rhombohedral) distortions of the underlying minimal surfaces, yet with the surfaces remaining minimal during the processes [108]. The inverse bicontinuous cubic phases are of particular relevance to the mechanism of membrane fusion, which is a ubiquitous process in cell membranes. The reason for this is that the fusion channel between bilayers is closely similar to the local structure of these cubic phases. Indeed, lamellar to cubic phase transitions in lyotropic liquid crystals must occur by a series of fusion events, and the bicontinuous cubic phase structures may be viewed as 3D lattices of such fusion pores. Figure 24b displays schematically the formation of a fusion pore via a stalk intermediate, which might also play a role in the final step of biological membrane fusion, which uses a variety of fusion proteins for approach and bending of the lipid bilayers, however.

The pressure-jump relaxation technique has also been applied to study the pressure dependence of the photocycle kinetics of bacteriorhodopsin (bR) from *Halobacterium salinarium* up pressures of 4 kbar [109]. The kinetics could be modelled by nine apparent rate constants, which were assigned to irreversible transitions of a single relaxation chain of nine kinetically distinguishable states P_1 to P_9. The kinetic states P_2 to P_6 comprise only the two spectral states L and M. From the pressure dependency of the bR photocycle kinetics, the volume differences ΔV_{LM} between these two spectral states could be determined, which range from $\Delta V_{LM} = -11.4$ mL mol^{-1} to $\Delta V_{LM} = 14.6$ mL mol^{-1} from P_2 to P_6. A model was developed that explains the dependence of ΔV_{LM} on the kinetic state by the electrostriction effect of charges that are formed or neutralized during the L/M equilibrium.

We conclude that pressure work on model membrane and lipid systems can yield a wealth of enlightening new information on the structure, energetics and phase behavior of these systems as well as on the transition kinetics between lipid mesophases.

4.3. Pressure effects on protein structure and stability

Since the discovery of high-pressure-induced protein denaturation by Bridgman in 1914 [110], it has been shown in numerous studies that the application of hydrostatic pressure results in the disruption of the native protein structure due to the decrease in the volume of the protein-solvent system upon unfolding. Pressure denaturation studies thus provide a fundamental thermodynamic parameter for protein unfolding, the ΔV^{o}, in addition to an alternative method for perturbing the folded state, and thus extracting its stability. A number of reviews on effects of pressure on proteins discuss these volume changes in greater detail [4,9-18]. Denaturation of proteins is usually studied at atmospheric pressure using high temperature, guanidinium hydrochloride or urea as denaturants. However, interpretation of the results obtained using such methods may be complicated by the facts that: 1) varying the temperature changes both the volume and the thermal energy of the system at the same time, and 2) the thermodynamic parameters of denaturation by guanidinium chloride or urea are influenced by the binding of these molecules to proteins. The use of pressure is also advantageous from a methodological point of view: The transition to native conditions (renaturation) is achieved simply by releasing the pressure. In general, the effects of pressure on proteins are reversible, and only seldom are they accompanied by aggregation or changes in covalent structure. The net volume change on denaturation comprises the effects of disruption of noncovalent bonds, changes in protein hydration and conformational changes. The incompressible volume of covalent bonds maintains a volume that does not change on unfolding, and the reduction in the net volume seems to be predominantly the result of the disappearance of solvent-inaccessible voids inside the protein [101].

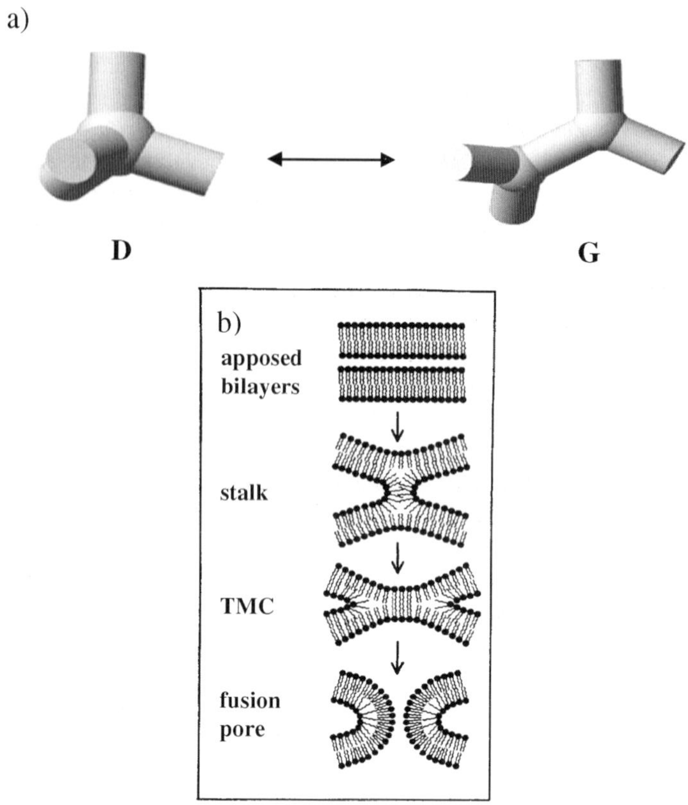

Fig. 24. a) Schematic illustration of the "stretching" of water channel junctions during the continuous transformation between the D and G cubic phases, which occur with no disruption of the bilayer topology. A junction of four water channels in the Q_{II}^{D} phase is converted into two three-way junctions in the Q_{II}^{G} phase. b) Possible mechanism of membrane fusion: the monolayers of two apposed lipid bilayers mix to form a stalk intermediate that expands radially to a trans monolayer contact (TMC), leading to rupture as a result of curvature and interstitial stresses and finally to the formation of a fusion pore.

The fundamental understanding of the process and kinetics of the folding of proteins to their native state has fascinated researchers for decades and remains one of the most interesting and challenging issues in modern biophysics. One experimental approach to understand the folding process is to characterize the nature of the barrier to folding or unfolding and the corresponding transition state. Also in this respect, pressure studies are of particular use. Moreover, pressure studies present an important advantage due to the positive activation volume for folding, the result of which is to slow down folding substantially, in turn allowing for relatively straightforward measurements of structural order parameters that are difficult or even impossible to measure on much faster timescales with all techniques.

4.3.1. Equilibrium studies of protein denaturation

As an example, we present data on the pressure-induced unfolding and refolding of staphylococcal nuclease (SNase). This protein has served as model for protein folding because it is small and has a well-known native structure. These studies were performed using synchrotron small-angle X-ray scattering (SAXS) and Fourier-transform infrared spectroscopy (FT-IR), which monitor changes in the tertiary and secondary structural properties of the protein upon pressurization or depressurization. SNase is a small protein of about 17.5 kDa containing 149 amino acids and no disulfide bonds. In the crystalline state the protein contains 26.2 % helices, 24.8 % β-sheets (barrel), 7.4 % extended chains, 24.8 % turns and loops, and 8.7 % unordered chains (8.1 % are uncertain). Analysis of the high pressure SAXS data reveal that over a pressure range from atmospheric pressure to approximately 3 kbar, the radius of gyration R_g of the protein increases from a value near 17 Å for native SNase two-fold to a value near 35 Å (Fig. 25). A large broadening of the pair distribution function $p(r)$ is observed over the same range, indicating a transition from a globular to an ellipsoidal or dumbbell-like structure [48-50,52]. Deconvolution of the FT-IR amide I' absorption band (Fig. 26) reveals a pressure-induced denaturation process over the same pressure range as the SAXS that is evidenced by an increase in disordered and turn structures and a drastic decrease in the content of β-sheets and α-helices (Fig. 27). Contrary to the temperature-induced unfolded state, the pressure-induced denatured state above 3 kbar retains some degree of β-like secondary structure and the molecules cannot be described as a fully extended random polypeptide coil, which is in accord with the SAXS results. Temperature-induced denaturation involves a further unfolding of the protein molecule which is indicated by a larger R_g-value of 45 Å, and significantly lower fractional intensities of IR-bands associated with secondary structure.

There are many indications now that the conformation of a protein denatured by pressure is more compact than that of a protein denatured by temperature or chemical agents. A growing body of experimental evidence shows that, according to the characteristic structural features (the presence of secondary structure and the absence of well-organized tertiary structure), pressure-denatured proteins often resemble "molten globules" type structures. This does not seem too surprising, as pressure is known to favor the formation of hydrogen bonds, which maintain the secondary-structure network, but is unfavorable for hydrophobic interactions, which are predominantly responsible for maintaining the tertiary structure of a protein. The idea is supported by theoretical results which suggest water penetration into the protein interior as a likely mechanism for pressure-denaturation of proteins due to a weakening of hydrophobic interactions, as opposed to the temperature-induced unfolding process [112]. Assuming the pressure-induced unfolding transition of SNase to occur essentially as a two-state process, analysis of the FT-IR pressure profiles yields a Gibbs free energy change for unfolding of ΔG^o = 17 kJ·mol^{-1} and a volume change for unfolding of $\Delta V^o = -80 \pm 20$ mL·mol^{-1} at ambient temperature and pressure.

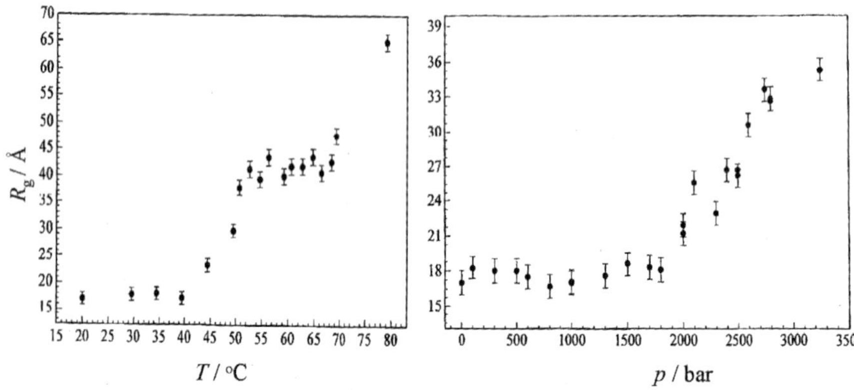

Fig. 25. Apparent radius of gyration R_g of SNase (1 % (w/w), pH 5.5) as a function of temperature and pressure (at T = 25 °C).

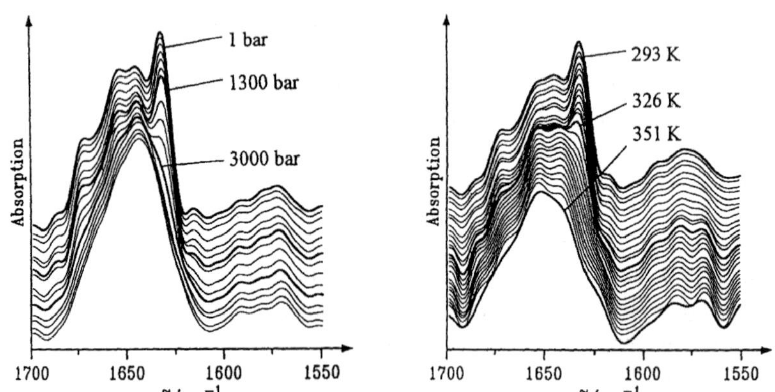

Fig. 26. Deconvoluted FT-IR spectra of SNase (5 % (w/w), pH 5.5) a) as a function of pressure at 25 °C and b) as a function of temperature (band assignment: 1611 cm^{-1} side chains, 1627 cm^{-1} β-sheets, 1651 cm^{-1} α-helices, 1641/1659/1666 cm^{-1} disordered structures/turns).

Recently, we have also characterized the temperature- and pressure-induced unfolding of SNase using high precision densitometric measurements [47]. At 45 °C we calculate a measured decrease in apparent molar volume due to pressure-induced unfolding of −55 cm^3 mol^{-1}. The threefold increase in compressibility upon unfolding reflects a transition to a partially unfolded state, which is consistent with our results obtained for the radius of gyration of the pressure-denatured state of SNase. Various experimental data indicate now that the underlying mechanism of pressure unfolding is the penetration of water into the protein matrix [47,113].

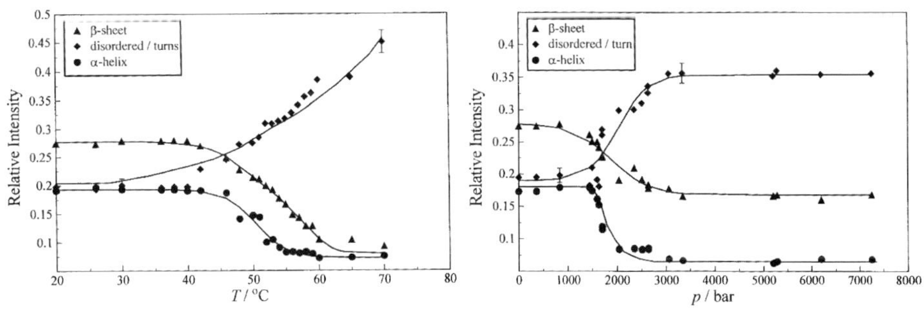

Fig. 27. Temperature and pressure effect on the areas of the IR bands associated with β-sheets, disordered/turn structures and α-helices of SNase at pH 5.5 and 25 °C.

The pressure midpoints at several temperatures obtained from the FT-IR and SAXS profiles are plotted as a *p*,*T*-phase diagram in Fig. 28a. It exhibits the elliptic-like curvature which is typical of many monomeric proteins (Fig. 28b) [4, 9-18, 113-115]. Interestingly, pressure-temperature stability diagrams of bacteriophages, and even bacterial cells, may be described by diagrams similar to that shown for proteins in Fig. 28b. There is also some evidence that sol-gel transitions of polysaccharides exhibit similar phase diagrams to those obtained for proteins [13], indicating that the role of water in playing an important role.

4.3.2. Theoretical calculation of the p,T-stability phase diagram of proteins

A complete thermodynamic description of the folding and unfolding reactions of proteins requires the characterization of the response of the protein structure also to pressure and will thus help in understanding protein stability and function. In general, the thermodynamic stability of a protein should be expressed most appropriately in multidimensional space as functions of temperature, pressure and solution conditions, giving an energy landscape as a multi-dimensional surface. When the solution conditions are held constant, the stability of the protein is a simultaneous function of temperature and pressure, giving the free energy difference between the native and denatured states as a three-dimensional surface on the temperature-pressure plane. Recently, we have calculated the free energy landscape of SNase as a function of temperature and pressure using all experimental input parameters available [116]). In the following, we elucidate what thermodynamic properties are required for calculating the stability diagram of a protein. The Gibbs free energy difference between the unfolded and native state is defined as:

$$\Delta G = G_{\text{unfolded}} - G_{\text{native}}. \tag{13}$$

As changes in ΔG are given by $\mathrm{d}\Delta G = \Delta V \mathrm{d}p - \Delta S \mathrm{d}T$, we obtain Eq. (14) upon integration of this equation from a chosen reference point T_0, p_0 to T, p [114,116,117]:

$$\Delta G = \Delta G_0 + \frac{\Delta\kappa'}{2}(p-p_0)^2 + \Delta\alpha'(p-p_0)(T-T_0) - \Delta C_p\left[T\left(\ln\frac{T}{T_0}-1\right)+T_0\right] + \Delta V_0(p-p_0) - \Delta S(T-T_0) \tag{14}$$

where Δ denotes the change of the corresponding parameter during unfolding, that is, the value in the unfolded, denatured state minus that in the native state; κ' is the compressibility factor, here defined as $\kappa' = (\partial V/\partial p)_T) = -V\kappa$, with κ being the isothermal compressibility defined in the usual way; α' is the thermal expansivity factor, here defined as $\alpha' = (\partial V/\partial T)_p = -(\partial S/\partial p)_T = V\alpha$, where α is the thermal expansion coefficient of the system in the usual definition; $C_p = (\partial H/\partial T)_p$ is the heat capacity. All other symbols have their usual meaning. Equation (14) is a second-order Taylor series of $\Delta G(T, p)$ expanded with respect to T and p around T_0, p_0. We have chosen p_0 = atmospheric pressure (1 bar = 10^5 Pa) and T_0 = 325 K, the unfolding temperature of SNase at ambient pressure.

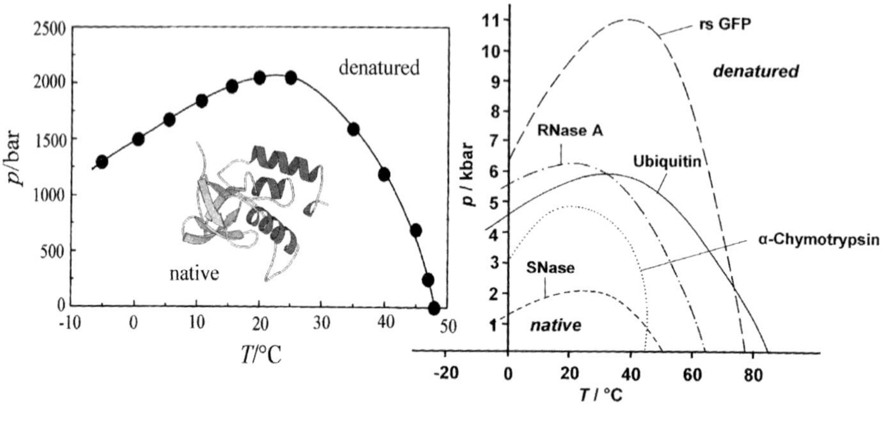

Fig. 28. *T,p*-stability diagram of a) SNase at pH 5.5 as obtained by SAXS, FT-IR and DSC measurements and b) of several other monomeric proteins.

The transition line, where the protein unfolds, is given by $\Delta G = 0$. The physically relevant solution of the curve in the T,p-plane has in fact an elliptical shape (Fig. 29). Taking only the first-order terms into account would give linear phase boundaries only. In the region where the protein is stable, i.e., in the native state, $\Delta G > 0$. As can be seen from Eq. (14), the shape of the elliptical phase boundary is defined by six thermodynamic parameters, ΔC_p, ΔV, ΔS, $\Delta\kappa'$, $\Delta\alpha'$, and the reference Gibbs free energy change of unfolding, ΔG_0. In our binary system (biomolecule + solvent), partial molar thermodynamic parameters of the solute should be considered, which are defined as the partial derivative of these parameters with respect to the number of moles of the protein. In the measurements, apparent thermodynamic quantities are determined, and it is assumed that the partial molar quantities of the solvent (water) in these dilute solutions are equal to the quantities of the pure solvent.

Truncating of the Taylor series at the second-order terms means that the second derivatives of the Gibbs free energy difference (ΔC_p, $\Delta\kappa'$, $\Delta\alpha'$) do not change significantly with temperature and pressure. If this assumption is not valid, an extended analysis is necessary, where the third order terms proportional to T^3, T^2p, $T\ p^2$ and p^3 are involved. As a consequence, the form of the ellipse remains but it gets distorted [114], in particular at high temperatures and pressures.

There are several specific points in the elliptical curve: The unfolding temperature at ambient pressure, T_u, the denaturation pressure p_u at room temperature, and the cold denaturation temperature, T_c. They are given by the following equations:

$$T_u = -\frac{\Delta S_0 T_0}{\Delta C_p} + \sqrt{\frac{\Delta S_0^2 T_0^2}{\Delta C_p^2} + 2\frac{\Delta G_0 T_0}{\Delta C_p}} + T_0 \tag{15}$$

$$p_u = -\frac{\Delta V_0}{\Delta\kappa'} + \sqrt{\frac{\Delta V_0^2}{\Delta\kappa'^2} - 2\frac{\Delta G_0}{\Delta C_p}} + p_0 \tag{16}$$

$$T_c = -\frac{\Delta S_0 T_0}{\Delta C_p} - \sqrt{\frac{\Delta S_0^2 T_0^2}{\Delta C_p^2} + 2\frac{\Delta G_0 T_0}{\Delta C_p}} + T_0. \tag{17}$$

At the maximum pressure and maximum temperature, where the native state is stable, the slope of the ellipse is zero and infinite at p_{max} and T_{max}, respectively. At $T = T_{max}$, $\Delta V = 0$, and $\Delta S = 0$ at $p = p_{max}$. A schematic picture of the elliptic phase diagram and of the lines for $\Delta V = 0$ and $\Delta S = 0$ is given in Fig. 29.

Knowing all input parameters from experimental measurements, the three-dimensional free-energy landscape can be calculated (Fig. 30a). The plot clearly shows that the protein is stable only (Gibbs free energy of unfolding $\Delta G > 0$) in a limited p,T- phase-space. The phase boundary for the unfolding transition (native state $\rightarrow$ unfolded state) is given by the condition $\Delta G = 0$. To demonstrate the shape of the stability curves, we show slices of the free energy in the p,T-plane. Figure 30b exhibits ΔG for −20, −10, 0 and 10 kJ mol^{-1}. In the figure, also the experimental data for the p,T-phase diagram of the protein as obtained by SAXS and FT-IR spectroscopy are superimposed. We also include DSC data for the heat and cold denaturation of the protein at ambient pressure. The agreement between the experimental data points and the theoretical curve for $\Delta G = 0$ is surprisingly good, thus indicating justification of the two-state assumption for the unfolding transition of SNase. Viewing the unfolding process in a more general energy landscape picture (folding funnel) [118-120] rather than in terms of such two discrete states, the states might represent phase-space coordinates in larger areas of the complex energy landscape of the protein. If these states have a rather well-defined average free energy, an effective two-state model may still be a reasonable approximation. Such an effective two-state model then allows for an evaluation of the whole set of thermodynamic quantities that determine the unfolding process and for an evaluation of the respective p,T-stability diagram. Our results for this small, monomeric protein conform relatively well to the type I scenario of the funnel model, i.e., a relatively smooth funnel topology. Large amplitude terms higher than second-order with respect pressure in Eq. (14) do not seem to be necessary for describing the gross features of the p,T-phase-space of this protein. Certainly, the temperature and pressure dependence of ΔC_p, $\Delta\alpha'$ and $\Delta\kappa'$ will have to be considered to get quantitative agreement with the experimental data over the whole temperature and pressure range. The shape of the p,T-plane, if "tongue-like" or "hillside-like", depends on the values of ΔC_p, $\Delta\kappa'$ and $\Delta\alpha'$ [114].

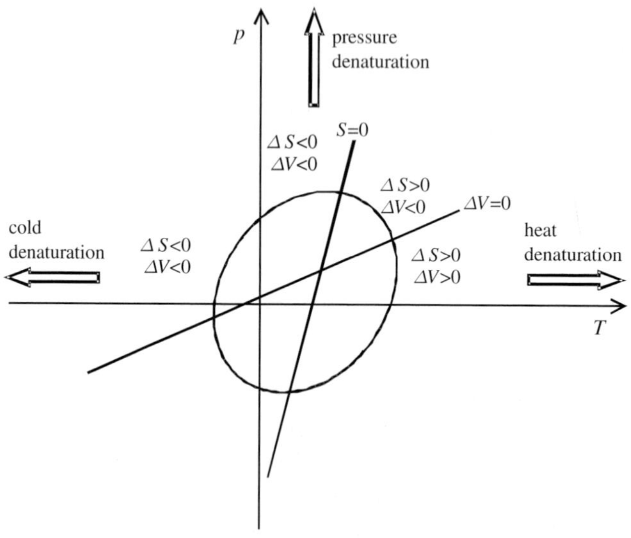

Fig. 29. Schematic representation of the elliptic phase diagram of monomeric proteins. The arrows show specific denaturation ways known as pressure, heat and cold denaturation. Relative positions of the $\Delta S = 0$ and $\Delta V = 0$ lines are also indicated.

The shape of the stability diagram certainly depends on the individual protein secondary structural composition and may be more complicated, in particular for larger proteins. Also additional regions in the phase diagram may appear, such as an extended region at high temperatures where aggregation occurs. One must also be aware of the fact that the unfolded states in the *p*,*T*-plane can be of considerably different structure, and that long-lived metastable states may occur.

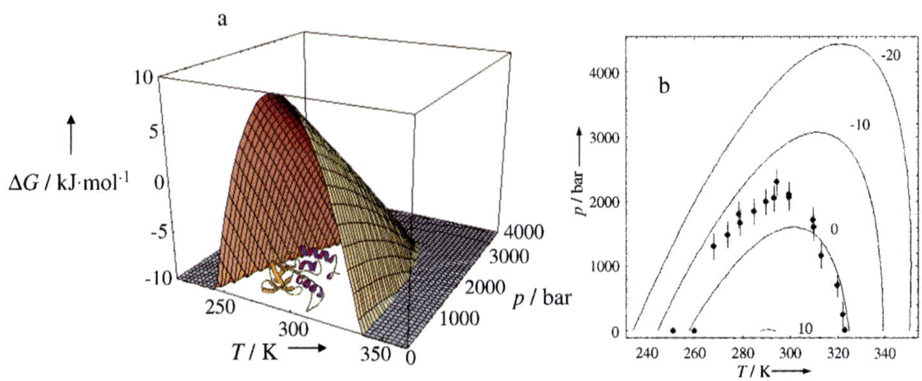

Fig. 30. a) Three-dimensional free energy landscape of SNase (pH 5.5) using experimentally thermodynamic parameters. The Gibbs free energy of unfolding, ΔG, is plotted from -10 to $+10$ kJ·mol^{-1}. $\Delta G < 0$ means that the protein unfolds in this phase-space of temperature and pressure. b) Contours of constant Gibbs free energy ΔG of unfolding of SNase on the pressure-temperature plane, obtained as slices of the three-dimensional free energy landscape shown in a). Points indicate measured values at different temperatures and pressure.

4.3.3. Cosolvent effects

It is well-known that the cytoplasm of the cell is relatively crowded and contains a rather complex aqueous solution including a variety of different salts and osmolytes. To date, still little is known about the pressure stability of proteins in the presence of particular reagents (in the present context, a cosolvent) in comparison to water or buffer (121, 122). Recently, we have explored the effect of different cosolvents on the high-pressure stability of a well-characterized monomeric protein, staphylococcal nuclease (SNase). Changes in the denaturation pressure, the volume change and the standard Gibbs free energy change of unfolding were obtained in the presence of different concentrations of polyhydric cosolvents (glycerol and sorbitol), sugars (sucrose), urea, chaotropic ($CaCl_2$) and kosmotropic ions (K_2SO_4) (Fig. 31). Additionally, FT-IR difference spectra were recorded and analyzed in an effort to detect conformational changes in the native and unfolded state of the protein in the presence of the different cosolvents [123]. Upon addition of the polyols glycerol, sorbitol and sucrose to SNase solutions, an increase in the denaturation pressure p_m is observed. The protection by these cosolvents against pressure-induced denaturation is a consequence of the increased ΔG^0(1 bar) value with increasing cosolvent concentrations. The ΔV^o seems to be almost independent of the concentration of these cosolvents, except for high glycerol concentrations. The unfolded state retains significant β-sheet character at high polyol concentrations, indicating that the protein is less unfolded under these conditions. With regard to temperature-induced denaturation, a different behavior of $\Delta V^o(c)$ is observed, which might be largely due to the strong temperature dependence of ΔV^o. We observed a marked decrease in $|\Delta V^o|$ in the presence of K_2SO_4, but without a significant change in ΔG^o.

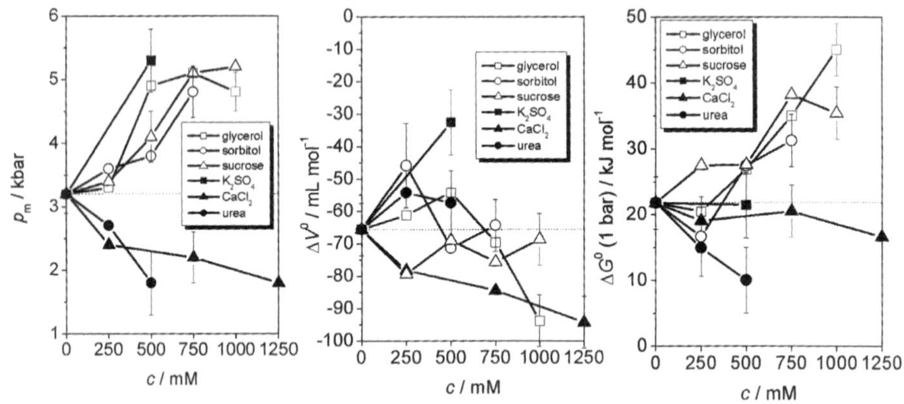

Fig. 31. Denaturation pressure (p_m), volume change of pressure-induced unfolding (ΔV^o), and standard Gibbs free energy change (ΔG^o), extrapolated to 1 bar, for the pressure-induced unfolding of SNase at different molar concentrations (c) of cosolvents.

In the presence of K_2SO_4, the pressure-denaturated state still contains a large population of ordered secondary structures, possibly resembling a molten globule kind of state. A similar structure is also seen for the corresponding temperature-induced unfolded state conformation. A more disordered structure in the pressure-denatured state, with regard to SNase without a cosolvent, is observed in the presence of $CaCl_2$, a strong chaotropic agent. Consequently, a linear increase in $|\Delta V^o|$, and a pressure-destabilization of SNase with increasing $CaCl_2$ concentration are found, which leads to a slight reduction in ΔG^o only. Upon addition of urea, SNase unfolds at lower p_m values and the effect of increasing urea concentrations on ΔV^o is small. As a result, ΔG^o decreases markedly with increasing urea concentrations. Urea induces a more disordered structure, i.e., the denaturated state contains fewer ordered secondary structure elements than SNase in pure buffer solution.

4.3.4. Kinetic studies of the un/refolding reaction of proteins

By crossing the phase boundary applying a p-jump, the folding and refolding kinetics can be studied [52, 54]. A rapid decrease of pressure for a solution of SNase at 25 °C from for example 4000 bar (denaturing conditions) to 800 bar (native conditions) results in a relatively rapid decrease in the value of the radius of gyration, R_g, from near 29 to 18 Å. The observed pressure-jump relaxation profile for the decrease in R_g fits well to a single exponential decay with a time constant τ of 4.5 s. In contrast, a positive pressure jump at 25 °C from 1000 bar (near-native conditions) to 3500 bar (fully denaturing conditions) results in a very slow relaxation of R_g from 19 to ~25 Å (Fig. 32). As for the negative pressure jumps, the positive pressure jump profile is well-fit by a single exponential function, with a much longer time constant of τ = 14 min. Further analysis of the SAXS data reveals a more detailed picture of structural changes during the folding/unfolding process (three-dimensional shape models in Fig. 32).

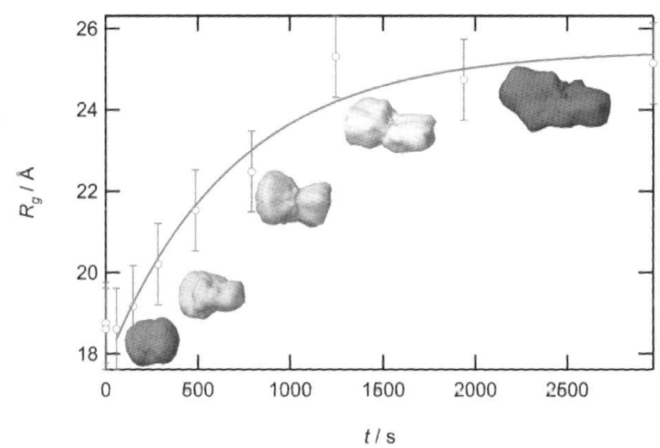

Fig. 32. Time evolution of the radius of gyration R_g (including the solvation shell) of SNase (1 % (w/w), pH 5.5) after a pressure jump from 1 to 3.5 kbar at T = 25 °C. Displayed is also the shape of the protein molecule in the course of the pressure-induced unfolding process.

Applying Eyring's transition state theory, we find from an analysis of the kinetic data that the activation volume for folding is large and positive (~57 mL mol^{-1}) and that for unfolding seems to be small and negative (~ −23 mL mol^{-1}). The volume of the protein solvent system in the transition state is thus significantly larger than in the unfolded state and somewhat smaller than in the folded state, so that the transition state lies closer to the folded than to the unfolded state in terms of system volumes. The positive activation volume of folding indicates that the transition state is accompanied by dehydration and chain collapse (with its accompanying packing defects). The positive activation volume for the folding process, which is responsible for the large increase in the relaxation time with pressure (allowing us to observe this process without resorting to ultrafast methods) may generally also be due to the fact that pressure acts by increasing the overall roughness of the free energy landscape.

The stage is now well set for further work addressing more complex questions, such as the study of the folding reaction of oligomers and protein complex formation as well as for studies of aggregation phenomena. Only a few studies have been performed in this direction so far [9-11, 18, 114, 124]. At pressures of 4-8 kbar, most small monomeric proteins unfold reversibly. Oligomeric proteins and multiprotein assemblies often dissociate into individual subunits already at pressures of 1-2 kbar (Fig. 33). The effect of pressure in promoting dissociation may be explained by the imperfect packing of atoms at the subunit interface, and negative volume changes resulting from the disruption of polar and ionic bonds in the intersubunit region. After dissociation by pressure, subunits may undergo further conformational changes.

We have discussed the fact that pressure destabilizes intermediates and leads to simplified folding kinetics in some cases, but is a powerful method for stabilizing folding intermediates in others. In several cases, high-resolution high-pressure NMR could also be used for detecting site-specific conformational fluctuations in proteins (for example, see Kremer et al., Akasaka, and Kamatari et al. in ref. [11]). The results suggest that both the rapid and slow conformational fluctuations are closely related to the dynamics of cavities and hydration.

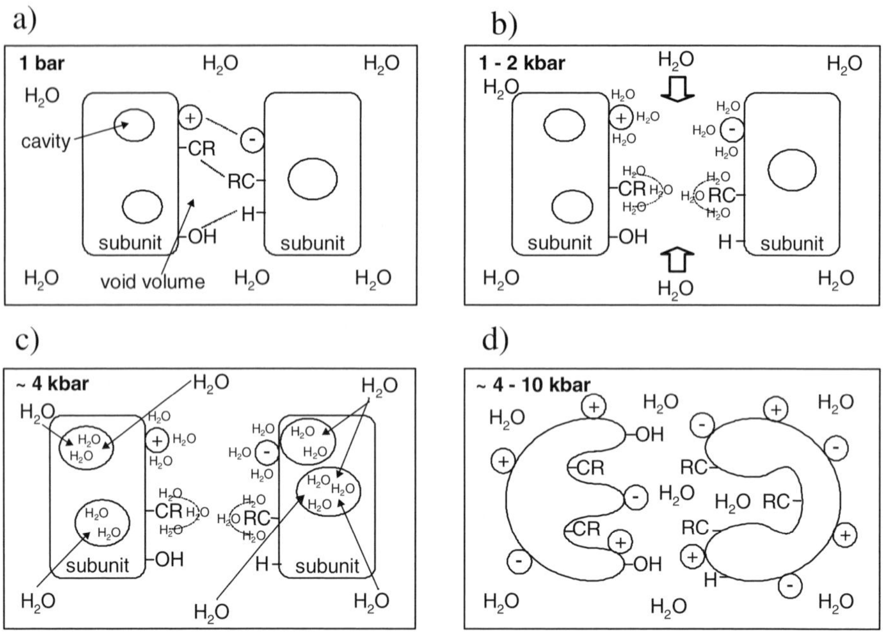

Fig. 33. Schematic representation of the effects of pressure on oligomeric proteins: a) native dimeric protein with cavities/voids; b) dissociation of the oligomer, hydration with electrostriction of polar/ionic groups, hydrophobic hydration of unpolar groups (-CR), release of void volume; c) weakening of hydrophobic interactions provides pathways for water to penetrate into the interior of the protein, swelling of the core - molten-globule like state; d) unfolding of subunits, disruption of the secondary/tertiary structure (hydration of residues not plotted here), loss of cavity volume within protein (adopted from ref. 139).

As described below, pressure studies might also lead to a better understanding of the interactions that lead to protein aggregation and will thus enhance our ability to design inhibitors and therapeutics for aggregation driven diseases.

4.3.5. Protein aggregation – amyloidogenesis

Since formation of the orderly-aggregated proteins - amyloids - was found to be associated with several neurodegenerative developments such as Alzheimer's, Huntington's, Creutzfeldt-Jakob's, and Parkinson's diseases, the problem of protein aggregation has begun to receive marked attention [125-127]. The phenomenon, initially thought to concern a handful of proteins, was later observed to be rather common, as amyloid β-fibrils even from otherwise stably-folded helical proteins were obtained [128, 129]. Mechanisms and thermodynamics lying behind protein aggregation, as well as the relation between aggregation and folding pathways remain still largely unclear. Recently, a hypothesis has been put forward that protein aggregation reflects a generic character of polypeptides as polymers and takes place whenever a polypeptide's main chain interactions are allowed to overrule specific native side-chain contacts in a folded protein. The presence of such stable tertiary contacts is a distinct feature of folded globular proteins discriminating them from "regular" polymers. Once these contacts are loosened, as it happens in a "molten globule" state, protein molecules become prone to aggregation. This is amply documented by numerous studies on amyloidogenesis under protein-destabilizing conditions such as low pH, point mutations, tertiary structure - disrupting trifluoroethanol, as well as high-pressure [130-138]. Pressure may affect the pathways of misfolding/aggregation and correct folding, as could be shown in several cases now. Formation of an amyloidogenic intermediate without further proceeding to aggregation is a unique property of pressure, which opens up the prospect to characterize the structure of the amyloidogenic form in detail. The isolation of these intermediates provides target for the development of antagonists capable of blocking protein misfolding and aggregation. Recently, we showed that under high hydrostatic pressure insulin forms amyloid of a different, a unique circular morphology (Fig. 34). The circular amyloid is accompanied by bent 20-100 nm long fibrils. The pressure-enhancement of a ring-like supramolecular fold suggests an anisotropic distribution of void volumes in regular amyloid fibres and the ability of high pressure to evoke drastic perturbations on an amyloidogenic pathway and may also help tune the conformation of amyloid templates.

In this chapter we clearly demonstrated that pressure dependent studies can help delineate the free energy landscape of proteins and hence help elucidate which features are essential in determining the uniqueness and stability of the native conformational state.

5. CONCLUSIONS

We conclude that pressure work on biomolecular systems can yield a wealth of enlightening new information on their structure, energetics, phase behavior and on their transition kinetics, and might promise fulfillment of the challenge set forth by W. Kauzmann when discussing thermodynamics of unfolding of proteins: "*Until more searching is done in the darkness of high-pressure studies, our understanding of the hydrophobic effect must be considered incomplete*" [140]. Furthermore, pressure may serve as valuable thermodynamic tweezers to study protein-protein interaction and aggregation. Ambitious goals, based on the rational modification of molecular structure-function relationships by pressure, still await a more-detailed understanding of the effects of pressure at a molecular level, however.

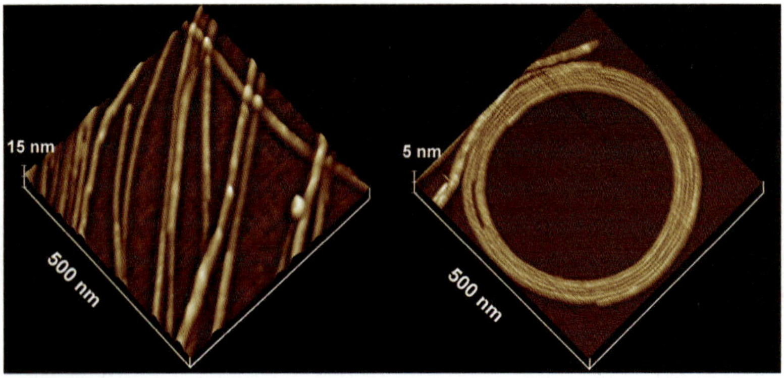

Fig. 34. AFM-images of insulin amyloid-fibres adsorbed on a mica surface. The left picture shows "normal" linear, temperature-induced (60 °C, incubation time: 1 day) fibres of "indefinite" length and a diameter ranging between 8 and 15 nm. In contrast, the image on the right shows that under high hydrostatic pressure conditions (60°C, 1500 bar, incubation time: 22 h), the protein is able to form fibres of a different topology: Amyloids of a unique circular morphology are observed here (width: 35-42 nm, the radius of 340-420 nm is the most abundant among the ring-shaped structures observed) [138].

ABBREVIATIONS

PC phosphatidylcholine, PE phosphatidylethanolamine, PS phosphatidylserine, FA fatty acid, LA lauric acid, MA myristic acid, PA palmitic acid, SA stearic acid, LC liquid-crystalline, TTC tetracaine, MO monoolein, ME monoelaidin, GD gramicidin D, DMPC 1,2-dimyristoyl-*sn*-glycero-3-phosphatidylcholine (di-$C_{14:0}$), DMPS 1,2-dimyristoyl-*sn*-glycero-3-phosphatidylserin (di-$C_{14:0}$), DTPE 1,2-ditetradecyl-*sn*-glycero-3-phosphatidylethanolamine (di-$C_{14:0}$), DPPC 1,2-dipalmitoyl-*sn*-glycero-3-phosphatidylcholine (di-$C_{16:0}$), DPPE 1,2-dipalmitoyl-*sn*-glycero-3-phosphatidylethanolamine (di-$C_{16:0}$); DSPC 1,2-distearoyl-*sn*-glycero-3-phosphatidylcholine (di-$C_{18:0}$), DOPC 1,2-dioleoyl-*sn*-glycero-3-phosphatidylcholine (di-$C_{18:1,cis}$), DOPE 1,2-dioleoyl-*sn*-glycero-3-phosphatidylethanolamine (di-$C_{18:1,cis}$), DEPC 1,2-dielaidoyl-*sn*-glycero-3-phosphatidylcholine (di-$C_{18:1,trans}$), POPC 1-palmitoyl-2-oleoyl-*sn*-glycero-3-phosphatidylcholine ($C_{16:0}$,$C_{18:1,cis}$), egg-PE egg-yolk phosphatidylethanolamine, SNase staphylococcal nuclease, DAC diamond anvil cell, DSC differential scanning calorimetry, SAXS small-angle X-ray scattering, SANS small-angle neutron scattering.

ACKNOWLEDGEMENTS

Financial support from the Deutsche Forschungsgemeinschaft (DFG) and the Fonds der Chemischen Industrie is gratefully acknowledged. I thank Drs. J. Seddon, R. Templer, J. Jonas, J. Silva, R. Vogel, K. Heremans, W. Dzwolak and C. Royer for many valuable discussions.

REFERENCES

[1] D. A. Wharton, Life at the Limits. Organisms in Extreme Environments, (Cambridge University Press, Cambridge, 2002).
[2] K. Hausmann and B. P. Kremer, (eds.), Extremophilc, (VCH, Wcinhcim,1995).
[3] P. Mentré and G. Hui Bon Hoa , Intern. Rev. Cytology, 201 (2001) 1.
[4] M. Groß and R. Jaenicke, Eur. J. Biochem., 221 (1994) 617.
[5] M. Groß, Exzentriker des Lebens, (Spektrum Akademischer Verlag, Heidelberg, 1997).
[6] F. Abe, C. Kato, and K. Horikoshi, Trends in Microbiol., 7 (1999) 447.
[7] D. H. Bartlett, Biochim. Biophys. Acta, 1595 (2002) 367.
[8] A. A. Yayanos, A. S. Dietz, and Van Boxtel R., Natl. Acad. Sci. USA, 78 (1981) 5212; C. Kato, A. Inoue, and K. Horikoshi, Trends Biotechnol., 14 (1996) 6.
[9] C. Balny, R. Hayashi, K. Heremans, and P. Masson, (eds.), High Pressure and Biotechnology, Colloque Inserm, Vol. 224, (John Libbey Eurotext, Montrouge, 1992).
[10] R. Winter and J. Jonas, (eds.), High Pressure Molecular Science, (Dordrecht, Kluwer Academic Publishers, NATO ASI E 358, 1999).
[11] R. Winter, (ed.), High Pressure Bioscience and Biotechnology II, (Springer-Verlag, Heidelberg, 2003).
[12] M. Kato and R. Hayashi, Biosci. Biotechnol. Biochem., 63 (1999) 1321.
[13] V. V. Mozhaev, K. Heremans, H. Frank, P. Masson, C. Balny, Tibtech, 12 (1994) 493.
[14] H. Ludwig, (ed.), Advances in High Pressure Bioscience and Biotechnology, (Springer-Verlag, Berlin, 1998).
[15] V. V. Mozhaev, K. Heremans, J. Frank, P. Masson, C. Balny, Struct. Func. Genet., 24 (1996) 81.
[16] R. Hayashi and C. Balny, (eds.), High Pressure Bioscience and Technology, (Elsevier Science, 1996).
[17] J. Jonas, (ed.), High Pressure NMR, (Springer-Verlag, Berlin, 1991).
[18] J. L. Silva and G. Weber, Annu. Rev. Phys. Chem., 44 (1993) 89.
[19] S. Janosch, E. Kinne-Saffran, R. K. H. Kinne, R. Winter, in: Advances in High Pressure Bioscience and Biotechnology II, R. Winter, (ed.), (Springer-Verlag, Heidelberg, 2003) p. 216.
[20] H. M. Ulmer, H. Herberhold, S. Fahsel, M. Gänzle, R. Winter, R. Vogel, Appl. Environm. Microbiol., 68 (2002) 1088.
[21] P. L. Chong, P. A. G. Fortes, D. M. Jameson, J. Biol. Chem., 260 (1985) 14484.
[22] F. Cornelius, Biochemistry, 40 (2001) 8842.
[23] K. Takahashi, T. Kubo, K. Kobayashi, J. Imanishi, M. Takigawa, Y. Arai, Y. Hirasawa, J. Orthoped. Res., 15 (1997) 150.
[24] A. A. Yayanos, A. S. Dietz,. R.van Boxtel, Science, 205 (1979) 808.
[25] A. A. Yayanos, Annu. Rev. Microbiol., 49 (1995) 777.
[26] M. H. Qureshi, C. Kato, K. Horikoshi, Extremophiles, 2 (1998) 93.
[27] F. Canganella, J. M. Gonzalez, M. Yanagibayashi, C. Kato, K. Horikoshi, Arch. Microbiol., 168 (1997) 1; F. Canganella, W. J. Jones, A. Gambacorta, G. Antranikian, Arch. Microbiol., 167 (1997) 233.
[28] N. L. Klyachko, A. V. Levashov, Curr. Opin. Colloid. Interf. Sci., 8 (2003) 179.
[29] R. Köhling, J. Woenckhaus, N. L. Klyachko, R. Winter, Langmuir, 18 (2002) 8626.
[30] P.-L. G. Chong and G. Weber, Biochemistry, 22 (1983) 5544.

[31] L. F. Braganza and D. L. Worcester, Biochemistry, 25 (1986) 2591; D. Worcester and B. Hammouda, Physica B, 241-243 (1998) 1175.
[32] P. T. T. Wong, D. J. Siminovitch, H. H. Mantsch, Biochim. Biophys. Acta, 947 (1988) 139.
[33] J. Jonas, C. –L. Cie, A. Jonas, P. J. Grandinetti, D. Campbell, A. Driscoll, Proc. Natl. Acad. Sci. USA, 85 (1988) 4115.
[34] R. Winter and W. C. Pilgrim, Ber. Bunsenges. Phys. Chem., 93 (1989) 708.
[35] A. Landwehr and R. Winter, Ber. Bunsenges. Phys. Chem., 98 (1994) 214.
[36] P. T. C. So, S. M. Gruner, E. Shyamsunder, Phys. Rev. Lett., 70 (1993) 3455.
[37] D. A. Driscoll, J. Jonas, A. Jonas, Chem. Phys. Lipids, 58 (1991) 97.
[38] O. Reis, R. Winter, T. W. Zerda, Biochim. Biophys. Acta, 1279 (1996) 5.
[39] C. Czeslik, R. Winter, G. Rapp, K. Bartels, Biophys. J., 68 (1995) 1423.
[40] A. Cheng, A. Mencke, M. Caffrey, J. Phys. Chem., 100 (1996) 299.
[41] S. Kaneshina, H. Ichimori, T. Hata, H. Matsuki, Biochim. Biophys. Acta, 1374 (1998) 1.
[42] H. Ichimori, H. Matsuki, T. Hata, S. Kaneshina, in: Advances in High Pressure Bioscience and Biotechnology, H. Ludwig, (ed.), (Springer-Verlag, Berlin 1998) pp. 161.
[43] S. Krishna Prasad, R. Shashidhar, B. P. Gaber, S. C. Chandrasekhar, Chem. Phys. Lipids, 43 (1987) 227.
[44] C. Czeslik, O. Reis, R. Winter, G. Rapp, Chem. Phys. Lipids, 91 (1998) 135.
[45] D. C. Fiech, B. B. Bonev, M. R. Morrow, Phys. Rev. E, 57 (1998) 3334.
[46] P. M. Duesing, J. M. Seddon, R. H. Templer, D. A. Mannock, Langmuir, 13 (1997) 2655.
[47] H. Seemannn, R. Winter, C. A. Royer, J. Mol. Biol. 307 (2001) 1091.
[48] G. Panick, R. Malessa, R. Winter, Biochemistry, 38 (1999) 6512.
[49] G. Panick, R. Malessa, R. Winter, G. Rapp, K. J. Frye, C. A. Royer, J. Mol. Biol., 275 (1998) 389.
[50] G. Panick, G. J. A. Vidugiris, R. Winter, C. A. Royer, Biochemistry, 38 (1999) 4157.
[51] M. W. Lassalle, H. Yamada, K. Akasaka, J. Mol. Biol., 298 (2000) 293.
[52] J. Woenckhaus, R. Köhling, P. Thiyagarajan, K. Littrell, S. Seifert, C. A. Royer, R. Winter, Biophys. J., 80 (2001) 1518.
[53] R. Winter, Biochim. Biophys. Acta,1595 (2002) 160; R. Winter, C. Czeslik, Z. Kristallogr., 215 (2000) 454; R. Winter, R. Köhling, J. Phys.: Condens. Matter, 16 (2004) S327.
[54] J. Woenckhaus, R. Köhling, R. Winter, P. Thiyagarajan, S. Finet, Rev. Sci. Instrum., 71 (2000) 3895.
[55] G. Büldt, R. Schlesinger, E. Pebay-Peyrola, H. J. Sass, in: Structure and Dynamics of Biomolecules: Neutron and Synchrotron Radiation for Condensed Matter Studies, E. Fanchon, E. Geissler, J.-L. Hodeau, J.-R. Regnard, P. A. Timmins, (eds), (Oxford University Press, Oxford, 2000) p. 251.
[56] P. Laggner, Topics in Current Chemistry, 145 (1988) 173.
[57] O. Kratky, P. Laggner, Encyclopedia of Physical Science and Technology, 14 (1987) (Academic Press, New York) 693.
[58] L. A. Feigin, D. I. Svergun, Structure analysis by small angle X-ray and neutron scattering, (Plenum Press, 1987).
[59] A. Guinier, A. Fournet, Small angle scattering of X-ray, (Wiley, New York, 1995).
[60] P. Lindner, T. Zemb, (eds), Neutron, X-ray and light scattering: introduction to an investigative tool for colloidal and polymeric system, (North-Holland, 1991).
[61] M. R. Arnold, H. R. Kalbitzer, W. Kremer, J Magn. Resonance, 61 (2003) 127.
[62] J. Eisenblätter, R. Winter, Magn. Res. Chem., 38 (2000) 662.
[63] R. Winter, Ann. Rep. NMR Spectr., 50 (2003) 163.
[64] M. –C. Lin, R. B. Macgregor Jr., Biochemistry, 35 (1996) 11846.
[65] R. B. MacGregor Jr., Biopolymers, 38 (1996) 321.
[66] D. N. Dubins, A. Lee, R. B. Macgregor Jr., T. V. Chalikian, J. Amer. Chem. Soc., 123 (2001) 9254.

[67] M. Giel-Pietraszuk, P. Salanski, J. Jurczak, J. Barciszewski, in: Advances in High Pressure Bioscience and Biotechnology II, R. Winter, (ed.), (Springer-Verlag, Heidelberg, 2003) p. 181.
[68] G. Cevc, (ed.), Phospholipids Handbook (Marcel Dekker, New York, 1993).
[69] R. Lipowski, E. Sackmann, (eds.), Structure and dynamics of membranes, Volumes 1A and 1B, (Elsevier, Amsterdam, 1995).
[70] J. M. Seddon, Biochim. Biophys. Acta, 1031 (1990) 1.
[71] G. Lindblom, L. Rilfors, Biochim. Biophys. Acta, 988 (1989) 221.
[72] M. W. Tate, E. F. Eikenberry, D. C. Turner, E. Shyamsunder, S. M. Gruner, Chem. Phys. Lipids, 57 (1991) 147.
[73] S. Hyde, S. Andersson, K. Larsson, Z. Blum, T. Landh, S. Lidin, B. W. Ninham, (eds.), The language of shape. The role of curvature in condensed matter: physics, chemistry and biology, (Elsevier, Amsterdam, 1997).
[74] V. Luzzati, Curr. Opin. Struct. Biol., 7 (1997) 661.
[75] T. Landh, FEBS Lett., 369 (1995) 13.
[76] H. Delacroix, T. Gulik-Krzywicki, J. Seddon, J. Mol. Biol., 258 (1996) 88.
[77] R. H. Templer, Langmuir, 11 (1995) 334.
[78] R. H. Templer, D. C. Turner, P. Harper, J. M. Seddon, J. Phys. II, 5 (1995) 1053.
[79] H. Chung, M. Caffrey, Nature, 368 (1994) 224.
[80] P. Mariani, B. Paci, P. Bösecke, C. Ferrero, M. Lorenzen, R. Caciuffo, Phys. Rev. E, 54 (1996) 5840.
[81] R. H. Templer, B. J. Khoo, J. M. Seddon, Langmuir, 14 (1998) 7427.
[82] R. H. Templer, J. M. Seddon, N. A. Warrender, A. Syrykh, Z. Huang, R. Winter, J. Erbes, J. Phys. Chem. B, 102 (1998) 7251.
[83] R. H. Templer, J. M. Seddon, P. M. Duesing, R. Winter, J. Erbes, J. Phys. Chem. B, 102 (1998) 7262.
[84] U. S. Schwarz, G. Gompper, Langmuir, 17 (2001) 2084.
[85] C. Czeslik, R. Malessa, R. Winter, G. Rapp, Nucl. Instr. Meth. Phys. Res. A, 368 (1996) 847.
[86] B.-S. Lee, S. A. Mabry, A. Jonas, J. Jonas, Chem. Phys. Lip., 78 (1995) 103.
[87] E. Shyamsunder, S. M. Gruner, M. W. Tate, D. C. Turner, P. T. C. So, C. P. S. Tilcock, Biochemistry, 27 (1988) 2332.
[88] J. Erbes, C. Czeslik, W. Hahn, R. Winter, M. Rappolt, G. Rapp, Ber. Bunsenges. Phys. Chem., 98 (1994) 1287.
[89] D. P. Siegel, W. J. Green, Y. Talmon, Biophys. J., 66 (1994) 402.
[90] D. P. Siegel, R. M. Epand, Biophys. J., 73 (1997) 3089.
[91] H. Chung, M. Caffrey, Biophys. J., 69 (1995) 1951; J. Briggs, H. Chung, M. Caffrey, J. Phys. II, 6 (1996) 723.
[92] O. Reis, R. Winter, Langmuir, 14 (1998) 2903.
[93] R. Winter, J. Erbes, R. H., Templer, J. M. Seddon, A. Syrykh, N. A. Warrender, G. Rapp, Phys. Chem. Chem. Phys., 1 (1999) 887.
[94] R. Winter, A. Gabke, C. Czeslik, P. Pfeifer, Phys. Rev. E, 60 (1999) 7354.
[95] S. Fahsel, E.-M. Pospiech, M. Zein, T. L. Hazlett, E. Gratton, R. Winter, Biophys. J., 83 (2002) 334.
[96] A. Landwehr, R. Winter, Ber. Bunsenges. Phys. Chem., 98 (1994)1585.
[97] P. L. –G. Chong, M. Zein, T. K. Khan, R. Winter, J. Phys. Chem. B., 107 (2003) 8694.
[98] H. Seeman, R. Winter, Zeitschr. Phys. Chem., 217 (2003) 831.
[99] C. Bernsdorff and R. Winter, Z. Phys. Chem., 193 (1996) 151.
[100] R. Winter, A. Landwehr, T. Brauns, J. Erbes, C. Czeslik, O. Reis, in: High Pressure Effects in Molecular Biophysics and Enzymology, J. L. Markley, D. B. Northrop, C. A. Royer (eds.), (Oxford University Press, 1996) p. 274.; O. Reis, A. Zenerino, R. Winter, in: Biological Macromolecular Dynamics, S. Cusack, H. Büttner, M. Ferrand, P. Langan, P. Timmins (eds.), (Adenine Press, Schenectady, 1997) p. 41.
[101] M. Zein and R. Winter, Phys. Chem. Chem. Phys., 2 (2000) 4545.

[102] N. Orr, E. Yavin, M. Shinitzky, D. S. Lester, Anal. Biochem., 191 (1990) 80.
[103] J. Erbes, R. Winter, G. Rapp, Ber. Bunsenges. Phys. Chem., 100 (1996) 1713.
[104] J. Erbes, A. Gabke, G. Rapp, R. Winter, Phys. Chem. Chem. Phys., 2 (2000) 151.
[105] M. Caffrey, J. Hogan, A. Mencke, Biophys. J., 60 (1991) 456.
[106] M. Steinhart, M. Kriechbaum, K. Pressl, H. Amenitsch, P. Laggner, S. Bernstorff, Rev. Sci. Instr., 70 (1999) 1540.
[107] A. M. Squires, R. H. Templer, J. M. Seddon, J. Woenckhaus, R. Winter, Langmuir, 18 (2002) 7384.
[108] J. M. Seddon, A. Squires, O. Ces, R. H. Templer, J. Woenckhaus, R. Winter, in: Self Assembly - The Future, B. H. Robinson, (ed.), (IOS Press, Amsterdam), in press.
[109] B. U. Klink, R. Winter, M. Engelhard, I. Chizhov, Biophys. J., 83 (2003) 3490.
[110] P. W. Bridgman, J. Biol. Chem., 19 (1914) 511.
[111] C. A. Royer, Biochim. Biophys. Acta, 1595 (2002), 201.
[112] G. Hummer, S. Garde, A. E. Garcia, M. E. Paulaitis, L. R. Pratt, Proc. Natl. Acad. Sic. USA, 95 (1998) 1552.
[113] H. Herberhold, S. Marchal, R. Lange, C. H. Scheyhing, R. F. Vogel, R. Winter, J. Mol. Biol., 330 (2003) 1153.
[114] K. Heremans and L. Smeller, Biochim. Biophys. Acta, 1386 (1998) 353.
[115] M. W. Lassalle, H. Yamada, K. Akasaka, J. Mol. Biol., 298 (2000) 293.
[116] R. Ravindra and R. Winter, Chem. Phys. Chem., 4 (2003) 359.
[117] S. A. Hawley, Biochemistry, 10 (1971) 2436.
[118] C. M. Dobson, A. Sali, M. Karplus, Angew. Chem., 110 (1998) 908.
[119] H. S. Chan and K. A. Dill, Proteins: Struct. Func. Genet., 30 (1998) 2.
[120] J. N. Onuchic, Z. Luthey-Schulten, P. G. Wolynes, Ann. Rev. Phys. Chem., 48 (1997) 545.
[121] K. J. Frye and C. A. Royer, Protein Sci., 6 (1997) 789.
[122] K. Ruan, C. Xu, Y. Yu, J. Li, R. Lange, N. Bec, C. Balny, Eur. J. Biochem., 268 (2001) 2742.
[123] H. Herberhold, C. A. Royer, R. Winter, Biochemistry, in press.
[124] G. Desai, G. Panik, M. Zein, R. Winter, C. A. Royer, J. Mol. Biol., 288 (1999) 461.
[125] J. D. Sipe and A.S. Cohen, J. Struct. Biol., 130 (2000) 88.
[126] G. S. Jackson, A. R. Clarke, Curr. Opin. Struct. Biol., 10 (2000) 69.
[127] L. C. Serpell, Biochim. Biophys. Acta, 1502 (2000) 16.
[128] J. Goers, S. E. Permyakov, E. A. Permyakov, V. N. Uversky, A. L. Fink, Biochemistry, 41 (2002) 12546.
[129] M. Fandrich, M. A. Fletcher, C. M. Dobson, Nature, 410 (2001) 165.
[130] A. D. Ferrão-Gonzales, S. O. Souto, J. L. Silva, D. Foguel, Proc. Natl. Acad. Sci. USA, 97 (2000) 6445.
[131] J. L. Silva, D. Foguel, C. A. Royer, Trends in Biochemical Sciences, 26 (2001) 612.
[132] F. Meersman, L. Smeller, K. Heremans, Biophys. J., 82 (2002) 2635.
[133] T. W. Randolph, M. Seefeld, J. F. Carpenter, Biochim. Biophys. Acta,1595 (2002) 224.
[134] A. D. Ferrão-Gonzales, L. Palmieri, M. Valory, J. L. Silva, H. Lashuel, J. W. Kelly, D. Foguel, J. Mol. Biol., 328 (2003) 963.
[135] J. Torrent, M. T. Alvarez-Martinez, F. Heitz, J.-P. Liautard, C. Balny, R. Lange, Biochemistry, 42 (2003) 1318.
[136] D. Foguel, M. C. Suarez, A. D. Ferrão-Gonzales, T. C. Porto, L. Palmieri, C. M. Einsiedler, L. R. Andrade, H. A. Lashuel, P. T. Lansbury, J. W. Kelly, J. L. Silva, Proc. Natl. Acad. Sci. USA, 100 (2003) 9831.
[137] D. Foguel, C. R. Robinson, P. C. de Sousa, J. L. Silva, A. S. Robinson, Biotechnol. Bioeng., 63 (1999) 552.
[138] R. Jansen, S. Grudzielanek, W. Dzwolak, R. Winter, J. Mol. Biol., in press.
[139] B. B. Boonyaratanakornit, C. B. Park, D. S. Clark, Biochim. Biophys. Acta, 1595 (2002) 235.
[140] W. Kauzmann, Nature, 325 (1987) 763.

Chemistry at Extreme Conditions
M. Riad Manaa (Editor)

Chapter 3

Molecules to Microbes: In-Situ Studies of Organic Systems Under Hydrothermal Conditions

Anurag Sharma,[a,b] George D. Cody,[b] James Scott,[b] and Russell J. Hemley[b]

[a] Department of Earth and Environmental Sciences, Rensselaer Polytechnic Institute, Troy, NY 12180

[b]Geophysical Laboratory, Carnegie Institution of Washington, 5251 Broad Branch Road, NW, Washington, DC 20015

1. INTRODUCTION

It has been proposed based on thermodynamic arguments that organic compounds synthesized abiotically within the subsurface of the earth may constitute a significant reservoir [1]. There remains, however, only one study that has actually identified natural hydrocarbons with non-biological origins [2]. Precise estimates of the extent of such a reservoir and its relationship to the larger carbon cycle is unknown. What has been shown is that the natural world provides abundant catalytic phases that are capable of promoting organic synthesis from simple substrates like CO_2 and H_2 [3-5]. The principal catalytic phases include a wide range of transition metal sulfides and oxides; the environment of synthesis exists within the deep subsurface and is associated with both past and presently active hydrothermal systems. One of the intriguing aspects of these subsurface organosynthetic reactions is that in many superficial ways, they mimic aspects of the extant anabolic reactions utilized by chemolithoautotrophic micro-organisms. It is therefore also possible that deep subsurface organosynthetic reactions could provide the primary production for sustaining complex subsurface microbial (e.g., methanogens, acetogens and select sulfate reducers) communities.

As it is virtually impossible to study the wide range of natural and active deep surface environments, we are left with the option of experimentally simulating such environments in order to place constraints on the subsurface capacity to synthesize organic compounds and sustain a "hot deep biosphere". With this in mind, studies were carried out in various organic-rich systems in a wide range of conditions with the aim of directly monitoring processes such as structural phase transformations, organic synthesis reactions, fluid-phase immiscibility, reactions kinetics and pathways. These studies have provided a fundamental basis for testing the response of microbial life to increased hydrostatic pressures and viability of sustaining biological activity within deep subsurface. These experiments constitute preliminary steps at establishing a unified approach to the study of geobiology at extreme conditions.

2. EXPERIMENTAL TECHNIQUES

Our experiments were initiated by the recognition of the opportunities and advantages of direct optical observations and spectroscopic measurements for conditions simulating deep subsurface. To this end, we have developed and modified diamond anvil cell (DAC) technology to address some of the key problems related to prebiotic chemistry, organic-mineral interaction dynamics and microbial adaptation at moderate pressures in aqueous organic-rich solutions. Most of these experiments were designed to provide direct information on the mechanisms, kinetics and processes involved in these reactions. The DAC was coupled with micro-Raman spectroscopy system that enabled direct optical observations as well as *in-situ* spectroscopy of the biogeochemical systems.

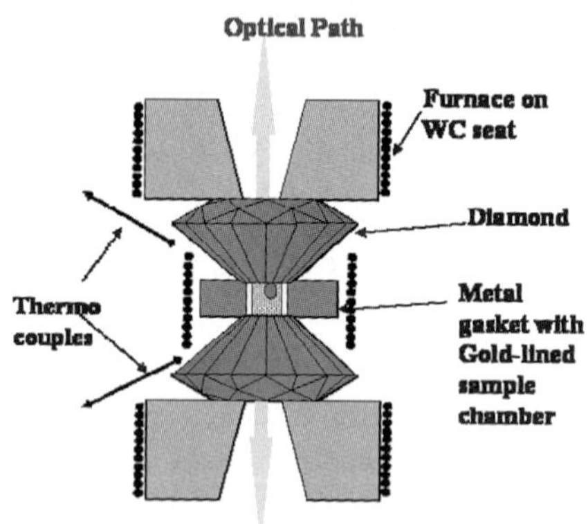

Figure 1. Schematic diagram of modified hydrothermal diamond cell (DAC) used in these studies. In some of the hydrothermal experiments, an additional resistance heater was mounted around the diamonds and sample to increase the heating rate and minimize the thermal gradients across the sample chamber.

2.1. Hydrothermal diamond cells

We used externally heated diamond anvil cells [6] for direct monitoring of reactions involved in the organic hydrothermal systems (e.g., methane – water, citric acid – water, formic acid – water) at high *P* and *T*. The DAC consists of two diamonds anvils that are mounted on tungsten carbide seats (Figure 1) surrounded by a pair of molybdenum wire resistance heaters capable of heating to 1000°C and wound around the tungsten carbide seats. In some experiments, an additional furnace was used around the diamonds for fast heating-rate response and a better control of thermal gradients. The diamonds as well as the heaters were held in place with ceramic cement capable of withstanding high thermal stress. The diamonds and the molybdenum heaters were protected from oxidation by introducing a mixture of 1% hydrogen and 99% argon into the cell during operation. A metal gasket (approximately 300 microns thick) with a 500 micron inner hole serves as the sample chamber, which provides an equivalent volume of 0.1 microliter.

In order to prevent the metal surface of sample chamber from acting as a catalyst in the hydrothermal reactions, we prepared the stainless steel and titanium gaskets with gold inserts (Figure 2), recognizing that for most of the reactions we were exploring gold exhibits no promoting capability [4]. A similar DAC set-up, but with a reduced optical working distance, was used for microbiological experiments at high pressures.

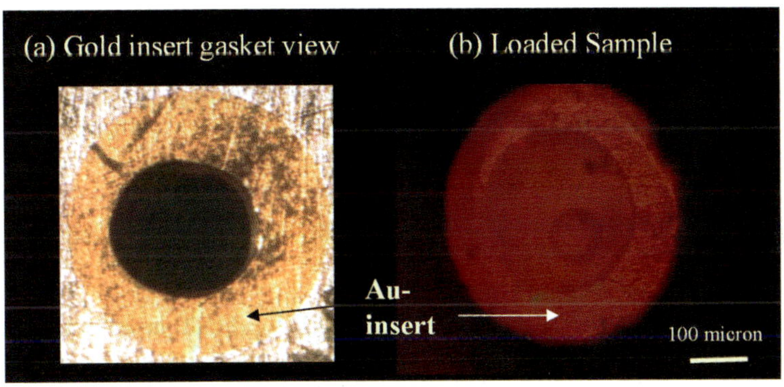

Figure 2. (a) Image showing the prepared gold-lined gasket. (b) Gasket loaded with citric acid –water solution in the DAC in reflected light.

2.2. Measurement of temperature and pressure

Sample temperatures were measured using a pair of chromel-alumel thermocouples cemented to the upper and lower diamonds on either side of the sample chamber. The thermocouples were calibrated against the melting points of $K_2Cr_2O_7$ (398 °C). The temperature uncertainties include thermocouple accuracy (±2 °C), controller variation (±1 °C), and thermal gradient across the sample. The furnace temperature was controlled by adjusting the input current regulated by a computer, which provided a temperature control to an accuracy of 0.1 °C. However, with the small thermal gradient considerations in the diamond cell, our temperatures are considered accurate to within ±1.0 °C.

Accurate pressure determination under these conditions has been the most difficult factor in such studies. In earlier studies of hydrothermal systems in the DAC [6], pressures in the cell were determined using the *P-V-T* properties of water [7]. This technique is based on the observation that, during heating of the cell from room temperature to some higher temperature, the volume of the liquid water phase expands. Consequently, the vapor bubble in the cell disappears at some temperature referred to as the homogenization temperature (T_h). At this temperature and beyond, the sample chamber contains liquid water or vapor (depending on the original size of the bubble) as well as any solids loaded into the cell. During heating beyond the homogenization temperature, the pressure is considered to change along, or close to, an isochore. However, due to stress related deformation of the gasket, a constant homogenization temperature is achieved only after several cycles of heating beyond T_h and then cooling to a temperature below T_h. Although limited to water-rich solutions, the technique is useful in obtaining a pressure approximation in the DAC at hydrothermal conditions. In our study, it was not possible to cycle the cell to obtain an isochoric sample chamber because the reactions of interest start irreversibly upon heating. To significantly reduce the stress-related deformation changes, the gasket was "pre-indented" with the diamond anvils to pressures of approximately 1 GPa (significantly higher than the intended experimental conditions). When tested with pure H_2O loaded in the cell and heated to the hydrothermal conditions, the pre-indented gaskets did not show any significant volume change to produce any pressure variation upon prolonged heating.

In order to obtain a direct and more accurate pressure determination, various internal pressure calibrants (e.g. quartz and ruby chips) are generally used. Internal calibrants, however, could not be used in the high temperature hydrothermal experiments due to interactions with the chemical system. In such cases, one of the more prevalent phases of the chemical system was calibrated as pressure indicator. For fluid-rich systems (methane-water), the pressures were also determined using the known phase equilibria of the methane hydrate decomposition and using shifts in the ruby fluorescence peak [8].

2.3. In-situ Raman spectroscopy

A TRIAX 550 spectrometer attached to an Andor –90°C cooled CCD detector was used for all spectroscopic measurements. Ar^+ laser lines at 488.0 nm and 514.3 nm were used. Reactions were monitored by time-resolved Raman spectroscopy, sequentially setup for two of three separate regions of interest within the spectral range. This enabled, for example, collecting information about the carboxylation/decarboxylation and hydration/dehydration processes by monitoring the various CO and CH vibration modes. This technique provided spectra in each region only after the collection of the spectra in other regions, and hence not favorable for faster kinetics. However, inclusion of OH and H_2 peaks gave a reasonably quantitative estimate on the extent of the hydrothermal reaction and valuable information for mass balance calculations (see further details in the experimental results for each system)

3. METHANE HYDRATES: PHASE TRANSFORMATION AND EQUILIBRIA

3.1. Overview

Methane hydrate is a hydrogen bonded network of water molecules that forms cage like cavities in which 'guest' gas molecules (methane and other species) reside. These phases are considered an important global energy resource because of the vast formations distributed globally in the deep regions of the continental shelves [9]. Recent experimental studies on these systems have found them stable at high pressures and exhibiting numerous phase transformations in *P*-*T* space. From a biochemical point of view, these compounds could provide a vast reservoir of biochemical energy trapped as a solid ice-like phase. If coupled with a suitable electron acceptor methane hydrates could be an enormous resource for deep-sea sediment microbial life (e.g. the anaerobic methane oxidizing syntrophic communities) [10]. With the likely presence of methane hydrates in planetary bodies such as Mars and Europa, these compounds may have significant biological and astrobiological importance.

However, a clearer understanding of the potential biological interaction of gas hydrates will benefit from a better understanding of the stability, structure and dynamics of these compounds under a range of environmental conditions.

To understand the high-pressure phase behavior of methane hydrate, in-situ observations were made in a hydrothermal diamond anvil cell under a microscope, and the identification of a new high-pressure phase (sII) was documented by Raman spectroscopy [11,12]. Raman spectroscopy has been used to identify the type of clathrate hydrate cage that contains methane [13,14] uniquely identifying the cages structure in sI, sII and sH gas hydrate phases. Our results show that several phase transitions can be observed, and some of these phase transitions project to lower pressure and temperature conditions, suggesting that they may impact the stability of gas hydrate phases in the deep ocean and in high-pressure engineering applications.

3.2. In-situ observations

Diamond cell experiments were performed between -40° and 60°C, and at pressures up to 900 MPa. To prepare samples, synthetic methane hydrate [11, 12] was loaded in the sample chamber of the cell. After the sample chamber is sealed, the total mass of the sample remains constant. The bulk density (and sample pressure) is then adjusted by changing the volume of the sample chamber, which is achieved by adjusting the distance between the two diamond-anvil faces. Once the volume of the sample chamber is fixed, the sample remains under isochoric conditions [6]. Thermal expansion of various parts of the cell, within the small temperature range of this study, has a negligible effect on the sample volume under these experimental conditions. Occasionally, samples were prepared by sealing Al_4C_3 and distilled water in the sample chamber, and CH_4 was generated at temperatures above 200 °C by the reaction

$$Al_4C_3(s) + 6H_2O(l) = 3\ CH_4(g) + 2Al_2O_3(s),$$

where s, l, and g represent solid, liquid, and gas, respectively. Even though there are no noticeable differences between results obtained in these two different loading methods, the Al_2O_3 powder generated in the reaction often hindered optical observations due to intense scattering.

As indicated earlier in the text, the pressure of the sample chamber was determined from the temperature at which the hydrate began to decompose with the formation of methane vapor (see Figure 5 for the *P-T* relations). This *P-T* point defines the density of the coexisting water.

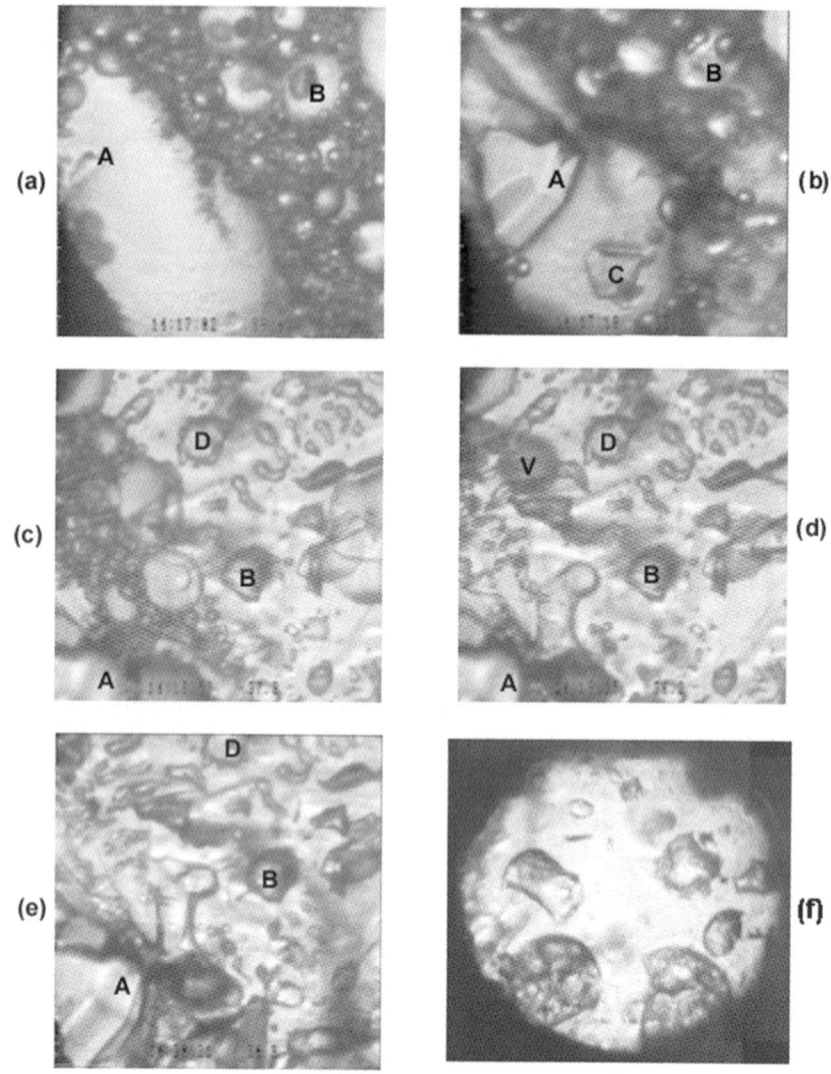

Figure 3. Images of a sample during a cooling cycle(a to e), and image of another sample (f) showing the optical relief of the new phase crystals. (a) The initial crystallization of the new methane hydrate phase (A) in coexisting water and methane vapor at 39.2 °C and 190 MPa with a crystal of the new phase (B). (b) The growth of new phase (A) with a second new phase (C) at 38.5 °C and 178 MPa in the presence of water and methane vapor. (c) The euhedral growth of the new phase (crystals B and D) in the presence of water and methane vapor at 37.5 °C and 163 MPa. Crystal A appears at the lower left corner. (d) The new phase (crystals A, B, D), and the last vapor bubble (V) in water at 35.9 °C and 142 MPa. (e) The coexistence of the new phase (crystals A, B, D, and many other high-relief crystals), sI methane hydrate phase (low-relief crystals), and water at 35.0 °C and 136 MPa. (f) In another sample, the new phase (high-relief crystals) with sI methane hydrate (background) and water at 25 °C and 125 MPa. The sample chamber is about 0.3 mm in diameter and 0.25 mm thick.

The sample pressure at lower temperatures, before the formation of ice, can be calculated from the isochore of H_2O of this particular density [7]. The bulk sample density was then increased by compressing the gasket with the two diamond anvils, and the new sample pressure, resulting from the reduction of sample volume, was determined in the same fashion. Typically, the sample pressure was high enough that methane hydrate was stable in the presence of water at room temperature.

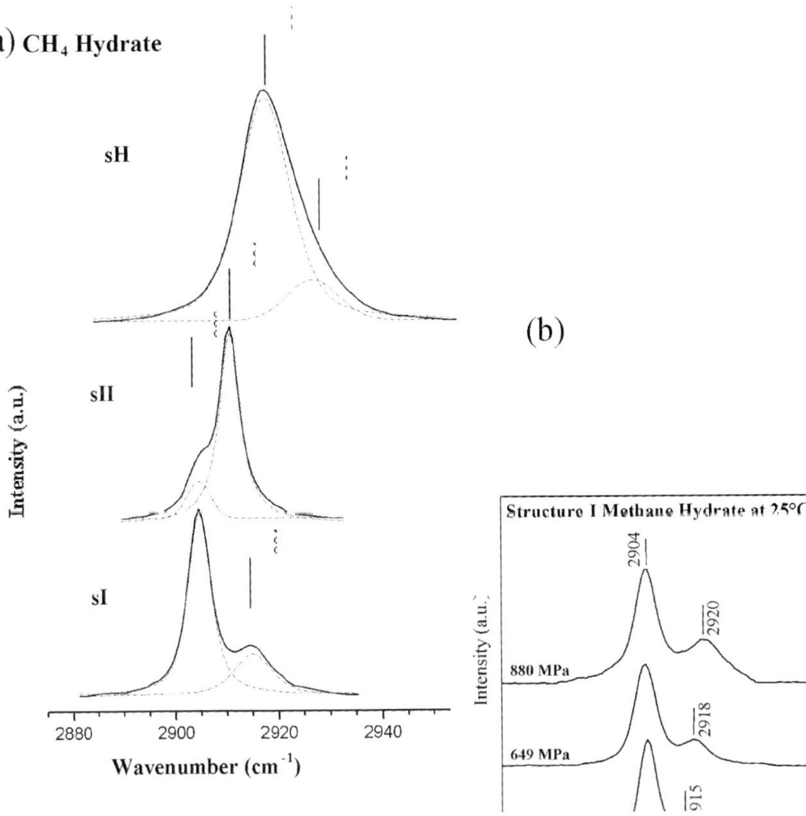

Figure 4. Comparison of the Raman spectrum of the CH_4 ν_1 band of sH methane hydrate at 25°C and 880 MPa (point E in Fig. 5) with the spectra of sI and sII methane hydrates at 25°C and 125 MPa (point H in Fig. 5); the intensity is given in arbitrary units. Also shown in dashed lines are the spectral deconvolutions of the Raman spectra. Spectra of (a; top) the new methane hydrate phase (sII) (high-relief crystals shown in Figure 3) and structure I methane hydrate (low-relief back ground crystals shown in Figure 3d) at 25 °C and 125 MPa; and (b; bottom) structure I methane hydrate at 25 °C and at pressures of 125, 649 and 880 MPa.

The growth of both the new high-pressure phase and structure I crystals during a cooling cycle is shown in Figure 3. Crystals of the new phase (crystals A, B, C, and D in Figure 3) have a higher optical relief, and a distinctly different Raman spectrum from that of structure I phase (Figure 4a). The pressures given in Figure 3 (a) to (d) were calculated from the data of

Nakano et al. [15] for the assemblage structure I methane hydrate-water- methane vapor and should be slightly higher than the actual pressures because of the presence of the new phase instead of structure I phase (see Figure 5). The pressure given in (e) was calculated from the equation of state of water [16] corresponding to a density of 1046.8 kg/m^3.

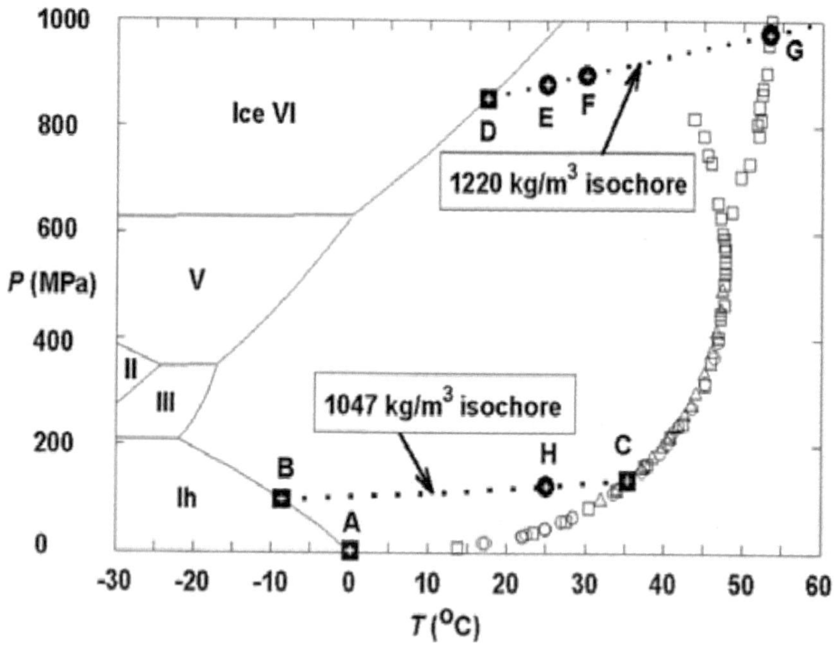

Figure 5. Previous experimental results and the *P–T* conditions for in situ observations in this study for the system CH_4–H_2O. The experimental data for the univariant *P–T* relations of the assemblage methane hydrate–water–methane vapor were taken from Marshall et al. (circles) [10], Dyadin et al. (squares) [11], and Nakano et al. (triangles) [12]. All symbols are for sI methane hydrate, except those squares branching out at higher *P–T* conditions. Points A, B, C and D (solid squares) indicate the *P–T* conditions for four invariant points, and they are for the following assemblages, respectively: sI methane hydrate–liquid water (lw)–ice Ih–methane vapor (v), sI and sII methane hydrates–lw–ice Ih, sI and sII methane hydrates–lw–v, and sI and sH methane hydrates–lw–ice VI. Points E, F and G (dots) are *P–T* points along the isochore of pure water for 1,220 kg/m^3 [14], and point H (dot) is a *P–T* point along the isochore of pure water for 1,047 kg/m^3. The former isochore was defined by the melting *P–T* condition of ice VI at point D (16.6 °C and 0.84 MPa) [15], and the latter isochore by that of ice Ih at point B (28.7 °C and 99 MPa; Chou et al., unpublished results). The latter isochore is also the univariant *P–T* conditions for the assemblage sI and sII methane hydrates–liquid water (Chou et al., unpublished results).

In some runs, structure I methane hydrate was found to metastably exist at higher pressures. Figure 4b shows the Raman spectra of structure I methane hydrate at 25 °C and at pressures of 125, 649 and 880 MPa. The shift of the methane peak for the small cavity from 2915 cm^{-1} to about 2921 cm^{-1}, as pressure increases from 125 to 880 MPa, is in agreement with the shift reported by Nakano et al. [15], except the effect of pressure on the peak position is not as strong. Apparently, the new high-pressure phase was not formed (or observed) in the experiments of Nakano et al. [15] and Hirai et al. [17], but has been reported in a more recent study by Kumazaki et al. [20].

3.3. Constraints on phase relations

Phase relations in the system CH_4-H_2O were determined by optical observations of the sample in the diamond cell during cooling and heating. In the presence of the new phase, structure I methane hydrate and water, ice Ih was formed during further cooling of the sample to about -40 °C. The invariant point, for the assemblage of the new phase-structure I methane hydrate-ice Ih-water (point B in Figure 5), was located by the melting temperature of ice Ih at -8.7(0.3)°C during warming of the sample [18]. In theory, in the presence of the new high-pressure phase and structure I methane hydrate, ice Ih melts at an invariant point (fixed *P* and *T*). However, in the presence of the new phase and structure I methane hydrate, ice Ih was observed to melt over a range of *P*-*T* conditions, indicating non-equilibrium conditions. Equilibrium is not expected to be achieved during our ice melting experiment under a heating rate of about 0.02 °C/s. Here, the invariant point was defined by the temperature at which the last crystal of ice Ih melts. On the basis of the equation of state of H_2O [18], the pressure of this invariant point is 99 MPa, and the density of water at this point is 1046.8 kg/m^3.

On further warming, the sample followed the *P*-*T* path shown as line BC in Figure 5, where the line ended at the invariant point C for the assemblage of the new phase-structure I methane hydrate-water-methane vapor. This invariant point was defined during warming by the temperature at which the first methane bubble appeared (35.3 (0.5) °C). Note that this observation is in agreement with the temperature at which the last methane vapor bubble almost disappeared at 35.9 °C during a cooling experiment (Figure 3d). The pressure of this invariant point, defined by the water density of 1046.8 kg/m^3 at 35.3 °C, is 136.7 MPa [16]. This pressure is very close to the values of 131.0 and 134.6 MPa, interpolated from the data of Marshall et al. [19] and Nakano et al. [15] respectively, for the assemblage structure I methane hydrate-water-methane vapor at 35.3 °C.

These studies have provided direct information on the phase boundaries associated with the sI, sII and sH phases. The formation of the sII phase in methane hydrates, as evidenced by the metastable persistence of sI phase, may be kinetically or compositionally controlled. Further investigations are required to obtain complete (i.e., compositionally dependent) phase relations in the methane hydrate system in this *P*-*T* range.

4. CITRIC ACID-WATER SYSTEM

4.1. Hydrothermal organic synthesis

Theoretical studies [21, 22] have postulated that emergence of autotrophic metabolic pathways, such as reductive citric acid cycle (RCC) may have occurred under high *P*-*T* hydrothermal conditions analogous to deep marine hydrothermal vents. Earlier experimental work established that citric acid decomposition under hydrothermal conditions leads to the formation of a number of products connected by a large number of reaction pathways [4, 23-25]. Using a high pressure hydrothermal apparatus, Cody *et al.* [21, 25] have detailed the effects of temperature as well as pressure in the in the hydrothermal decomposition of citric acid and have followed the reaction pathways by following quench product analysis (Figure 6). Key questions, such as the controls of the phase behavior of CO_2 in this system and its effect on the pathways remain unanswered. In order to better understand the kinetics as well as the effects of pressure and fluid phase behavior on the reaction pathways, direct observations on this system at high P and T are required.

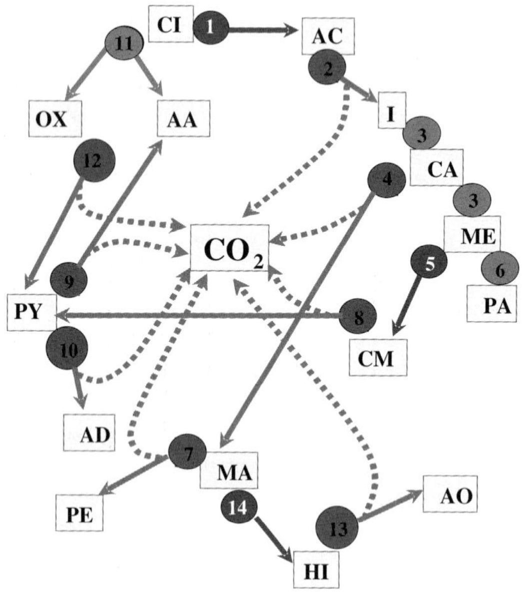

Figure 6. A compilation of the important decomposition pathways for citric acid based on Cody et al. [25]. Each reaction is numbered to correspond with discussions in the text. A solid line encloses each distinct pathway, described in detail within the text. The end products of each pathway are acetone, acetic acid and propene with CO_2. The various reactions within this reductive citric acid cycle are:

1. Citric (CI) <==> Aconitic (AC) + H_2O
2. Aconitic (AC) ==> Itaconic (I) + CO_2
3. Citraconic (CA) <==> Itaconic (I) <==> Mesaconic (ME)
4. Itaconic (I) ==> Methylacrylic (MA) + CO_2
5. Unsaturated dibasic acids (I, CA, ME) + H_2O ==> Citramalic (CM)
6. Itaconic (I) <==> Paraconic (PA)
7. Methylacrylic (MA) ==> Propene (PE) + CO_2
8. Citramalic (CM) ==> Pyruvic (PY) + CO_2 + H_2
9. Pyruvic (PY) ==> Acetic (AA) + CO_2 + H_2
10. Pyruvic (PY) ==> Acetaldehyde (AD) + CO_2
11.Citric (CI) ==> Oxaloacetic (OX) + Acetic (AA)
12. Oxaloacetic (OX) ==> Pyruvic (PY) + CO_2
13. Hydroxy Isobutanoic (HI) ==> Acetone (AO) + CO_2 + H_2
14. Methylacrylic (MA) + H_2O <==> Hydroxy Isobutanoic (HI)

This study was undertaken to develop techniques for direct monitoring of organic hydrothermal synthesis reactions in DACs combined with direct microscopic and spectroscopic observations. Such direct observation provided additional insight into the mechanism, kinetics and phase behavior (miscibility characteristics) of the fluid-rich system. Described below are some results on the direct monitoring of citric acid-H_2O system at high *P* and *T* with DAC Raman spectroscopy combined with quench product gas chromatographic analysis.

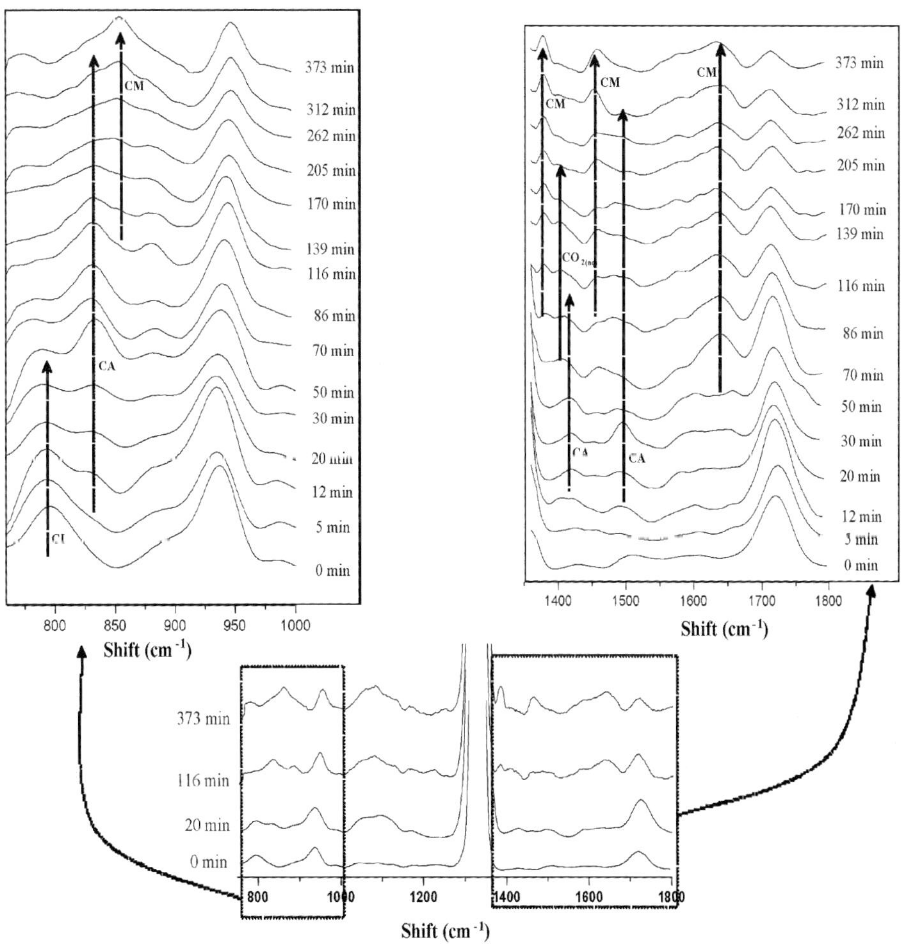

Figure 7. Compilation of Raman spectra with time in the C=O stretching region . Expanded view showing changes in the (a) C-OH vibrational frequencies in the ~800 cm^{-1} range, (b) C-O and C=O vibrational frequencies near 1380cm^{-1} and 1640cm^{-1}.

4.2. In-situ Raman spectroscopy

Typical hydrothermal experiments were setup by loading a dilute citric acid solution in a Au-lined stainless steel gasket enclosed between the two diamond anvils in the optical cell. The sample was abruptly heated to and stabilized at 200°C. In-situ Raman spectroscopy measurements were continuously made from the initiation of the heating run. Each of the Raman spectra was collected for 60 seconds using a 1500 grooves/mm grating for high resolution. By switching the position of the spectrometer after every 60 seconds, the spectra were collected first between the 750 – 1800 cm^{-1} (to monitor the C=O, C-O vibrations) and subsequently in the 2825 – 3640 cm^{-1} range (to monitor the C-H and OH stretching vibrations). This process of spectroscopic observation was repeated continuously throughout the experiment, which typically lasted for 4-6 hrs. Critical differences in the Raman bands of these carboxylic acids in the C=O and C-H stretching regions were employed for monitoring the progress of the citric acid decomposition reaction and evolution of the various byproduct phases (Figure 7). Additionally, progressive concentration changes of the carboxylic acids during the course of the experiment was estimated using the ratio of C-H vibrational peak for the carboxylic acids (~2960 cm^{-1}) with the maximum intensity peak of the O-H stretching band of water (~3400 cm^{-1}).

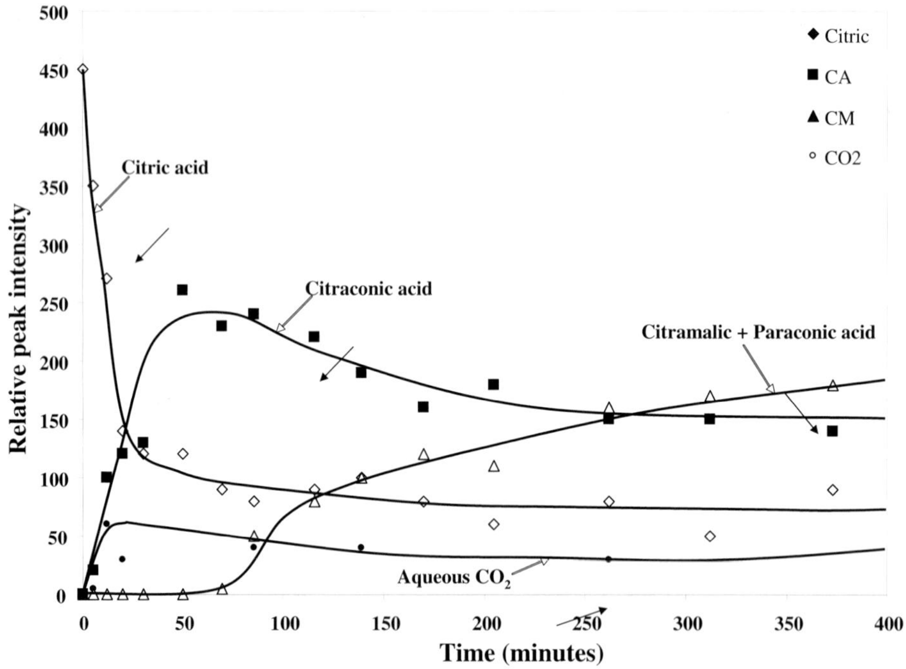

Figure 8. Kinetics data for organic synthesis reactions in the citric acid-water system at 200°C and 200 MPa, the results show decay of citric acid coupled with synthesis of various carboxylic acids (CA: citraconic acid; CM : citramalic and paraconic acids) . Note that because of the phase separation of CO_2 from the aqueous phase, the data here reflect the solubility of CO_2 in the aqueous phase at these conditions.

Monitoring the citric acid decomposition reaction using Raman spectroscopy provided a direct record of the chemical evolution of the system as reflected by progressive changes in the various stretching modes. A compilation of the Raman spectra in the C=O stretching range (Figure 7) shows the development of additional peaks near the 850 cm^{-1}, 1380 cm^{-1} and 1640 cm^{-1}, indicative of formation of by-product phases. On the other hand, the C-H vibration band exhibits a continuous change in the shape and intensity of 2900 cm^{-1} C-H stretch peak, reflecting dehydration of citric acid to aconitic acid, which is rapidly followed by decarboxylation of aconitic acid to form itaconic acid. Using the changes in the fitted peak area ratios, the rate of evolution of the hydrothermal system is shown in Figure 8.

Within the first 30 minutes of the initiation of the heating run, an additional C-OH deformation peak developed at 834 cm^{-1}, close to the preexisting 794 cm^{-1} band (Figure 7). This is consistent with the formation of a second hydrated phase (citramalic acid) formed from the hydration of itaconic acid. Although this reaction is persistent throughout the experiment, the presence of aconitic is difficult to detect after the initial stages of the reaction. This is because the rate of decarboxylation of aconitic is much faster than the rate of dehydration of the citric acid, which reduces the concentrations of aconitic in solution. Subsequently, this change in the C-OH frequency is followed by the development of additional peaks in the 1640 cm^{-1} range corresponding to the C=O stretching modes of the unsaturated dicarboxylic acids citraconic, itaconic and mesaconic acid. Due to the combination of the slightly different C=O frequencies of the citraconic, itaconic and mesaconic phases, the resultant peak is very broad and it is difficult to distinguish each dicarboxylic acid group individually. The change in the C=O peaks is supplemented by a simultaneous evolution of a 1380 cm^{-1} band, corresponding to the formation of CO_2 from the decarboxylation process (Figure 7). This change in the vibrational spectrum with continued heating further enhances the intensities of peaks corresponding to the CO_2 and citric decarboxylation by-products.

4.3. Chromatographic analysis

The quenched reaction products (after esterification) were analyzed using a Hewlett-Packard 6890 Series Gas Chromatograph (GC) interfaced with a 5972 Series quadrupole mass spectrometer (MS). To enable the GC-MS analysis on this exceedingly small volume sample (~0.1 microliter), an elaborate and careful sample extraction technique was developed. Immediately after the experiment, the diamond cell containing the sample was stored in a freezer at –20 °C. Before extracting the sample, the cell was immersed in liquid nitrogen, completely solidifying the sample. The cell was then carefully opened and the gasket containing the frozen sample was transferred into a glass vial for standard esterification procedure with BF_3-propanol for analysis.

Consistent with the previous experimental studies [4, 25], the analysis shows the presence of citric acid along with itaconic, citraconic, mesaconic, citramalic, as well as paraconic acid (Figure 9). In addition to these components methyl succinic and hydroaconitic acids show a dominant presence. The presence of these saturated compounds indicates that water reacted with the steel gasket to produce significant H_2; the stainless steel gasket is likely to provide an excellent surface site for catalytic hydrogenation. Although the sample was initially shielded from the steel gasket by a gold insert (Figure 2), there evidently was some interaction between the fluids and the steel gasket, which helps stabilize the hydroaconitic acid. In a previous attempt, we had visually observed the breach of the gold insert at high temperatures. The GC-MS analysis of the sample showed methyl succinic and hydroaconitic acid as the dominant species. Although some citric acid has persisted, only small amounts of citraconic and itaconic acids were detected. These results highlight the difficulties in controlling reaction conditions. Extreme care must be taken during DAC experiments and careful consideration must be made for the potential of fluid leaking across the inert gasket to the reactive steel gasket.

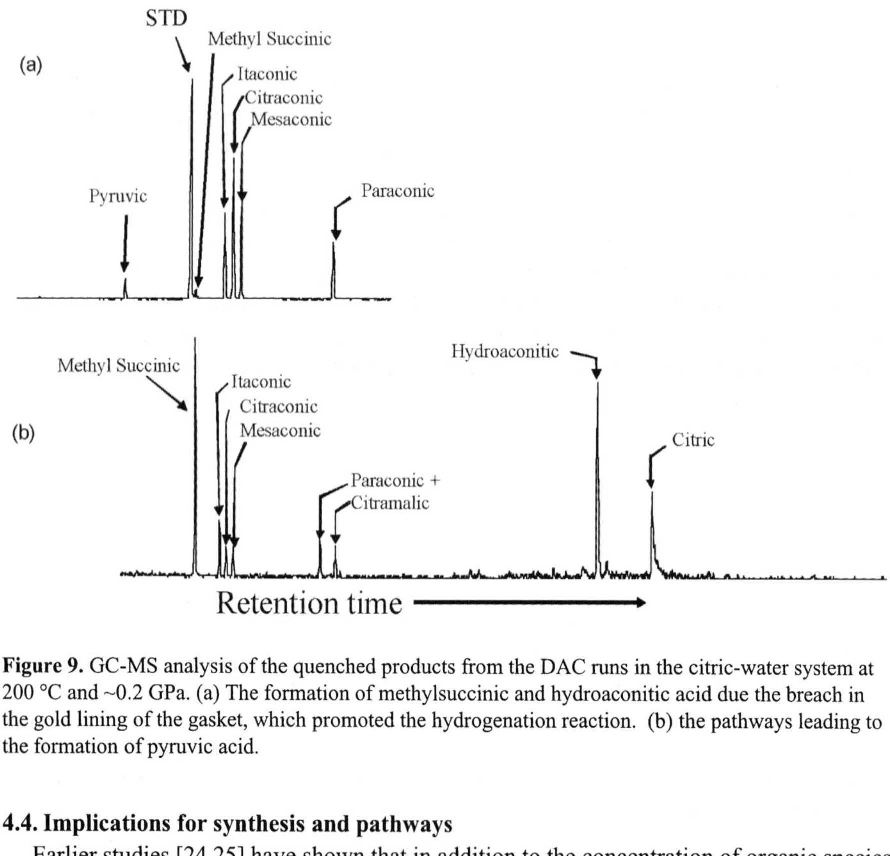

Figure 9. GC-MS analysis of the quenched products from the DAC runs in the citric-water system at 200 °C and ~0.2 GPa. (a) The formation of methylsuccinic and hydroaconitic acid due the breach in the gold lining of the gasket, which promoted the hydrogenation reaction. (b) the pathways leading to the formation of pyruvic acid.

4.4. Implications for synthesis and pathways

Earlier studies [24,25] have shown that in addition to the concentration of organic species, temperature is the most significant variable controlling the kinetics of the citric acid-H_2O system. Cody *et al.* [25] further show that for a given citric acid concentration, the system exhibits a very complex behavior at high *P* and *T*. These studies have described the behavior of the system under hydrothermal conditions in detail and have formed the basis for interpretation of these diamond cell experiments. In this study, we have used in-situ Raman spectroscopy to monitor the reaction pathways and phase behavior in the decarboxylation reactions as well as derived kinetic information from these measurements.

The differences in the C=O vibrational frequencies of the citric acid and its decarboxylation byproducts (CE, I, ME) has been used to extract reaction kinetic information (Figure 8). Figure 7 (a) shows a plot of the relative variation of the ratios of the 1653 cm^{-1} and 1740 cm^{-1} peaks with the progression of the reaction. These calculations indicate that the reaction rapidly approaches equilibrium at these hydrothermal conditions. This observation is consistent with those of Cody et al. for this system at similar conditions [25]. Additional information regarding the progression of the dehydration reactions can be inferred from the changes in the ratios of the CH (2930 cm^{-1}) and the OH (3430 cm^{-1}) peaks as shown in Figure 7. As would be expected, this aspect of the citric acid decomposition competing with the various decarboxylation reactions would be much slower to reach the steady state. However, direct interpretation of the CH/OH peak variation is difficult due to the combined effects of the OH stretching vibrations of the various organic species added during the ongoing reaction. The decarboxylation process was monitored using the O-C-O symmetric stretching vibration intensity at 1330 cm^{-1}, as shown in Figure 7. This variation indicates a very slow initiation of the decarboxylation reactions and does not exhibit steady state even after the equilibration of the citric acid decomposition as inferred form Figure 7. To explain this anomaly it is important to understand the mode of interaction of CO_2 with the organic fluid solution. According to the data of Barnes and Greenwood [26], for CO_2 - H_2O solutions, the CO_2 phase at these experimental conditions of 200 MPa and 200 °C should phase separate. Visual in-situ examination of the sample at these hydrothermal conditions shows no indication of any immiscibility in the sample until after a significant formation of citramalic and paraconic molecules.

5. FORMIC ACID – WATER – FeS SYSTEM

5.1. Metal sulfides as catalysts

In all extant organisms, transition metal sulfide clusters play a crucial catalytic role in biological energy conversion systems [27]. The potential connection between this essential role and the predominance of mineral sulfides in hydrothermal and volcanic vents has led some to speculate that life may have emerged from such environments [28, 29]. Such a connection is further supported by the molecular phylogeny showing early lineages for such bacteria and archea [30]. Recent experiments designed to replicate aspects of the primitive hydrothermal vent solution chemistry have revealed the intrinsic potential for the synthesis of alkyl thiols [3] from reactions involving iron sulfides in the presence of CO. The formation of acetate [31] has been demonstrated in aqueous solutions containing methyl thiol and CO in the presence of iron and nickel sulfides. Huber and Wächtershäuser [31] proposed that iron and nickel sulfides can mimic the functionality of acetyl coenzyme A (CoA) synthase, the iron- and nickel-containing enzyme complex present in chemoautotrophic anaerobic organisms [31]. Cody et al [3] demonstrated the formation of pyruvate, a key molecule for entry into metabolic reaction pathways, under hydrothermal conditions. More recently Cody et al. [4] showed that many transition metal sulfides are capable of promoting reactions that mimic the activity of the acetyl-CoA synthesase complex.

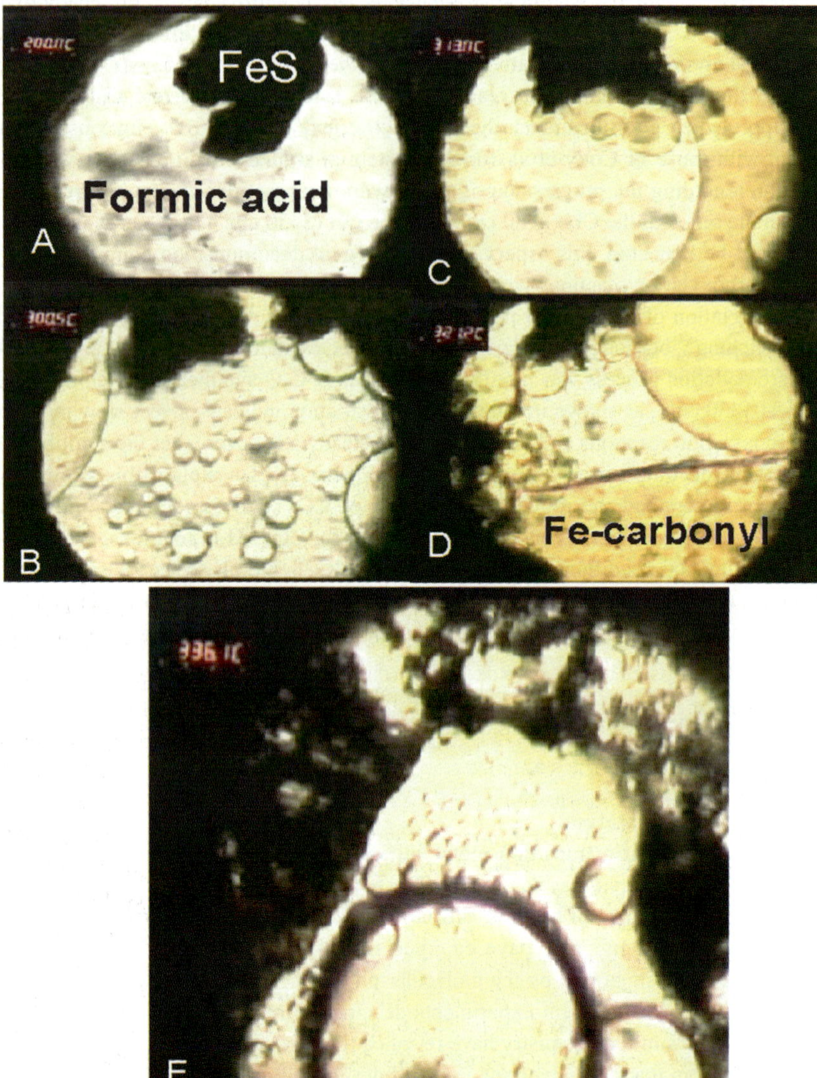

Figure 10. Hydrothermal reactions in the formic-rich system under hydrothermal conditions at ~300°C and ~0.4 GPa. (A-D) Formic acid –water with FeS single crystal at hydrothermal conditions. The sequence follows the reaction: $FeS + HCOOH = Fe(CO)_5 + H_2O + S^0$. Starting with (A) FeS crystals with formic, shows subsequent formation of the orange colored Fe-carbonyl phase, with nanoparticulate cluster of elemental sulfur (D). (E) Formic acid-FeS-alkyl thiol reaction. This shows the formation of Fe-bearing organometallic complexes.

5.2. High *P-T* experiments

The precise mechanism for the synthesis of these biorelevent organics provides an obvious challenge. Insight into the chemical system at *P* and *T* can be gained by monitoring the evolution of the system under hydrothermal conditions. With this in mind, hydrothermal reactions with formic acid were performed in the presence of FeS and subsequently an alkane thiol. Formic acid hydrothermal decomposition provides a stoichiometric source of CO_2, CO, H_2O, and H_2, governed by the thermodynamics and fast kinetics of the water-gas shift reaction, $CO_2 + H_2 = CO + H_2O$. In the absence of metal sulfide catalyst, at high temperature, $CO_2 + H_2$ is strongly thermodynamically favored, as we observe in the DAC (Figure 10). In the presence of a synthetic FeS single crystal within the formic acid – water system at ~300°C and ~0.5 GPa (Figure 10), we observed rapid formation of iron pentacarbonyl in aqueous solution. This reaction resulted in a CO dominated reaction solution with all the iron from FeS combined in $Fe(CO)_5$ and formation of noncrystalline S^o aggregates. Experiments conducted with the addition of alkyl thiol to the formic-water-FeS system (Figure 10) produced iron-rich carbonylated organic complexes such as $Fe_2(RS)_2(CO)_6$. These organometallic compounds may have promoted the synthesis of pyruvatic acid via a double carbonyl insertion reaction [3]. The significance of the formation of pyruvate in such systems is that it is a high-energy organic molecule capable of fueling a broad range of metabolic synthetic reactions as well as an excellent source molecule to drive biological energy conversion for ATP.

6. HIGH-PRESSURE MICROBIOLOGY

6.1. Monitoring viability under extreme conditions

Microbial communities are known to adapt to a wide range of pressures, temperatures, salinities, pH, and oxidation states. Although the chemical and physical conditions in these extreme environments are reasonably well constrained, the consequence of these physical parameters on the physiology of microbial communities is not well understood. Significant attention has been focused on the effects of high and low temperature on physiology [32, 33] There is some evidence that elevated pressure may also manifest interesting effects on cellular physiology [34, 35]. For example, recent studies report that elevated pressure may lead to enzyme inactivation, compromise cell-membrane integrity, and suppress protein interactions with various substrates [34–37]. Whereas the cumulative impact of these pressure-induced effects on microbial metabolism and physiology is an inhibition in growth rate and cellular division in microorganisms [38–40], exactly how these factors affect intact cells is not well understood [35, 41] The high-pressure studies on biological systems have been either on individual biomolecules related to high pressure biophysics studies or based on "indirect" measurements of cell growth related to microbiological studies in stressed environments. A generalized notion of pressure effects on life is based on pressure producing a corresponding change in volume.

Biophysical studies have defined tentative boundary conditions to which the biomolecules can sustain structural and functional integrity. Studies have shown that pressure affects the structure of membranes, biopolymers, and multimacromolecular assemblies because of actions on noncovalent bonds. Differtial alterations in the hydrophobic interactions within proteins [43-45], micelles, and lipids [46] are observed due to increased pressure.

Studies of microbiology under extreme conditions started with ZoBell's [40] early discovery of barophiles (or piezophiles) in the 1950s. Further breakthroughs on the characterization of extemophiles were made in devising high pressure culturing techniques [47]. Since the discovery of piezophilic and piezotolerant organisms, much interest has been focused on effects of hydrostatic pressure on a variety of deep-sea and surface-dwelling organisms [48 – 50]. These 'extremophiles' are considered specialized organisms that have adapted for living at deep ocean high pressure (and also temperature) conditions. Studies on the various biomolecular components in the extremophiles have provided insight into the mechanisms of life's adaptation to extreme conditions. Early studies by Landau [51] also studied effects of pressure on non-barophiles, such as *Eschericia coli*. This and subsequent studies [52] showed a remarkable reduction in viability of microorganisms with increased pressure. Combined with the fact that pressure causes no dramatic change in the covalent bonding on compounds, a wide ranging food preservation applications emerged in the 1990's. Numerous recent studies [53-56] showing microbial inactivation at high pressures have made high pressure biotechnology a viable option.

However, these studies have been limited to conventional high pressure techniques that required extraction and subsequent growth of microbes for testing viability. Although for moderate pressures (up to 100 MPa), continuous culture reactors have been used [47] high-pressure microbial inactivation experiments [53-56] pressurization is done in pressure steel vessels with samples enclosed in a plastic/teflon bag. Upon depressurization, the cells are extracted and suspensions are serially diluted in a nutrient rich medium and plated onto filter-sterilized sodium pyruvate added to the molten agar. Colonies are then counted after an incubation of 48 hours [54]. This test for viability with such highly stressed microorganisms, although crucial [57], is not likely to provide a realistic picture. The microorganisms that may have survived at high pressures will most likely not have the same growth rates and may even be further stressed from increased nutrients in the 'new' post-pressure growth environment. Direct observations using fluorescence/bioluminescence measurements or flow cytometry would provide a more realistic constraint on the survivability of microbes under extreme pressures.

We have made direct optical observations and measurements of microbial activity at various pressures. As in the above experiments, we have used diamond anvil cells in combination with micro-Raman spectroscopy and optical microscopy to directly monitor their viability and metabolic activity at extreme conditions [58]. The following is an overview of these direct observations of microbial activity under extreme pressures and their implications for adaptive mechanisms of life (as we know it) on this planet.

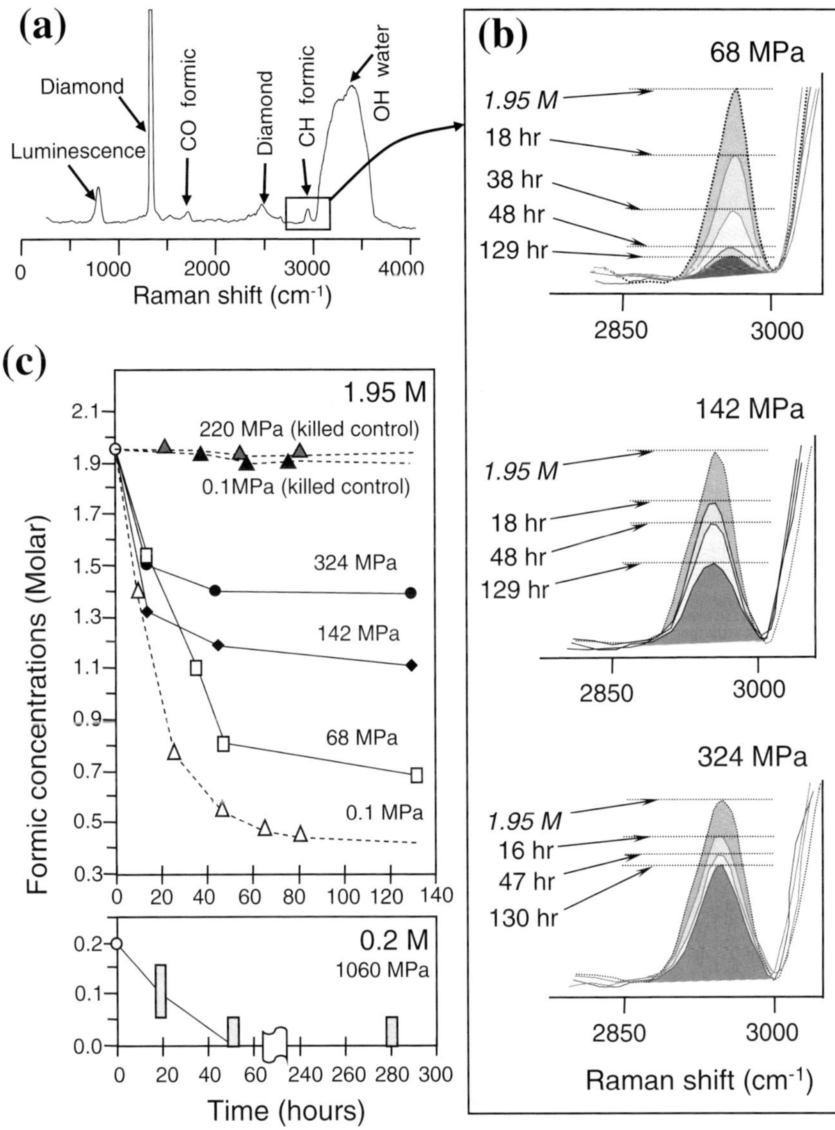

Figure 11. Raman spectra of the formic-biological system (**A**) shown with the vibration peaks of formic and diamond anvils used in this study. The outlined boxed region is shown at higher resolution (**B**) to quantify the successive decrease in the peak intensity of the C-H stretch of formic acid at pressures of 68, 142, and 324 MPa. The equivalent formate concentrations (**C**), corresponding to each peak height change, are based on comparisons with a known calibration curve. All experiments were performed at 25°C, with diamond anvil cells with gold-lined sample chambers. Pressures were estimated using Raman shifts in quartz used as an internal calibrant.

6.2. Experimental overview

For these biological experiments, we further modified the DACs by reducing the optical working distance for high magnification microscopic observations by decreasing the thickness of carbide supports for the diamonds and enlarging the opening angle of the diamond anvil cell. This enabled direct visualization of microbes with direct optical magnification exceeding 800X. For this study we used two pressure sensitive species of bacteria, *Shewanella oneidensis* MR1 (formerly *Shewanella putrefaciens* MR1), a metal-reducing, facultative anaerobe [62] and *Escherichia coli* strain MG1655, were utilized in this study because of their ability to oxidize formate. Although both MR1 and MG1655 are non-piezophilic, the genus *Shewanella* appears to have some species predisposed to pressure tolerance [63]. The inocula was grown overnight (16 to 24 hrs) to stationary phase at 30°C in Luria Bertani broth (LB) [64]. The cells were washed in 50 mM potassium phosphate buffer (pH 7.4) 3 times and re-suspended in either a saturated solution of 2 M formate and 0.2 M sodium fumarate, 0.1 grams per liter of yeast, 0.2 M sodium formate, 0.2 sodium fumarate and 10 mM potassium phosphate (pH 7.4) or LB medium. The bacteria were immediately aseptically loaded in a Au-lined sample chamber and sealed between the diamond anvils.

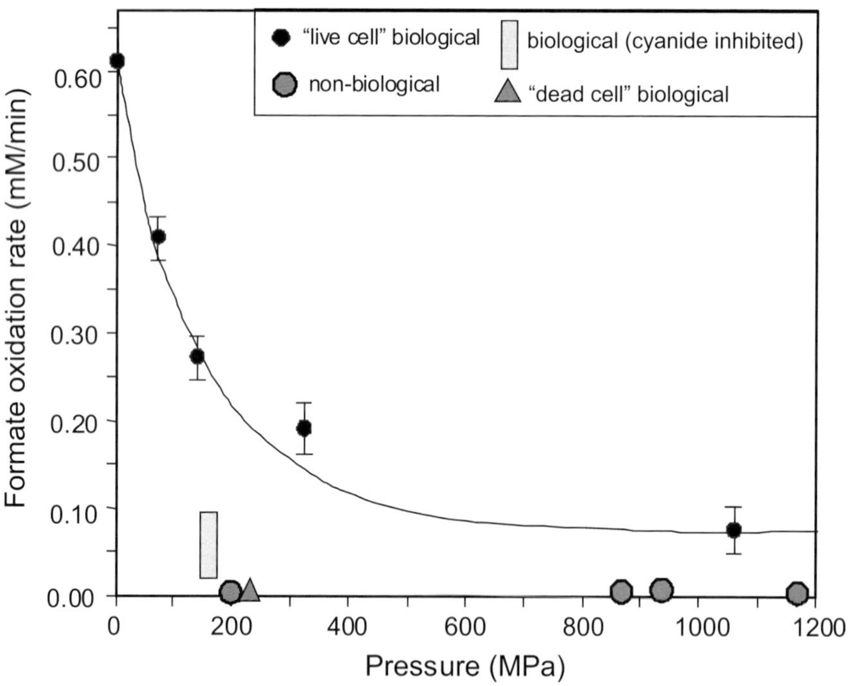

Figure 12. Shown are formate oxidation rates for experiments at high pressures for the first 48 hours. Whereas biological formate oxidation rates exhibit an inversely exponential relationship with pressure, the cyanide-inhibited system shows a lower formate oxidation rate.

We used the rate of microbial formate oxidation as a probe of metabolic activity. The utilization of formate by microorganisms is considered a fundamental metabolic process in anaerobic environments [59] that can be easily monitored via molecular spectroscopy. We monitored in situ microbial formate oxidation for *Shewanella oneidensis* MR1 to pressures of 1060 MPa, at room temperature (25°C), using Raman spectroscopy in diamond anvil cells (Figure 11). Raman spectroscopy of the formic bearing solutions gave us a control on the extent of metabolic activity. Using the ratio of C-H stretching vibration peak for formic (2950 cm^{-1}) to the O-H stretching vibration peak for water (~3500 cm^{-1}), we were able to monitor the metabolic formic oxidation rate at various pressures. Pressures in the diamond anvil cells were determined using Raman peak shifts in quartz [60, 61] added as an internal calibrant, and are accurate to within 20 MPa. The pressures were typically increased very quickly (within 5-30 seconds) and allowed to equilibrate (for 5-60 minutes) before making measurements.

The Raman data revealed rapid oxidation of formate determined from the reduction in the peak height of C-H vibration mode (Figure 11). Microscopic observations to pressures of 1060 MPa confirmed continued cell viability (intact and motile) over the duration of more than 30 days. As anticipated, formate oxidation rates were significantly reduced upon the addition of sodium cyanide (0.2 M). Cyanide acts as an inhibitor of respiration by acting in competition as an alternate electron acceptor. This suggests that the oxidation of formate is coupled to the biological respiration. Additional experiments were done with heat-inactivated MR1 cells, inactivated at 50°C for ~1 hr until lysis was observed. The spectroscopic analysis of the heat inactivated cells did not exhibit any detectable chemical changes (Figure 12). Both cell-free and dead cell controls indicate that formate oxidation was exclusively a result of biological activity. The rate of formate oxidation, determined based on the spectral data of first 48 hours of the experiment, shows an exponential decrease. The extrapolation of the formate oxidation rates, as indicated from the Raman data (Figure 12), could reflect an experimental 'limits' to the viability of life.

It is also possible that the decrease in the formate oxidation rates with increasing pressure in the living cells may simply reflect a decrease in the cell numbers. To constrain the cell numbers, we first attempted light scattering based experiments within the DAC. However, the presence of cell debris and cahngesin the cell morphology, precluded from any direct interpretation. We therefore obtained the cell numbers by direct counts at high pressure (1.4 GPa), analyzing 100 to 150 frames selected randomly through the sample. At t = 0 hr, the bacteria were observed through the entire depth of the sample chamber (*E. coli*: 200-micron thickness, 300-micron diameter; *Shewanella* MR1: 200-micron thickness, 500-micron diameter), while after 30 hrs only one distinct layer of cells was observed due to sedimentation. The total cell numbers were then calculated by integrating the average of each count over the entire volume of the sample chamber [total cell number = average cell number (sample chamber volume/area of each frame). For *E. coli*: at 0 hour, average cells per frame (n = 11) = 44 (std = 9.26; variance = 21%); at 30 hrs, average cells per frame (n = 11) = 60 (std = 9.3; variance = 15%). For *Shewanella* MR1: at 0 hour, average cells per frame (n = 12) = 56 (std = 10.3; variance = 18%); at 30 hours, average cells per frame (n = 12) = 43 (std = 8.1; variance = 19%).

Based on these estimates, we determined by direct cell counts on both *E. coli* [ranging from 10^6 cells (at 0 hours) to 10^4 cells (after 30 hours)] and *Shewanella* MR1 [ranging from 3 X 10^6 cells (at 0 hours) to 10^4 cells (after 30 hours) (*20*)]. Distinct changes in cellular morphology were also observed, including a significant increase in cell size and polar invaginations, both clear indications of arrested cellular division (*25, 26*). Upon a subsequent increase of pressure to 1680 MPa, the veins of organic-rich fluids developed a "stringy" texture as well as an increase in fluorescence background in Raman spectra.

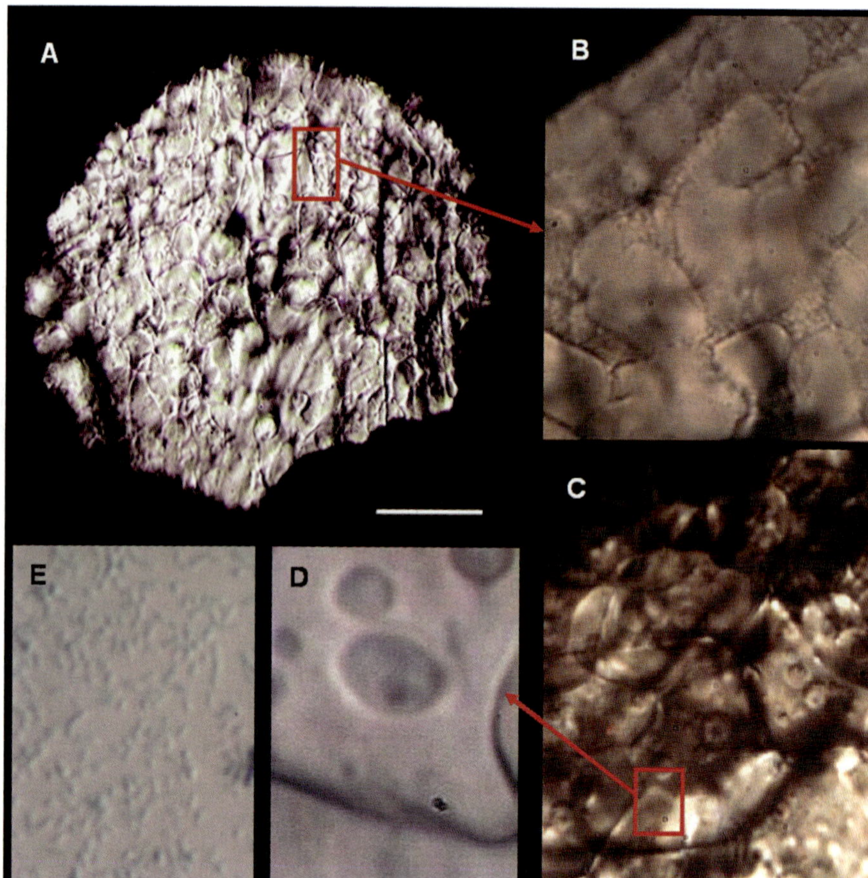

Figure 13. Microbial activity and viability in ice-VI at 1.4 GPa. (A) Complete view of the sample chamber with ice-VI phase with bacteria. Clearly visible is the vein-like structure surrounding the ice phase. (B) Close-up view showing the ice-VI crystal boundary surrounded by clusters of bacteria (*Shewanella* MR1) within organic-rich veins. (C) After approx. 1 hour, textural changes in the ice occur, defined by the formation of organic-rich inclusions within ice crystals. (D) Close-up view of the fluid inclusions containing clusters of motile bacteria. (E) Upon decompression (to <100 MPa), the viable bacteria (dyed with methylene blue) are seen clustered at the diamond surface but are still observed to be motile.

6.3. In-situ observations above 1 GPa

We extended these experiments to pressures where ice-VI becomes stable and explored how the pressure-sensitive bacteria [41,65], *Shewanella* MR1 and *Escherichia coli*, respond to this change of state, in a nutrient-rich medium. Optical observations at pressures ranging from 40 to 1200 MPa, lower than ice-VI formation, show a gradual decrease in cell motility, an increase in cellular adhesion to surfaces, decrease in cell numbers and reduction in metabolic activity (Figure 13). We added methylene blue (0.05 mM) as a dye for microbes that would also indicate metabolic activity. This is because under reduced conditions the dye changes color from blue-violet (oxidized) to colorless (reduced) [66], because it is used as an alternate electron acceptor by *Shewanella* MR1 [67] and *E. coli* [68]. Furthermore, methylene blue in its reduced form during anaerobiosis is an indication of flavoprotein activity in the bacteria. Furthermore, methylene blue, when added as an electron acceptor to test for respiration at elevated pressures, remained reduced throughout the experiment, indicating that bacterial respiration continued under these conditions.

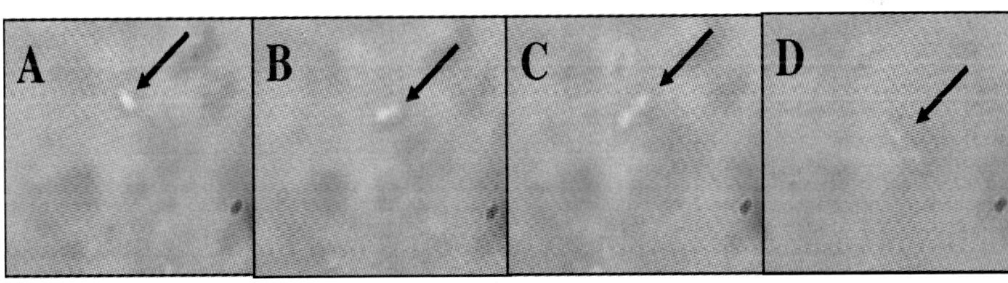

Figure 14. A series of time-lapse (approx 20 minutes between frames) images (A-D) showing a single *Shewanella* MR1 cell dividing (growth?) at 0.1GPa pressure after decompression from 1.4 GPa. The observed cells are on top of the anvil surface of the DAC. Subsequent extraction and growth of cultures confirm cell viability and show a longer 'lag' period before exponential growth.

At ~1250 MPa, the formation of ice-VI was instantaneous. Initially, mosaics of ice-VI crystals were observed separated by thin "veins" of organic-rich fluid containing the bacteria (Figure 13). However, over time the ice melted partially along the interstitial veins and developed numerous organic-rich fluid inclusions containing microbial clusters (Figure 13). These textural changes in the ice, observed to pressures of 1600 MPa, might either be the consequence of the phase separation associated with the formation of ice in any organic-rich medium, or the result of the microorganisms protecting ice-sensitive cellular components [69, 70].

Upon decompression from ~1400 MPa pressures (microbes associated with ice-VI), we observed a small fraction of cells surviving the pressure excursion. Observations of cell replication within the DAC (Figure 14) at decompressed conditions of ~100MPa suggest viable cell-replication. This is further confirmed by culturing and growth of extracted bacteria, which show a long lag time for the start of growth phase. It is possible that these stressed microorganisms are unable to grow on the agar plates, but may do so at some non-ambient condition. Study is currently underway to characterize the cell membrane and the growth characteristics of microbes under pressure.

6.4. Implications for adaptation under pressure

Adaptability of *Shewanella oneidensis* MR1 and *Escherichia coli* in these experiments indicates that microorganisms can continue to metabolize substrate at pressures far beyond those previously reported [34, 35, 41]. Although an evolutionary component to the adaptation of microbial communities to temperature and salinity is well known [71], whether there might be any evolutionary component for pressure adaptation is still in question. *Shewanella* MR1 belongs to a genus that contains a number of piezophiles; however, *E. coli* clearly does not. Despite this, there is evidence that exposure of *E. coli* to pressures up to 800 MPa selects a population of cells less sensitive to pressure inactivation [71]. Furthermore, it is well known that the increase in pressure tolerance is also associated with heat tolerance [71].

Our observations on pressure-sensitive bacteria may indicate that piezo-tolerance of these microorganisms could be a protective mechanism in response to stress [72-74]. Whether this is a short-term or an evolutionary adaptation remains unanswered. These results imply that pressure may not be a significant impediment to life. The maximum pressure explored in this work is equivalent to a depth of ~50 km in the Earth, or ~160 km in a hypothetical ocean. The more modest pressures encountered at the depths of thick ice caps and the deep crustal subsurface thus may not be a limiting factor for the existence of life. Our observations suggest that deep (water/ice) layers of Europa, Callisto, or Ganymede and subduction zones on Earth might provide viable settings for life unhindered by high pressure [75, 76].

7. SUMMARY

These experiments represent an initial step in a new in-situ approach to understanding biogeochemical processes within the deep subsurface. We have adapted recently developed high-pressure tools to obtain direct information about complex biological processes under extreme conditions from experiments on simple systems, key biochemical reactions, and ultimately organisms themselves. With increased realization of the significant contributions from the deep subsurface to near surface processes (such as the global carbon cycle), such experimental studies are needed to obtain crucial information on biochemistry and biological activity in these largely inaccessible regions. Results obtained to date indicate the synthesis of organic compounds, new mineral-fluid organic interactions and the feasibility of biological activity at very high pressures. As such, these findings should expand our understanding of a broad range of chemical phenomena under extreme conditions.

ACKNOWLEDGMENTS

We wish to thank the late H. S. Yoder, Jr. for his great support of the work described above, and for many stimulating discussions. We are also grateful to H. K. Mao, I. -M. Chou, W. A. Bassett, R. M. Hazen, M. L. Fogel, A. F. Goncharov, V. V. Struzhkin, S. Gramsch and W. T. Huntress, Jr. for discussions and help in various aspects of this study. This work was supported by the NASA Astrobiology Institute (Cooperative Agreement NNA04C09A), the National Science Foundation, and the Carnegie Institution of Washington.

REFERENCES

[1] T. Gold, Proc. Natl. Acad. Sci. U.S.A. 89 (1992) 6045.
[2] B. Sherwood Lollar et al., Nature 416 (2002) 522.
[3] G. D. Cody et al., Science 289 (2000) 1337
[4] G. D. Cody et al., Geochim. Cosmochim. Acta 68 (2004) 2185.
[5] D. I. Foustoukos,W. E. Seyfried Jr., Science 304 (2004) 1002.
[6] W. A.Bassett et al. 1993, Rev. Sci. Instr., 64 (1993) 2340.
[7] L. Haar, J. S. Gallagher, and G. S. Kell, NBS/NRC Steam Table (1984).
[8] H. K. Mao et al., J. Geophys. Res. 91 (1986) 4673.
[9] W. S. Holbrook et al., Science 273 (1996) 1840.
[10] V. J. Orphan et al., Science 293 (2001) 484.
[11] I. -M. Chou et al. Proc. Natl Acad. Sci. USA 97 (2000) 13484.
[12] I. -M. Chou et al., J Phys Chem A, 105 (2001) 4664.
[13] E. D. Sloan, Jr., Clathrate Hydrates of Natural Gases (Dekker, New York), 2nd Ed.(1998).
[14] S. Subramanian et al., Chem. Eng. Sci. 55 (2000) 1981.
[15] S. Nakano et al., J. Chem. Eng. Data 44 (1999) 254.
[16] A. Saul and W. Wagner J. Phys. Chem. Ref. Data 18 (1989)1 537.
[17] H. Hirai et al., J. Phys. Chem. B 104 (2000) 1429.
[18] W. Wagner et al., J. Phys. Chem. Ref. Data 23 (1994) 515.
[19] D. R. Marshall et al., Chem. Eng. J. 10 (1964) 202.
[20] T. Kumazaki et al., Chem Phy Lett 388 (2004) 18.
[21] G. Wächtershäuser, Prog. Biophys. Mol. Biol. 58 (1992) 85.
[22] M. J. Russell and A. J. Hall, J. Geol. Soc. London 154 (1997) 377.
[23] M. Sakai, Bull. Chem. Soc. Japan 49 (1976) 219.
[24] Carlsson et al., Ind. Eng. Chem. Res. 33 (1994) 1989.
[25] G. D. Cody et al., Geochim Cosmo Acta 65 (2001) 3557.
[26] H. J. Greenwood and H.L.Barnes, GSA Memoir, 97 (1966) 385.
[27] R. Cammack, Adv. Inorg. Chem. 38 (1992) 281.
[28] G. Wächtershäuser, Proc. Natl. Acad. Sci. U.S.A. 87 (1990) 200.
[29] M. J. Russell and A.J. Hall, J. Geol. Soc.London, 154 (1997) 377.
[30] C. Woese et al., Proc. Natl. Acad. Sci. U.S.A. 87 (1990) 4576.
[31] C. Huber and G. Wächtershäuser, Science 276 (1997) 245.
[32] T. D. Brock, Science 158 (1967) 1012.
[33] J. W. Deming and J. A. Baross, Geochim. Cosmochim. Acta 57 (1993) 3219.
[34] D. H. Bartlett, Sci. Progress 76 (1992) 479.
[35] F. Abe, C. Kato, K. Horikoshi, Trends Microbiol. 7 (1999) 447.
[36] C. A. Royer, High Pressure Res. 19 (2000) 213.
[37] Y. Yano, A. Nakayama, K. Ishihara, H. Saito, Appl. Environ. Microbiol. 64 (1998) 479.
[38] C. E. Zobell, in Ken Sugawara Vol. Y. Miyake, T. Koyoma, Eds. (Maruzen, Tokyo, 1964) pp.83-116.
[39] C. O. Wirsen, S. J. Molyneaux, Appl. Environ. Microbiol. 65 (1999) 5314.
[40] C. E. Zobell, Science 115 (1952) 907.
[41] F. Abe, K. Horikoshi, Trends Biotechnol. 19 (2001) 102.
[42] A. A. Yayanos, in Extremophiles, K. Horikoshi, W. D. Grant, Eds. (Wiley, New York, 1998).
[43] J. L. Silva, D. Foguel and C. Royer, Trends Biochem. Sci. 26 (2001) 612.
[44] V. V Mozhaev et al., Proteins Struct Func. Gene. 24 (1996) 81.
[45] M. Gross and R. Jaenicke, Eur J Biochem. 221 (1994) 617.
[46] D. H. Bartlett and K. A. Bidle, in Enigmatic Microorganisms and Life in Extreme Environments (ed. Seckbach, J.) 503–512 (Kluwer, Dordrecht, 1999).
[47] H. W. Jannasch et al., Appl. Env. Microbio 62 (1996) 1593.
[48] A. A. Yayanos, Annu. Rev. Microbiol. 49 (1995) 777.
[49] C. Kato and D. H. Bartlett, Extremophiles 1 (1997) 111.

[50] F. Abe and K. Horikoshi, Trends Biotechnol. 19 (2001) 102.
[51] J. V. Landau Biochim. Biophys. Acta 149 (1967) 506.
[52] K. J. A. Hauben et al. Appl. Env. Microbio 63 (1997) 945.
[53] D. H. Bartlett et al. Nature 342 (1989) 572.
[54] C. Garcia-Graells et al. Appl. Env. Microbio 64 (1998) 1566.
[55] R. Pagán and B. MacKey Appl. Env. Microbio 66 (2000) 2829.
[56] A. D. Molina-Garcia et al., Meat Science 66 (2004) 709.
[57] A. A. Yayanos, Science 297 (2002) 295.
[58] A. Sharma et al., Science 295 (2002) 1514.
[59] J. G. Ferry, FEMS Microbiol. Rev. 7 (1990) 377.
[60] K. J. Dean, W. F. Sherman, G. R. Wilkinson, Spectrochimica Acta 38A (1982) 1105.
[61] R. J. Hemley, in High Pressure Research in Mineral Physics, M. H. Manghnani, Y. Syono, Eds, 347-359 (1987).
[62] C. R. Meyers, K. H. Nealson, Science 240 (1988) 1319.
[63] C. Kato, et al., Env. Stressors and Gene Responses, K. B. Storey, J. Storey, Eds, 277 (2000).
[64] J. Sambrook, E. Fritsch, T. Maniatis, Molecular Cloning: A Lab Manual (Cold Spring Harbor Laboratory Press, Cold Spring Harbor, NY 1989).
[65] K. J. A. Hauben et al., Appl. Environ. Microbiol. 63 (1997) 945.
[66] K. M. Jones, in Data for Biochemical Research, R. M. C. Dawson, D. C. Elliott, W. H. Elliott, K. M. Jones, Eds. (Oxford Univ. Press, London, 1969).
[67] J. H. Scott and K. H. Nealson, J. Bacteriol. 176 (1994) 3408.
[68] H. G. Enoch, R. L. Lester, J. Biol. Chem. 250 (1975) 6693.
[69] J. F. Carpenter, T. N. Hansen, Proc. Natl. Acad. Sci.U.S.A. 89 (1992) 8953.
[70] J. C. Priscu et al., Science 280 (1998) 2095.
[71] C. R. Woese, O. Kandler, M. L. Wheelis, Proc. Natl.Acad. Sci. U.S.A. 87 (1987) 4576.
[72] K. J. A. Hauben et al., Appl. Environ. Microbiol. 63 (1997) 945.
[73] T. J. Welch, A. Farewell, F. C. Neidhardt, D. H. Bartlett, J. Bacteriol. 175 (1993) 7170.
[74] K. Liberek et al., Proc. Natl. Acad. Sci. U.S.A. 89 (1992) 3516.
[75] C. F. Chyba, C. B. Phillips, Proc. Natl. Acad. Sci. U.S.A. 98 (2001) 801.
[76] M. H. Carr et al., Nature 391 (2000) 363.

Chemistry at Extreme Conditions
M. Riad Manaa (Editor)

Chapter 4

Application of High Pressure in Inorganic and Bioinorganic Chemistry

Rudi van Eldik and Colin D. Hubbard

Institute for Inorganic Chemistry, University of Erlangen-Nürnberg, Egerlandstraße 1, 91058 Erlangen, Germany

1. INTRODUCTION

The principal focus in this chapter is on application of pressure to inorganic and bioinorganic reactions in solution and an examination of the consequences of the application. The examination is most often in the form of a kinetic characterisation and interpretation of kinetic and thermodynamic parameters that results in mechanistic conclusions. Normally the reaction is not irreversibly changed by pressure, and therefore unlike in many other fields of endeavour of exploitation of the pressure variable, our interest is not in additionally characterising the system after decompression. Following decompression, the system will have been restored to equilibrium at ambient pressure or the reaction will have been transformed from reactants into products.

It should be clear from the outset that this contribution is mostly restricted to liquid phase chemistry. It is recognised that in gas phase chemistry the pressure variable in controlling product yield is widely applied. Reactions where pressure is used to improve yields in liquid phase or solid phase inorganic chemistry are not covered in this chapter. Information on the spectroscopic characterisation of solids under pressure may be accessed from specialised literature. [1] Likewise, while there is fascinating progress in the area of inorganic reactions conducted in supercritical fluids, [2(a)] for example, the reactions of transition metal noble gas complexes in supercritical noble gas solutions, [2(b)] and these are indeed reactions where the system is under pressure, this subject requires a separate treatment. Theoretical and computational investigations relating to high pressures and especially to kinetics at elevated pressures will be included where they are complementary to experimental studies.

It is obvious that an "extreme" condition can mean different things to different scientific sub-disciplines. For example, high pressures in protein folding experiments may often approach 1000 – 1200 MPa, [3(a)] although such high pressures are not necessarily obligatory in this field of research. [3(b)] In mechanistic inorganic and bioinorganic solution chemistry, pressures are more typically in the range up to 200 MPa. These may not seem extreme in the context of the subject, but this is an upper limit of 2000 atmospheres (about 2 kbar), and technical issues are not trivial. In fact operating at higher pressures may provide further valuable information about the reacting system (*vide infra*); however, particularly in cases of reactions in non-aqueous solutions, effects on bulk solvent properties begin to be exhibited and this can complicate what otherwise might be relatively, kinetic simplicity. [4] Furthermore, in macromolecular containing systems other non-kinetic events may be initiated above the 200 MPa threshold guideline. Conversely for the mechanistic information usually sought, it is not necessary to resort to higher pressures. These would require an associated increase in technical sophistication, and expense.

1.1. Background and literature

The fundamental relationships regarding the effect of pressure on the kinetics of reactions have been known and introduced in standard textbooks for some time, [5] and indeed many examples of organic reactions studied as a function of pressure were reported. [6] Only some three decades ago did some in the inorganic chemistry field embrace the application of pressure as a powerful mechanistic tool. [7] There is a partial explanation for this in that many of the inorganic reactions at issue were of coordination complexes of transition metals and were sufficiently rapid to require specialised instrumentation for kinetic studies at ambient pressure. This instrumentation had to be modified to enable measurements to be made at pressures up to 200 MPa. Many technical issues were adequately resolved, and the field has blossomed, particularly for reactions involving solvent exchange, ligand substitution and electron transfer processes. The relevant experimental methods are cited below. The last few years have seen the appearance of many authoritative reviews and books on various aspects of effects of pressure, especially kinetics at elevated pressure. Therefore, for this chapter we have concentrated mostly on research articles published in the past two years, and readers may consult the relatively recent literature cited throughout for the historical background for each particular topic. Although not exclusively devoted to high-pressure kinetics, three chapters of reference [8] deal respectively with mechanisms of solvent exchange, ligand substitution and those for reactions of nitric oxide with biologically relevant metal centres. Chapters within reference [9] contain coverage of methods and reactions relevant to this contribution, and a comprehensive account of many reaction classes has appeared, [10] and exhaustive compilations of derived parameters from high-pressure kinetics, with accompanying narrative, have been reported. [11] More specifically focused reviews are those directed toward the effects of pressure on relevant photochemical reactions [12] and electron transfer reactions in general [13] and also to self-exchange electron transfer reactions. [14] Although the emphasis here is on mechanistic characterisation from an analysis of kinetics results obtained over a range of pressure, there are other applications of high pressure in inorganic chemistry that will be cited including electron spin-state conversion and piezochromism and in organometallic chemistry where in some cases transition metal compounds serve as catalysts. Clearly of parallel interest is the application of pressure in organic chemistry both from the perspectives of preparative chemistry (yield/stereochemical preference) and in mechanistic chemistry. These subjects can be followed in literature relevant to a period up to about ten years ago [15] and more recently [9], and compilations of kinetic parameters and mechanistic conclusions from them [11] are available. Suitable detailed and more general reviews may be consulted. [16]

1.2. Basic principles of kinetics at high pressure

The origin of the relationship between the rate constant (k) of a reaction in solution and the hydrostatic pressure (P) has been introduced frequently in the literature and this aspect will not be repeated here. It suffices to provide the relationship itself:

$$(\delta ln(k))/\delta P)_T \quad = \quad -(\Delta V^{\neq}/RT) \tag{1}$$

where $\Delta V^{\neq}$ is the volume of activation defined as the difference between the partial molar volumes of the transition state and the reactant(s), and R is the ideal gas constant and the absolute temperature is T. Providing the volume of activation is not dependent on pressure then:

$$ln(k_p) \quad = \quad ln(k_0) - \Delta V^{\neq}/RT \tag{2}$$

where k_p and k_0 are then rate constants at pressures P and 0, respectively, and $\Delta V^{\neq}$ is the volume of activation. Hence the volume of activation can be obtained from a plot of $ln(k_p)$ versus P providing the plot is linear. If this is not the case, the initial slope of the plot in the lower pressure range is taken to calculate $\Delta V^{\neq}$. Thus, a positive value is obtained from a rate constant that decreases with increasing pressure and a negative value is obtained when a pressure increase causes rate acceleration. In the event that the volume of activation is not independent of pressure, one or more terms are added to the latter equation. This could be the term $\Delta\beta^{\neq}P^2/2RT$, where $\Delta\beta^{\neq}$ is the compressibility coefficient of activation. In almost all examples of reactions reported herein the focus of the investigation was mechanistic, and therefore the pressure was limited to the range where the volume of activation is independent of pressure. Various treatments of experimental data are available [4, 16, 17] for investigators interested in the pressure dependence of the volume of activation. In the case of multi-step reactions the situation can become more complicated as the measured volume of activation can represent a composite value of volume changes associated with each rate or equilibrium constant in the rate law. Therefore, experimental design in terms of selected experimental conditions to minimise such complications should be undertaken.

A mechanistic classification of inorganic reactions has been explained fully [18(a)] and the relevance to high pressure kinetics has been presented. The volume of activation consists of intrinsic changes (bond making or bond breaking) as the transition state is reached, and other volume changes that arise from solvation changes. When the charge on species changes, or changes in bond polarity occur, these can influence adjacent solvent molecules and volume changes due to solvent electrostriction can occur. The volume of activation can be a powerful indicator of mechanism particularly if the latter of the two contributions to its magnitude is absent. Thus in the case of solvent exchange reactions, i.e. where a solvent molecule bound to a reactant species exchanges with a bulk solvent molecule, there is no net reaction change and the partial molar volumes of reactant and product are identical. In addition, no change in electrostriction occurs. A consequence of this is that, in principle, the volume of activation is a relatively straightforward diagnostic indicator of intrinsic volume changes directly related to the mechanism of the reaction.

Ligand substitution (of which solvent exchange is one form) has been classified as A (associative), D (dissociative) or I (interchange) with the latter having sub-classifications I_a and I_d. The first involves a reaction proceeding through an intermediate of increased coordination number, whereas a D mechanism involves formation of an intermediate of decreased coordination number. An I mechanism is one in which an outer-sphere precursor complex is formed with greater sensitivity to the nature of the entering ligand (I_a) or the departing ligand (I_d). The connection between these mechanistic classifications and perhaps the more familiar S_N2(lim), S_N2, S_N1 and S_N1(lim) mechanisms has been made. [18(a)] It has also been pointed out that there are subtle differences between various interpretations of the original Langford-Gray mechanistic classification nomenclature. [18(b), 19]

It should be noted that high pressure kinetics studies are frequently able to yield more mechanistic insight than can be obtained from activation parameters determined from studying the temperature dependence of kinetic parameters. The entropy of activation, $\Delta S^{\neq}$, it has been argued frequently, is not precisely determined as the method used to obtain it involves considerable extrapolation of the data acquired. It may also be argued that from the point of view of mechanistic analysis that since entropy change involves essentially both energy and molecular order components it does not easily complement volume of activation data even if it were precisely obtained. Furthermore, in the case of solvent exchange reactions it has been proposed that interactions beyond the first coordination sphere of the species concerned can contribute to the magnitudes of both $\Delta S^{\neq}$ and $\Delta H^{\neq}$, thus adversely affecting their mechanistic value in these reactions. [20]

For reactions where solvational changes are anticipated, and these can be interpreted in terms of increasing or decreasing solvent electrostriction, each reaction must be assessed on its own merit, and combining intrinsic with solvation volume changes in understanding the underlying mechanism is a challenge. The experimental approach to obtain insight into the relative contributions will depend on the specific properties of the system.

A complete exposition of the thermodynamic relationships involving the pressure variable has been presented. [4] It is pertinent to distil from that presentation that the temperature dependence of $\Delta V^{\neq}$ is not substantial, and when experimentally feasible the kinetics measurements leading to calculation of $\Delta V^{\neq}$ are invariably carried out at 298 K. Minimal effects on the value of $\Delta V^{\neq}$ would ensue if the reactions were carried out within 20 K on either side of 298 K, and therefore have no significant effect on the mechanistic value. However, if the rate law is a composite of several terms, then the temperature dependence of the various contributions to the overall $\Delta V^{\neq}$ could conceivably combine to yield non-trivial differences in the overall temperature dependence of the latter.

Finally and fairly obviously it should be mentioned that equilibrium thermodynamic properties of the reaction system may be afforded if the kinetics of the reaction can be monitored in both the forward and reverse directions. This leads to the value of the equilibrium constant as a function of pressure and the reaction volume, the difference in volume between the partial molar volumes of products and reactants. This information can also be established if the equilibrium constant can be obtained directly as a function of pressure. Precise density measurements of reactant and product solutions in experimentally amenable systems may be used to determine the partial molar volumes of reactants and products, and then the volume scale and the volume profile for the reaction can be set on an absolute rather than only a relative scale. Desirable properties required for the last approach are plentiful supply of substances, high solubility of species in the reaction solvent and prolonged stability of each species in solution. Abbreviations, symbols and formulae in the following sections are most usually those used in the original literature, and therefore at times these will not always be completely uniform or necessarily standard.

2. EXPERIMENTAL METHODS

The technique used for monitoring a given reaction at elevated pressures is invariably the same as that used at ambient pressure. By far the most widely used techniques, but not exclusively, for monitoring inorganic reactions in solution are UV/Visible spectrophotometry and nuclear magnetic resonance (NMR) spectroscopy. [4, 8, 10, 11] In certain specialised areas of research, high pressure infrared (IR) spectroscopy has been a valuable but mostly qualitative method in species characterisation; the method has been described [21] and reviewed. [22]. A very recent application in biochemistry has been reported. [3(a)] Choice of method for a given reaction is dictated by the characteristic properties of the reaction and the particular instrumentation selection depends on the range of half-lives likely to be encountered. General and quite comprehensive accounts have been provided in several sources [9, 10, 23, 24] and reports of instrumental methods relating particularly to high pressure NMR (HPNMR) spectroscopy [25], [26] and to UV/Visible/near IR spectroscopic measurements have appeared. [27] A pioneering example of a high pressure NMR spectroscopy probe for catalysis studies utilising the carbon-13 nucleus was reported initially in 1981. [28] A more comprehensive description was provided subsequently. [29] Redox chemistry in appropriate examples has been investigated by high pressure electrochemical methods. [30] At the time of preparation we were not aware of major developments regarding instruments relating to high pressure applications in solution phase inorganic chemistry but refinements, improvements and extension of the pressure range are ongoing. [27] Other high pressure techniques have been described; some are not of widespread use in inorganic reactions, but are of parallel technical interest. Fourier transform infrared (FTIR) [3], [31] and fluorescence spectroscopies [31] have not been extensively used in inorganic reactions, but are used in the important area of protein folding investigations. The activation volumes of folding and refolding of a protein in the presence of denaturants have been determined using pressure-jump and high-pressure stopped-flow techniques. [32] A high pressure jump method in conjunction with small angle synchotron x-ray scattering technique has also been applied in protein folding studies. [33] The potential of high pressure Raman spectroscopy for protein folding investigations has been indicated. [34] A high pressure vessel for elastic and inelastic x-ray diffraction experiments for liquids over a wide temperature range has been described. [35] Characterisation of kinetics of protein colloidal micellar dissociation has been carried out using high pressure turbidity and high pressure light scattering measurements. [36] Flow and so-called jump methods will be described in more detail below.

Having presented the various physical and spectroscopic methods of indirectly monitoring reaction species concentrations or detection of intermediates during system compression, we now turn to a wider examination of the practical aspects involved in the various experimental methods, and also their incremental development. Although electrochemical methods have been successfully applied, in the vast majority of literature relating to high pressure kinetics of inorganic and bioinorganic reactions, the monitoring techniques used are UV/Visible spectrophotometry or NMR spectroscopy of various appropriate nuclei. This is the case in most of the reports in Sections 3, 4 and 5 below.

In the former, the initiation method of reaction may differ, i.e. a straightforward thermal reaction, a photo-initiated process or a pulse radiolytic initiation, [37] and the instrumentation used is very dependent on reaction half-life. For reactions with a half-life of more than 10 to 20 min a spectrophotometer that houses a pressurisable cuvette may be used, and all other functions are the same as for ambient pressure. The pressure transmitting cuvette, known as a pill-box, is immersed in an autoclave possessing optical windows and containing optically transparent water or heptane as the pressurising fluid. [38] By its design the pill-box cuvette allows pressure to be transmitted through compression of two movable closely fitting cylindrical components. Following its filling by a syringe needle technique, one cylindrical component is rotated through 180° to seal the cuvette. Under pressure the window ends of the cylinder move closer together as a consequence of the compression of the sample solution, a factor that must be duly accounted for in terms of light absorption and the optical path length, particularly for highly compressible organic solvents. The pressure generating system typically consists of a hydraulic pump used to generate pressure on oil, and this pressure is then transmitted by a separator to the pressurising medium. Normal practice would be to monitor reaction kinetics at several pressures during both a compression and decompression cycle at one or more single wavelengths. In suitable cases the spectrum of a chemical equilibrium mixture can be recorded as a function of pressure to establish the pressure dependence of the equilibrium constant and hence obtain the reaction volume.

For reactions exhibiting a half-life in the range of a couple of minutes down to a few milliseconds, the stopped-flow (sf) method, pioneered for use at ambient pressure by Gibson and colleagues, [39] is widely available commercially in different configurations. The sf has been adapted for high pressure applications (hpsf) (usually up to about 150 MPa). [40 – 44] Most hpsfs are locally designed and constructed and have individual characteristics, although a commercial version is available. [45] The sf method involves flowing two or more reactant solutions from syringes into a special mixer, and upon arresting the flow the reaction kinetics are monitored. Frequently the monitoring method is UV/Visible spectrophotometry, although an hpsf instrument utilising fluorescence detection has been reported. [46] In principle circular dichroism, polarimetry or conductivity monitoring of reactions could be incorporated. In the hpsf method the sf unit is usually immersed in the pressurising liquid, requiring introduction of reactant solutions before subsequent pressurisation. In some designs many replicate runs are possible with one loading of the system meaning that kinetics at several pressures can be observed before decompression. A disadvantage is that ambient pressure sfs have a dead time of the order of 2 ms, but this is usually in the range of 20 to 50 ms for hpsf.

When a reaction is more rapid than the time range of the sf method (at temperatures close to ambient), and can only be studied in aqueous solution, then the low temperature (down to about 200 K) sf is not an option. However, the reaction may be amenable to a kinetic study by a relaxation method. The latter was mostly developed by Eigen and coworkers, and involves a perturbation (single pulse – temperature, pressure, electrical field; or periodic pulse – ultrasound) to a chemical reaction system at equilibrium. [47, 48], The response to the perturbation can be analysed in terms of the kinetics of equilibrium restoration. The scope of exploitation of relaxation methods was also lucidly presented. [49] By far the preponderance of reports relating to inorganic reactions have involved "joule heating", yielding a temperature- jump of a few degrees in about 10 μs. An early report of inorganic reactions studied by the temperature-jump method established kinetic parameters for $Ni(aq)^{2+}$ reacting with glycine and imidazole. [50] Early reports of modifying a temperature-jump instrument for high pressure measurements can be attributed to Caldin, Tregloan and coworkers. [51] Systems studied in aqueous medium include transition metal complex equilibria, [52, 53], binding of small molecules to heme and to proteins, [54] and in non-aqueous media for proton transfer especially in cases where proton quantum mechanical tunnelling was anticipated. [55] In non-aqueous media joule heating cannot be used and a laser was used to initiate perturbation. [51] These early studies [52, 53], were influential in confirming mechanisms postulated as I_d for formation of some transition metal complexes from ambient pressure kinetic studies. [56] Later work has emphasised transition metal complex reactions, [57] and particularly in the case of copper(II) where variation of the coordination number from six to five by virtue of converting the geometry to trigonal bipyramidal employing tetradentate tripodal ligands, has a dramatic effect on the kinetic lability of the coordinated water. [58] Various aspects of the coordination geometry of aquated Cu(II) and Cu(II) complexes have recently been discussed based on combinations of x-ray crystallographic studies, EXAFS studies, molecular dynamics simulations and neutron diffraction studies, and differences of interpretation and opinion have emerged. [59]

The use of HPNMR in species characterisation has been cited above. It should be emphasised that this application involves gases under high pressure, whereas mechanistic studies that will be reported in some detail below involve only liquid solutions. Furthermore, many of the latter are of solvent exchange on fully or partly solvated metal ions where there is no net reaction. Clearly constructing an nmr probe for HPNMR is not a trivial exercise and has only been carried out in a few University laboratories. Developments of the technical aspects of HPNMR can be followed. [28, 29, 60 – 65(a)] Of particular note is the very recent report of a new HPNMR probe designed for a narrow bore magnet system (see Figures 1 to 3 for more details). [65(b)] The advantages of rapid sample introduction and exchange, and improved spectral resolution were emphasised. Furthermore this new design eliminated the need to modify the spectrometer magnet to perform high pressure measurements. Detailed descriptions of the design and construction are beyond the scope of this chapter. Reference [26] is a valuable update of the state of technical development in one of the principal practitioner's laboratories. It must be recognised and emphasised that the primary data acquired in nmr experiments, the results from which will be described in section 3.1 below, require considerable treatment before kinetic parameters can be generated. This is a situation that is very different from the more direct conversion of primary data to kinetic parameters in UV/Visible spectrophotometry, usually with relatively straightforward software.

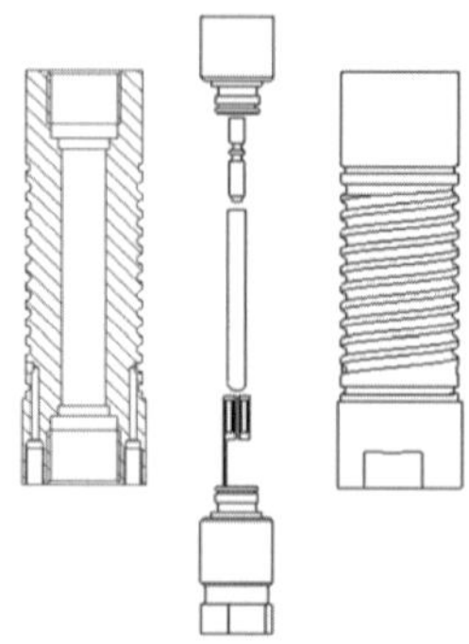

Figure 1. Cross sectional view of the high pressure autoclave with top and bottom plug, sample tube, macor plug and sample coil.

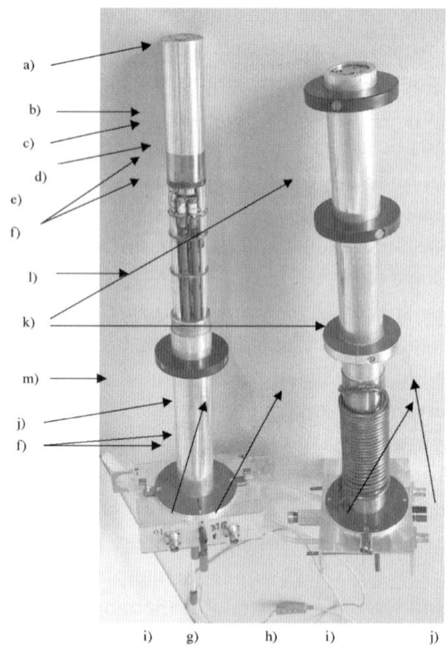

Figure 2. Photograph of two narrowbore probeheads : a) aluminum jacket sealing the double helix used for thermostating, b) high pressure vessel, c) platform carrying the autoclave, d) capacitors, e) capacitor platforms, f) tuning rods, g) high pressure connector, h) thermocouple, i) BNC connector, j) Pt-100 connector, k) copper tubing, l) widebore adapter.

Photolysis of photosensitive reaction systems can give rise to an equilibrium perturbation with subsequent observation of the relaxation process, or it can give rise to an overall reaction. The technique of flash photolysis whereby very rapid reactions can be studied was developed several decades earlier. [66] The photolysis method has been thoroughly modernised and examples of the application of pressure to provide an additional dimension to mechanistic studies have been reported. [67] An authoritative account of photoreactions of transition metal complexes, including the effect of pressure upon them, has been given and references are given to experimental aspects. [68] Methods for studying redox reactions, and heterogeneous electron transfer reactions have been developed; [30, 69, 70] these are highly specialised and available in only a very few laboratories. High pressure pulse radiolysis chemistry has been reviewed, [37] and the special cell developed for the relevant reactions has been described. [71]

It should be clear that applying high pressure in studying inorganic and bioinorganic reactions is not a simple matter, since almost all the equipment and instrumentation is constructed in investigators' laboratories and workshops. Furthermore, an expensive instrument may be required to be dedicated to the high pressure mode. Safety issues are almost standard laboratory practice providing only liquids are under pressure, since a component failure would result in pressure loss, liquid leakage and no further consequences except if toxic substances are involved then adherence to pertinent safety practice would be required. When compressed gases are involved in the pressurising system or the reaction itself involves gases under pressure the safety regime must be much more rigorous and appropriate precautions and training of personnel must be undertaken.

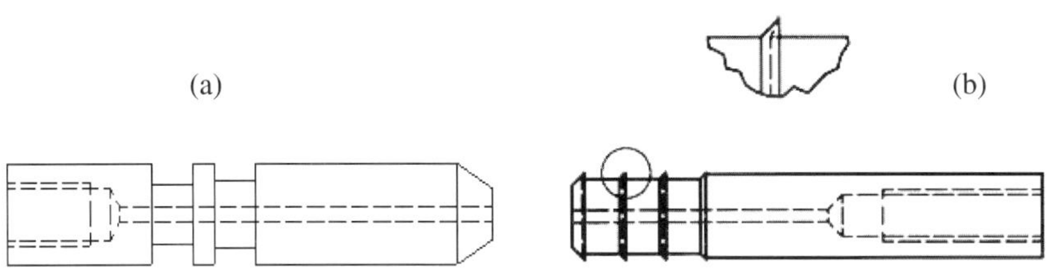

Figure 3. (a) Macor plug (two o-ring sealing), (b) PTFE – plug.

3. INORGANIC REACTIONS

3.1. Solvent exchange

Solvated metal cations have exclusively been studied and frequently the solvent is water. In many cases the number of water or other solvent molecules coordinated to the cation is known. Why is solvent exchange of a solvent molecule between a coordinated solvent molecule and one of the bulk solvent important to investigate? In principle, any solvated metal cation is of interest since the chemical reactivity of a solvated ion in solution is intimately connected with its stability and lability of the coordinated solvent molecules. Thus, knowledge of the solvent exchange characteristics is of fundamental importance in understanding the reactions of solvated or aquated metal ions with other ligands. [18, 72] It is the most fundamental substitution process that can occur on a solvated metal ion and in fact in many cases controls the nature of ligand substitution reactions on such metal cations. Obviously differences in reactivity can arise if some of the coordination sites are occupied by other donor molecules (spectator ligands) than solvent molecules,

$$M(H_2O)_x{}^{n+} \quad + \quad H_2O^* \quad \rightarrow \quad M(H_2O)_{x-1}(H_2O^*)^{n+} \quad + \quad H_2O\,.$$

Thus the kinetics of the reaction in water are of interest, where for s, p and d block elements whose cations are amenable to study, x is usually 4 or 6 and n is usually 2 or 3. The water molecule bearing the asterisk is merely indicated as the one from bulk solvent that replaces one in the first coordination sphere. The values of x are normally the same for other unidentate solvents.

$\Delta V^{\neq}$ for solvent exchange is a particularly valuable parameter as it relates only to an intrinsic volume change upon reaching the transition state, uncomplicated by electrostriction changes. Mechanistic conclusions are predicated on the assumption that the volume of activation represents events involving water or other solvent exchange, and there are no volume changes in the system associated with the non-exchanging ligands. This issue is addressed further in section 6. For solvent exchange processes there is no reaction volume since it is a symmetric reaction. The past two decades have witnessed a tremendous growth in knowledge of the rate constants for water and solvent exchange derived from the temperature and pressure dependence of nmr spectroscopic signals. Until about 15 years ago transition metal cations were the subject of most investigations reflecting the interest in the lability, or lack thereof, and in the variety of oxidation states of these ions, in some cases, that had attracted the attention of coordination chemists. Increasingly studies devoted to f block elements have also been reported. Volumes of activation for solvent exchange reported in the period 1987 to 1996 have been presented in tabular form accompanied by an explanatory narrative. [11] Results of more recent studies have been included in comprehensive review articles principally for water exchange, [73] and for other solvents as well. [74] The experimental and technical aspects concerning obtaining these results have already been cited. [26] One remarkable feature emerging is the wide range of lability; the mean life time of a coordinated water molecule on Ir^{3+}(aq) is about 10^{10} s, whereas ions of Eu(II) and Cu(II) are extremely labile (lifetimes of about 10^{-10} s). It has been pointed out in earlier nmr studies of the temperature dependence of solvent exchange, specifically acetonitrile exchange on Ni^{2+}, monitoring either the ^{1}H or ^{14}N nuclei, that $\Delta H^{\neq}$ ranged from 40 to 67 kJ mol^{-1} and $\Delta S^{\neq}$ ranged from –33 to +43 J mol^{-1} K^{-1}. [18(b)] Thus mechanistic interpretations on the basis of $\Delta S^{\neq}$ data are subjected to large uncertainties as shown by this range of experimental values. This observation may have its explanation residing in the difficulty in converting primary data into kinetic parameters.

The topic of solvent exchange has not escaped the scrutiny of theoretical chemists. The seemingly accepted mechanism for water exchange on first row divalent transition metal ions based on experimental findings, [75, 76] was challenged. [77-79] Experiments had been interpreted to indicate that an I_a mechanism prevailed for the ions in order of increasing atomic number up to and including Mn^{2+}, whereas for Fe^{2+}, Co^{2+} and Ni^{2+} the mechanism of exchange was I_d. The mechanistic changeover could be rationalised on the basis of changes in ionic radii and d-orbital occupancy along the first row transition metals. Theoretical work purported to show that all first row divalent transition metal ions exchanged water by an I_d mechanism. [77-79] Further calculations yielded arguments in favour of the mechanistic changeover. [80-82] However, except for highlighting this controversy, results that have been thoroughly interpreted, documented and reported as late as 2001 and early 2002 will not be repeated here. The remaining part of this section will be devoted to very recent results and these mostly refer to water exchange on lanthanide aqua-ions and lanthanide complex ions. Nevertheless, some background information will be needed in a few cases to set the stage for the more recent results.

There is intrinsic interest in water exchange on lanthanide ions. The scale of the investigations has been driven by the need to establish basic chemical properties of hydrated lanthanide ions, or partly hydrated lanthanide complex ions, in relation to their application as magnetic resonance imaging (MRI) agents. Whereas the coordination numbers of many s, p or d- block elements are known reliably, originally such numbers were less certain for the lanthanide ions. However, different experimental techniques have shown that the lighter trivalent cations (La^{3+} - Nd^{3+}) are predominantly nine-coordinate, whereas water coordination of Pm^{3+} to Eu^{3+} is characterised by equilibria between eight- and nine-coordinate states, and Gd^{3+} to Lu^{3+} are predominantly eight-coordinate. This variation is consistent with partial molar volumes of these ions obtained in aqueous solution. [74, 83, 84] Activation volumes for water exchange on eight-coordinate Gd^{3+} to Yb^{3+} are modestly negative and indicative of an I_a mechanism. [85-87] The influence of non-exchanging (spectator) ligands on water exchange rates and in ligand substitution kinetics has been investigated thoroughly in many systems and this aspect dates back decades. The influence of such ligands on water exchange of gadolinium(III) ions has been studied extensively since Gd^{3+} species have potential use as contrast agents in MRI. Three properties that are desirable, and perhaps essential, for Gd(III) species to be of value as MRI contrast agents are that the complex should be thermodynamically stable, i.e. have a high stability constant, the coordinated ligand does not occupy all the coordination sites, thus allowing for at least one water molecule to remain coordinated (available for exchange), and that the presence of the coordinated ligand does not dramatically retard the water exchange rate (relative to that of Gd^{3+}(aq)). The reasons for each of these features are, in turn, limiting the presence of toxic Gd^{3+}(aq), retaining water molecules to monitor water exchange kinetics and acquiring, by fast water exchange, high proton relaxivity. The last of these factors is required for useful MRI contrast agent reasons.

The kinetic and activation parameters have been determined for water exchange on several poly(aminocarboxylate) ligands coordinated to Gd^{3+} together with either one or two coordinated water molecules. In these systems the metal ion is nine-coordinate and the exchange rate constant is reduced by two or three orders of magnitude. [88] Distinctly positive $\Delta V^{\neq}$ values were obtained (D-activation); these could be explained by the steric crowding of the metal ion and the complex ion rigidity relative to the fully aquated ion.

Europium(II), isoelectronic with Gd^{3+}, according to an EXAFS study has a coordination number of 7.2. [89]. The aqua Eu(II) ion is extremely labile, k_{ex} at 298 K is 5 x 10^9 s^{-1}, and the volume of activation (-11.7 $cm^3\ mol^{-1}$) is more negative than that for trivalent lanthanide ions, and therefore the mechanism is close to being fully associative. [90] (It should be noted here that this value of the volume of activation was obtained at 298 K, as were the values to be reported later in this chapter; however, it has already been pointed out that this parameter is not particularly temperature sensitive). Since the EXAFS study indicated an equilibrium between seven- and eight-coordinate states, the reason for the low activation barrier is apparent. [89] In this report [89] an analysis, incorporating results from several other sources, provides a comprehensive interpretation of mechanistic changes within these species - aquated Eu(II) ions, Eu(II) complexes, aquated Gd(III) ions and Gd(III) complexes. Arguments therein, based on the metal – oxygen bond distances and steric crowding allow an understanding of the relative values of water exchange rate constants, volumes of activation and the consequent mechanism variation. The relevant parameters are conveniently tabulated and therefore the report serves as mini-review of information available.

The marked loss of water exchange lability observed upon complexation of Gd^{3+} with poly(aminocarboxylate) ligands is not shared by similar complexes of Eu(II) which can be explained partly by the significantly lower charge density on the latter complex ions. Volumes of activation for water exchange on complexes of Eu(II) were less positive than for comparable complexes of Gd(III) reflecting less steric crowding. [88] Appropriate manipulation of the composition of the ligand in nine-coordinate monohydrated Gd(III) poly(aminocarboxylate) complexes has resulted in acceleration of the water exchange rate by two orders of magnitude compared with that of a complex in an earlier report. [91] Modification of the ligand caused steric compression around the water binding site; rapid water exchange is a favourable property for a Gd(III) MRI contrast agent, as noted earlier. [92] Variable pressure ^{17}O nmr experiments yielded in one case a distinctly positive $\Delta V^{\neq}$ indicating an I_d mechanism, but in another case a value for $\Delta V^{\neq}$ close to zero indicated an interchange mechanism. [91] A much more modest influence on the rate of water exchange is brought about by varying the substituent on bis(alkylamide) derivatives of a Gd(III) DTPA complex, and the activation volume was independent of the size of the substituents on the amide nitrogen atoms. [93] H_5DTPA is diethylenetriamine-N,N,N′,N′′,N′′-pentaacetic acid. The substituents ranged from CH_3 up to n-hexyl while $\Delta V^{\neq}$ varied only between +6.3 and +7.3 $cm^3\ mol^{-1}$, which indicated that the inner sphere of the complex was unaffected by the nature and magnitude of the amide substituents.

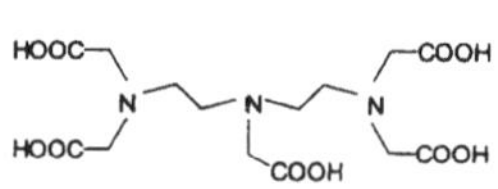

H_5DTPA

The cryptand 2.2.2 (4,7,13,16,21,24-hexaoxa-1,10-diazabicyclo[8.8.8]hexacosane) supplies eight donor atoms in coordinating to Eu(II) and in addition there are two coordinated water molecules, yielding what is reported to be the first ten-coordinate Eu(II) complex described. The coordination is preserved in aqueous solution, the water exchange rate is more favourable, i. e. more rapid, for a potential MRI agent, than it is for monohydrated Eu(II) complexes. An activation volume of +0.93 $cm^3\ mol^{-1}$ is indicative of an interchange mechanism, implying the incoming water molecule participates in the rate-determining step. [94] A further study of polydentate complexes of Eu(II) involved the two macrocyclic ligands DOTA and TETA. H_4DOTA is 1,4,7,10-tetraazacyclododecane-1,4,7,10-tetraacetic acid and H_4TETA is 1,4,8,11-tetraazacyclotetradecane-1,4,8,11-tetraacetic acid. The two complexes $[Eu(DOTA)(H_2O)]^{2-}$ and $[Eu(TETA)]^{2-}$ have been thoroughly characterised in terms of thermodynamic and redox stability, proton relaxivity, electron spin relaxation and in the former complex water exchange kinetics. Water exchange is extremely rapid; $k_{ex} = 2.5 \times 10^9$ s^{-1} at 298 K, and variable pressure ^{17}O nmr spectroscopy yielded a near zero activation volume (+0.1 cm^3 mol^{-1}) pointing to an interchange mechanism. This rapidity of water exchange could be rationalised on the basis of two factors: the low charge density on Eu(II), and a longer than expected Eu-O(water) distance (2.85 X), which lead to the interchange mechanism. [95]

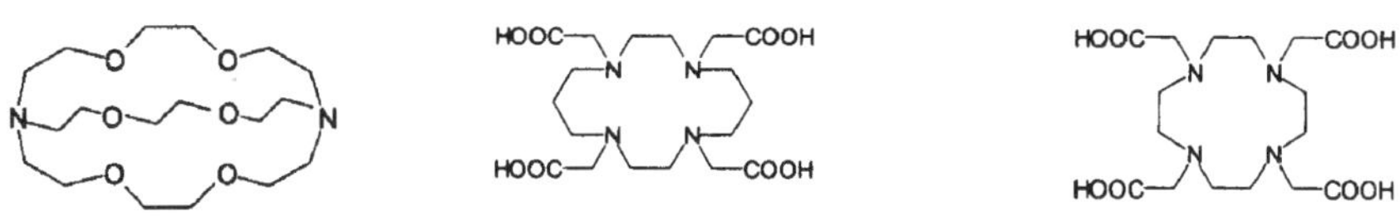

2.2.2 **H_4DOTA** **H_4TETA**

Among further efforts to design and develop MRI contrast agents, beyond the success of optimised chelates of Gd(III), binuclear poly(aminocarboxylate) complexes of Y, Eu, Gd and Tb in oxidation state three have been studied by a range (nmr, UV/Visible, epr and luminescence) of spectroscopies, as well as by molecular dynamics simulations. [96] The purpose of this type of study is to develop systems that possess the properties of slowing down tumbling, thus increasing relaxivity and increasing the water exchange rate of the complex ion. The application of pressure complements the overall mechanistic understanding relating to these properties. These binuclear chelates can undergo an isomerisation and the pertinent equilibrium is,

$$[Ln_2(OHEC)(H_2O)_2]^{2-} \rightarrow [Ln_2(OHEC)]^{2-} + 2H_2O$$

where Ln = Y, Eu, Gd, Tb and OHEC = octaazacyclohexacosane-1,4,7,10,14,17,20,23-octaacetate, in which the metal centre changes its coordination number from nine to eight. The isomerisation equilibrium has been characterised (K, ΔH^o, ΔS^o, ΔV^o) for the Eu(III) complex; the reaction volume is +3.2 cm^3 mol^{-1}. Isomerisation is slow, k(isomerisation) = 73.0 s^{-1} at 298 K and $\Delta V^{\neq}$ is $+7.9 \pm 0.7$ cm^3 mol^{-1}. Water exchange in the $[Gd_2(OHEC)(H_2O)_2]^{2-}$ complex is slow (relative to water exchange on some Gd(III) mononuclear complexes and compared with the aquated Gd(III) ion), and the mechanism is considered to be a dissociative interchange ($\Delta V^{\neq} = +7.3 \pm 0.3$ cm^3 mol^{-1}). An analysis of the consequences of these findings is presented in a comprehensive manner. [96]

The Gd(III) ion, Gd-2, has a stability constant K ($\log_{10}K$ = 24.9), and therefore fulfils the criterion of a low *in vivo* concentration of free Gd^{3+}. [97] Water exchange is very rapid and proceeds by an associative interchange mechanism, as determined by variable pressure (0 – 200 MPa) ^{17}O nmr spectroscopy, with $\Delta V^{\neq} = -5 \pm 1$ cm^3 mol^{-1}. The latter is a similar value to

$\Delta V^{\neq}$ for water exchange on $[Gd(H_2O)_8]^{3+}$, viz. -3.3 cm^3 mol^{-1}. Based on these findings hydroxypyridonate complexes of Gd(III) (of which Gd-2 is one) were proposed to be a promising new class of MRI agents.

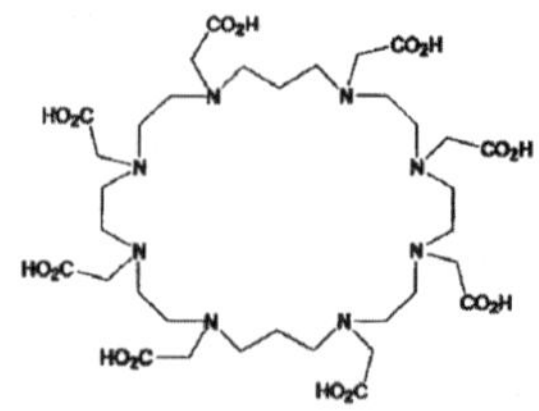

H_8OHEC

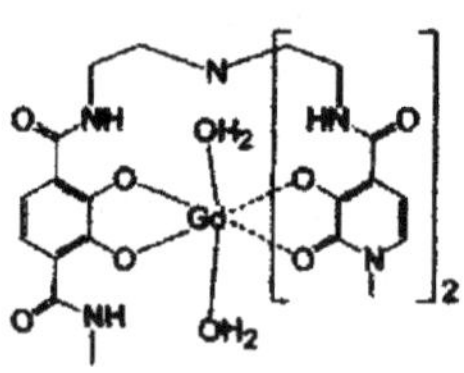

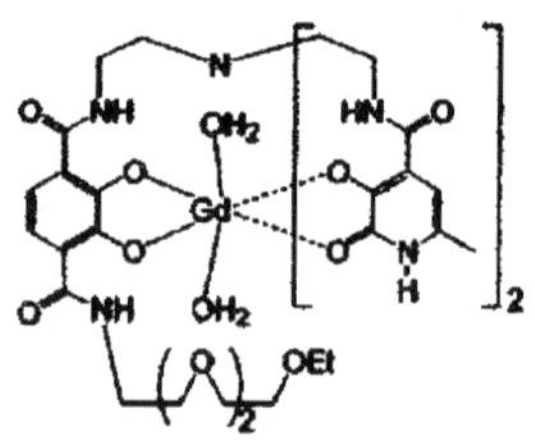

Gd-1 **Gd-2**

A further mechanistic update to that provided [89] was given following another structural XAFS investigation of Eu^{2+} and Sr^{2+} poly(aminocarboxylates) in the solid state and in solution. [98] The $DOTA^{4-}$, $DTPA^{5-}$ and $ODDA^{2-}$ complexes of both metal ions were studied, where $DOTA^{4-}$ is 1,4,7,10-tetraazacyclododecane-1,4,7,10-tetraacetate, $ODDA^{2-}$ is 1,4,10,13-tetraoxa-7,16-diazacyclooctadecane-7,16-diacetate, and the formula of DTPA has been given above. Mechanistic differences, and different water exchange rate constants could be understood in terms of charge densities about the central metal ion and the metal-oxygen bond distance by inference from these structural studies and also comparing and contrasting properties with similar Gd(III) species.

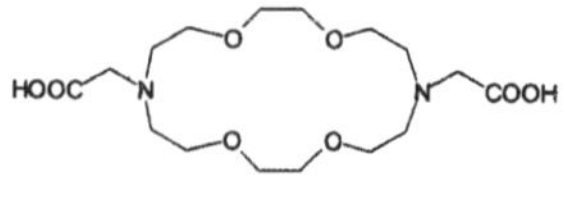

H_2ODDA

Although the preponderance of recent reports centres upon some inner transition metal ions (lanthanides), one prominent investigation features water exchange on a first row organo-transition metal compound. Organometallic aqua ions in which a transition metal ion complexes a π-cyclic or carbonyl ligand(s) with only water as the co-ligands nearly almost always contain a metal of the second or third row of transition elements. Hence efforts to generate and characterise such a complex of a first row transition metal are expected, and are a logical consequence. A further motivating consideration is that a comparison or contrast of the properties of such a complex with its 4d and 5d partners in the Group would be possible. The tricarbonyltriaqua manganese complex, $[(CO)_3Mn(H_2O)_3]^+$, although previously prepared and its properties postulated, has now been more thoroughly characterised. [99] Stable species in an oxidation state of +1 are quite rare for aqueous chemistry and for σ-coordinating ligands specifically for Mn, Tc and Re. It was proposed that in this case the CO groups stabilise the t_{2g} orbitals by back bonding thus stabilising the monovalent species. The exchange of water on the manganese compound was studied by ^{17}O nmr spectroscopy, yielding a value of k_{ex} at 298 K of $19 \pm 4\ s^{-1}$, and from application of pressure, $\Delta V^{\neq}$ of -4.5 $cm^3\ mol^{-1}$ was determined, indicative of an I_a process. The exchange of water on the corresponding rhenium compound is four orders of magnitude slower. Only a preliminary value for the rate of water exchange on the similar Tc compound has been established, but the rate constant will probably fit between those of the other two compounds. Thus the Mn analogue is the fastest of the triad; at present there are virtually no reliable data for water exchange on similar triads for comparative purposes. Perhaps surprisingly water exchange on the rhenium compound has been shown to operate by an I_d mechanism. The substitution reactions of $[(CO)_3Re(H_2O)_3]^+$ (see below for details), exhibited a mechanistic changeover: I_d (hard donor substituting ligands) changing to I_a (for a soft donor). Naturally it will be of undoubted interest to determine the ligand substitution mechanism(s) of $[(CO)_3Mn(H_2O)_3]^+$, and to examine thoroughly and systematically the properties of the Group triad.

3.2. Ligand substitution

In addition to ligand substitution in classical inorganic reactions, this sub-section will embrace some examples where ligand substitution occurs in what strictly speaking are organometallic reactions. A comprehensive account of ligand substitution reactions at both ambient and elevated pressures that brings the subject up to date to the beginning of 2002 has been provided. [100] Consequently, attention will be directed here mostly to reactions studied as a function of pressure, in the intervening period. All of the complexes cited in this section involve transition metal centres, and the order of coverage is by increasing atomic number of the metal.

3.2.1. Iron

Hepta-coordinate 3d metal complexes are relatively rare and solution reactions of such complexes have not been investigated systematically. Iron (III) forms a seven-coordinate complex with 2,6-diacetylpyridine-bis-(semioxamazide) (H_2dapsox).

The ligand provides five donor atoms in an equatorial plane with two axial water molecules completing a pentagonal pyramid. Beside the intrinsic interest in studying substitution reactions (of water) of this complex, iron complexes of this type have relevance to the enzyme superoxide dismutase. [101] The kinetics of substitution of $[Fe(dapsox)(H_2O)_2]^+$ by thiocyanate ion have been studied as a function of SCN^- concentration, pH, temperature and pressure, in aqueous medium. [102] Analysis of the kinetics results showed that there were two stages in the overall substitution process. An I_a mechanism was proposed for reaction of the diaqua complex, an I_d mechanism was operative for the aquathiocyanate complex, and an I_d or possibly a D mechanism was suggested for the reaction of SCN^- with the aquahydroxy forms of the complex. It was noted that the somewhat unexpected case of the I_a mechanism could be rationalised by the lack of steric hindrance in the axial positions since the pentadentate ligand is ideally planar. In methanol the kinetics of SCN^- substitution were analysed with the invocation of a limiting D mechanism for both steps. [103] This, it was argued was a consequence of both the weakly coordinating power of methanol and the proposal that one of the coordinating methanol molecules is actually present as a methoxy ligand.

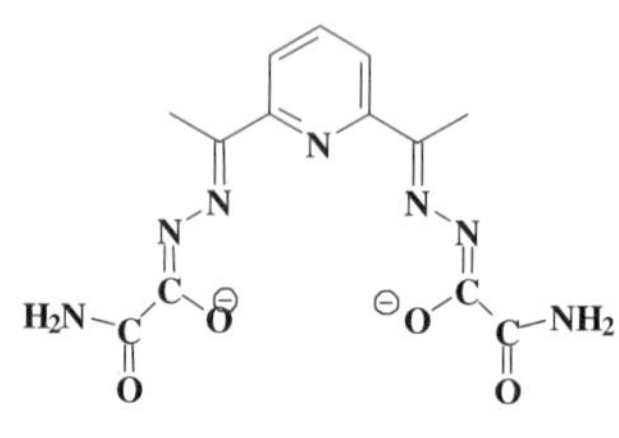

dapsox^{2-}

3.2.2. Cobalt

Transition metal complexes of benzosemiquinonediimine (s-BQDI) are important in terms of redox properties and in relation to biologically relevant catechol and benzoquinone derivatives.

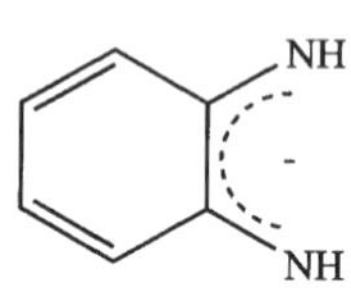

s-BQDI

The Co(III) complex, $[Co^{III}(s\text{-}BQDI)_2(Ph_3P)]ClO_4$ is square pyramidal in the solid state with the PPh_3 group occupying an axial position. [104] A coordinating solvent molecule such as CH_3OH supplies a solvent molecule to occupy the sixth coordination position in solution. In acetonitrile comparable Co(III) complexes with the PF_6^-, I^- or SCN^- counter ion have been described as potentially five-coordinate. Equilibria between five- and six-coordinate species of such complexes is a topic of interest with respect to ligand substitution and the mechanisms thereof, and particularly to alkyl cobalamines and cobinamides, as will be discussed below. The effects of pressure and temperature variation upon the kinetics of axial ligand substitution of S (S = CH_3OH, CH_3CN) in $[Co^{III}(s\text{-}BQDI)_2(Ph_3P)S]^+$ by imidazole and 4-dimethylaminopyridine have been investigated. [105] Reaction between imidazole and the metal complex in methanol is characterised by rapid loss of a solvent molecule, followed by a rate-determining associative ligand exchange process. Since the progress of formation of the imidazole complex can be followed, and furthermore the reaction of Ph_3P with the imidazole complex can also be monitored (the reverse reaction) both as a function of pressure, then a volume profile could be established (see Figure 4). In acetonitrile the pressure dependence of the rate constant for the substitution reaction (imidazole for Ph_3P) and other kinetic information resulted in the proposal of an associative interchange mechanism thus precluding the formation of a five-coordinate intermediate. The significant difference in rate constant, $k_{2nd\ order}(CH_3CN) >> k_{2nd\ order}(CH_3OH)$ for reaction of imidazole suggested that hydrogen bonding deactivates the attacking ligand in methanol, or that the stronger donor property of acetonitrile labilises the trans position thereby inducing rapid displacement of PPh_3.

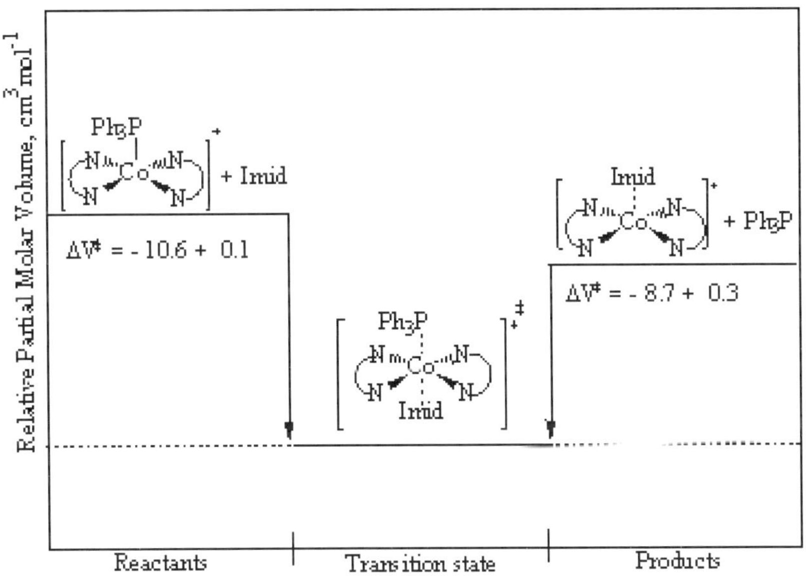

Figure 4. Volume profile for the reaction:$[Co^{III}(s\text{-}BQDI)_2(Ph_3P)]^+$ + Imid ≡ $[Co^{III}(s\text{-}BQDI)_2(imid)]^+$ + Ph_3P.

In a parallel study the kinetics of substitution of coordinated water or CH_3OH by three pyridine nucleophiles (pyridine itself and the 4-CN- and 4-NH_2-derivatives) in *trans*-$RCo(Hdmg)_2(H_2O)$ or *trans*-$RCo(Hdmg)_2(CH_3OH)$, (R = $C_6H_5CH_2$ or CF_3CH_2 and dmg is dimethylglyoxime) were studied as a function of nucleophile concentration, temperature and pressure. [106] From analysis of the derived activation parameters it was concluded that the substitution mechanism varies with R substituent and solvent, *viz.* a limiting D mechanism for R = $C_6H_5CH_2$, reflecting a stronger donor property than when CF_3CH_2 is the substituent. The substrate containing the latter substituent underwent water substitution by an I_d mechanism.

The mechanism also varies in that the trend in $\Delta V^{\neq}$ values depended on the basicity of the nucleophile with progressively increasing positive $\Delta V^{\neq}$ values (4-CN- > H- > 4-NH_2-).

A series of closely related reports examined the equilibria between five- and six-coordinate species and possible adduct formation in Co^{III}-corrinoids, [107] the thermodynamic and kinetic properties for what is termed the base-on/base-off equilibration of alkylcobalamins, [108] and the kinetics and thermodynamics of parallel equilibria of alkyl-13-epicobalamins, [109] respectively. The cyanation reaction of alkyl-13-epicobalamins has also been addressed in the last study. In the first report the pressure dependence of the UV/Visible spectra of the five-coordinate (yellow)/six-coordinate (red) equilibrium for both methylcobalamin and vinylcobinamide was obtained. Water is the ligand that converts the five- to a six-coordinate species. From the UV/Visible spectra the equilibrium constant as a function of pressure could be obtained and hence the reaction volume derived. The latter was identical for both equilibria (ΔV = -12.5 ± 1.2 and -12.5 ± 1.0 cm^3 mol^{-1} for the methyl and vinyl complexes, respectively), and remarkably close to the numerical value of -13 or +13 cm^3 mol^{-1} advocated and generally accepted for limiting A or D mechanisms for the displacement of a water molecule. [11, 110] Thus, the reaction volume is a measure of the reversible loss of a water molecule from six- to a five-coordinate state. Initial experiments with both vinylcobinamide (five-coordinate) and dicyanocobinamide (six-coordinate) showed that adducts were formed with a selection of various added solutes, consistent with their place in the lyotropic series. [107] It was suggested that the solute species are probably located within the hydrophobic cylinder surrounding the cobalt ion. The base-on/base-off equilibration of a series of alkylcobalamins (XCbl) has been studied to investigate the influence of the electronic nature of X. An important characteristic of cobalamins (see Figure 5) is the equilibrium in which the axially coordinated dimethylbenzimidazole moiety (DMBz) is displaced by water (the base-off form) and protonated. (The DMBz group is substituted in the biological cycle.) A combination of equilibrium and kinetic measurements as a function of temperature, pressure, and, in the kinetics studies of the equilibration base-on/base-off as a function of acid concentration indicated, after a complex results analysis, that the base-off form of different XCbl species has different coordination. [108] AdoCbl and ethylcobalamin are mainly five-coordinate, whereas the CF_3-, CN- and H_2O-cobalamins have little base-off form and exist mainly as six-coordinate species. When X = CH_3, CH_2Br and CH_2CF_3 there is an equilibrium mixture of the five- and six-coordinate species. Analysis of the acid dependent measurements provided evidence for an acid catalysed dechelation (of DMBz) pathway. The properties of structurally modified analogues of AdoCbl may shed light on the chemistry of the unmodified coenzyme (see structure given in Figure 5). In this regard two derivatives ($NCCH_2$ and CN^-) of the epimer obtained by epimerisation at C13, in which the e-propionamide side chain adopts an upwardly axial configuration, have been the subject of an extensive study of the cyanation reaction. Kinetics of the cyanation reaction have been conducted with variation in the CN^- concentration as well as the temperature and pressure of measurement. [109] It was shown that epimerisation has a sensitive influence on the kinetics of displacement of the DMBz moiety first by water and then by the cyanide ion, and the rate of substitution is reduced compared with an alkyl cobalamin. Displacement of DMBz shows saturation kinetics at high cyanide ion concentration, and the activation parameters, $\Delta S^{\neq}$ and $\Delta V^{\neq}$ for the limiting rate constant are +77 ± 4 and +82 ± 4 J mol^{-1} K^{-1} and +13.3 ± 1.0 and +14.8 ± 1.2 cm^3 mol^{-1} for the $NCCH_2$ and CN^- epimers, respectively. These values were interpreted as evidence of a limiting D mechanism.

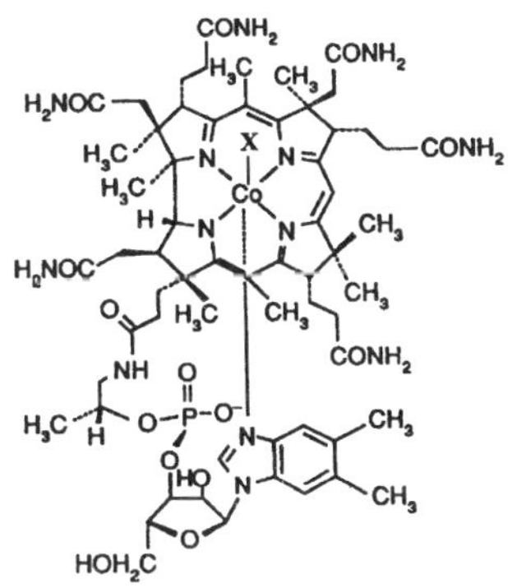

Figure 5. Structure of cobalamin, where X = CN^- for cyanocobalamin (vitamin B_{12}), X = H_2O for aquacobalamin (vitamin B_{12a}) and X = 5'-deoxyadenosyl for coenzyme B_{12}.

Similarly a limiting D-mechanism had been proposed for the kinetics of displacement of DMBz by CN^- in XCbl, where X = β-$NCCH_2$ and CN^- to form $NCCH_2(CN)Cbl$ and $(CN)_2Cbl$, respectively. This was also based upon saturation kinetics at high cyanide concentration and significantly positive entropies and volumes of activation. This report also contained a formulation in terms of a general mechanism for this type of reaction that included earlier results and results of reinvestigations. [111]

In order to establish a consistent, comprehensive view of substitution reactions in Co(III) compounds that have analogy with the more direct bio-relevant compounds, the kinetics of substitution of reactions of various cobaloximes have been investigated with a mechanistic goal. [112] Again application of pressure, in addition to temperature variation, to the reacting systems provided mechanistic insight. In one study the kinetics of axial water substitution by cysteine in six different cobaloximes, in aqueous solution, have been investigated. The cobaloxime system was again *trans*-$RCo(Hdmg)_2(H_2O)$ with R = cyclo-C_5H_9, CH_3CH_2, CH_3, $C_6H_5CH_2$, C_6H_5 and CF_3CH_2. While five-coordinate species have been inferred from results of studies on coenzyme B_{12} in aqueous medium and on alkylcobaloximes in non-coordinating solvents, the evidence (largely on the basis of the magnitude of $\Delta V^{\neq}$) in the reported investigation was that the mechanism is dissociative interchange, i.e. the absence of five-coordinate species. The absence of a linear isokinetic plot implied a mechanistic variation with the compound containing the CF_3CH_2 derivative operating by an interchange mechanism ($\Delta V^{\neq}$ = 0 ± 1 cm^3 mol^{-1}). It was also concluded that all the activation parameters are influenced by a combination of electronic and steric effects.

The pressure dependencies of kinetic parameters have been vital in finally bringing significant clarity to the mechanism of substitution reactions of cobalamins (see structure given in Figure 5). In particular the effect of various alkyl substituents in the *trans* position of the Co(III) centre on the kinetic, thermodynamic and ground state properties has been investigated. [113] The specific cobalamins at issue in these studies were cyanocobalamin (vitamin B_{12}) and aquacobalamin (vitamin B_{12a}) and the complex formed when the cyano or water ligand in these species is replaced by 3'-deoxyadenosyl. Complete details of the experimental results, the derived kinetic parameters and mechanistic conclusions from a substantial body of research have been reported. [113] In addition to using $\Delta V^{\neq}$ in mechanistic diagnosis, the effect of pressure on the UV/Visible spectra of various cobalamins demonstrated the key role of the alkyl group in controlling the equilibrium between five- and six-coordinate species, relating to these species potential participation in ligand substitution reactions. It has become clear that the nature of the cobalt-carbon bond, and the character of the substituent supplying the carbon atom to that bond are vitally important. It has also been shown by applying UV/Visible and ^{1}H nmr spectroscopies, and including the determination of $\Delta V^{\neq}$ (= -4.5 cm^3 mol^{-1}) for the reaction of co-enzyme B_{12} with cyanide ion, that the rate-determining step involves solvent assisted heterolytic cleavage of the Co-carbon bond. [114]

3.2.3. Ruthenium

The potential inhibition of sulfhydryl proteases such as papain and bromelein by $[Ru^{III}(EDTA)(H_2O)]^-$ has been examined. [115] Sulfhydryl or thiol proteases have been implicated in protein degradation in several medical disorders. Since inhibition was in fact recorded, this finding prompted an investigation of the reaction of the ruthenium complex with some amino acids. [115] This investigation took the form of potentiometric and kinetic studies; the coordinated water molecule in $[Ru(EDTA)(H_2O)]^-$ is labile and therefore mixed ligand complexes can be formed. The kinetic data for the reaction of glycine, cysteine and S-methylcysteine indicated a two-step reaction that could be interpreted as an initial, rapid amino acid concentration-dependent complex formation step followed by a slower ring-closure step. At the pH of measurement it is believed that the ring-closure step occurs when the protonated amine group of the amino acid deprotonates and coordinates to the Ru^{III} centre accompanied by displacement of one of the carboxylate arms of EDTA. The activation parameters for the first step ($\Delta S^{\neq}$ = +31 ± 5 and -28 ± 7 J mol^{-1} K^{-1} and $\Delta V^{\neq}$ = -3.3 ± 0.2 and -4.7 ± 0.4 cm^3 mol^{-1}) for glycine and cysteine, respectively, were indicative of the operation of an associative interchange mechanism. This is consistent with the mechanism proposed for similar reactions of Ru^{III}(EDTA) type complexes with various thiol ligands. [116] It was also of interest that the stability constant for the 1:1 complex of the three amino acids was highest for cysteine and the latter also bound more efficiently as evidenced by the largest rate constant.

3.2.4. Rhodium

Kinetic measurements of substitution on $[Rh(OH_2)_6]^{3+}$ ions by Cl^- or Br^- were analysed and led to a D mechanism being proposed. [117], [118], [119] Substitution of a second water molecule led to the product *trans*-$[Rh(OH_2)_4(X)_2]^+$, where X = Cl^- or Br^-, was also concluded to proceed by a D mechanism. However, high pressure ^{17}O nmr spectroscopy kinetic studies yielded $\Delta V^{\neq}$ values of -4.2 cm^3 mol^{-1} for water exchange on $[Rh(OH_2)_6]^{3+}$ [120] and of -5.7 cm^3 mol^{-1} for water exchange on $[Ir(OH_2)_6]^{3+}$. [121] There would appear to be a conflict here as the latter values for water exchange would suggest an I_a process, and in many ligand-for-water substitution reactions the mechanism mirrors this process. The energetics of water replacement on $[Rh(OH_2)_6]^{3+}$ were modelled using an *ab initio* quantum mechanical treatment. [122] These calculations did not favour a D mechanism. A variable pressure spectrophotometric study of the complexation of bromide by the aquarhodium(III) ion, in which the range of acid concentration was extended beyond that employed in earlier studies, yielded values of $\Delta V^{\neq} = -3.3 \pm 1.0$ cm^3 mol^{-1} for complexation by the hexaaqua ion and $\Delta V^{\neq} = +7.7 \pm 1.0$ cm^3 mol^{-1} for the pentaaquahydroxy species. [123] Both values were obtained after corrections for estimates of the pressure dependence of the outer-sphere association constant. The former value, although not very substantially, supports an I_a mechanism, and the latter an I_d mechanism for substitution at the lower charged $[Rh(OH_2)_5(OH)]^{2+}$ species. The fact that the rate constants for water exchange and Br^- or Cl^- anation on $[Rh(OH_2)_6]^{3+}$ were similar reflects on the fact that these ligands are not discriminated by $[Rh(OH_2)_6]^{3+}$ within an I_a mechanism. The monohydroxy species exhibited a weaker interaction with the departing molecule in the transition state which can be attributed to charge neutralisation and desolvation. Arguments developed [123] were extended to discuss ligand substitution on other divalent and trivalent transition metal ions as well as on $[Ga(OH_2)_6]^{3+}$ whose kinetics of water exchange had been reported earlier. Further theoretical work regarding the mechanism of water exchange on $Rh(H_2O)_6^{3+}$ and on $Ir(H_2O)_6^{3+}$ will be presented in section 6.

3.2.5. Palladium

Reasons for studying complexes of palladium(II) in the context of anti-tumour drug design revolve around the lack of efficacy of platinum(II) complexes in certain circumstances. The structural and equilibrium properties of palladium complexes are similar to the corresponding platinum complexes but the reactivity of the former complexes is generally orders of magnitude higher. In an extension to earlier investigations, and with emphasis here on sulphur containing ligands bound to Pd(II), the complex-formation equilibria and stoichiometry of reactions between $[Pd(SMC)(H_2O)_2]^+$ (SMC = S-methyl-L-cysteine) and inosine, inosine-5′-monophosphate, guanosine-5′-monophosphate, L-glycine and 1,1-cyclobutanedicarboxlate have been studied. [124] The choice of substituting ligand was driven partly with the view of generating basic information on DNA constituents in such equilibria. In addition the kinetics of the complex-formation reactions were studied as a function of temperature and pressure. Potentiometric measurements were obtained for the complex-formation equilibria. Various fitting procedures were adopted and these gave rise to species distribution profiles over the pH range 1 to 13. An investigation of the kinetics of complex formation of the first three cited ligands showed that the reaction was characterised by two steps, both in the stopped-flow spectrophotometer time range, but sufficiently widely separated in time allowing facile extraction of the rate constants. [124] Both reaction steps exhibited a linear dependence on the concentration of incoming nucleophile and the plots of observed rate constant versus nucleophile concentration had significant intercepts indicating reversible reactions. Each of the two steps denotes the progressive substitution of the coordinated water molecules by the nucleophile. The faster step was attributed to the displacement of the water molecule in the *trans* position to the S-coordination site. This is a consequence of the strong *trans* labilising effect of coordinated sulphur. Inosine reacts much more rapidly than the phosphate containing components since the phosphate groups were considered to undergo a pre-association with the metal complex, a process that retards the actual substitution reaction. This pattern of reactivity is in accord with reactions of the same ligands with $[Pd(met)(H_2O)_2]^+$ and $[Pd(pic)(H_2O)_2]^{2+}$, where met = methionine and pic = 2-picolylamine. [125], [126] Both forward reactions were characterised by markedly negative activation entropies indicative of associative character of the substitution processes. This was supported by the pressure dependence of the rate constants for the second substitution; negative volumes of activation were obtained. (The former step was too rapid for studying the rate constant as a function of pressure). The overall substitution pattern is similar to that found for Pd(II) complexes that are chelated through nitrogen atoms, and it can therefore be concluded that the sulphur donor ligand initiates a faster rate for the first substitution process.

3.2.6. Rhenium

The stability, electrochemical, photochemical and spectroscopic properties of tricarbonylrhenium complexes are favourable with respect to application in the field of photochemistry. Rhenium compounds have also been investigated regarding their potential application in medicinal chemistry. These aspects of rhenium chemistry and related studies on "half-sandwich" organic-aqua complexes of other metals (Ru, Os, Rh, Ir) in which the organic ligand may labilise the *trans* water molecules, stimulated interest in the rate and mechanism of water exchange on tricarbonyl *fac*-$[(CO)_3Re(H_2O)_3]^+$. The kinetics of water substitution in this complex by various ligands were studied with the view of examining the sensitivity to the nucleophilicity of the entering ligand as shown in the following reaction. [127]

$$[(CO)_3Re(H_2O)_3]^+ + L \rightarrow [(CO)_3Re(H_2O)_2L]^+ + H_2O$$

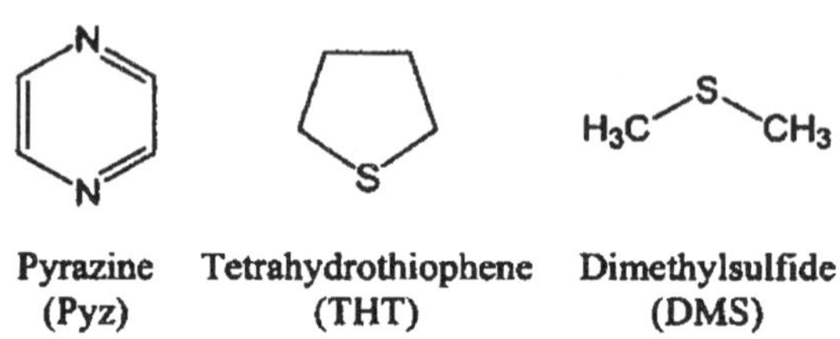

However, unambiguous determination of the mechanism of substitution of water by pyrazine, tetrahydrothiophene and dimethylsulphoxide was only possible when variable pressure kinetic studies were performed. [128] The calculated interchange rate constants were not significantly different from the water exchange rate constant, but it could be shown from a detailed analysis of the volume profiles (see Figure 6) that for hard nucleophiles, with O- and N-donating atoms the I_d mechanism was operative, whereas a soft S-donor nucleophile having a greater affinity for the soft rhenium centre leads via an I_a mechanism to an associative transition state.

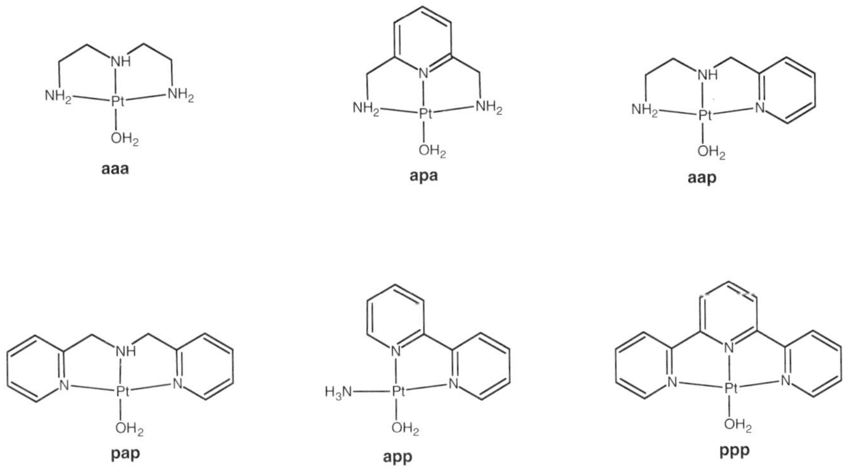

Figure 6. Schematic representation of the complexes used in a systematic s tudy of the influence of the π-acceptor effects on a series of Pt(II) complexes.

3.2.7. Platinum

A multitude of reactions of platinum compounds continues to receive attention. In particular ligand substitution reactions have been widely studied as some of these reactions have biological and pharmacological relevance. Examples of reactions in which pressure has been shown to be a valuable experimental variable, are now presented. The first is the latest [129] in a series of reports investigating complex-formation reactions of $[Pt(NNN)(H_2O)]^{2+}$ with different nucleophiles. [130] In this case NNN represents a terdentate ligand 2,2′:6′,2′′-terpyridine, (terpy) and is used to examine the reaction of $[Pt(terpy)(H_2O)]^{2+}$ with thiol ligands, L, specifically L-cysteine, DL-penicillamine and glutathione, and also with thiourea (TU). This study has importance in relation to platinum anti-tumour compounds.

$$[Pt(terpy)(H_2O)]^{2+} + L \rightarrow [Pt(terpy)(L)]^{2+} + H_2O \quad k_1/k_{-1}$$

The reactions were followed using excess L, and plots of k_{obs} versus [L] could be used to show that the reactions proceed virtually to completion, from which the second-order rate constants were obtained. The reactivity order was DL-penicillamine < L-cysteine < glutathione < thiourea, a trend not unexpected and consistent with the σ-donor properties of the entering ligand in an associative mode of activation. Some reactivity detail revealed that steric effects are also implicated, for both the non-exchanging terpyridine and the entering ligand. From the second-order forward rate constants obtained as a function of pressure the volumes of activation were determined. These and the entropies of activation were distinctly negative. This is compatible with expectation for an associative process in which a five-coordinate intermediate was proposed. The values of $\Delta V^{\neq}$ are in qualitative agreement with those for the corresponding reactions of Pd(II), [124, 131] but more negative. Since structural studies have shown that the metal to ligand bond distances are very similar in both the Pt(II) and Pd(II) complexes, [132] it can be argued that in the transition state there is less orbital overlap and a less compact structure for the latter complexes.

A combined theoretical and experimental study of the electronic tuning of the lability of Pt(II) complexes through π-acceptor effects has been reported. [133(a)] This is a recurrent theme in attempting to understand all the factors regarding the properties of the bound ligands, exchanging ligands in substitution processes both in terms of σ and π acceptor or donor character in the lability of Pt(II) complexes. Another aspect of the motivation attached to such an investigation was the desire to design Pt(II) complexes with appropriate kinetic and thermodynamic properties to possess required efficacy in a particular biological environment. This subject will also be addressed further below. [133(b), (c)] A series of Pt(II) complexes containing three nitrogen donor ligands and one coordinated water molecule were prepared and the lability of the coordinated water molecule investigated in order to assess the influence of the π-acceptor effect upon the lability of these complexes (see Figure 7). The kinetics of water substitution by TU, N,N′-dimethylthiourea (DMTU), N,N,N′′,N′′-tetramethylthiourea (TMTU), I^- and SCN^- were studied as a function of nucleophile concentration, temperature and pressure. Analysis of the results indicated that a π-acceptor ligand moiety in the *cis* position is more influential than π-acceptor property in the *trans* position to the leaving water molecule, and furthermore it was found that the two π–acceptor effects were multiplicative, and the reaction mechanism was associative. The results from density functional theory (DFT) calculations are presented in section 6 below.

A comprehensive study of several cyclometallated analogues of platinum(II) terpyridine/chloro complexes involved crystal structure determinations, substitution kinetics in methanol as solvent with the nucleophiles Br^- and I^- and in two cases, [Pt-N-N-C)Cl] and [Pt(N-C-N)Cl] with (TU), (DMTU), and (TMTU), where N-N-CH is 6-phenyl-2,2′-bipyridine and N-CH-N is 1,3-di(2-pyridyl)benzene (see Figure 8). [133(b)] One principal objective of this study was an examination of the relationship between σ-donor and π-acceptor effects in Pt(II) chemistry. The much higher reactivity observed for the N-C-N compound was reflected in a 20 kJ mol^{-1} lower activation enthalpy than for the N-N-C compound; this could be explained by the influence of the *trans* phenyl group in destabilising the ground state, thus facilitating cleavage of the Pt-to-leaving group bond. The observed activation enthalpy sequence with, for example the N-N-C compound was TMTU < DMTU ≈ TU, whereas for the N-C-N compound the relationship was TMTU < DMTU < TU. This sequence was not in accord with reactivity trends, and therefore it was concluded differences in activation entropies were responsible for the observed steric retardations. The volumes of activation were predicted by the authors to be, and were found to be negative for the three thiourea ligands reacting with the N-N-C and N-C-N compounds. This arises from the volume contraction in the transition state because of bond formation by the entering nucleophile, a contraction which is partially offset by the geometry changing from square planar to trigonal bipyramidal (the five-coordinate intermediate state). In the case of the N-C-N compound the platinum-ligand bonds of the entering and leaving ligands are longer due to the *trans* influence, and thus the observed $\Delta V^{\neq}$ values were less negative for the N-C-N reaction than for the N-N-C reaction where the corresponding bonds are shorter. On the basis of results in this report and in the copious literature on Pt(II) substitution reactions it was concluded that the magnitude and direction of the *cis* effect (as compared to the *trans* effect of the same ligand) depends on the σ-donor strength, π-acceptor ability and the steric property of the *cis* ligand. [130] The subtle interplay among these factors and in relation to the relative magnitudes of these factors, (*cis*/*trans*) and the consequences for reactivity were discussed. It was also intimated how consideration of such factors could be important in the design of Pt(II) complexes for anti-tumour drugs, C-H activation and homogeneous catalysis.

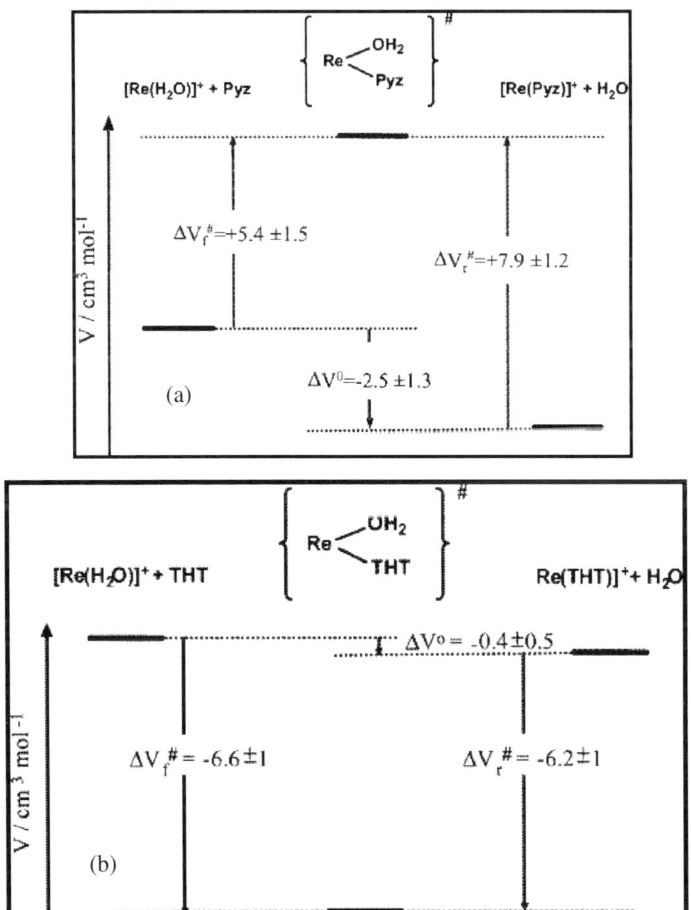

Figure 7. Volume profiles for the reactions $[(CO)_3Re(H_2O)_3]^+ + L \rightarrow [(CO)_3Re(H_2O)_2L]^+ + H_2O$, where L = Pyz and THT, respectively. Figure 7a) is typical for an I_d mechanism, whereas Figure 7b) is typical for an I_a mechanism.

An investigation related to the one just described had the objectives of exploring the role of chelate substituents and the *cis* σ-effect on the rate of ligand substitution at Pt(N-N-N) and Pt(N-N-C) centres. [133(c)] The terdentate (non-substituting) ligands were terpyridine itself, 4′-Ph-terpy, 4′-(2′′′-CF_3-Ph)-terpy, and 4′-(2′′′-CF_3-Ph)-6-Ph-2,2′-bipy, and the chloro ligand in the complexes [Pt(N-N-X)Cl] (X= N or C). The rates of substitution of the chloro ligand by TU, DMTU and TMTU in these complexes were monitored as a function of incoming nucleophile concentration, temperature and pressure by stopped-flow spectrophotometry. The kinetics were straightforward: the rate law established that the reaction is first-order in both platinum complex and in nucleophile concentrations. These facts and the negative activation entropies and volumes supported an associative substitution mechanism. Substitution on the terpyridine parent ligand did not affect the kinetic parameters significantly. When a carbon σ-donor was in the *cis* position the reaction was slower than when a N σ-donor was present, indicating that this is a different situation from the effect of a Pt-C bond in the *trans* position.

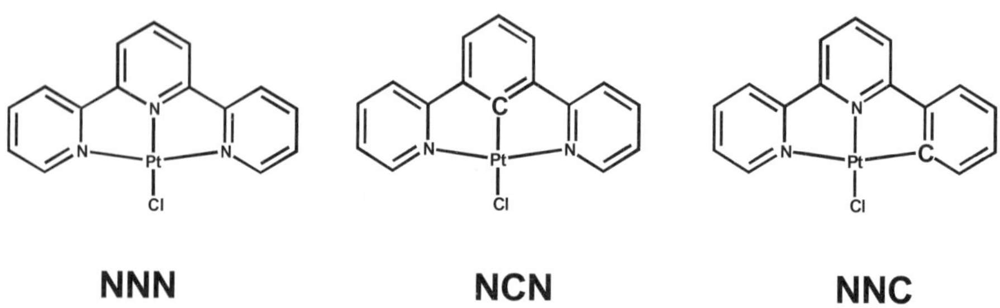

Figure 8. Schematic representation of the complexes used in a systematic study of the influence of π-acceptor and σ-donor effects on a series of Pt(II) complexes.

The activation volume for C-H activation for benzene substitution at a Pt(II) centre has been determined and analysed. [134] In this investigation a platinum(II) monomethyl cation (the three other positions around the platinum centre were occupied by an unsaturated bidentate arene-containing ligand and a solvent molecule) reacts with benzene in CF_3CH_2OH/H_2O mixtures to yield a product containing the phenyl cation. An important aspect of this investigation was to resolve the mechanism of the mode of entry of the hydrocarbon into the Pt(II) coordination sphere. Many ligand substitution reactions of square planar d^8 metal complexes follow an associative mechanism, but in this case there would be doubt since the entering hydrocarbon is a very weak nucleophile. The pressure dependence of the kinetic parameters yielded a composite value of the volume of activation. Fortunately this could be dissected and a value of -9.5 $cm^3\ mol^{-1}$ emerged as the volume of activation for the activation of benzene by the solvento-(CF_3CH_2OH)Pt(II) complex, which together with the negative entropy of activation supports the operation of an associative mechanism.

In another study, determination of the activation volume has provided a powerful experimental criterion for understanding the mechanism of substitution with concomitant isomerisation in cyclometallated dimethylhalo-platinum(IV) complexes. [135] In the course of studying the substitution reactions of several *cis*-($CH_3/S(CH_3)_2$) arranged cyclometallated dimethylhaloplatinum(IV) complexes, whereby PPh_3 substitutes the $S(CH_3)_2$ group, an intramolecular isomerisation process was revealed in two cases. It was possible to monitor the latter by low temperature ^1H and ^{31}P nmr spectroscopies. The process could be described as a very energetically demanding turnstile twist-type reorganisation of the molecule, after a significant degree of dissociation of the ligand. The thermal activation parameters were about 100 kJ mol^{-1} for $\Delta H^{\neq}$ and +88 and +49 J mol^{-1} K^{-1} for $\Delta S^{\neq}$. The pressure derived activation parameters were +15 and +20 cm^3 mol^{-1}. These reported values indicate that the process involves a substantial increase in volume in forming the transition state. In other cases complexities arose from various *mer*/*fac* ratio dispositions.

3.3. Redox reactions

Understanding electron transfer in solution is a goal not restricted to inorganic chemistry and is of widespread importance in biochemistry with the attendant significance in a medical context. Fundamental studies on relatively uncomplicated reactions have been widely pursued. The availability of high quality experimental results both at ambient and elevated pressures, whether for spontaneous self-exchange or non-symmetrical redox reactions, has enabled investigators to test the predictions of Marcus-Hush theory under these conditions. [136, 137] In addition, studies of redox reactions in electrochemical cells both at ambient and high pressure have advanced knowledge of electron transfer in heterogeneous systems. A considerable fraction of the reports in the recent period at issue involve compounds or complexes of iron.

Kinetic studies of redox properties of organic compounds in the presence of metal complexes can assist in understanding electron transfer mechanisms in organic molecules. N-substituted phenothiazines are important in biochemistry and pharmacology. The kinetic and equilibrium parameters for a series of phenothiazines in electron transfer reactions with hexaaquairon(III) were established reliably about 25 years ago. [138]

$$\text{R-(phenothiazine)} + Fe(aq)^{3+} \rightarrow \text{R-(phenothiazine)}^{+} + Fe(aq)^{2+}$$

High pressure stopped-flow experiments upon the reaction of hexaaquairon(III) and promazine (R = $CH_3CH_2CH_2N(CH_3)_2$, see reaction given below) yielded $\Delta V^{\neq}$ values of -6.3 and -12.5 cm^3 mol^{-1} for the forward and reverse reactions, respectively. [139] The derived reaction volume of 6.2 cm^3 mol^{-1} has been confirmed by calculation from the pressure dependence of the equilibrium constant obtained from the variation of the UV/Visible spectra as a function of pressure. The volume profile (see Figure 9) displays the fact of a compact transition state, and the positive reaction volume was suggested to arise from charge dilution in proceeding from the reactants to $Fe(aq)^{2+}$ and the promazine cation radical.

It was also indicated that the compact nature of the transition state could arise, in part, from a reduction in volume as a precursor contact pair is formed, with this factor being greater for the reverse than for the forward reaction. Markedly negative entropies of activation in both directions, determined earlier, [138] indicated that electrostriction increase is important particularly for the back reaction. Intrinsic volume changes were not considered to contribute significantly as any bond length changes in promazine during the reaction are small or negligible. From application of Marcus-Hush theory it was concluded that oxidation of promazine controls volume changes associated with the forward reaction, whereas oxidation of iron(II) controls volume changes associated with the reverse reaction, reasonably firm conclusions not usually possible in this field.

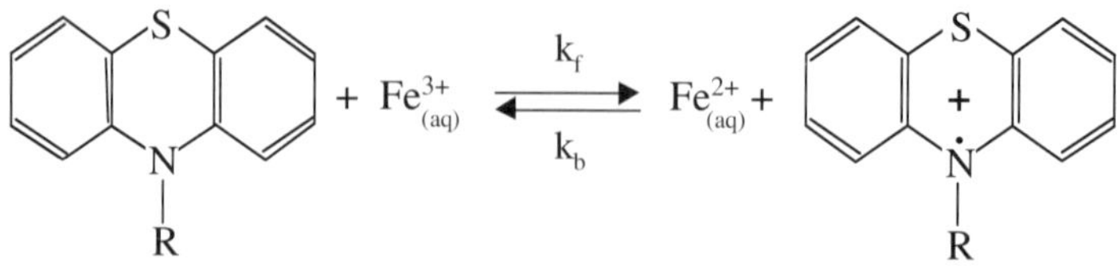

$R = CH_2CH_2CH_2N(CH_3)_2$

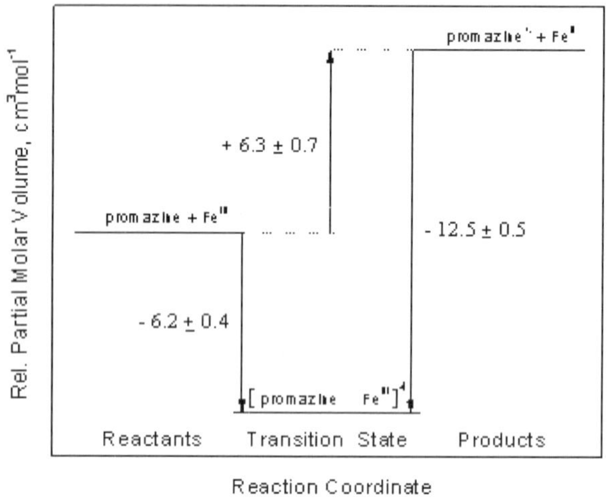

Figure 9. Volume profile for the reaction: promazine + $Fe^{3+}_{(aq)}$ $\leftrightarrow$ promazine$^{+\cdot}$ + $Fe^{2+}_{(aq)}$.

There has been considerable debate regarding the mechanism and modelling of the self-exchange reaction of metallocenes. Considerable clarity has been brought to this topic by the recent investigation of the reaction between $[(\eta^5\text{-}C_5(CH_3)_5)_2Fe^+]$ ($DmFc^+$), (hexafluorophosphate or tetrafluoroborate counter ions) and $[(\eta^5\text{-}C_5(CH_3)_5)_2Fe]$ ($DmFc^0$) in deuterated acetone, dichloromethane and acetonitrile. [140(a)] Values of the exchange rate constant, k_{ex}, were obtained from ^{1}H nmr spectroscopy at ambient and elevated pressure, although practical limitations rendered only semi-quantitative results in the last solvent. Only very marginal effects of ion-pairing were encountered. Values of $\Delta V^{\neq}$ were distinctly negative (-8.5 to -8.9 $cm^3\,mol^{-1}$ in $(CD_3)_2CO$, and -6.3 to -7.2 $cm^3\,mol^{-1}$ in CD_2Cl_2). The small variations arise when different counter ions or concentrations are used. The various contributions to $\Delta V^{\neq}$ in terms of an extension of the Marcus-Hush theory, with particular reference to non-aqueous solvents were carefully evaluated. It was concluded that the results were entirely consistent with a two sphere activation model and it was not necessary to invoke solvent dynamics in the explanation. Earlier reports had proposed that solvent dynamics were involved in electron self-exchange, [141] but this high pressure study was able to marshal evidence against this argument. Perhaps the importance of this topic to a broad range of inorganic chemists could be gauged by the fact that an art work depiction of the article [140(a)] was chosen as the front cover of the issue of the journal.

In a related study, the volume of reaction, the volume of activation for diffusion ($\Delta V_{diff}^{\neq}$) (from the variation of D, the diffusion coefficient, with pressure) and the volume of activation, obtained from the standard electrode reaction rate constant at various pressures ($\Delta V_{el}^{\neq}$), were determined for the decamethylferrocene ($DmFc^{+/0}$) system, as defined above, [140(a)] in several non-aqueous solvents. [140(b)] The choice of the decamethylated ferrocene couple was partly dictated by the large size and low charge of the positive ion form offering minimisation of complicating Coulombic interactions, such as ion-pairing. The unmethylated $Fc^{+/0}$ couple is less stable in the rigour of the electrochemical experimental conditions required, and the higher charge density of the positive ion would increase the likelihood of the unwanted involvement of ion-pairing, although limited data were reported for the latter system for comparison. A further objective was to compare the pressure effects on the corresponding outer-sphere bimolecular electron transfer (self-exchange) rate constants in homogeneous solution [140(a)] with those on the non-aqueous electrode kinetics. The reaction volumes, ΔV_{cell} could be expressed as a linear function of the term Φ, (the Drude-Nernst relation, $\Phi = (1/\varepsilon)(\delta \ln \varepsilon/\delta P)_T$, where ε is the solvent dielectric constant). $\Delta V_{diff}^{\neq}$ (for diffusion) ranged from +7 to +17 $cm^3\,mol^{-1}$ for the $DmFc^{+/0}$ system and generally increased with increasing solvent viscosity. The values of $\Delta V_{el}^{\neq}$ for the electrode reaction of both $DmFc^{+/0}$ and $Fc^{+/0}$ were all positive and within experimental uncertainty correspond to the relevant values of $\Delta V_{diff}^{\neq}$; this is in distinct contrast to the negative $\Delta V_{ex}^{\neq}$ values for the electron self-exchange reactions in homogeneous solution. It was concluded that the electrode reactions but not the electron self-exchange reaction were subject to solvent dynamical control. This report included an extensive discussion of the unsatisfactory relationship of the radii of the electro-active species according to the Drude-Nernst and Stokes-Einstein equations relative to those related to crystallographic radii, a topic to be consulted in the original literature. [140(b)]

An examination of the effect of cations on electron transfer between two anionic species has been stimulated by finding that catalysis by alkali metal ions occurs in some anion-anion redox reactions. [142] Such catalysis has been observed in both self-exchange and in non-symmetrical reactions. A model was presented to account for the catalysis; this was derived from a reaction with a large positive volume of activation. [143] The latter was accounted for by desolvation from the cation prior to it catalysing electron transfer between two anions. Other authors have proceeded by calculating an ion-triplet (anion-cation-anion) association constant and suggested that minimisation of repulsion could be implicated, but pointed out that often the cation is present in much higher concentration than the reacting partners and hence more than one cation may participate. [144] Thus far, apart from the study cited above, there have been few investigations aimed at acquiring and interpreting the volume of activation for the catalysed and uncatalysed pathways. Sometimes efforts have been thwarted by the fact that the extent of catalysis and further reaction acceleration upon pressure application has put measurement inconveniently outside of the time range of the hpsf instrument. [142(f)] However, an elegant recent study of the reaction between ferricyanide, $[Fe(CN)_6]^{3-}$, and ferrocyanide, $[Fe(CN)_6]^{4-}$, ions using principally ^{13}C nmr spectroscopy has made a valuable contribution to our understanding. [145] The counter ion was K^+, an ion associated with significant catalysis. The uncatalysed pathway was studied by sequestering the K^+ ions by two macrocycles, the cryptand, crypt-2.2.2 and the crown ether, 18-crown-6; the K^+-independent self-exchange kinetic parameters were obtained and hence the volume of activation was determined. The value of -11.3 cm^3 mol^{-1} was independent of the sequestering agent. The magnitudes of both the exchange rate constant and $\Delta V^{\neq}$ were accounted for satisfactorily by Marcus-Hush type theory implying that the outer-sphere electron transfer is adiabatic. The catalytic effect of K^+ or other catalytically active alkali metal ion was greater than can be accounted for solely by reduction of the Coulombic work terms by ion pairing or by the electrostatic effect of a bridging K^+ ion on the solvent reorganisation component of the free energy. A volume of activation value close to zero for the catalysed pathway, i.e. 9 to 10 cm^3 mol^{-1} more positive than for the uncatalysed pathway, appears to reflect a partial de-aquation of the K^+ ion in reaching the transition state. Apparent requirement of partial or complete de-aquation of an aqua catalytic ion explains why the strongly complexed K^+ by cryptand or crown ether did not exhibit catalytic activity. It has become clear that the catalysis of anion-anion electron transfer will be most effective if the electron transfer path provided by a bridging cation is as short as possible. This postulate could be made following an examination of collated results from catalysis by alkali metal ions and by R_4N^+ ions, where R can be changed to vary the radius, and by consideration of hydration enthalpies of the cationic species. The reaction was not uncomplicated by ion-pairing and this was treated at length in this report. It was intimated that the cation catalysed pathway has some of the attributes of an inner-sphere mechanism, and because the cation provides a positive potential between the anions it is a more effective catalyst in electron transfer than is an anion in cation-cation reactions.

The importance of reactions of cobalt(I) complexes with proton donors in the context of understanding the photo- and electrocatalytic reduction of water to H_2 and of carbon dioxide to HCO_2^- has been highlighted. [146] In a series of reports the thermodynamics and kinetics of proton binding to the cobalt(I) complex of the macrocycle, 5,7,7,12,14,14,-hexamethyl-1,4,8,11-tetraazacyclotetradeca-4,11-diene (= L) were established. [147] Pulse–radiolysis techniques were used to generate the $Co^I(L)^+$ complex from the corresponding Co(II) complex whereupon the reaction with proton donors (HA) to form the hydride complex in aqueous medium could be monitored.

$$Co^I(L)^+ + HA + H_2O \rightarrow \textit{trans}\text{-}Co^{III}(L)(H)(H_2O)^{2+} + A^-$$

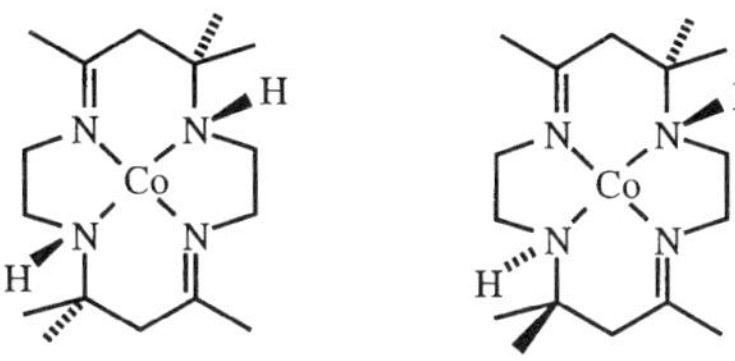

rac **L** *meso* **L**

The rate constants for these reactions decreased with increasing pK_a of the donor acid consistent with a reaction occurring by proton transfer. However, the mechanism required elucidation and a systematic study of the reaction kinetics over a pressure range has recently been undertaken. [148] Increasing pressure accelerated the reactions, with more pronounced acceleration occurring for acids of lower pK_a (for example phenol, $pK_a = 10$) resulting in negative $\Delta V^{\neq}$ values progressively higher in magnitude for the weaker acids. In the latter case the observed $\Delta V^{\neq}$ correlated with the reaction volumes for the ionisation of the weak acids with a slope of 0.44 (see Figure 10). This indicated that in volume terms the bond dissociation of the weak acid molecule bound to the metal centre proceeds approximately halfway at the transition state along the reaction coordinate. This partial cleavage of the H-A bond in the transition state $[Co^{I}---H^{\delta+}---A^{\delta-}]^{\neq}$ is accompanied by an increase in electrostriction, whereas the stronger acids do not form comparable species and therefore have significantly less negative volumes of activation. The reaction involved charge transfer to oxidise formally Co(I) to Co(III) and reduce H^+ to H^-. Since earlier studies had shown that the bond distances to the ligand from the cobalt centre were virtually independent of cobalt oxidation state, [149] it was safe to rule out any significant contribution to volume changes from this potential source.

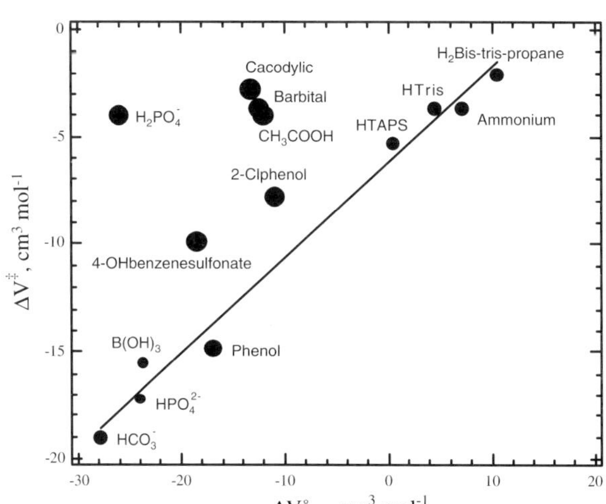

Figure 10. Plot of $\Delta V^{\neq}$ for the reaction of CoL^+ with HA versus ΔV°_{ion} of HA. The size of each marker is correlated to the magnitude of the protonation rate constants. The slope of the plotted line is 0.44.

Copper(I) complexes catalyse a variety of organic reactions, many of which are of synthetic or commercial importance. [150] Halide abstraction from an aryl or alkyl halide by the Cu^I catalyst is thought to be the rate-determining step. Complexing Cu^I with various ligands has been proposed as a method by which the Cu^I species has reasonable solubility, compared to Cu^I halides, in solution and also limits disproportionation if Cu^+(aq) were used; that is, the Cu^I species would promote more efficient catalysis. [151] Other considerations regarding ligand suitability were discussed and based on a kinetics study of Cu^I with Cl_3COO^- (TCA) in the presence of fumarate ($HOOCCH=HCCOO^-$) a multi-step general mechanism was proposed. [152] A further investigation applying two specially designed ligands that stabilise Cu^I and also using CH_3CN and NH_3 in reactions with TCA was initiated with a view to improve understanding of the catalytic mechanism. [153] In some cases Cu^{II} species were starting materials and Cu^I species were generated by pulse-radiolysis. A combination of stopped-flow spectrophotometry and pulse-radiolysis methods was used to obtain kinetics parameters, and these results were interpreted in an assessment of ligand properties favourable for catalysis. The pressure dependence of the rate of reaction of Cu^+(aq) with TCA in the presence of

$$Cu^+(aq) + CCl_3COO^- \rightarrow CuCl^+ + {}^{\cdot}CCl_2COO^-$$

fumarate was examined and yielded a volume of activation of -9.9 ± 1.0 cm^3 mol^{-1}. Since this value contains a contribution of -4.5 ± 1.0 cm^3 mol^{-1} for the reaction volume, and thus for the reaction in question, the volume of activation for the rate-determining redox reaction step is -14.4 ± 2.0 cm^3 mol^{-1} and thus demonstrated that the mechanism is associative.

It has been appreciated for some time that while guidelines can be established for the extent to which mechanistic features contribute to intrinsic volume changes, less is known definitively about magnitudes of electrostrictive contributions to volume changes. Indeed it is not clear over what distance from a charge change site do electrostrictive effects operate. Some progress has been made in that a relationship has been developed between the reaction volume in a redox reaction and the difference of the square of the charges on the oxidised and reduced forms of complexes. [70(a)] This correlation could be made because the study involved systems in which there were effectively no intrinsic changes involved. It was suggested that electrostrictive effects, though they may be the major component in the overall reaction volume in some reactions, their influence may be quite local. [154] This proposition has been explored in a recent investigation of the reduction, electrochemically in aqueous media of mononuclear and bridged ruthenium complexes in which differently charged complexes were introduced as bridging partners. [155] The $Ru^{III}(edta)(H_2O)]^-$ complex was subject to high pressure electrochemistry at different pH values to test the effects of ionisation in the region adjacent to the redox centre. Bridged dinuclear complexes such as $[(edta)Ru^{III}(\mu\text{-}CN)Fe^{II}(CN)_5]^{5-}$ and $[(NH_3)_5Ru^{III}(\mu\text{-}CN)Fe^{II}(CN)_5]^-$ were employed; the two redox centres in close proximity can separately undergo reversible electrochemical change. Whether there is an effect of electronic communication via the bridging ligand on the effective charge of the redox active metal centre was probed. Cyclic voltammograms obtained at pressures up to 150 MPa were treated by fitting procedures to obtain the relevant reduction potentials at each pressure, and thence the applicable reaction volumes. The results could be analysed to conclude that electrostriction effects exhibit a highly localised character. The model employing the difference in the square of the charges on the oxidised and reduced forms enabled the intrinsic and electrostrictive volume effects to be separated. As discussed above for cyanometallate couples, [145] the anion-anion reaction was dominated by catalysis of the counter ion. For pathways catalysed by alkali metal ions, the volumes of activation, $\Delta V_{ex}^{\neq}$, in several cases examined were positive, and the catalytic power was found to increase in the order for the positive ions of Li < Na < K < Rb < Cs. This was a trend that could be related to the order of increasing heats of hydration of these ions.

In order to determine whether these and consistent findings for the corresponding electrode reactions related to the specific properties of cyanometallates, cation and pressure effects upon the electrode kinetics of the polyoxometallates, 12-tungstocobaltate ($CoW_{12}O_{40}^{n-}$, n = 4 - 8; CoW_{12}), and 12-tungstophosphate ($PW_{12}O_{40}^{n-}$, n = 3 - 5; PW_{12}) were investigated. [156] In addition to the stated purpose of this investigation these polyoxometallate species are of interest since they are important as catalysts and multivalent redox reagents. Specifically the effects of supporting electrolytes and pressure on the electrode reactions of the aqueous $CoW_{12}O_{40}^{5-/6-}$ couple were studied and the results reported. More limited results were obtained for a parallel investigation of the $PW_{12}O_{40}^{3-/4-}$ and $PW_{12}O_{40}^{4-/5-}$ couples. The volumes of activation for diffusion, $\Delta V_{diff}^{\neq}$(average value -0.9 ± 1.1 $cm^3\ mol^{-1}$) and the electrochemical cell reaction volumes for the cobaltate couple did not exhibit any significant dependence on the electrolyte identity or concentration. The two phosphate couples yielded electrochemical cell reaction volumes of -14 and -26 $cm^3\ mol^{-1}$ for the 3-/4- and 4-/5- couples respectively, results which suggested a dependence on (z^2) (z = ionic charge number), a dependence that could be predicted from the Born-Drude-Nernst theory of solvent electrostriction. However, such dependence was not followed by the cobaltate system or by other anion-anion couples. For aqueous electrode reactions of CoW_{12} the rate constants k_{el} showed specific alkali metal cation catalysis, ($Na^+ < K^+ < Rb^+ < Cs^+$) and $\Delta V_{el}^{\neq}$ is usually positive in the presence of supporting electrolytes. In fact for the higher atomic number cations the values of $\Delta V_{el}^{\neq}$ were in the range +10 to +15 $cm^3\ mol^{-1}$, values that were compatible with their partial dehydration in facilitating electron transfer catalysis and indicative of the absence of solvent dynamical control.

3.4. Reactions of nitric oxide

It is convenient for readers with a special interest in nitric oxide (NO) chemistry that the reactions of NO are grouped together, although sections on ligand substitution, or redox reactions or bioinorganic reaction could accommodate them, as appropriate. In the last decade or so the chemistry of NO has undergone a renaissance partly owing to its direct or implied involvement in many important biochemical processes. [157] The focus here is on reactions at metal centres, notably transition metal centres. Readers are urged to consult the authoritative review of nitric oxide mechanisms, [158] and more succinct accounts. [159] In this section we will present recent literature where progress has been made from interpretation of high pressure kinetics measurements.

A detailed study of the kinetics of the reaction between nitric oxide and aquapentacyanoferrate(III) resulting in the generation of nitroprusside, $[Fe(CN)_5NO]^{2-}$ (NP) was undertaken to elucidate the reaction mechanism. [160] It prevailed that the mechanism was different from that for typical ligand substitution reactions involving $[Fe^{III}(CN)_5H_2O]^{2-}$ for which the mechanism had been proposed earlier to have a dissociative character. [161] Formation of NP was characterised by $\Delta V^{\neq} = -13.9 \pm 0.5$ cm^3 mol^{-1} and $\Delta S^{\neq} = -82 \pm 4$ J mol^{-1} K^{-1} which certainly at face value do not comport with a dissociative mechanism. By considering all the available evidence it was proposed that the iron(III) reactant complex undergoes a rate-determining outer-sphere electron transfer producing the corresponding iron(II) complex, whereupon the coordinated water molecule is rapidly replaced by HONO or NO_2^- depending on pH, but with the final product being $[Fe(CN)_5NO]^{2-}$. It would therefore follow that the negative activation parameters arose as a consequence of significant electrostriction following reduction. The proposition that the aquapentacyanoferrate(II) ion is an intermediate was probed by employing pyrazine or the thiocyanate ion as scavengers. Analysis of the ambient pressure kinetics results and changes in the UV/Visible spectrum confirmed this proposal. Further support for the presence of an iron(II) intermediate was provided by kinetics results and changes in spectra for reactions between NO and other $Fe^{III}(CN)_5L]^n$ species (L = pyridine, NCS^-, NO_2^-, CN^-).

The nitroprusside ion, $[Fe(CN)_5NO]^{2-}$ has been widely used pharmacologically for the provision of NO. The related iron(II) complex ion is also of considerable interest but kinetic studies involving formation or dissociation of the latter can be fraught with the experimental uncertainty of knowing whether NO solutions are free from (oxidising) impurities. It has been shown recently from ambient and high pressure kinetic studies that the formation of $[Fe(CN)_5NO]^{3-}$ from the pentacyanoaqua complex ion and NO is characterised by a markedly positive $\Delta V^{\neq}$ value (+17.4 ± 0.3 cm^3 mol^{-1}) and a positive $\Delta S^{\neq}$ value (+34 ± 4 J mol^{-1} K^{-1}). [162] Clearly this indicates the reaction proceeds by a dissociative mechanism, and indeed the volume change was larger than predicted (+13 cm^3 mol^{-1}) for dissociation from an octahedral hexaaqua complex, [110], [163] and thought to be a consequence of the highly solvated nature of the complex ion. There was no evidence of volume reduction within the iron complex ion as the five-coordinate intermediate is formed. The reaction product $[Fe(CN)_5NO]^{3-}$ is not particularly stable but the kinetics of the reverse reaction could be studied by confining the system to only the desired reaction by the presence of sodium cyanide. The release of NO was also accompanied by a positive $\Delta V^{\neq}$, a value of +7.1 ± 0.2 cm^3 mol^{-1}, also leading to the conclusion of a dissociative mechanism, and thereby allowing the calculation of the reaction volume of +10.3 ± 0.5 cm^{3}mol^{-1} (see volume profile in Figure 11). This latter value is more positive than the +4.8 ± 0.6 cm^3 mol^{-1} reaction volume value for reaction of NO with hexaaquairon(II) ion, but in the latter reaction there is a formal change in oxidation state as $[Fe(H_2O)_6]^{2+}$ yields, with NO, $[Fe^{III}(H_2O)_5NO^-]^{2+}$, and this has been advanced in part as the reason for the difference (see volume profile in Figure 12).

A mechanistic study of the reaction of ruthenium ammine complexes with NO was reported almost four decades earlier. Although $[Ru(NH_3)_6]^{3+}$ is very inert, an acidic solution of this complex ion was observed to undergo a rapid reaction in the presence of NO to form $[Ru(NH_3)_5(NO)]^{3+}$. [164] An electrophilic substitution process was invoked, with NO considered to function as an electrophile and the product proposed to be Ru^{IV}-NO^-. The conclusion from a subsequent study was that a bond making mechanism operates for this reaction in acidic solution. [165] Curiosity regarding the rapidity of the reactions of these complexes with NO and the widespread interest in NO reactions prompted a renewed investigation in which the reaction of $[Ru(NH_3)_5X]^{(3-n)+}$ with NO (X = Cl^-, NH_3, H_2O) in acidic aqueous solution was monitored as function of NO concentration, temperature and pressure. [166]

$$[Ru^{III}(NH_3)_5X]^{(3-n)+} + NO \rightarrow [Ru^{III}(NH_3)_5(NO)]^{3+} + X^{n-}$$

The reaction was first-order in the concentration of each reactant and there was no evidence for a reverse reaction step. Entropies of activation for the three reactions were in the range of -110 to -130 J mol^{-1} K^{-1}. The volumes of activation were -13.6 ± 0.3 and -18.0 ± 0.5 cm^3 mol^{-1} for the substitution of the NH_3 and Cl^- groups, respectively (determination of this parameter for the substitution of H_2O was not possible). The authors presented several possible mechanisms for consideration to explain the rapid reactions and the magnitudes of the activation parameters. It was eventually concluded that the most compatible mechanism consistent with the results and product species characterisation was a unique combination of associative ligand binding and concerted electron transfer to yield the stable ruthenium(II) nitrosyl complex.

$$[Ru^{III}(NH_3)_5X]^{(3-n)+} + NO \rightarrow [Ru^{II}(NH_3)_5(NO^+)]^{3+} + X^{n-}$$

The faster reaction of the aqua complex could be accounted for partially, it was argued, because ligand displacement by an entering nucleophile could be expected to depend to some extent on the lability of the leaving group even when the reaction follows an associative mechanism.

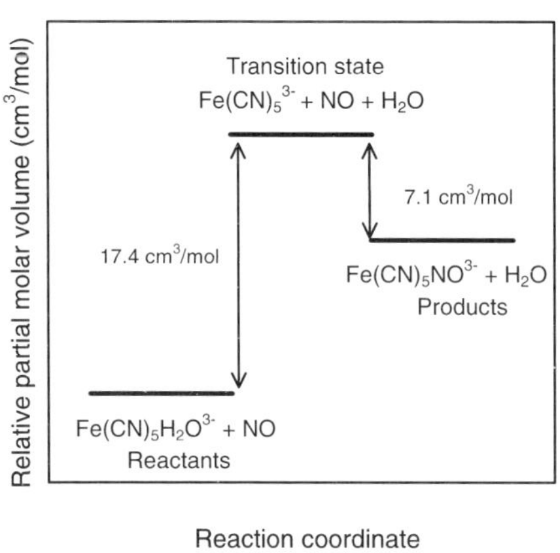

Figure 11. Volume profile for the overall reaction: $[Fe^{II}(CN)_5H_2O]^{3-} + NO \equiv [Fe^{II}(CN)_5NO]^{3-} + H_2O$.

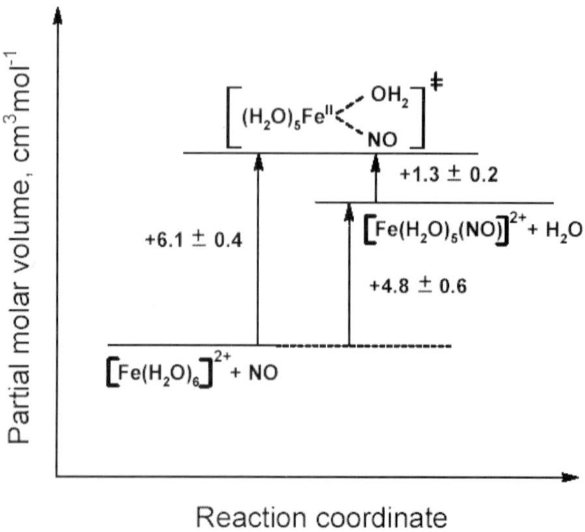

Figure 12. Volume profile for the reaction: $[Fe^{II}(H_2O)_6]^{2+} + NO \leftrightarrow [Fe^{III}(H_2O)_5(NO^-)]^{2+} + H_2O$.

4. BIOINORGANIC REACTIONS

One initial objective in studying electron transfer reactions between a metal complex and a biologically active molecule is to determine the reaction mechanism, i. e. is it inner-sphere or outer-sphere? Another aspect is to design the system to be studied in a way that assists in making this distinction. If oppositely charged reactants are employed then, in principle, it may be possible to form an encounter complex with sufficient strength of ion-pairing that saturation kinetics would be observed, which then allows the rate constant of the subsequent electron transfer to be obtained. When saturation kinetics are not observed then only a composite term of kinetic or equilibrium parameters is obtained. With these considerations in mind, the kinetics of reduction of cytochrome c by $[Fe(edta)(H_2O)]^{2-}$ were studied comprehensively since various properties of the reaction participants combined to offer delineation of the kinetic steps and to distinguish possibly between inner- and outer-sphere mechanisms. [167] In an earlier investigation of the reaction

$$[Fe^{II}(edta)(H_2O)]^{2-} + [\text{cytochrome } c^{III}]^{7+} \rightarrow [Fe^{III}(edta)(H_2O)]^{-} + [\text{cytochrome } c^{II}]^{6+},$$

the kinetic data were analysed in terms of an outer-sphere mechanism. [168] In the interim, many other bioinorganic electron transfer systems have been investigated. Given that it has been shown that the hepta-coordinate reactant iron complex has a labile water molecule [169] it could be speculated that the latter could participate in effective formation of an inner-sphere complex with a suitable potential binding site on the protein partner. In addition, availability of appropriate high pressure instrumentation (hpsf) provided an additional tool in the repertoire for reaction mechanistic insight. It prevailed that saturation kinetics were not observed and plots of the forward pseudo first-order rate constants versus excess iron(II)(edta) complex concentration were linear with a small but non-negligible intercept. The reverse reaction was confirmed directly to be a very slow process. Thus the second-order rate constant, $k = Kk_{ET}$, where K is the precursor complex-formation constant, and k_{ET} is the rate constant for the mutual reduction/oxidation of the iron(II) complex and cytochrome c reactant respectively, remains as a composite value and the two constants cannot be determined separately from the available kinetic data. On the basis of calculations from Marcus-Hush theory and other arguments it could be shown that the reactant iron(II) complex ion does not experience the overall charge on the protein surface, rather only a small fraction of it, signifying that localised interactions are responsible for the electron transfer process. The small enthalpy of activation values for the forward reaction, 31 ± 1 and 26 ± 1 kJ mol^{-1} at two different pH values were consistent with other values for outer sphere electron transfer, and the large negative $\Delta S^{\neq}$ values (-107 ± 4 and -128 ± 4 J mol^{-1} K^{-1}) suggested a highly structured transition state. However, it was suggested that the magnitudes of these parameters were not inconsistent with an inner-sphere mechanism. The volume of activation was determined to be -8 ± 1 cm^3 mol^{-1}, and the reaction volume could be obtained from cyclic voltammograms generated as a function of pressure, following data treatment, and was found to be +1.7 ± 1.0 cm^3 mol^{-1}. The latter results from a combination of a volume increase due to oxidation of the iron(II)edta complex ion and a volume decrease due to the reduction of cyt c^{III}. The compact nature of the transition state could have arisen from an effective formation of an inner-sphere precursor species. The volume changes are illustrated in the volume profile given in Figure 13.

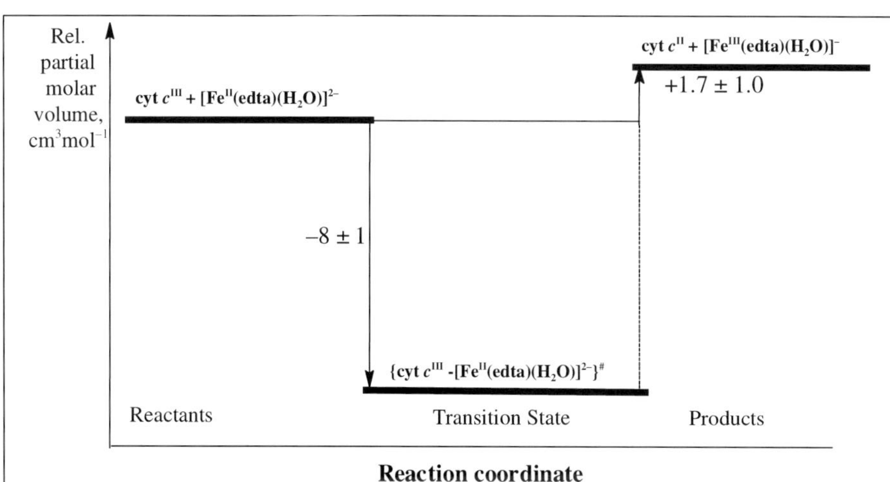

Figure 13. Volume profile for the reduction of cyt $c^{III}]^{7+}$ by $[Fe^{II}(edta)H_2O]^{2-}$.

In contrast to other systems in which the mechanism (inner-sphere or outer-sphere electron transfer) may be a matter of debate, the reaction between the protein horse heart cytochrome c with anionic $Cu^{II/I}$ complexes was adjudged to proceed by an outer-sphere mechanism. [170] The copper(II) complex bis(5,6-bis(4-suphonatophenyl)-3-(2-pyridyl)-1,2,4-triazine)Cu(II), (the ligand is commonly known as ferrozine) possesses square pyramidal geometry with the two bidentate ligands in the equatorial plane, and the fifth axial position is occupied by a water molecule, in aqueous solution.

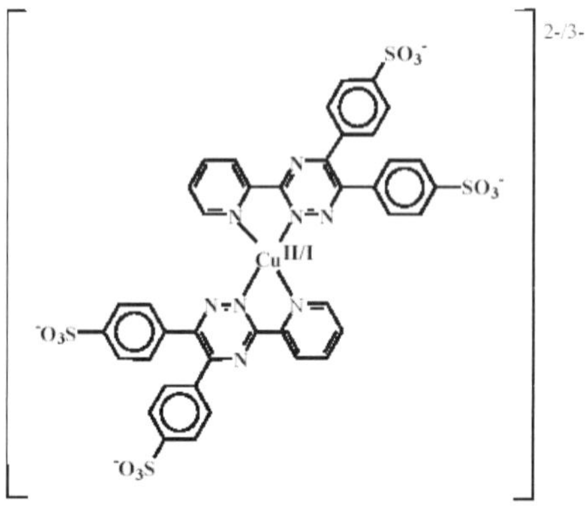

The Cu(I) complex with two ferrozine ligands has a tetrahedral structure, based on UV/Visible spectra and electrochemical properties. In the reaction of cyt c^{II} with the copper(II) complex, in the neutral pH region, a plot of the observed rate constant versus excess copper(II) complex concentration showed saturation kinetics. This made possible extraction of the precursor equilibrium constant, $K_{OS} = 7.7 \times 10^3$ M^{-1} and the electron transfer rate constant, $k_{ET} = 6.2$ s^{-1} at 288 K from the plot at pH 7.4 and an ionic strength of 0.2 mol dm^{-3} ($LiNO_3$). From an Eyring plot the values of ΔH_{OS} and ΔS_{OS} for the precursor complex formation, respectively -4 ± 8 kJ mol^{-1} and +91 ± 28 J mol^{-1} K^{-1}, were obtained. These were interpreted as a small non-specific interaction but an entropy driven outer-sphere complex formation, whereas the values for $\Delta H^{\neq}_{ET}$ of 85 ± 4 kJ mol^{-1} and $\Delta S^{\neq}_{ET} = -61 \pm 13$ J mol^{-1} K^{-1} were reported to be indicative of a significant enthalpic requirement and of a structurally constrained process (see free energy profile in Figure 14). A study of the pressure dependence of the reaction kinetics permitted separation: $\Delta V_{OS} = +0.8 \pm 1.3$ and $\Delta V^{\neq}_{ET} = +8.0 \pm 0.7$ cm^3 mol^{-1}. A series of estimates of various contributions to the overall reaction volume led to a value of +30 cm^3 mol^{-1}, implying that the transition state is "early". A significant fraction of the reaction volume arises from the fact that the Cu(II) complex upon being reduced converts its coordination number from five to four with loss of the coordinated water molecule; the usually accepted estimate for this process as discussed earlier, is 13 cm^3 mol^{-1} (see volume profile in Figure 15). Therefore, it was concluded that water loss occurred following transition state formation. It was also argued that the transition state for the actual redox processes in both directions was about half way between reactants and products on a volume basis. This is a proposition that has been more clearly demonstrated for reactions of cytochrome c with a series of pentaammine ruthenium complexes where no change in coordination number occurs. [171] The complete data set in terms of free energy and volume changes associated with the electron transfer process enabled for the first time the construction of a combined three dimensional reaction profile as illustrated in Figure 16. In this profile free energy changes can be directly correlated with volume changes along the reaction coordinate.

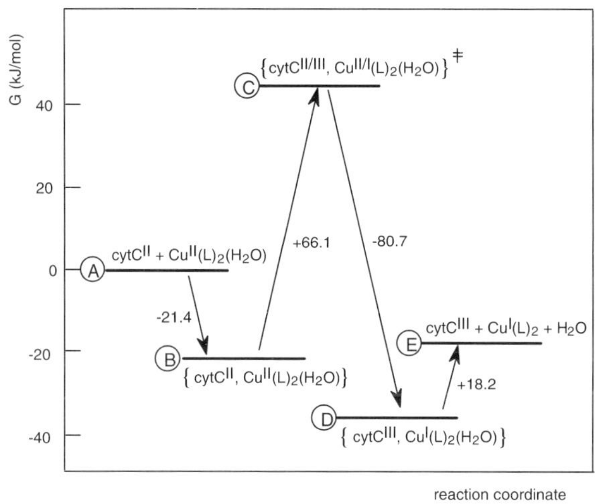

Figure 14. Free energy profile for the oxidation of cytochrome *c* by the Cu(II) complex measured with respect to the free energy of the reactants.

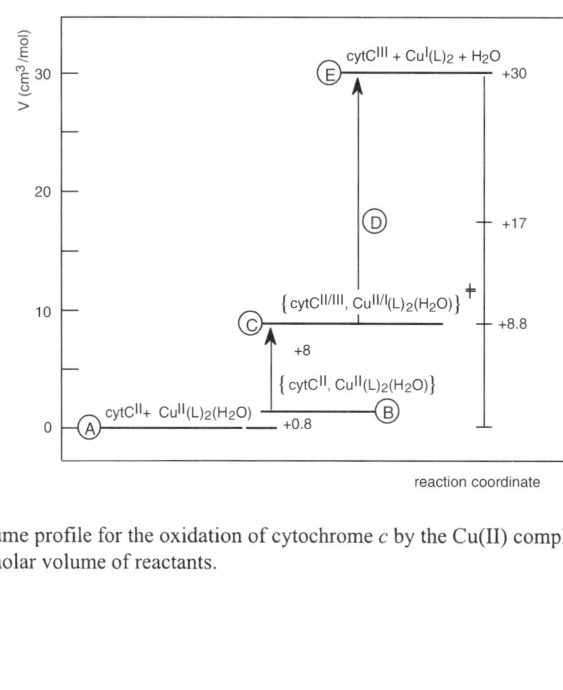

Figure 15. Volume profile for the oxidation of cytochrome *c* by the Cu(II) complex, measured with respect to the molar volume of reactants.

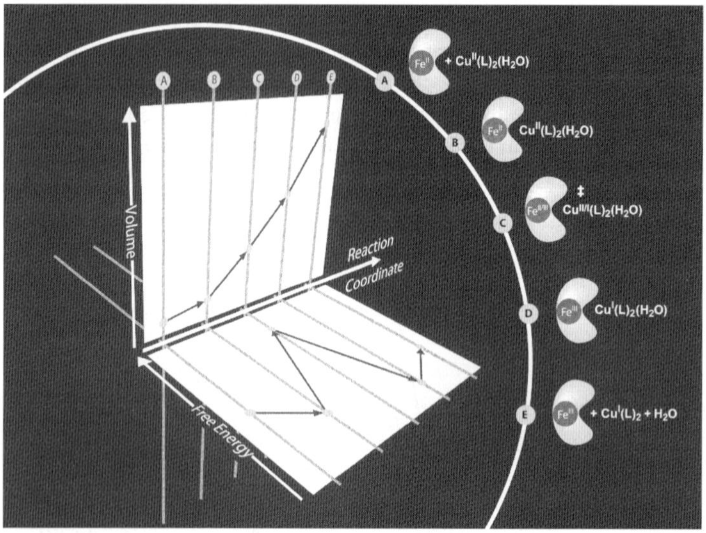

Figure 16. Combined energy-volume profile for the oxidation of cytochrome *c* by the Cu(II) complex.

The electron exchange kinetics within cytochrome c itself have been determined by employing fast scanning cyclic voltammetry, using specially modified gold electrodes. [172] Further, by determining kinetic parameters as a function of pressure the activation volume was obtained. The value, +6.1 ± 0.5 $cm^3 mol^{-1}$, from the pressure dependence of the heterogeneous rate constant is similar to that for the homogeneous cyt c self-exchange process predicted from the Marcus-Stranks cross reaction treatment. [173] It was acknowledged that the agreement could be coincidental, but nevertheless this novel experimental method confirmed reasonably well the hypothesis of the approach. Pressure increases the viscosity of the protein globule but not the bulk solvent and the results supported the adiabatic "protein friction" mechanism [174] rather than the extended non-adiabatic charge transfer model. [175] It was pointed out that there was very satisfactory agreement with earlier results in which the viscosity had been varied directly, adding credence to the arguments presented on the protein friction mechanism.

Cytochrome P450 enzymes are a widely distributed group of hemoproteins involved in a broad range of vital physiological processes. A variety of hydrophobic compounds undergoes catalysed oxidation by these enzymes; in general an incipient active species is formed following heterolytic O-O cleavage of a heme bound O_2. Reports of research investigations into several aspects of cytochrome P450's activities and structure are abundant, particularly in the past three decades. Of critical interest in the context of this article has been the ligand arrangement around the heme iron and whether bound water at this site is displaced by a substrate such as camphor accompanied by the low- to high-spin transition of the iron(II) centre. [176] Efforts to understand the bioinorganic chemistry of this system have been pursued by studying the kinetics of binding of small molecules such as CO and NO at the heme centre. In a study of the dynamics of bound water in the heme domain of a cytochrome

P450, application of pressure caused a high- to low-spin shift and subsequently a P450 to P420 transition, in the presence of palmitic acid. Reaction volumes for the low-to high-spin transition of substrate-free cytochromes (+20 to +23 cm^3 mol^{-1}) were considered to be consistent with the displacement of one water molecule. Volume changes for the spin transitions of substrate bound P450s showed a linear relationship with the ΔG^o values of the spin transition suggesting a common mechanism. [177] The kinetics of reaction of ferrous cytochrome P450 with carbon monoxide have been studied as a function of temperature and pressure. [178] The rate constant and the volume of activation for the "on" reaction varied with the nature of the bound substrate. Arguments to explain these differences have been presented and the topic of the high pressure kinetics of CO binding to iron-proteins has been debated for some time. [54, 178 (d)]

Unlike CO, nitric oxide can bind to both ferrous and ferric forms of cytochrome P450. Since it has been suggested that P450 enzymes might be a target for NO *in vivo* it seemed incumbent upon investigators to examine the kinetics and deduce the mechanism for interaction of NO with cytochrome P450. A recent report of the reaction of ferric cytochrome P450 with NO, in the presence and absence of a substrate, in this case camphor, employed stopped-flow and flash photolysis methods at different temperatures and pressures. [179] An analysis of the measurements indicated that the mechanisms for the forward and reverse reactions are very different for the substrate-free and substrate-bound reactions. The following kinetics results and derived parameters permitted the drawing of volume profiles for the reactions of NO with a cytochrome P450 in both the presence and absence of the substrate (see volume profiles given in Figures 17 and 18). Reaction of NO with substrate-free cytochrome $P450_{cam}$ was characterised kinetically by two steps, readily separable in time, with the enthalpy barrier for both steps being about 90 kJ mol^{-1}. Large positive $\Delta S^{\neq}$ values (+169 and +128 J mol^{-1} K^{-1}) for the fast and slow steps, respectively, were derived and these values were accompanied by large positive values of $\Delta V^{\neq}$ (+28 (fast) and +30 (slow) cm^3 mol^{-1}). Dissociation of NO from $P450_{cam}$(NO) occurs in a single step also characterised by large positive $\Delta S^{\neq}$ (+155 $Jmol^{-1}K^{-1}$) and $\Delta V^{\neq}$ (+31 cm^3 mol^{-1}) values. The released NO is rapidly scavenged by an excess of the $[Ru^{III}(edta)H_2O]^-$ complex. The two phase kinetics for the forward reaction were interpreted as arising from two conformational substates in cytochrome $P450_{cam}$; the existence of multiple substates had been established and characterised by ftir spectroscopy. [178(i)], [180] The substates are thought to possess different hydrogen bonding networks from differently packed water molecules in the heme pocket. This has consequences for the binding of NO to the heme iron, whereupon rearrangement of other water molecules in the heme pocket accompanied dissociation of the coordinated water molecule, but at different rates. It is not clear whether the NO bound product is a mixture of the two substates. A mechanistic scheme was advanced and this proposed that the dissociation of a water molecule gave rise to a five-coordinate high-spin intermediate before NO binds. This low-spin to high-spin process contributes a further +12 to +15 cm^3 mol^{-1} (six-coordinate, low-spin aqua complex to five-coordinate high-spin intermediate) to that of about +13 cm^3 mol^{-1} accepted (*vide supra*) for dissociation of a coordinated water molecule. This is a compatible total volume change with the observed value of $\Delta V^{\neq} = +28 \pm 2$ cm^3 mol^{-1} (faster step). The bond formation step gives rise to a linearly bound diamagnetic complex, formally Fe^{II}-NO^+. Upon reversing the reaction, breakage of the iron-nitrosyl bond is accompanied by formal oxidation back to Fe(III) and solvent reorganisation around the Fe^{II}-NO^+ species and reformation of the five-coordinate state is associated with a change from low-spin to high-spin. Together these mechanistic features militate in favour of large positive values of $\Delta S^{\neq}$ and $\Delta V^{\neq}$ for the reverse reaction, as observed. The volume profile (see Figure 15), using the $\Delta V^{\neq}_{on}$ from the faster step therefore exhibited a close to zero reaction volume; a similar finding was noted for reaction between NO and the ferric hemoprotein, metmyoglobin. [181]

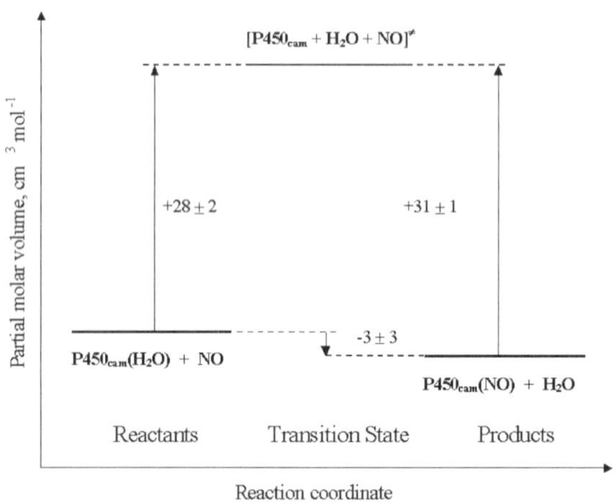

Figure 17. Volume profile for the reversible binding of NO to substrate-free cytochrome $P450_{cam}$.

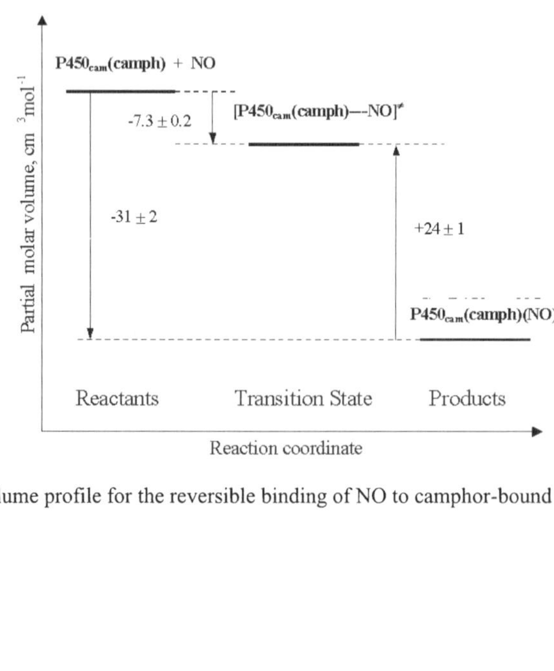

Figure 18. Volume profile for the reversible binding of NO to camphor-bound cytochrome $P450_{cam}$.

An investigation of the kinetics and mechanism of the binding of NO to substrate bound (camphor) cytochrome $P450_{cam}$ revealed significant differences in both rates and coordination states. [179] However, explanations were readily forthcoming. First, binding of the substrate 1 R-camphor to the enzyme that possesses a low-spin six coordinate iron(III) centre caused a change to virtually 100 % of the high-spin form, with coordinated water being expelled resulting in a five-coordinate heme-iron(III) centre. Therefore the rate of ligation by NO is not limited by the rate of water molecule displacement, and the Fe-NO bond formation is an order of magnitude faster. This is in distinct contrast to CO binding to the iron(II) enzyme where CO binds two orders of magnitude slower to the substrate-bound form than to the substrate-free form; this was explained by invoking an argument based on solvent compressibility of solvent molecules in the heme pocket of cytochrome $P450_{cam}$. [178(d)] A subtle interplay between the availability of a coordination site, rigidity of the active site, and differences in the reorganisation of spin multiplicity during reaction with NO can explain the differences in kinetic parameters *vis a vis*, ferric-, ferrous-enzymes, substrate-free, substrate-bound variations, and with model porphyrin complexes. Stopped-flow spectrophotometry and laser flash photolysis techniques provided reassuring agreement for $\Delta S^{\neq}$ and $\Delta V^{\neq}$ for the substrate-bound case (-74 J mol^{-1} K^{-1} and -6.9 cm^3 mol^{-1}, average values of these two parameters, respectively). Although the reaction of NO is very rapid, an analysis showed that a diffusion controlled reaction could be ruled out, and the cyt $P450_{cam}$ (NO) species was formed in a rate-determining step following encounter complex formation. The volume of activation for the reverse reaction was +24 cm^3 mol^{-1}, a value consistent with a mechanism in which the iron-nitrosyl bond is broken. The Fe-NO cleavage is accompanied by charge transfer from the metal to the nitrosyl ligand, since the initial complex containing Fe^{II}-NO^+ character was transformed to Fe^{III} as NO was released, and the low-spin state of iron returned to a high-spin state. Solvational changes also occurred, and all these factors contributed to the large positive $\Delta V^{\neq}$. Thus the volume profile is radically different for the reaction of NO with camphor-bound cytochrome $P450_{cam}$ and the reaction volume is -31 cm^3 mol^{-1}, and for the on (forward) reaction the transition state was considered to be "early", i.e. mainly involving bond formation.

5. OTHER RELEVANT REACTIONS AND EFFECTS OF PRESSURE

Phosphine functionalised ferrocenes are important in homogeneous catalysis, and chiral derivatives are of particular interest for asymmetric catalysis. Such compounds containing two planar chiral units may exhibit *rac* and *meso* isomers. Conversion of one isomer to the other would require one of the ferrocenyl rings to flip over and coordinate to the iron atom by the other face. See the Scheme given below.

Until very recently this type of conversion had not been observed for phosphine functionalised ferrocenes, although under certain conditions (solvent, acid, photochemical intiation, for example) similar conversions have been reported for related systems. [182] The bis-planar chiral ferrocenyldiphosphine bis(1-(diphenylphosphino)-η-indenyl)iron(II) was found to undergo an isomerisation from the *meso* isomer to the *rac* isomer in tetrahydrofuran solvent at room temperature. [183] Among attempts to understand the mechanism of this isomerisation the reaction was studied by adding other solvents, adding salts, varying the temperature and pressure of the reaction, and by judicious isotopic substitution within the reactant species. The rate of isomerisation was retarded by the presence of non-coordinating solvents, accelerated by the presence of salts such as LiCl and $LiClO_4$, and was accelerated by temperature and pressure increases. The retardation of the reaction as increasing quantities of chloroform were added indicated that a coordinating solvent (THF) facilitates the reaction and is, in fact vital as the isomerisation does not proceed in 100 % chloroform. Reaction acceleration by the presence of both LiCl and $LiClO_4$ enabled participation by Cl^- as a nucleophile to be ruled out, but implied that a polar or salt-like intermediate occurred. Both at ambient and elevated temperatures and pressures the kinetics of the isomerisation could be followed by ^{31}P nmr spectroscopy, leading to the activation parameters $\Delta H^{\neq} = 57 \pm 4$ kJ mol^{-1}, $\Delta S^{\neq} = -145 \pm 15$ J mol^{-1} K^{-1}, and $\Delta V^{\neq} = -12.9 \pm 0.8$ cm^3 mol^{-1}. A rigorous appraisal of mechanistic possibilities was presented including an examination of the fate of the substituted deuterium atoms. It was concluded that the mechanism that fits the kinetics results and activation parameters involved an associative solvent-mediated ring-slipping process resulting in dechelation of the idenide and coordination of the phosphine in the key intermediate species. This is followed by coordination of the idenide by the other face and formation of the other isomer. The significance of the ring-slipping process and racemisation in homogeneous asymmetric catalysis was pondered, although the process should not be important for analogous cyclopentadienyl systems.

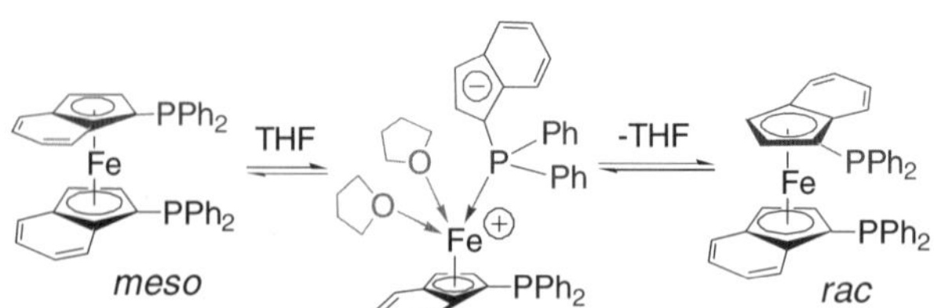

Preferred Ring-slippage Isomerization Process.

The possibility of radical coupling of the superoxide ion and nitric oxide to form peroxynitrous acid (ONOOH) has stimulated interest in the chemical reactivity of this latter compound since it may be associated with undesirable consequences, i.e. diseases associated with oxidative stress. Despite what may be termed exhaustive studies of the pressure dependence of the reaction of conversion of peroxynitrous acid to nitric acid, the volumes of activation obtained span a range of values that have prevented an unequivocal proposition of the reaction mechanism. [184] Since this and related reactions are so important physiologically, a detailed mechanistic understanding has been sought by many investigators. One suggested mechanism involves rotation around the NO bond followed by intramolecular OH transfer, a process that would be predicted to yield a moderate increase in volume upon reaching the transition state. A second mechanism has been proposed in which homolysis of the O-O bond produces free nitrogen dioxide and hydroxy radicals resulting in a more distinctly positive value of $\Delta V^{\neq}$. [185] Pulse radiolysis experiments gave rise to values of $\Delta V^{\neq}$ of $+10 \pm 1$ cm^3 mol^{-1} [186] and high pressure stopped-flow studies generated an average value of +6.7 ± 0.9 $cm^3 mol^{-1}$, although recently the latter method has yielded a $\Delta V^{\neq}$ of +9.7 ± 1.4 $cm^3 mol^{-1}$. [184] The higher values would be consistent with the homolysis mechanism, yet conversely the former mechanism would be compatible with the lower values. A myriad range of explanations can be offered for this dichotomy, including different instrumental methods, different preparations of peroxynitrite and different media. Even for a given set of experiments the error range on individual points (i.e. rate constant at a given pressure) was greater than usually found by either technique leading to $\Delta V^{\neq}$ values with uncertainty limits, in some cases, approaching 20 %. Thus quite unusually, the high pressure approach did not provide the key decisive mechanistic advantage.

In the introduction of this contribution it was noted that coverage of gas phase or heterogeneous reactions was not to be included. However, it would be remiss not to refer briefly to both the important mechanistic understanding of catalytically important organometallic reactions, studied under pressure and technical developments that have occurred. Rhodium carbonyl clusters have been investigated as potentially effective catalysts in conversion of CO and syngas mixtures in hydroformylation reactions. There is some uncertainty regarding the particular form of the cluster that is the actual catalytic agent. Attempts to resolve this matter have included use of gas pressures up to 100 MPa and application of hpnmr spectroscopy, monitoring ^{13}C to deduce aspects of the reaction mechanism. [187] High pressure ir spectroscopy was used to complement and substantiate the mechanistic conclusions. [187] Recent technical aspects with respect to *in situ* hpnmr spectroscopy in this context have been presented thoroughly. [28, 29, 188] The copolymerisation of styrene with carbon monoxide catalysed by a palladium(II) complex has been studied by hpnmr spectroscopy, for example, [189] and a report of the delineation of the steps in the carbomethoxy cycle for the carboalkoxylation of ethene by a palladium–diphosphane catalyst described related palladium-catalytic chemistry. [190]

Few inorganic compounds display significant solvatochromism. Indeed only a small selection of compounds is likely to possess suitable solubility characteristics in a range of polar and non-polar solvents to permit appropriate experiments to be conducted. Ternary iron(II)-diimine-cyanide complexes possess reasonably favourable properties in this context however, and have been used to establish correlations with the Reichardt E_T [191] parameter or with the solvent acceptor number. Some of these iron(II) complexes have been amenable to piezochromic determination and correlations between piezochromism expressed as $\delta\nu/\delta P$, where ν is an electronic absorption wavenumber, and solvatochromic shift have been presented. [192] Raman, luminescence or x-ray absorption spectroscopic (spin-state crossover) studies of inorganic solid state compounds at high pressures have been reported, where the pressure is usually applied using a diamond anvil cell, and these references and others [193-195] provide suitable literature starting points. The pressure variation of vibrational modes from the Raman spectrum of the ammonia-borane complex, NH_3BH_3, (also obtained using a diamond anvil cell) has been analysed in an investigation to provide evidence of the dihydrogen bond. [196] An apparatus for high pressure combinatorial screening of homogeneous catalysts for the hydrogenation of CO_2 has been reported. [197] The particular significance of this study is that it enabled large arrays of potential catalysts to be tested simultaneously and with a non-instrumental, i.e. visual dye assay method to detect product yield, in the first instance. A further important aspect was the highly desirable goal of finding a catalyst not containing a metal of the "precious" set (Ru, Rh, Pd, Ir, Pt). Pressures up to 20 MPa were applied and some of the results obtained using the apparatus have been described. [197] Quite recently moderate pressures (0 to 10 MPa) have been used in experiments to determine solubilities of CO in ionic liquids. [198] The purpose was to develop useful information about such properties that are relevant to catalytic systems that depend on gases such as CO, for example the $RhH(CO)(PPh_3)_3$ catalysed hydroformylation of 5-hexene2-one. Sapphire nmr tubes were used in the ^{13}CO nmr spectroscopic measurements. In another aspect of considering possible catalysts in hydroformylation reactions the pentacarbonyl species $Mo(CO)_5(Sol)$ (Sol = solvent) has been proposed as an intermediate, prompting an investigation of its reaction with CO to form the hexacarbonyl species. [199] By using a special high pressure / variable temperature flow cell a flash photolysis study led to the kinetic characterisation of the regeneration of $Mo(CO)_6$. [199] An interchange mechanism was proposed. The pressure range of CO (0 - 2 MPa) was not sufficient to determine the activation volume, but nevertheless again this study illustrates the success of a combination of innovative equipment development (in the pressure regime) and a well designed chemistry approach.

6. THEORETICAL STUDIES

Helm and Merbach have summarised efforts aimed at calculating exchange mechanisms for water exchange on first row, second and third row transition metal cations and on lanthanide ions. [26] The limitations within the calculations and where there is consistency with parameters obtained from ambient and high pressure kinetics results have been pointed out. An example of a particular theoretical investigation will be cited below.

Attention has also been drawn to consideration of water exchange mechanisms on the isoelectronic ruthenium(II) and rhodium(III) hexaaqua ions. [121] On the basis mainly of activation volumes the mechanisms were assigned as I_d [200] and I_a [201], respectively. These assignments stimulated further discussion. [201], [202] Accordingly, quantum mechanically based calculations that included consideration of hydration effects were undertaken. [122] Detailed calculations and arguments led to the conclusion that these aqua-ions do indeed exchange water by different mechanisms. Obviously the charges on the ions are different, but a pivotal factor was the estimation of the relative metal-oxygen bond strengths. The calculations showed that for a fully dissociative (D) mechanism to occur, the activation energy for the Ru^{II} ion was only about half that needed in the case of the Rh^{III} ion, where 137 kJ mol^{-1} would be needed. The Rh^{III}-O bonds are thus considerably stronger than the Ru^{II}-O bonds, a pattern sustained in M---O bonds of the respective transition states for an interchange mechanism. Hence because of the strong Rh^{III}-O bonds, water exchange on $Rh(H_2O)_6^{3+}$ proceeds via the I_a (close to I) pathway, while the same reaction of $Ru(H_2O)_6^{2+}$, which has considerably weaker bonds was said to follow an I_d or D mechanism. The findings of this theoretical investigation have been confirmed for the mechanism of water exchange on $Rh(H_2O)_6^{3+}$ by a later experimental study, [123] treated at length in section 3.2.

One way to combat any suspicion that experimental values of)$V^{\neq}$ for water exchange are equivocal with respect to mechanism assignment because they may contain a significant but unknown contribution from extension or compression of the bonds to the central metal ion of the non-exchanging ligands, is to perform calculations leading to volume changes for attainment of various hypothetical transition states. If these calculations are sufficiently refined and the corresponding energy profiles $\Delta H^{\neq}$, $\Delta G^{\neq}$ are also calculated then mechanistic assignment by comparison with experimental parameters can be made, and the volume contribution from possible movement of non-exchanging ligands can be assessed. Such an approach has been successful in confirming the mechanisms of water exchange for the rhodium(III) and iridium(III) hexaaqua ions. [203] Both volume and energy profiles have been computed for two distinct water exchange mechanisms (D and I_a) using methods that included hydration effects (see Figure 19). The calculated energy of activation for water exchange by an I_a mechanism on $Ir(H_2O)_6^{3+}$ was 128 kJ mol^{-1} (experimental values of $\Delta H^{\neq}$ and $\Delta G^{\neq}$ were 131 and 130 kJ mol^{-1}, respectively, at 298K) whereas the activation energy for exchange by a D mechanism was determined to be close to 160 kJ mol^{-1}. Volumes of activation were calculated to be -3.9 and -3.5 cm^3 mol^{-1} for the hexaaqua ions of Rh^{3+} and Ir^{3+}, respectively. (Experimental values obtained earlier were -4.2 and -5.7 cm^3 mol^{-1}, respectively). In the case of the iridium(III) ion calculations for the D mechanism showed a shortening of 0.025 X for the Ir-O bonds of the spectator water molecules. This translated to a volume change of -1.8 cm^3 mol^{-1}, and together with the calculated value of +7.3 cm^3 mol^{-1} for the bond breakage of the departing water molecule, yielded a value of $\Delta V^{\neq}_{calc}$ for the hypothetical D mechanism of +5.5 cm^3 mol^{-1}. Further calculations showed that a volume decrease of 0.9 cm^3 mol^{-1} occurred owing to, surprisingly, a shortening of the Ir-O bonds of non-exchanging water molecules in an I_a mechanism, resulting in a volume value -2.4 cm^3 mol^{-1} for entry in the first coordination sphere of the seventh water molecule *cis* to the departing molecule. These findings led to firm conclusions that both ions exchange coordinated with bulk water molecules by an I_a mechanism albeit with a modest degree of associative character, although such a mechanism would not be predicted for species with a t_{2g}^6 electronic configuration. Energy profiles and profiles of the sum of all M-O bond length changes for water exchange on the two trivalent ions together with those for the $Ru(H_2O)_6^{2+}$ ion are shown in Figure 20. From this report it can be deduced that volume changes associated with non-exchanging ligands do not contribute significantly to measured volumes of activation, for these hexaaqua ions.

The theoretical part of a combined theoretical and experimental study of the electronic tuning of the lability of Pt(II) complexes through π-acceptor effects permitted the conclusions presented below. [133(a)] First we will reiterate the experimental aspects. A series of Pt(II) complexes containing three nitrogen donor ligands and one coordinated water molecule were prepared, and the lability of the coordinated water molecule investigated in order to assess the influence of the π-acceptor effect upon the lability of these complexes. The kinetics of water substitution by TU, DMTU, TMTU, I^- and SCN^- were studied under various conditions (details in section 3.2). Analysis of the results indicated that a π-acceptor ligand moiety in the *cis* position is more influential than π-acceptor property in the *trans* position to the leaving water molecule, and furthermore it was found that the two π–acceptor effects were multiplicative, and the reaction mechanism was associative. Density functional theory (DFT) calculations showed that by the addition of π–acceptor ligands to the metal the positive charge on the metal centre increased, and the energy separation of the frontier molecular orbitals of the ground state Pt(II) complexes decreased. The calculations supported the experimentally observed additional increase in water lability when two π–accepting rings were adjacent to each other, an effect that was attributed to electronic communication between the two pyridine rings of the coordinated ligand. It was also reported that the results demonstrated that the pK_a of the coordinated water molecule was controlled by the π-accepting property of the chelate system and that reflected the electron density around the platinum centre.

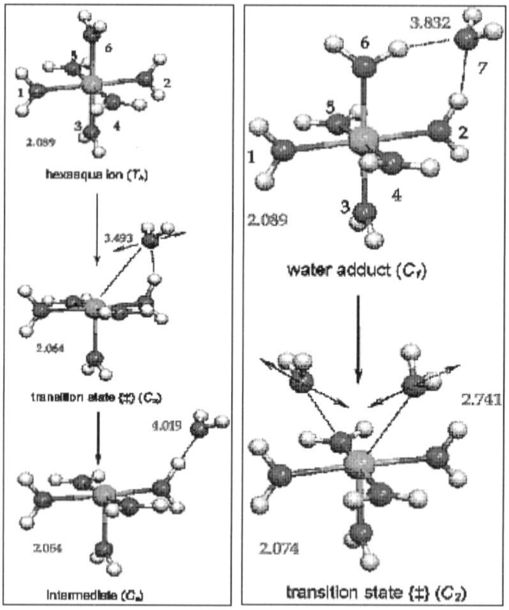

Figure 19. Theoretical mechanistic details for water exchange reactions on $[Ir(H_2O)_6]^{3+}$ following a limiting D (left) and an I_a (right) mechanism, respectively.

In Section 3.1, summaries of several experimental studies regarding water exchange on ions of lanthanide elements were provided. Beside the purpose of uncovering the inherent mechanism it was emphasised that understanding these properties had the additional value of examining the potential application of aqua-complexes of the certain lanthanide ions, specifically of Gd(III), as MRI contrast agents. A key desirable property of such complexes is an enhanced water proton relaxation rate (relaxivity) which arises from protons of water molecules directly coordinated to the metal ion that exchange with bulk water molecules, but also from an outer-sphere relaxation due to dipolar interactions through space with surrounding water molecules. In order to obtain a better understanding of events in the first coordination sphere, how intramolecular motions can influence the water exchange rate and the reasons for rapid water exchange, a combined x-ray crytallographic and molecular dynamics simulations investigation has been undertaken. [204] Specifically $[Gd(egta)(H_2O)]^-$

($egta^{4-}$ = 3,12-bis(carboxymethyl)-6,9-dioxa-3,12-diazatetradecanedioate(4-)) was subject to molecular dynamics simulations, and crystal structures were used as a basis for setting the conditions of simulations of this chelate species, and of the corresponding de-aquated complex, $[Gd(egta)]^-$. The volume change for loss of the water molecule could be calculated as +7.2 $cm^3\ mol^{-1}$, which compares reasonably with the previously published experimental value of +10.5 $cm^3\ mol^{-1}$. The simulations addressed changes in the conformation of the complex with flips of some dihedral angles, very rapid changes in the symmetry orientation of the coordination polyhedron and steric constraints of the ligand on the inner-sphere water molecule, factors which could be related to the water exchange rate. These features of the calculations are an invaluable outcome toward understanding proton relaxivity and the molecular mechanisms extant in the experimental parameters, and can provide further insight regarding appropriate design of MRI agents.

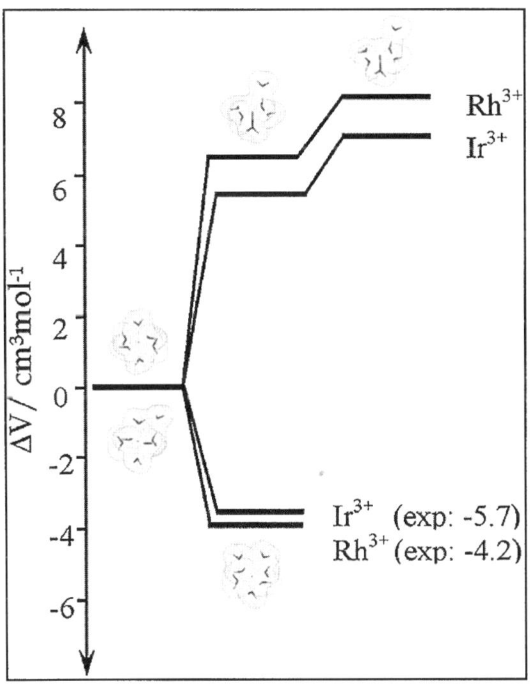

Figure 20. Comparison of theoretical and experimental volume profiles for water exchange reactions on $[Rh(H_2O)_6]^{3+}$ and $[Ir(H_2O)_6]^{3+}$ according to limiting D (top) and I_a (bottom) mechanisms. The experimental volumes of activation favor the operation of an associative interchange (I_a) mechanism in both cases.

7. CONCLUDING REMARKS

It is evident that application of pressure has wide scope to enhance our knowledge in fields of several scientific disciplines (chemistry, biochemistry, geochemistry, materials science, physics, atmospheric processes, engineering for example). Within chemistry, exploitation of the pressure variable has provided a range of additional information in synthetic and mechanistic investigations in inorganic, organic, organometallic catalytic and polymer chemistry. In this contribution we have highlighted some of the more recent advances in inorganic and bioinorganic chemistry with particular emphasis on mechanistic application in solution systems. Almost ten years ago [205] we wrote that "Considering the variety of chemistry currently studied and likely to be studied in the future, the next decade promises to generate new mechanistic challenges to rival the exciting progress over the past ten years". We have endeavoured to illustrate that the (inorganic high pressure) field has shown remarkable growth, and that the last few years of the past decade have witnessed the revelation of some very innovative and exciting chemistry understanding and perhaps surpasses rather than rivals progress over the previous decade.

ACKNOWLEDGEMENTS

It is a pleasure to acknowledge the fruitful collaborations with many students and scientists who appear as co-authors in publications of the authors of this chapter. We gratefully acknowledge financial support from the Deutsche Forschungsgemeinschaft through SFB 583 "Redox-active metal complexes", and the Fonds der Chemischen Industrie.

REFERENCES

[1] For example, G. Demazeau, J. Phys.: Condensed Matter, 14, (2002) 11031; J. V. Badding, Ann. Rev. Mater. Sci., 28, (1998) 631.

[2] (a) J. Hyde, W. Leitner and M. Poliakoff, in "High Pressure Chemistry. Synthetic, Mechanistic and Supercritical Applications" R. van Eldik and F.-G. Klärner, Eds., (Wiley-VCH, Weinheim, 2002). (b) O. S. Jina, X. Z. Sun and M. W. George, Dalton Trans., 1773 (2003).

[3] (a) L. Smeller, F. Meersman, J. Fidy and K. Heremans, Biochemistry, 42, (2003) 553; (b) G. Pappenberger, C. Saudan, M. Becker, A. E. Merbach and T. Kiefhaber, Proc. Nat. Acad. Sci., USA, 97 (2000) 17.

[4] R. van Eldik in "Inorganic High Pressure Chemistry. Kinetics and Mechanism", R van Eldik, Ed., (Elsevier, 1986) Chapter 1.

[5] For example, K. J. Laidler, "Chemical Kinetics" 2nd edition, (McGraw-Hill, 1965) p. 231.

[6] T. Asano and W. J. le Noble, Chem. Rev., 78 (1978) 407.

[7] D. R. Stranks, Pure and Applied Chem., 38 (1974) 303.

[8] (a) R. van Eldik, C. Dücker-Benfer and F. Thaler, Adv. Inorg. Chem., 49 (2000) 1; (b) Adv. Inorg. Chem., Vol. 54, R. van Eldik and C. D. Hubbard, Eds., (Academic Press 2003), Chapters 1, 2 and 4.

[9] "High Pressure Chemistry. Synthetic, Mechanistic and Supercritical Applications", R van Eldik and F.-G. Klärner, Eds., (Wiley-VCH, Weinheim, 2002).

[10] R. van Eldik and C. D. Hubbard, S. Afr. J. Chem., 53 (2000) 139.

[11] A. Drljaca, C. D. Hubbard, R. van Eldik, T. Asano, M. V. Basilevsky and W. J. le Noble, Chem. Rev., 98 (1998) 2167.

[12] G. Stochel and R. van Eldik, Coord. Chem. Rev., 159, 153 (1997); R. van Eldik and P. C. Ford, Adv. Photochem., 24 (1998) 61.

[13] J. Macyk and R. van Eldik, Biochem. Biophys. Acta, 1595 (2002) 283.

[14] T. W. Swaddle in Ref. [9], Chapter 5.

[15] "Chemistry under Extreme or Non-Classical Conditions", R. van Eldik and C. D. Hubbard, Eds., (Wiley-Spektrum, New York-Heidelberg, 1997), Chapters 3 and 4.

[16] G. Jenner, Angew. Chem. Internat. Ed., 14 (1975) 137; G. Jenner, J. Phys. Org. Chem., 15 (2002) 1.
[17] C. A. Eckert, Annu. Rev. Phys. Chem., 23 (1972) 239.
[18] "Inorganic Reaction Mechanisms", M. L. Tobe and J. Burgess, (Addison-Wesley-Longmans, Harlow, 1999); (a) Chapter 2, (b) Chapter 7.
[19] "Ligand Substitution Reactions", C. H. Langford and H. B. Gray, (Benjamin, New York, 1965).
[20] Chapter 1 in reference [8b].
[21] R. Whyman, K. R. Hunt, R. W. Page and S. Rigby, J. Phys. E, 17 (1984) 559.
[22] "Laboratory Methods in Vibrational Spectroscopy", H. A Willis, J. H. Van der Maas and R. G. J. Miller, (Wiley, New York, 1987), p. 289.
[23] R. van Eldik and C. D. Hubbard, Instrum. Sci. Technol., 22 (1995) 1.
[24] "High Pressure Techniques in Chemistry and Physics. A Practical Approach" W. B. Holzapfel and N. S. Isaacs, Eds., (Oxford University Press, 1997).
[25] W. E. Price and H.-D. Lüdemann, Chapter 5 in reference [24].
[26] L. Helm and A. E. Merbach, J. Chem. Soc., Dalton Trans., 633 (2002).
[27] A. Zahl, P. Igel, M. Weller, D. Koshtariya, M. S. A. Hamza and R. van Eldik, Rev. Sci. Instrum., 74 (2003) 3758.
[28] B. T. Heaton, J. Jonas, T. Eguchi and G. A. Hoffman, J. Chem. Soc., Chem Commun., 331 (1981).
[29] B. T. Heaton, L. Strona, J. Jonas, T. Eguchi and G. A. Hoffman, J. Chem. Soc., Dalton Trans., 1159 (1982).
[30] For example: (a) H. Doine, T. W. Whitcombe and T. W. Swaddle, Can. J. Chem., 70 (1992) 82; (b) J. I. Sachinidis, R. D. Shalders and P. A. Tregloan, J. Electroanal. Chem Interfacial Electrochem., 327 (1992) 219.
[31] For example, H. Heberhold, S. Marchal, R. Lange, C. H. Scheyhing, R. F. Vogel and R. Winter, J. Mol. Biol., 330 (2003) 1153.
[32] M. H. Jacob, C. Saudan, G. Holterman, A. Martin, D. Perl, A. E. Merbach and F. X. Schmid, J. Mol. Biol., 318 (2002) 837 (2002).
[33] J. Woenckhaus, R. Kohling, R. Winter, P. Thiyagarajan and S. Finet, Rev. Sci. Instrum., 71 (2000) 3895.
[34] N. W. A. Uden, H. Hubel, D. A. Faux, D. J. Dunstan and C. A. Royer, High Pressure Research, 23 (2003) 206.
[35] S. Hosokawa and W.-C. Pilgrim, Rev. Sci. Instrum., 72 (2001) 1721.
[36] H. Rollema, D. Keenan, S. A. Galema, K. K. de Kruif and C. D. Hubbard, manuscript in preparation.
[37] R. van Eldik and D. Meyerstein, Acc. Chem. Res., 33 (2000) 207.
[38] (a) W. J. le Noble and R. Schlott, Rev. Sci. Instrum., 47 (1976) 770; (b) Reference [4]; (c) D. T. Richens, Y. Ducommun and A. E. Merbach, J. Am. Chem. Soc., 109 (1987) 603.
[39] Q. H. Gibson, Disc. Faraday Soc., 17 (1954) 137; B. Chance, R. H. Eisenhardt, Q. H. Gibson and K. K. Lonberg-Holm, Eds., "Rapid Mixing and Sampling Techniques in Biochemistry", (Academic Press, New York, 1964); Q. H. Gibson and L. Milnes, Biochem. J., 91 (1964) 161.
[40] K. Heremans, J. Snauwaert and J. Rijkenberg, Rev. Sci. Instrum., 51 (1980) 806.
[41] (a) R. van Eldik, D. A. Palmer, R. Schmidt and H. Kelm, Inorg. Chim. Acta, 50 (1981) 131; (b) R. van Eldik, W. Gaede, S. Wieland, J. Kraft, M. Spitzer and D. A. Palmer, Rev. Sci. Instrum., 64 (1993) 1355.
[42] S. Funahashi, K. Ishihara and M. Tanaka, Inorg. Chem., 20 (1981) 5 ; K. Ishigara, S. Funahashi and M. Tanaka, Rev. Sci. Instrum., 53 (1982) 1231.
[43] P. J. Nichols, Y. Ducommun and A. E. Merbach, Inorg. Chem., 22 (1983) 3993.
[44] C. Balny, J. L. Saldana and N. Dahan, Anal. Biochem., 139 (1984) 178.
[45] High-Tech Scientific, Brunel Road, Salisbury, SP2 7PU, UK.
[46] P. Bugnon, G. Laurenczy, Y. Ducommun, P.-Y. Sauvageat, A. E. Merbach, R. Ith, R. Tschanz, M. Deludda, R. Bergbauer and E. Grell. Anal. Chem., 68 (1996) 3045.

[47] M. Eigen, Disc. Faraday Soc., 17 (1954) 194; M. Eigen and L. De Maeyer, in "Investigation of Rates and Mechanisms of Reactions", S. L Friess, E. S. Lewis and A. Weissberger, (Interscience 1963), Part II, Chapter 18.
[48] G. Czerlinski and M. Eigen, Z. Elektrochem., 63, 652 (1959); G. Czerlinski, Rev, Sci. Instrum., 33 (1962) 1184.
[49] E. F. Caldin, "Fast Reactions in Solution", (Blackwell Scientific, Oxford, 1964).
[50] G. G. Hammes and J. I. Steinfeld, J. Am. Chem. Soc., 84 (1962) 4639.
[51] E. F. Caldin, M. W. Grant, B. B. Hasinoff and P. A. Tregloan. J. Phys. E., Sci. Inst., 6 (1973) 349.
[52] E. F. Caldin, M. W. Grant and B. B. Hasinoff, J. Chem. Soc., Faraday Trans., I 68 (1972) 2247.
[53] M. W. Grant, J. Chem. Soc., Faraday Trans., I 69 (1973) 560.
[54] E. F. Caldin and B. B. Hasinoff, J. Chem. Soc., Faraday Trans. I, 515 (1975); B. B. Hasinoff, Biochemistry, 13 (1974) 3111.
[55] C. D. Hubbard, C. J. Wilson and E. F. Caldin, J. Am. Chem. Soc., 98 (1976) 1870.
[56] R. H. Holyer, C. D. Hubbard, S. F. A. Kettle and R. G. Wilkins, Inorg. Chem., 4 (1965) 929; M. Eigen and R. G. Wilkins, Adv. Chem. Series, American Chemical Society (1965).
[57] R. Doss and R. van Eldik, Inorg. Chem., 21 (1982) 3993.
[58] (a) D. H. Powell, A. E. Merbach, I. Fabian, S. Schindler and R. van Eldik, Inorg. Chem., 33 (1994) 4468; (b) F. Thaler, C. D. Hubbard, F. W. Heinemann, R. van Eldik, S. Schindler, I. Fabian, A. Dittler-Klingermann, F. E. Hahn and C. Orvig, Inorg. Chem., 37 (1998) 4022; (c) A. Neubrand, F. Thaler, M. Koerner, A. Zahl, C. D. Hubbard and R. van Eldik, J. Chem. Soc., Dalton Trans., 957 (2002).
[59] (a) A. Pasquarello, I. Petri, P. S. Salmon, O. Parisel, R. Car, E. Toth, D. H. Powell, H. E. Fischer, L. Helm and A. E. Merbach, Science 291 (2001) 856; (b) I. Persson, P. Persson, M. Sandström and A.-S. Ullström, J. Chem. Soc., Dalton Trans., 1256 (2002).
[60] J. Jonas, in "High Pressure Chemistry", H. Kelm, Ed., (Reidel, Dordrecht, 1978) p. 65.
[61] Reference [4]; A. E. Merbach in "High Pressure Chemistry and Biochemistry", R. van Eldik and J. Jonas, Eds., (Reidel, Dordrecht, 1987).
[62] T. W. Swaddle in reference 4, page 273.
[63] H.-D. Lüdemann, Polish J. Chem., 70 (1996) 387.
[64] M. M. Hoffman and M. S. Conradi, Rev. Sci. Instrum., 68 (1997) 159.
[65] (a) A. Zahl, A. Neubrand, S. Aygen and R. van Eldik, Rev. Sci. Instrum., 65 (1994) 882; (b) A. Zahl, P. Igel, M. Weller and R. van Eldik, Rev. Sci. Instrum., in press, (2004).
[66] G. Porter in "Investigations of Rates and Mechanisms of Reactions" 2nd Edition, S. L. Friess, E. S. Lewis and A. Weissberger Eds., (Interscience, 1963), Part II, Chapter 19.
[67] A. Wanat, M. Wolak, L. Urzel, M. Brindell, R .van Eldik and G. Stochel, Coord. Chem. Rev., 229 (2002) 37.
[68] P. C. Ford and L. E. Laverman, Chapter 6 in reference [9].
[69] T. Fu and T. W. Swaddle, Chem. Commun., 1171 (1996).
[70] (a) J. I. Sachinidis, R. D. Shalders and P. A. Tregloan, Inorg. Chem., 33 (1994) 6180; (b) J. I. Sachinidis, R. D. Shalders and P. A. Tregloan, Inorg. Chem., 35 (1996) 2497.
[71] R. van Eldik and C. D. Hubbard, Chapter 1 in Reference [9].
[72] J. Burgess, "Metal Ions in Solution", (Ellis Horwood, Chichester, 1978).
[73] L. Helm and A. E. Merbach, Chapter 4 in Reference [9].
[74] F. A. Dunand, L. Helm and A. E. Merbach, Chapter 1 in Reference [8].
[75] Y. Ducommun, K. E. Newman and A. E. Merbach, Inorg. Chem., 19 (1980) 3696.
[76] Y. Ducommun, D. Zbinden and A. E. Merbach, Helv. Chim. Acta, 65 (1982) 1385.
[77] S. K. Kang. B. Lam, T. A. Albright and J. F. O'Brian, New J. Chem., 15 (1991) 757.
[78] R. Åkesson, L. G. M. Pettersson, M. Sandström, P. E. M. Siegbahn and U. Wahlgren, J. Phys. Chem., 97 (1993) 3765.
[79] R. Åkesson, L. G. M. Pettersson, M. Sandström and U. Wahlgren, J. Am. Chem. Soc., 116 (1994) 8691; *ibid* 8705.

[80] F. P. Rotzinger, J. Am. Chem. Soc., 118 (1996) 6760.
[81] F. P. Rotzinger, J. Am. Chem. Soc., 119 (1997) 5230.
[82] M. Hartman, T. Clark and R. van Eldik, J. Am. Chem. Soc., 119 (1997) 5867.
[83] F. H. Spedding, L. E. Shiers, M. A. Brown, J. L. Derer, D. L. Swanson and A. Habenschuss, J. Chem. Eng. Data, 20 (1975) 81.
[84] F. H. Spedding, P. F. Cullen and A. Habenschuss, J. Phys. Chem., 78 (1974) 1106.
[85] C. Cossy, L. Helm and A. E. Merbach, Inorg. Chem., 27 (1988) 1973.
[86] C. Cossy, L. Helm and A. E. Merbach, Inorg. Chem., 28 (1988) 2699.
[87] K. Mieskei, D. H. Powell, L. Helm, E. Brucher and A. E. Merbach, Magn. Reson. Chem., 31 (1993) 1011.
[88] F. A. Dunand, L. Helm and A. E. Merbach, Chapter 1 of Reference [8]. Table XVIII and accompanying narrative.
[89] G. Moreau, L. Helm, J. Purans and A. E. Merbach, J. Phys. Chem. A, 106 (2002) 3034.
[90] P. Caravan, E. Toth, A. Rochenbauer and A. E. Merbach, J. Am. Chem. Soc., 121 (1999)10403.
[91] S. Laus, R. Ruloff, E. Toth and A. E. Merbach, Chem. Eur. J., 9 (2003) 3555.
[92] R. Ruloff, E. Toth, R. Scopelliti, R. Tripier, H. Handel and A. E. Merbach, Chèm. Commun, 2630 (2002).
[93] F. Botteman, G. M. Nicolle, L. van der Elst, S. Laurent, R. N. Muller and A. E. Merbach, Eur. J. Inorg. Chem., 2686 (2002).
[94] L. Burai, R. Scopelliti and E. Tóth, Chem. Commun., 2366 (2002).
[95] L. Burai, E. T∴th, G. Moreau, A. Sour, R. Scopelliti and A. E. Merbach, Chem. Eur. J., 9 (2003) 1394.
[96] G. M. Nicholle, F. Yerly, D. Imbert, U. Böttger, J.-C. Bünzli and A. E. Merbach, Chem. Eur. J., 9 (2003) 5453.
[97] M. K. Thompson, M. Botta, G. Nicholle, L. Helm, S. Aime, A. E. Merbach and K. N. Raymond, J. Am. Chem. Soc., 125 (2003) 14274.
[98] G. Moreau, L. Burai, L. Helm, J. Purans and A. E. Merbach, J. Phys. Chem. A, 107 (2003) 758.
[99] U. Prinz, U. Koelle, S. Ulrich, A. E. Merbach, O. Maas and K. Hegetschweiler, Inorg. Chem., 43 (2004) 2387.
[100] J. Burgess and C. D. Hubbard in Adv. Inorg. Chem., 54, (Academic Press 2003), R. van Eldik and C. D. Hubbard, Eds., Chapter 2.
[101] D. Zhang, D. H. Busch, P. L. Lennon, R. H. Weiss, W. L. Neumann and D. P. Riley, Inorg. Chem., 37 (1998) 956.
[102] I. Ivanovic-Burmazovic, M. S. A. Hamza and R. van Eldik, Inorg. Chem., 41 (2002) 5150.
[103] I. Ivanovic-Burmazovic, M. S. A. Hamza and R. van Eldik, Inorg. Chem. Commun., 5 (2002) 937.
[104] S. Nèmeth, L. I. Simándi, G. Argay and A. Kálmán, Inorg. Chim. Acta, 166 (1989) 31.
[105] B. M. Alzoubi, M. S. A. Hamza, C. Dücker-Benfer and R. van Eldik, Eur. J. Inorg. Chem., 2972 (2003).
[106] B. M. Alzoubi, M. S. A. Hamza, A. Felluga, L. Randaccio, G. Tauzher and R. van Eldik, Eur. J. Inorg. Chem., 653 (2004).
[107] M. S. A. Hamza, R. van Eldik, P. L. S. Harper, J. M. Pratt and E. A. Betterton, Eur. J. Inorg. Chem., 580 (2002).
[108] M. S. A. Hamza, X. Zou, K. L. Brown and R. van Eldik, Eur. J. Inorg. Chem., 268 (2003).
[109] M. S. A. Hamza, X. Zou, K. L. Brown and R. van Eldik, Dalton Trans., 2986 (2003).
[110] T. W. Swaddle, Rev. Phys. Chem. Jpn., 50 (1980) 230.
[111] M. S. A. Hamza, X. Zou, K. L. Brown and R. van Eldik, Dalton Trans., 3832 (2003).
[112] M. S. A. Hamza, A. Felluga, L. Randaccio, G. Tauzher and R. van Eldik, Dalton Trans., 287 (2004).
[113] M. S. A. Hamza and R. van Eldik, Dalton Trans., 1, (2004) and references *loc cit.*
[114] M. S. A. Hamza, A. G. Cregan, N. E. Brasch and R. van Eldik, Dalton Trans., 596 (2003).

[115] D. Chatterjee, M. S. A. Hamza, M. M. Shoukry, A. Mitra, S. Deshmukh and R. van Eldik, Dalton Trans., 203 (2003).
[116] H. C. Bajaj and R. van Eldik, Inorg. Chem., 28 (1989) 1980; H. C. Bajaj and R. van Eldik, Inorg. Chem., 29 (1990) 2855.
[117] W. Plumb and G. M. Harris, Inorg. Chem., 3 (1964) 542.
[118] K. Swaminathan and G. M. Harris, J. Am. Chem. Soc., 88 (1966) 4411.
[119] R. J. Buchacek and G. M. Harris, Inorg. Chem., 15 (1976) 926.
[120] G. Laurenczy, I. Rapaport, D. Zbinden and A. E. Merbach, Magn. Reson. Chem., 29 (1991) S45.
[121] A. Cusanelli, U. Frey, D. T. Richens and A. E. Merbach, J. Am. Chem. Soc., 118 (1996) 5265.
[122] D. deVito, H. Sidorenkova, F. P. Rotzinger, J. Weber and A. E. Merbach, Inorg. Chem., 39 (2000) 5547.
[123] S. C. Galbraith, C. R. Robson and D. T. Richens, Dalton Trans., 4335 (2002).
[124] Z. D. Bugarcic, M. M. Shoukry and R. van Eldik, J. Chem. Soc., Dalton Trans., 3945 (2002).
[125] T. Rau, M. M. Shoukry and R. van Eldik, Inorg. Chem., 36 (1997)1454.
[126] F. F. Prinsloo, J. J. Pienaar and R. van Eldik, J. Chem. Soc., Dalton Trans., 3581 (1995).
[127] B. Salignac, P. V. Grundler, S. Cayemittes U. Frey, R. Scopelliti, A. E. Merbach, R. Hedinger, K. Hegetschweiler, R. Alberto, U. Prinz, G. Raabe, U. Kölle and S. Hall, Inorg. Chem., 42 (2003) 3516.
[128] P. V. Grundler, B. Salignac, S. Cayemittes, R. Alberto and A. E. Merbach, Inorg. Chem., 43 (2004) 865.
[129] Z. D. Bugarcic, G. Liehr and R. van Eldik, J. Chem. Soc., Dalton Trans., 2825 (2002).
[130] D. Jaganyi, A. Hofmann and R. van Eldik, Angew. Chem., Int. Ed., 40 (2001) 1680.
[131] Z. D. Bugarcic, G. Liehr and R. van Eldik, J. Chem. Soc., Dalton Trans., 951 (2002).
[132] P. Casten, F. Duhan, S. Wimmer, F. L. Wimmer, J. Chem. Soc., Dalton Trans., 2679 (1990); K. W. Jennette, J. T. Gill, J. A. Sadownick, S. J. Lippard, J. Am. Chem. Soc., 98 (1976) 6159; J. A. Bailey, M. G. Hill, R. E. Marsh, V. M. Miskowski, W. P Schaefer, H. B. Gray, Inorg. Chem., 34, (1995) 4591.
[133] (a) A. Hofmann, D. Jaganyi, O. Q. Munro, G. Liehr and R. van Eldik, Inorg. Chem., 42 (2003) 1688; (b) A. Hofmann, L. Dahlenburg and R. van Eldik, Inorg. Chem., 42 (2003) 6528; (c) D. Jaganyi, D. Reddy, J. A. Gertenbach, A. Hofmann and R. van Eldik, Dalton Trans., 299 (2004).
[134] J. Procelewska, A. Zahl, R. van Eldik, H. A. Zhong, J. A. Labinger and J. E. Bercaw, Inorg. Chem., 41 (2002) 2808.
[135] M. Font-Bardia, C. Gallego, G. Gonz<lez, M. Martinez, A. E. Merbach and X. Solans, Dalton Trans., 1106 (2003).
[136] R. A. Marcus, J. Phys. Chem., 72 (1968) 891, and references *loc cit.*
[137] N. S. Hush, Coord. Chem. Rev., 64 (1985) 135, and references *loc cit.*
[138] E. Pelizzetti and E. Mentasti, Inorg. Chem., 18, (1979) 583.
[139] J. Wisniewska and R. van Eldik, Inorg. Chem., 41 (2002) 3802.
[140] (a) A. Zahl, R.van Eldik, M. Matsumoto, T. W. Swaddle, Inorg. Chem., 42 (2003) 3718, and references *loc cit.*; (b) M. Matsumoto and T. W. Swaddle, Inorg. Chem., 43 (2004) 2724.
[141] R. M. Nielson, G. E. McManis, M. N. Golovin and M. J. Weaver, J. Phys. Chem., 92 (1988) 3441; R. M. Nielson, G. E. McManis, L. K. Safford and M. J. Weaver, J. Phys. Chem., 93 (1989) 2152; T. Gennet, D. F. Milner and M. J. Weaver, J. Phys. Chem., 89 (1985) 2789.
[142] For example, (a) H. Bruin, S. Nigam and J. F. Holzwarth, Faraday Discuss., Roy. Soc. Chem., 74 (1982) 129; (b) L. Spiccia and T. W. Swaddle, Inorg. Chem., 26 (1987) 2265; (c) A. Rampi-Scandola, F. Scandola and A. Indelli, J. Chem. Soc., Faraday Trans I, (1985) 2967; (d) J. M. Leal, P. L. Domingo, B. Garcia and S. Iheas, J. Chem. Soc., Faraday Trans., (1985) 3571; (e) B. Goyal, S. Solanki, S. Arora, A. Prakash, R. N. Mehrotra, J. Chem. Soc., Dalton Trans., (1996) 3109; (f) C. D. Hubbard, A. Gerhard, R. van Eldik, J. Chem. Soc., Dalton Trans., (2001) 1069.
[143] P. D. Metelski and T. W. Swaddle, Inorg. Chem., 38 (1999) 301.
[144] W. H. Mulder, T. P. Dasgupta and G. Stedman, Inorg. React. Mech., 1 (2000) 257.
[145] A. Zahl, R. van Eldik and T. W. Swaddle, Inorg. Chem., 41 (2002) 757.
[146] N. Sutin, C. Creutz and E. Fujita, Comments Inorg. Chem., 19 (1997) 67.

[147] A. M. Tait, M. Z. Hoffman and E. Hayon, J. Am. Chem. Soc., 98 (1976) 86; C. Creutz, H. A. Schwartz, J. F. Wishart, E. Fujita and N. Sutin, J. Am. Chem. Soc., 113 (1991) 3361; C. Creutz, H. A. Schwartz, J. F. Wishart, E. Fujita and N. Sutin, J. Am. Chem. Soc., 111 (1989) 1153.
[148] E. Fujita, J. F. Wishart and R. van Eldik, Inorg. Chem., 41 (2002) 1579.
[149] E. Fujita and R. van Eldik, Inorg. Chem., 37 (1998) 360.
[150] For example, B. Cornils and W. A. Herrmann, Applied Homogeneous Catalysis with Organometallic Compounds, (VCH, Weinheim, 1996).
[151] F. A. Cotton and G. Wilkinson, Advanced Inorganic Chemistry, 4th Edition, (John Wiley and Sons, New York, 1980).
[152] N. Navon, H. Cohen and D. Meyerstein, Inorg. Chem., 36 (1997) 3781.
[153] N. Navon, A. Burg, H. Cohen, R. van Eldik and D. Meyerstein, Eur. J. Inorg. Chem., 423 (2002).
[154] T. W. Swaddle and P. A. Tregloan, Coord. Chem. Rev., 187 (1999) 255.
[155] H. C. Bajaj, P. A. Tregloan and R. van Eldik, Inorg. Chem., 43 (2004) 1429.
[156] M. Matsumoto, N. L. Neuman and T. W. Swaddle, Inorg. Chem., 43 (2004) 1153.
[157] A. R. Butler and R. Nicholson, "Life, Death, and Nitric Oxide", (Royal Society of Chemistry, 2003).
[158] P. C. Ford, L. E. Laverman and I. M. Lorkovic, Chapter 4 in Reference [8(b)].
[159] M. Wolak and R. van Eldik, Coord. Chem Rev., 230 (2002) 263.
[160] F. Roncarli, J. A. Olabe and R. van Eldik, Inorg. Chem., 41 (2002) 5417.
[161] G. Stochel and R. van Eldik, Inorg. Chim. Acta, 190 (1991) 55.
[162] F. Roncarli, J. A. Olabe and R. van Eldik, Inorg. Chem., 42 (2003) 419.
[163] T. W. Swaddle, Inorg. Chem., 22 (1983) 2663.
[164] J. N. Armor, H. A. Scheidegger and H. Taube, J. Am. Chem. Soc., 90 (1968) 5928.
[165] S. Pell and J. N. Armor, J. Am. Chem. Soc., 95 (1973) 7625.
[166] A. Czap and R. van Eldik, Dalton Trans., 665 (2003).
[167] J. Macyk and R. van Eldik, Dalton Trans., 2704 (2003).
[168] H. L. Hodges, R. A. Holwerda and H. B. Gray, J. Am. Chem. Soc., 96 (1976) 3132.
[169] T. Schneppensieper, A. Wanat, G. Stochel and R. van Eldik, Inorg. Chem., 41 (2002) 2565.
[170] M. Körner, P. A. Tregloan and R. van Eldik, Dalton Trans., 2710 (2003).
[171] M. Meier, J. Sun, J. F. Wishart and R. van Eldik, Inorg. Chem., 35 (1996) 1564.
[172] T. D. Dolidze, D. E. Khoshtariya, D. H. Waldeck, J. Macyk and R. van Eldik, J. Phys. Chem. B, 107 (2003) 7172.
[173] M. Meier and R. van Eldik, Inorg. Chim. Acta, 242 (1996) 185.
[174] D. E. Khoshtariya, Biofizika, 31 (1986) 391.
[175] Y. Fu, A. S. Cole and T. W. Swaddle, J. Am. Chem. Soc., 121 (1999) 10410; H. Doine and T. W. Swaddle, Can. J. Chem., 66 (1988) 2763; T. W. Swaddle , Chapter 5 in reference [9].
[176] B. W. Griffin and J. A. Peterson, J. Biol. Chem., 250 (1975) 6445.
[177] D. R. Davydov, G. Hui Bon Hoa and J. A Peterson, Biochemistry, 38 (1999) 751.
[178] (a) M. Kato, R. Makino and T. Iizuka, Biochim. Biophys. Acta, 1246 (1995) 178; (b) M. Unno, K. Ishimori, Y. Ishimura and I. Morishima, Biochemistry, 33 (1994) 9762; (c) A. P. Koley, R. C. Robinson and F. K. Friedmann, Biochimie, 78 (1996) 706; (d) C. Jung, N. Bec and R. Lange, Eur. J. Biochem., 269 (2002) 2989; (e) C. Jung, Biochim. Biophys. Acta, 1595 (2002) 309; (f) C. Jung, G. Hui Bon Hoa, D. R. Davydov, E. Gill and K. Heremans, Eur. J. Biochem., 2343 (1995) 600; (g) H. Schulze, G. Hui Bon Hoa and C. Jung, Biochim. Biophys. Acta, 1338 (1997) 77; (h) H. Schulze, O. Ristau and C. Jung, Biochim. Biophys. Acta, 1183 (1994) 491; (i) C. Jung, G. Hui Bon Hoa, K.-L. Schoeder, M. Simon and J. P. Doucet, Biochemistry, 31 (1992) 12855; G. Stochel and R. van Eldik, Coord. Chem. Rev., 189 (1999) 329.
[179] A. Franke, G. Stochel, C. Jung and R. van Eldik, J. Am. Chem. Soc., 126 (2004) 4181.
[180] C. Jung, H. Schulze and E. Deprez, Biochemistry, 35 (1996) 15038; C. Jung, O. Ristan, H. Schulze and S. G. Sligar, Eur. J. Biochem., 235 (1996) 660; C. Jung, J. Mol. Recognition, 13 (2000) 325.

[181] L. E. Laverman, A. Wanat, J. Oszajca, G. Stochel, P. C. Ford and R. van Eldik, J. Am. Chem. Soc., 123 (2001) 285.
[182] A literature survey is contained in Reference [183].
[183] O. J. Curnow, G. M. Fern, M. L. Hamilton, A. Zahl and R. van Eldik, Organometallics, 23 (2004) 429.
[184] R. Kissner, C. Thomas, M. S. A. Hamza, R. van Eldik and W. H. Koppenol, J. Phys. Chem. A, 107 (2003) 11261.
[185] References *loc cit.* in Reference [184].
[186] S. V. Lymar, R. F. Khairutdinov and J. K. Hurst, Inorg. Chem., 42 (2003) 5259.
[187] D. T. Brown, T. Eguchi, B. T. Heaton, J. A. Iggo and R. Whyman, J. Chem. Soc., Dalton Trans., 677 (1991).
[188] J. A. Iggo, D. Shirley and N. C. Tong, New J. Chem., 22 (1998) 1043.
[189] J. A. Iggo, Y. Kawashima, J. Liu, T. Hiyama and K. Mozaki, Organometallics, 22 (2003) 5418.
[190] J. Liu, B. T. Heaton, J. A. Iggo and R. Whyman, Angew. Chem. Int. Ed., 43 (2004) 90.
[191] S. Spange, M. Lauterbach, A.-K. Gyra and C. Reichardt, Liebigs Ann. Chem., 323 (1991).
[192] A. Alousy, N. J. Blundell, J. Burgess, C. D. Hubbard and R. van Eldik, Transition Met. Chem., 27 (2002) 244, and references *loc cit.*
[193] C. Piquer, F. Grandjean, O. Mathon, S. Pascarelli, D. L. Reger, C. A. Little and G. J. Long, Inorg. Chem., 42 (2003) 982.
[194] J. K. Grey, I. S. Butler and C. Reber, Inorg. Chem., 42 (2003) 6503.
[195] High Pressure Research: An International Journal; Taylor and Francis. Volume 23 (2003), and earlier and later volumes.
[196] S. Trudel and D. F. R. Gilson, Inorg. Chem., 42 (2003) 2814.
[197] C.-C. Tai, T. Chang, B. Roller and P. G. Jessop, Inorg. Chem., 42 (2003) 7340.
[198] C. A. Ohlin, P. J. Dyson and G. Laurenczy, Chem. Commun., 1070 (2004).
[199] C. Kayran, M. Richards and P. C. Ford, Inorg. Chim. Acta, 357 (2004) 143.
[200] N. Aebischer, G. Laurenczy, A. Ludi and A. E. Merbach, Inorg. Chem., 32 (1993) 2810.
[201] G. Laurenczy, L. Rapaport, D. Zbinden and A. E. Merbach, Magn. Reson. Chem., 29 (1991) S45.
[202] D. T. Richens, "The Chemistry of Aquaions", (John Wiley, Chichester, 1997) p. 465.
[203] D. DeVito, J. Weber and A. E. Merbach, Inorg. Chem., 43 (2004) 858.
[204] F. Yerly, K. L. Hardcastle, L. Helm, S. Aime, M. Botta and A. E. Merbach, Chem. Eur. J., 8 (2002) 1031.
[205] Chapter 2 of reference [15].

Chemistry at Extreme Conditions
M. Riad Manaa (Editor)

Chapter 5

High Pressure Materials Research: Novel Extended Phases of Molecular Triatomics

Choong-Shik Yoo

Lawrence Livermore National Laboratory, Livermore, California 94551, USA

1. HIGH PRESSURE MATERIALS RESEARCH

Application of high pressure significantly alters the interatomic distance and thus the nature of intermolecular interaction, chemical bonding, molecular configuration, crystal structure, and stability of solids [1]. With modern advances in high-pressure technologies [2], it is feasible to achieve a large (often up to a several-fold) compression of lattice, at which condition the material can be easily forced into a new physical and chemical configuration [3]. The high-pressure thus offers enhanced opportunities to discover new phases, both stable and metastable ones, and to tune exotic properties in a wide-range of atomistic length scale, substantially greater than (often being several orders of) those achieved by other thermal (varying temperatures) and chemical (varying composition or making alloys) means.

Simple molecular solids like H_2, C, CO_2, N_2, O_2, H_2O, CO, NH_3, and CH_4 are bounded by strong covalent intramolecular bonds, yet relatively weak intermolecular bonds of van der Waals and/or hydrogen bonds. The weak intermolecular bonds make these solids highly compressible (i.e., low bulk moduli typically less than 10 GPa), while the strong covalent bonds make them chemically inert at least initially at low pressures. Carbon-carbon single bonds, carbon-oxygen double bonds and nitrogen-nitrogen triple bonds, for example, are among the strongest. These molecular forms are, thus, often considered to remain stable in an extended region of high pressures and high temperatures. High stabilities of these covalent molecules are also the basis of which their mixtures are often presumed to be the major detonation products of energetic materials as well as the major constituents of giant planets. However, their physical/chemical stabilities are not truly understood at those extreme pressure-temperature conditions. In fact, an increasing amount of experimental evidences contradict the assumed stability of these materials at high pressures and temperatures.

Figure 1 illustrates the principle Hugoniots of simple molecules like O_2, CO, N_2, and CO_2 [4]. Clearly, all these materials exhibit the cusps on their Hugoniots at the pressure range between 20 and 40 GPa. At these pressures, these materials could heat up to several thousand degrees because of their high compressibilities. The calculated shock temperature of carbon dioxide, for instance, is about 4500 K at 40 GPa. The presence of such a distinctive cusp on the Hugoniot is surely an indication for chemical reaction or phase change. In fact, many previous statistical mechanical calculations have shown that these materials undergo strong chemical changes such as the decomposition of CO_2 and CO to the elementary products like carbon and oxygen and the dissociation of N_2 and O_2 to diatomic/monatomic ionic products. Note that the Hugoniots of unreacted CO and N_2 are nearly identical, attributing to their isoelectronic characteristics resulting in same initial density and similar nonbonded atom-atom potential. The previous diamond-anvil cell studies of these materials [5-7] also found very similar phase diagrams with many isostructural polymorphs. Later in the chapter, we shall also see a similar parallelism existing in the phase diagrams of isoelectronic triatomics CO_2 and N_2O.

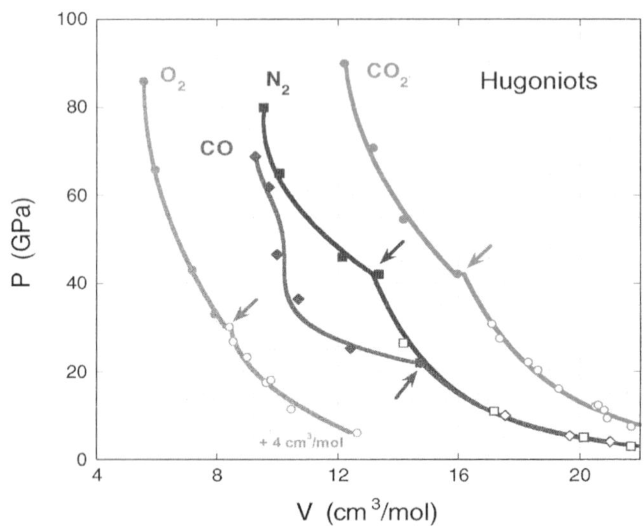

Figure 1. Hugoniots of selected simple molecular solids, reproduced from reference [4]. Each of these materials exhibits a cusp (indicated by arrow), a strong indication of chemical and/or physical change. Open and closed symbols, respectively, represent unreacted and reacted part of the Hugoniots.

There are numerous examples, also indicating the increase of chemical instability of unsaturated molecular bonds at high static pressures. The examples include many recent discoveries: covalently bonded nonmolecular phases of nitrogen [8, 9], carbon dioxide [10], cyanogen [11], carbon monoxide [12], charge transferred ionic solids of nitrous dioxide [13, 14], oxygen [15], hydrogen [16], metallic phases of oxygen [17, 18, 19], iodine [20, 21], xenon [22, 23], hydrogen bonded extended solids of symmetric ice [24, 25], hydrogen cyanide [26], dissociative products of methane [27, 28], and aromatic compounds [29]. These fundamental changes in chemical bonding of simple molecular solids may or may not occur reversibly upon the reversal of pressure and temperature, offering the opportunity to understand the materials metastability. These transformations and the associated changes in thermodynamic, mechanical, electronic and magnetic properties are also fundamental to understand the state of matters in the deep interiors of Earth and other planets, and the chemistry behind high energetic detonation and combustion.

Modern advances in theoretical and computational methodologies now make possible to explain or even predict novel structures and properties in a relatively wide range of length scales on the basis of thermodynamic stability. These theoretical calculations have been successful, not only to explain the details of materials discovered in experiments such as crystal structures, stabilities, properties, and transition dynamics [30-32], but also to predict new often highly unusual phases that might exist at the extreme conditions. To list a few recent predictions includes super ionic phases of H_2O and NH_3 [33], superconducting metallic H_2 [34], and nonmolecular H_2 fluid [35]. The development or realization of these predicted, potentially useful materials is, however, controlled by the stability of solids as well as the metastablility. In fact, many nanoparticles and surface structures are engineered based on knowledge of their metastability. This makes the experimental confirmation of material in a given stability (or metastability) field the priority in materials research. Furthermore, theoretical calculations using the first principles of physics and chemistry are computationally intensive (because of the intrinsic N^3 dependence) and rapidly become challenges as the system gets large and/or the transition takes long, even with the most powerful computational tool available today. In this regard, a close dialog of theory and experiments is the most powerful way to address fundamental issues in high-pressure materials research.

1.1. Fundamental principles of high pressure chemistry

The recent discoveries of nonmolecular phases of simple molecular solids [8-29] demonstrate the proof-of-the-principles for producing exotic phases by application of high pressure. More importantly, such a transition from a molecular solid to a denser covalently bonded framework structure indicates the fundamental principle of high-pressure chemistry. This occurs because electron kinetic energy has higher density dependence ($\rho^{2/3}$) than that of electrostatic potential energy ($\rho^{1/3}$). As a result, electrons localized within intramolecular bonds become increasingly less stable as density (or pressure) increases and the intermolecular potential becomes highly repulsive (Fig. 2). At high enough pressures, it will essentially lead to physical and chemical changes of molecular solids and modification of their chemical bonds to more delocalized states such as polymeric and metallic solids. This perhaps is the reason for which many unsaturated molecular bonds become unstable above 10 GPa and network structures are ubiquitous at high pressures as found in the crystal structures of diamond, *c*-BN, β-C_3N_4 [36], symmetric-H_2O [24, 25], and CO_2 [37].

Three mechanisms may occur at high pressures to delocalize electrons and soften repulsive potentials: (i) the pressure-induced ionization creating attractive electrostatic coulomb interaction, (ii) polymerization delocalizing intramolecular electrons between neighboring molecules, and (iii) metallization completely delocalizing electrons through conduction bands. These processes represent the collective properties of solids, strongly dependent on the intermolecular separation. Therefore, it is likely that these processes occur with increasing pressure as molecular phases $\rightarrow$ ionic species $\rightarrow$ polymeric phases $\rightarrow$ metallic phases, in a way to produce the configuration with more itinerant electrons.

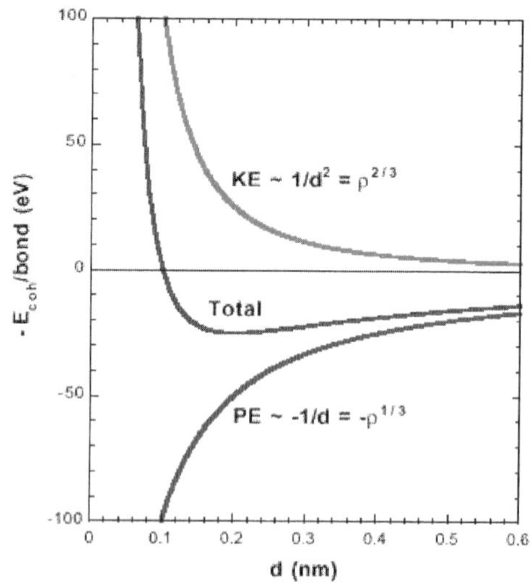

Figure 2. A conceptual representation of intermolecular energy change as a function of intermolecular distances. The nature of intermolecular potential becomes highly repulsive at a short distance or high density.

Because of large modification in chemical bonding associated with the molecular-to-nonmolecular phase transition, one might expect large activation energies in the reverse process and thus the nonmolecular product to be metastable even at the ambient condition. Furthermore, these types of extended molecular solids, particularly made of low-Z first and second row elements, are entirely a new class of novel materials that may exhibit interesting properties such as super-hardness [37], optical nonlinearity [10], superconductivity [34, 38-39], and high energy density [40], to name a few. Previous theoretical calculations [40], for example, have predicted that polymeric nitrogen may contain a dramatically enhanced energy density (Energy/Volume) equal to about three times that of HMX (one of the most powerful conventional high explosives available today). Metallic H_2 has been predicted to be a high Tc superconductor [34], as are many other low-Z molecular solids like B, Li and S [38, 39] found to be.

1.2. Generalized phase diagram of simple molecule

Figure 3 illustrates several chemical and physical changes of molecular solids occurring at high pressures and high temperatures. At high pressures of 100 GPa, electrons develop huge kinetic energy (Fig. 2) and, thereby, the core and valance electrons can strongly mix with valence electrons of its own or nearby molecules. Such a core swelling and/or a valence mixing create an excellent environment for simple molecules to chemically transform into nonmolecular phases such as polymeric and metallic solids. At high pressures of 100 GPa, the mechanical energy ($P\Delta V$) of the molecular system often exceeds an eV, comparable to those of most chemical bonds and certainly enough to induce chemistry acquiring bond scissions. The products are controlled by collective behaviors of molecules, leading to strongly associated phases probably in a pressure range of 10-50 GPa, multi-dimensional polymeric products at around 50 and 100 GPa, and eventually band-gap closing molecular and atomic metals typically above 100 GPa. At sufficiently high pressures of ~1 TPa, most solids will lose their periodic integrities [41] and the system with simple or no core electrons (*e.g.* H_2 and He) may even convert into a bare nuclei.

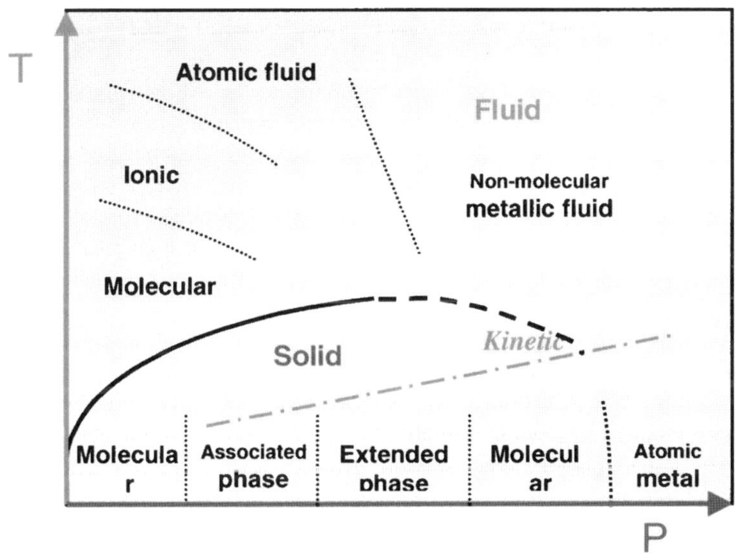

Figure 3. A conceptual generalized physical/chemical phase diagram of solids at high pressures and temperatures, illustrating the melting maximum and phase boundaries in both solid and fluids.

The materials at high temperatures, on the other hand, often transform into an open structure like bcc because of a large increase of entropy [41]. The melting transition is another example of electron delocalization in a simple electron-gas model [42]. In fact, at extremely high pressures where the matter is composed of bare nuclei, one can expect the melting to occur at zero K [43]. This would result in a melting maximum and a close loop of melting curve as illustrated in Fig. 3. Further increasing temperatures well above the melt will eventually ionize, dissociate or even decompose molecules into elemental atoms [3, 4, 44, 45]. Such a temperature-induced ionization would eventually produce a conducting state of matter if the pressure were sufficiently high [46, 47]. This means that the molecular-to-nonmolecular and/or insulator-to-metallic transitions would also form a close loop in the pressure-temperature phase diagram. These close loops of melting and molecular-to-nonmolecular phase lines should intersect at a triple point of intermediate high pressures and temperatures [35, 48]. Therefore, the combined effect of high pressure and high temperature will provide a way of probing a delicate balance between mechanical ($P\Delta V$) and thermal

($T\Delta S$) energies or between pressure-induced electron delocalization and temperature-induced electron ionization, reflected on stabilities of phases and the phase boundaries. These pressure-temperature induced changes are unique, establishing an entirely different set of periodic behaviors in crystal structure and electronic and magnetic properties not found in the conventional periodic table. In return, this is what makes the "*Mbar chemistry*" unique from any ambient-pressure combinatorial chemistry based on variation of chemical composition and temperature. New opportunities to discover interesting phenomena and exotic materials exist in both liquids and solids at high pressures.

2. EXPERIMENTAL TOOLS FOR HIGH PRESSURE RESEARCH

Studies of high-density molecular solids and fluids at the extreme pressure-temperature conditions where molecular solids transform into nonmolecular polymeric and metallic phases are very challenging, because of the difficulties associated with achieving such formidable high pressure-temperature conditions, the absence of *in-situ* structural probe for a minute sample inevitable in static high pressures, and the transient nature of species encountered in dynamic high pressure conditions. With recent developments of high pressure-temperature membrane diamond-anvil cells coupled with micro-probing diagnostic methods available at third-generation synchrotron x-ray sources [49] and modern laser systems [50], these challenges on one hand are rapidly becoming more attainable for static experiments. There are also rapid growing efforts of utilizing a large volume press in high-pressure materials research, made of WC anvils, sintered-diamond anvils, Møssanite anvils [51], Sapphire anvils, or CVD grown large volume anvils [52]. Gas gun, laser, and magnetic drivers, on the other hand, can also be used in high-pressure materials research to investigate the dynamic aspect of material behaviors at high pressures and temperatures. While these dynamic experiments are typically performed to exploit the materials on the Hugoniot states, the method can be modified to provide variable loading that can range from near isentropic all the way to the Hugoniot [53-55] and to utilize modern diagnostic developments capable of probing transient events such as a ps-time resolved x-ray diffraction and a sub-ps laser probes [56].

Shock and static high pressures are complimentary in many aspects including thermal conditions, kinetics, states of stress, rates of loading, etc., all of which could have different implications for materials applications. Because of these differences, the materials behave very differently under shock and static conditions. For example, the materials at shock compressions favor a martensitic transformation than a reconstructive one [57]. Shock-compressed liquid is often found at the *P, T-* conditions well above its melt curve, due to the kinetics associated with forming long-range ordered solids [58]. Large crystals can be grown in static conditions, whereas shock wave typically results in nanocrystals or amorphous materials. Shock-induced reactions are often dissociative, whereas the static reactions are typically associative [59]. The shear-band interaction is a typical mechanism for the reactions in shock-compressed solids, whereas such an interaction is absent in static conditions [60]. Clearly, complementary information from shock- and static- high pressures experiments is critical to gain insight of materials transformation at high pressures and temperatures.

3. EXAMPLES OF TRIATOMIC MOLECULAR SOLIDS

There are numerous theoretical and experimental results demonstrating that simple molecular solids transform into nonmolecular phases at high pressures and temperatures, ranging from monatomic molecular solids such as sulfur [61], phosphorous [62] and carbon [63] to diatomic molecular solids such as nitrogen [8, 9, 40], carbon monoxide [12] and iodine [20, 21], to triatomic molecules such as ice [24, 25], carbon dioxide [10, 31, 37] and carbon disulfide [64, 65] to polyatomics such as methane [27, 28] and cyanogen [11], and aromatic compounds [29]. In this section, we will limit our discussion within a few molecular triatomics: first to review the transformations in two isoelectronic linear triatomics, carbon dioxide and nitrous dioxide, and then to discuss their periodic analogies to carbon disulfide and silicone dioxide.

3.1. Carbon dioxide: CO_2

Carbon dioxide is a good example of material with a richness of high-pressure polymorphs and a great diversity in intermolecular interactions, chemical bonding and crystal structures. The phase diagram of carbon dioxide (Fig. 4) summarizes the physical and chemical changes and their crystal structures (Fig. 5) at high pressures and temperatures. Early high-pressures studies [60, 66-68, 70] established the existence of two molecular solid phases: a cubic (*Pa3*) phase I, and an orthorhombic (*Cmca*) phase III, both stabilized by quadruple interactions between the linear molecules [71]. Recent diamond-anvil cell studies [37, 72-74] have discovered three additional phases whose chemical bondings and crystal structures are very different from those of molecular solids. New phases discovered include tetrahedral bonded polymeric phase V ($P2_12_12_1$) like SiO_2-tridymite, bent phase IV ($P4_12_12$ or *Pbcn*) like SiO_2-cristobalite or a post-stishovite α- PbO_2, and strongly associated pseudo-six-folded phase II ($P4_2/mnm$ or *Pnnm*) like SiO_2-stishovite (or its orthorhombic distortion to a $CaCl_2$-like structure). The evidence of the sixth phase VI [74] has also been reported but its crystal structure and stability field is not well understood. It is also known that carbon dioxide molecules undergo strong chemical changes under shock compression evident from a cusp in shock Hugoniot near 40 GPa and 4500 K (see Fig. 1). Though no chemical change was observed in pure carbon dioxide at high temperatures (at least up to 3000 K) below 30 GPa, an interesting ionic form of carbon dioxide dimer, $CO^{2+}CO_3^{2-}$, was produced by laser heating carbon particles in oxygen to above 2000K at around 10 GPa [75].

3.1.1. Molecular phase I and III

Carbon dioxide molecule is the simplest form of linear molecular triatomics abundant in nature. At ambient temperatures, it crystallizes into cubic (*Pa-3*) phase I, known as "dry ice", at around 1.5 GPa and then to orthorhombic phase III (*Cmca*) above 12 GPa (see Figs. 4 and 5). Both of these structures commonly appear in many other molecular solids [76, 77], for which stabilities have been well understood in terms of the intermolecular quadrupole-quadrupole interaction. In these phases at relatively low pressures below 15 GPa, the nearest intermolecular separation is in a range of 3.0 to 2.5 Å, typically 2 - 2.5 times of the intramolecular C=O bond distance ranging 1.35 – 1.30 Å (all depending on pressure). These values are typical for molecular solids [1].

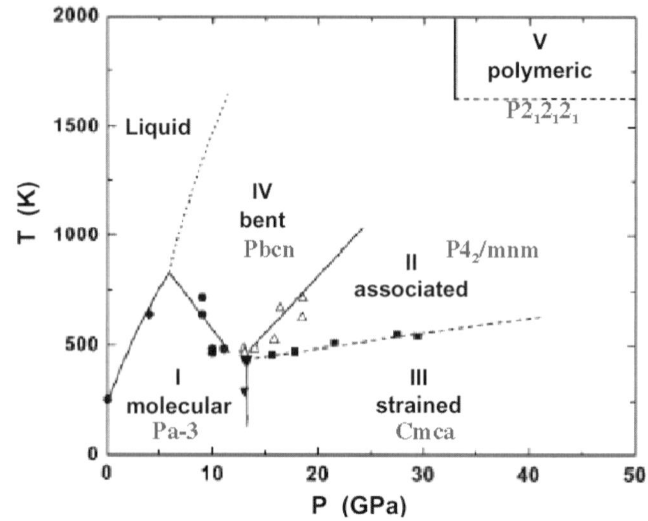

Figure 4. Phase diagram of carbon dioxide with five polymorphs with 50 GPa and 2000 K. All high temperature phases, II, IV and V, can be stabilized at the ambient temperature over an entire stability range of phase III. This may suggest that phase III is metastability, frozen in only through compression of phase I, and result in four phase boundaries of I through IV being accidentally degenerated at a single thermodynamic point. This phase diagram indicates that pure molecular solid like I is stable only within a limited range of pressure and temperature (less than 10-20 GPa and 500 K) and transforms into non-molecular extended phase V through intermediate phases like II, IV and to some extent highly strained phase III at high pressures.

Figure 5 (next page). Crystal structure of carbon dioxide polymorphs: (a) a cubic (Pa-3) phase I with four molecules per unit cell. In this structure, carbon atoms at the face centered positions and the molecular axis aligned to the great diagonal direction. (b) an orthorhombic (Cmca) phase III with four molecules per unit cell, a layer structure with all carbons at the face centered positions and all molecules are on the ab-plane. (c) a tetragonal ($P4_2/mnm$) structure with pseudo-six folded carbon atoms with two elongated intramolecular bonded oxygens and four collapsed intermolecular bonded oxygen atoms in the four nearest neighbor molecules. Because of a short oxygen-oxygen contact distance, this phase exhibits an orthorhombic distortion (Pnnm) and dynamic disorder. (d) an orthorhombic (Pbcn) structure with four molecules per unit cell with bent molecular configurations. This phase also shows elongated intramolecular bonds and collapsed intermolecular bonds. (5) an orthorhombic ($P2_12_12_1$) structure with eight molecules per unit cell. In this structure, all carbon atoms are four fold coordinated with carbon-oxygen single bonds.

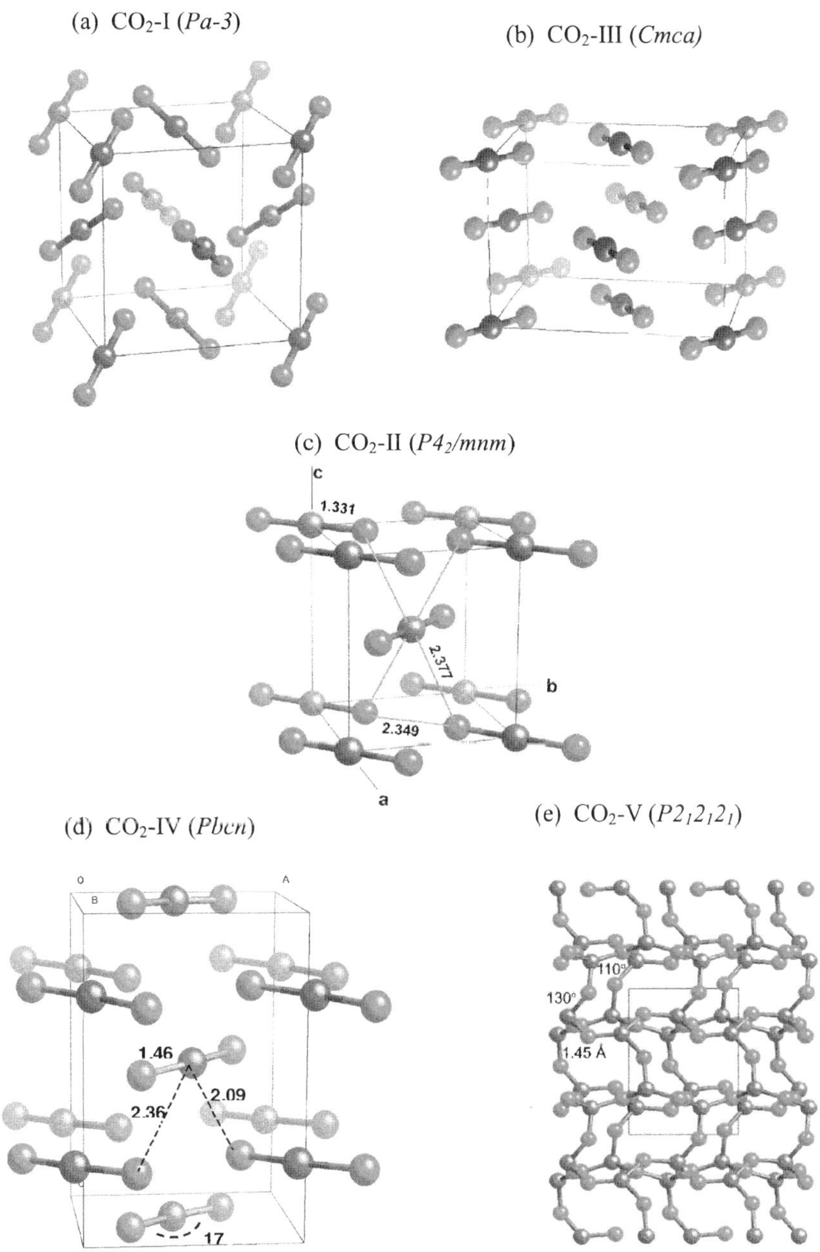
(a) CO2-I (Pa-3)
(b) CO2-III (Cmca)
(c) CO2-II (P42/mnm)
c
1.331
2.377
b
2.349
a
(d) CO2-IV (Pbcn)
0
A
B
1.46
2.36
2.09
17
(e) CO2-V (P212121)
110°
130°
1.45 Å

The crystal structures of these two molecular phases are similar. All carbon atoms are at the face centered positions. Carbon dioxide molecules in phase I are aligned along the great diagonal direction, whereas those in phase III are aligned approximately along the face diagonal within the ab-plane. As a result, the I→III phase transition is associated with only a minor change in molecular rearrangement; that is, a slight tilt of CO_2 molecules from the great diagonal to the face diagonal without any apparent discrete change in their specific volumes [9]. This martensitic nature makes the I→III phase transition sluggish at ambient temperature, and both phases coexist over an extended pressure range between 12 and 22 GPa. The extended metastability of cubic CO_2-I to 22 GPa also reflects its small energy difference from that of CO_2-III in this pressure range, and a presence of small lattice strain would prolong the stability of CO_2-I well above its stability field as was observed.

There is, however, a subtle but important difference between the two phases. Note that the molecular axis of carbon dioxide is slightly tilted from the exact diagonal direction at 51.7 degree. As a result, oxygen atom in phase III faces approximately the center of C=O bonds, not the carbon atoms of nearest neighbor molecules. Therefore, one may consider the *Cmca* phase as a "paired" layer structure. Such a pairing of molecules in the *Cmca* structure has an important consequence at high pressures (above 20 GPa), converting this phase III to a non-typical molecular solid. It develops high strains in the lattice, evident from its characteristic texture and the ability to support a large pressure gradient (~100 GPa/mm at 30 GPa). It also has unusually high bulk modulus of 80 GPa [78] (comparable to that of Si - 87 GPa [79]). Therefore, it is a possibility that molecular phase III is not stable in this pressure range, but the kinetic barrier may preclude any further transformation at the ambient temperature. In fact, this conjecture is supported by its transformation at high temperatures to nonmolecular phase V above 40 GPa and to intermediate phases II and IV above 20 GPa. Further convincing is the fact that all of these high-temperature phases II, IV and V can be quenched in an entire stability field of CO_2-III.

3.1.2. Nonmolecular extended phase V

Laser heating the phase III transforms into an extended nonmolecular solid, phase V, above 40 GPa and 1800 K [10]. The vibration spectrum of this phase shows a strong C-O-C stretch mode at around 800 cm^{-1} at 40 GPa, clearly indicating that it is an extended covalent solid made of carbon-oxygen single bonds. Though it occurs above 1800 K, the transition appears to have no strong dependence on temperature. Thus, it is likely that the experimentally observed phase boundary be a kinetic barrier. In fact, the first principles calculation at 0K suggests that such a molecular-to-nonmolecular phase transition would take place above 40 GPa. The phase V can be quenched at the ambient temperature as long as the pressure retains above 10 GPa. Below 10 GPa, it depolymerizes into the phase I, although the remnant of polymeric phase V can be seen at substantially lower pressures down to 1 GPa where CO_2 liquidifies or sublimes.

Determining the crystal structure of phase V has been challenging for several reasons, including (i) its coexistence with other phases due to an incomplete transformation of phase III and/or the metastability of other high temperature phase IV and II, (ii) the presence of large lattice distortion and (iii) highly preferred orientation. Nevertheless, the x-ray data indicate that the crystal structure is similar to that of trydimite ($P2_12_12_1$) [37]. In this structure of CO_2-V, each carbon atom is tetrahedrally bound to four oxygen atoms. These CO_4 tetrahedral units share their corner oxygens to form six-fold distorted holohedral rings with alternating tetrahedral apices pointing up and down the ab-plane. The apices of tetrahedra are then connected through oxygen atoms along the c-axis. This interconnected layer structure of tetrahedra results in the C-O-C angle 130 (±10°), which is substantially smaller than those of SiO_2-tridymites, 174°-180° [80].

It is well known that in SiO_2 there is very little energy difference for various polymorphs of tridymite. In addition, there often exists a substantial distortion in the oxygen sublattice of SiO_2-tridymite. In fact, recent theoretical calculations have shown that there is a little difference among those candidate structures of CO_2-V, including α, β-quartz, *m*-chacopalite, trydimite, coesite, etc. However, contrary to a wide range of Si-O-Si bond angles in SiO_2 from near 180° in tridymite to 145° in quartz [81], all C-O-C bond angles in CO_2-V were estimated to be about 130 degrees. Such rigidity in the C-O-C bond angle results in a relatively large distortion in the six-fold holohedra along the ab-plane of CO_2-V. It in turn reflects the fact that oxygen atoms in CO_2-V are more tightly bound than in SiO_2 and results in a higher covalence and bulk modulus for CO_2-V than for any SiO_2 polymorphs.

The synthesis of "polymorphic carbon dioxide" resembling SiO_2 glass has long been a challenge in chemistry for many reasons such as high strength, high thermal conductivity, wide band gap, high chemical stability, etc. The high-pressure synthesis of polymeric phase V clearly demonstrates the very existence of CO_2 polymer and, more importantly, reveals several interesting properties. It is an optically nonlinear solid, converting infrared light into green light with a high conversion efficient unparallel to any of conventional nonlinear crystals [10]. It also has an extremely low compressibility, nearly the same as *c*-BN (Table I), and it is thus likely to be super hard [37]. The recovery of this phase V at the ambient condition, however, remains to be a challenge to date.

Table I. The stiffness of carbon dioxide phases in comparison with other covalent materials, showing extremely low compressibilities of nonmolecular carbon dioxide phases.

Soilds	ρ (g/cm³)	B_o (GPa)
Diamond	3.50	450
cubic-BN	3.48	369
CO_2-V	2.90	362
Al_2O_3	4.00	239
hcp-Fe	8.29	165
CO_2-II	2.45	130
Silicon	2.33	98
CO_2-III	2.22	87
CO_2-I	1.75	6
δ-N_2	1.03	3

3.1.3. Intermediate phases II and IV

At 19 GPa, CO_2-III transforms to a new phase, CO_2-II, above 500 K and then to CO_2-IV above 650 K [82]. These transformations are apparent from distinct changes in both visual appearance and Raman spectrum as represented in Fig. 6. The Raman spectrum of quenched CO_2-IV exhibits a triplet bending mode ν_2 (O=C=O) near 650 cm^{-1}, suggesting a broken inversion symmetry because of molecular bending in this phase.

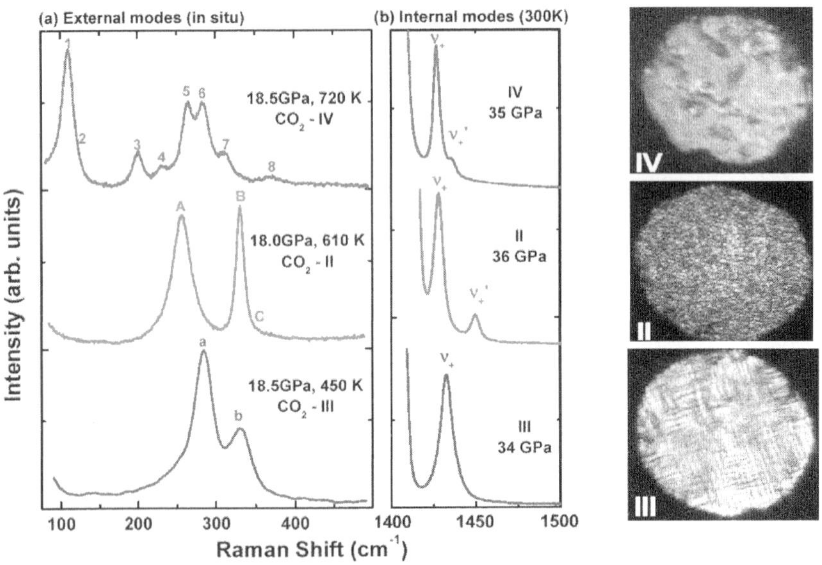

Figure 6. Characteristic visual appearances and Raman spectra of carbon dioxide phases at high pressures and temperatures. The microphotographs were taken at 18.5 GPa as temperature increases to 450, 610 and 720 K for each phase. Note that the large separation of the ν_1 mode of phase II indicates a strong association of CO_2 molecules.

CO_2-II crystallizes into a stishovite-like structure ($P4_2/mnm$), where carbon atoms are pseudo-six fold coordinated with oxygen atoms: two bonded oxygen atoms at the elongated C=O distance ~1.33 Å and four nonbonded oxygen atoms of nearest molecules at about 2.34 Å. Note that the intermolecular distance is even less than twice the intramolecular distance. Based on the elongated intramolecular bond distance and the collapsed intermolecular distance, the phase II should be considered as an intermediate phase between molecular and nonmolecular solids [73]. Strong molecular association of carbon dioxide molecules in this highly distorted octahedral structure in turn results in a high bulk modulus near B_o = 130 GPa (Table II) and a large separation of symmetric ν_1 vibration (see two bands at around 1450 cm^{-1} of phase II at 36 GPa in Fig. 6). Furthermore, this is a layer structure with an extremely short oxygen-oxygen contact distance, 2.35 Å, in the ab-plane, resulting in a tetragonal-to-orthorhombic (*Pnnm*, *CaCl₂*-like) distortion and the dynamic disorder evident in both Raman and x-ray data.

The crystal structure of CO_2-IV can be interpreted in terms of two plausible models: the $P4_12_12$ (α-SiO_2 cristobalite) and the *Pbcn* (α-PbO_2, post-stishovite). Carbon dioxide molecules are bent slightly more in the *Pbcn* phase (<C-O-C = 160 degree) than in the $P4_12_12$ (171 degree). In fact, the Raman spectrum of phase IV [83] clearly includes the ν_2 bending modes, typically forbidden in a linear molecule. Thus, it too suggests the bending of CO_2 molecules. Because of dipole interactions, the intermolecular interactions are expected to be even stronger in this phase than that of phase II. For both models, the intramolecular C-O bond is increased to 1.5 (±0.1) Å at the distance of C-O single bond, whereas the intermolecular C..O distance is reduced to around 2.1 (±0.2) Å. The bending and elongation of the molecular units suggest that phase IV is also an intermediate state between the molecular and non-molecular extended phases. Therefore, it is apparent that the molecular-to-nonmolecular transformation in carbon dioxide occurs gradually via intermediate phases IV and II.

Both phases II and IV can be quenched at ambient temperatures over the entire stability field of phase III. This result indicates that the phase III is likely a metastable phase. Interesting changes have also observed from the quenched phase IV in particular. The 650 cm^{-1} bending modes rapidly soften with increasing pressure, indicating an enhanced intermolecular interaction among neighboring bent CO_2 molecules. At 80 GPa, the low-frequency vibron collapses into high-frequency phonons, and CO_2-IV becomes an extended amorphous solid, another form of nonmolecular CO_2. Upon laser heating the quenched phase IV, CO_2-IV transforms into polymeric CO_2-V above 30 GPa. Note that the IV→V transition pressure of 30 GPa is substantially lower than that of the III→V transition of 40 GPa. It is probably due to the bent configuration of CO_2-IV, which lowers the activation barrier of the polymerization.

3.1.4. Ionic solids

At relatively low pressures below 10 GPa carbon dioxide remains purely molecular. Carbon-oxygen double bonds are highly stable and no transformation has been observed to 3000 K in phase I. On the other hand, there has been experimental evidence for which the direct elementary reaction of carbon and oxygen at about 2000 K and 9 GPa yields a nearly transparent ionic product of carbon dioxide dimer [75]. The fact that the ionic carbon dioxide dimer does not form directly from molecular carbon dioxide implies an existence of a large activation barrier for the dimerization pathway. However, once formed at high P and T, the dimer can be quenched to ambient temperature at high pressures.

The Raman spectrum of quenched products (Fig. 7) consists of the symmetric stretching of excess β-O_2 at 1585 cm^{-1}, two Fermi-resonance bands of CO_2 at 1270 and 1400 cm^{-1}, and three new additional sharp bands at 734, 1079, and 2242 cm^{-1}. The systematic of the latter three bands are very similar to those of nitrosonium nitrate $NO^+NO_3^-$, an ionic dimer of nitrogen dioxide. This similarity suggests that the products also include a species with carbonates and carbosonium, $CO^{2+}CO_3^{2-}$. The vibrations of carbonate ions appear at 713 and 1082 cm^{-1} in $CaCO_3$ [84], and the CO vibration appears about 2150 cm^{-1} at 5 GPa [85]. Electronic structure calculations for CO^{2+} [86, 87] suggest that there are several low lying states of CO^{2+}, whose vibrational frequencies vary between 1000 and 2000 cm^{-1}. The yield of ionic dimer in this sample is small; other samples show nearly complete conversion (*vida infra*, the 9 GPa trace in Fig. 6).

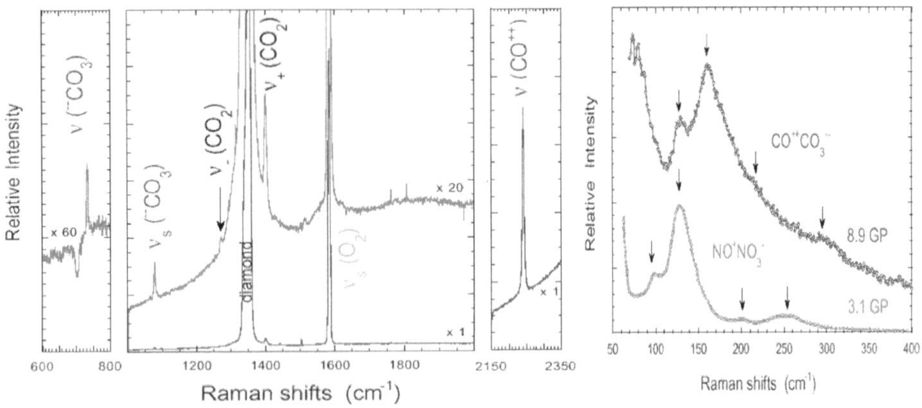

Figure 7. Raman spectra of $CO^{2+}CO_3^{2-}$ quenched from the elementary reactions of C and O_2 at 9 GPa and 2000 K, in (a) internal and (b) external vibrational regions.

Figure 7 also compares the lattice phonons of the C-O_2 products with those of $NO^+NO_3^-$ and shows striking similarities in the number of bands, band shapes, widths and intensities. Differences in the peak positions can be attributed to different pressures, force constants, and reduced masses. Similarities of vibrons and lattice phonons between $NO^+NO_3^-$ and the C-O_2 reaction products imply that they have similar molecular configurations, $CO^{2+}CO_3^{2-}$, and similar crystal structures. The crystal structure of $CO^{2+}CO_3^{2-}$ has not been determined as yet; that of $NO^+NO_3^-$ has been determined to be aragonite-like structure (see below).

3.1.5. Dissociative solids

Shock-compressed carbon dioxide exhibits a strong slope change in the Hugoniot (recall Fig. 1), a clear indication of chemical reaction, at around 40 GPa and estimated temperature of 4500 K [1]. The previous theoretical calculation has confirmed that it is indeed due to chemical dissociation of carbon dioxide to elementary products such as diamond and oxygen. In recent diamond-anvil cell experiments [74], the similar dissociative products, lonstaleite diamond and oxygen; have also been observed from the quenched products after laser-heating of CO_2 samples at 67 GPa. The transition temperatures were estimated to be about 2500 K at 35 GPa, substantially lower than the estimated shock transition temperature 4500 K.

3.2. Nitrous oxide: an electronic analog

CO_2 and N_2O are isoelectronic, linear triatomics with similar molecular weights, melting temperatures and quadrupole moments. Although N_2O has no inversion symmetry, it has been shown to resonate between two bonding configurations with opposing dipole moments: $^-N{=}N^+{=}O$ and $N{=}N^+{=}O^-$ [88]. As a result, the net dipole moment of nitrous oxide is negligible compared to its substantial quadrupole moment [89, 90] at relatively low pressures. Therefore, one can find a close parallelism between the phase diagrams of N_2O in Fig. 8 [91] and CO_2 discussed above (Fig. 4). However, note that such a close phase parallelism is maintained mainly in a molecular regime. Upon breaking or weakening of N=O (C=O) bonds at high pressures and temperatures, the different nature of carbon and nitrogen enhances the ionic character in N_2O phases and eventually leads to ionization of N_2O, instead of the polymerization as seen in CO_2.

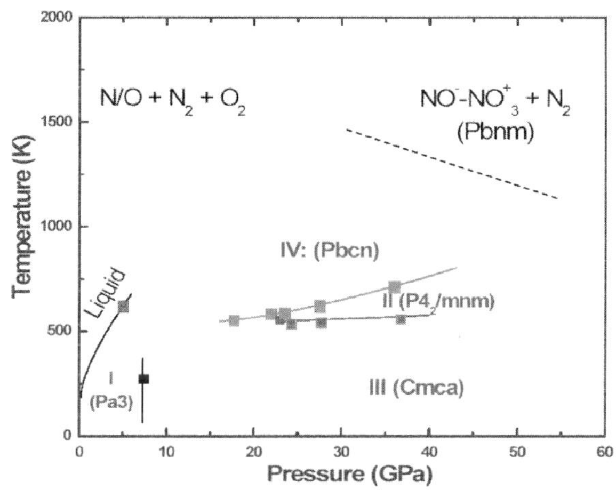

Figure 8. Phase diagram of nitrous oxide.

3.2.1. *The phase diagram of N_2O*

In the absence of dipole ordering, N_2O molecules are oriented randomly and crystallize in the same configurations as CO_2 [92] as phase I (*Pa3*) at 1 GPa and phase III (*Cmca*) above ~5 GPa. At high temperatures, there exist two additional N_2O phases (labeled II and IV in Fig. 8). Above ~600 K, phase II stabilizes above 23 GPa in a relatively narrow temperature range 10-30 K. Phase IV, on the other hand, is obtained by heating either phase III below 23 GPa, or phase II at higher pressures. These transformations can be readily observed by abrupt changes in the visual appearance of the sample and in the Raman spectrum, similarly to the case of CO_2 phases.

While the one-to-one phase analogy between CO_2 and N_2O is maintained at relatively low pressures and temperatures, these materials develop significant differences with increasing pressure and temperature. For example, at ambient temperatures, N_2O-III remains stable to at least 135 GPa, whereas CO_2-III becomes highly unusual above 20 GPa and becomes unstable above 40 GPa with respect to its polymeric phase V. The crystal structure of high-temperature phase IV also exhibits a subtle but important difference between N_2O and CO_2. That is, the center nitrogen atoms in N_2O-IV occupy the face-centered-cubic (*fcc*) sites; whereas the carbon atoms in CO_2-IV deviate from the *fcc* packing and form zigzag chains. This difference results in a perfect layer structure and a relatively large bending angle 132° in N_2O, but in a large buckling of CO_2 layers and a substantially smaller bending angle 170° in CO_2. This divergence is due to the difference in ionicity (or covalency) between N_2O and CO_2. Such a difference in ionicity, in turn, leads to further significant divergence in their chemistry at high pressures and temperatures. That is, N_2O disproportionate into ionic $NO^+NO_3^-$ and N_2 [13, 14] whereas CO_2 polymerizes into an extended covalent solid above 35 GPa and 2000 K [10, 72].

3.2.2. Ionization and dissociation

Laser-heating of N_2O-III at lower pressures, 10-20 GPa, to about 2000 K (Fig. 9) produces a more complicated set of products, consisting of δ-N_2 (the doublet near 2365 cm^{-1}), β-O_2 (doublet at 1650 cm^{-1}), $NO^+NO_3^-$ (730, 835, 1095, and 2250 cm^{-1}), nitrogen-oxygen products (features between 750 cm^{-1} and 1070 cm^{-1}). Subsequent heating of the ionic product also results in a similarly complex set of dissociation products. Therefore, it is apparent that the ionic phase $NO^+NO_3^-$ further dissociates into N_2 and O_2. In fact, the ionization is always accompanied by dissociation when N_2O-III is laser-heated at pressures below 30 GPa, whereas no evidence for further dissociation of $NO^+NO_3^-$ was observed to 3370 K at higher pressures. These results suggest that the dissociation temperature of the ionic dimer increases with increasing pressure. Note also in Fig. 9c the splitting of the oxygen stretching mode, indicating that the oxygen is dissolved in δ-N_2 (Pm3n) [93]. This splitting is due to a vibration-vibration resonance transfer between molecules at two types of molecular sites of δ-N_2 [94]. The relative intensities of the components of the doublet vary with pressure and composition [21], on which basis we estimate the oxygen content of the δ-N_2 in Fig. 9c to be about 10 %.

Both the ionic and dissociative reaction products (Fig. 9b and 9c, respectively) are quenchable at room temperature at all pressures studied (between 10 and 55 GPa). The ionic phase ($NO^+NO_3^-$) is stable in a wide pressure region to 55 GPa, the maximum pressure applied. No reverse transition of $NO^+NO_3^-$ to the molecular phases of N_2O_4, N_2O, or NO_2 was observed even below the N_2O-I/III transition pressure 4-5 GPa. This is contrarily to the high pressure-temperature phases of CO_2 (phases II, III, IV, and V and its ionic dimer phase [95]), all of which transform back to the phase I (*Pa3*) near 11 GPa. Note that the splitting of the ν_4 mode above 35 GPa probably resulted from the anisotropic strain developed in the lattice.

3.2.3. Novel ionic crystal

A rigorous determination of the crystal structure of $NO^+NO_3^-$ has not been made to date, because of several experimental challenges such as the coexistence of by-products like δ-N_2 (*Pm3n*) and N_2O-III (Cmca) and their highly preferred orientation. Nevertheless, based on the Le Bail fit, the x-ray diffraction data of $NO^+NO_3^-$ can be explained in terms of an orthorhombic cell with a plausible space group of either *Pnma* or *Pn2₁a*. Note that the Pnma structure is analogous to the aragonite, $CaCO_3$, as occurred in other nitrates such as KNO_3 or NH_3NO_3 [96].

$NO^+NO_3^-$ is an extremely high density ionic solid with the density ~2.7 g/cc at the ambient condition. Figure 10 compares room-temperature isotherms for N_2O-III and $NO^+NO_3^-$ to 55 GPa with those of CO_2-III and CO_2-V. At pressures below 15 GPa, N_2O-III is relatively soft initially (B_o= 10.9 GPa), as is typical of molecular solids, e.g., N_2O-I (B_o=7.9) and CO_2-I (B_o=6.2). At higher pressures, N_2O-III rapidly stiffens and its compression curve becomes essentially identical to that of CO_2-III. The ionic phase of $NO^+NO_3^-$, however, behaves quite differently from the non-molecular phase CO_2-V. $NO^+NO_3^-$ is substantially softer (B_o=45.0 GPa) than polymeric CO_2-V (B_o=362 GPa) [37]. As a result, $NO^+NO_3^-$ becomes denser than CO_2-V above 12 GPa despite its lower density at the ambient pressure. The higher density of $NO^+NO_3^-$ than CO_2-V at high pressures probably reflects a more efficient packing of the ion pairs in $NO^+NO_3^-$. It is probably due to relatively strong attractive columbic interaction of the ion pairs, in contrast to very stiff covalent bonds of CO_2-V with its low-coordination structure. This result is also consistent with the higher number of nearest neighbors in $NO^+NO_3^-$ than CO_2-V; for example, each nitrosonium (NO^+) ion has six nearest nitrate ions, whereas each carbon atoms in CO_2-V has only four nearest oxygen atoms.

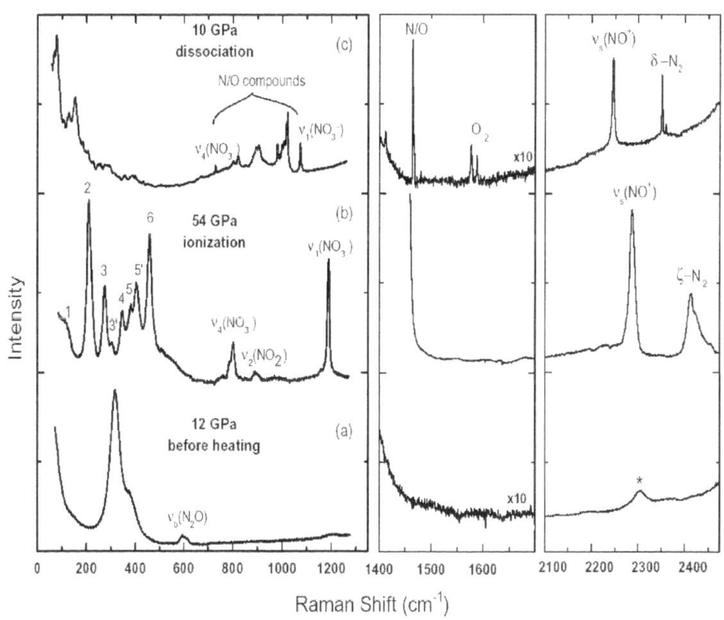

Figure 9. Raman spectra of N_2O-IIIL (a) before and (b, c) after laser-heating, showing the pressure-induced reactions to (b) the ionization products of $NO^+NO_3^-$ and η-N_2 at 54 GPa and (c) the dissociation products δ-N_2 containing dissolved O_2, N/O-compounds, and $NO^+NO_3^-$ at 10 GPa.

The inset of Fig. 10 compares the molar volumes of N_2O-III with the ionic and dissociative products. Above 5 GPa, the mixture of $NO^+NO_3^-$ and N_2 [97] has smallest molar volume and thus is favored over both N_2O-III and the dissociative mixture of N_2 and O_2 [98]. On the other hand, the molar volume of the N_2 and O_2 mixture becomes smaller than that of β-N_2O above 56 GPa and, based on the extrapolation, of the mixture of $NO^+NO_3^-$ above 130 GPa. This result thus suggests that the ionization is primarily driven by densification at high pressures, whereas the dissociation observed between 10 and 30 GPa results from the combined effects of densification and entropy increase at high pressures and temperatures. This conclusion is also consistent with the presumption that the dissociation would require higher temperature than the ionization.

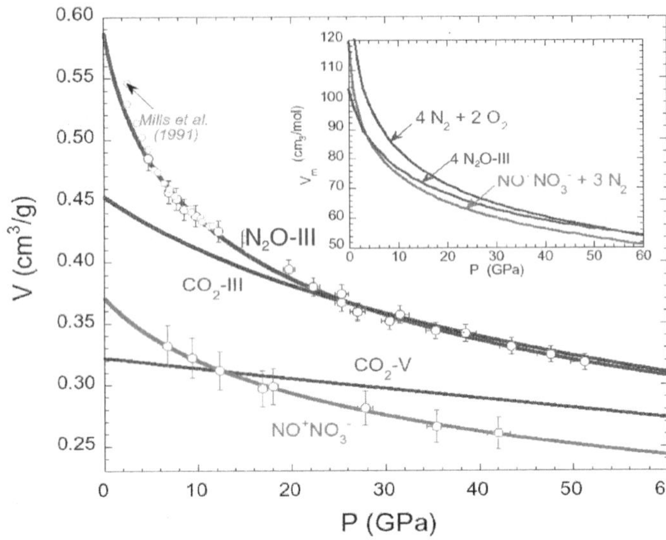

Figure 10. Pressure-volume plots of $NO^+NO_3^-$ in comparison with N_2O and CO_2 phases. (Inset) Molar volumes of N_2O in comparison with ionic and dissociated products.

3.3. Carbon disulfide: a periodic analog

Carbon disulfide is also another example of CO_2 analog, a centro-symmetric linear triatomic molecule with a similar valence electronic structure. The phase diagram of carbon disulfide is shown in Fig. 11. At room temperature, CS_2 molecules crystallizes into an orthorhombic (*Cmca*) structure at ~0.5 GPa [99]. While this structure is identical to those of CO_2-III and N_2O-III, it is interesting to note that the cubic *Pa-3* structure seen in CO_2-I and N_2O-I is absent in the phase diagram of CS_2. Nevertheless, it is still consistent with the periodic structural variation with pressure; for example, the absence of the graphite structure in silicon and the second-row CO_2 transforming to the structures of the third-row compound SiO_2 at high pressures. Furthermore, as in the cases of CO_2-III, the CS_2 molecules in the *Cmca* phase behave cooperatively and lead to strong chemical reactions at high pressures and high temperatures. In fact, there are many other examples showing strong collective behaviors in the *Cmca*, including X_2 [100], H_2-III [101], Li_2 [102], etc.

Bridgman was the first to report the chemical transformation of carbon disulfide to a black polymer under static high pressure-temperature condition (~5.5 GPa and 450 K) [103]. However, Agnew and coworkers [104] later found that the chemistry of CS_2 is actually substantially more complicated at high pressures and temperatures as illustrated in Fig. 11. Several reaction zones were identified, all of which contain the mixtures of multimer products of carbon dioxide. Note that at the ambient temperature carbon disulfide transforms into a dimeric product above 9 GPa, signifying the dimeric pairing of the *Cmca* structure.

The chemical reactions of CS2 have also been studied under various (single [105], double [106], and multiple [107]) shock conditions, which follow different thermal paths as illustrated in Fig. 11. For example, CS_2 molecules under single shock conditions decompose to carbon and sulfur [105], whereas they behave collectively under multiple shock conditions [108, 109]. These studies found that the primary effect of pressure and temperature on CS_2 is the π-electron delocalization. Such a π-electron delocalization induces the molecular bending, evident from the appearance of "T-band" in absorption [110], which could be a precursor to the chemical reaction. Based on the cooperative behavior of CS_2 and the absorption spectral changes, the reaction was suggested to be an associative one to CS_2 multimer and/or Bridgman black polymer, similar to those observed under the static conditions. The collective behavior and polymerization of CS_2 molecules are in a way analogous to those of CO_2.

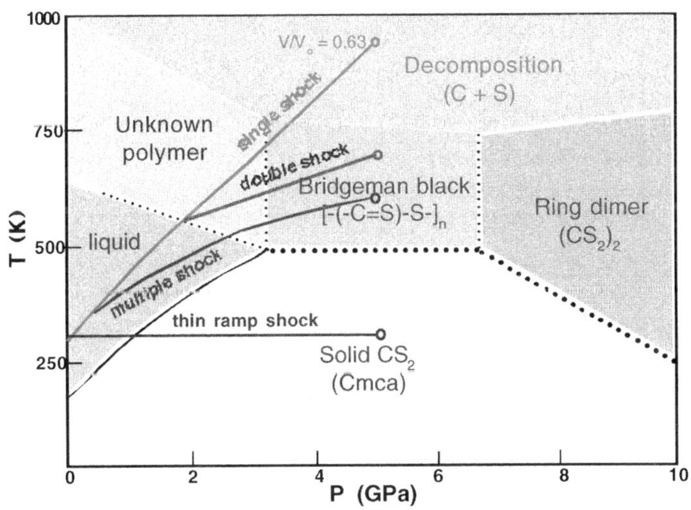

Figure 11. Phase diagram of carbon disulfide, showing several reaction zones at high pressures and temperatures. The pressure-temperature conditions of various shock wave experiments are also reproduced to highlight the similarity observed in reaction products between shock and static high-pressure experiments.

3.4. Silicon dioxide: a periodic analog

Despite a huge difference in the electronic structure, the crystal structures of CO_2 phases exhibit a great degree of similarity with those of many SiO_2 polymorphs as shown in Fig. 12. The examples include stishovite-like CO_2-II ($P4_2/mnm$) [73], β-cristobalite-like CO_2-IV ($P4_12_12$ *or Pbcn*) [72], tridymite-like CO_2-V ($P2_12_12_1$) [37], and even α-cristobalite-like CO_2-I (*Pa-3*). While the structures of CO_2 phases are (or close to) isostructural to those of SiO_2 polymorphs, the nature of chemical bonding is clearly different between CO_2 and SiO_2

polymorphs. With increasing pressure and temperature, the intermolecular bonding in CO_2 phases, for example, increases from nearly non-bonding in the phase I to approximately a half-bonding in phases II and IV and to a full covalent bonding in phase V.

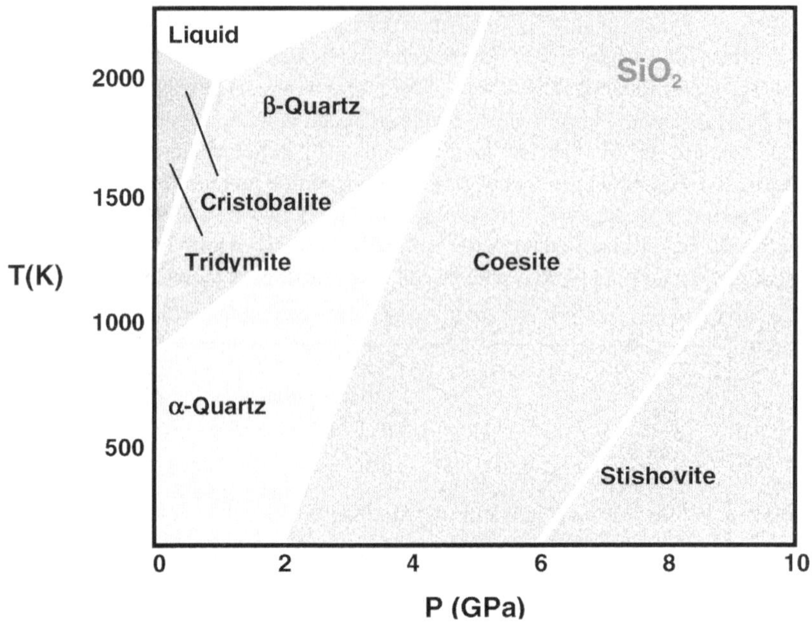

Figure 12. Phase diagram of SiO_2.

The crystal structure and transformation of CO_2-II also exhibits subtle difference from stishovite: (1) because of high covalency in C-O bonds [111-113], the O-C-O and O-C-O angles are more rigid and favor 110-130 degrees, which contrasts with a wider range of angles, 90 to 180 degrees, observed in various SiO_2 polymorphs [114, 115]. (2) There are no nearby *d*-bands in carbon, which makes it difficult to stabilize nonbonding electrons of oxygen atoms at pressures below 100 GPa. As a result, the transition of CO_2-II to a "perfect" six-folded extended phase is limited at these intermediate pressures. Instead, the lattice develops various distortions like the tetragonal-to-orthorhombic distortion and bending of linear molecules, which precede a transition to a four-fold carbon dioxide-V phase. (3) Finally, CO_2-II appears at lower pressures than four-fold CO_2-V, whereas six-fold stishovite appears at substantially higher pressures than four-fold quartz and coesite. Clearly, it reflects the intermediate nature of CO_2-II between molecular and extended solids.

4. CONCLUDING REMARKS

While the pressure-induced electron delocalization explains qualitatively the molecular-to-nonmolecular phase transition, the detailed mechanisms are substantially more complex because of the existence of intermediate phases, metastability, kinetics, and lattice strain. For all triatomics molecules discussed above, the molecular phases are only stable within a limited range of pressures (<10 GPa) and temperatures (<500 K) and that, at higher pressures and temperatures, they transform into molecular configurations with more itinerant electrons. For example, molecular CO_2 phase I and III first transform to pseudo-six fold coordinated phase II (*$P4_2/mnm$* or *Pnnm*) and bent phase IV (*Pbcn* or *$P4_12_12$*), both with elongated molecular bonds and compressed intermolecular distances, and finally to a fully extended phase V (*$P2_12_12_1$*), made of four-fold coordinated carbon atoms. These experimental results suggest that the electron delocalization occurs gradually, *via* intermediate phases (II, IV and III to an extent) to a fully extended covalent solid (V). The formation of intermediary phases lowers the barrier to breaking the strong C−O molecular double bonds to form singly bonded tetrahedral CO_4 structures.

Similarly, molecular N_2O phase I and III transform to the intermediary phase II (*$P4_2/mnm$*) and IV (*Pbcn*) and then disproportionates to an ionic product of $NO^+NO_3^-$ and N_2. The increase of ionicity in N_2O leads to the ionic solid, that again occurs gradually *via* dipole ordered N_2O-III and IV phases.

However, theoretical descriptions of such a gradual electron delocalization through intermediate phases have been challenges. Recent total energy calculations of CO_2 [116], for example, assert a different picture that the high-pressure, "intermediate" phases may be strictly molecular and have entirely different phase stabilities. This calculation, however, fails to account for the stability of the bent phase IV (*Pbcn*) and, instead, suggests that a molecular *Cmca* structure (experimentally found at the ambient temperature) occupies the entire stability field of phase IV (experimentally found only at high temperatures). This description advocates an extended stability domain for molecular CO_2 and seems to imply that the molecular-to-nonmolecular transition occurs rather *abruptly* at the phase boundary between phases III and V.

While the existence and stability of bent configuration in CO_2, N_2O, and CS_2 are apparent [72, 83, 91, 107], theoretical descriptions of such phases also face challenges [116] and result in substantially higher energy (several eV) than their linear configuration. Nevertheless, it is not surprising, considering the fact that the bent configuration is stabilized only by their collective behaviors at high pressures. Without strong molecular association (seen in the phase II and III and, to an extent, the paired layer phase III at high pressures), the bent configuration may simply represent an excited state of these triatomics, which is bent and several eV higher than the linear ground states. Furthermore, considering that transformation kinetics plays an important role in determining the phase stability of both N_2O and CO_2 (and very likely other molecular compounds), any calculation aiming to predict their phase stabilities *must* include molecular dynamics simulations of a large number of structural configurations and reaction paths in addition to the minimum energy calculations.

ACKNOWLEDGEMENT

The experimental works reported in this article have previously been done in collaboration with Drs. V. Iota, J. Park and H. Cynn at the LLNL; Prof. M.F. Nicol at UNLV; Prof. Y. Gupta at the WSU. This work has been supported by the LDRD and PDRP programs at

Lawrence Livermore National Laboratory, University of California under the auspices of the U.S. Department of Energy under Contract No. W-7405-ENG-48.

REFERENCES

[1] L. Pauling, Nature of the Chemical Bond (Cornell University Press, Itaca, N.Y.1940).
[2] A. Jayarama, Rev. Mod. Phys., 55 (1983) 65.
[3] R.J. Hemley, Ann. Rev. Phys. Chem., 51 (2000) 763.
[4] W.J. Nellis, A.C. Mitchell, F. H. Ree, M. Ross, N.C. Holmes, R.J. Trainor, J. Chem. Phys., 95 (1991) 5269; W.J. Nellis, F.H. Ree, M. vsn Thiel, and A.C. Mitchell, J. Chem. Phys., 75 (1981) 3055; W.J. Nellis, N.C. Holmes, A.C. Michell, and M. van Thiel, Phys. Rev. Lett., 53 (1984) 1661; W.J. Nellis and A.C. Mitchell, J. Chem. Phys., 73 (1980) 6137.
[5] R.L. Mills, B. Olinger, and D.T. Cromer, J. Chem. Phys., 84 (1986) 2837.
[6] D.T. Cromer, R.L. Mills, D. Schiferl, and L.A. Schwalbe, Acta Crystalogr. B, 37 (1981) 8.
[7] H. Schneider, W. Hafner and A. Wokaun. J. Chem. Phys., 96 (1992) 8046.
[8] M.L. Erements, R.J. Hemley, H.K. Mao, E. Gregoryanz, Nature, 411 (2001) 170.
[9] A.F. Goncharov, E. Gregoryanz, H.K. Mao, Z. Liu, R.J. Hemley, Phys. Rev. Lett., 85 (2000) 1262.
[10] V. Iota, C.S. Yoo, H. Cynn, Science, 283 (1998) 1510.
[11] C.S. Yoo and M. F. Nicol, J. Phys. Chem., 90 (1986) 6726; *ibid* 90 (1986) 6732.
[12] A. Katz, D. Schiferl, R.L. Mills, J. Phys. Chem., 88 (1984) 3176.
[13] M. Somayazulu, A. Madduri, A.F. Goncharov, O. Tschauner, P.F. McMillan, Phys. Rev. Lett., 87 (2001) 135504.
[14] C.S. Yoo, V. Iota, H. Cynn, M. Nicol, J.H. Park, T. Le Bihan, and M. Mezouar. J. Phys. Chem. B, 107 (2003) 5922.
[15] M. Nicol and K. Syasson, Phys. Rev. B28, (1983) 1201.
[16] R.J. Hemley, Z.G. Soos, M. Hanfland, and H.K. Mao, Nature, 369 (1994) 384.
[17] A.F. Goncharov, E. Gregoryanz, R.J. Hemley, and H. K. Mao, Phys. Rev. B68, (2003) 100102.
[18] F.A. Gorelli, L. Ulivi, M. Santoro, and R. Bini, Phys. Rev. Lett., 83 (1999) 4093.
[19] S. Desgreniers, Y.K. Vohra, A.L. Ruoff, J. Phys. Chem., 94 (1990) 1117.
[20] K. Takemura, S. Minomura, O. Shinomura, Phys. Rev. Lett., 49 (1982) 1772.
[21] K. Takemura, K. Sato, H. Fujihisa, O. Mitsuko, Nature, 423 (2003) 971.
[22] R. Reichlin, M.Ross, S. Marin and K.A. Goettel, Phys. Rev. Lett., 56 (1986) 2858.
[23] K.A. Goettel, J.H. Eggert, and I.F. Silvera, Phys. Rev. Lett., 62 (1989) 665.
[24] K. Aoki, H. Yamawaki, M. Sakashita, H. Fujihita, Phys. Rev. B54 (1996) 15673.
[25] Ph. Pruzan, I.C. Chervin, B. Canny, J. Chem. Phys., 99 (1993) 9842.
[26] K. Aoki, S. Usuba, M. Yoshida, Y. Kakudate, K. Tanaka, and S. Fujiwara, J. Chem. Phys., 89 (1988) 529.
[27] M. Ross, Nature, 292 (1981) 435.
[28] L.R. Benedetti, J.H. Nguyen, W.A. Caldwell, H. Liu, M. Kruger, R. Jeanloz, Science, 286 (1999) 100.
[29] L. Ciabini, M. Santoro, R. Bini, V. Schettino, J. Chem. Phys., 116 (2002) 2928.
[30] F. Ancilotto, G.L. Chiarotti, S. Scandolo, E. Tosatti, Science, 275 (1997)1288.
[31] S. Serre, C. Cacvaxxoni, G.L. Chiarotti, S. Scandolo, E. Tosatti, Science, 284 (1999) 788.
[32] J. Dong, J.K. Tomfohr, O.F. Sankey, Phys. Rev. B62, (2000) 14685.
[33] C. Cavazzoni, G.L. Chiarotti, S. Scandolo, E. Tosatti, M. Bernasconi, M. Parrinello, Science, 283 (1999) 44.
[34] C.F. Richardson and N.W. Ashcroft, Phys. Rev. Lett., 78 (1997) 118.
[35] S. Scandolo, Proc. Natl. Acad. Sci. U.S.A., 100 (2003) 3051.
[36] A.Y. Liu and M.L. Cohen, Science, 245 (1989) 842.
[37] C. S. Yoo, H. Cynn, F. Gygi, G. Galli, V. Iota, M. Nicol, S. Carlson, D. Hauserman, and C. Mailhiot, Phys. Rev. Lett., 83 (1999) 5527.

[38] M. Eremets, V.V. Struzhkin, H.-K. Mao, R.J. Hemley, Science, 293 (2001) 272.
[39] V.V. Struzhkin, M. Eremetsm W. Gan, H.-K. Mao, R.J. Hemley, Science, 298 (2002) 1213.
[40] C. Mailhiot, L.H. Yang, and A.K. McMahan, Phys. Rev. B46 (1992) 14419.
[41] D. A. Young, Phase Diagrams of the Elements. (UC Press, CA, 1991).
[42] D.M. Ceperley and B.J. Alder, Phys. Rev. Lett., 45 (1980) 566.
[43] D.A. Kirzhnits, Usp. Fiz. Nauk 104, 489 (1971); Sov. Phys. Usp., 14 (1972) 512.
[44] L.B. Da Silva et al., Phys. Rev. Lett., 78 (1997) 483.
[45] M.D. Knudson, D.L. Hanson, J.E. Bailey, C.A. Hall, and J. R. Asay, Phys. Rev. Lett., 87 (2001) 225501.
[46] R. Chau, A.C. Mitchell, R.W. Minich, and W.J. Nellis, Phys. Rev. Lett., 90 (2003) 245501.
[47] S.T. Weir, A.C. Mitchell, W.J. Nellis, Phys. Rev. Lett., 76 (1996) 1860.
[48] Bonev, submitted (2004).
[49] G. Shen, M.L. Rivers, Y. Wang, and S. R. Sutton, Rev. Sci. Instrum., 72 (2001) 1273.
[50] F. Datchi, R. LeToullec and P. Loubeyre, J. Appl. Phys., 81 (1997) 3333.
[51] J.-A. Xu and H.-K. Mao, Science, 290 (2000) 783.
[52] W.L. Mao, et al., Appl. Phys. Lett., 83 (2003) 5190.
[53] C.S. Yoo and Y.M. Gupta, J. Chem. Phys., 93 (1990) 2082.
[54] C.A. Hall et al., Rev. Sci. Instrum., 72 (2001) 1.
[55] G.E. Duvall and R.A. Graham, Rev. Mod. Phys., 49 (1974) 523.
[56] A.M. Lindenberg, et al., Phys. Rev. Lett., 84 (1999) 111.
[57] D.J. Erskine and W.J. Nellis, Nature, 349 (1991) 317 (1991).
[58] C.S. Yoo, J.J. Furrer, G.E. Duvall, S.F. Agnew, and B.I. Swanson, J. Phys. Chem., 91 (1987) 6577.
[59] J.J. Dick, R.N. Mulford, W.J. Spencer, D.R. Pettit, E. Garcia, D.C. Shaw, J. Appl. Phys., 70 (1991) 3572; J.J. Dick, J. Phys. Chem., 97 (1993) 6195.
[60] C.S. Yoo and W.J. Nellis, Science 254, 1489 (1991); C.S. Yoo, W.J. Nellis, M.L. Sattler, and R.G. Musket, Appl. Phys. Lett., 61 (1992) 273.
[61] V.V. Struzhkin, R.J. Hemley, H. K. Mao, Y. A. Timofeev, 390, 382 (1997).
[62] Y. Katayama, T. Mizutani, W. Utsumi, O. Shimonura, M. Yamakata, K. Funakoshi, Nature, 403 (2000) 170.
[63] F.P. Bundy and J.S. Kasper, J. Chem. Phys. 46, 3437 (1967); F.P. Bundy, Physica 156A (1989) 169.
[64] P. W. Bridgman, Proc. Am. Acad. Arts Sci., 74 (1942) 399.
[65] S.F. Agnew, R.E. Mischke. B.I. Swansen, J. Phys. Chem., 92 (1988) 4201; S.F. Agnew, B.I. Swansen, D.G. Eckhart, in Shock Wave in Condensed Matter, Y. M. Gupta, Ed., (Plenum, New York, 1986) p 221.
[66] P.W. Bridgman, Proc. Am. Arts. Sci., 72 (1938) 207.
[67] R.C. Hanson and L.H. Jones, J. Chem. Phys., 75 (1981) 1102.
[68] B. Olinger, J. Chem. Phys., 77 (1982) 6255.
[69] R.C. Hanson, J. Phys. Chem., 89 (1985) 4499.
[70] K. Aoki et al., Science, 263 (1994) 356.
[71] B. Kuchta and R. Etters, Phys. Rev. B47, (1993) 14691.
[72] J.-H. Park, C.S. Yoo, V. Iota, H. Cynn, M.F. Nicol, and T. Le Bihan, Phys. Rev. B 68, (2003) 014107.
[73] C. S. Yoo, H. Kohlmann, H. Cynn, M. F. Nicol, V. Iota, and T. Le Bihan, Phys. Rev. B 65, (2002) 104103.
[74] O. Tschauner, H.-K. Mao, and R.J. Hemley, Phys. Rev. Lett., 87 (2001) 75701.
[75] C.S. Yoo, Science and Technology of High Pressure, M. H. Manghnani, M.F. Nicol, Eds., vol. 1 (University Oress, Hyberadad, India, 2000) p 86.
[76] C.A. English, J.A. Venables, Proc. R. Soc., London A, 340 (1974) 57.
[77] J.A. Venables and C.A. English, Acta Crystallogr. B30, (1974) 929.
[78] C.S. Yoo et al., Phys. Rev. Lett., 83 (1999) 5527.

[79] E. Knittle, Handbook of Physical Constants, T. Ahrens, Ed., (AGU, Washington D.C., 1995) p98-142.
[80] H. Graetsch and O.W. Florke, Z. Kristallogr., 195 (1991) 31.
[81] R.F. de Dombal and M.A. Carpenter, Eur. J. Mineral, 5 (1998) 171.
[82] V. Iota and C.S. Yoo, Phys. Rev. Lett., 86 (2001) 5922.
[83] C.S. Yoo, V. Iota, and H. Cynn, Phys. Rev. Lett., 86 (2001) 447.
[84] A. Anderson, The Raman Effect, vol. 2, (Mercel Dekker, Inc., New York, 1973) p 911.
[85] A. Katz, D. Schiferl, M.J. Mill, J. Phys. Chem., 88 (1984) 3176.
[86] P. Lablanquie et al., Phys. Rev. A, 40 (1989) 5673.
[87] N. Correia, A. Flores-Riveros, K. Helenelund, L. Asplund, U. Gelius, J. Chem. Phys., 83 (1985) 2035.
[88] L. Pauling, Nature of the Chemical Bond (Cornell University, Ithaca, 1940).
[89] B. Kuchta and R.D. Etters, J. Chem. Phys., 95 (1991) 5399.
[90] D.E. Stogryn and A. P. Stogryn, Mol. Phys., 11 (1966) 371.
[91] V. Iota, J.-H. Park and C.S. Yoo, Phys. Rev. B69, (2004) 064106.
[92] R.L. Mills, B. Olinger, D.T. Cromer, and R.L. LeSar, J. Chem. Phys., 95 (1991) 5392.
[93] B. Baer and M.J. Nicol, J. Phys. Chem., 94 (1990) 1073.
[94] D.T. Cromer, R.L Mills, D. Schiferl, L. Schwalbe, Acta Crystallogr., B37 (1981) 8.
[95] C.S. Yoo, Science and Technology of High Pressure, M. H. Manghnani, M.F. Nicol, Eds., vol.1, (University Oress, Hyberadad, India, 2000) p 86.
[96] M. Somayazulu, A. Madduri, A.F. Goncharov, O. Tschauner, P.F. McMillan, H.K. Mao, R.J. Hemley, Phys. Rev. Lett., 87 (2001) 135504.
[97] R.W.G. Wyckoff, Crystal Structures, 2nd ed., vol.2, (John Wiley & Sons, New York, 1964) Ch. VII.
[98] M. Hanfland, M. Lorenzen, C. Wassilew-Reul, F. Zontone, Rev. High Press. Sci. Technol., 7 (1998) 787; also, see the ESRF Web page by M. Hanfland, 2000.
[99] S.F. Agnew, R.E. Mischke, and B.I. Swansen, J. Phys. Chem., 92 (1988) 4201.
[100] Y. Fujii, et al., Phys. Rev. Lett., 63 (1989) 536.
[101] J. B. Neaton and N.W. Ashcroft, Nature, 400, (1999) 141.
[102] M. Hanfland, K. Syassen, N.E. Christensen and D.L. Novikov, Nature, 408 (2000) 174.
[103] P.W. Bridgman, Proc. Am. Acad. Arts Sci., 74, (1942) 399.
[104] S.F. Agnew, R.E. Mischke, and B. I. Swanson, J. Phys. Chem., 92 (1988) 4201.
[105] R.D. Dick, J. Chem. Phys., 52 (1970) 6021.
[106] S.A. Sheffield and G.E. Duvall, J. Chem. Phys., 79 (1983) 1981; *ibid*, 81 (1984) 3048.
[107] C.S. Yoo, G.E. Duvall, J.J. Furrer, and R. Granholm, J. Phys. Chem., 93 (1983) 3012.
[108] C. S. Yoo and Y. M. Gupta, J. Phys. Chem., 94 (1990) 2857.
[109] C. S. Yoo and Y. M. Gupta, J. Chem. Phys., 93 (1990) 2082.
[110] Ch. Jungen, D.M. Malm, Can. J. Phys., 51 (1973) 1471.
[111] C. S. Yoo, H. Cynn, F. Gygi, G. Galli, V. Iota, M. Nicol, S. Carlson, D. Hauserman, and C. Mailhiot, Phys. Rev. Lett., 83 (1999) 5527.
[112] B. Holm, R. Ahuju, A. Belomoslike, and B. Johansson, Phys. Rev. Lett., 85 (2000) 1258.
[113] J. Dong, J.K. Tomfohr and O.F. Sankey, K. Leinenweber, M. Somayazulu, and P.F. McMillan, Phys. Rev. B61, (2000) 5967.
[114] R.M. Hazen, L. W. Finger, R.J. Hemley, and H.K. Mao, Solid State Commun., 72, (1989) 507.
[115] H. Gratsch and O.W. Förke, Z. Kristallogr., 195 (1991) 31.
[116] S.A. Bonev, F. Gygi, T. Ogitsu, and G. Galli, Phys. Rev. Lett., 91 (2003) 065501.

Chemistry at Extreme Conditions
M. Riad Manaa (Editor)

Chapter 6

Nitrogen-Containing Molecular Systems at High Pressures and Temperature

Yang Song[a], Russell J. Hemley[a], Ho-kwang Mao[a] and Dudley R. Herschbach[b]

[a]Geophysical Laboratory, Carnegie Institution of Washington, 5251 Broad Branch Rd. NW, Washington DC 20015, USA

[b]Department of Chemistry and Chemical Biology, Harvard University, 12 Oxford St., Cambridge, MA 02138, USA

1. INTRODUCTION

Key aspects of pressure as a governing chemical parameter appear in a simple heuristic model: an atom or molecule imprisoned in an infinite-walled box of shrinking volume. Venerable prototype examples include the hydrogen atom in a spherical box [1] and the hydrogen molecule in a spheroidal box [2]. A major effect of increasing compression, simulated by such models [1-5], is the marked increase in electronic kinetic energy. This results both from the momentum increase required by the uncertainty principle and from repulsive forces, induced by the Pauli exclusion principle, that arise from overlap of electron clouds as neighboring molecules crowd together. Figure 1 illustrates the dominant role of these factors. As a consequence, high pressure typically weakens chemical bonds and thereby fosters rearrangements to produce new phases or molecular species.

This chapter briefly surveys current experimental capabilities for subjecting molecules to extreme conditions of pressure and temperature and the use of Raman and infrared spectroscopy and x-ray diffraction to characterize the solid-state chemistry of some simple nitrogen molecular systems. We consider chiefly phase transformations and vibrational dynamics of nitrogen bearing systems, including polynotrigen species at high pressure. There has been considerable progress in the study of framework nitrides at high pressures (e.g., Refs. [6, 7]), but this is beyond the scope of this chapter. Our discussion of nitrogen-bearing molecular systems complements other recent reviews [8, 9] in exemplifying how even simple molecules can be endowed at high P-T with unorthodox properties and incarnations.

2. EXPERIMENTAL CAPABILITIES

Static high pressure experiments employing diamond anvil cells (DAC) now can routinely attain pressures from the kilobar (0.1 MPa) to the multimegabar (>100 GPa) range. The pair of opposing anvils are formed from brilliant-cut single crystal diamonds with small culet faces. A variety of gasket materials, sandwiched between the anvils, can be used to form the sample chamber [10]. The transparency of diamonds over a wide wavelength range enables use of Raman scattering and infrared spectroscopy to examine the pressure dependence of molecular vibrations. The availability of 2nd and 3rd generation synchrotron radiation facilities, which provide extremely high photon flux and brilliance tunable over a broad energy range, has greatly facilitated *in situ* investigation of novel structures formed under pressure, particularly infrared spectroscopy and x-ray diffraction studies.

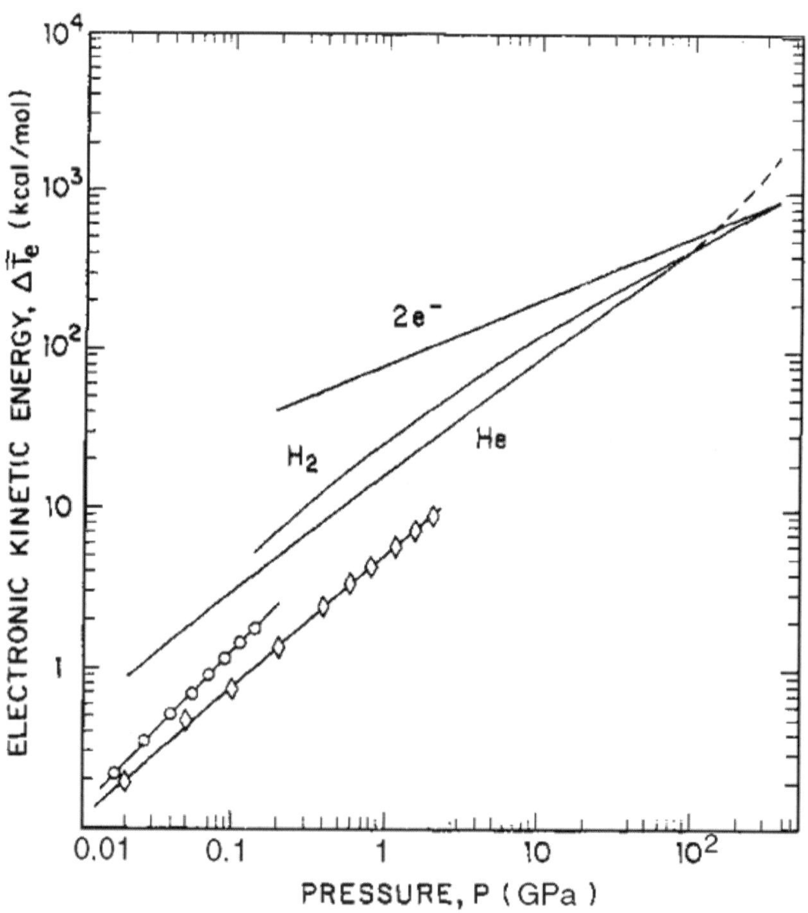

Fig. 1. Pressure-induced change in average electronic kinetic energy, ΔT_e, for He in a spherical box and for H_2 and for two noninteracting electrons in a spheroidal box. Points for He (diamonds) and H_2 (circles) derived from experimental data (adapted from Ref. [2]).

In our experiments, we typically used R_1 ruby fluorescence [11] to measure the pressure according to the relation,

$$P\ (GPa) = A/B[\{(1+\Delta)/\lambda_0)\}^B - 1] \qquad (1)$$

where A=1904 GPa, B= 7.665, λ_0 = 694.28 nm at 298 K and $\Delta\lambda$ is the difference between the wavelength of the R_1 line at pressure P and that at atmospheric pressure. Under the quasi-hydrostatic conditions of these experiments, the accuracy of this ruby pressure scale is ± 0.5%, from self-consistent Brillouin scattering / x-ray diffraction calibration [12]. Since ruby fluorescence decreases and broadens significantly with increasing temperature, the use of Eq.(1) is limited to below ~700 K. Alternatively, one can Sm:YAG for high temperature pressure calibration ; below 820 K, the pressure induced frequency shifts for both the Y1 and Y2 lines of Sm:YAG have no obvious temperature dependence [13] . Thus the pressure can be determined by P(GPa)=-0.12204(ω^{Y1}-16187.2) or P(GPa)=-0.15188(ω^{Y2}-16232.2), where ω^{Y1} and ω^{Y2} are the observed frequencies.

2.1. Raman spectroscopy

The availability of single grating imaging spectrometers with holographic optics, together with very sensitive CCD detectors has improved the measurements of Raman Spectra with DACs. We typically used the strongest lines of a Coherent Innova Ar^+ laser, at 488.0 and 514.5 nm as the excitation source, with power of 0.1-0.5 Watt (much less at the sample). The spectral and spatial purity of the laser beam is selected by a band pass filter and improved by a spatial filter. The collimated laser beam is typically focused on the sample in a backscattering geometry. A neutral or notch filter beamsplitter and the same long working distance objective lens collects the scattered light, which is sent through notch filters to remove the unwanted Rayleigh scattering, the traverses a pinhole before entering the slit of the spectrograph. The resolution achieved using a 460 mm focal length f/5.3 imaging spectrograph (ISA HR460) equipped with an 1800 grooves/mm grating is ± 0.1 cm^{-1}. The wavelength calibration, done using Ne lines, has an uncertainty of ± 1 cm^{-1}. Additional details of the technique are given in Ref. [14].

2.2. Infrared spectroscopy

The synchrotron IR beamline U2A at the National Synchrotron Light Source (NSLS) of Brookhaven National Laboratory (BNL) enables high quality measurements of near- to far-infrared spectra in DACs [14]. Briefly, the synchrotron light is collected in a 40 x 40 mrad solid angle and collimated to a 1.5″ diameter beam before entering a Bruker IFS 66V vacuum Fourier transform spectrometer. The beam is then sent to one of the three IR microscopes. The spectrometer is equipped with a number of beam splitters and detectors including a silicon bolometer and MCT. In addition, a grating spectrograph with a CCD array detector can be used with an Ar^+ laser, Ti-sapphire laser, and standard lamps for Raman, fluorescence, absorption, and reflectivity measurements in the visible range. Altogether, the system provides spectral coverage from 50-20,000 cm^{-1}.

2.3. X-ray diffraction

X-ray diffraction techniques are essential for the characterization of high pressure phases. Energy dispersive x-ray diffraction measurements were carried out at the superconducting wiggler beamline X17C at the National Synchrotron Radiation Source (NSLS) at Brookhaven National Laboratory (BNL) [15]. The high photon flux in the energy range of 20-100 keV readily penetrates diamond windows. The beamline instrumentation and operation procedures of the primary beam x-ray optics as well as the diffracted beam collimation and sample positioners are designed to optimize spatial and temporal resolution, signal-to-noise ratio, and stability for minute samples. In the energy dispersive x-ray diffraction (EDXD) configuration, the analysis is based on the Bragg relation, $E\ (\text{keV}) = 6.1993/d(\text{Å})\sin\theta$. The measurements are carried out at fixed 2θ angles. In addition, the cell can be oscillated along ω and χ to minimize the effects of preferred orientation. A germanium solid state detector is used to collect the diffraction signal which is processed by a multi-channel analyzer in the energy range of 5-80 keV. The diffraction angle is typically calibrated with diffraction lines of gold at ambient conditions.

For angle dispersive x-ray diffraction we used the High Pressure Collaborating Access Team (HPCAT, sector 16) insertion device diffraction beamline at the Advanced Photon Source (APS) of Argonne National Laboratory (ANL). The x-ray beam in the energy range of 24-35 keV is dispersed by a double-crystal monochromator in an upstream enclosure. In the current setup we chose a single wavelength, such as 0.3699 or 0.4084 Ångstroms, with diffraction configurations calibrated by collecting the pattern of CeO_2 standard. The beam was focused to 10 μm (horizontal) by 14 μm (vertical) by two platinum coated 300 mm long electrode bimorph mirrors then guided by a pinhole mounted at the end of 3mm diameter tubes before incident into the diamond anvil cell. A high-resolution MAR345 imaging plate was used as detector and the typical exposure time is 30-120 seconds for each pattern. The two-dimensional diffraction rings were then converted to one-dimensional angle dispersive diffraction pattern using FIT2D program.

2.4. Variable temperature

Combining the variable of temperature with that of pressure leads to new phenomena. Samples at high pressure may be brought up to 1000 K by a variety of resistive (furnace) heating techniques with thermocouples. To prevent oxidation of the heating wires and the metal yoke holding the anvils, a reducing gas of 1% H_2 in Ar was continuously supplied to the furnace. Higher temperatures, up to >6000 K, can be obtained by directing a laser beam into the DAC, using either cw- or pulsed-lasers (e.g., CO_2, Nd-YAG or Nd-YLF). In the present studies, we chiefly used a focused beam of diameter 30-50 microns from a cw CO_2 laser (10.6 μm); heating power of less than 50W readily provides temperatures in the range 1000-2000 K. Pyrometry is used to determine the temperature, by collecting the black body thermal radiation emitted from the sample into a CCD equipped spectrograph. For low temperature studies, the DAC was placed in a cryostat cooled by liquid nitrogen or liquid helium. The cryostat was equipped with specially designed optical windows which enable Raman and infrared spectroscopy as well as x-ray diffraction to be carried out *in situ*.

3. HIGH-PRESSURE STUDIES OF NITROGEN-CONTAINING COMPOUNDS

Table I gives an overview of selected nitrogen systems thus far studied at high pressures and high or low temperatures by means of *in situ* DAC experiments. For each system, we list the initial molecular species loaded in the DAC, the range of pressures and temperatures examined, the data reported (R, IR, Opt, Dif, EC and NMR denote Raman, infrared, optical spectra, x-ray or neutron diffraction, electrical conductivity and magnetic nuclear resonance). We also indicate the phases observed, whether involving the initial molecular species (denoted by M) or a different species (denoted "nonmolecular," NM, if not specifically identified) that becomes prominent at high P and T. The references cited include the original research reports that provide experimental data and most theoretical papers pertinent to these studies.

Our commentary will not attempt to abstract for each system specifics of the analysis of spectra and other data. Rather, we present some vignettes chosen to exemplify current capabilities or to illustrate either typical or atypical responses to compressing, heating, or cooling nitrogen species. The intent is threefold: to provide researchers in the field with a comprehensive listing of literature references; to conduct nonspecialists on a circumspect tour of a growing family album; and to serve both audiences by emphasizing inferences about electronic structure aspects, the fundamental chemical perspective sought in DAC experiments.

3.1. Nitrogen: diatomic, polyatomic, and polymeric

As an archetypal homonuclear diatomic molecule with a very strong triple bond, nitrogen particularly invites the study of pressure-induced transformations, expected to produce delocalization of electronic shells and eventual molecular dissociation. Among the several solid molecular nitrogen phases are two at low pressure (α and γ) that represent alternate ways of packing quadrupoles; a disordered, plastic phase (β) that solidifies from the supercritical fluid; three phases at higher pressures with nonquadrupolar-type ordering (δ, ε, ζ); and two other distinct phases (ι, θ), only recently discovered [66], which have exceptionally large regions of stability and metastability. These two phases extend far into regions long thought to belong solely to previously known phases. The newly found phases also can both be quenched to room temperature. The important general conclusion is that the definitive determination of the equilibrium phase relations even of molecular nitrogen is more complex than previously thought due to the presence of substantial transformation barriers between different classes of structures [66]. These structures include potential "polynitrogen" phases in which there are covalent linkages between polyatomic nitrogen molecules or molecular ions. These polyatomic molecular ions (N_3^-, N_5^+) are discussed in detail in later sections. Like polymeric nitrogen, such compounds should be highly energetic materials because single or double bonds are much weaker than the triple bond; thus decomposition to molecular nitrogen would be highly exothermic. Although thermodynamically very metastable, once formed polynitrogen might prove kinetically reluctant to decompose at low temperature, so offer a powerful, environmentally benign fuel for rocket propulsion. Several theoretical calculations conclude that metastable polynitrogen could exist at ambient pressure [19, 20].

Table I. Survey of high P-T studies of nitrogen-bearing molecular systems

Initial Speices	Pressure range, GPa	Temperature range, K	Measurements[a]	Phases	High P-T species[b]	Experimental references	Theoretical references
N_2	0-140 125-240	10-6000 30-500	R, IR, Opt, Dif R, IR, Opt, Dif, EC	α, β, γ, δ, ε, ζ, θ, ι, η	M, NM or PL	[35-61], [62-67]	[16-18], [19-25]
NO	0-1.5	109-300	R, IR, Dif	Monomer, Dimer	N_2O_2, N_2O, N_2O_3, N_2O_4	[68-76]	[26, 27]
N_2O	0-135	20-1000	R, IR, Dif	α, β, II, IV	M, $NONO_3$ (I)	[77-82]	[28, 29]
N_2O_4	0-27	20-1000	R, IR, Dif	α, β, γ	M, $NONO_3$ (I)	[83-93]	
HCC-CN NC-CN	0-1.5 0-10	248-300 290-350	R, IR, Opt, Dif R, IR, Dif	P21/m I, II, III	$(R\text{-}CN)_x$, $[(R\text{-}CN)2]_m$, p-DISN[c]	[94, 95], [96-99]	
NaN_3	0-50 50-160	120-3300	R, IR, Opt, Dif, EC	α, β, I, II, III, IV, V, VI	M, NM or PL	[100-106], [107-113]	[30-34]

a Experimental measurements are denoted by R (Raman), IR (Infrared), Opt (Optical), Dif (x-ray or neutron diffraction), EC (electrical conductivity) and NMR (nuclear magnetic resonance).

b High P-T species are denoted by M (molecular), NM (non-molecular), PL (Polymeric), I (ionic) or Dc (decomposed).

c p-DISN represents poly (2, 3- diiminosuccinonitrile). See Ref. [97].

For nitrogen, destabilization of the triple bond should occur at a sufficiently high pressure, leading to a nonmolecular phase (NM) formed by a network of nitrogen-nitrogen single bonds. Recent x-ray diffraction measurements have been reported that are consistent with the formation of the gauche structure [66, 67]. A theoretical calculation by McMahan and LeSar in 1985 [21] and subsequent calculations [22-24] predicted such a transition from the molecular nitrogen solid (M) to various threefold coordinated NM phases below 100 GPa. Shock-wave experiments in 1986 indeed found evidence for a NM transition in the liquid phase at 30 GPa and 6000 K [62]. For solid nitrogen, however, dissociation of nitrogen molecules by compression was not experimentally confirmed until 2000 and did not occur until pressures above 150 GPa at room temperature, but exhibited a large hysteresis at low temperature, suggesting an equilibrium transition near 100 GPa [63-65]. This work is reviewed in detail elsewhere (Goncharov and Gregoryanz chapter, [8]).

3.2. Nitrogen oxides: prevalence of $NO^+NO_3^-$

A naïve chemical notion led to the first high pressure study of NO, carried out in 1985 at Los Alamos [73]. Interest in polysulfur nitride, $(SN)_x$, a conducting polymer or "non-metallic metal," raised the question whether nitric oxide, isoelectronic to SN, might polymerize under high pressure [114]. However, it was found that a pressure of only 1.5 GPa at 176 K was sufficient to induce NO to undergo a facile disproportionation reaction, to form N_2O + N_2O_4. No evidence for a polymeric form of nitric oxide appeared up to the highest pressure reached in the Los Alamos experiment, 14 GPa. In the same study, As the bond strengths of nitric oxide and carbon monoxide are unusually high, corresponding to bond orders of 2.5 and 3, such facile transformations offered a dramatic demonstration that high pressure can drastically reduce chemical activation energies, in effect acting like a powerful catalyst.

In general, nitrogen oxides are among those molecules whose reactivity are tremendously altered by high pressures in a broad temperature range. These molecules are stable at ambient conditions but are exceptionally reactive under high pressure conditions, such that a wealth of pressure induced chemical reactions accompanied by the formation of new phases and species are observed. In particular, the most intriguing observation is that at high pressures many nitrogen oxides transform to a remarkable stable ionic isomer, nitrosonium nitrate ($NO^+NO_3^-$). Although this ionic species and its molecular precursor, N_2O_4, have been the subject of several earlier studies [86, 87, 115-118], the mechanisms governing the transformation between the two forms had remained unclear.

Under ambient pressure and low temperatures, there are several means to produce $NO^+NO_3^-$. The formation of $NO^+NO_3^-$ was first detected in IR spectra of oxidized NO [115]. Subsequent experiments on N_2O_4 established that the ionic form could be spontaneously produced when N_2O_4 is trapped in a Ne matrix [116] or by temperature-induced autoionization of N_2O_4 [87]. The transformation of molecular N_2O_4 to ionic $NO^+NO_3^-$ under high pressures was first observed by Jones and co-workers [86, 117]. They discovered that laser irradiation of cubic Im3 α-N_2O_4 results in the formation of β-N_2O_4 with unknown non-cubic structure at 1.16 GPa at room temperature. Under pressures of 1.5-3.0 GPa at room temperature, β-N_2O_4 exhibits a reversible phase transition to the ionic form of $NO^+NO_3^-$ with a large hysteresis. Recently, a different way to synthesize $NO^+NO_3^-$ under pressure was reported. [76, 81] At 10-30 GPa laser heating to 1000-2000 K transforms N_2O via the reaction $4N_2O \rightarrow NO^+NO_3^- + 3\ N_2$ (i.e., thermal instead of photo induced reaction) .

In summary, $NO^+NO_3^-$ can be formed via totally different paths with different starting nitrogen oxides. However, unambiguous interpretation of the structure, stability and transformation mechanism involving this peculiar ionic species is unavailable. Therefore, comprehensive experimental means, including x-ray diffraction measurement, Raman and Infrared spectroscopy were employed to understand the fundamental properties of $NO^+NO_3^-$ synthesized via the last path mentioned above. While the diffraction measurements established the P-V equation of state, the optical spectra, especially the low-temperature Raman data, elucidate important aspects of the transformation, thermodynamic properties, and stability diagram of $NO^+NO_3^-$.

3.2.1. Raman spectra and phase transitions

Here we chose the thermal disproportionation of N_2O at high pressures to synthesize $NO^+NO_3^-$. Raman spectrum of $NO^+NO_3^-$ was typically measured to check the degree of completion of reaction and transformation upon laser heating of pressurized N_2O. Between 10 GPa and 40 GPa, heating of N_2O results in an inhomogeneous dark mass within the cell that appears fabric-like to the eye. The formation of $NO^+NO_3^-$ and N_2 can be confirmed by characteristic peaks (e.g. at 13 GPa) at 740 cm^{-1} (ν_4), 827 cm^{-1}(ν_2), 1096 cm^{-1}(ν_1), 2253 cm^{-1}(ν_{NO+}), 2362 cm^{-1} and 2383 cm^{-1}(N_2) as well as abundant lattice modes below 400 cm^{-1}. On decompression, the major characteristic modes shift to lower frequencies. In addition, the number of resolvable Raman peaks is reduced as a result of peak broadening. This trend is consistent with that $NO^+NO_3^-$ becomes more disordered on decompression [76]. When the pressure drops below 1 GPa, the Raman spectrum shows significant changes, in particular a broad peak near 280 cm^{-1} is enhanced and the strong NO^+ peak disappears. This accords with the low temperature results as discussed below, which suggest a phase transformation of $NO^+NO_3^-$ occurs at low pressure.

Raman spectra were collected and examined as a function of pressure to detect possible phase transitions. On decompression from 40 to 1 GPa, changes in the room temperature Raman spectra appear to be steady and continuous. Moreover, the evolution of the peak positions with pressure is linear for ν_4, ν_2, ν_1, ν_{NO+} as well as the major lattice modes. Therefore, low-temperature Raman spectra were collected in the expectation that peaks would be better resolved and thereby respond more sensitively to pressure. Using liquid nitrogen as cryogen, we maintain the system at constant temperature 80 K. Under such conditions, the Raman spectra were measured as a function of pressure between 14 GPa and ambient and are plotted in Fig. 2. At such a low temperature, both the lattice and internal modes exhibit sharp profile and significant pressure shifts. These changes in the Raman spectrum can be interpreted as evidence for a phase transition in $NO^+NO_3^-$.

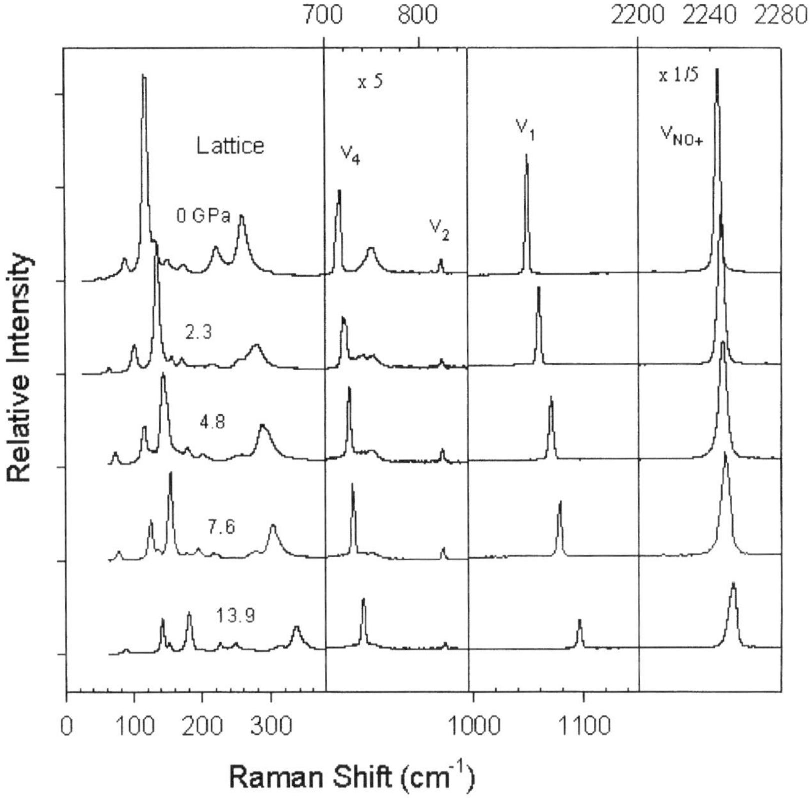

Fig. 2. Raman spectra of $NO^+NO_3^-$ in the regions 50-380; 700 860; 1000-1150; and 2200-2280 cm^{-1} measured near 80 K and five pressures. The measurements were performed at successive steps of decompression, starting from 13.9 GPa. Due to the low intensity of ν_4 and ν_2 and high intensity of ν_{NO+}, the spectra in the 700-850 cm^{-1} and 2200-2280 cm^{-1} regions are scaled by 5 and 1/5 respectively. (from Ref. [80])

The peak positions of Raman modes in the lattice mode region observed at 80 K on as a function of pressure are depicted in Figure 3. These bands exhibit a distinct change in $(d\nu/dP)_T$ at about 5 GPa, indicating a transition occurs near that pressure. The behavior of the second highest frequency mode changes most markedly at 5 GPa. In addition, nine modes can be identified at high pressures while at low pressures (<5 GPa), only seven are discernible. The pressure dependence of higher frequency intramolecular modes (not shown) shows changes at 5 GPa, but this is much less pronounced. The evolution of these Raman bands in the lattice-mode region was also examined on compression in a separate run. There are slight differences in slopes for several modes between compression and decompression (most likely due to different stress conditions), but a distinct change at 5 GPa on compression is still prominent, indicating the transition is reversible and has very little hysteresis.

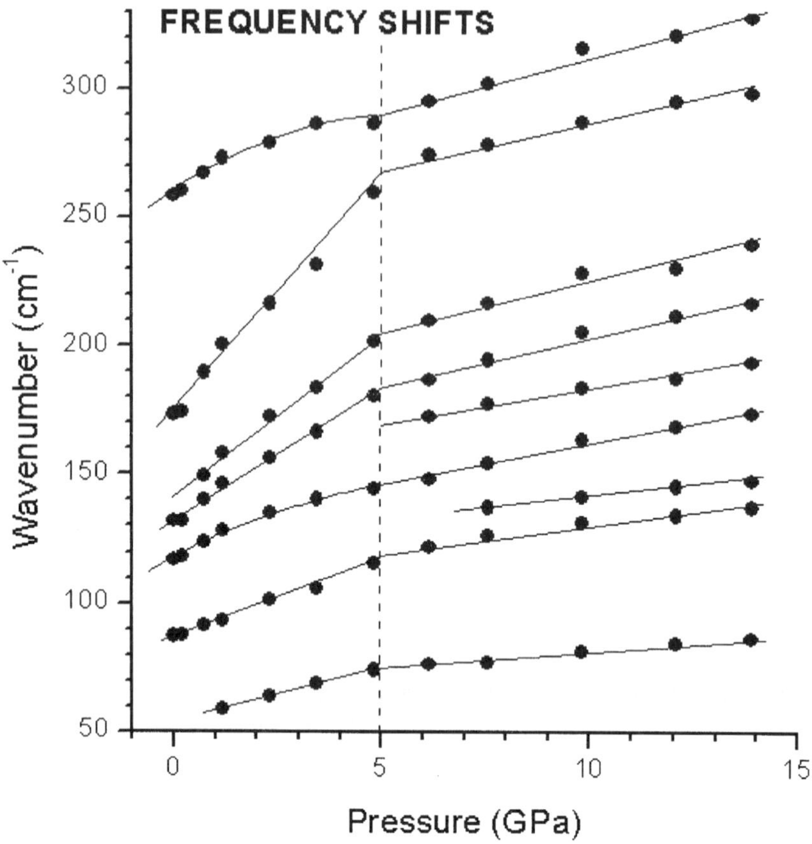

Fig. 3. Variation with pressure of Raman shifts in the lattice region (50-350 cm^{-1}) up to 15 GPa at 80 K on decompression. The lines are guides to the eye. The vertical dashed line at about 5 GPa indicates the approximate phase boundary. (from Ref. [80])

Upon decompression to ambient pressure, these low-temperature Raman measurements also established that $NO^+NO_3^-$ can be recovered, since all the principal low- and high-frequency modes associated with $NO^+NO_3^-$ persist down to ambient pressure. The accompanying N_2 formed by the laser heating of N_2O escaped, as evidenced by the disappearance of the peaks at about 2350 cm^{-1}, and $NO^+NO_3^-$ appears to be the sole product. The sample was warmed at ambient pressure and Raman spectra of $NO^+NO_3^-$ measured at five temperatures ranging from 80 K to 215 K. It is found that the highest temperature at which the $NO^+NO_3^-$ retains its identity is about 180 K. When the sample was heated to 190 K at ambient pressure, significant changes in the Raman spectrum were observed. New vibrational bands with spectroscopic features of molecular species such as NO_2 or its dimer N_2O_4 are observed [86, 87, 91]. Therefore it can be concluded that at 190 K, ionic $NO^+NO_3^-$ has mostly converted to molecular N_2O_4 in bulk, and the Raman spectrum arises from a mixture of the two phases. Further heating of the sample to 215 K resulted in loss of sample due to evaporation and disappearance of the Raman feature.

The transformation from molecular N_2O_4 to ionic $NO^+NO_3^-$ has been investigated since the early 1980's. Temperature-induced autoionization of solid N_2O_4 condensed on the surface of a copper mirror was observed by Bolduan *et al.* [87]. They discovered that when the N_2O_4 was heated to 180 K, autoionization to ionic $NO^+NO_3^-$ occurs and the product remains stable from 15 to 180 K. The transformation temperature in our current experiments, which is between 180 K and 190 K, is in excellent agreement with their result. However, our experiments correspond to the opposite transformation, from ionic $NO^+NO_3^-$ to molecular N_2O_4. The combination of results of the present study with previous observations provides convincing information about the thermodynamic properties of ionic $NO^+NO_3^-$ and molecular N_2O_4, specifically that 1) $NO^+NO_3^-$ is the more stable phase at low temperatures and ambient pressure; 2) the transformation from either side involves a thermochemical barrier characterized by a temperature of about 180 K at ambient pressure; and 3) the transformation is reversible.

3.2.2. *IR spectra and ionicity*

Synchrotron based IR spectroscopy provides appealing advantages in probing detailed vibrational structures of novel materials formed at high pressures, especially with far-IR capacity. In Figure 4 the IR spectra of $NO^+NO_3^-$ obtained at room temperature and pressures ranging from 32.5 to 0.6 GPa are compared. Between 32 and 10 GPa, the IR bands in both the lattice mode and internal vibration regions evolve smoothly with pressure. The pressure dependence of the major IR-active modes is close to linear, and fairly gradual. The smooth evolution of the major IR modes indicates a single phase of $NO^+NO_3^-$ persists in this broad pressure region, consistent with x-ray diffraction measurements [79]. However, when the pressure is reduced further, a significant change in the absorption is observed. For example, at about 3 GPa, the IR spectrum exhibits a significant red shift of the lattice mode, accompanied by abundant new IR bands at 600-800 cm^{-1}, and the disappearance of the peak at 1800 cm^{-1}, although the major internal modes (ν_4, ν_2 and ν_1) are preserved. These changes strongly suggest that a new phase of $NO^+NO_3^-$ occurs in the low-pressure region. On further release of pressure to below 1 GPa (Fig 4), the absorption pattern again changed dramatically, including the loss of characteristic modes of ionic $NO^+NO_3^-$. It can be concluded that below 1 GPa, $NO^+NO_3^-$ has transformed to molecular N_2O_4, since the major peaks match the active modes of N_2O_4 unambiguously [85, 88, 90, 91]. The three strongest peaks centered at 740, 1250 and 1722 cm^{-1} can be assigned as ν_{12} (B_{3u}, NO_2 deformation), ν_{11} (B_{3u}, NO_2 symmetric stretch) and ν_9 (B_{2u}, NO_2 asymmetric stretch) of ordered or disordered N_2O_4 [88, 90, 91, 119].

The characteristic absorption peaks observed in the present study, such as ν_{NO+} (2264 cm^{-1}), ν_1 (1130 cm^{-1}, NO_3^- symmetric stretch) and ν_3 (1403 cm^{-1}, NO_3^- asymmetric stretch), are in excellent accord with the previous IR investigations of $NO^+NO_3^-$ conducted at atmospheric pressure and at low temperatures [88, 90, 91]. However, additional peaks were observed in both low and high frequency IR spectra that were not observed in the previous atmospheric pressure measurements. As an ionic form of molecular N_2O_4, nitrosonium nitrate had previously been produced by thermal or photolytic autoionization [87, 115-117]. The N_2O_4 molecule is planar with D_{2h} symmetry but has an unstable, asymmetric isomer ONO-NO_2 (denoted D″) which is believed to be the precursor of $NO^+NO_3^-$. However, Givan *et al.* [91] reported observing a direct transformation of N_2O_4 to $NO^+NO_3^-$ at atmospheric pressure without precursors or subsequent induction by visible light. Therefore, the nature of this conversion process, speculated as arising from either intra- or intermolecular mechanisms, still poses an interesting question.

The present work provides the first far-IR data at any pressure on this material, providing important insight into the lattice dynamics. Accurate assignments of the spectra require factor group-analysis [120] based on detailed information on the crystal structures involved. X-ray diffraction studies have provided determination of the unit-cell symmetry and constraints on possible space group, but no information on the atom coordinates needed for detailed symmetry assignments. Alternatively, we can compare the spectra with those of materials that appear related, such as KNO_3, whose IR and Raman spectra have been studied both experimentally and theoretically [119, 121, 122]. The analogic analysis is based on that both materials have an aragonite structure with four ion pairs per cell and both ion pair has the same C_{2v} symmetry [119, 122]. The predicted ambient pressure IR-active modes involving the vibrations between the ion pairs of KNO_3 ($NaNO_3$) are ν_4(A1): 234 (322) cm^{-1}, ν_9(B2): 187(255) cm^{-1} and ν_6 (B1): 73 (108) cm^{-1} [121]. There are 18 predicted active Raman modes in the lattice region of phase II KNO_3, as compared with fewer experimentally observed and reported in the region of 53-165 cm^{-1}. Of these Raman modes, those below 100 cm^{-1} are believed to be rotational (or librational) modes of the nitrate ion while those above 100 cm^{-1} are due to translational modes [119, 122]. Considering the above, we believe all peaks observed at 180-360 cm^{-1} in the far-IR spectra of $NONO_3$ are most likely due to the vibrations between the NO^+ and NO_3^- ion pair, while rotational modes may occur at lower frequencies (not resolved in the present study). Further x-ray and spectroscopic studies are required to address these issues unambiguously.

One of the interesting questions regarding nitrosonium nitrate is the degree of ionicity, or symmetry breaking charge transfer in the material, and how this changes with pressure. Our previous work suggested evidence for an increase in charge transfer with increasing pressure based on the behavior of the high frequency mid-IR bands [76]. The far-IR contains much more definitive information, because of some uncertainty in the assignment of the higher frequency bands, but this region was not explored in the previous study. Clearly, the overall intensity of the bands in the far-IR region increases with pressure, indicating that the ionicity of $NO^+NO_3^-$ is enhanced by pressure.

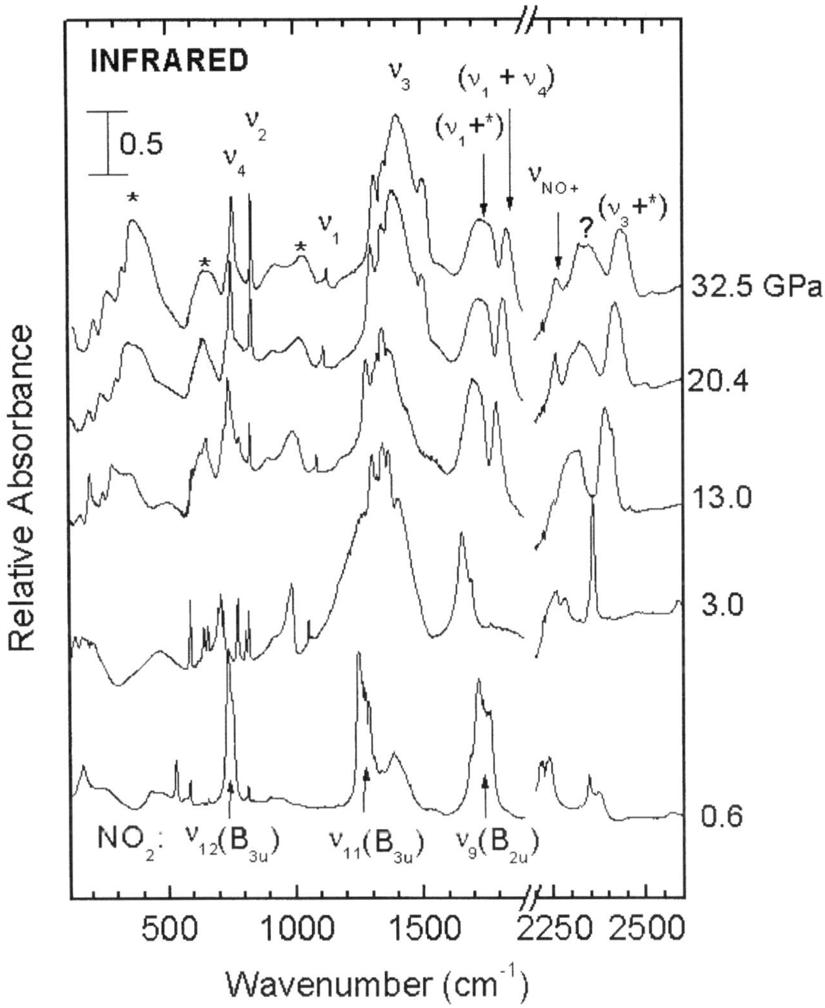

Fig. 4. IR spectra of $NO^+NO_3^-$ in the range 100-2500 cm^{-1} measured at room temperature for five pressures (indicated in GPa on right hand ordinate). The absorbance has been normalized with respect to the beam current of the synchrotron light source. The sample thickness was about 23 μm. The region 1900-2200 cm^{-1} is omitted because of interfering absorptions from the Type IIa diamonds used as anvils. Asterisks (*) indicate lattice modes or combinations. (from Ref. [80])

3.2.3. X-ray diffraction and equations of state

Figure 5 shows typical diffraction patterns for pressures of 9.9, 21.4 and 32.2 GPa. The assignments of the major peaks arising from $NO^+NO_3^-$ diffraction are labeled. By tracing the major peaks, the diffraction patterns can be consistently assigned at all pressures. The conversion from energy to d-spacings was based on the well-known equation [15]. In the previous x-ray diffraction study carried out at a single pressure of 21 GPa [76], diffraction due to ε-N_2 was identified and the remaining peaks of $NO^+NO_3^-$ were indexed to give an orthorhombic unit cell with cell parameters a=5.658 Å, b=7.324 Å and c =6.202 Å. The systematic extinctions in the diffraction pattern suggested space groups *Pmcn* or $P2_1cn$, similar to the aragonite phase of $CaCO_3$ and KNO_3. In the analysis of our new data, the indices of the peaks observed at all pressures likewise suggest the point group of *mm*2 with a primitive cell. The coincidence of the Raman and IR lines indicates a non-centrosymmetric cell and therefore $P2_1cn$. The unit-cell volume was calculated from the orthorhombic cell parameters and the molecular volume was determined assuming each cell containing four molecules (Z=4). The room-temperature P-V relations for $NO^+NO_3^-$ are plotted in Fig. 6. Also shown is the data point from the new refinement of the earlier study at 21 GPa with cell parameters a=5.66(2) Å, b=6.47(2) Å and c=5.39(1) Å, which gives a denser structure than previously reported [76]. Recently, Yoo *et al.* [81] re-examined the high pressure structure of $NO^+NO_3^-$ and reported a higher density for this material. Their results are also plotted in Figure 6.

The compression data for $NO^+NO_3^-$ were fit to a Birch-Murnaghan [128] and Vinet [129] equation of state (EOS). The bulk modulus and its derivative are determined to be K_0 = 45.2 (±0.9) GPa, K_0'=3.18 (±0.90) and K_0 = 42.3±1.0 GPa and K_0'=3.52±0.60, respectively. The results of both fits are shown in the figure. In addition, both of these values are consistent with the later reported K_0 = 45.0 GPa by Yoo *et al.* [81]. Figure 6 also shows the room-temperature *P-V* relations for O_2, N_2 and N_2O. Also shown is the volume of the assemblage of one N_2 and two O_2 molecules, an equivalent stoichiometric assemblage of $NO^+NO_3^-$. The zero-pressure molecular volumes for N_2 and O_2 are 45 Å^3 and 40.2 Å^3 [124]. The molecular volume for $NO^+NO_3^-$ is 66.25 Å^3 as compared to the volume of the assemblage $N_2 + 2O_2$ of 125.4 Å^3. The unit cell volume for molecular N_2O_4 (isomer of $NO^+NO_3^-$) has been determined by single crystal diffraction at -40 °C and ambient pressure [83]. The results gave 78.18 Å^3/ molecule, 18% larger than the fitted zero-pressure volume for $NO^+NO_3^-$ obtained here. Moreover, $NO^+NO_3^-$ is denser than the assemblage at all pressures studied here. The cell volume of O_2 determined by single crystal x-ray diffraction measurements by Johnson *et al.* [130] are plotted in Fig. 6. The results indicate that $NO^+NO_3^-$ is a denser phase than the N_2 and O_2 assemblage. The *P-V* relation of N_2O derived from x-ray diffraction by Mills *et al.*[127] is also plotted. It can be seen that $NO^+NO_3^-$ is also denser than N_2O + 2/3 O_2.

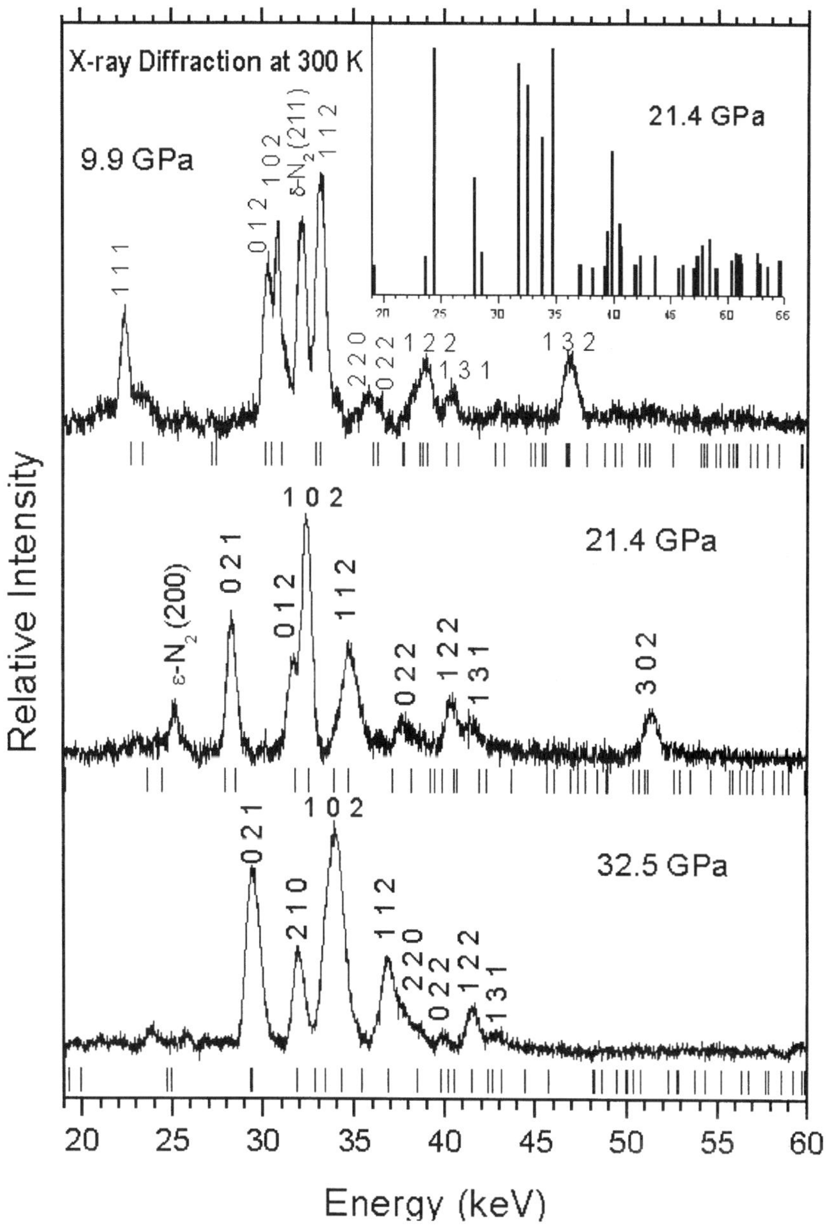

Fig. 5. Energy dispersive x-ray diffraction pattern of $NO^+NO_3^-$ measured at (a) 9.9 GPa, (b) 21.4 GPa and (c) 32.2 GPa and room temperature. Background has been subtracted. The energy calibration was obtained from a gold external standard diffraction pattern and the pattern has been background subtracted. The 2θ used was 8.99°. The calculated d-spacings are indicated below each diffraction pattern. The calculated intensity profile for the energy–dispersive x-ray diffraction pattern at 21.4 GPa is shown in the inset. (from Ref. [79])

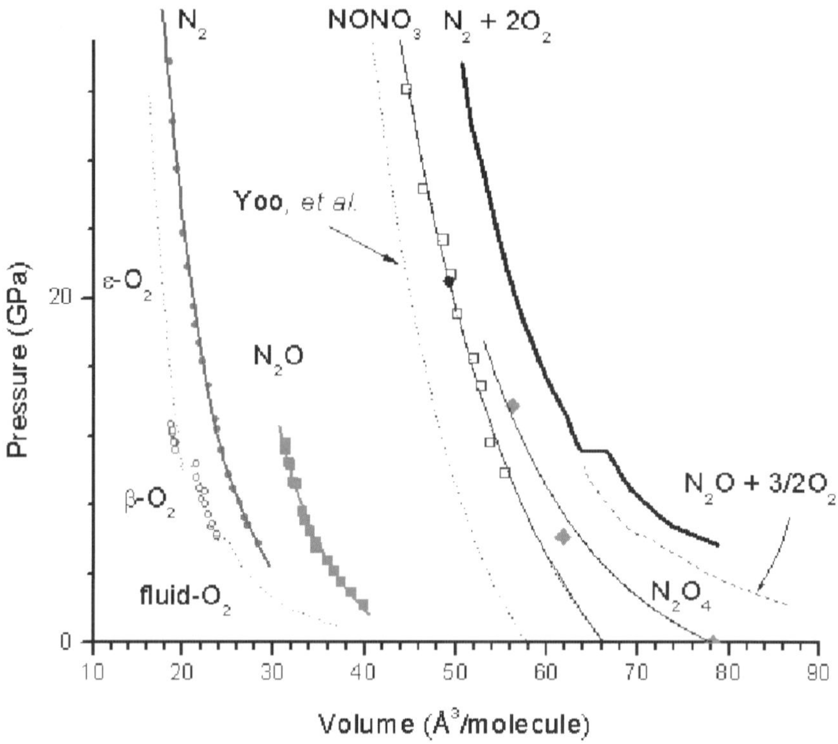

Fig. 6. Pressure-volume relations for $NO^+NO_3^-$ and other molecular systems. $NO^+NO_3^-$ determined from the present energy-dispersive x-ray diffraction (□) and that from previous angle-dispersive x-ray diffraction with refined cell parameters (•), and that from C.S. Yoo *et al.* (···) (Ref. [81]), compared with a third-order Birch-Murnaghan (—) and Vinet *et al.* EOS fits (-.-). For O_2 (···) data, below 5.5 GPa are for fluid O_2 (Ref. [123]); above 5.5 GPa for the solid (Ref. [124]). Experimental data for O_2 (o) at several pressures performed from Ref. [125] are also plotted. For N_2 (•), experimentally determined EOS is from Ref. [126], for N_2O (■) from Ref. [127]. Volumes for N_2O_4 (♦) determined in the present study is fitted by the Birch-Murnaghan equation of state (—) tentatively. Also shown are the corresponding volumes of stoichometrically equivalent assemblages of $N_2 + 2O_2$ (—) and N_2O+ 3/2 O_2 (---).

3.2.4. Stability diagram

The density of $NO^+NO_3^-$ established by the equation of state gives important insight into the stability, thermodynamic properties, and the reaction mechanisms related to $NO^+NO_3^-$. Previous observations of the formation of $NO^+NO_3^-$ were either by temperature-induced transformation at ambient pressure or by photon-induced autoionization of molecular N_2O_4 at low pressures [86, 87, 116, 117]. However, the symmetry-breaking transformation (or chemical reaction) from N_2O to $NO^+NO_3^-$ can be interpreted as being driven by the higher density of the product $NO^+NO_3^-$ + N_2. Upon heating N_2O breaks down via two competing channels. Below 10 GPa, heating N_2O results in its dissociation into N_2 and O_2, while above 10 GPa, laser heating of the sample predominantly forms $NO^+NO_3^-$. The blocking of the dissociation channel by high pressure strongly indicates that $NO^+NO_3^-$ is a more stable phase with lower free energy at high pressures. This observation, together with the density comparison, suggests that heating a mixture of N_2 and O_2 under pressure will directly produce $NO^+NO_3^-$, a result that has been confirmed [131]. Kinetic factors associated with these reaction channels should be investigated further. These results provide evidence that at high pressures $NO^+NO_3^-$ is a stable phase both at room temperature and high temperatures.

These observations provide a basis for extending the stability diagram of N_2O to high pressures and temperatures and provide useful information for understanding the formation of $NO^+NO_3^-$ from other species. At room temperature and ambient pressure, N_2O is a colorless gas and becomes fluid at 184 K, subsequently solidifying at 182 K. At low pressures and room temperature, N_2O forms the α- phase (*Pa*3) below 4 GPa and β-phase (*Cmca*) above 5 GPa. At intermediate pressures, x-ray diffraction measurements indicate the coexistence of the two phases [127]. It is reported that the transition pressure between α and β has no significant temperature dependence. The melting point of N_2O was measured by Clusius *et al.*[132] to 0.025 GPa, and was extrapolated by Mills *et al.* [127]. In our study we also explored the melting curve at several other pressures using the resistance heating method. The melting was confirmed by both visual observation and Raman spectroscopy. We have refitted the melting curve using a Simon type equation on both current measurements and those from Ref. [132] (Fig 7). On heating, N_2O transforms to $NO^+NO_3^-$ and N_2 irreversibly at high pressure, it can dissociate into nitrogen and oxygen upon heating at other pressures. No attempt was made to study the reaction yield as a function of pressure and temperature. Several parallel heating experiments conducted at different pressures up to 40 GPa indicate that the transformation is complete and $NO^+NO_3^-$ is stable up to 2000°C. The region where N_2O transforms to $NO^+NO_3^-$ is shown schematically in Fig. 7. We note that molecular N_2O was found to be stable up to 40 GPa and below 300 K (i.e., without heating) [126]. Additional transformations at intermediate P-T conditions to form additional phases of molecular N_2O have been reported [82].

The crystal structure of $NO^+NO_3^-$ at 21 GPa appears to be orthorhombic with four molecules per unit cell. By analogy to related ionic materials, a possible space group is $P2_1cn$ [76]. In the present study, cell parameters were found to evolve smoothly over the entire pressure range from 9.9 to 32.2 GPa. This is consistent with IR and Raman measurements [80], which likewise indicate that no major phase transitions occur in this pressure range. However, these spectroscopic data do reveal the presence of a transition below 10 GPa. The lowest pressure at which we observed x-ray diffraction at room temperature was 6.3 GPa. The diffraction pattern at this pressure differs significantly from those at higher pressures, such that the cell parameters are not consistent with the same orthorhombic structure. At 2.7 GPa, the diffraction peaks have become too weak to clearly identify, even when the sample was exposed to x-rays for a prolonged period. This weakening of the diffraction peaks may indicate that $NO^+NO_3^-$ at this pressure has an amorphous or disordered structure. Notably, it has been reported that at atmospheric pressure, $NO^+NO_3^-$ is predominantly in an amorphous phase [88, 90, 91]. We suggest that $NO^+NO_3^-$ transforms at room temperature from the orthorhombic structure to a disordered form on decompression from 9.8 GPa to 2.7 Gpa. This transition may be gradual, with intermediate ordered or partially ordered structures in between, making the boundary difficult to determine.

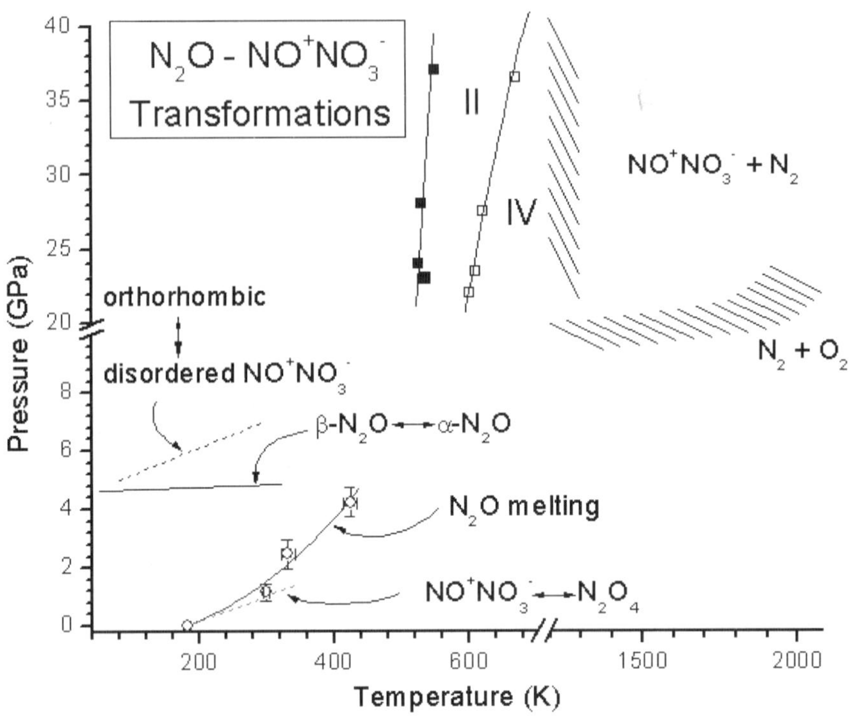

Fig. 7. Schematic phase and reaction diagram of $NO^+NO_3^-$ and N_2O. The boundary (—) between the α-Pa3 and β-Cmca phases of N_2O is from Ref. [127]. The phase boundary between α and fluid is indicated by open circles (present study) and solid line (fitted by Simon-type equation on data from both current measurements and Ref. [132]). The approximate P-T regime for the transformation of N_2O into $NO^+NO_3^-$ or N_2 and O_2 is indicated by shaded lines (\\\\\\). The stability fields of other high P-T phases of N_2O reported by Iota et al. are also shown (■ and □ with — as eye guide). The boundary (···) between ionic $NO^+NO_3^-$ and molecular N_2O_4 is from the spectroscopic measurements (Ref. [80]). The approximate boundary (···) for the transformation (which may be gradual) between the orthorhombic $NO^+NO_3^-$ and disordered $NO^+NO_3^-$ was estimated from behavior of x-ray diffraction patterns observed at low temperatures and room temperature (see text).

3.3. Nitrogen oxide: molecular N_2O_4 revisited

Extensive investigation on ionic $NO^+NO_3^-$ under high pressures provided essential information on the structure, dynamics and transformation of this species and its molecular isomer N_2O_4. In this section we come back to look at molecular N_2O_4 under high pressures and high temperatures since there are still fundamental questions that remain unanswered, such as 1) What is the higher-pressure behavior of this compound? Up to now there has been just one high-pressure study [86] on this molecular solid, extending up only 7.6 GPa. 2) What is the corresponding high-temperature behavior? Previous studies focused chiefly on the low-temperature region (< 300 K) [87, 88]. 3) What are the crystal structures of the materials under pressure? As yet no x-ray diffraction measurements at high pressures have been reported. In addressing these questions, we document a new transition associated with pressure-induced change in molecular geometry. Further experiments using Raman spectroscopy combined with CO_2 laser heating identified the ionic phase of N_2O_4 in the high P-T region.

3.3.1. Vibrational spectroscopy

NO_2 (or N_2O_4) was loaded cryogenically into a DAC to various pressures before warming up to room temperature. Then Raman spectra were collected on compression. Significantly different Raman features are observed when the pressure is increased from 8.8 to 12.3 GPa (see Figure 8). These features, indicating a new phase of N_2O_4, include enhanced structure in the lattice mode region 210 cm^{-1} to 370 cm^{-1}, splitting and broadening of the peak at 730 cm^{-1} (ν_8 B_{1g}, NO_2 wagging mode) and the appearance of new peaks at 1104 $cm^{-1}$and 2208 cm^{-1}. We designate the new phase as γ (β-N_2O_4 was first observed [86] by laser irradiation of α-N_2O_4 at 1.2 GPa). In the IR spectra, such pressure-induced changes are also consistent with a phase transition between 8 and 11 GPa, although the changes are much less pronounced than those seen in the Raman spectra.

To probe the high temperature regime, we performed heating experiments at several high pressures using the CO_2 infrared laser. This spectrum exhibited vibrational modes and a lattice profile characteristic of the ionic species $NO^+NO_3^-$ [76, 79, 80]. Previously, the transformation from molecular N_2O_4 to ionic $NO^+NO_3^-$ had been observed only at ambient pressure when induced by heating or at pressures below 3 GPa by laser irradiation. Our studies seem to be the first to observe such a transformation at high pressure and high temperature.

The Raman data indicate that γ–N_2O_4 has a molecular structure closer to α–N_2O_4 than to $NO^+NO_3^-$, with some but less pronounced ionic character and lower symmetry than D_{2h}. Even when the pressure was increased up to 18 GPa, the prominent peak at 2208 cm^{-1} did not shift noticeably. This peak, the most distinctive for γ–N_2O_4, is appreciably lower in frequency but correlates with the stretching mode of the NO^+ moiety seen in the ionic isomer $NONO_3$.

The pressure and temperature induced phase transitions reported here are strongly associated with kinetic factors and path dependent. When the N_2O_4 was loaded cryogenically to high pressure (> 6 GPa) before warming, γ–N_2O_4 is readily accessible as described above. However, when the initial loading pressure is not sufficiently high, for example, only raised to 2 GPa before warming, the γ–N_2O_4 is suppressed even when very high pressure is subsequently applied. Figure 8 shows a Raman spectrum obtained under the latter conditions; even for a final pressure of 13.0 GPa, the two most characteristic vibrational modes for the γ–phase (i.e., 1104 cm^{-1} and 2208 cm^{-1}) are not observed in the Raman spectrum (despite similar laser power and exposure time). On the other hand, compressing the γ–N_2O_4 to higher pressures (> 20 GPa) without heating results only in enhanced intensity of the characteristic peak at 2208 cm^{-1}, without typical pressure-induced frequency shift (see Figure 8 inset), indicating that pressure is insufficient to induce complete transformation to the ionic phase. It is also found that at different pressures, the extent of transformation induced by heating varies. For example, at 8.3 GPa the conversion to γ–N_2O_4 is incomplete, as judged from the intensity of Raman active modes associated with residual α–N_2O_4 (Fig. 8).

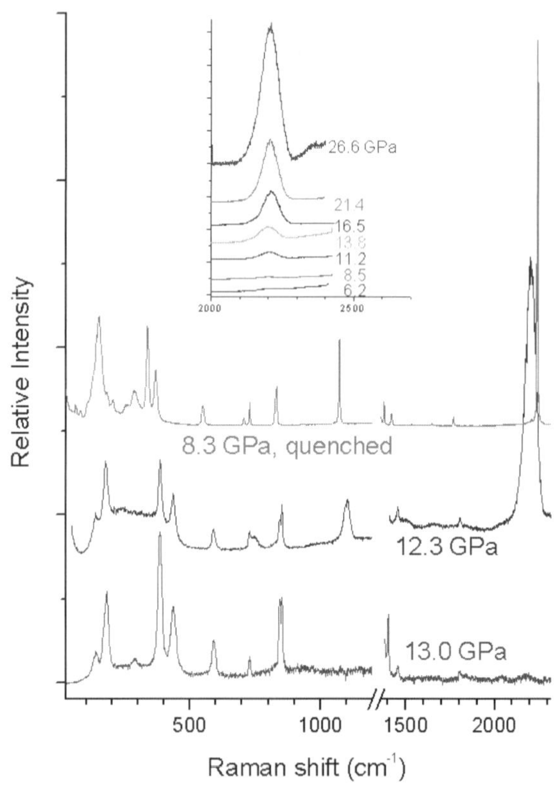

Fig. 8. Raman spectra of N_2O_4 measured for different conditions. Top spectrum for sample heated at 8.3 GPa and quenched to room temperature. Middle spectrum for sample loaded to high pressure (> 6GPa) at low temperature, then warmed to room temperature and pressurized to 12.3 GPa. Bottom spectrum for sample loaded to low pressure (< 2 GPa) at low temperature, then warmed to room temperature and pressurized to 13.0 GPa. The inset shows the evolution of the characteristic peak at 2208 cm^{-1} on compression under the same conditions under which the middle spectrum was obtained.

3.3.2. X-ray diffraction

The crystal structure of N_2O_4 at high pressure is important for establishing the transformation mechanism and the equation of state. Previous x-ray diffraction at ambient pressure and -40°C found that solid N_2O_4 crystallizes in a cubic unit cell with space group *Im3* (T_h^5) and six molecules per unit cell.[83] The cell parameter a =7.77 Å yields a cell volume of 469.1 Å^3 and a molecular volume of 78.2 Å^3. On the basis of Raman measurements, Agnew *et al.*[86] suggested that high-pressure N_2O_4 is identical to the low-temperature cubic *Im3* phase. Since the x-ray diffraction pattern we obtained at 6.2 GPa can be consistently indexed by the same *Im3* (T_h^5) space group, our study supports their argument and provides experimental evidence that N_2O_4 can persist from ambient pressure up to at least 12 GPa. Figure 9 displays x-ray diffraction patterns collected at 6.2 GPa and 13.8 GPa. The patterns conform well with indices for a cubic unit cell. The space group *Im3* is assumed as a starting point, as indicated by the previous ambient pressure and low-temperature x-ray diffraction study of single-crystal N_2O_4 [83], although in our studies N_2O_4 is in a high-pressure phase.

As our Raman spectra indicate pressure-induced phase transitions above 12 GPa, we expected to find x-ray patterns differing from that at ambient pressure. However, the pattern we obtained at 13.8 GPa showed only smooth d-spacing shifts with pressure, with the ambient pressure d-spacings preserved, and thus indexed by the same space group. The analysis assumes that each unit cell contains six molecules. To check the plausibility of that assumption, we compared in Fig. 6 the P-V equations of state for oxygen and nitrogen with that estimated from the unit cell volumes. The molecular volume of N_2O_4 is seen to be bounded by that of $NONO_3$ and the assemblage N_2 + $2O_2$, in excellent agreement with previous observation that $NONO_3$ is denser than molecular N_2O_4 [79].

3.3.3. Transformation mechanisms

The main evidence for a transition near 12 GPa stems from the appearance of new peaks and altered lattice profiles in our Raman spectra. In our heating experiment, the two strong peaks at 1097 cm^{-1} and 2250 cm^{-1} at 15.3 GPa can be assigned to characteristic stretching modes of NO_3^- and NO^+, respectively. At 12.3 GPa, these peaks were likewise prominent in previous Raman studies at ambient pressure and low temperature [87, 88] wherein spectral features induced by radiation indicated formation of isomeric forms of N_2O_4. The Raman spectrum we observed at 8.8 GPa indicates the molecular geometry at this pressure has D_{2h} symmetry since the spectrum exhibits the corresponding active Raman modes, as seen in the N_2O_4 spectrum at low temperature and ambient pressure. On compression, the appearance of new peaks associated with the principal NO^+ and NO_3^- modes indicates that the molecular symmetry is no longer D_{2h}. This conclusion is reinforced by the broad structure that emerges in the lattice region. Formation of the ionic $NONO_3$ species implies a totally broken symmetry as established in previous studies [76, 79, 80]. The high-pressure phase (designated as γ- N_2O_4) thus could be interpreted as intermediate between the α-phase of molecular N_2O_4 with D_{2h} symmetry and the ionic $NONO_3$ with an orthorhombic structure [76]. The molecular units within the γ phase could be close to the D or D' type isomers of

N_2O_4 known at ambient pressure [87, 88]. Figure 10 summarizes the pressure-induced transitions between phases with different molecular geometries.

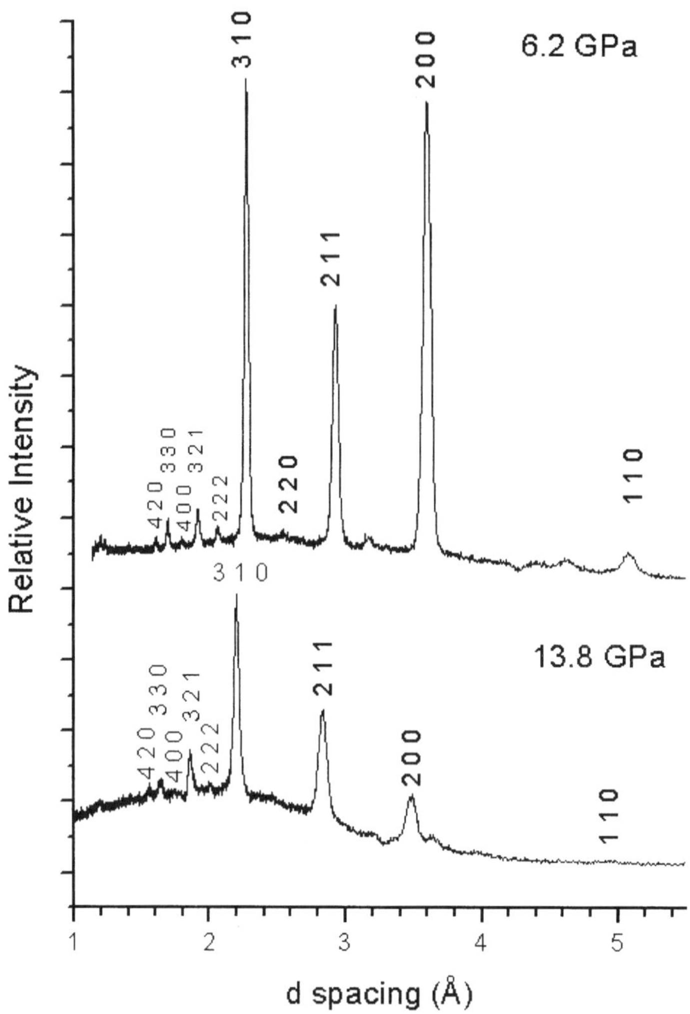

Fig. 9. X-ray diffraction patterns for N_2O_4 obtained at 6.2 GPa and 13.8 GPa. The respective 2θ angles used were 9.00° and 13.00°. The weak peaks are most likely associated with diffractions from other species of impurity, such as gasket material. (from Ref. [93])

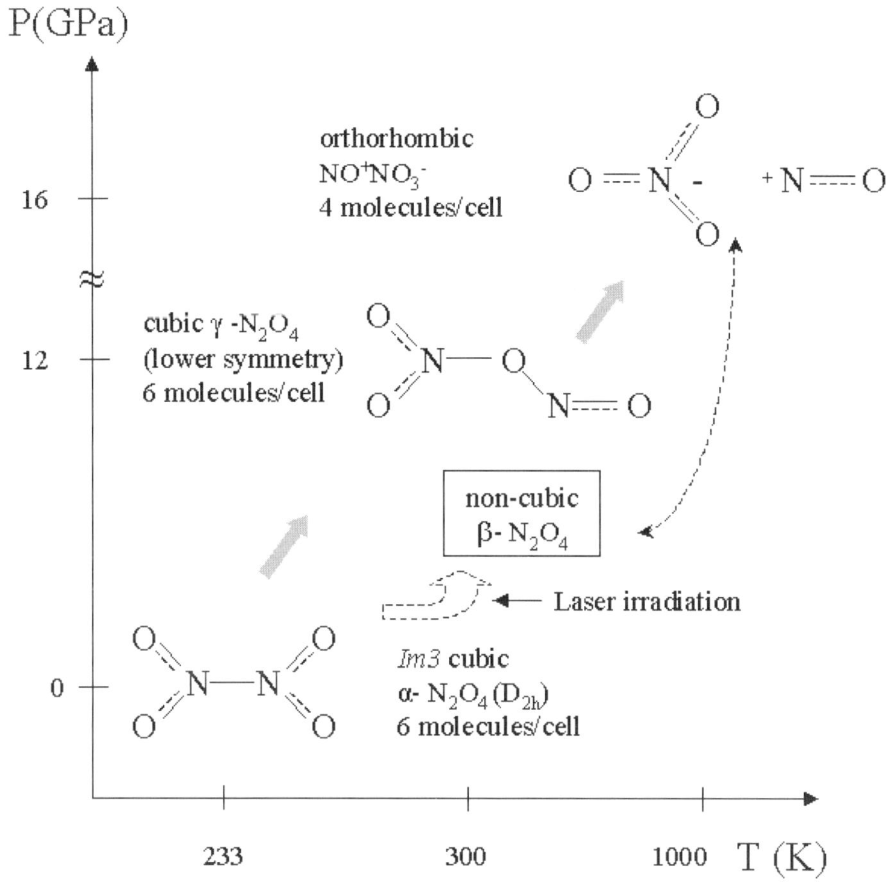

Fig. 10. Schematic diagram of phase transitions of N_2O_4 in the P-T region of the present study. Gray arrows indicate phase transitions induced by high pressures and temperature. The arrow with dashed edge indicates the photon-induced transition. The long dashed line with double arrows indicates a reversible transformation between the β and ionic phases. (from Ref. [93])

3.4. Further pursuit of polynitrogen

3.4.1. Pentanitrogen hexafluoroantimonate (N_5SbF_6)

Polynitrogen has been of particular interest due to its promising potential to serve as high energy density material [32]. However, the currently known nitrogen species are limited:

only a few species, such as N_2, N_3^- and N_4 are experimentally accessible [100, 133]. The recent synthesis of N_5^+ salt by Christe *et al.* [134] has given great impetus to efforts to discover larger polynitrogen species. The N_5^+ salt was first synthesized as a hexafluoroarsenate as $N_5^+AsF_6^-$ [134] and later as more stable species, $N_5^+SbF_6^-$ and $N_5^+Sb_2F_{11}^-$ [135]. The remarkably stable N_5^+ cation has a V-shape geometry; all the nine vibrational frequencies corresponding to this geometry have been observed.

The experimental discovery of N_5^+ salt stimulated a number of theoretical studies. Bartlett and colleagues [136] studied the stability of $N_5^+N_5^-$ salt. It represents a potential solid nitrogen rocket fuel that would be much more efficient than currently rocket propellant. Unfortunately, due to the unavailability of N_5^-, a direct experimental test of this ion pair is not feasible. The stability of another ion pair, $N_5^+N_3^-$, has been investigated by Kortus *et al.* [137]. Unlike $N_5^+N_5^-$ for which a stability minimum was predicted as two-ion-pair clusters, the N_5^+ N_3^- ion pair can spontaneous isomerize to azidopentazole with lower energy and the latter will decay to molecular N_2 spontaneously [137]. More recently, using *ab initio* molecular orbital theory, Dixon *et al.* [138] predicted that neither the $N_5^+N_5^-$ nor the $N_5^+N_3^-$ ion pair are stable; both should decompose spontaneously into N_3 radicals and N_2.

These theoretical studies prompted us to make preliminary studies of the structure and stability of N_5^+ salt at high pressures as well as to examine the possibility of forming an ion pair with azide anion [139]. We loaded $N_5^+SbF_6^-$ or a mixture with NaN_3 into the DACs and carried out Raman and IR measurements at different pressures and at room temperature. Fig 11 shows the Raman spectra of pure $N_5^+SbF_6^-$ at pressures of 1.4 to 29.1 GPa at room temperature, in the region of 2200 to 2400 cm^{-1} where the characteristic fundamentals of ν_7 and ν_1 for N_5^+ appear. Since most low frequency modes are associated with SbF_6^- anion, their evolutions on pressure are not of particular interest in this study. At modest pressures such as 1.4 GPa, the ν_7 mode occurs at 2217 cm^{-1} with less intensity than the sharp mode of ν_1 at 2274 cm^{-1}. The small peak at 2342 cm^{-1} could be associated with N_2 from the unavoidable spontaneous decomposition of the salt during loading. As pressure is increased, the ν_1 mode loses intensity while the intensity for the ν_7 mode remains more or less the same. When pressure is increased to 15 GPa (not shown here), a new mode starts to evolve and become prominent at 2294 cm^{-1} at 17.8 GPa. Upon further compression, this mode dominates the high frequency region as observed at 20.7, 23.1 and 29.1 GPa while the old ν_7 and ν_1 evolve into broad peaks at these high pressures. The occurrence of the new peak could indicate a major change of the "V" shaped geometry of N_5^+ predicted at ambient pressure.

Under high pressure, intermolecular distances are reduced such that the interaction between adjacent ions could play a more prominent role that is responsible for this new mode. Since the x-ray diffraction measurements on N_5^+ are only available so far in the $Sb_2F_{11}^-$ salt, direct *in situ* diffraction measurement on $N_5^+SbF_6^-$ are therefore required to confirm the high pressure geometry for both N_5^+ cation and SbF_6^- anion. Nevertheless, a phase transition can be proposed around 13-15 GPa. To demonstrate this more clearly, we plot the Raman shifts of ν_1, ν_7 and ν_8 as a function of pressure in Fig 12. The Raman mode at 2294 cm^{-1} appearing at 17.8 GPa can be attributed to a new mode instead of continuation of ν_7 due to the significantly different origins. The slight discontinuity at about 13 GPa also suggests the occurrence of a new high pressure phase.

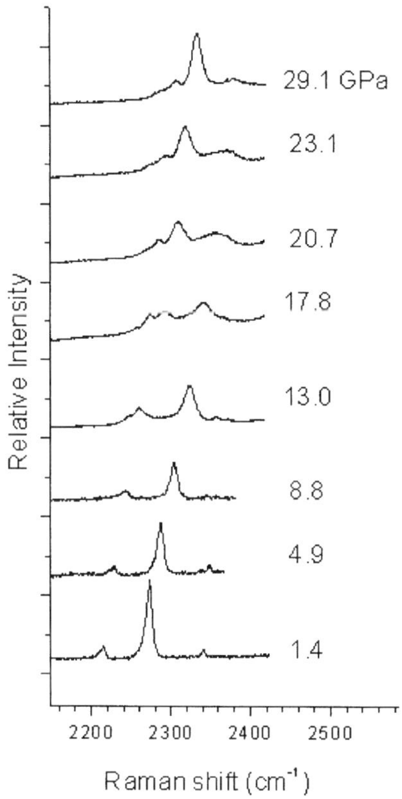

Fig. 11. Raman spectra of N_5^+ at high pressures from 1.4 to 29.1 GPa in the region of 2200-2400 cm^{-1}. (from Ref. [139])

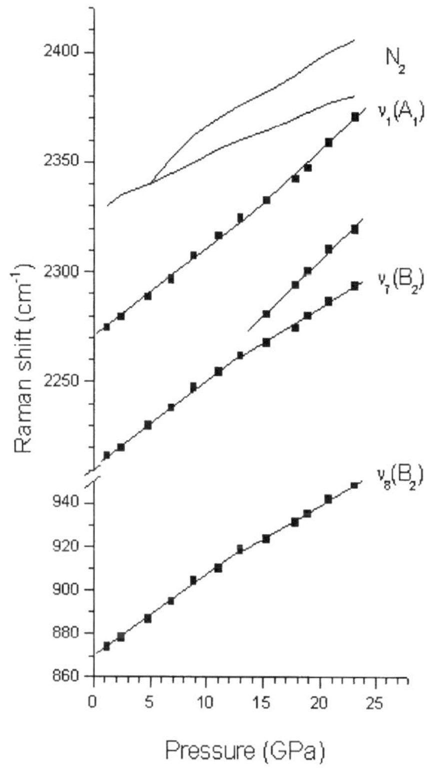

Fig. 12. Raman shifts of ν_1, ν_7 and ν_8 modes (■) of N_5^+ vs. pressure. The solid lines across are for eye guidance. The nitrogen vibrons are also plotted for comparison. (from Ref. [139])

Under ambient conditions, the stability of N_5^+ was an essential requisite for synthesis of polynitrogen salts. The first example, $N_5^+AsF_6^-$ salt, tended to explode, whereas $N_5^+SbF_6^-$ and $N_5^+Sb_2F_{11}^-$ proved more stable. Studies of the stability N_5^+ salts at higher pressures and temperatures may aid discovery of other polynitrogen species. It is found that if $N_5^+SbF_6^-$ is pressurized to 4.5 GPa and heated, only when the temperature exceeds 205 °C, does the sample start to decompose, as evidenced by the appearance of N_2 vibration at 2333 cm^{-1}. However, prolonged heating at this P-T condition did not result in the complete decomposition of the salt. In contrast, heating the sample at a lower pressure, such as 2.7 GPa to the same temperature induced significantly more complete transformation, evidenced by the depletion of ν_7 mode as well as the appearance of new modes near 690 cm^{-1} and in the lattice region. These heating experiments demonstrated that thermal stability of N_5^+ salt is enhanced by pressure. This suggests that synthesis of other polynitrogen species might be favorable at high pressures.

The availability of both the N_5^+ and N_3^- salts provide a straightforward way to examine possibility of forming the ionic pair considered in theoretical studies [137]. On compression of a mixture of $N_5^+SbF_6^-$ and $Na^+N_3^-$ to 40 GPa and ambient temperature, we found that the two salts remain "inert" and exhibit their individual high-pressure behavior as two independent single phases. The lack of evidence of a chemical reaction is consistent with experimental results reported by Dixon *et al.* [138]. They found that these two salts can be mixed as dry powers at room temperature without sign of reaction, in contrast to the violent reaction between CsN_3 and N_5SbF_6 [138]. In order to promote reaction, we employed resistant heating at various pressures. When the mixture of N_5SbF_6 and NaN_3 was pressurized to 6.4 GPa and heated to 483 K (210 °C), extensive reaction occurred, as evidenced by the significant depletion of the ν_7 Raman mode at 2278 cm^{-1}. These heating experiments further established that chemical stability of N_5^+ is also enhanced at high pressures [139].

3.4.2. Sodium azide (NaN_3)

To form polynitrogen from N_2 involves breaking the strong triple bond (954 kJ/mol), likely only possible at ultrahigh pressures or temperatures. The azide anion, $(N=N=N)^-$, offers a more amenable precursor, by virtue of its quasi-double bond order and lower bond energy (418 kJ/mol).

Recently, Eremets *et al.* [113] performed a study of NaN_3 at pressures up to 120 GPa by Raman spectroscopy. Upon compression, at least 3 phases are observed. First transition occurs at less than 1 GPa, corresponding to the pressure-induced β to α transition. Then around 15-17 GPa, significant change in the lattice pattern and the appearance of IR active modes ν_2 and ν_3 of azide anion indicate the transition to a new phase, denoted as phase I. Starting from 50 GPa, another new phase seems to develop as new Raman peaks appear, accompanied by darkening of sample. When pressure is increased to 80 GPa, the sample becomes completely opaque. At the highest pressure of 120 GPa, all the Raman features smeared out, indicating the possible formation of the network of nitrogen or polynitrogen. On decompression from 120 GPa, another four different phases are accessed irreversibly.

Despite the observation of profuse spectroscopic changes on compression and decompression of NaN_3, the *in situ* high-pressure structures of the various phases remain unknown. The small sample size required by high pressure studies and large hysteresis due to the strain and stress make difficult *in situ* diffraction measurements. At high temperature conditions, however, the transformation from an azide to a polynitrogen phase may be more readily characterized as the requisite transformation pressures may be much lower and less subject to hysteresis. Here we describe the heating experiments on pure sodium azide and a mixture with boron at high pressures and indeed find that heating enables transformation to new phases of NaN_3 to occur at much lower pressures [140]. Most important, we use angle dispersive x-ray diffraction measurements to directly characterize the newly observed high-pressure structures of azide.

For pure azide, prominent Raman bands are observed mainly in two regions [101, 103]: the lattice region (<500 cm^{-1}) and the symmetric stretch mode (~1450 cm^{-1}) although at higher pressures the 1900 cm^{-1} mode becomes noticeable. Raman spectra were obtained for heated pure NaN_3 at 6.6 GPa, 9.2 GPa, 14.0 GPa and 46.6 GPa. The typical observation upon heating over such a broad pressure range is that the both prominent Raman bands are significantly diminished. The doublet lattice modes at low pressures (<15 GPa) are converted to a very broad band centered at a lower frequency or totally featureless, while the 1450 cm^{-1} mode is completely depleted in all cases. The peak at 1900 cm^{-1} visible at 46.6 GPa also disappears upon heating. No other new features are observed in the Raman spectra of heated pure azide.

However, striking new features are observed when azide is heated in the mixture with metals or amorphous boron. Figure 13 compares the Raman spectra of a heated mixture of azide with boron at 9.6 GPa and 22.2 GPa with spectra obtained before heating. At 9.6 GPa, heating of the mixture produces a very strong and narrow peak at 1940 cm^{-1}. Other features include the appearance a new lattice mode near the doublet although this mode is sometime visible before heating upon loading and compression. The relative Raman intensity of both the lattice mode and symmetric stretching mode decreased upon heating but these remain fairly strong even after prolonged heating. The characteristic peak at 1940 cm^{-1} can be observed upon heating at several pressures up to about 20 GPa, but not observable at higher pressures such as 22.2 GPa (Fig 13). The decompression of the heated mixtures to low enough pressures such as 1.6 GPa results in the transition to low pressure or ambient azide structures (α- and β-phases) [102, 110] again indicating the transformation is reversible.

The structures of sodium azide have been extensively studied at ambient and low pressures [30, 106, 107, 141, 142]. At room temperature and ambient pressure, NaN_3 is a highly ionic structure with one formula unit per primitive cell or three molecules in the hexagonal cell with space group of $R\bar{3}m$ (D_{3d}^5), which is designate as β-azide [102]. By lowering temperature below 20 °C, or increasing pressure, this transforms to a monoclinic phase (α-azide) with a distorted rhombohedral cell and space group $C_{2/m}$ [110]. In this structure, the sodium and azide ions each form two dimensional arrays with N_3^- axis perpendicular to the layers. Figure 14 shows typical diffraction patterns of mixture of azide and amorphous boron under different conditions. For reference, the bottom plot is diffraction pattern of the mixture compressed to 9.8 GPa without heating. When the mixture of azide and boron is heated at 9.6 GPa and then quenched to room temperature, the diffraction pattern becomes dramatically different, indicating a new structure. This is also confirmed by the Raman spectra, with the appearance of the characteristic peak at 1940 cm^{-1}. In another experiment, when the mixture is pressurized to 22.2 GPa then heated, the diffraction pattern displays another profile different from that for the unheated sample, and with broadened peaks. To facilitate comparison, in Fig. 14 we also plot the diffraction pattern obtained when the heated sample was decompressed to 8.5 GPa.

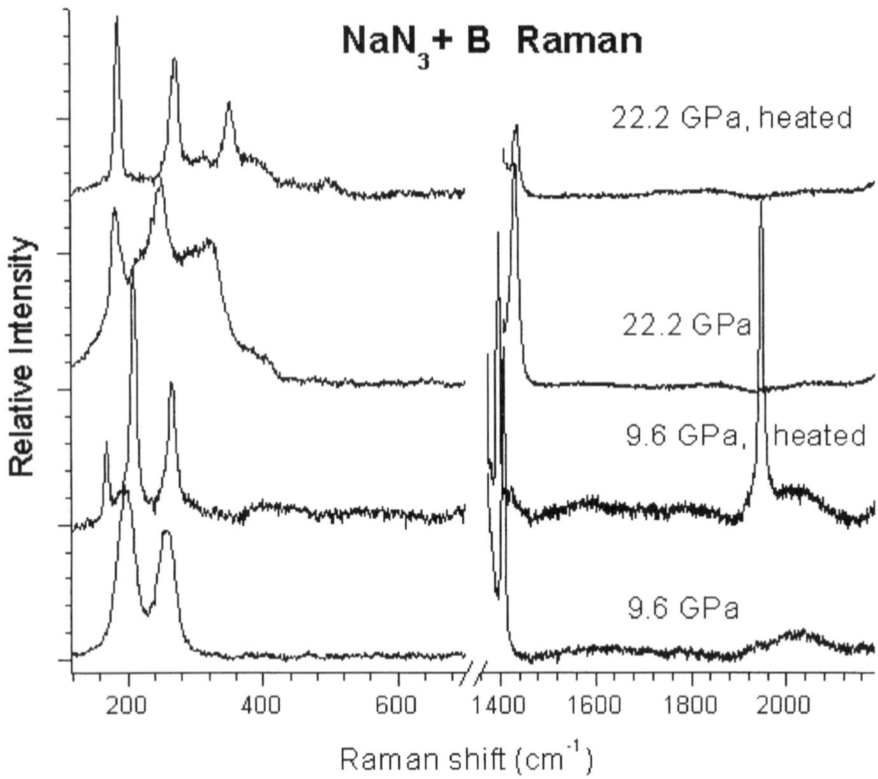

Fig. 13. Raman spectra of a mixture of azide and amorphous boron at 9.6 GPa and 22.2 GPa before and after heating by a CO_2 infrared laser. The region of 1250-1350 is obscured by the T_{2g} mode of the diamond anvil. (from Ref. [140])

The preliminary analysis of diffraction data for the pure azide at 9.8 GPa without heating indicates it remains a possible monoclinic structure with space group $C_{2/m}$. The cell parameters are determined to be a=5.635(8) Å, b=3.419 (6) Å and c=4.936(8) Å, β=99.5(1)° and V=93.8 $Å^3$. Compared with the monoclinic structure at ambient pressure and low temperature, for which the cell parameters are a=6.1654 Å, b=3.6350 Å and c=5.2634 Å, β= 107.543°, the high-pressure phase seems to remain monoclinic structure but with isotropic compression of three axes with compression ratio of 91%, 94% and 94% respectively. This indicates that high-pressure phase at 9.8 GPa has the same or a similar structure as ambient low temperature phase, although the unit cell may be a further distorted monoclinic structure characterized by a different β angle.

The diffraction pattern of heated sample at 9.6 GPa displays more sharply resolved peaks indicating a totally different structure than the unheated sample at the nearly the same pressure, 9.8 GPa. This phase could have even lower symmetry or the diffraction pattern originates from a mixture of phases. Due to the broad profile of the diffraction patterns at 22.2 GPa and its decompressed sample at 8.5 GPa, unambiguous analysis is feasible. The angle dispersive x-ray diffraction measurements here are the first attempt to determine high P-T structures of sodium azide. Additional systematic measurements on compression should permit structure refinement and evaluation of the equation of state to aid interpretation of the new structures indicated by the Raman spectra [140].

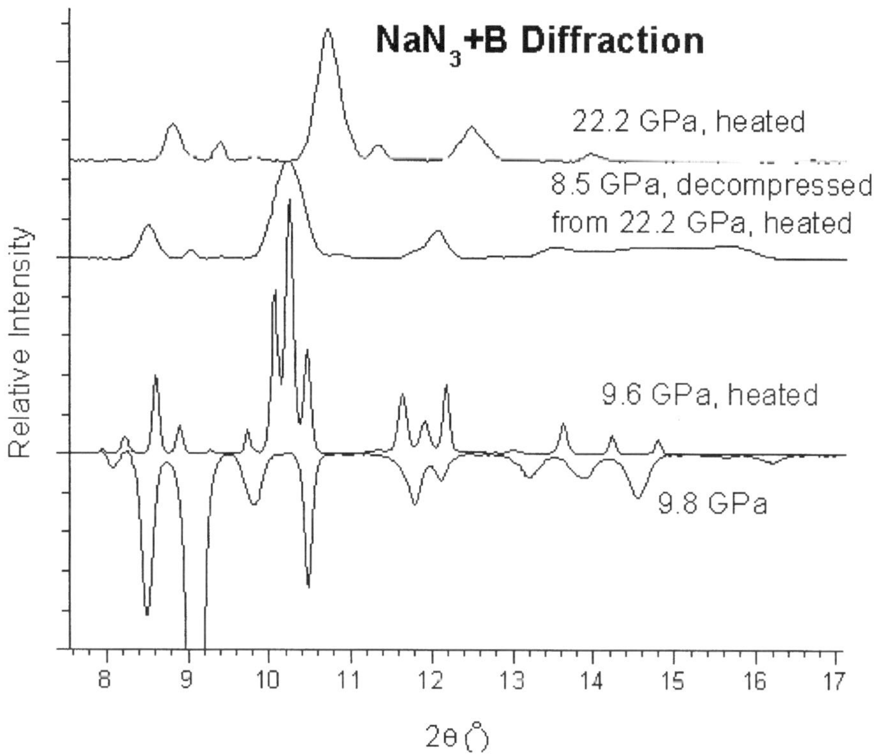

Fig. 14. Angle dispersive x-ray diffraction patterns for mixture of azide with amorphous boron collected at different conditions. Bottom flipped pattern is from the sample compressed to 9.8 GPa without heating. Due to the strong intensity at $2\theta=9.094^{\circ}$, this pattern is flipped for convenience of comparison with other patterns. The pattern at ground level is from the sample heated at 9.6 GPa. The top pattern is from the sample compressed to 22.2 GPa followed by CO_2 laser heating, and the pattern immediately below it is collected after decompression from 22.2 GPa to 8.5 GPa of the same sample. (from Ref. [140])

4. THEMATIC PERSPECTIVES AND PROSPECTS

As illustrated in Fig. 1, compressing molecules "loosens" the electronic structure. When the neighboring electron clouds crowd in, the consequent repulsions markedly attenuate the otherwise major role of attraction of valence electrons to the nuclear framework of the molecule. Thereby, compression experiments can markedly alter a wide range of chemical interactions and "dial up" behavior not acceptable to uncrowded molecules. From a thermodynamic perspective [9], the shrinkage in volume available to reaction products causes the familiar disparity between strong intramolecular bonds and weak intermolecular interactions to fade away. At sufficiently high compression, the intramolecular and intermolecular forces become comparable, making free energy changes become favorable for some unorthodox pathways but unfavorable for certain ordinarily elite processes.

Systematic explorations of the chemical domains made accessible by DAC techniques are still in their infancy. Most of the experiments on nitrogen systems compiled in Table I were done during the past five years. The scope of *in situ* experiments has been much enhanced by the availability of advanced synchrotron radiation sources. However, as yet for most of the systems of Table I crystal structures have not been determined and the cell parameters need to be refined. Inelastic scattering and other analytical techniques now feasible for DAC use [9, 93, 143] , also should be added to the standard repertoire.

We note in particular several inviting opportunities for high pressure kinetic studies. The membrane-type DACs [144-146] drive compression of the diamond anvils by the expansion of a stainless steel diaphragm when it is filled with an inert gas. This enables fine tuning and steady scanning of the pressure exerted on the sample by the anvils [95, 98]. Applied to phase transitions or chemical reactions, DAC techniques can be used to determine the pressure dependence of activation energies, fundamental information entirely lacking at present. Another versatile means for DAC kinetic experiments, not yet exploited, would measure relaxation rates following perturbation of an equilibrium by a sudden pressure or temperature jump. Recent work even puts in prospect subpicosecond x-ray diffraction capable of following the kinetics of structural changes. These kinetics experiments are best accomplished using diaphragm or piezoelectric driven DACs [147]. Dramatic improvements in chemical vapor deposition technique now allow the production of large, high-quality single-crystal diamond anvils [148, 149]. This gives the prospect of considerably enlarging the DAC sample volume, thereby enhancing the prospects for studies of chemical dynamics, including experiments using neutron scattering and nuclear magnetic resonance in the megabar pressure range. In company with these anticipated experimental advances, we welcome the growing theoretical interest in high pressure processes. The value of symbiotic interactions between theory and experiment is well exemplified in the case of polynitrogen. As more molecular systems and properties under compression are explored, the mutual challenges, needs and opportunities for prediction and interpretation will expand in scope and variety. Long ago, in perhaps the first theoretical study of effects of entrapment in a box of shrinking volume, this was demonstrated in a compelling way by Edgar Allan Poe [150].

ACKNOWLEDGEMENTS

We are grateful to our colleagues, Z. Liu, M. Somayazulu, J. Hu, Q. Guo, J. Shu, O. Tschauner, A. F. Goncharov, V.V. Struzhkin, P. Dera, C. Prewitt, J. Lin and O. Degtyareva for experimental assistance and helpful discussions. We also thank K. Christe and W. Wilson for samples and helpful discussions. Our high pressure studies of nitrogen compounds have been supported by LLNL (subcontract to Harvard), AFSOR, DARPA, NSF and DOE.

REFERENCES

[1] S. R. d. Groot and C. A. t. Seldam, Physica, 12 (1946) 669.
[2] R. LeSar and D. R. Herschbach, J. Phys. Chem., 85 (1981) 2798.
[3] T. Pang, Phys. Rev. A, 49 (1994) 1709.
[4] S.A. Cruz, J. Soullard and E. G. Gamaly, Phys. Rev. A, 60 (1999) 2207.
[5] S. Mateos-Cortes, E. Ley-Koo and S. A. Cruz, Int. J. Quantum Chem., 86 (2002) 376.
[6] A. Zerr, G. Miehe and R. Riedel, Nat. Mater., 3 (2003) 185.
[7] E. Gregoryanz, C. Sanloup, M. Somayazulu, J. Badro, G. Fiquet, H. K. Mao and R. J. Hemley, Nat. Mater., 3, 294 (2004).
[8] R. J. Hemley, Annu. Rev. Phys. Chem., 51 (2000) 763.
[9] V. Schettino and R. Bini, Phys. Chem. Chem. Phys., 5 (2003) 1951.
[10] R. J. Hemley and H. K. Mao, in High Pressure Phenomena. (IOS Press/Società Italiana di Fisica, Amsterdam, 2002) pp. 3-40.
[11] H. K. Mao, J. Xu and P. M. Bell, J. Geophy. Res., 91 (1986) 4673.
[12] C. S. Zha, H. K. Mao and R. J. Hemley, P. Natl. Acad. Sci., 97 (2000) 13494.
[13] N. J. Hess and D. Schiferl, J. Appl. Phys., 71 (1992) 2082.
[14] A. F. Goncharov, V. V. Struzhkin, R. J. Hemley, H. K. Mao and Z. Liu, in Science and Technology of High Pressure, M.H. Manghnani, W. Nellis and M. Nicol, Eds., (Universities Press, Hyderabad, India, 2000) pp. 90-95.
[15] J. Z. Hu, Q. Z. Guo and R. J. Hemley, in Science and Technology of High Pressure, M.H. Manghnani, W. Nellis and M. Nicol, Eds., (Universities Press, Hyderabad, India, 2000) pp. 1039-42.
[16] J. Belak, R. LeSar and R. D. Etters, J. Chem. Phys., 92 (1990) 5430.
[17] S. Nosé and M. L. Klein, Phys. Rev. Lett., 50 (1983)1207.
[18] R.D. Etters, V. Chandrasekharan, E. Uzan and K. Kobashi, Phys. Rev. B, 33 (1986) 8615.
[19] M. R. Manaa, Chem. Phys. Lett., 331, (2000) 262.
[20] M. R. Manaa, Theoret. Comput. Chem., 13 (2003) 71.
[21] A. K. McMahan and R. LeSar, Phys. Rev. Lett., 54 (1985) 1929.
[22] R. M. Martin and R. J. Needs, Phys. Rev. B, 34 (1986) 5082.
[23] C. Mailhiot, L. H. Yang and A. K. McMahan, Phys. Rev. B, 46 (1992) 14419.
[24] L. Mitas and R. M. Martin, Phys. Rev. Lett., 72 (1994) 2438.
[25] T. W. Barbee, Phys. Rev. B, 48 (1993) 9327.
[26] J. E. Williams and J. N. Murrell, J. Am. Chem. Soc., 93 (1971) 7149.
[27] P. N. Skancke and J. E. Boggs, Chem. Phys. Letts., 21 (1973) 316.
[28] B. Kuchta and R. D. Etters, J. Chem. Phys., 95 (1991) 5399.
[29] B. Kuchta and R. D. Etters, Phy. Rev.B, 45 (1992) 5072.
[30] L. B. Kanney, N. S. Gillis and J. C. Raich, J. Chem. Phys., 67 (1977) 81.
[31] P. Botschwina, J. Chem. Phys., 85 (1986) 4591.
[32] W. J. Lauderdale, J. F. Stanton and R. J. Bartlett, J. Phys. Chem., 96 (1992) 1173.
[33] P. Zielinski and C. Marzluf, J. Chem. Phys., 96 (1992) 1735.
[34] M. M. Ossowski, J. R. Hardy and R. W. Smith, Phys. Rev. B, 60 (1999) 15094.
[35] R. Bini, L. Ulivi, J. Kreutz and H. Jodl, J. Chem. Phys., 112 (2000) 8522.
[36] D. A. Young, C.–S. Zha, R. Boehler, J. Yen, M. Nicol, A. S. Zinn, D. Schiferl, S. Kinkead, R. C. Hanson and D. A. Pinnick, Phys. Rev. B, 35 (1987) 5353.
[37] S. Zinn, D. Schiferl and M. F. Nicol, J. Chem. Phys., 87 (1987) 1267.
[38] W. L. Vos and J. A. Schouten, J. Chem. Phys., 91 (1989) 6302.
[39] S. C. Schmidt, D. Schiferl, A. S. Zinn, D. D. Ragan and D. S. Moore, J. Appl. Phys., 69 (1991) 2793.
[40] C. A. Swenson, J. Chem. Phys., 23 (1955) 1963.
[41] R. L. Mills and A. F. Schuch, Phys. Rev. Lett., 23 (1969) 1154.
[42] F. Schuch and R. L. Mills, J. Chem. Phys., 52 (1970) 6000.
[43] J. R. Brookeman and T. A. Scott, J. Low. Temp. Phys., 12 (1973) 491.

[44] W. E. Streib, T. H. Jordan and W. N. Lipscomb, J. Chem. Phys., 37 (1962) 2962.
[45] R. LeSar, S. A. Ekberg, L. H. Jones, R. L. Mills, L. A. Schwalbe and D. Schiferl, Solid State Comm., 32 (1979) 131.
[46] S. Buchsbaum, R. L. Mills and D. Schiferl, J. Phys. Chem., 88 (1984) 2522.
[47] D. T. Cromer, R. L. Mills, D. Schiferl and L. A. Schwalbe, Acta Crystallogr. B, 37 (1981) 8.
[48] D. Schiferl, S. Buchsbaum and R. L. Mills, J. Phys. Chem., 89 (1985) 2324.
[49] R. Reichlin, D. Schiferl, S. Martin, C. Vanderborgh and R. L. Mills, Phys. Rev. Lett., 55 (1985) 1464.
[50] R. L. Mills, B. Olinger and D. T. Cromer, J. Chem. Phys., 84 (1986) 2837.
[51] H. Olijnyk, J. Chem. Phys., 93 (1990) 8968.
[52] H. Schneider, W. Haefner, A. Wokaun and H. Olijnyk, J. Chem. Phys., 96 (1992) 8046.
[53] M. I. M. Scheerboom and J. A. Schouten, Phys. Rev. Lett., 71 (1993) 2252.
[54] M. I. M. Scheerboom and J. A. Schouten, J. Chem. Phys., 105 (1996) 2553.
[55] R. Bini, M. Jordan, L. Ulivi and H. J. Jodl, J. Chem. Phys., 108 (1998) 6849.
[56] A. Mulder, J. P. J. Michels and J. A. Schouten, J. Chem. Phys., 105 (1996) 3235.
[57] A. Mulder, J. P. J. Michels and J. A. Schouten, Phys. Rev. B, 57 (1998) 7571.
[58] T. Westerhoff, A. Wittig and R. Feile, Phys. Rev. B, 54 (1996) 14.
[59] H. Olijnyk and A. P. Jephcoat, Phys. Rev. Lett., 83 (1999) 332.
[60] A. P. Jephcoat, R. J. Hemley, H. K. Mao and D. E. Cox, Bull. Am. Phys. Soc., 33 (1988) 522.
[61] A. F. Goncharov, E.. Gregoryanz, H.K. Mao and R. J. Hemley, Fizika Nizkikh Temperatur, 27 (2001) 1170.
[62] H.B. Radousky, W.J. Nellis, M. Ross, D. C. Hamilton and A. C. Mitchell, Phys. Rev. Lett., 57 (1986) 2419.
[63] A. F. Goncharov, E. Gregoryanz, H.K. Mao, Z. Liu and R. J. Hemley, Phys. Rev. Lett., 85 (2000) 1262.
[64] M. Eremets, R. J. Hemley, H. K. Mao and E. Gregoryanz, Nature, 411 (2001) 170.
[65] E. Gregoryanz, A. F. Goncharov, R. J. Hemley and H. K. Mao, Phys. Rev. B, 64 (2001) 052103.
[66] E. Gregoryanz, A. F. Goncharov, R. J. Hemley, H.K. Mao, M. Somayazulu and G. Shen, Phys. Rev. B, 66 (2002) 224108.
[67] M. I. Eremets, A. G. Gavriliuk, I. A. Trojan, D. A. Dzivenko and R. Boehler, Nat. Mater., in press.
[68] W. J. Dulmage, E. A. Meyers and W. N. Lipscomb, J. Chem. Phys., 19 (1951) 1432.
[69] W. J. Dulmage, E. A. Meyers and W. N. Lipscomb, Acta Cryst., 6 (1953) 760.
[70] C. E. Dinerman and G. E. Ewing, J. Chem. Phys., 54 (1971) 3660.
[71] W. N. Lipscomb, J. Chem. Phys., 54 (1971) 3659.
[72] A. Anderson and B. Lassier-Govers, Chem. Phys. Letts., 50 (1977) 124.
[73] S. F. Agnew, B. I. Swanson, L. H. Jones and R. L. Mills, J. Phys. Chem., 89 (1985) 1678.
[74] P. Brechignac, S. E. Benedictis, N. Halberstadt, B. J. Whitaker and S. Avrillier, J. Chem. Phys., 83 (1985) 2064.
[75] J. M. Fernández, G. Tejeda, A. Ramos, B. J. Howard and S. Montero, J. Mol. Spectr., 194 (1999) 278.
[76] M. Somayazulu, A. F. Goncharov, O. Tschauner, P. F. McMillan, H. K. Mao and R. J. Hemley, Phys. Rev. Lett., 87 (2001) 135504.
[77] H. Olijnyk, H. Daufer, M. Rubly, H.-J. Jodl and H. D. Hochheimer, J. Chem. Phys., 93 (1990) 45.
[78] R. L. Mills, B. Olinger, D. T. Cromer and R. LeSar, J. Chem. Phys., 95 (1991) 5392.
[79] Y. Song, M. Somayazulu, H. K. Mao, R. J. Hemley and D. R. Herschbach, J. Chem. Phys., 118 (2003) 8350.
[80] Y. Song, R. J. Hemley, Z. Liu, M. Somayazulu, H. K. Mao and D. R. Herschbach, J. Chem. Phys., 119 (2003) 2232.
[81] C. S. Yoo, V. Iota, H. Cynn, M. Nicol, J. H. Park, T. L. Bihan and M. Mezouar, J. Phys. Chem. B, 107 (2003) 5922.

[82] V. Iota, J.-H. Park and C. S. Yoo, Phy. Rev.B, 69 (2004) 064106.
[83] R. W. G. Wyckoff: Crystal Structures, (Wiley, New York, 1963).
[84] R. V. Louis and B. Crawford, J. Chem. Phys., 42 (1965) 857.
[85] C. H. Bibart and G. E. Ewing, J. Chem. Phys., 61 (1974) 1284.
[86] S. F. Agnew, B. I. Swanson, L. H. Jones, R. L. Mills and D. Schiferl, J. Chem. Phys., 87 (1983) 5065.
[87] F. Bolduan, H. J. Jodl and A. Loewenschuss, J. Chem. Phys., 80 (1984) 1739.
[88] A. Givan and A. Loewenschuss, J. Chem. Phys., 90 (1989) 6135.
[89] A. Givan and A. Loewenschuss, J. Chem. Phys., 91 (1989) 5126.
[90] A. Givan and A. Loewenschuss, J. Chem. Phys., 93 (1990) 7592.
[91] A. Givan and A. Loewenschuss, J. Chem. Phys., 94 (1991) 7562.
[92] D. A. Pinnick, S. F. Agnew and B. I. Swanson, J. Phys. Chem., 96 (1992) 7092.
[93] Y. Song, R. J. Hemley, Z. Liu, M. Somayazulu, H. K. Mao and D. R. Herschbach, Chem. Phys. Lett., 382 (2003) 686.
[94] F. V. Shallcross and G. B. Carpenter, Acta Cryst., 11 (1958) 490.
[95] K. Aoki, Y. Kakudate, M. Yoshida, S. Usuba and S. Fujiwara, J. Chem. Phys., 91 (1999) 778.
[96] A. S. Parker and R. E. Hughes, Acta Cryst., 16 (1963) 734.
[97] C. S. Yoo and M. Nicol, J. Phys. Chem., 90 (1986) 6726.
[98] C. S. Yoo and M. Nicol, J. Phys. Chem., 90 (1986) 6732.
[99] A. S. Parks and R. E. Hughes, Acta Crystallogr., 16 (1963) 734.
[100] T. Curtius, Ber. Dtsch. Chem. Ges., 23 (1890) 3023.
[101] J. I. Bryant, J. Chem. Phys., 45 (1966) 689.
[102] G. E. Pringle and D. E. Noakes, Acta Cryst., B24 (1968) 262.
[103] J. I. Bryant and R. L. Brooks, J. Chem. Phys., 54 (1971) 5315.
[104] Z. Iqbal and M. L. Malhotra, J. Chem. Phys., 57 (1972) 2637.
[105] Z. Iqbal, J. Chem. Phys., 59 (1973) 1769.
[106] G. J. Simonis and C. E. Hathaway, Phys. Rev. B, 10 (1974) 4419.
[107] Z. Iqbal and C. W. Christoe, J. Chem. Phys., 62 (1975) 3246.
[108] C. S. Choi and E. Prince, J. Chem. Phys., 64 (1976) 4510.
[109] N. E. Massa, S. S. Mitra, H. Prask, R. S. Songh and S. F. Trevino, J. Chem. Phys., 67 (1977) 173.
[110] S. R. Aghdaee and A. I. M. Rae, Acta Cryst., B40 (1984) 214-18.
[111] S. M. Peiris and T. P. Russell, in Science and Technology of High Pressure, M.H. Manghnani, W. Nellis and M. Nicol., Eds., (Universities Press, Hyderabad, India, 2000) pp. 667.
[112] S. M. Peiris and T. P. Russell, J. Phys. Chem. A, 107 (2003) 944.
[113] M. I. Eremets, M. Y. Popov, I. A. Trojan, V. N. Denisov, R. Boehler and R. J. Hemley, J. Chem. Phys., 120 (2004) 10618.
[114] D. R. Herschbach, Ann. Rev. Phys. Chem., 51 (2000) 1.
[115] L. Parts and J. T. Miller, J. Chem. Phys., 43 (1965) 136.
[116] F. Bolduan and H. J. Jodl, Chem. Phys. Lett., 85 (1982) 283.
[117] L. H. Jones, B. I. Swanson and S. F. Agnew, J. Chem. Phys., 82 (1985) 4389.
[118] M. Somayazulu, A. F. Goncharov, O. Tschauner, P. F. McMillan, H. K. Mao and R. J. Hemley, Phys. Rev. Lett., 87 (2001) 135504.
[119] M. Balkanski, M. K. Teng and M. Nusimovici, Phys. Rev., 176 (1968) 1098.
[120] G. Turrell, Infrared and Raman Spectra of Crystals, (Academic Press, London and New York, 1972).
[121] W.-J. Lo, M.-Y. Shen, C.-H. Yu and Y.-P. Lee, J. Mol. Spectrosc., 183 (1997) 119.
[122] D. Liu, F. G. Ullman and J. R. Hardy, Phys. Rev. B, 45 (1992) 2142.
[123] E. H. Abramson, L. J. Slutsky, M. D. Harrell and J. M. Brown, J. Chem. Phys., 110 (1999) 10493.
[124] S. Desgreniers and K. E. Brister, in High Pressure Science and Technology. (World Scientific Publishers, Singapore, 1996) pp. 363-65.

[125] B. Olinger, R. L. Mills and J. R.B. Roof, J. Chem. Phys., 81 (1984) 5068.
[126] H. Olijnyk, H. Daufer, M. Rubly, H.-J. Jodl and H. D. Hochheimer, J. Chem. Phys., 93 (1990) 45.
[127] R. L. Mills, B. Olinger, D. T. Cromer and R. LeSar, J. Chem. Phys., 95 (1991) 5392.
[128] F. Birch, J. Geophys. Res., 95 (1978) 1257.
[129] P. Vinet, J. Ferrante, J. H. Rose and J. R. Smith, J. Geophy. Res., 92 (1987) 9319.
[130] S. Johnson, M. Nicol and D. Schiferl, J. Appl. Cryst., 26 (1993) 320.
[131] Y. Ma, H. K. Mao and R. J. Hemley, unpublished.
[132] K. Clusius, U. Piesbergen and E. Varde, Helv. Chim. Acta, 18 (1960) 1290.
[133] F. Cacace, G. D. Petris and A. Troiani, Science, 295 (2002) 480.
[134] K. O. Christe, W.W. Wilson, J. A. Sheehy and J.A. Boatz, Angew. Chem., Int. Ed, 38 (1999) 2004.
[135] A. Vij, W. W. Wilson, V. Vij, F. S. Tham, J. A. Sheehy and K. O. Christe, J. Am. Chem. Soc., 123 (2001) 6308.
[136] S. Fau, K. J. Wilson and R. J. Bartlett, J. Phys. Chem. A, 106 (2002) 463.
[137] J. Kortus, M. R. Peterson and S. L. Richardson, Chem. Phys. Lett., 340 (2001) 565.
[138] D. A. Dixon, D. Feller, K. O. Christe, W. W. Wilson, A. Vij, V. Vij, H. D. B. Jenkins, R. M. Olson and M. S. Gordon, J. Am. Chem. Soc., 126 (2004) 834-43.
[139] Y. Song, R. J. Hemley, D. R. Herschbach, H. K. Mao, W. W. Wilson and K. Christe, in preparation.
[140] Y. Song, R. J. Hemley, D. R. Herschbach, P. Dera, O. Degtyareva and H. K. Mao, in preparation.
[141] Z. Iqbal, J. Chem. Phys., 59 (1973) 1769.
[142] M. M. Ossowski, J. R. Hardy and R. W. Smith, Phys. Rev. B, 60 (1999) 15094.
[143] Y. Meng, H. K. Mao, P. J. Eng, T. P. Trainor, M. Newville, M. Y. Hu, C. C. Kao, J. F. Shu, D. Hausermann and R. J. Hemley, Nat. Mater., 3 (2004) 111.
[144] R. LeToullec, J. P. Pinceaux and P. Loubeyre, High Pressure Res., 1 (1988) 77.
[145] J. H. Eggert, H. K. Mao and R. J. Hemley, Phys. Rev. Lett., 70 (1993) 2301.
[146] W. B. Daniels and M. G. Ryschkewitsch, Rev. Sci. Instrum., 54 (1983) 115.
[147] C. S. Yoo, Private Communication.
[148] C. S. Yan, Y. K. Vohra, H. K. Mao and R. J. Hemley, Proc. Nat'l Acad. Sci., 99 (2002) 12523.
[149] C. S. Yan, H. K. Mao, W. Li, J. Qian, Y. S. Zhao and R. J. Hemley, Phys. Status Solidi. A, 201 (2004) R25.
[150] E. A. Poe, in The Gift. (Carey and Hart, Philadelphia, 1842) pp. 133-51.

Chemistry at Extreme Conditions
M. Riad Manaa (Editor)

Chapter 7

Aqueous Chemistry in the Diamond Anvil Cell up to and Beyond the Critical Point of Water

William A. Bassett[a], I-Ming Chou[b], Alan J. Anderson[c], Robert Mayanovic[d]

[a]Department of Earth and Atmospheric Sciences, Cornell University, Ithaca, NY 14853, USA

[b] MS 954, US Geological Survey, Reston, VA 20192, USA

[c] Department of Earth Sciences, St. Francis Xavier University, Antigonish, Nova Scotia, B2G 2W5 Canada

[d] Department of Physics, Astronomy, and Materials Science, Southwest Missouri State University, Springfield, MO 65804, USA

1. INTRODUCTION

The hydrothermal diamond anvil cell (HDAC) has been developed for the study of fluids and their interactions with other phases. It is capable of pressures up to 10 GPa and temperatures from -190°C to 1200°C. It has found application in studies of equations of state of fluids, reactions between fluids and solids as well as fluids and melts, hydration and dehydration of hydrous solids under P_{H2O}, fractionation of species between fluids and solids as well as fluids and melts, the effect of P_{H2O} on melting of silicates, structures of ions and clathrates in solution, preservation of hosts of fluid inclusions at high temperatures, and reactions in clathrates and other organic materials. Visual, spectroscopic, and X-ray methods are used to analyze samples by taking advantage of the exceptional transparency of the diamond anvils. Examples of successful applications of the HDAC include the equation of state (EOS) of water, stability of the various stages of hydration of montmorillonite and calcium carbonate, leaching of elements from zircon, the effect of P_{H2O} on the melting of albite, speciation and structures of Sc, Fe, Cu, Zn, Y, La, Yb, and Br in solution, stability of methane hydrates and $Ca(OH)_2$, identifying a new H_2O ice form and sII of methane hydrate. The description of diamond cell configuration, analytical methods, and examples of applications provide evidence of the utility of the technique for many studies of fluids at temperatures and pressures up to and beyond the critical point of water.

Diamond anvil cells (DAC) have been used extensively since their invention in the late 1950's [1]. In 1964 Van Valkenburg [2] used liquids as pressure media and in 1971 Van Valkenburg et al. [3] investigated a reaction between liquid water and a solid at ambient temperature. In 1993 Bassett et al. [4] designed a diamond anvil cell specifically for the study of fluid samples at simultaneous high temperatures and high pressures. This cell consisted of electric heater windings around the seats that support the diamond anvils. Care was taken to avoid distortions of the mechanical parts of the cell during heating and to provide the most uniform and constant temperature, pressure, and volume.

Because water is one of the most important fluids in nature and its study was the principal reason for making the modifications, this cell was called the Hydrothermal Diamond Anvil Cell (HDAC) (Fig. 1). Examining the EOS of water with the new instrument was an important first step in using the HDAC. This was not only because of the inherent interest in the subject but also because of the opportunity offered by the EOS of water for making accurate pressure determinations in samples having water as a major component [5].

There were a number of important objectives beyond analyzing the EOS of water. Because water at elevated pressures and temperatures plays such an important role in the earth's interior, the HDAC was developed primarily for conducting experiments that would yield data useful for a better understanding of geologic processes. Although that was the major motivation behind the development of the HDAC, the potential usefulness of its applications in other fields was obvious.

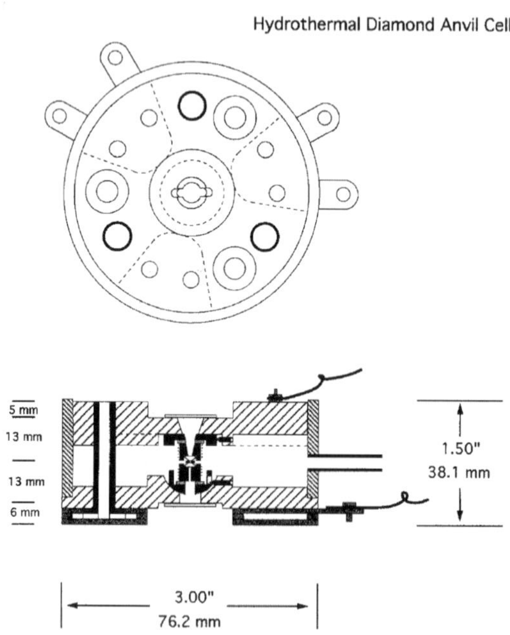

Figure 1. Diagram of the hydrothermal diamond anvil cell (HDAC). Three screws placed well away from the heat source pull the platens together. Three posts also similarly placed well away from the heat source serve to guide the upper platen. Bellville springs under the heads of the diver screws minimize the effect of differential expansion of diamond cell parts during heating and cooling.

2. THE HYDROTHERMAL DIAMOND ANVIL CELL (HDAC)

The HDAC (Fig. 1) derived from the design described by Merrill and Bassett [6] is modified to accommodate heaters and to keep important parts of the cell at low enough temperatures to avoid distortions due to differential expansion. The screws that provide the force and the posts that serve as guides are located well away from the heaters. The heaters are formed by winding resistance wire, typically molybdenum or chromel, around the tungsten carbide (WC) seats that support the diamond anvils (Fig. 2). A ceramic cement coating on the WC and covering the wires provides electrical insulation. Another important objective was to provide as uniform and constant a temperature as possible at the sample by minimizing the solid angle through which radiative heat loss could occur. Because diamond is an excellent thermal conductor, thermocouples (type K) located under ceramic cement in contact with the diamond anvils yield temperatures that are within a few degrees Celsius of the sample temperature even at temperatures as high as 1000°C. A metal gasket, typically made of rhenium or stainless steel, has a hole drilled in it for forming a sample chamber when squeezed between the diamond anvils. A fluid sample sealed within this small chamber can be observed through the anvils with a microscope. Likewise, X-rays, infrared, and other portions of the electromagnetic spectrum can pass through the anvils. For some studies, a depression is milled in the face of one of the diamonds and X-rays can have access to the sample by passing through the sides of the diamond anvil. The diamond anvils, WC supports, heating wire, and other parts are protected from oxidation by means of a slightly reducing gas such as argon or helium mixed with 1% to 4% hydrogen. Springs under the heads of the driver screws even out the forces and accommodate the small changes in dimensions produced by modest temperature changes during heating. Hollow posts and sometimes hollow driver screws can provide for air cooling.

An important aspect of the HDAC is its ability to maintain a nearly constant sample volume during changes in pressure and temperature. It was found that Re gaskets are sufficiently stiff so that they can maintain constant sample volume under certain conditions. Typically, initial changes in temperature and pressure lead to a small reduction of volume as the metal relaxes and extrudes slightly from between the anvils. However, the volume remains essentially constant during decrease of temperature and pressure. Subsequent increases in temperature and pressure cause far less change in volume, and on the third cycle the volume change is usually negligible.

A laser interferometry method was developed for monitoring the volume of the sample chamber during variation in temperature and pressure. Interference fringes are produced by reflection of laser light from the upper and lower anvil faces. They remain unchanged if the separation of the faces and the refractive index of the sample remain constant. In most experiments an isochore is followed from the homogenization point, as described below, to a point at which an observation is made; therefore, the refractive index remains constant. Images of the outline of the gasket hole indicate that in general there is no change in the diameter of the hole if the separation of the anvil faces remains constant. The laser interferometry method developed for this purpose also has found application for detecting changes in refractive index and dimensions of solid samples, particularly during phase transformations.

In addition to high temperatures, the HDAC can be used for studying samples at temperatures down to –190 °C by introducing cold gas or liquid nitrogen into the space surrounding the diamond anvils. This is best done in a glove bag to avoid condensation of moisture on the HDAC surfaces.

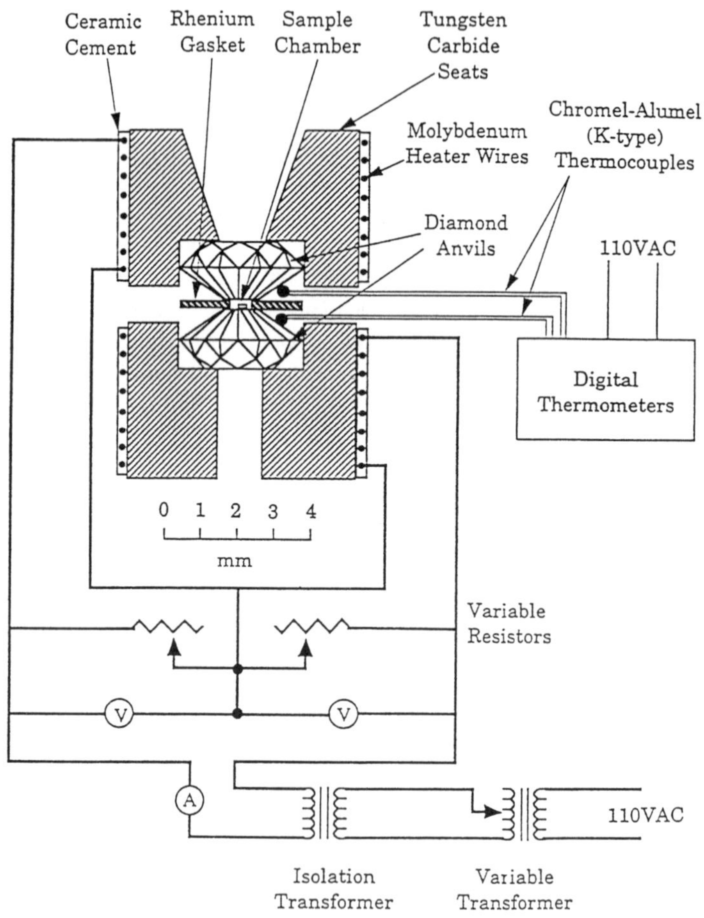

Figure 2. Sample, diamond anvils, heaters, and electric circuits for heating, controlling, and measuring the temperature of the sample in the HDAC.

3. SAMPLE PREPARATION

Loading a liquid into a hole in a gasket only a few tenths of a millimeter in diameter can be difficult. The more volatile the liquid, the more difficult it is to load the sample. If the hole is large (>0.4 mm) and the liquid adheres to diamond, a drop can be hung from the upper anvil before closing the cell. If, however, the hole is <0.4 mm, the surface tension attending a liquid such as water may prevent filling of the hole. In this case the most successful technique consists of introducing the liquid through a capillary tube directly into the hole, removing the tube, and closing the anvil faces on the gasket to seal the sample in the chamber. Because this needs to be done rapidly, the upper anvil is typically suspended just above the sample chamber so it can be lowered quickly once the liquid is in place. Samples such as CO_2 and NH_4 whose liquid stability fields lie above atmospheric pressure can be loaded in the solid state and then allowed to melt under pressure.

4. PRESSURE AND TEMPERATURE MEASUREMENT

For calibration, the temperature at the sample location can be determined by visually observing the melting of salts of known melting temperatures. The salts that have proved to be most reliable for this purpose are NaCl (800.5 °C) and $NaNO_3$ (306.8 °C). The α-β phase transition in quartz (574°C) can be used as well. The laser interferometry method described above can be used to accurately pinpoint the transition in an optically perfect platelet of quartz.

A number of methods for measuring pressure at ambient temperature in the DAC have been developed since its invention. These include known phase transitions [7], lattice parameters determined by X ray diffraction [8], wavelength shift of ruby fluorescence [9], and shift of Raman spectra in diamond [7]. Full EOS information on well characterized materials such as NaCl, MgO, and Au have provided a means for determining pressure at high temperatures. These methods have proven to be very valuable over pressure ranges of gigapascals (GPa) but offer inadequate precision over ranges of megapascals (MPa).

For a more accurate method of pressure measurement over a range of MPa, Shen et al. [5] turned to the EOS of H_2O (Fig. 3). The dependence of its structure and density, ρ, on pressure, P, and temperature, T, is a topic that has engaged the interests and efforts of many investigators over the years as it is basic to the understanding of the properties and behavior of H_2O. The ability of the HDAC to maintain a constant sample volume as determined by laser interferometry, was central to this effort. When sample volume can be held constant, the P-T-ρ relationship offers a means of determining pressure over the range of a few hundred MPa that is roughly an order of magnitude more precise than other methods of pressure determination such as X-ray diffraction, ruby fluorescence, and Raman spectroscopy.

Several published equations of state for H_2O were available when the first HDAC was made. Of these, Haar et al. [10] and Saul and Wagner [11] were considered to be the most comprehensive and accurate. One of the first applications of the HDAC [5] was undertaken to compare EOS data of water with the published α-β phase transition boundary in quartz [12]. The conclusion of this study was that the EOS published by Haar et al. [10] could provide pressures accurate to ±1% for the pressure range up to about 1500 MPa. However, other studies such as that of Brodholt and Wood [13] find evidence that the EOS of Saul and Wagner [11] is more accurate. Consequently, a claim of a precision of ±1% is perhaps more appropriate than an accuracy of ±1%. A comparison of various EOS can be found in Haselton et al. [14]. The EOS of Haar et al. [10] has been adopted as the basis for pressure determination in most HDAC studies involving water. When an aqueous solution of sufficient concentration to significantly affect the EOS was studied, its EOS was determined independently using the α-β quartz boundary [12] for calibration, using the method of Shen et al. [5]. This new EOS was then used in the same way as the EOS of water for pressure determination.

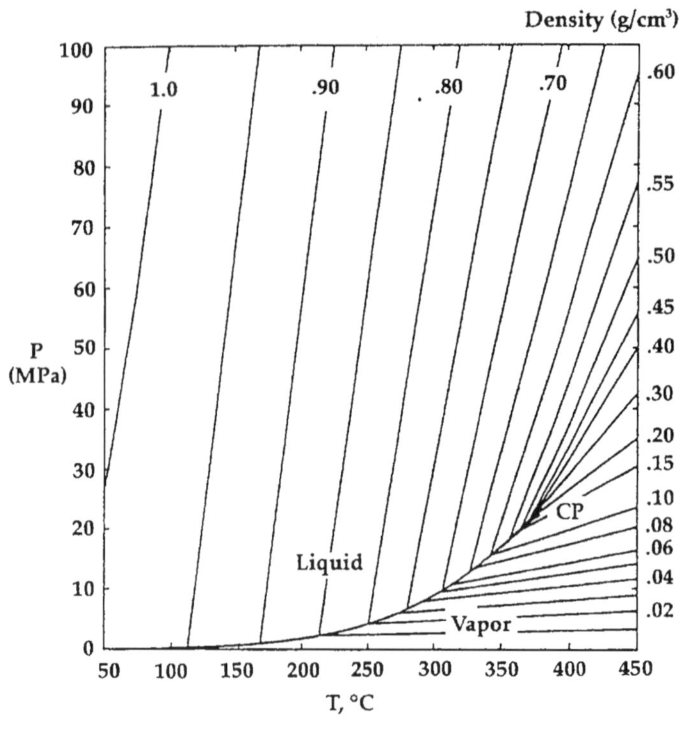

Figure 3. Pressure-temperature-density diagram for liquid water. LV represents the liquid-vapor boundary; CP represents the critical point. The isochors are labeled by density in g/cm^3.

Pressure determination by fluid EOS requires accurate EOS data, constant sample volume, and accurate temperature measurement. Calculating pressure involves two steps: determination of ρ from measured homogenization temperature (T_h) along the liquid-vapor (LV) boundary and then using the EOS to solve for P at a point of interest using the T and ρ measured at that point. In practice this consists of loading a liquid sample with a vapor bubble and heating the sample until the bubble shrinks and vanishes at T_h. With further heating the sample follows an isochore of ρ based on T_h. P is then calculated from the measured ρ and T using the EOS (Fig. 3). To determine the pressure of a liquid sample with ρ greater than at the solid-liquid-vapor triple point (ρ_{SLV}), the temperature is lowered or the pressure is increased until solid and liquid are observed to coexist. The temperature is then raised until the solid melts. This temperature (T_m) can then be used to identify the density, ρ, of the isochore that the sample follows as temperature is further increased. As with liquid having densities less than ρ_{SLV}, the pressure can be found using the EOS, ρ, and T at any point on the P-T-ρ surface [14, 15].

Phase transitions in some solids can be used for measuring pressure at high temperatures. Chou has suggested $BaTiO_3$, $Pb_3(PO_4)_2$, and $PbTiO_3$ as suitable pressure calibrants [7]. The following criteria were the basis for these choices: (1) they are ferroelectric phase transitions that are rapid and can be determined visually; (2) their phase boundaries plotted in P-T space intersect isochores of common geologic fluids at high angles; (3) they are inert in most pressure media over a wide P-T range. A doubly polished platelet is placed in the sample chamber along with the sample. $BaTiO_3$ and $Pb_3(PO_4)_2$ have phase transitions that are visible because of the disappearance and reappearance of twin boundaries. $PbTiO_3$ has a phase transition that can be detected by the movements of phase fronts. Schmidt and Ziemann [16] have proposed Raman spectroscopy of quartz as a pressure sensor.

5. ANALYTICAL TECHNIQUES

The extraordinary properties of diamond including hardness, high thermal conductivity, and transparency over large portions of the electromagnetic spectrum make it an ideal material for anvils that also serve as windows. All of the analytical techniques described here take advantage of these properties.

5.1. Visual

Visual observation using a microscope is valuable both for monitoring a sample and as an analytical tool. Phases such as liquid, vapor, and solid are clearly identifiable. Even different solid phases usually have distinctive appearances that allow them to be distinguished from each other. In many cases, visual observation is all that is required for determining phase relationships of a chosen system over a range of P and T. Not only can relative sizes of the phases be used to measure their relative abundances, but optical properties such as refractive index can be used as an indication of changes in crystal structure (e.g., quartz) and composition (e.g., albite melt). Visual observation in conjunction with other analytical techniques is important as well. For instance, when fluorescent diamonds are selected for use in studies using an intense X-ray beam, visual observation can provide information about the location of the X-ray beam with respect to the sample.

When a TV camera is installed on the microscope, a video cassette recorder can be used to make a permanent record of observations. This is especially valuable when the temperature and time are superimposed on the image of the sample allowing review as a function of time and temperature. This can be done a frame at a time if necessary.

5.2. X-ray Diffraction (XRD)

X-ray diffraction patterns of samples under pressure in DACs have been an important source of information since the invention of the DAC [1]. The earliest patterns were collected on photographic film or by diffractometer using angle dispersion. More recently, imaging plates and charge couple devices (CCD) have largely replaced photographic film for obtaining angle-dispersion patterns, especially with the use of synchrotron radiation. With the advent of improved energy-discriminating solid-state detectors, energy dispersion techniques became a very attractive way to collect X-ray diffraction data because of the rapidity with which patterns could be collected, especially when used with the intense chromatic (white) radiation available at synchrotron sources. Both polycrystalline samples and single crystal samples are analyzed routinely by these methods.

The placement of the heaters in the HDAC allows openings to be large enough for 2θ dispersion up to 30° when the X-rays pass through both anvils. In hydrothermal studies, X-ray diffraction has played an important role in determining changes in crystal structure caused by reactions with the surrounding fluid at elevated P and T. Gasket materials that are transparent to X-rays, e.g., beryllium and boron-epoxy make it possible to obtain diffraction patterns by passing X-rays through the gasket. In some cases such as beryllium, however, the gasket material may be highly reactive with solutions and therefore unsuitable on the basis of chemistry.

5.3. X-ray Fluorescence (XRF)

Excitation of X-ray emission from portions of a sample inside the HDAC can be used for obtaining chemical analyses of individual phases at high P-T conditions. This is especially valuable for determining partitioning of species between phases and for studying the leaching of species from a solid or melt subjected to high T at high P_{H2O}. The finely focused, intense X-ray beam available at synchrotron sources along with the use of visually fluorescent diamond anvils has greatly enhanced the ability of the experimenter to select the phases to be analyzed and thus improved the usefulness of this technique. Background due to rayleigh and compton scattering is minimized when the detector is located at 90° from the X-ray beam in the plane of polarization. Achieving this requires that either the impinging X-rays or the emitted X-rays or both must pass through the gasket or through the side of one of the anvils. When the sample is contained within a depression in the face of one of the anvils, X-rays need pass only through single-crystal diamond. Anvils with such sample-containing depressions have represented an important advance in application of X-ray fluorescence to chemical analyses of phases in DAC samples.

5.4. X-ray Absorption Fine Structure (XAFS)

X-ray absorption spectra of samples at high P and T in the HDAC can be obtained by transmission or fluorescence. When transmission is used, the X-ray beam passes through a detector, enters though one of the anvils, passes through the sample, exits through the other anvil, and enters a second detector. The ratio of intensities between the two detectors as a function of energy yields an absorption spectrum for the sample. When the elements being analyzed by XAFS have absorption edges with energies below ~10 keV, the diamond anvils impose a severe loss due to absorption and scattering. This can be minimized by using a pair of anvils with holes drilled to within about 0.15 mm of the sample (Fig. 4). Such anvils have been used successfully for analyzing first-row transition elements [17, 18]. Sharp pseudo-absorption peaks are produced when strong diffraction by the anvils deflects X-rays with wavelengths that satisfy the Bragg equation for sets of planes in the single-crystal diamonds. These peaks can be shifted by slightly rotating the DAC with respect to the X-ray beam. When spectra collected with the DAC in two different orientations are combined, these pseudo-absorption peaks can be eliminated [17, 18].

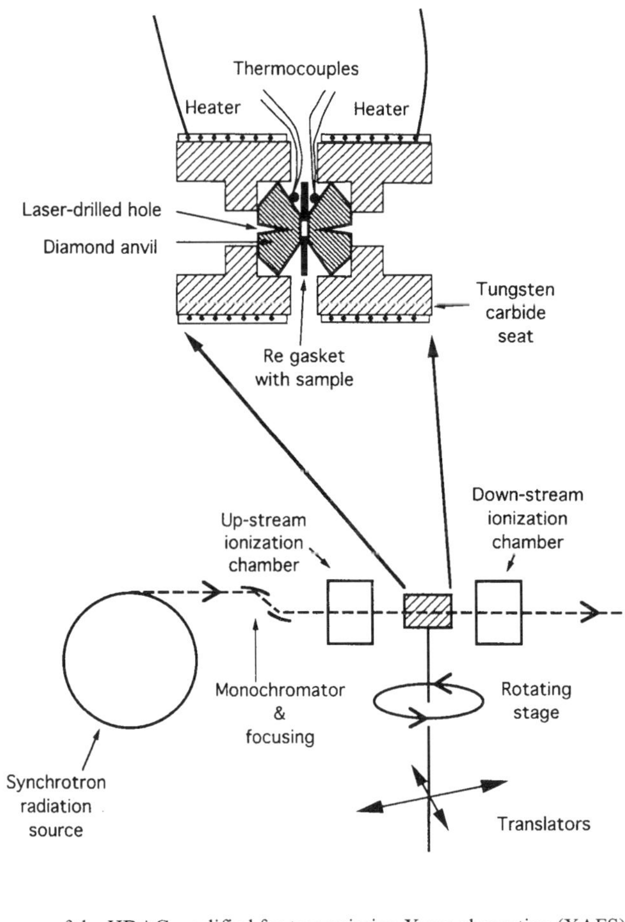

Figure 4. Diagram of the HDAC modified for transmission X-ray absorption (XAFS) studies of elements with low absorption-edge energies. Holes drilled using a laser to within 0.15 mm of the anvil faces reduce the attenuation of the X-ray beam by diamond for elements with absorption energies less than 10 keV.

The second method for obtaining XAFS spectra is by using fluorescence emission as an indication of absorption by an element. The method can use the same geometry as described in the section on XRF and shown in Figure 5. However, in this case the impinging monochromatic X-ray beam is scanned through the desired range of energies while the detector is set to record fluorescence radiation within a narrow energy window bracketing the most intense emission peak from the element. The fluorescence XAFS method has several advantages. 1. It is more sensitive because the signal is directly related to absorption by the sample rather than being a small effect superimposed on a large signal. 2. Pseudo-absorption effects caused by diffraction in the anvils are reduced because the impinging X-ray beam passes through only one diamond. The fluorescence emission signal is not subject to the same effect because it has only one energy. 3. The minimizing of background due to rayleigh and compton scattering by utilizing the geometry described under XRF improves the sensitivity. 4. Anvil faces can be ion milled or laser machined with greater precision making it possible to better control the amount of diamond the X-rays must traverse. As with the transmission method, remaining pseudo-absorption peaks can be eliminated by combining spectra collected with the HDAC in two different orientations. Diamond anvils that fluoresce to produce visible light (Fig.5b) offer a valuable method for aiming the X-ray beam through the sample.

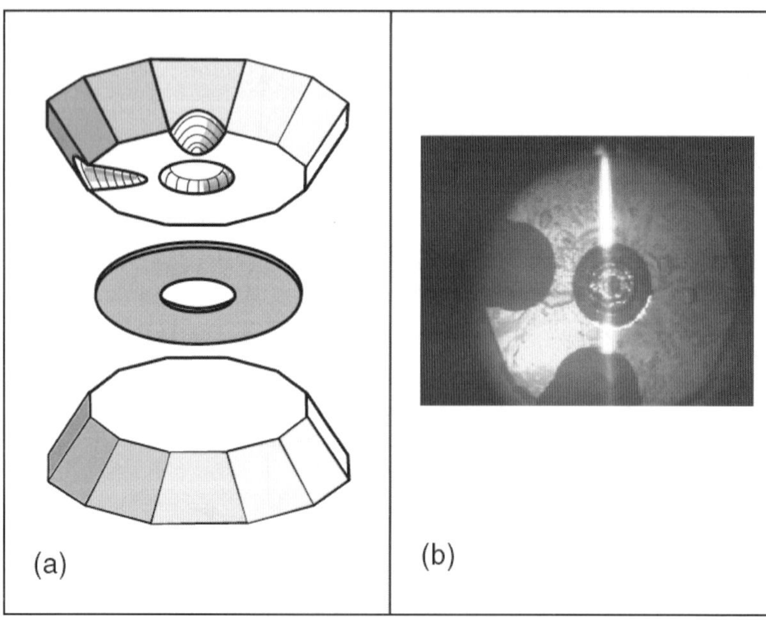

Figure 5. (a) A recess 0.3 mm in diameter and 0.05-0.1 mm deep, machined by laser or ion milling into one of the diamond anvils, contains the fluid sample. Grooves in the same anvil provide access for the incident and fluorescence X-rays to minimize attenuation by diamond in their paths. A Re gasket 0.05 mm thik placed between the anvils forms a seal. (b) Visible fluorescence (light streak) is produced by the incident X-ray beam as it passes through the anvil and sample chamber. Note the small amount of diamond that the incident beam must pass through.

5.5. Raman spectroscopy

Raman spectroscopy of samples in DACs has been a powerful method for studying samples subjected to high P and T. A highly focused laser beam can be directed at any portion and therefore any phase within a sample in a DAC. Phases have characteristic Raman frequency shifts due to the vibrational modes of bonded atoms within them. These frequency shifts can be used for identification purposes or to study the response of bonding frequencies to changes in P and T [19].

5.6. Infrared absorption

Infrared absorption, like Raman, is sensitive to vibrational frequencies exhibited by the molecular constituents of the sample. It can be used for identification of phases [20], to measure the abundance of a particular molecular species [21], or to determine the response of vibrational frequencies to changes in P and T of a sample.

5.7. Luminescence

In situ transformations in organic samples can be observed in real time in the DAC by using a microscope equipped with a CCD video monitoring system and luminescence spectrometer. A high-pressure mercury lamp is used to excite fluorescence in the sample. The emitted light from the sample is split with one part used for visual observation and the other part for analysis made possible by use of a grating spectrometer and a photomultiplier detector [22].

6. EXAMPLES

To date the HDAC has been used with a variety of analytical techniques to investigate properties of fluids, reactions between fluids and solids, and the melting of solids under H_2O pressure.

6.1. EOS of Water

Accurate information on the EOS of water is basic to all research on hydrothermal studies. The approach taken by Shen et al. [5] is described in Section 4 on accurate measurement of pressure in the HDAC.

6.2. Montmorillonite

Montmorillonite is a clay mineral with H_2O molecules bound to alkali and alkali earth ions between silicate sheets. Although its dehydration in air had been the subject of numerous studies, its dehydration under P_{H2O} was not investigated until the HDAC was developed. Dehydration is reversible when it takes place under P_{H2O}. It can therefore be treated as an equilibrium phase transition or reaction and thermodynamic principles can be applied. Furthermore, reactions under P_{H2O} are more pertinent to the understanding of geologic processes than dehydration alone. Three samples, Na-, Mg-, and Ca-montmorillonites, were prepared from the same natural material by exchanging the interlayer cations [23, 24]. All three were found to retain their interlayer water to higher temperatures under P_{H2O} than had previously been expected. X-ray diffraction measurements of interlayer distances showed discrete dehydration events and a clear dependence of dehydration temperature on the charge and size of the ions. Application of the clapeyron equation to the P-T boundaries indicated that, for this reaction, change of configurational entropy, ΔS, plays a much greater role than change of volume, ΔV.

6.3. Ikaite

Ikaite, a hydrous form of calcium carbonate, $CaCO_3 \cdot 6H_2O$ occurs naturally at mineral springs in very cold waters [25]. Prior to the development of the HDAC, ikaite's stability field and properties were studied using the DAC at ambient temperature [3]. Shahar et al. [15] have extended the known stability field of ikaite to 2 GPa and 110 °C. Their results show that the boundary separating ikaite from aragonite + water in P-T space is curved like most melt curves. This observation is consistent with the greater compressibility of a liquid than a solid.

6.4. Silicate melts and solutions under hydrothermal conditions

Shen & Keppler [26] melted glass with a composition close to that of albite ($NaAlSi_3O_8$) under hydrothermal conditions in a HDAC. A sample consisting of nearly equal parts of the silicate and water remained as two visually distinct phases up to 763 °C at 1.45 GPa. As temperature increased from 763 °C to 766 °C, the meniscus between the melt and the aqueous solution vanished and the two phases became a single fluid. This phenomenon is critical behavior similar to the critical behavior between vapor and liquid of a single substance. In a similar experiment Shen & Keppler [21] used infrared absorption (IR) spectroscopy to measure the relative abundances of H_2O and OH^- in a melt of similar composition coexisting with an aqueous solution. Their results showed that the ratio of H_2O:OH^- in the melt decreased with increasing temperature indicating that water becomes more strongly bonded to the silicate as it is converted to OH^-.

Audetat and Keppler [27] have determined the viscosities of silicate-rich aqueous fluids by measuring the rates at which spheres of different densities drop through the fluids inside a HDAC. They have found that the viscosity of these fluids increases rather slowly with increasing silicate content. For instance, a fluid with 20 wt % silicate at 800 °C has approximately the viscosity of water at ambient P-T conditions and a fluid with 50 wt % silicate has approximately the viscosity of olive oil at ambient conditions. Viscosity continues to increase nearly linearly up to ~80 wt % silicate beyond which it rises exponentially toward the value of dry silicate melt. They found quite similar behavior for albite, leucite, and pectolite compositions.

H_2O plays an important role in virtually all magmas. Using the HDAC Chou et al. [28] studied melting in the system $Ca(OH)_2$-$CaCO_3$-H_2O along six isochores and found eutectic melting temperatures ranging from 610 °C at 120.4 MPa to 583 °C at 720.3 MPa. These results agree within 3 °C with those obtained by high pressure differential thermal analysis. Chou & Anderson [29] studied melting in petalite-quartz. Anderson & Chou [28] studied crystallization in the system $NaAlSi_3O_8$-SiO_2-$LiAlSiO_4$-H_2O. Sowerby & Keppler [31] investigated the effect of F, B, and Na in the albite-H_2O system. Bureau & Keppler [32] examined melting behavior in silicate-H_2O systems. Veksler et al. [33] used the HDAC to find evidence of three coexisting immiscible fluids in synthetic granite pegmatite doped with B, P, and F. Sowerby & Keppler [32] studied water speciation in rhyolite melt by in situ infrared spectroscopy using a hydrothermal diamond cell.

6.5. Solubility and leaching under hydrothermal conditions

Schmidt & Rickers [35] have developed a novel way for determining mineral solubilities at high pressures and temperatures using X-ray fluorescence analysis of samples in a modified HDAC. A shallow recess in the face of one of the anvils contains the fluid portion of their sample. To one side of the sample chamber a small pocket cut into the gasket holds the solid portion of the sample where it is shielded from the direct X-ray beam entering the sample chamber through one of the anvils. X-ray fluorescence emissions excited by the X-ray beam as it passes through the fluid then pass out through the anvil at right angles to the beam and are recorded by a detector. This X-ray fluorescence analysis of the fluid that is in direct contact with the solid mineral fragment provides data that can be used to determine the mineral's solubility over a range of pressures and temperatures. The authors expect to increase the sensitivity of the technique by better than an order of magnitude using a third generation synchrotron source and reducing the thickness of diamond traversed by the X-rays.

Zircon, which is commonly used in U-Pb age dating, can lose radiogenic Pb by fluid leaching. Domains within the zircon that are variably affected by alpha-decay damage are more likely to suffer Pb loss . To better characterize this phenomenon, Anderson et al. [36] conducted in situ x-ray fluorescence analyses of dilute KF aqueous solution during its interactions with a zircon grain from the Georgeville granite, Nova Scotia, at elevated temperatures and pressures in the HDAC. The absorption of the incident and secondary X-rays by surrounding diamond was reduced by employing a diamond anvil with laser-milled sample chamber and grooves in the surface of the anvil (Fig. 5a). X-rays had to pass through only 80 microns of diamond between the grooves and the sample chamber. The analyses showed that the concentration of Pb in the dilute KF aqueous solution increased from 0 to 37 ppm over a period of 150 minutes at 300 °C and 110 MPa. During the run, the Pb concentration of the aqueous fluid increased at a rate of 0.15 ppm/min for the first hour, followed thereafter by an increase to 0.42 ppm/min. Variation in the rate of Pb leaching is most likely due to the dissolution of nanometer size inclusions of galena (PbS) encountered by the fluid as it penetrated the nearly amorphous zircon grain. Similar results were obtained when a dilute HCl solution was used.

6.6. Fluid inclusion studies

Stretching or decrepitation can occur during microthermometric analysis of fluid inclusions, especially when the host solid is soft or the fluid pressure in the inclusion is unusually high. Following the suggestion of Chou et al. [37], Schmidt et al. [38] and Darling and Bassett [39] used the HDAC to apply external pressure to samples containing synthetic and natural fluid inclusions in quartz. A thin platelet containing the fluid inclusions was immersed in water and subjected to pressure in the sample chamber of the HDAC. The coexisting phases in the fluid inclusions could be visually observed through the diamond anvils as temperature was increased. This technique enabled making successful measurements to much higher temperatures without loss of fluid from the inclusions. These authors, however, did observe significant change in fluid inclusion volume as a function of external pressure. This observation suggests that the normal practice of using molar volume data of the host during such analyses is questionable and needs further investigation.

6.7. Radiation-induced small Cu particle cluster formation in aqueous $CuCl_2$

When high-energy (X-ray) synchrotron radiation is used, the photons can induce oxidation and/or reduction in a solution, thus altering the solution's chemistry. Jayanetti et al. [40] used the fluorescence method to collect XAFS spectra of a $CuCl_2$ aqueous solution containing 55 ppm Cu. Their results showed the formation and growth of copper clusters. Nearest neighbor distances in these clusters increased from 2.48 ± 0.02 Å to 2.55 ± 0.01 Å during irradiation indicating an approach to the lattice dimensions of bulk copper. At the same time, the Debye-Waller factor of the copper clusters was observed to increase by ~50% - 55%. These observations are consistent with the coarsening of a mixture of small clusters 5 - 10 Å across and large clusters of essentially bulk copper. During this process, the nearest neighbor coordination number was observed to increase in a manner consistent with the decrease in the surface to volume ratio. The authors [40] concluded that the copper ions in solution were reduced to the metallic state by reacting with hydrated electrons produced as a result of radiolysis of water by the incident X-ray photons. Cluster growth probably proceeded through agglomeration of copper atoms that may have reacted initially to form dimers. This study illustrates the importance of considering chemical changes that can occur in a sample as a result of impinging high-energy photons.

6.8. Hydration structure of aqueous La^{3+} up to 300 °C and 160 MPa

Using the fluorescence method at the La L_3-edge, Anderson et al. [41] collected XAFS spectra of an aqueous solution of lanthanum nitrate containing 0.007 *m* La over a range of temperatures from 25 to 300 °C and pressures up to 160 MPa. Their results indicated that each La^{3+} ion has a hydration number of 9 and that the solvating waters surround the ion in a tricapped trigonal prismatic arrangement. As temperature is increased from 25 to 300 °C, the bond distance between the equatorial plane oxygens and the La^{3+} ion increases from 2.59 ± 0.02 Å to 2.79 ± 0.04 Å. The bond distance between the oxygens in the tricapped trigonal prismatic sites and the La^{3+} ion decreases by 0.13 ± 0.03 Å, in the same temperature range. This study illustrates the usefulness of XAFS studies on samples in the HDAC for understanding rare earth elements in solution.

6.9. Structure of Yb^{3+} aqua ion and chloro complexes in aqueous solutions up to 500 °C and 270 Mpa

Mayanovic et al. [42] used fluorescence XAFS at the Yb L_3-edge to obtain spectra of the Yb^{3+} ion in nitrate and chloride aqueous solutions. Over a range of temperatures from 25 to 500 °C and pressures up to 270 MPa they found that the Yb-O distance in the Yb^{3+} aqua ion in the 0.006M Yb / 0.16M HNO_3 solution decreases at a rate of 0.02 Å/100 °C, while the number of oxygens decreases from 8.3 ± 0.6 to 4.8 ± 0.7. No nitrate complexes were found. Yb^{3+}

persists as an aqua ion up to 150 °C in the 0.006M $YbCl_2$ / 0.017M HCl aqueous solution. In the 300 to 500 °C range, chloro complexes form, probably of the type $Yb(H_2O)_{\delta-n}Cl_n^{+3-n}$ ($\delta \approx$ 7). The Yb-Cl distance of the chloro ytterbium(III) complexes decreases at a rate of about 0.02 Å/100 °C, while the number of chlorides increases from 0.5 ± 0.3 to 1.8 ± 0.2 in the 300 to 500 °C temperature range. On the other hand, the Yb-O distance decreases at a rate of 0.007 Å/100 °C, while the number of oxygens drops from 8.3 ± 0.5 to 5.1 ± 0.3 in going from 25 to 500 °C in the same solution.

6.10. Zinc halide solutions and evidence for hydrogen bond breaking in aqueous solutions near the critical point

Mayanovic et al. [43] using transmission XAFS (Fig. 4) obtained XAFS spectra indicating that the $ZnCl_4^{2-}$ complex is predominant in 1*m* $ZnCl_2$/6*m* NaCl solution, while the $ZnCl_2(H_2O)_2$ complex is predominant in the 2*m* $ZnCl_2$ aqueous solution, at all temperatures and pressures. The Zn-Cl bond length in both types of chlorozinc complexes was found to decrease at a rate of about 0.01 Å/100 °C.

Later, Mayanovic et al. [44] obtained XAFS spectra of 1*m* $ZnBr_2$ /6*m* NaBr aqueous solution using both Zn K- and Br K-absorption edges. Analysis shows that between 25 to 500 °C, a 63% reduction of waters occurs in the solvation shell of $ZnBr_4^{2-}$, the predominant complex at all pressure-temperature points investigated. A similar reduction in the hydration shell of the Br^- aqua ion was found. They conclude that the water-anion and water-water bond breaking mechanisms occurring at high temperatures are essentially the same. This is consistent with the hydration waters being weakly hydrogen bonded to halide anions in electrolytic solutions.

6.11. Transformations in methane hydrates

Chou et al. [45, 46] made a detailed study of pure methane hydrates in a HDAC by visual, Raman, and X-ray microprobe techniques. Their results revealed two previously unknown high-pressure structures, sII and sH. At 250 MPa sII has a cubic unit cell of a = 17.158 Å. At 600 MPa sH has a hexagonal unit cell of a = 11.980 Å and c = 9.992 Å. The compositions of these two phases are still not known. They conclude that within deep hydrate bearing sediments underlying continental margins, and in the presence of other gases in the structure, the sII phase is likely to dominate over the sI phase.

6.12. New H_2O ice form

Chou et al. [47] report a previously unknown solid phase of H_2O recognized by its peculiar growth patterns, distinctive pressure-temperature melting relationships and Raman vibrational spectra. Morphologies of ice crystals and their P-T melting relationships were directly observed in a HDAC for H_2O bulk densities between 1.203 and 1.257 g/cm^3 between -10 °C and 50 °C. Under these conditions, four different ice forms were observed to melt: two stable phases, ice V and ice VI, and two metastable phases, ice IV and the new ice phase. The Raman spectra and crystal morphology of the new phase are consistent with a distorted anisotropic structure with some similarities to ice VI.

6.13. Organic material

Techniques to monitor organic transformations in the HDAC were developed by W.-L. Huang and G.A. Otten in the labs of the Exxon Production Research Company [7, 20, 22, 48]. These include visual observation, fluorescence (visual spectrum), and micro-fourier transform infrared spectroscopy (micro-FTIR). These techniques have made it possible to run experiments under hydrous and high P-T conditions and to observe relative abundances of organic species during pyrolysis of kerogen, cracking of crude oil, and gas generation during pyrolysis of coal. These data have provided new valuable information about the nature of the reactions taking place as well as their rates.

6.14. Biological applications of the HDAC

Sharma et al. [49] subjected *Shewanella oneidensis* strain MR1 and *Escherichia coli* strain MG1655 to pressures of 68 to 1680 MPa. At pressures of 1200 to 1600 MPa they observed living bacteria residing in fluid inclusions in ice VI crystals. These bacteria continued to be viable at high pressure and after release to ambient pressure (0.1 MPa). Their evidence of microbial viability and activity at these extreme pressures extends by an order of magnitude the range of habitable conditions. The maximum pressure explored by Sharma et al. [49] is equivalent to a depth of ~50 km beneath rock, or ~160 km beneath water/ice. Their results suggest that deep layers of water/ice on Europa, Callisto, or Ganymede, and Mars as well as subduction zones on Earth might provide viable settings for life unhindered by high pressures.

7. FUTURE DEVELOPMENTS

One of the most frustrating aspects of present fluid experiments in the MPa pressure range is the need to have EOS data for the fluid in order to accurately determine pressure at elevated temperature. Known phase transitions such as α-β quartz, $BaTiO_3$, $Pb_3(PO_4)_2$, and $PbTiO_3$ can provide pressure determinations only along transition boundaries. While these are useful for calibrating the EOS of a given fluid, they do not provide a pressure-measuring technique that can be applied at any set of P-T conditions. There is a serious need for a pressure-measuring technique that is independent of the sample composition, does not interfere with sample chemistry, and can be used at any point in P-T space. Some device like a micro-aneroid or bimetal strip may someday help solve this problem.

Another problem results from the need for gaskets that are simultaneously stiff and inert. An interesting solution to this is to line the sample chamber with a metal that is more inert such as gold or platinum. The extruded gold or platinum not only provides an inert surface but also provides a good seal for fluid samples. This technique has been applied successfully in a number of cases and is becoming a popular method for handling samples. Using the HDAC can be rather demanding, especially for some of the applications. However, as it is more widely used, new developments are making its operation more versatile, routine, and reliable. Interfacing with some techniques has led to changes in dimensions, configurations, and anvil designs. The addition of new techniques to those already used with DACs will lead to the development of new configurations. Because of the simplicity and versatility of the DAC principle, experimenters should never hesitate to attempt such modifications. Further modifications are certain to benefit future users.

REFERENCES

[1] C.E. Weir, E.R. Lippincott, A. Van Valkenburg, E.N. Bunting, J. Res. Nat'l. Bur. Standards (U.S.), 63A (1959) 55.
[2] Van Valkenburg, Diamond Research, (1964) 17.
[3] Van Valkenburg, H.K. Mao, P.M. Bell, Carnegie Institution of Washington, Year Book, 70 (1971) 237.
[4] W.A. Bassett, A.H. Shen, M. Bucknum, I.-M. Chou, Rev. Sci. Instrum., 64 (1993) 2340.
[5] A.H. Shen, W.A. Bassett, I.-M. Chou, Am. Mineral., 78 (1993) 694.
[6] L. Merrill, W.A. Bassett, Acta Cryst., B31 (1975) 343.
[7] W.A. Bassett, T.-C. Wu, I.-M. Chou, H.T. Haselton, Jr., J. Frantz, B. Mysen, W.-L. Huang, S.K. Sharma, and D. Schiferl, in Mineral Spectroscopy: a tribute to Roger G. Burns, The Geochemical Society, Special Publication, No. 5 (1996) 261.
[8] D.L. Decker, J. Appl. Phys. 42 (1971) 3239.
[9] H. K. Mao, J. Xu, P. M. Bell, J. Geophs. Res., 91 (1986) 4673.
[10] L. Haar, J.S. Gallagher, G.S. Kell, NBS/NRC steam tables: Thermodynamic and transport properties and computer programs for vapor and liquid states of water in SI units. 320 p. (Hemisphere, Washington, DC, 1984).
[11] Saul, and W. Wagner, J. Phys. Chem. Ref. Data, 18 (1989) 1537.
[12] P.W. Mirwald, H.-J. Massonne, J. Geophys. Res., 85 (1980) 6983.
[13] J.P. Brodholt, and B.J. Wood, J. Geophys. Res., 98 (1993) 519.
[14] H.T. Haselton, Jr., I.-M. Chou, A.H. Shen, W.A. Bassett, Am. Mineral., 80 (1995) 1302.
[15] Shahar, W.A. Bassett, H.K. Mao, I.-M. Chou, and W. Mao, in preparation.
[16] C. Schmidt and M.A. Ziemann, Am. Mineral., 85 (2000) 1725.
[17] W.A. Bassett, A.J. Anderson, R.A. Mayanovic, I.-M. Chou , Chemical Geology, 167 (2000) 3.
[18] W.A. Bassett, A.J. Anderson, R.A. Mayanovic, I.-M. Chou, Z. Kristallogr., 215 (2000) 711
[19] T.R. Lettieri, E.M. Brody, W.A. Bassett, Solid State Communications, 26 (1978) 235.
[20] W.-L. Huang and G.A. Otten, Organic Geochemistry, 29 (1998) 1119.
[21] A.H. Shen and H. Keppler, Am. Mineral., 80 (1995) 1335.
[22] W.-L. Huang and G.A. Otten, Organic Geochemistry, 32 (2001) 817.
[23] W.-L. Huang, W.A. Bassett, T.-C. Wu, Am. Mineral., 79 (1994) 683.
[24] T.-C. Wu, W.A. Bassett, W.-L. Huang, S. Guggenheim, A.F. Koster van Groos, Am. Min., 82 (1997) 70.
[25] H. Pauly, Naturens Verden, Copenhagen, (1963) 168.
[26] A.H. Shen and H. Keppler, Nature, 358 (1997) 710.
[27] A. Audetat and H. Keppler, Science, 303 (2002) 513.
[28] I.-M. Chou, W.A. Bassett, T.B. Bai, Am. Mineral., 80 (1995) 865.
[29] I.-M. Chou and A.J. Anderson, Mineral. Mag., 62A (1998) 327.
[30] A.J. Anderson and I.-M. Chou, in Joint Annual Meeting of Geological Association of Canada and Mineralogical Association of Canada, 24 (1999) 2.
[31] J.R. Sowerby and H. Keppler, Contrib. Mineral. Petrol., 143 (2002), 32.
[32] H. Bureau and H. Keppler, Earth Planet. Sci. Lett., 165 (1999) 187.
[33] I.V. Veksler, R. Thomas, C. Schmidt, Am. Mineral., 87 (2002) 775.
[34] J.R. Sowerby and H. Keppler, Am. Mineral., 84 (1999) 1843.
[35] Schmidt and K. Rickers, Am. Mineral., 88 (2003) 288.
[36] A.J. Anderson, R.A. Mayanovic, W.A. Bassett, I.-M. Chou, M. Newville, S.R. Sutton, in preparation.

[37] I.-M. Chou, A.H. Shen, W.A. Bassett, in “Fluid Inclusions in Minerals: Methods and Applications”, B. De Vivo and M.L. Frezzotti, Eds., (Short course of the working group, IMA, "Inclusions in Minerals" 1994) 215.
[38] Schmidt, I.-M. Chou, R.J. Bodnar, W.A. Bassett, Am. Mineral., 83 (1998) 995.
[39] R.S. Darling and W.A. Bassett, Am. Mineral., 87 (2002) 67.
[40] S. Jayanetti, R.A. Mayanovic, A.J. Anderson, W.A. Bassett, I.-M. Chou, J. Chem. Phys., 115 (2001) 954.
[41] A.J. Anderson, S. Jayanetti, R.A. Mayanovic, W.A. Bassett, I.-M. Chou, Am. Mineral., 87 (2002) 262.
[42] R.A. Mayanovic, S. Jayanetti, A.J. Anderson, W.A. Bassett, I.-M. Chou, J. Phys. Chem. A, 106 (2002) 6591.
[43] R.A. Mayanovic, A.J. Anderson, W.A. Bassett, I.-M. Chou, J. Synchrotron Rad., 6 (1999) 195.
[44] R.A. Mayanovic, A.J. Anderson, W.A. Bassett, I.-M. Chou, Chem. Phys. Lett., 336 (2001) 212.
[45] I.-M. Chou, A. Sharma, R.C. Burruss, J. Shu, H.K. Mao, R.J. Hemley, A.F. Goncharov, L.A. Stern, S.H. Kirby, S.H., Proceed. Nat. Acad. Sci., 97 (2000) 13484.
[46] I.-M.Chou, A. Sharma, R.C. Burruss, R.J. Hemley, A.F. Goncharov, L.A. Stern, S.H. Kirby, Physical Chemistry A, 105 (2001) 4664.
[47] I.-M. Chou, J.G. Blank, A.F. Goncharov, H.K. Mao, R.J. Hemley, Science, 281 (1998) 809.
[48] W.-L. Huang, Organic Geochemistry, 24 (1996) 95.
[49] Sharma, J.H. Scott, G.D. Cody, M.L. Fogel, R.M. Hazen, R.J. Hemley, W.T. Huntress Science, 295 (2002) 1514.

Chemistry at Extreme Conditions
M. Riad Manaa (Editor)

Chapter 8

Solid Nitrogen at Extreme Conditions of High Pressure and Temperature

Alexander Goncharov[a,b] and Eugene Gregoryanz[a]

[a]Geophysical Laboratory, Carnegie Institution of Washington, 5251 Broad Branch Road, Washington, D.C., 20015, USA.

[b]Lawrence Livermore National Laboratory, University of California, 7000 East Avenue, Livermore, CA 94551, USA.

ABSTRACT

We review the phase diagram of nitrogen in a wide pressure and temperature range. Recent optical and x-ray diffraction studies at pressures up to 300 GPa and temperatures in excess of 1000 K have provided a wealth of information on the transformation of molecular nitrogen to a nonmolecular (polymeric) semiconducting and two new molecular phases. These newly found phases have very large stability (metastability) range. Moreover, two new molecular phases have considerably different orientational order from the previously known phases. In the iota phase (unlike most of other known molecular phases), N_2 molecules are orientationally equivalent. The nitrogen molecules in the theta phase might be associated into larger aggregates, which is in line with theoretical predictions on polyatomic nitrogen.

1. INTRODUCTION

The evolution of molecular solids under pressure constitutes an important problem in condensed-matter physics [1]. Under compression, delocalization of electronic shells and eventual molecular dissociation is expected, leading to the formation of a framework or close-packed structures. However, this process may not necessarily be simple and direct, because of large barriers of transformation between states with different types of bonding and molecular structures with various types of orientational order, including possible associated and charge-transfer intermediate states. Nitrogen is an archetypal homonuclear diatomic molecule with a very strong triple bond. It was expected to undergo the transition to a network structure related to a destabilization of its triple bond in the pressure range accessible by modern experimental techniques. This prediction is based on the results of shock-wave experiments, which provided evidence for the transition to a nonmolecular phase in the liquid phase at 30 GPa and 6000 K [2], as well as theoretical calculations for the solid [3-5]. The latter predictions involved transitions to various three-fold coordinated phases below 100 GPa. It has been suggested that the observed transition pressure can be substantially higher because of a large potential barrier to dissociation (*e.g.*, Refs. [4, 5]).

For transformation to framework structures, information about the type and character of the ordering is important, since it affects the energy barriers associated with the transition. Solid nitrogen is an important system for such studies because of the stability and simple electronic structure of the isolated molecule. Moreover, solid nitrogen has been well extensively studied theoretically, and accurate experimental data provide an important test of condensed matter theory [4-8].

The phase diagram of nitrogen (Fig. 1) is complex at moderate pressures and temperatures [9, 10] and until recently has been little studied over a wider range. The melting curve of N_2 has been only measured to about 22 GPa [11-13]. Higher P-T data are available from shock wave CARS study [14], which refer to a liquid state. Solid nitrogen has a wealth of molecular phases that differ in the types of orientational ordering and crystal structures formed [9, 10, 12, 13, 15-33]. The ordering of the low pressure-temperature α and γ phases is controlled by quadrupole-quadrupole interactions, whereas at higher pressures a class of molecular structures (δ, δ_{loc}, ε, ζ) stabilized by additional anisotropic intermolecular interactions is found [10]. The δ phase (Fig. 2) is proposed to have a disordered cubic structure (space group *Pm3n*) [22] with two different types of molecules exhibiting sphere - and disk- like disorder and giving rise to two classes of vibron bands ν_1 and ν_2. With increasing pressure and/or decreasing temperature, nitrogen molecules exhibit orientational order through a sequence of the phase transformations (to δ_{loc} and then ε) as determined by vibrational spectroscopy and x-ray diffraction [27-36]. The structure of the ε phase is rhombohedral (space group of $R\bar{3}c$), which can be viewed as a distortion of cubic [25].

Theoretical calculations have been contradictory in their predictions of stable molecular ordered phases [6-8]. In particular, Monte Carlo (MC) [6] and Molecular Dynamics (MD) [7] simulations based on proposed intermolecular potential favor tetragonal structures, while *ab initio* calculations [8] predicted the stability of $R\bar{3}c$ phase in agreement with experiment. Recent MC calculations [31, 32] pointed out the importance of using accurate atom-atom potentials (especially anisotropic terms) to obtain adequate results.

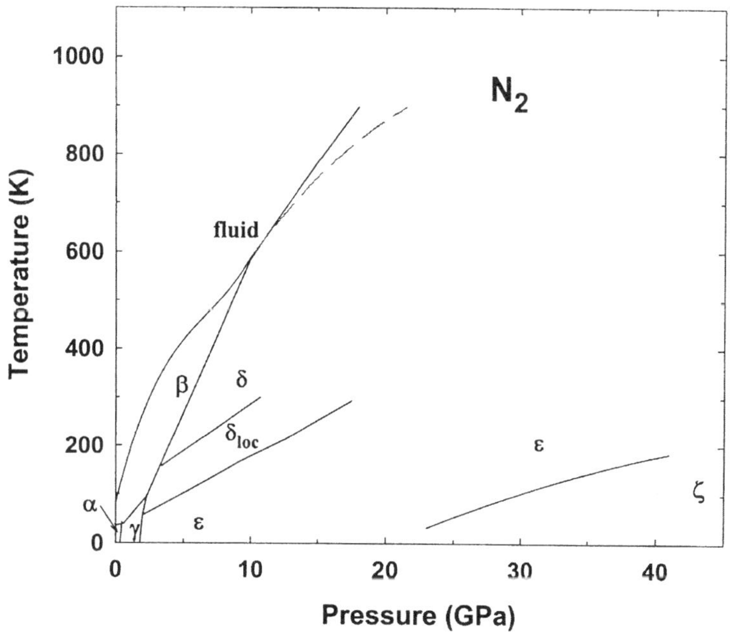

Fig.1. A portion of the phase diagram of nitrogen at moderate pressure and temperature conditions [9,11,12,25]. Dashed gray line and thin solid line at high temperature correspond to the melting curve from Refs. [12] and [11], respectively.

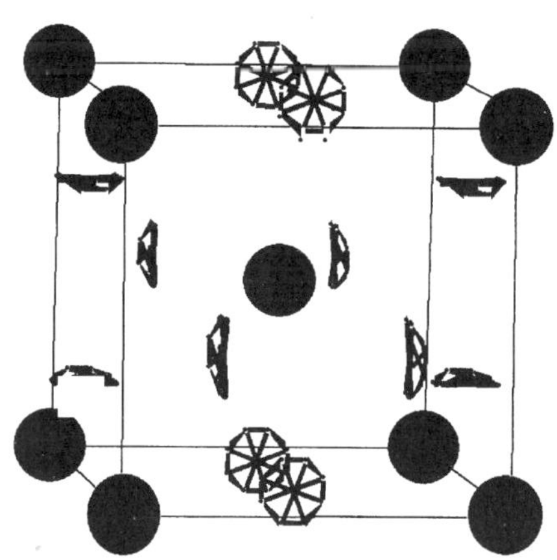

Fig. 2. The unit cell of δ–N_2 [22]. Black spheres are N_2 molecules in sphere-like positions (2*a* site symmetry), black "wheels" are N_2 molecules in disk-like positions (48*l* site symmetry).

At 20-25 GPa and 10 K a transformation has been found by Raman spectroscopy [23] and confirmed by later Raman and IR measurements [9]. No x-ray data were available until recently for this phase (called ζ). According to vibrational spectroscopy data, its structure has a strong similarity to ε and δ phases. It has been inferred to have *R3c* space group. The ε - ζ phase boundary has been extended recently to 180 K and 40 GPa [9]. At low temperatures, a significant region of metastability is reported [9], making difficult to clarify the mechanism of the transformation.

At room temperature, the vibrational spectroscopy data are still contradictory about a number and nature of the transformations above 20 GPa. A sequence of new phases has been reported on the basis of several splitting of the Raman vibron modes [24], including one just above 20 GPa [27]. In contrast, x-ray studies indicate the stability of ε phase to 50 GPa [26, 33] in agreement with latter Raman study [34]. A change in x-ray diffraction pattern was observed above 60 GPa [36], but interpretation requires additional measurements. Recent Raman and IR measurements to 42 GPa show clear correspondence between the number of observed lattice and vibron modes and group-theoretical predictions for the ε phase [9].

Here we review our recent combined optical, Raman, synchrotron IR and x-ray diffraction high-pressure studies of nitrogen up to 270 GPa between 10 and 300 K and temperatures above 1000 K at pressures up to 150 GPa. [37-40].

2. EXPERIMENTAL

Pressure was generated with a diamond anvil cell (DAC) employing beveled anvils with central flats ranging from 20 to 100 μm and flat diamonds with 200-500 μm culets. Two types of DAC were used: modified (to match a continuous flow He cryostat) Mao-Bell cell for operations at room and low temperatures [41] and a Mao-Bell high-T external heating cell [42]. The latter one is equipped with two heaters and thermocouples. Four experiments were performed at RT aiming to highest pressure and the final pressures varied from 180 to 268 GPa. For low-temperature measurements we used a continuous-flow He cryostat, which allowed infrared and *in situ* Raman/ fluorescence measurements. More details about our IR/Raman/fluorescence setup at the NSLS are published elsewhere [41].

To explore high P-T transformations of nitrogen, we performed more than 20 experiments ranging from 15 to 1050 K and up to 150 GPa using in situ Raman spectroscopy and also synchrotron infrared (IR) spectroscopy and x-ray synchrotron diffraction on temperature-quenched samples. For Raman spectroscopy we used 514.5 and 487.9 nm and tunable red lines of a Ti: Sapphire laser as excitation sources. The temperature was measured to within ±1 K below 600 K and ±5 K above 600 K. To determine the pressure, we used in situ fluorescence measurements of ruby and Sm: yttrium aluminum garnet chips loaded in the sample chamber. Energy dispersive x-ray diffraction was carried out at the beam line X17C of the National Synchrotron Light Source (NSLS), using a focused white beam. To reduce possible texture problems the sample was rocked in v and x over an angular range allowed by the diamond backing plates. The majority of the diffraction data, however, were collected at the 13-ID beam line of GeoSoilEnviroCARS (GSECARS) at the Advanced Photon Source (APS). A focused, monochromatic beam at 0.4246 Å (29.2 keV) was used, and the data were recorded on a MAR charge coupled device calibrated with a CeO_2 standard. The sample was rocked in v by ±10° to average over as many crystallites as possible.

3. ORIENTATION ORDER IN MOLECULAR PHASES

Fig. 3 shows Raman spectra as a function of pressure. As was discussed above, the splitting of the vibron in δ– N_2 (not shown, see e.g. Refs. [20]) and ε-N_2 to ν_1 and ν_2 components manifests the presence of sites of different site symmetry (disk 48*l* and sphere-like 2*a*). With pressure, the ν_2 mode further splits successively to several components. The first splitting corresponds to the delta-epsilon transition at 18 GPa. The second component is first seen as a weak shoulder on the high-frequency side of the major component. The splitting occurs due to the reduction of symmetry (monoclinic distortion), so differently polarized vibrations have different symmetry and may have different energies. An additional splitting of the main component is related to phase transformations due to a sequential orientational ordering of molecules, as will be discussed later. Additional and very important information about orientational ordering can be obtained from the spectra of the translational (phonons) and rotational (librons) vibrations (see below).

Qualitatively observed similar phenomena are observed at low temperatures, but the transformations are more pronounced and occur at lower pressures. The sequence of Raman spectra measured as a function of pressure at low temperatures is shown in Fig. 4. At 17.5 GPa we observe two Raman peaks in the vibron region -slightly broadened ν_2 (lower frequency) and ν_1. This is in excellent agreement with the experiments for the ε phase (see also Refs. [9, 23]). Increasing pressure splits the ν_2 peak, so three components can be seen. This splitting becomes obvious at the highest pressure (44 GPa), while at 24-38 GPa peak fitting is required to reveal the two components (e.g. the spectrum at 24.8 GPa in Fig. 4). Only a slight broadening of the ν_1 peak is observed as the pressure is increased. Lattice modes also change at 18-25 GPa, revealing splitting of the major bands and appearance of new bands (Fig. 5) in agreement with previous studies for ε-N_2 [9, 23].

Infrared spectra at 17.5 GPa (Fig. 4b) show a very weak absorption in a spectral range of the ν_2 vibron. The absorbance increases substantially at 18-25 GPa, so a doublet of IR vibrons is clearly visible at higher pressure. As for the Raman bands, a moderate broadening of the IR peaks is observed as the pressure is increased.

The pressure dependence of the observed Raman frequencies is shown in Fig. 6. Corresponding data from Ref. [9] are also shown for comparison. Earlier results [23] for the ν_2 multiplet are very close to those reported in Ref. [9] and are not shown for clarity. For the ν_1 band, our low-temperature data are very close to those measured at the room temperature (see also Ref. [32]).

The substantial difference in the pressure dependence of the frequency of the Raman ν_2 multiplet will be discussed later. The pressure dependences of the Raman and infrared frequencies are compared in the inset of Fig. 6. We find that Raman and infrared frequencies do not coincide (cf. Ref. [9]) suggesting a non-centrosymmetric space group. Unlike the Raman data, our infrared frequencies are in good agreement with those reported in Ref. [9] in the pressure range overlapped by the two studies (to 25.2 GPa).

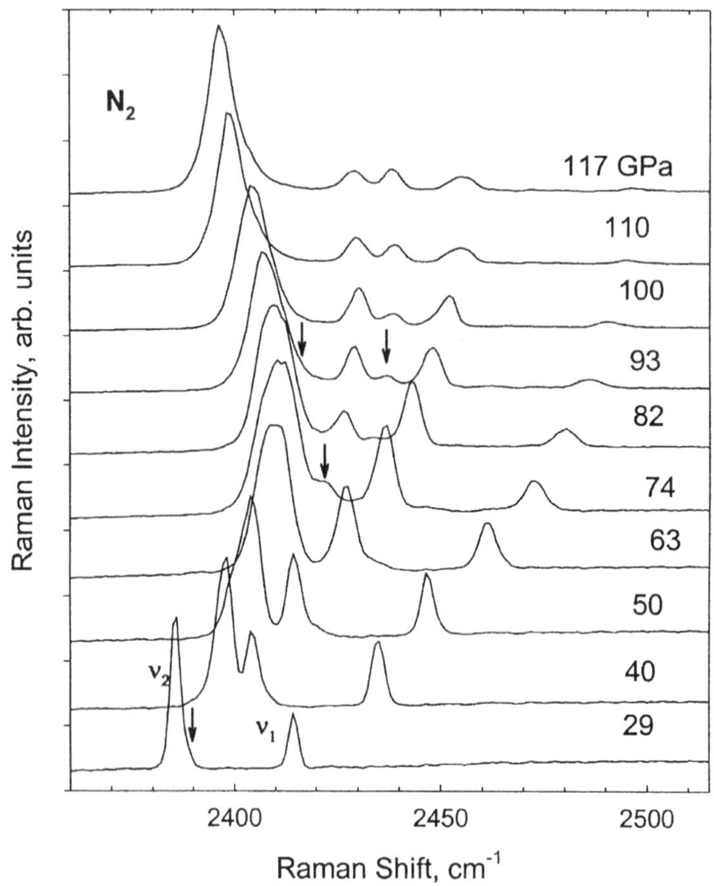

Fig. 3. A sequence of Raman spectra of N_2 as pressure increases at room temperature. Vertical arrows indicate new bands, which appear as pressure increases.

At room temperature, Raman spectra of N_2 measured above 60 GPa (Fig. 3) contain more vibron bands (four) that are allowed by group theory for the R $\overline{3}$ c symmetry of ε–N_2 (three) [23]. Increasing of pressure through the 60 GPa range gives rise to a new Raman peak (ν_{2b}), designated by arrow in Fig. 3 (see also Refs. [24, 35]). A similar observation was made when cooling down from approximately the same starting point [36]. At this pressure (and room temperature) we also observed an increase in intensity of the infrared vibron [3]. The splitting of the lattice modes increases with pressure with multiplets evolving into distinct bands.

Raman spectra to 140 GPa (Fig. 3) reveal an abundance of vibrational modes in agreement with previously reported results [24, 27, 35]. At moderate pressures, our data are in agreement with those reported in Refs. [24, 27, 35]. As in Refs. [24, 27, 35], we observe branching of the Raman-active vibrons at around 90 GPa followed by an increase in their frequency separation. Intensity redistribution is also observed as the lowest frequency vibron ν_2 gathers most of the intensity at the highest pressures. As in Refs. [24, 35, 45], we also observe a softening of the low-energy Raman vibron, which was initially interpreted as an indication of approaching dissociation of N_2 molecules. In contrast, we find a monotonic increase in frequency (Fig. 5) of all lattice modes and redistribution of their intensities with pressure, so that high-frequency bands dominate at the highest pressures.

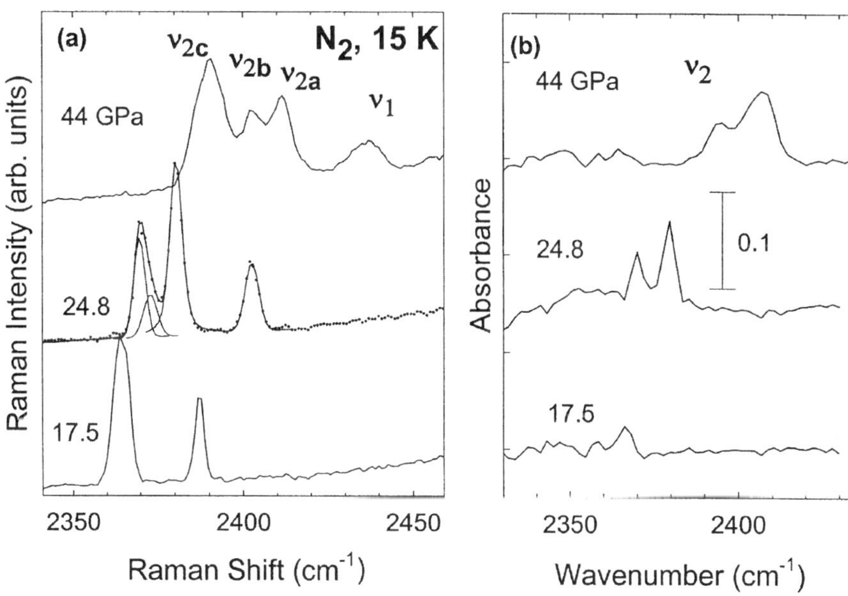

Fig. 4. Raman (a) and IR (b) spectra of nitrogen through the low-temperature transition to the ζ phase. The spectra are shifted in vertically for clarity. The Raman spectrum at 24.8 GPa (points) is shown along with the fitting to multiple components (Voigt profiles).

IR absorption spectra (see Fig. 7) reveal a multiplet of modes in the vibron spectral range and also a newly discovered mode of lattice vibrations (see below). The pressure dependencies of IR vibron frequencies match well the extrapolation of the lower pressure data reported in Refs. [44, 45]. The pressure dependence of the lattice mode frequency is very close to that of the high-frequency Raman lattice mode (Fig. 8). As in the case of Raman vibrons, the IR vibrons show branching with pressure, so that up to 5 vibrons can be observed at high pressure, all of which originate from the ν_2 multiplet. The Raman and IR vibrons probe different points of the large Brillouin zone and become accessible because of its folding. The lowest-frequency Raman vibron corresponds to the lowest frequency of the Brillouin zone and represents the case where all molecules on the faces of the unit cell (e.g.

Ref. 27) vibrate in phase [37]. The other vibrons (Raman and IR) involve different out-of-phase vibrations and form a compact group close to the frequency of the uncoupled N-N stretch. The splitting between the highest and lowest frequency vibrons approximately quantifies the Brillouin zone bandwidth, which increases dramatically with pressure [37]. The frequency of the uncoupled N-N stretch tentatively determined near the position of the density maximum of the observed vibrons (Fig. 7) levels off in the 100-150 GPa pressure range. Thus, the turnover and softening with pressure of the lower-frequency Raman vibron is naturally explained by an increase in the vibrational splitting, which in turn is caused by an increase in the intermolecular interactions. The observed redistribution of intensities between lattice and vibron modes can be considered a consequence of mixing of vibrational states arising from the difference in crystallographic positions at the unit cell face and origin. This means that nitrogen molecules are sitting in quasi-homogeneous sites prior to the transition to a nonmolecular state unlike another possible scenario involving formation of heterostructures [46].

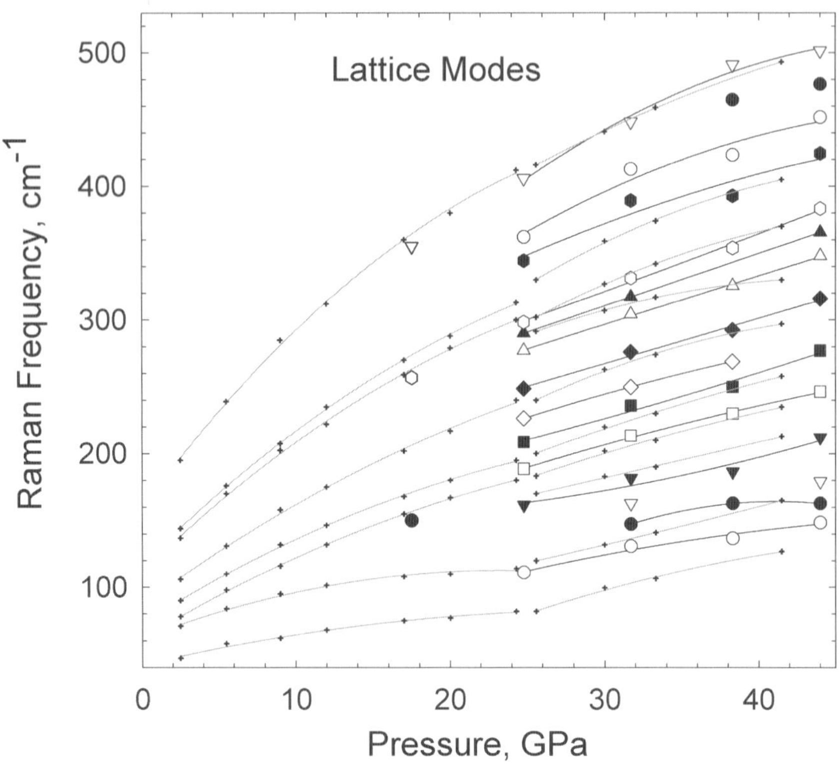

Fig. 5. Raman frequencies of the lattice modes at low temperatures as a function of pressure. Large symbols and solid lines are our data. Dots and grey lines (square fittings) are data from the Ref. [9].

At pressures above 90 GPa and room temperature a new Raman vibron band (ν_{2a}) become visible (see also, Ref. [35]). At this point the infrared absorption spectra of vibron modes also show a change: more IR modes become visible and the overall intensity increases (Fig. 7). Phase transition has been suggested in Ref. [24] on the basis of similar observations, but it was not confirmed in more recent publication [35]. The data presented here and also the results of the recent unpublished x-ray study (Sanloup et al., unpublished) clearly indicate a phase transformation at 100-110 GPa. This transition may be related to a next (probably ultimate) stage of orientation ordering of N_2 molecules.

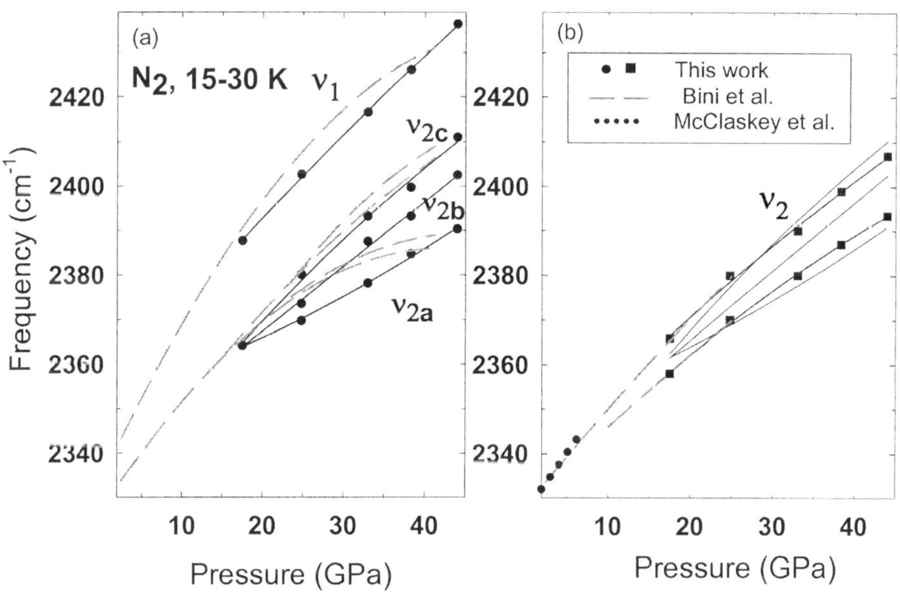

Fig. 6. Raman (a) and IR (b) vibrational frequencies as a function of pressure at 15 K. Filled circles are the Raman data from this work. The solid thick solid lines are the guides to the eye. The dashed lines are the Raman data from Ref. [9]. The panel (b) shows the comparison between Raman and infrared frequencies. Filled squares are the infrared data from this work. The solid thick solid lines are the guides to the eye. Thin solid lines are the Raman data from this work.

The splitting of the vibron bands and change in the lattice mode spectrum observed at low temperature indicate a phase transformation related to orientational ordering of the nitrogen molecules. The phase diagram of Ref. [9] suggests that these changes correspond to the $\varepsilon \rightarrow \zeta$ transition. Qualitatively, our data and those presented in Refs. [9, 23] show similar trends, but detailed comparison shows different Raman spectra for the high-pressure phase (Fig. 6). We believe that the disagreement arises from the use of different experimental procedures and the nature of the high-pressure phase (or phases). In contrast to experiments reported in Refs. [9,

23], we changed pressure at low temperature. It is useful to note that when infrared spectra were measured in a manner similar to ours [9], the results from both studies agree very well (inset to Fig. 3). The evidence that the properties of the high-pressure phase depend on the thermodynamic path, suggests that this phase is not thermodynamically stable (i.e. metastable at the indicated P-T conditions). This is supported by observations of a large hysteresis of the transition at low temperatures [9]. An alternative (but related) explanation is that different properties of the high-pressure phase arise from relatively large pressure inhomogeneities in our experiment (since we changed pressure at low temperature) as indicated by broadening of Raman and infrared bands at higher pressures (Fig. 4).

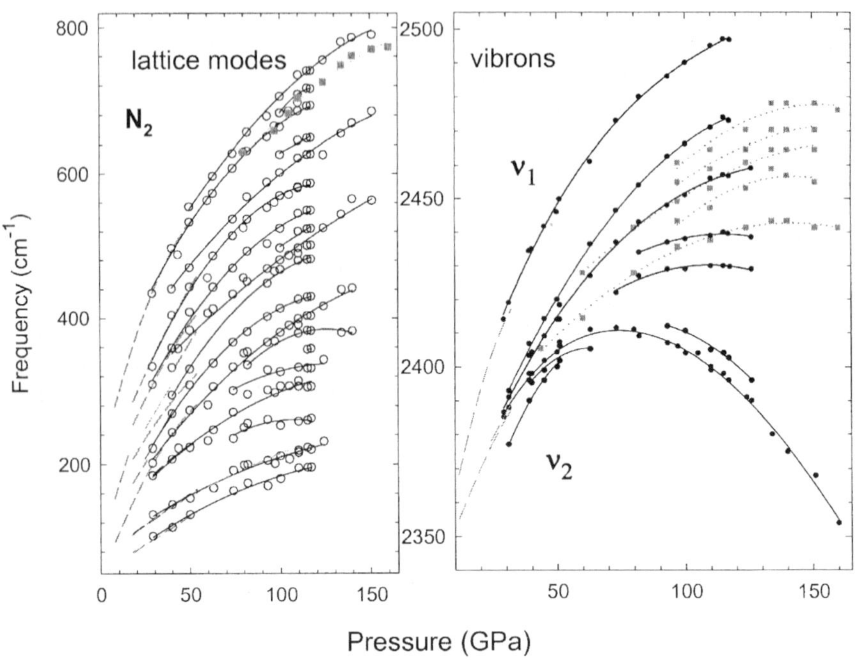

Fig. 7. Raman and infrared lattice and vibrational frequencies of N_2 at high pressures and room temperature. Circles – Raman data, squares – IR data. Solid line and dotted grey lines are guides to the eye for these data, respectively. Grey dashed lines are from Ref [27].

The changes in Raman and infrared spectra above 60 GPa at room temperature are very similar to those observed at low temperatures. Moreover, the reported transition boundary [9] extrapolated to high pressure and temperature matches this room-temperature point. According to the observed Raman and infrared spectra, the vibrational properties of the high-pressure phase are very similar at room and low temperature. Thus, we will consider it to be the same phase (ζ).

In view of the absence of sufficient x-ray data for the ζ phase, we can only speculate on its crystal structure. The number of the vibron modes (in either our experiment or those reported in Ref. 35) exceeds that predicted for the *R3c* structure based on the space group theory proposed in Ref. [23]. According to Ref. [35], the increase of number of vibron modes is due to the increase in the number of different site symmetries occupied by N_2 molecules. Following this idea, up to 5 different site symmetry positions should be invoked to explain the observed number of Raman vibron peaks above 60 GPa, which does not seems plausible. A critical examination of the spectra of Ref. [35] shows that this number can probably be reduced to 3 according to the number of observed distinct peaks in the Raman excitations of the guest molecules. Thus, it seems natural to propose that the branching of vibron modes is related to sequential lifting of degeneracy of the ν_2 term of the cubic δ phase. In the first stage (δ to ε transition), the ν_2 band splits into A_{1g} and E_g components by the crystal field. In the second one (ε to ζ transition), the symmetry is further reduced (to orthorhombic or monoclinic), with doubly degenerate level splitting into two components. Additional splitting (vibrational or Davydov-type) of these major components could be caused by intermolecular interactions. This is related to a possible increase in the number of molecules in the unit cell as well as associated symmetry lowering. High-quality diffraction data are required to examine these hypotheses.

We find that the properties of the high-pressure, low-temperature phase of nitrogen obtained by "cold" compression are different from those for the phase quenched from high temperature. This suggests that the ζ phase is metastable and/or transitions to it are sensitive to nonhydrostatic effects. Raman and infrared spectra of ε–N_2 above 40 GPa and ζ–N_2 are not compatible with $R\bar{3}c$ and *R3c* symmetries proposed in Ref. [23] because the number of vibron bands is larger than predicted for the standard structures based on these space groups. This increase in the number of bands is probably related to the additional lowering of the symmetry and multiplication of the size of the unit cell.

The present vibrational spectroscopy data provide additional constraints on the structure and properties of the high-pressure phases. They also suggest that known phases are not necessarily thermodynamically stable in the P-T region in which they can be observed. As for other molecular crystals, sluggish kinetics can complicate the determination of the true thermodynamic phase diagram (see e.g. Ref. [47]). Further theoretical and experimental effort is necessary to obtain a better understanding of the phase diagram of nitrogen at these high-pressure conditions.

4. NEW CLASSES OF MOLECULAR PHASES

When compressed at 300 K nitrogen transforms from the δ to ζ phase around 60 GPa (see Fig. 11). When heated at pressures higher than 60 GPa the material first back transforms from ζ to ε along a boundary that we find to be on the extension of the line established in Ref. 9 at lower temperatures. At 95 GPa when the temperature reaches >600 K, the transition to θ nitrogen takes place. The transition can be observed visually since ε(ζ)-N2 normally shows substantial grain boundaries, while after the transition to the q phase, the sample looks uniform and translucent. In most cases the transition happens instantaneously and goes to completion within seconds as determined by Raman spectroscopy. If ε-N_2 is heated at even lower pressures (e.g., 65–70 GPa), it transforms above 750 K to ι-N_2. It is also possible to access the ι phase from θ: we observed the transformation from the θ to ι phase on pressure release at 850 K at 69 GPa (see below).

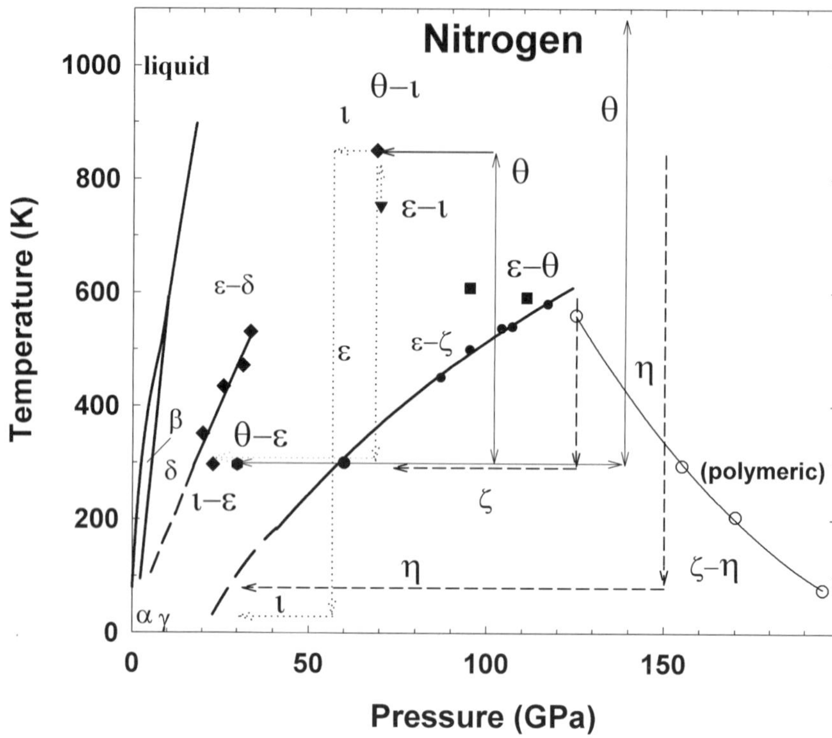

Fig. 8. Phase and reaction diagram of nitrogen at high pressures and temperatures. Solid thick lines are thermodynamic boundaries. Solid circles show the transitions between δ– ε and ε– ζ phases investigated in this work. Filled symbols (ι—diamonds, θ—squares) show the *P-T* points at which new phases were reached or back transformed to the known phases. The arrows show thermodynamic paths (schematic) used to reach θ (solid, thin lines) and ι (dotted, thin lines) phases and paths taken to investigate their stability. The transformation to nonmolecular η-nitrogen is shown by the open circles (this work and Ref. 48) and thin solid line, which is only a guide to the eye; the thin, dashed arrows are paths to investigate the stability. This region should be treated as a kinetic boundary. Phase boundaries at low *P-T* (open squares) are from Ref. 9 and the melting curve is from Ref. 11. The phase boundaries for the α, γ, and δ_{loc} phases are not shown.

Figure 9 shows the Raman spectra of the ι and θ quenched to room temperature. In order to have spectra of all phases discussed here at similar conditions we present them at room temperature. The high-temperature spectra of ι and θ are very similar to ones measured upon quenching the sample to 300 K. All IR measurements were performed at 300 K. The Raman and IR spectra of the ζ phase (obtained by compression at 300 K) from which these phases were formed are shown for comparison at 69 GPa (Fig. 9) and 97 GPa (Fig. 10).

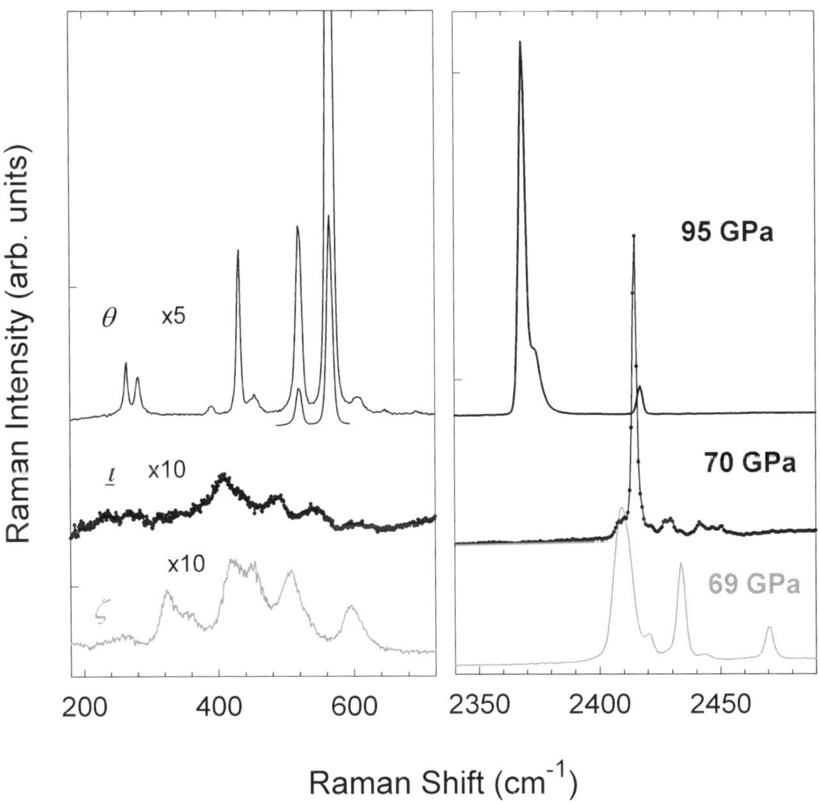

Fig. 9. Representative Raman spectra of θ and ι phases measured at 95 and 70 GPa and 297 K after quenching from high temperature. The spectra of the ζ phase used as a starting material are shown for comparison at the same temperature and at 69 GPa.

Raman spectra of both θ and ι exhibit vibron excitations (Fig. 9, right panel), although their number and frequencies differ from those of all other known molecular phases. The changes in the low-energy region of the spectra (Fig. 9, left panel) for θ phase are very pronounced. The lattice modes of θ nitrogen are very sharp and high in intensity compared to either ι or ζ (also ε). This is a clear indication that molecular ordering in θ is essentially complete, whereas other molecular phases may still possess some degree of static or dynamic orientational disorder. The lattice modes of ι-N_2 are also different from those in ζ and ε: they are extremely weak and broad, suggesting that this phase is not completely orientationally ordered.

Figure 10 shows infrared-absorption spectra of new phases together with that of the ζ phase. Again, number, frequencies, and intensities of vibron excitations (Fig. 10, right panel) are different from other known phases. In the case of ι-N_2 the spectrum differs mainly in positions of the absorption bands while the IR vibron mode of the θ phase has a much larger oscillator strength compared to other N_2 phases cf. H_2 in phase III (Ref. [49]) and ε-O_2 (Ref. [50]). There is a more pronounced Raman and IR softening of the vibron bands of θ nitrogen compared to the other modifications (ι, ζ, and ε). This observation together with the presence of a strong IR vibron is consistent with the existence of charge transfer e.g., see Refs. [49–51] related to the formation of lattice-induced dipole moments or association of molecules [50–52].

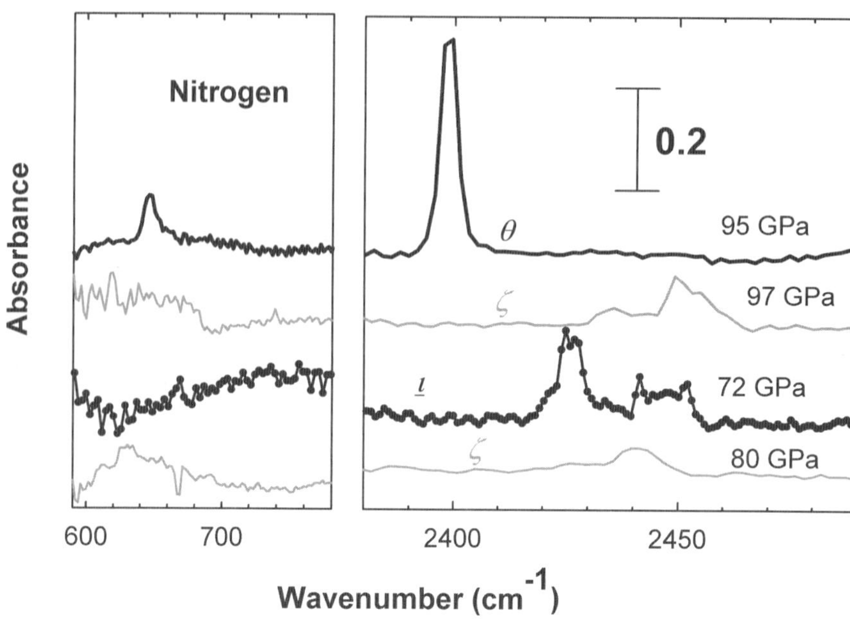

Fig. 10. Infrared modes of θ and ι phases measured at 95 and 70 GPa and 297 K after quenching from high temperature. The spectra of the ζ phase used as a starting material were measured at 97 GPa and 80 GPa are shown for comparison at the same temperature.

The pressure dependence of the Raman-active vibron modes (Figs. 11(a) and 11(b)) was studied on unloading at 300 K in new phases. ι–N_2 exhibits typical behavior for such molecular crystals: branching of vibrational modes and increasing of separation between them with pressure due to increasing intermolecular interactions. All of the vibrational modes originate from the same center, which is close to the frequency of the ν_2 disk-like molecules in ε-N_2. Thus, the structure of the ι phase is characterized by the presence of just one type of site symmetry for the molecules and the large number of vibrational modes arises from a unit cell having a minimum of eight molecules. For the θ phase, two different site symmetries are present. The higher frequency $\nu_{1\theta}$ gives rise to three Raman bands and one in the IR, while the lower frequency $\nu_{2\theta}$ correlates with only one Raman band. Figures 11(a) and 11(b) also show that the spectra have several cases of frequency coincidence of Raman and IR vibron modes, which excludes an inversion center for both structures. Surprisingly, we find that the θ and ι phases have a very large domain of pressure and temperature metastability. The pressure dependence of the Raman spectra (Fig. 11) was studied in a wide pressure range at 300 K in both phases. Most data were taken on pressure release, although experiments carried out on different conditions did not reveal that any measurable differences in frequencies of the θ and ι phases depended on the sample history. Performing x-ray analyses on low-Z materials in the 100 GPa pressure range and high temperatures is difficult due to the low scattering efficiency and small openings in the diamond backing plates, leading to low signal/noise ratios. Nevertheless, we performed synchrotron x-ray diffraction to confirm the existence of two new structures. First, we found good agreement with previously reported results for the low-pressure phases [26, 33]. Only a few reflections could be observed above 50 GPa because of strong sample texture. The highly textured nature of the sample could result in substantial changes in intensities of diffraction peaks from an ideal powder and even prevent observation of some of them. For the θ-phase results presented here, we combined the energy- and angle-dispersive measurements for three samples with presumably different preferred orientations of crystallites. No major changes in the x-ray diffraction patterns were observed at 60 GPa and room temperature, corresponding to the ε– ζ transition (see also Ref. 36). This observation is consistent with vibrational spectroscopy, which shows only moderate changes identified as a further distortion of the cubic unit cell of the δ–phase [24, 39]. In contrast, the x-ray-diffraction patterns of the samples after ε→θ and ε→ι transformations differ substantially from those of the ε and ζ phases, and from each other. Tentative indexing of the peaks of θ nitrogen gives an orthorhombic unit cell e.g., with lattice parameters a=6.797(1), b=7.756(5), and c=3.761(1) Å at 95 GPa). The systematic absences, lack of inversion center and presence of high-symmetry sites (see above) are consistent with space groups $Pma2$, $Pmn2_1$, $Pmc2_1$, $Pnc2$, and $P2_12_12$. The a/c ratio is close to $\sqrt{3}$, which suggests that the lattice is derived from a hexagonal structure. Extrapolation of the equation of state of ε-N_2 measured to 40 GPa (Ref. 26) shows that volume for this phase is about 14 $Å^3$/molecule at 95 GPa, which gives an upper bound assuming a pressure-induced (density driven) transition. Comparison with the experimentally determined unit-cell volume (198 $Å^3$) suggests 16 molecules in the unit cell, giving 12.4 $Å^3$/molecule in the θ phase and an 11% volume collapse at the ε– θ transition. The number of molecules is in agreement with vibrational spectroscopy data, although it is possible to describe the vibrational spectra with a smaller number (up to 8).

In order to better understand the P-T provenance of new phases and their relation with other polymorphs of nitrogen (Fig. 8), we pursued extensive observations in different parts of the phase diagram. The new phases are found to persist over a wide P-T range. As noted above, both phases could be quenched to room temperature. On subsequent heating, the θ–phase remained stable when heated above 1000 K between 95 and 135 GPa and does not transform to the nonmolecular hphase (shown by arrow in Fig. 8). But on pressure release it transforms to ι-N_2 at 69 GPa at 850 K. In view of the relatively high temperature of this transformation and its absence at room temperature, this observation implies that the transformation point is close to the $\theta - \iota$ equilibrium line (see Fig. 8). At room temperature, θ-nitrogen remains metastable as low as 30 GPa on unloading. Similarly, ι-nitrogen remains metastable to 23 GPa; at these pressures both phases transform to ε-N_2 on unloading. ι-N_2 was found to be stable at low temperatures (down to 10 K) at pressures as low as 30 GPa. We note that η nitrogen so far has been accessed only from ζ-N_2 (see Fig. 8). The apparent kinetic boundary (open circles in Fig. 8) that separates these phases can be treated as a line of instability of ζ-N_2. Likewise, the ι and θ phases have been reached only from the ε phase, though they probably can be formed from δ as well. We observed that on further increase of pressure and temperature, the θ phase does not transform to the nonmolecular η phase (to at least 135 GPa and 1050 K). We suggest that it might instead transform to a (perhaps different) nonmolecular crystalline phase on compression. This nonmolecular phase may not be easily reached by compression at 300 K because of a kinetic barrier separating it from ζ-N_2.

The above results provide important insights into the behavior of solid nitrogen at high pressures and temperatures. The ι-phase appears to represent a different kind of lattice consisting of disk-like molecules, presumably packed more efficiently compared to the mixed disk- and sphere-like δ-family structures. The θ phase is more complex. Its striking vibrational properties indicate that it is characterized by strong intermolecular interactions, perhaps with some analogy to H_2-phase III [51] ε-O_2 [50] or CO_2-II [53]. If the interactions are strong enough, the phase may be related to theoretically predicted polyatomic species [54] but this requires further investigation. Our data show that the new phases are thermodynamically stable high-pressure phases since they are formed irrespective of thermodynamic path. Indeed, our data indicate that ζ-N_2 may be metastable in much of the P-T range over which it is observed, since it is typically obtained only as a result of compression at $T<500$ K of the ε-N_2 (see above). An important general conclusion of this work is that the definitive determination of the equilibrium phase relations of nitrogen is more complex than previously thought due the presence of substantial transformation barriers between different classes of structures. These structures include the well-known phases based on weakly interacting N_2 molecules [10] the recently observed nonmolecular phase and the strongly interacting molecular phases documented here (see also Ref. [48]).

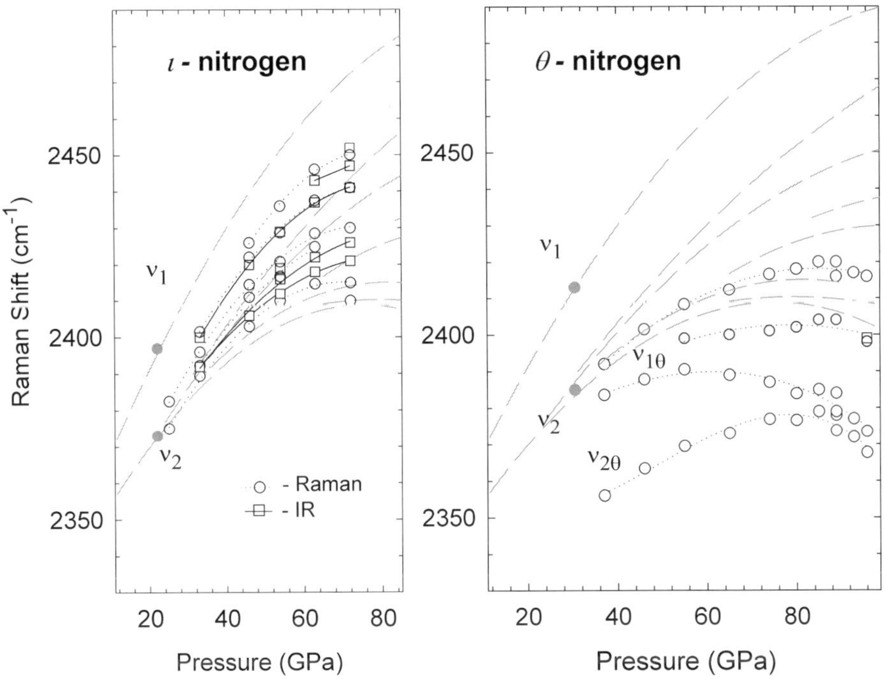

Fig. 11. (a) Raman (open circles) and infrared (open squares) frequencies of vibron modes as a function of pressure for ι phase. (b) Raman (open circles) and infrared (open square) frequencies of θ phase. All measurements were done on the pressure release at 300 K. Filled circles correspond to the vibron frequencies after the transformation to the ε phase from θ and ι phases. Gray dashed lines are data for the ε and ζ phases from Ref. 35.

5. POLYMERIC NITROGEN

Before this study had begun, Raman measurements of nitrogen have been carried out to 130 GPa [24] and 180 GPa [43]. The lowest-frequency vibron has been observed in both studies to the highest pressure reached, and the persistence of this vibron was interpreted as the existence of molecular phase to those pressures. Also, visual observations [24, 43] and visible transmission measurements in Ref. [24] reveal color changes at 130-180 GPa but no quantitative characterization has been done.

In contrast to Refs. [24, 43], we used only a minute amount of ruby to determine pressure, so the majority of the sample was available for optical measurements. We observe that the Raman and IR vibrons (Fig. 12) lose their intensities in 140-160 GPa pressure range and completely disappear at higher pressures. This is also observed for the Raman and IR lattice modes. It may be argued that the disappearance of the Raman modes is attributed to the presence of a luminescence background (quite moderate with red excitation) and an increase of the visible sample absorption (see below). However, IR intensities are totally independent of these factors because the sample remained transparent in the mid-IR spectral range.

Fig. 13 presents the results of visible and IR absorption measurements of nitrogen at elevated pressures. Below 140 GPa nitrogen is transparent in the entire (except the absorption on relatively weak vibrational excitations) measured spectral range (600-20000 cm^{-1}). At 150 GPa a wide absorption edge appears in the visible part of the spectra, at which point the sample becomes yellow, and then totally opaque (Fig. 13) at 160 GPa. This transformation substantially affected the measurement of IR spectra because of increased absorption in the near-IR range. Nevertheless, we measured IR absorption spectra to the highest pressure reached in the experiment (about 170 GPa). Inspecting the spectra at different pressures above 150 GPa one can easily infer that, to a first approximation, they can be obtained by simple scaling, indicating an increase of the abundance of the new phase with increasing pressure. This matches closely the vibrational spectroscopic observations of a gradual disappearance of all excitations in the molecular phase (see above) between 140 and 170 GPa. The Raman and IR spectra of the new phase show a rather broad, weak Raman band at 640 cm^{-1} as well as a broader IR band at 1450 cm^{-1} (Fig. 12). Their intensities seem to increase gradually with pressure, concomitant with a decrease in the intensity of molecular excitations, implying the coexistence of two phases between 140 GPa and 170 GPa. The complete change in vibrational excitations and appearance of the low-energy band gap provide evidence for the transformation from the molecular phase to a nonmolecular phase with a narrow gap. Theoretical calculations predict a transformation to threefold-coordinated cubic or distorted cubic structures [3-5] associated with a substantial volume discontinuity (25-33%). We can rule out with a high probability the simple cubic high-pressure phase [3], which must be metallic. Other calculations indicated semiconducting phases [3-5], which agree with our experimental data (see also Ref. [48]).

Theoretical calculations [5, 55] predict at least a two-fold decrease in the frequency of the N-N stretch as a result of the transformation from a triple to a single-bonded molecule. Our IR measurements reveal a broad band in this spectral range (Fig. 12). A Raman band at 640 cm^{-1} was also observed, which could be identified as a bending mode or correspond to a peak in the phonon density of states (see below). Comparison of our data with the calculations of Barbee (Ref. [55]) shows even more striking qualitative correspondence between experiment and theory assuming the "cubic gauche" [5] high-pressure structure (i.e., possible appearance of the second weak IR peak at 900 cm^{-1} (Fig. 12). A variety of IR and Raman modes are predicted for this phase, and most of them are degenerate. Those modes can split under nonhydrostatic conditions or further lowering of the symmetry Ref. [55], which can explain the observed large linewidth of Raman and IR excitations. Alternatively, the high-pressure phase could be a fine mixture of different tree-coordinate structures with very close total energies *e.g.* arsenic, black phosphorus, "cubic gauche") and the chain-like structure [5, 56] with a high concentration of stacking faults which would make it appear amorphous at the scale of vibrational spectroscopy (see below).

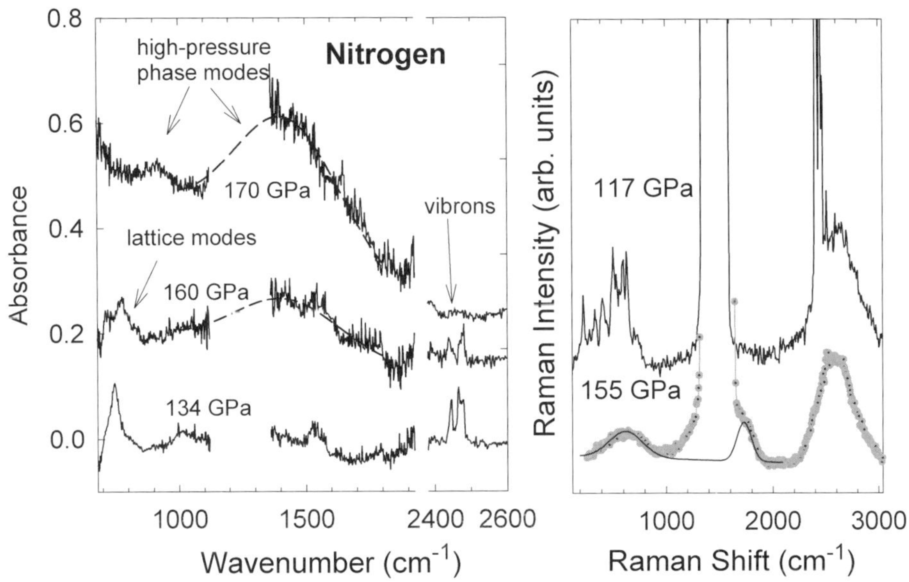

Fig. 12. (a) IR absorption spectra in the vibrational spectral range through the transition to a polymeric phase. Dashed lines are guide to the eye. No data could be obtained in the 1120-1370 cm^{-1} and 1920-2380 cm^{-1} spectral ranges because of strong absorption by the diamonds. (b) Raman spectra through the transition. A strong band at 1350 cm^{-1} and a broad band at 2600 cm^{-1} are due to the first and the second order Raman signal of the diamond anvils. The solid line is a guide to the eye.

The analysis of the shape of the absorption edge of the high-pressure phase (Fig. 13) shows the existence of two spectral ranges with different types of energy dependence on the absorption coefficient. At high values of absorption it follows the empirical Tauc relation [57] in the case of parabolic band edges (Fig. 13(b)), while at smaller absorption a so-called Urbach or exponential absorption tail [58, 59] is observed (Fig. 13(c)). The existence of this kind of absorption edge is normally related to amorphous semiconductors. The optical absorption gap determined from our experiment is 0.6-0.7 eV and it decreases with pressure (see below). The slope of the Urbach tail, which can be considered as a measure of a random microfield [59] is found to be Γ=2.6 eV^{-1} at 160 GPa. This is very close to what one would expect for an amorphous phase with a coordination of 2.5 [59].

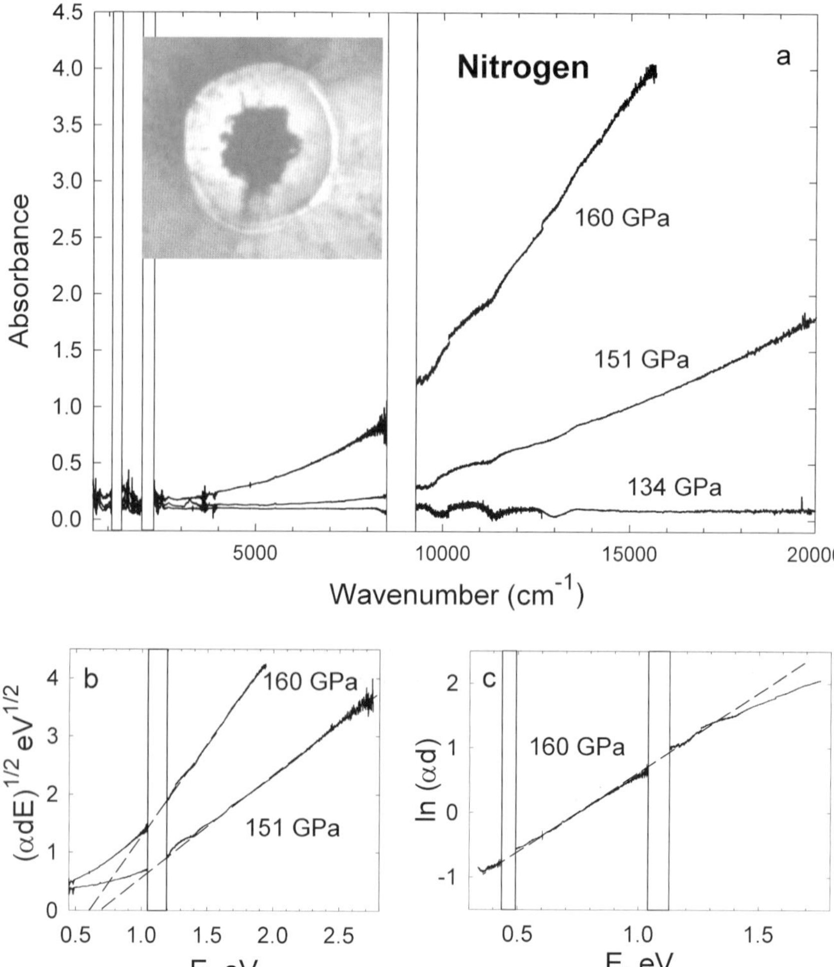

Fig. 13 (a) Optical absorption spectra in a wide spectral range at different pressures (Inset: the microphotography of the opaque sample at 160 GPa (reflected and transmitted light)). (b) Absorption data at 151 and 160 GPa plotted in photon energy (E) versus $(\alpha d\mathrm{E})^{1/2}$ (α-absorption coefficient, d-sample thickness) coordinates illustrating a direct allowed gap absorption law at high energies. Extrapolations of those dependencies to zero absorption give an estimate of the direct gap value. (c) Absorption data at 160 GPa plotted in E versus $\ln(\alpha d)$ coordinates illustrating Urbach-like absorption law at low energies. The frequency ranges of strong diamond absorption are shown by empty boxes. Absorbance A and absorption coefficient α are related by expression $\alpha d = A \ln 10$.

Figure 14 shows representative IR and visible transmission spectra demonstrating the effect of temperature on the semiconducting optical edge characteristic of the η phase. The spectra presented correspond to conditions when no molecular phase is present in the sample as determined from IR vibrational spectroscopy. No variation of the shape and position of the band gap can be detected from transmission spectra at different temperatures and constant pressure of 200 GPa. Figure 14(b) shows that the low-energy portion of the spectra plotted on a logarithmic scale (Urbach plot) has a constant slope (Γ) in a 10–200 K range. This also agrees with 300 K data thus showing that Γ is not temperature dependent. Similar spectra have been reported for amorphous phosphorus at zero pressure [60]. This is typical for solid amorphous semiconductors [59] because the random microfield is caused by static disorder in the system as opposed to crystalline materials [58] where the vibrations generate a temperature-dependent dynamical disorder. Determination of the band gap from our data is a complicated issue, because there is no characteristic feature of the spectra which can be associated with the band gap ~e.g., Ref. [61]. This is especially important for our measurements, since we essentially deal with samples of various thickness (which is a function of anvil geometry and pressure). As the result, visual observations of the sample above 230 GPa showed that it is red or yellowish in transmission and black in reflection, which is consistent with the semiconducting state. The color of the sample (compare with the observations of dark nitrogen in Refs. 24, 37, 43, 48) may be explained by its thickness of the order of 1 μm) compared to the samples brought to 150 GPa being up to 5 μm). At the highest pressure (268 GPa), visible transmission spectra clearly show the presence of the fundamental absorption edge characteristic of semiconductors (Fig. 14(a)). This result is in agreement with direct electrical measurements performed to 240 GPa [48]. The high-energy absorption edge, which can be observed in this case, corresponds to electronic transitions between extended states (unlike Urbach absorption, which is presumably caused by transitions from localized to extended states). Extrapolation of the absorption spectra plotted as $(h\nu\alpha)^{0.5}$ versus $h\nu$ gives the value of optical gap [57]. These values at different pressures are shown in Fig. 15. Note that data from different experiments agree, despite the different sample thickness and the fact that some of the data are taken on pressure release in a metastable pressure region (see below). We observed a monotonic redshift of the band gap with pressure (see also Fig. 14(a)). The pressure dependence of the band gap is sublinear mainly due to contribution from the points obtained on decompression. The extrapolation of the band gap values gives metallization at pressures slightly above 300 GPa. Linear extrapolation of this curve to higher pressures (not taking into account points obtained upon decompression) gives a value of 280 GPa.

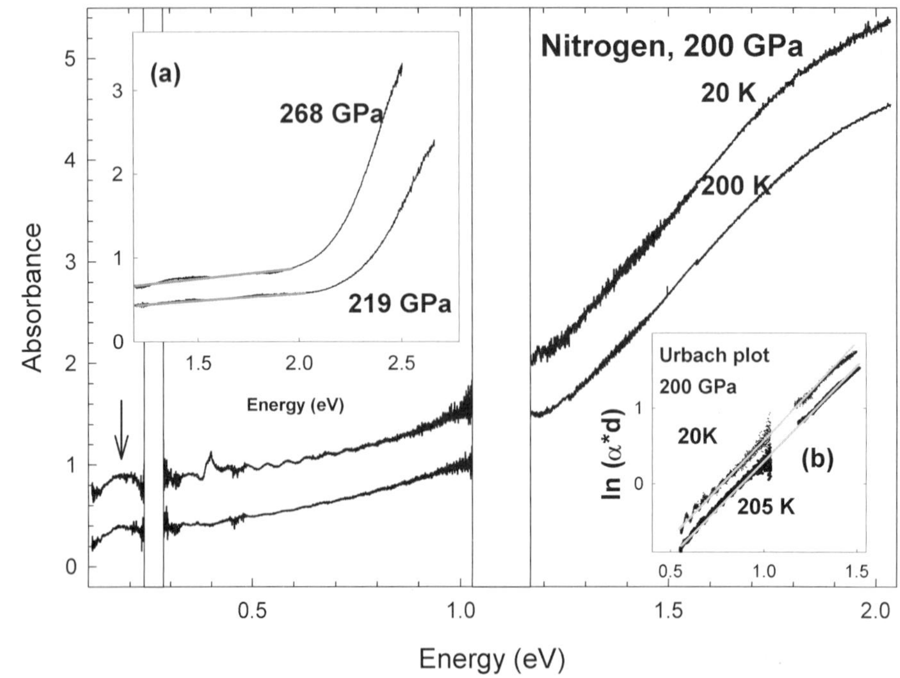

Fig. 14. Transmission spectra of polymeric nitrogen as a function of temperature. Spectra are shifted vertically for clarity. The characteristic peak of the η phase is marked by a vertical arrow. Inset (a) shows the pressure dependence of the absorption spectra of nitrogen at very high pressures and room temperature. Gray lines represent the Tauc fits to the spectra in an appropriate spectral range. The determination of the energy gap from these measurements is obscured by additional losses caused by the presence of a fine ruby powder in the chamber. The high-energy absorption edge is most probably due to stress-induced absorption of diamond anvils (Ref. 62). (b) Urbach plots at 200 GPa and different temperatures (shifted vertically). Gray lines are guides to the eye.

We now present temperature measurements of the vibrational properties of the η phase. Type II diamonds were used for mid-IR measurements to avoid interference with the characteristic absorption of the sample. The representative absorption spectra at different temperatures (see Fig. 14) clearly show the presence of a broad 1700 cm^{-1} IR band (compare with Fig. 12). Its presence was also observed in the sample heated to 495 K at 117 GPa (see below). The position of the band and its damping (if fitted as one band) does not depend on pressure and temperature within the error bars. The Raman spectrum of the η phase obtained on heating (see below) does not show any trace of the molecular phase (see Fig. 12(b)). Careful examination of the spectrum in this case showed a weak broad band at 640 cm^{-1} and a shoulder near 1750 cm^{-1} (both indicated by arrows in Fig. 12(b)). For an amorphous state, the vibrational spectrum would closely resemble a density of phonon states [63] with the maxima corresponding roughly to the zone boundary acoustic and optic vibrations of an underlying structure [3-5, 55], which is consistent with our observations. The only lattice dynamics calculations for hypothetical high-pressure crystalline phases of nitrogen are available for the cubic gauche phase [55] and calculated phonon frequencies are in a qualitative agreement with our measurements (Fig. 16). The vibrational spectroscopy and band gap structure indicate the absence of a long-range order. The material can still possess some short-range order, for example, related to pyramidal coordination of nitrogen atoms. The absence of long-range order can be due to structural flexibility because each atom forms bonds with only three other atoms out of six nearest neighbors [3-5].

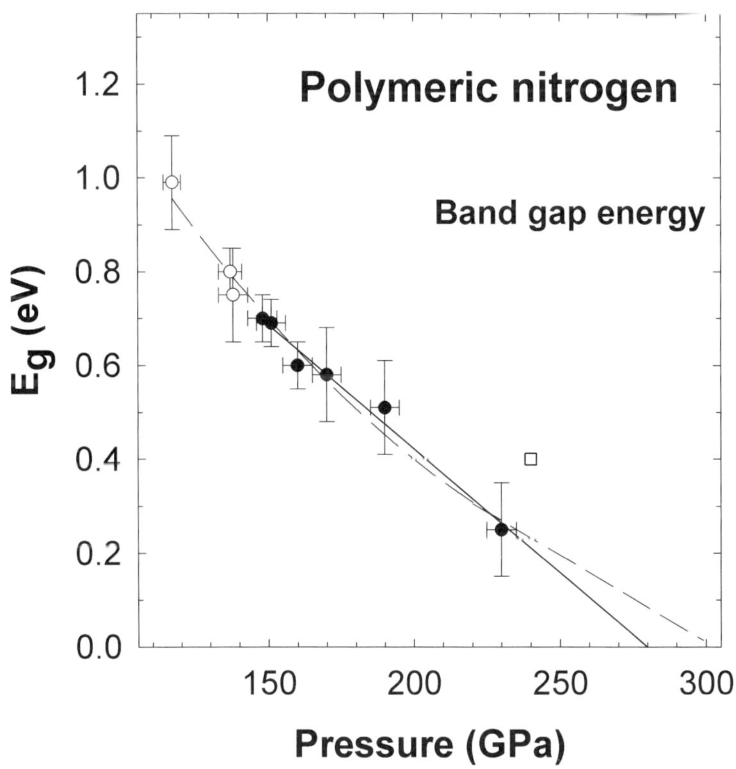

Fig. 15. Band gap of the η phase as a function of pressure. Solid circles represent increasing pressure and open circles decreasing pressure. Linear and quadratic extrapolations are shown as dashed lines. Square corresponds to the electrical conductivity measurements from the Ref. [48].

We probed the forward and reverse transformations of the molecular to the η phase in different regions of *P-T* space (Fig. 8) We used IR transmission spectra as diagnostics of the degree of transformation to the nonmolecular phase. The absence of IR bands corresponding to vibrons and lattice modes of the molecular phase was used as a criterion. Since both the molecular and nonmolecular phases are transparent in the mid-IR, the amount of the phase present is simply *proportional* to the amplitude of the corresponding IR peaks. This is unlike the situation with Raman spectra, which are attenuated by absorption of the η phase. We examined the transformation at 205 K and elevated pressures and found that it starts at 155 GPa and is completed at 185 GPa. This is shifted to higher pressures compared to our 300 K data and is in agreement with the trend reported in Ref. [48]. The sample has been cooled down to 10 K at 200 GPa and warmed up after subsequent release of pressure at 130–150 GPa. IR and visible transmission spectra and Raman spectra clearly showed the persistence of the η phase without any reverse transformation down to 120 GPa. At this point the pressure dropped to 87 GPa and the sample transformed instantaneously back to a transparent phase (called ζ' here). The molecular nature of this phase is confirmed by its Raman spectrum although the positions of the vibron lines do not correspond to those observed on pressure increase [38]. This means that the amorphous phase back-transforms to a molecular phase, which differs from the one observed on upstroke. On further release of pressure (to 60 GPa) we observe the Raman spectra, which are similar to those of ε phase in positions and intensities of vibron peaks.

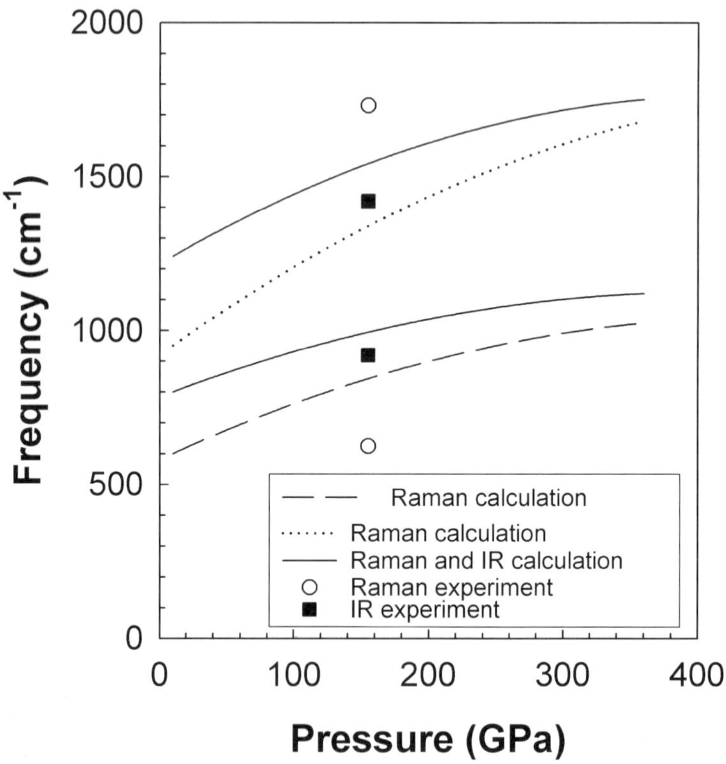

Fig. 16. Comparison of the measured (symbols) and calculated (lines) vibrational frequencies for the cubic gauche structure [55].

In the heating experiment we first exposed the sample to 495 K at 117 GPa. The effect of temperature caused a gradual transformation (starting at 450 K) similar to that observed at 300 and 200 K. A large increase of fluorescence typically precedes transformation to the η phase. Once the transformation is completed (see below), this fluorescence disappears (Fig. 12(b)). The comparison of Raman modes revealed more than a tenfold decrease of intensity in the Raman vibrons and no observable lattice modes. Quenching of the sample to room temperature did not change the color and visible absorption spectra. Surprisingly, the infrared spectra revealed the presence of molecular vibrons, indicating an incomplete transformation (about 30% of the nonmolecular phase judging from the infrared activity). During the second heating the sample was completely transformed to the η phase. Then the pressure was dropped to 105 GPa at 460 K, causing an instantaneous reverse transformation to a transparent molecular phase. The spectral positions of the bands and their number do not correspond to those observed at this pressure on compression but are similar to those obtained during the unloading at 300 K (see above). Increasing pressure to 135 GPa at 510 K drove the direct transformation into the η phase again. Finally, we tested the stability of the η with respect to the transformation to another phase (e.g., crystalline *cf* phase [5, 55]). Raman measurements at 155 GPa and temperatures to 850 K show no sign of any transformation. Figure 8 summarizes our data for the phase diagram of nitrogen obtained in a course of extensive *P-T* measurements. Substantial hysteresis is observed for the transformation from and back to the molecular phase, so the observed curves should be treated as kinetic boundaries. For a direct transformation, our data are in good agreement with the results of visual observations of Ref. 48. Our high-temperature data show that the hysteresis becomes quite small at temperatures above 500 K. There is large hysteresis at lower temperature such that the molecular ζ phase can be metastably retained beyond the ζ– ζ boundary (above approximately 100 GPa; see also Ref. 48). Thus, observation of another molecular phase (ζ') in this *P-T* conditions means that this phase is either kinetically favored or thermodynamically stable with respect to the ζ phase.

If the potential barrier between two crystalline phases is high (molecular dissociation is required in our case), the transition may be preempted by a transformation to a metastable phase, which may be amorphous [64]. This defines an intrinsic stability limit (e.g., spinodal) for the diatomic molecular state of nitrogen. In view of the amorphous component of the higher-pressure phase, the transition may be considered as a type of pressure-induced amorphization. As such, the transformation boundary could track the metastable extension of the melting line of the molecular phase, and if so, it should have a negative slope (consistent with negative ΔV and positive ΔS for a transition to dense amorphous state [64]. Alternatively, one can view this in terms of an intrinsic (elastic or dynamical) instability of the structure of the molecular solid. In this sense, the behavior of the material parallels other amorphizing systems that undergo coordination changes (see Ref. [65]).

In conclusion, we present optical evidence for a transition of molecular nitrogen to a nonmolecular state. The transition occurs on compression when the ratio of inter-to-intramolecular force constants reaches 0.1 [37]. This is small compared to the highest ratio reached for hydrogen (in its molecular phase). It suggests that the destabilization of the triple-bonded nitrogen molecule is the driving force of the nonmolecular transition. Vibrational and optical spectroscopic data indicate that the high-pressure phase is a narrow-gap, disordered, and single-bonded phase. The amorphous nature of the high-pressure phase may represent the common case of a transition in a field of deep metastability [66].

ACKNOWLEDGEMENT

This work was supported by the Carnegie/DOE Alliance Center (CDAC), which is supported by the DOE/NNSA, and by NSF-DMR, NASA, and the W. M. Keck Foundation. Work at Lawrence Livermore National Laboratory was performed under the auspices of the University of California under DOE Contract No. W-7405-Eng-48.

REFERENCES

[1] R. J. Hemley and N. W. Ashcroft, Phys. Today, 51 (1998) 26.
[2] H. B. Radousky et al., Phys. Rev. Lett., 57 (1986) 2419.
[3] A. K. McMahan and R. LeSar, Phys. Rev. Lett., 54 (1985) 1929.
[4] R. M. Martin and R. J. Needs, Phys. Rev. B, 34 (1986) 5082.
[5] C. Mailhiot, L. H. Yang, and A. K. McMahan, Phys. Rev. B, 46 (1992) 14419.
[6] J. Belak, R. LeSar, and R. D. Etters, J. Chem. Phys., 92 (1990) 5430.
[7] S. Nosé and M. L. Klein, Phys. Rev. Lett., 50 (1983) 1207.
[8] R. D. Etters, V. Chandrasekharan, E. Uzan, and K. Kobashi, Phys. Rev. B, 33 (1983) 8615.
[9] R. Bini, L. Ulivi, J. Kreutz, and H. Jodl, J. Chem. Phys., 112 (2000) 8522.
[10] V. G. Manzhelii, Y. A. Freiman, Physics of cryocrystals, American Institute of Physics, (College Park MD, 1997).
[11] D. A. Young et al., Phys. Rev. B, 35 (1987) 5353.
[12] S. Zinn, D. Schiferl, and M. F. Nicol, J. Chem. Phys., 87 (1987) 1267.
[13] W. L. Vos and J. A. Schouten, J. Chem. Phys., 91 (1989) 6302.
[14] S. C. Schmidt, D. Schiferl, A. S. Zinn, D. D. Ragan, and D. S. Moore J. Appl. Phys., 69 (1991) 2793.
[15] C. A. Swenson, J. Chem. Phys., 23 (1955) 1963.
[16] R. L. Mills and A. F. Schuch, Phys. Rev. Lett., 23 (1969) 1154.
[17] F. Schuch and R. L. Mills, J. Chem. Phys., 52 (1970) 6000.
[18] J. R. Brookeman and T. A. Scott, J. Low. Temp. Phys., 12 (1973) 491.
[19] W. E. Streib, T. H. Jordan, and W. N. Lipscomb, J. Chem. Phys., 37 (1962) 2962.
[20] R. LeSar, S. A. Ekberg, L. H. Jones, R. L. Mills, L. A. Schwalbe, and D. Schiferl, Solid State Comm., 32 (1979) 131.
[21] S. Buchsbaum, R. L. Mills, and D. Schiferl, J. Phys. Chem., 88 (1984) 2522.
[22] D. T. Cromer, R. L. Mills, D. Schiferl, and L. A. Schwalbe, Acta Crystallogr. B, 37 (1981) 8.
[23] D. Schiferl, S. Buchsbaum, and R. L. Mills, J. Phys. Chem., 89 (1985) 2324.
[24] R. Reichlin, D. Schiferl, S. Martin, C. Vanderborgh, and R. L. Mills, Phys. Rev. Lett., 55 (1985) 1464.
[25] R. L. Mills, B. Olinger, and D. T. Cromer, J. Chem. Phys., 84 (1986) 2837.
[26] H. Olijnyk, J. Chem. Phys., 93 (1990) 8968.
[27] H. Schneider, W. Haefner, A. Wokaun, and H. Olijnyk, J. Chem. Phys., 96 (1992) 8046.
[28] M. I. M. Scheerboom and J. A. Schouten, Phys. Rev. Lett., 71 (1993) 2252.
[29] M. I. M. Scheerboom and J. A. Schouten, J. Chem. Phys., 105 (1996) 2553.
[30] R. Bini, M. Jordan, L. Ulivi, and H. J. Jodl, J. Chem. Phys., 108 (1998) 6849.
[31] A. Mulder, J. P. J. Michels, and J. A. Schouten, J. Chem. Phys., 105 (1996) 3235.
[32] A. Mulder, J. P. J. Michels, and J. A. Schouten, Phys. Rev. B, 57 (1998) 7571.
[33] M. Hanfland, M. Lorenzen, C. Wassilew-Reul, and F. Zontone, in Abstracts of the International Conference on High Pressure Science and Technology, (Kyoto, Japan, 1997) p. 130.
[34] T. Westerhoff, A. Wittig, and R. Feile, Phys. Rev. B, 54 (1996) 14.

[35] H. Olijnyk and A. P. Jephcoat, Phys. Rev. Lett., 83 (1999) 332.
[36] A. P. Jephcoat, R. J. Hemley, H. K. Mao, and D. E. Cox, Bull. Am. Phys. Soc., 33 (1988) 522.
[37] A. F. Goncharov, E. Gregoryanz, H.-k. Mao, Z. Liu, and R. J. Hemley. Phys. Rev. Lett., 85 (2000) 1262.
[38] E. Gregoryanz, A. F. Goncharov, R. J. Hemley, and H-k. Mao. Phys. Rev. B, 64 (2001) 052103.
[39] A. F. Goncharov, E.. Gregoryanz, H-k. Mao, and R.J. Hemley. Fizika Nizkikh Temperatur, 27 (2001) 1170.
[40] E. Gregoryanz, A. F. Goncharov, R. J. Hemley, H-K. Mao, M. Somayazulu, and G. Shen, Phys. Rev. B, 66 (2002) 224108.
[41] A. F. Goncharov, V. V. Struzhkin, R. J. Hemley, H. K. Mao, and Z. Liu, in: Science and Technology of High Pressure, Vol. 1, M. H. Manghnani, W. J. Nellis and M. F. Nicol, Eds., (Universities Press, Hyderabad, India, Honolulu, Hawaii, 1999) p. 90 95.
[42] Y. Fei, in Mineral Spectroscopy: A Tribute to Roger G. Burns, M.D. Dyar, C. McCammon and M. W. Schafer, Eds.,(Geochemical Society, Houston, 1966) p. 243.
[43] P. M. Bell, H. K. Mao, R. J. Hemley, Physica , 139\&140B (1986) 16.
[44] M. D. McCluskey, L. Hsu, L. Wang, and E. E. Haller, Phys. Rev. B, 54 (1996) 8962.
[45] R. Bini, M. Jordan, L. Ulivi, H. J. Jodl, J. Chem. Phys., 106 (1998) 6849.
[46] D. Hohl, V. Natoli, D. M. Ceperley, and R. M. Martin, Phys. Rev. Lett., 71 (1993) 541.
[47] R. Jeanloz, J. Geophys. Res., 92 (1987) 10352.
[48] M. Eremets, R. J. Hemley, H. K. Mao, and E. Gregoryanz, Nature, 411 (2001) 170.
[49] M. Hanfland, R. J. Hemley, and H. K. Mao, Phys. Rev. Lett., 70 (1993) 3760.
[50] F. Gorelli, L. Ulivi, M. Santoro, and B. Bini, Phys. Rev. Lett., 83 (1999) 4093.
[51] J. Hemley, Z. Soos, M. Hanfland, and H. K. Mao, Nature, 369 (1994) 384.
[52] J. Kohanoff, S. Scandolo, S. Gironcoli, and E. Tosatti, Phys. Rev. Lett., 83 (1999) 4097; I. I. Mazin, R. J. Hemley, A. F. Goncharov, M. Hanfland, and H. K. Mao, ibid. 78 (1997) 1066.
[53] V. Iota and C. S. Yoo, Phys. Rev. Lett., 86 (2001) 5922.
[54] R. Bartlett, Chem. Ind., 4 (2000) 140.
[55] T. W. Barbee III, Phys. Rev. B, 48 (1993) 9327.
[56] M. M. G. Alemany, J. L. Martins, Phys. Rev. B, 68 (2003) 024110.
[57] J. Tauc, R. Grigorovici, and A. Vancu, Phys. Status Solidi, 15 (1966) 627.
[58] F. Urbach, Phys Rev., 92 (1953) 1324.
[59] N. F. Mott and E. A. Davis, Electronic Processes in Non-Crystalline Materials, 2nd Ed., Clarendon Press, Oxford (1979).
[60] L. J. Pilione, R. J. Pomian, and J. S. Lannin, Solid State Commun., 39 (1981) 933.
[61] S. Knief, Phys. Rev. B, 59 (1999) 12940.
[62] Y. K. Vohra, in Recent Trends in High Pressure Research, A. K. Singh, ed., (Oxford & IBH, Calcutta, 1991) p. 349.
[63] M.H. Brodski in: Light scattering in solids, Topics in Applied Physics, vol. 8, M. Cardona , ed., (Springer-Verlag, New York 1983).
[64] For recent reviews see S. M. Sharma and S. K. Sikka, Prog. Mat. Sci., 40 (1996) 1; P. Richet and P. Gillet, Eur. J. Mineral., 9 (1997) 907.
[65] R. J. Hemley, A. Jephcoat, H. Mao, L. Ming, and M. Manghnani, Nature , 334 (1988) 52; J. Badro, J.-L. Barrat, and P. Gillet, Europhys. Lett., 42 (1998) 643. This includes the likelihood that material produced on compression is heterogeneous (i.e., partly crystalline); see R. J. Hemley, J. Badro, and D. M. Teter, in Physics Meets Mineralogy, H. Aoki, Y. Syomo, and R. Hemley, Eds., (Cambridge University Press, Cambridge, England, 2000) p. 173.
[66] E. G. Ponyatovsky and O. I. Barkalov, Mater. Sci. Rep., 8 (1992) 1471.

Chemistry at Extreme Conditions
M. Riad Manaa (Editor)

Chapter 9

Non-Equilibrium Molecular Dynamics Studies of Shock and Detonation Processes in Energetic Materials

Brad Lee Holian[a], Timothy C. Germann[b], Alejandro Strachan[a], and Jean-Bernard Maillet[c]

[a]Theoretical Division, Los Alamos National Laboratory, Los Alamos, NM 87545 USA

[b]Applied Physics Division, Los Alamos National Laboratory, Los Alamos, NM 87545 USA

[a]Commissariat à l'Energie Atomique, DAM Ile-de-France, DPTA BP12, 91680 Bruyères-le-Châtel, FRANCE

1. INTRODUCTION

Detonation involves a complex interplay of chemical reactions and energy transfer between lattice and internal molecular modes, leading to a steady shock wave at the front of the detonation and a self-similar release wave of expanding reaction products at the rear [1-3]. Attempts to relate macroscopic observables, such as the detonation velocity and pressure, to the underlying reaction thermodynamics began with Chapman (1899) and Jouget (1905, 1917). The field was advanced significantly with the independent development of ZND theory by Zeldovich (1940), von Neumann (1942), and Doering (1943) during World War II. The rapidity and violence of detonations have hindered direct experimental attempts to probe the underlying molecular processes, although modern advances in ultrafast shock wave spectroscopy [4] and interferometry [5] are beginning to provide insight. Despite such advances, atomistic computer experiments, namely non-equilibrium molecular dynamics (NEMD) simulations, provide unmatched spatial and temporal resolution enabling the atomic-level characterization of the fundamental processes that govern the properties of energetic materials. Furthermore, NEMD is particularly well suited to the short time and length scales of shock and detonation processes. Examples of such studies are the shock-induced plasticity [6-8] and phase transformations [9] that may occur in metallic solids.

We note that in molecular-dynamics (MD) simulations we make no approximations other than the ones implied in the interatomic potentials and the fact that the dynamics of the atoms is purely classical (no quantum effects on the atomic motion). For example, no approximation is made as to what type of chemical reaction can or can not occur; complex phenomena such as pressure effects, multi-molecular reactions, and relaxation are explicitly described in NEMD. In this sense, the simulations presented here provide a full-physics, full-chemistry description of energetic materials.

In this chapter we review our recent efforts towards understanding many of the salient features of detonation using NEMD simulations. We will focus on large-scale NEMD simulations using a model interatomic potential (denoted REBO) to study generic, but complex, detonation phenomena and the use of a new, computationally more intensive, potential (denoted ReaxFF) that accurately describes a real nitramine energetic material.

2. A SHORT HISTORY OF REACTIVE POTENTIALS

The fundamental input required for molecular-dynamics (MD) simulations is the interatomic potential energy surface upon which the classical dynamics of the nuclei takes place. Typical high explosives, including TNT, TATB, RDX, and HMX, are complicated organic (C, H, N, O) molecules [10]. Developing accurate force fields capable of describing these molecules at high pressures and temperatures, including the subsequent reaction chemistry and products, is an extremely challenging task, which has only been recently attempted [11-13]. Moreover, the complicated energy transfer and sequence of reactions (occurring within the so-called "reaction zone") for these energetic materials are thought to take place on a microsecond (10^{-6} s) timescale. This is far longer than is presently possible using NEMD simulations that have to resolve individual sub-picosecond atomic vibrations with integration timesteps on the order of a femtosecond (10^{-15} s). As a result, only the initial chemical reaction events in such materials can be directly simulated (see below). However, if one were to consider an energetic material with a simpler reaction chemistry pathway, it should have a shorter reaction zone, which could be entirely contained within an NEMD simulation cell. Likely candidates for probing generic behavior are diatomic molecular fluids or solids—either actual explosives such as NO, or model systems. In this chapter, we will review our recent efforts along both of these paths, namely, model systems to capture general phenomena and sophisticated reactive potentials capable of describing the initial response of specific energetic materials.

2.1. Reactive bond-order potentials

Molecular dynamics studies of diatomic model detonations were first carried out by Karo and Hardy in 1977 [14]. They were soon followed by other groups [15, 16]. These early studies employed "predissociative" potentials, in which the reactant dimer molecules are metastable and can dissociate exothermically. More realistic models, combining an endothermic dissociation of reactants with an exothermic formation of product molecules, were introduced by White and colleagues at the Naval Research Laboratory and U.S. Naval Academy, first using a LEPS (London-Eyring-Polanyi-Sato) three-body potential for nitric oxide [17], and later a Tersoff-type bond-order potential [18] for a generic AB model, loosely based on NO [19, 20].

We have chosen to study the latter, reactive empirical bond-order (REBO) potential, introduced by Brenner et al. [20]. The binding energy of an *N*-atom system is given by

$$E_b = \sum_{1 \leq i < N} \sum_{i < j \leq N} f_c(r_{ij})\left[V_R(r_{ij}) - \overline{B}_{ij} \cdot V_A(r_{ij})\right] + V_{vdW}(r_{ij}) \tag{1}$$

consisting of a molecular bonding term (including a cutoff function $f_c(r)$ which smoothly goes to zero at a finite distance) and a weak, longer-range van der Waals interaction (also vanishing at a finite distance) to stabilize the molecular solid, a herringbone lattice in two dimensions (2D) and the Pa3 (α-N_2) structure in 3D. The key to the REBO potential is the bond-order function [18], $\overline{B}_{ij} \equiv (b_{ij} + b_{ji})/2$, which varies from 0 to 1 depending on the local environment around atoms i and j (see Fig. 1), hence weakening the i-j bond as either b_{ij} or b_{ji} decreases. For an isolated diatomic molecule, $\overline{B}_{ij} = 1$ and $V_R(r) - V_A(r)$ is a Morse potential with bond length r_e and well depth D_e, separated into repulsive and attractive components. As the local density around atom i increases, b_{ij} decreases and weakens the attraction to neighbor atom j. For instance, if atom i has two neighbors, j and k, with $r_{ij} < r_{ik}$, then $0 < b_{ik} < b_{ij} < 1$, i.e., the *i-j* and *i-k* attractions are both reduced, but more so for the more distant pair (*i-k*) which is effectively "screened" by the *i-j* interaction. The effect of this is to introduce a preferential valence of 1 for each atom (the triplet *ijk* is a transition state). The precise form and parameters used for f_c, V_R, V_A, b_{ij}, and V_{vdW} are given in Table I (as corrected in a subsequent erratum by the authors) of Ref. [20]. Parameters worth noting here are the molecular bond energies, $D_e^{AB} = 2.0$ eV for reactants, and $D_e^{AA} = D_e^{BB} = 5.0$ eV for products, resulting in an exothermicity of 6.0 eV for the reaction 2 AB $\rightarrow A_2 + B_2$. The van der Waals interaction is three orders of magnitude smaller, a Lennard-Jones 6-12 potential (splined to zero at both short and long distances) with well depth $\varepsilon = 5.0$ meV. Although a systematic study has not been carried out, slightly different parameter sets can lead to qualitatively different behavior, including dissociative phase transformations [19, 21-23].

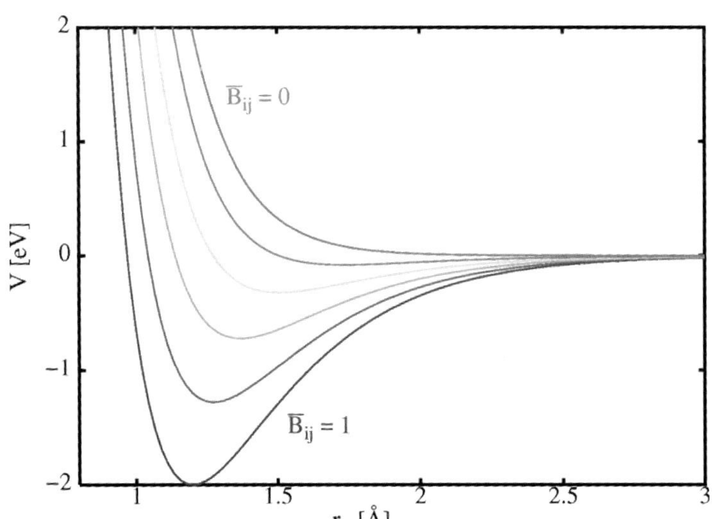

Figure 1. Weakening of the Morse interaction by the bond-order term, which varies from 1 for an isolated molecule towards 0 for a bond that is "screened" by intervening neighbor atoms.

The short-range nature of the REBO potential makes it amenable to implementation in a parallel domain-decomposition molecular-dynamics code, such as SPaSM (Scalable Parallel Short-range Molecular dynamics) [24, 25] at Los Alamos National Laboratory. Since neighbor lists are not normally stored in such a code, during each force (or energy) evaluation an initial pairwise interaction (message-passing) loop is required to evaluate the bond-order function $b$ij for each pair of neighbors, and a second pass to compute the total force (or energy). This is similar to the complexity of the embedded atom method (EAM) potential widely used for metals, where an initial pass to compute the electron density at each atom is required. The computational complexity is significantly greater to calculate torsional forces and/or equilibrate partial charges for long-range Coulombic interactions within such a domain-decomposition code; such interactions are present in more sophisticated REBO-related hydrocarbon potentials [11, 13], as we describe next.

2.2. Transferable Reactive Force Fields: ReaxFF

The ab initio-based reactive potential ReaxFF was developed at Caltech by van Duin, Goddard and collaborators [13], and is currently being extended and improved. The original version was designed with hydrocarbons in mind, but it was soon realized that ReaxFF could accurately describe complex processes (including chemical reactions) in a wide range of materials, from nitramine energetic materials [26] to metallic and covalent solids, their oxides and alloys [27, 28]. Quantum mechanical (QM) methods (such as quantum Monte Carlo, density functional theory, and Hartree-Fock theory) describe atomic interactions from first principles, and usually provide a very accurate description of atomistic processes. Manaa and collaborators have shown that first principles MD can provide detailed information regarding the chemistry of energetic materials [29]; they studied the thermal decomposition of HMX at conditions of temperature and density close to CJ conditions and found very fast chemistry (on the scale of a few picoseconds). Unfortunately, QM methods are computationally too intensive to directly simulate with today's computer power many processes relevant to the energetic materials community. ReaxFF provides a computationally more efficient alternative to ab initio methods, although, as with any classical force field, one has to be very careful with the accuracy of the description, especially for processes that occur far away from the region of configurations where the potential was parameterized or tested.

It is not our intention here to provide a detailed description of each energy term in ReaxFF (we refer the reader to the original publications [13, 27] for that); rather, we intend to describe the key concepts that enable an accurate classical description of intrinsically quantum phenomena.

2.2.1. Key concepts and total energy expression

ReaxFF describes the total energy of an atomistic system with three main terms: i) covalent (bonds, angles, torsions, etc.), ii) electrostatics with environment-dependent charges, and iii) van der Waals interactions. Covalent interactions are based on the concept of partial bond orders that are calculated solely from atomic positions (no pre-determined connectivities). Once the bond order between every pair of atoms is known, bond energies, angles, and torsions are determined. The second key concept in reactive force fields (also used in the latest versions of the REBO family) is the environment-dependent charges to compute electrostatic energy.

2.2.2. Covalent interactions: bond order calculations

ReaxFF defines and calculates σ, π, and ππ bond orders between atoms to describe single, double, and triple bonds. The calculation of these bond orders is done in two steps. First, a pairwise approximation to the bond orders (BO') between every pair of atoms i and j is calculated in terms of the distance between them (r_{ij}):

$$BO_{ij}' = BO_{ij}'^{\sigma} + BO_{ij}'^{\pi} + BO_{ij}'^{\pi\pi} = \exp[p^{\sigma}(r_{ij}/r^{\sigma})^{q\sigma}] + \exp[p^{\pi}(r_{ij}/r^{\pi})^{q\pi}] + \exp[p^{\pi\pi}(r_{ij}/r^{\pi\pi})^{q\pi\pi}] \quad (2)$$

These two-body bond orders are corrected for over-coordination of atoms sharing a bond. If the bond order functions are parameterised to decrease with bond distance slowly enough to correctly describe bond dissociation, atoms sharing a common neighbour have, in many circumstances, a non-zero bond order. These so-called 1-3 bond orders make the potential not only computationally more intensive (since more angle and torsion terms need to be accounted for) but also less intuitive and harder to parameterise. Bonds orders between two over-coordinated atoms are reduced in a non-linear manner (weak bonds are affected the most). Figure 2(b) shows the bond order correction in action for an ethane molecule where the C-C bond has been reduced to 1 Å, both C and H atoms are over-coordinated before the correction (see Fig. 2(a)). The bond order correction [13] removes all the 1-3 bond orders, without affecting the strong C-H bonds; it completely removes the over-coordination in H atoms, and substantially removes the over-coordination in the C atoms.

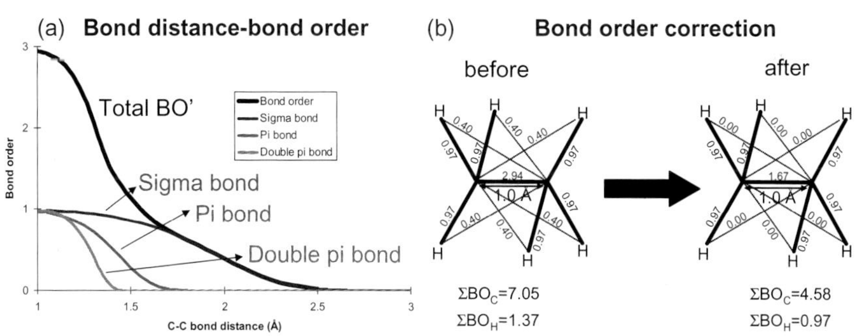

Figure 2. (a) Bond order vs bond distance functions for C-C bonds; we show sigma, pi, and double pi bonds. (b) Bond order correction. Effect of bond order correction on ethane molecule where the C-C bond was reduced to 1 Å.

Once the bond orders have been calculated, they are used to compute bond energies, angles, and torsions. These terms are also used in non-reactive force fields, but their use in reactive simulations requires some modifications. All covalent terms are pre-multiplied by the bond orders involved; this ensures that whenever a bond is broken, all the terms involving it vanish smoothly. Also, the equilibrium angle in covalent-angle terms depends upon the bond orders involved. Finally since there are no pre-determined connectivities in reactive force fields, a term limiting the number of bonds an atom can form is also required (this is obviously not necessary in non-reactive force fields). ReaxFF handles this via a penalty to over-coordination—an energy term that depends on the total bond order of each atom and increases steeply when it becomes larger than the atomic valence (4 for C, 1 for H, etc.).

ReaxFF also includes other terms—covalent terms to describe the effect of lone pairs, conjugation, under-coordination, etc. The reader is referred to the original publications for a description of these terms [13, 27][13].

2.2.3. Electrostatic interactions with self-consistent, variable charges

A force field that intends to describe atoms in different environments with the same set of rules (e.g., Al in fcc metal or Al_2O_3 oxide) cannot use fixed charges to describe electrostatic interactions. There are various methods currently in use to describe charge transfer between atoms; see for example, Refs. [30, 31]. They all compute environment-dependent, self-consistent charges by minimizing an approximation to the total electrostatic energy with respect to atomic charges, while maintaining charge neutrality and keeping atomic positions fixed. The total electrostatic energy consists of the electrostatic interactions plus an atomic term, usually taken as a second-order polynomial that depends on electronegativity and hardness [32]. In ReaxFF electrostatic interactions are computed between all pairs of atoms (no exclusions) and the form of the interaction takes into account the atomic overlaps:

$$E_{\text{elec}}(r_{ij}) = C\, q_i q_j / (r_{ij}^3 + \gamma^3)^{1/3}, \tag{3}$$

where γ represents the overlap radius between atoms i and j. $E_{\text{elec}}(r_{ij})$ converges to the Coulomb law for large r_{ij}; for small separation distances, it describes the finite size of atoms via a shielded potential.

2.2.4. Van der Waals interactions

Van der Waals (vdW) interactions are also computed between all pairs of atoms (including bonded ones). As in electrostatics, a shielded Morse potential is used [13]; this avoids extremely large repulsions at normal bonding distances that would be obtained with the usual vdW interactions.

2.2.5. Force field optimization

The parameters of ReaxFF are based on a large number of QM calculations (over 40 reactions and over 1600 equilibrated molecules were included to describe nitramines, for example), which are designed to characterize the atomic interactions under various environments—likely and unlikely (high energy)—that an atom can encounter. The training set contains bond-breaking and compression curves for all possible bonds, angle-bending, and torsion data for all possible cases [13, 26-28]. Also, zero temperature equations of state (energy- and pressure- volume curves) for various crystals (stable and unstable ones) are used in the parameterization of ReaxFF [27, 28].

To illustrate the capability of ReaxFF for describing complex chemical reactions, Fig. 3 shows ReaxFF calculations (full lines with filled symbols) of the energetics of the 15 intermediate species and 15 transition states involved in the three most important pathways for uni-molecular decomposition of RDX [sequential HONO elimination (circles), homolytic cleavage of an NN bond (NO_2 elimination) and subsequent decomposition (diamonds), and concerted decomposition (triangles)]. This is compared with the results from extensive ab initio QM calculations [38] (dashed lines with open symbols in Fig. 3), showing that ReaxFF describes accurately both the relative stability of the intermediate fragments and the transition states, providing an accurate description of the complex chemistry of RDX.

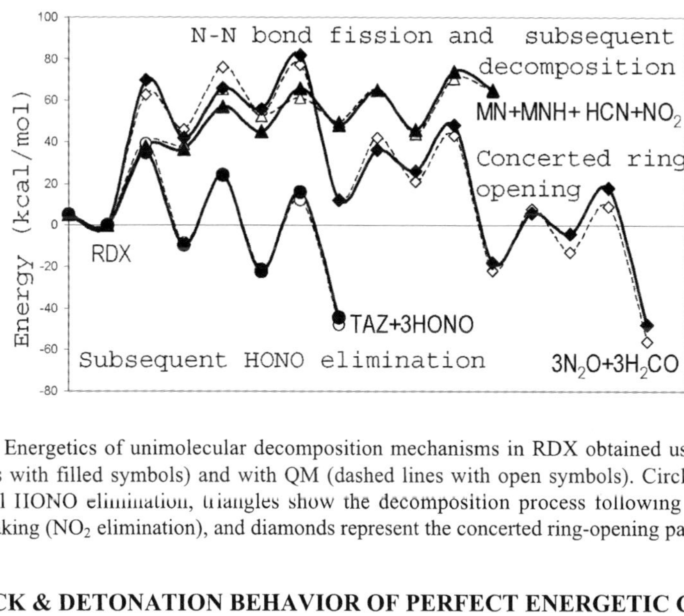

Figure 3. Energetics of unimolecular decomposition mechanisms in RDX obtained using the ReaxFF (full lines with filled symbols) and with QM (dashed lines with open symbols). Circles represent the sequential HONO elimination, triangles show the decomposition process following homolytic N-N bond breaking (NO_2 elimination), and diamonds represent the concerted ring-opening pathway.

3. SHOCK & DETONATION BEHAVIOR OF PERFECT ENERGETIC CRYSTALS

3.1. Non-Equilibrium Molecular Dynamics (NEMD) Shockwave Simulations

Two different methods for introducing a shock wave (which may or may not subsequently turn into a detonation wave) into the target AB material have been used. The first employs a thin, inert A_2 solid flyer plate, which is launched at a velocity $+2u_p$ towards the target, resulting in an initial particle velocity of $+u_p$ into the sample (and one of $-u_p$ into the flyer plate, due to the virtually perfect impedance match of A_2 and AB). If chemical reactions have not begun by the time the rarefaction (release-wave) fan from the rear end of the flyer plate overtakes the initial unreactive shock (a time proportional to the flyer plate thickness), then the shock wave decays and eventually fails to propagate.

On the other hand, if detonation occurs, a leading von Neumann pressure spike is observed, followed by a drop in pressure P through the reaction zone to the C-J point. Behind the steady reaction zone is a self-similar Taylor release wave of expanding gaseous products, as pressure drops back to $P = 0$ (see Fig. 4, and [23] for a good discussion of the AB detonation structure). The steady-state detonation velocity, D, is independent of initiation conditions (flyer plate thickness and velocity), as expected [20]. By "steady," we mean that a profile of density, pressure, or energy, as a function of coordinate x in the shock direction and time t can be scaled by $x - Dt$; "self-similar" profiles scale like x/t.

The second approach enforces a finite-pressure rear boundary condition, i.e., a supported shock wave, by sending the AB target at a velocity $-u_p$ towards a few layers of a rigid A_2 solid. Since the approaching AB molecules "see" the piston A_2 molecules as soon as they are within the potential cutoff distance, this provides a smoother version of the sudden "momentum mirror" which specularly reflects atoms reaching the $z = 0$ boundary [7]. (Using the perfect momentum mirror can result in peculiar behavior; in 2D at low velocities, it collapses the herringbone lattice into an amorphous or even melted structure, rather than the expected elastic response.)

Whichever boundary condition is chosen, the unsupported flyer plate or supported piston, periodic boundary conditions in the transverse dimension(s) are typically used, emulating the center of a large sample which does not "see" release effects from free surfaces on the timescale of shock passage. If desired, free lateral boundaries can be used to study critical diameter effects (width effects in 2D strips [33]), as described later. During the simulation, instantaneous snapshots of longitudinal profiles of properties, such as pressure, temperature, or density, are periodically computed (see Fig. 4); it is frequently useful to also make movies in which atoms are colored according to their bonded neighbors (no neighbors = free radical; neighbor of opposite type = reactant AB; neighbor of same type = product A_2 or B_2; or multiply bonded = transient reaction complex).

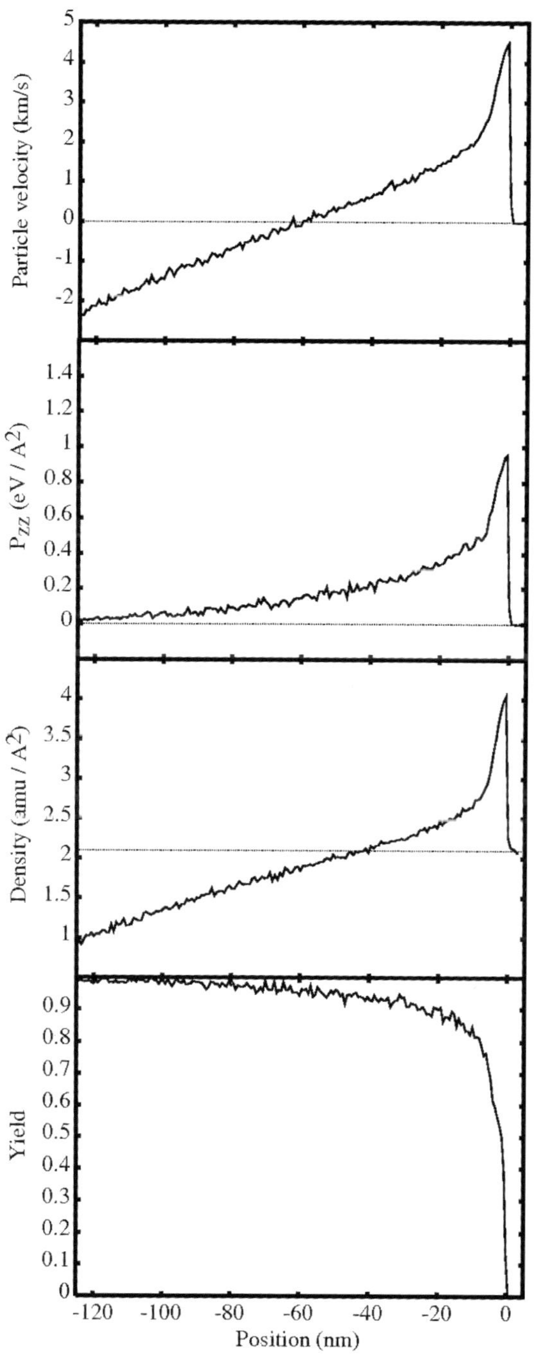

Figure 4. Unsupported detonation profiles in a 2D AB sample (from [34]). The (periodic) cross-section is 124 nm, and the entire sample contains 234,200 atoms. Yield is defined as the fraction of atoms in product (A2 or B2) molecules.

3.2. REBO AB Shock Hugoniot

Using the supported piston to impose a final particle velocity u_p as just described, the set of final shocked states (commonly referred to as the Hugoniot) can be measured from a series of simulations at various u_p. These can be plotted in various planes; three common choices are P-V, P-u_p, and u_s-u_p, the last of which is shown in Fig. 5 for the AB solid in two and three dimensions. Shock loading in two different orientations of the 2D herringbone lattice is shown; the resulting Hugoniots are virtually identical except for some minor differences in the elastic regime. (The detonation properties using a thin flyer plate impactor for these orientations are also indistinguishable, with only a slight orientation dependence in the initiation behavior [35].)

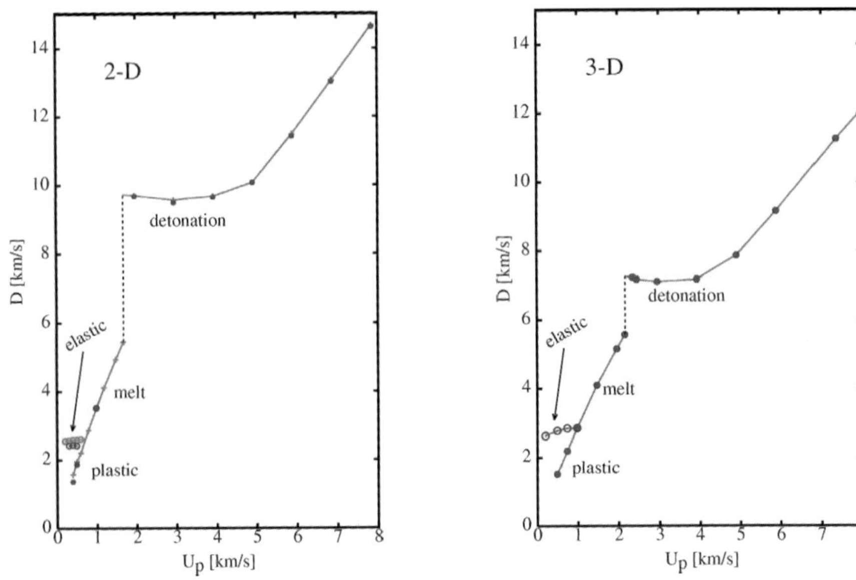

Figure 5. Shock velocity – particle velocity Hugoniots for the REBO AB energetic solid. Left: 2D herringbone lattice in two different crystal orientations (red and blue symbols), rotated 90° from each other. Right: 3D α-N_2 structure.

Qualitatively, 2D and 3D crystals exhibit similar Hugoniots, with several distinct regimes which we now discuss, in order of increasing shock strength. At the slowest piston velocities, a purely elastic shock wave is possible; upon release the solid returns to its original structure with no residual heating. A slight, but definite, orientation dependence in this regime is evident in Fig. 5, due to the anisotropic sound velocities in the 2D herringbone lattice. As the shock strength increases, the uniaxial compression also increases, eventually reaching a point where plastic deformation is nucleated and builds up into a slower, trailing shock front (the Hugoniot Elastic Limit, or HEL). Such split shock waves, with an elastic precursor ahead of a lagging plastic wave, are common and have been observed in MD simulations of fcc metals [8], the bcc-to-hcp transformation of solid iron [9], and for a different REBO parameterization [19] which undergoes a high-pressure dissociative phase transformation [21]. As u_p is further increased, the plastic wave speed u_s increases and eventually overtakes the elastic wave, resulting in a single overdriven plastic wave. Due to the very weak intermolecular van der Waals attraction between AB molecules, the solid-liquid phase boundary is crossed soon thereafter, around $u_p \sim 1.0$ km/s in 2D. A slight change in the u_s-u_p slope, imperceptible in Fig. 5, likely occurs at the transition point. (Experimentally, it is also quite difficult to identify the melt boundary by the u_s-u_p Hugoniot; a discontinuity in the (rarefaction) sound speed is the preferred method for locating the melting transition [36]).

Around $u_p \sim 1.6$ km/s (in 2D), corresponding to a compression of $V/V_0 \sim 0.75$, the detonation threshold is reached, and the Hugoniot jumps from the reactant to the product branch. For a narrow region above this threshold (see Fig. 6, for $u_p = 1.96$ km/s), one can observe at early times an unreactive melt wave propagating at the velocity expected from extrapolation of the equilibrium Hugoniot melt branch in Fig. 5. At some later time, ignition occurs in the shock-compressed and heated liquid, in the region that has been in that state the longest (i.e., near the impact plane). A so-called "superdetonation" wave in the compressed liquid subsequently overtakes the melt wave, slowing down at that point (due to the reduced density of the unshocked AB crystal) into a steady-state detonation wave [37]. This detonation velocity ($D \sim 9.7$ km/s in 2D) is noticeably greater than that measured in unsupported flyer plate simulations ($D \sim 9.25$ km/s). (In 3D, the measured supported and unsupported detonation velocities are much closer: 7.2 km/s and 7.14 km/s, respectively.) In ZND theory, the detonation velocities (and underlying reaction zone structures) would be identical for both cases, so there is some hint of a non-ideal ZND behavior here, particularly in 2D. Likely related to this is the small fraction (10-20%, see Fig. 4) of reactions which occur in the Taylor expansion wave of unsupported detonations; these reactions will occur more quickly in supported detonations since the pressure and density do not fall off to zero, but are limited by the imposed particle velocity. Although these reactions cannot contribute energy directly to the detonation, since they occur beyond the sonic point, perhaps they alter the reaction zone properties enough to account for the faster velocity of supported detonations. Further investigation of this question is clearly needed. In any event, as u_p is further increased, it eventually exceeds the normal particle velocity at the CJ point, resulting in an overdriven detonation whose velocity increases with u_p.

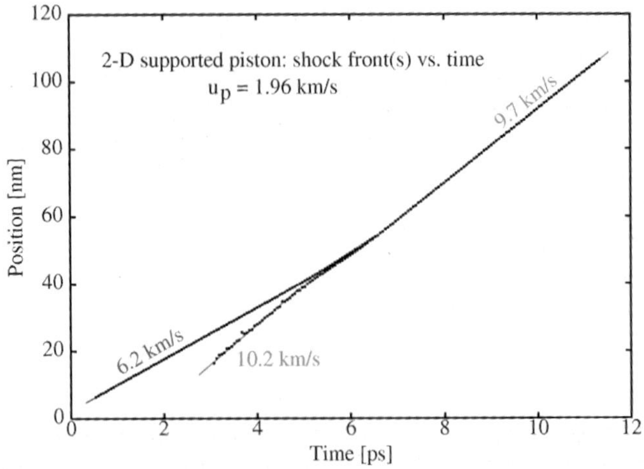

Figure 6. Delayed homogeneous initiation in a 2D AB perfect crystal supported shock wave. Ignition takes place in the shock-compressed and heated AB liquid, resulting in a superdetonation wave (10.2 km/s) which overtakes the nonreactive melt wave (6.2 km/s) to result in a steady-state detonation wave ($D = 9.7$ km/s).

3.3. Failure diameter in cylindrical samples

As is apparent from Fig. 4, the reaction zone in 2D samples is extremely short, on the order of 10 nm. This short reaction zone leads to a critical width for detonation that is small enough to be accessible to direct molecular dynamics simulation. Indeed, the propagation of a reactive wave through a 2D strip using free boundary conditions on the edges was found to reach a steady state provided the strip is large enough (diameter greater than 13 nm) [33]. For smaller strips, rarefaction waves coming from the sides relieve the pressure and quench the reactive wave. We have performed similar simulations in 3D using cylindrical samples of solid crystalline material, with radii ranging between 60 and 150 Å and containing between one and two million atoms. A shock wave, which rapidly transforms to an (unsupported) detonation wave, was initiated by a flyer plate composed of four slices of A_2 molecular crystal traveling at 8.84 km/s. The use of free boundaries allows the expansion of detonation products, as shown in Fig. 7.

Rarefaction waves travel from the edges toward the center of the cylinder, causing the reaction wave to travel more slowly at the edges, thus curving the front. As a consequence, the velocity of the wave depends on the radius of the cylinder, as depicted in Figure 8.

In the case of small cylinders (R < 100 Å), the reactive wave slows down and finally vanishes. For cylinder radii greater than 100 Å, the reactive wave is found to be steady; its velocity is given by the slope of the curves presented in Fig. 8 (see Table 1). For comparison, a 3D simulation using periodic boundary conditions was used to calculate the planar unsupported detonation velocity (CJ velocity).

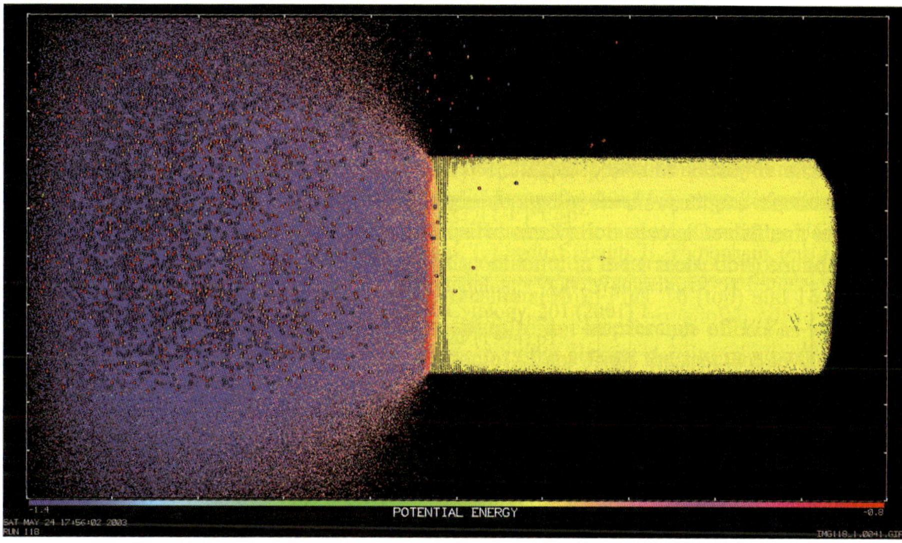

Figure 7. Instantaneous configuration of a detonation wave propagating in a 3D cylinder of radius 150 Å. Atoms are colored according to their potential energy (yellow for unreacted AB material, blue for A_2 and B_2 detonation products).

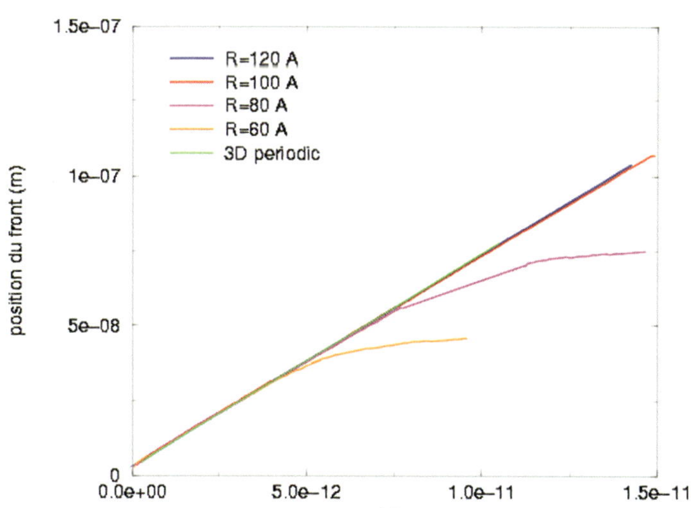

Figure 8. Shock front position versus time for a series of 3D cylinder simulations.

Table 1: Detonation velocity as a function of the cylinder radius.

Radius (Å)	Velocity (m/s)
100	6990
120	7095
3D Periodic (CJ)	7140

The failure diameter is deduced from Fig. 8, and is approximately equal to 100 Å. All the results presented so far: the existence of a failure diameter, the curvature of the front and the variation of the detonation velocity with the cylinder radius, are consistent with continuum theory. Nevertheless, the necessary condition to get a steady detonation wave is the existence of a sonic point that isolates the reaction zone from the rarefaction waves coming from the back. This is investigated in the case of a planar detonation wave (using periodic boundary conditions in 3D).

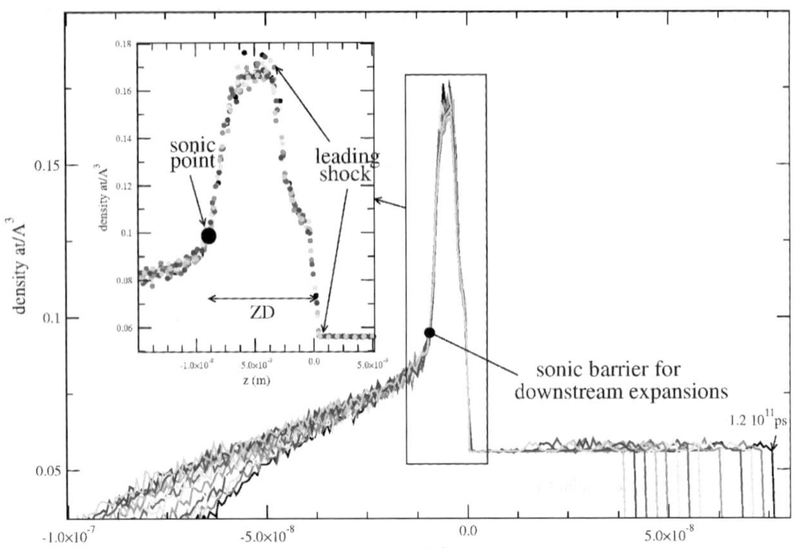

Figure 9. Density profiles measured at different times of the simulation and plotted in a shock front reference frame.

The density profiles computed at different times of the simulation are plotted in Fig. 9, in a shock front reference frame. Each profile exhibits a peak with a similar shape followed by a decrease, which we associate to the von Neumann peak and ZND reaction zone. The constant shape (versus time) of this peak accounts for the steadiness of the reactive wave. Behind the reaction zone, the density profiles decrease almost linearly, reflecting the self-similar Taylor wave expansion of the detonation products. These two regimes (steady reaction zone and spreading self-similar rarefaction waves) are separated by the sonic line (on which the particle velocity equals the local sound velocity), such that any perturbations behind this point are unable to reach the reaction zone.

3.4. Initial chemical reaction events in RDX perfect crystals under shock loading

In contrast with the AB system described above, RDX and most other energetic materials have long reaction times—fractions of a microsecond—and extended reaction zone lengths, on the order of a millimeter. Due to the size of the reaction zone and the complexity of the interatomic potentials necessary to describe real nitramines, steady-state NEMD simulations of detonation are beyond current and near-future capabilities, both in computation time and computer memory requirements. Keeping these limitations in mind, we use NEMD to study the *initial* chemical events in RDX under shock loading. In Section 5 we will describe equilibrium MD simulations to study phenomena at longer time-scales.

We simulate shock propagation in RDX perfect crystals via high-velocity, symmetric impacts using MD with interactions given by ReaxFF [26]. Figure 10 shows snapshots from an MD run with particle velocity u_p = 5 km/s at different times. The top frame shows the initial configuration, consisting of two 3D slabs with periodic boundary conditions in the two directions perpendicular to the free surfaces. After equilibration, each atom is assigned the desired impact velocity in addition to its thermal velocity, and the dynamics of the system is simulated with constant-energy MD. This setup generates two symmetric shockwaves propagating in opposite directions in the two slabs.

We find that for sufficiently strong shocks ($u_p > 3$ km/s) the RDX molecules decompose and react to form a variety of small molecules on a very short time-scale (< 3 picoseconds). For lower velocities, only NO_2 is formed, in agreement with photoelectron spectroscopy (XPS) analysis on samples shocked with sub-detonation strengths [38]. Figure 11(a) shows the time evolution of the population of several small molecules for $u_p = 3$ km/s. We see that NO_2 is the only small molecule formed and many RDX molecules remain intact. It is interesting to note that NO_2 is the only product in weak shocks even though the gas phase energy barrier for HONO elimination is essentially the same (Fig. 3). The analysis of our simulations for $u_p = 4$ km/s shows that, together with NO_2, significant amounts of N_2, NO and OH are produced on a very short timescale [see Fig. 11(b)]. The products predicted by our simulations are consistent with those observed experimentally at much longer times.

The MD trajectories contain the clues that should allow sense to be made from the complex chemistry of energetic materials. Thus each time a molecule is formed we can go backward in time to see which the key reactions were. Such analysis indicates complex (multi-molecular) processes; for $u_p = 4$ km/s, we find (at 4 ps) a large percentage of the final molecules were formed by multimolecular processes: 80% for HONO, 60% for OH, 20% for N_2, 10% for NO and 5% for NO_2.

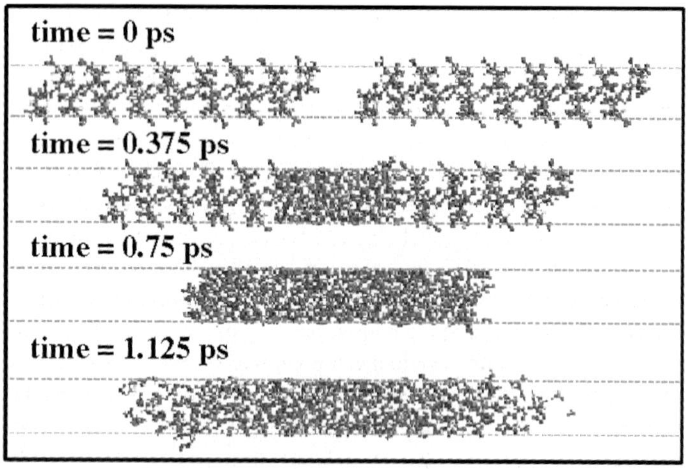

Figure 10. Snapshots of an MD shock simulation of RDX using ReaxFF with $u_p = 5$ km/s.

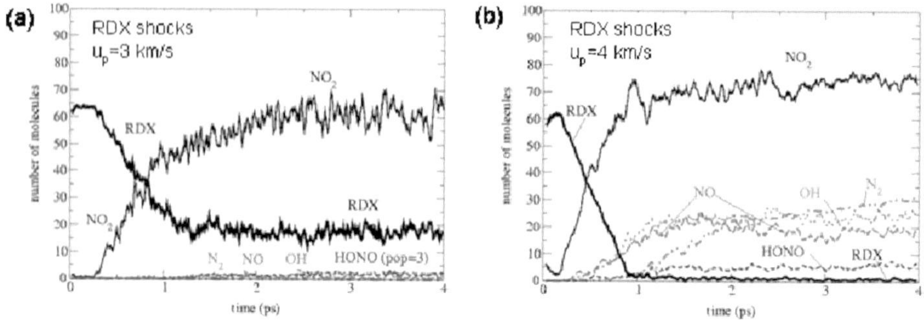

Figure 11. Time evolution of several small molecules for shocks with (a) $u_p = 3$ km/s and (b) $u_p = 4$ km/s.

4. SHOCK AND DETONATION BEHAVIOR OF DEFECTIVE CRYSTALS

4.1. How do hotspots happen? The view from the atomic scale

Impacts upon perfect crystals of high explosives have to be pretty violent to get them to detonate; introduce defects into them, and the job is much, much easier. Porosity in the form of voids has been assumed the principal source of hot spots, or heterogeneous nuclei for chemical reaction, as a shock wave passes by [39, 40]. We have investigated the interaction of shock waves with voids at the atomistic level, to see just how the hot spots are formed [41].

In Section 3, we showed that large-scale molecular-dynamics (MD) simulations could be used to study the effect of impacts upon perfect crystals of high-explosive diatomic molecules whose interactions are modeled by the reactive empirical bond-order (REBO) potential. We showed that perfect crystal shock simulations lead to detonation above a threshold impact velocity, with characteristics that satisfy the ZND theory of detonations. To see if the threshold for initiation of chemical reaction can be lowered, we also introduced a variety of defects into our samples.

The first thing that we found was that defects had to be a minimum size to have any effect whatsoever. A vacancy, or even a divacancy, can be passed over by a shock wave without so much as a hiccup. However, beyond a certain size, a void (or vacancy cluster) can produce not only a warp in the shock front, but serious heating upon its collapse. Such an overheat can subsequently lead to initiation of chemical reaction originating at the void, even for a shock wave that would not have been strong enough to produce initiation in a perfect crystal (see Fig.12). For materials that don't react chemically, such hotspots can still be nucleation sites for other shock-induced phenomena, such as plastic flow or polymorphic phase transformations.

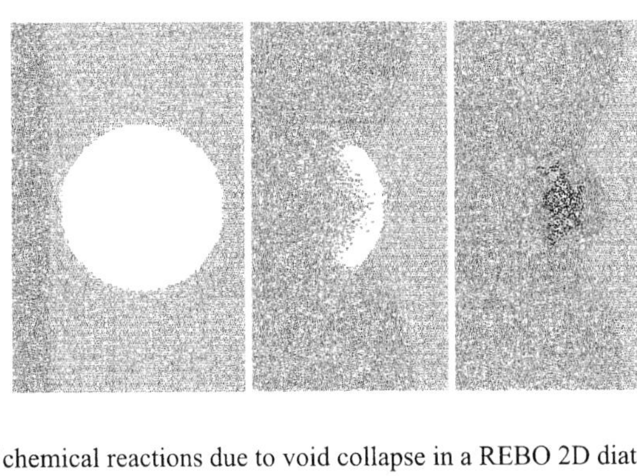

Figure 12. Initiation of chemical reactions due to void collapse in a REBO 2D diatomic (AB) solid.

Besides shock strength and void size, a third factor to be considered is the shape of the void. Elliptical voids [34] illustrate the importance of focusing effects; collapse along the long axis of the void is more effective at initiating chemical reactions than along the short axis, for otherwise equivalent (in shape and size) voids. White and co-workers [23] simplified the issue of shape by considering planar gaps of varying width. They found that if the gap width was too small, no initiation occurred, unless the impact strength was above the perfect-crystal limit. No matter how wide the gap, there was a minimal velocity required to initiate chemical reactions. (Whether these reactions are sufficient to lead to a self-sustained detonation wave is a stronger condition, with higher threshold strength.) The planar gap results are intriguing; there is still a substantial threshold dependence for gaps significantly longer than the range of the potential (7.32 Å). Initial speculation [23] centered on the velocity doubling of atoms ejected into the crack, but this was discounted because reactions did not occur immediately upon impact with the downstream wall, but only much later. Another idea was that wider gaps would permit sufficient rotational tumbling of the ejected AB molecules as they traversed the gap, so that for gaps sufficiently wide, a more favorable collisional orientation (e.g., direct impact of the A atoms of two AB molecules) might occur. This hypothesis was also disproved upon closer inspection.

In order to simplify the ignition problem even further, we studied the thermal overheating that occurs in a completely *unreactive* system, containing planar gaps of varying widths, using the traditional Lennard-Jones pair potential in a 2D monatomic solid [41]. Again, we observed that significant overshoot in temperature occurred only when the initial impact was strong enough to eject material from the upstream side of the gap, corresponding to a shock temperature sufficient to nearly melt the material. Moreover, the gap had to be wide enough before overheating occurred at the downstream side of the gap.

In light of these observations from our MD simulations, we have proposed a simple model [41] describing the role of shockwave interactions with microscopic voids that leads to significant heating, sufficient to thermally initiate chemical reactions in solid explosives, or phase transitions in metals. The key ingredients to this dramatic overshoot in temperature are shown in Fig. 13. The dependencies on both shock strength (piston velocity u_p) and one-dimensional gap width l, which we observed in atomistic simulations of a two-dimensional unreactive Lennard-Jones solid, for the thermal overshoot ΔT was well predicted by our straightforward model:

$$k\Delta T \approx \frac{3}{4d} m u_s u_p [1 - \exp(-l/l_0)], \tag{4}$$

where u_s is the shock velocity, l_0 is a characteristic width inversely proportional to u_p, and d is the number of spatial dimensions. This simple 1D model demonstrates how hot spots can be generated under shockwave conditions.

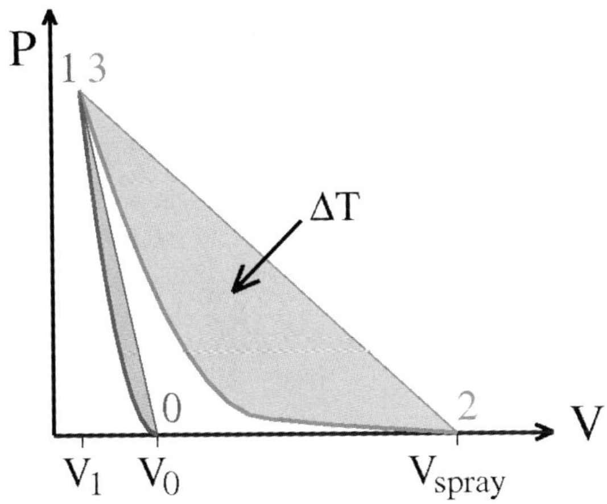

Figure 13. Cartoon of planar gap collapse: (1) initial shock from specific volume $V_0 = 1/\rho_0$ to V_1 with temperature rise T_1; (2) adiabatic expansion (and cooling) of ejecta spray to a mean specific volume $V_{spray} \leq 2V_0$; (3) recompression at the downstream side of the gap to V_1, with a temperature rise ΔT.

4.2. Initial chemical events in RDX crystals with planar gaps

We have shown in Section 3.4 that NEMD simulations using ReaxFF allow one to study the initial chemical events induced by shockwaves in RDX. We have also seen that the interaction of voids in the crystals with shocks can lead to large local heating and that this localization of energy can initiate chemical reactions in the AB system. We now turn to analyze the effect of small voids (planar gaps) on the initial chemical events in RDX.

As in Section 3.4 and in Ref. [26] we simulate the impact between two 2D periodic slabs, but this time one of them contains a gap of length l_{gap}. We impose free boundary conditions in the direction of the shock and periodic boundary conditions in the other two directions. The gap simulates the center of an ellipsoid void with two long axes perpendicular to the shock direction. After thermalization, we assigned the desired relative velocity to each slab on top of thermal velocities and followed the dynamics with constant-energy MD. When the shockwave reaches the gap, molecules are ejected from the upstream surface and expand into the void. The ejected material then collides against the remainder of the slab and re-compresses. The pressure-volume work during this recompression causes local heating. Here we will focus on $u_p = 3$ km/s, since for perfect crystals this velocity separates two different regimes: for shocks with $u_p > 3$ km/s a variety of small molecules are formed, while for lower velocities only small fractions of NO_2 are observed [26].

Figure 14 shows profiles (along the shock direction) of the local particle velocity and local temperature. At time $t = 0$, we see the imposed relative velocities and initial temperature (10 K). At $t = 0.5$ ps the slabs have collided and two symmetric shock waves are propagating: the velocity of the shocked material is zero and its temperature is between 1500 and 2000 K. From $t = 0.75$ ps to 1 ps, the shock on the right slab has already reached the gap, and molecules near the upstream surface are accelerated into the void. Local velocities relative to the down-stream, un-shocked, portion of the right slab reach $2u_p$. The ejecta from the free surface hits the remainder of the slab at time $1 < t < 1.1$ ps, leading to high local temperatures (see temperature profiles for $t = 1.1$, 1.25 and 1.5 ps). We can see from Fig. 14(b) that the local temperature caused by the shock is much higher in the regions surrounding the gap than that in the perfect slab.

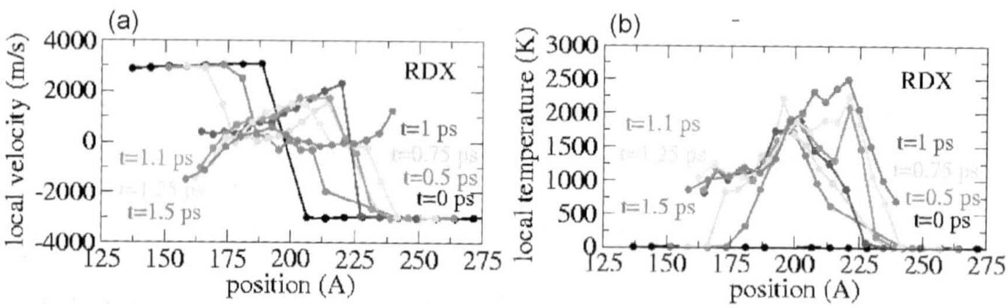

Figure 14. Particle velocity (a) and local temperature (b) profiles at different times. Particle velocity 3 km/s, gap width 20 Å.

In order to study the local heating caused by the interaction of the shock wave and the gap, as well as its dependence on gap width, we plot (see Fig. 15) the time evolution of the local temperature in the final region traversed by the shock (8 molecules in the downstream half of the right slab) for l_{gap} = 5, 10, and 20 Å. We also show that time evolution of the local temperature of the corresponding (mirror) region in the perfect slab; the temperatures of the regions in the perfect slab are independent of the gap width and are displayed as black lines. We can see that the initial heating of the regions by the gap is similar to that of the ones in the perfect slab but they reach a lower temperature (independent of gap width) because some of the shock energy is used to accelerate them into the void. When the ejected (expanding) molecules collapse with the upstream part of the right slab (at a time that depends on void width) they re-compress and heat up to temperatures much higher (~1000 K for l_{gap} = 20 Å) than that attained in the perfect slab. In agreement with the hot spot model described in the previous subsection, the ReaxFF MD simulations show that heating increases with gap width.

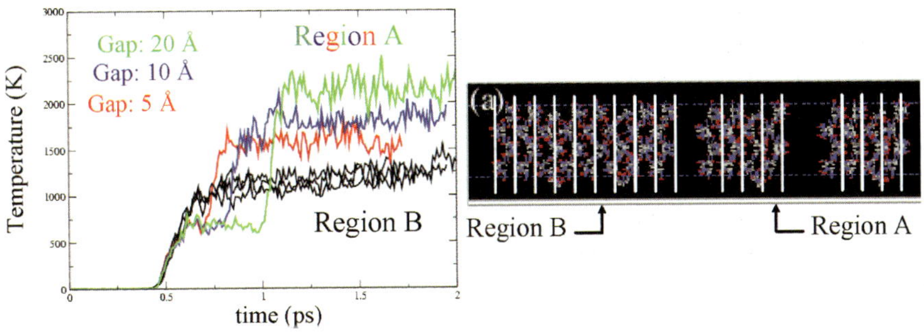

Figure 15. Time evolution of local temperature of the last region (see right panel) the shock goes through in the downstream half of the right slab for l_{gap} = 5, 10, and 20 Å. We also show that time evolution of the local temperature of the corresponding (mirror) region in the perfect slab (black lines).

Once we have seen local heating due to void-shock interaction, we can ask whether the overheating achieved with these void widths is sufficient to facilitate any chemistry. Figure 16 shows the profiles of the population along the shock propagation direction of four small molecules [NO_2, NO, OH and N_2] at times t = 1.25 and 2 ps. Red and blue circles below the profiles show the atomic positions at 2 ps, differentiating the atoms belonging to the left slab and to both portions of the defective slab. It is clear from the figure that the population profiles are very asymmetric, with a larger quantity of small product molecules formed in the defective crystal. The quantities of NO and OH produced on the defective half are comparable to those obtained in perfect crystals at a higher impact velocity: u_p = 4 km/s [26].

Much larger simulations will be necessary to establish if the enhancement of the initial chemistry observed here is enough to generate a self-sustained detonation and consequently increase the sensitivity of the material, or if instead the chemical reactions will die out.

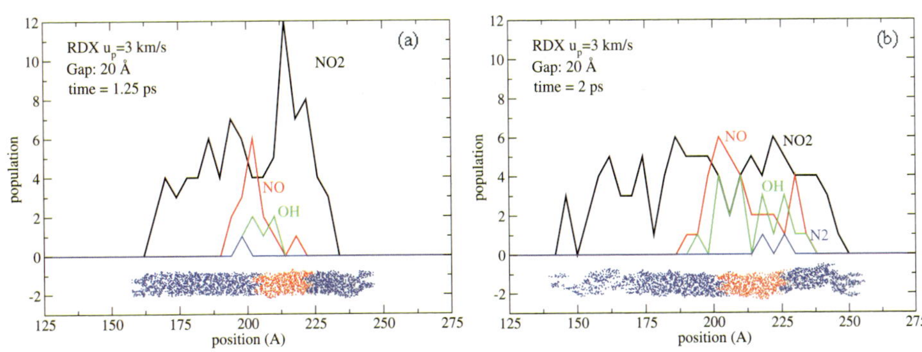

Figure 16. Profile of the population of several species for times t = 1.25 ps (a) and 2 ps (b). RDX shock u_p = 3 km/s, gap width 20 Å.

5. THERMAL DECOMPOSITION OF RDX

Equilibrium MD simulations can provide valuable information about the thermal decomposition of energetic materials and can also enable the exploration of phenomena with time-scales much longer than in shockwaves. As an example, we studied the decomposition and subsequent reactions of RDX under various temperatures (between $T = 1200$ K and $T = 3000$ K) and densities (at low density, 0.21 g/cm^3; near normal density, 1.68 g/cm^3; and under compression, 2.11 g/cm^3), using MD with RDX interactions given by the reactive potential ReaxFF.

5.1. Reactive MD simulations

The initial state of the simulations consisted of RDX perfect crystals using simulation cells containing 8 molecules (one unit cell, 168 atoms) and 3D periodic conditions. After relaxing the atomic positions at each density with low temperature MD, we studied the time evolution of the system at the desired temperature with isothermal isochoric (NVT ensemble) MD simulations (using a Berendsen thermostat; the relaxation time-scale associated with the coupling between the thermostat and the atomistic system was 200 femtoseconds).

While one unit cell (8 molecules) is very small for most applications, we think it is enough to characterize the overall time evolution of the chemistry. In order to test system size dependence of our results, we performed some simulations with larger cells (2x2x2 unit cells, 64 RDX molecules and 1344 atoms). Both reaction time-scales and equilibrium distribution of products are in excellent agreement with the smaller simulations. Using a single unit cell enabled us to explore lower temperatures where several nanoseconds-long simulations were necessary to reach an equilibrium distribution of products.

5.2. Energetics of decomposition

Figure 17 shows the time evolution of the potential energy of RDX for temperatures from $T = 1200$ to 3000 K for the three densities studied. We find that higher densities lead to faster chemistry, and the temporal evolution of the potential energy can be described rather well with a simple exponential function, from which we extract an overall characteristic time of reaction (the solid lines in Fig. 17 shows the exponential functions).

Our simulations reveal that the characteristic times are density dependent and show Arrhenius temperature dependence; Fig. 18 shows the characteristic times (in logarithmic scale) as a function of inverse temperature (the so-called Arrhenius plot). The red line corresponds to MD results under compression (2.11 g/cm^3), the black line to 1.68 g/cm^3, and the green represents the theoretical results at low density (0.21 g/cm^3). The blue line represents the behavior of HMX (rather than RDX) obtained from a wide variety of experimental ignition times [42]. Figure 18 shows that our first-principles-based calculations are in reasonable *quantitative* agreement with experiments. The Arrhenius behavior shown by both MD and experiment can be attributed to a single rate-limiting step in the sequence of reactions that lead to decomposition. From the MD simulations we obtain an activation energy for this limiting step that increases as density decreases: for 2.11 g/cm^3 and 1.68 g/cm^3 we obtain ~23 kcal/mol, compared to 26.6 kcal/mol for 0.21 g/cm^3. The theoretical numbers are somewhat smaller than the experimental activation energy for HMX (rather than RDX), namely, 35.6 kcal/mol.

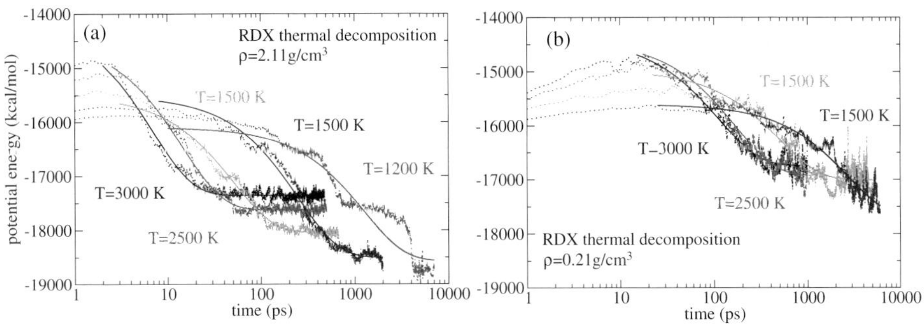

Figure 17. Time evolution of potential energy (in kcal/mol per unit cell) for temperatures from T = 1200 to 3000 K under compression (a) and expansion (b). Solid lines represent exponential fits to MD data.

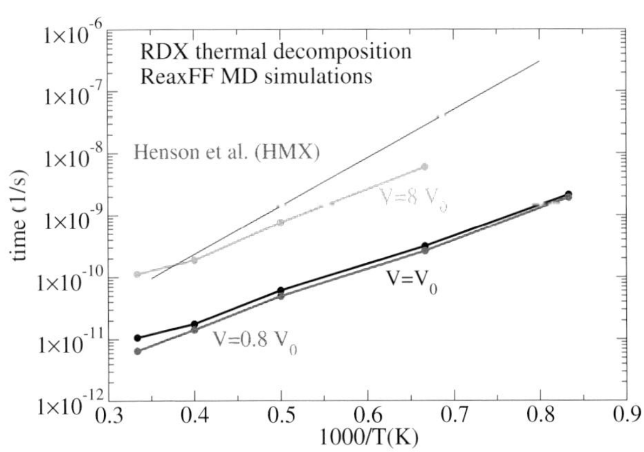

Figure 18. Characteristic time vs. inverse temperature from ReaxFF MD simulations for three densities: 2.11 g/cm^3, 1.68 g/cm^3, and 0.21 g/cm^3. We also show the Arrhenius behavior obtained from experimental ignition times [42] in HMX.

5.3. Time evolution of products

In Fig. 19 we show the time evolution of the population of key products N_2, H_2O, CO, and CO_2, and the intermediate NO_2; the time is scaled with the corresponding characteristic time obtained from the energy evolution. We find that NO_2 elimination is the first chemical reaction in all cases. Time-scales of H_2O and N_2 formation are similar to one another and to the overall characteristic time; they form during the early stages of the process. The formation and final concentration of CO and CO_2 show marked density dependence. We see from Fig. 19 that at low density, virtually all C atoms form CO molecules, while we find no CO_2. Figure 20(a) shows a snapshot of the equilibrium state. All atoms form small molecules: apart from the ones shown in Fig. 5.3, we find O_2 with asymptotic populations of ~4. As the density increases, C atoms start clustering into a condensed phase aggregate. Consequently, the gas phase is poor in C, leading to an increase in the population of CO_2 molecules at the expense of CO. For $\rho = 2.11$ g/cm^3, we find almost no CO molecules: all the C atoms either form CO_2 molecules or belong to the condensed phase aggregate. Fig. 20(b) shows a snapshot of an equilibrium configuration for $\rho = 2.11$ g/cm^3, which shows a small carbon cluster.

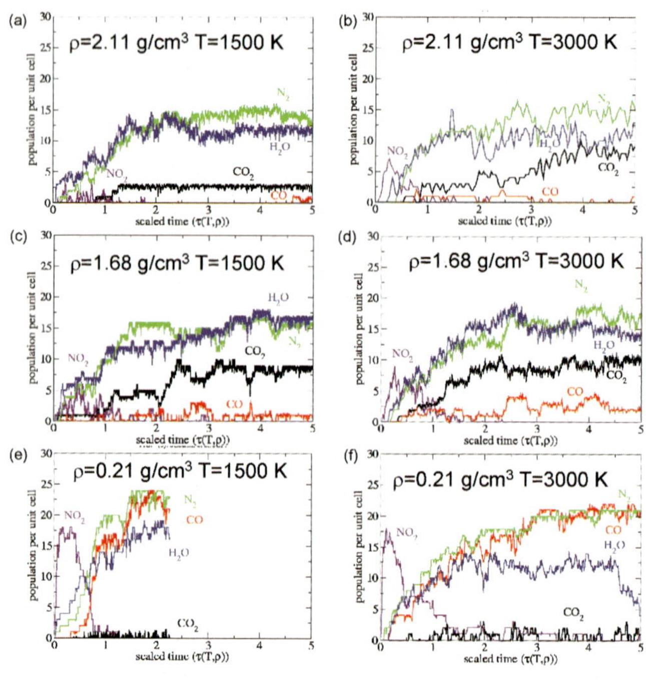

Figure 19. Time evolution of products N_2 (green), H_2O (blue), CO (red), and CO_2 (black), and intermediate NO_2 (purple), for temperatures 1500 K (left) and 3000 K (right) and three densities: 2.11 g/cm^3 (top), 1.68 g/cm^3 (middle), and 0.21 g/cm^3 (bottom).

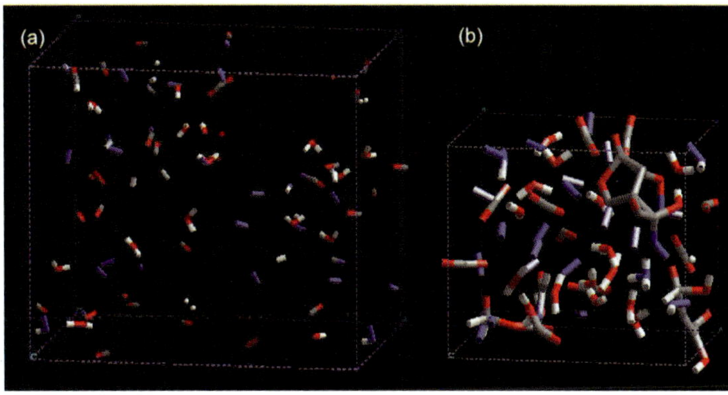

Figure 20. Snapshots from reactive MD simulations of equilibrium configurations at temperature 3000 K and density (a) ρ=0.21 g/cm^3 and (b) ρ=2.11 g/cm^3. A small carbon cluster can be seen for ρ = 2.11 g/cm^3 in the top right corner of the figure.

The molecular state of the material at CJ conditions, a critical question to the energetic materials community, is very hard to determine experimentally. Our results show that under conditions near the C-J point (ρ = 2.11 g/cm^3 and T = 3000 K), a significant amount of the C atoms cluster into a condensed phase. (The size and time-scales of our simulations only allow for the initial stages of carbon clustering to be observed.) Furthermore, at these conditions, we can clearly identify molecules with relatively long mean lifetimes (several bond vibrations): the mean lifetime of N_2 molecules is hundreds of picosenconds, and those of H_2O and CO_2 are of the order of tens of picoseconds.

6. SUMMARY AND CONCLUSIONS

We anticipate that MD simulations employing both simple model systems, such as the REBO AB potential, and more realistic potentials, such as ReaxFF, for describing actual energetic materials, will play an increasingly important role in the understanding of the molecular level mechanisms of detonation and decomposition of energetic materials. Such a fundamental understanding is a key step towards the development of physics-based, predictive models of energetic materials. Once validated, these models will play an important role in the design of new-generation materials with improved safety, reliability and performance tailored for specific applications.

Interatomic potentials that can accurately describe real materials (such as ReaxFF) provide valuable information on energetic materials, such as: i) the initial chemical events under shock loading, ii) decomposition and reactions under thermal loading, and iii) characterization of the C-J state. Unfortunately, the simulation size required to capture the entire reaction zone for most compounds of interest (such as HMX, RDX, and TATB) via NEMD is forbiddingly high.

Simple models potentials, such as REBO, allow the study of generic, but complex, detonation phenomena such as the one shown in Fig. 21. In this 2D simulation, two input strips, each 25 nm wide (approximately twice the critical width [33]) are simultaneously ignited at $t = 0$; 11 ps later they enter a main charge and send out curved detonation waves, with unreacted regions due to gradual corner-turning. (Note also the reactions initiated by high-velocity reaction products, particularly in the region equidistant between the two input strips.) The detonation waves expand and eventually collide, resulting at later times in a steady, curved detonation front. Although this is a model compound with an extremely short reaction zone, the resulting detonation physics is generic, and can provide a challenging microscopic benchmark for numerical hydrodynamic models of detonation.

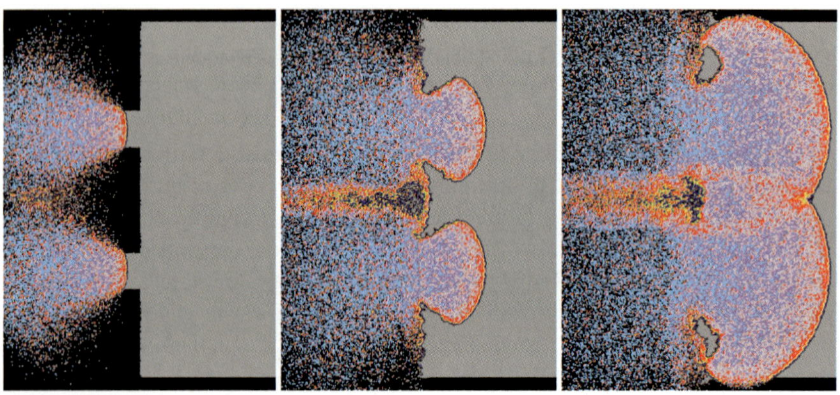

Figure 21. Corner-turning and colliding detonation waves in a 2D AB sample, at 10.2, 15.3, and 22.9 ps (left to right). The input strips are 25 nm wide and run for 100 nm before entering the main charge, with a height of 249 nm and length (not completely shown) of 519 nm, resulting in a total simulation of 1.63 million atoms. The color-coding, from gray to light blue, represents the average reaction progress, or chemical yield, in an Eulerian region.

ACKNOWLEDGEMENTS

We would like to thank our colleagues, David Funk, Bill Goddard, Niels Jensen, Ed Kober, David Moore, Laurent Soulard, C. Matignon, Adri van Duin, and Carter White, for numerous discussions and suggestions relating to this work. Work carried out at Los Alamos National Laboratory has been supported by the U.S. Department of Energy under contract no. W-7405-ENG-36, through the Advanced Simulation and Computing (ASC) Materials & Physics Modelling program, and the High Explosive Reaction Chemistry via Ultrafast Laser Excited Spectroscopies (HERCULES) project.

REFERENCES

[1] W.C. Davis, Sci. Am., 256 (1987) 106.

[2] W. Fickett and W.C. Davis, Detonation (University of California Press, Berkeley, 1979).

[3] A.N. Dremin, Toward Detonation Theory (Springer-Verlag, New York, 1999).

[4] D. Dlott, Annu. Rev. Phys. Chem., 50 (1999) 251.

[5] S.D. McGrane, D.S. Moore, and D.J. Funk, J. Appl. Phys., 93 (2003) 5063.

[6] B.L. Holian, Phys. Rev. A, 37 (1988) 2562.

[7] B.L. Holian and P.S. Lomdahl, Science, 280 (1998) 2085.

[8] T.C. Germann, B.L. Holian, P.S. Lomdahl, and R. Ravelo, Phys. Rev. Lett., 84 (2000) 5351.

[9] K. Kadau, T.C. Germann, P.S. Lomdahl, and B.L. Holian, Science, 296 (2002) 5573.

[10] J.A. Zukas and W.P. Walters, Eds., Explosive Effects and Applications (Springer-Verlag, New York, 1998).

[11] G.D. Smith and R.K. Bharadwaj, J. Phys. Chem. B, 103 (1999) 3570.

[12] S.J. Stuart, A.B. Tutein, and J.A. Harrison, J. Chem. Phys., 112 (2000) 6472.

[13] A.C.T. van Duin, S. Dasgupta, F. Lorant, and W.A. Goddard III, J. Phys. Chem. A, 105 (2001) 9396.

[14] A.M. Karo and J.R. Hardy, Int. J. Quantum Chem. 12(S) (1977) 333; A.M. Karo, J.R. Hardy, and F.E. Walker, Acta Astron., 5 (1978) 1041.

[15] D.H. Tsai and S.F. Trevino, J. Chem. Phys., 81 (1984) 5636.

[16] P. Maffre and M. Peyrard, Phys. Rev. B, 45 (1992) 9551.

[17] M.L. Elert, D.M. Deaven, D.W. Brenner, and C.T. White, Phys. Rev. B, 39 (1989) 1453.

[18] J. Tersoff, Phys. Rev. B, 37 (1988) 6991.

[19] D.H. Robertson, D.W. Brenner, and C.T. White, Phys. Rev. Lett., 67 (1991) 3132.

[20] D.W. Brenner, D.H. Robertson, M.L. Elert, and C.T. White, Phys. Rev. Lett., 70 (1993) 2174; 76 (1996) 2202(E).

[21] C.T. White, D.H. Robertson, M.L. Elert, J.W. Mintmire, and D.W. Brenner, Mat. Res. Soc. Symp. Proc., 296 (1993) 123.

[22] B.M. Rice, W. Mattson, J. Grosh, and S.F. Trevino, Phys. Rev. E, 53 (1996) 623.

[23] C.T. White, D.R. Swanson, and D.H. Robertson, in Chemical Dynamics in Extreme Environments, R.A. Dressler, ed. (World Scientific, Singapore, 2001) pp. 547-592.

[24] P.S. Lomdahl, P. Tamayo, N. Grønbech-Jensen, and D.M. Beazley, in Proceedings of Supercomputing 93, G.S. Ansell, ed. (IEEE Computer Society Press, Los Alamitos, CA, 1993) pp. 520-527.

[25] D.M. Beazley and P.S. Lomdahl, Par. Comput., 20 (1994) 173-195.

[26] A. Strachan, A.C.T. van Duin, D. Chakraborty, S. Dasgupta, and W.A. Goddard III, Phys. Rev. Lett., 91 (2003) 098301.

[27] A.C.T. van Duin, A. Strachan, S. Stewman, Q. Zhang, X. Xu, and W.A. Goddard III, J. Phys. Chem. A, 107 (2003) 3803.

[28] Q. Zhang, T. Çagin, A. van Duin, W.A. Goddard III, Y. Qi, and L.G. Hector, Jr., Phys. Rev. B, 69 (2004) 045423.

[29] M.R. Manaa, L.E. Fried, C.F. Melius, M. Elstner, and T.J. Frauenheim, J. Phys. Chem A, 106 (2002) 9024.

[30] W.J. Mortier, S.K. Ghosh, and S. Shankar, J. Am. Chem. Soc., 108 (1986) 4315.

[31] K. Rappe and W.A. Goddard III, J. Phys. Chem., 95 (1991) 3358.

[32] D. Chakraborty, R.P. Muller, S. Dasgupta, and W.A. Goddard III, J. Phys. Chem. A, 104 (2000) 2261.

[33] C.T. White, D.H. Robertson, D.R. Swanson, and M.L. Elert, in Shock Compression of Condensed Matter, M.D. Furnish, L.C. Chhabildas, and R.S. Hixson, eds. (American Institute of Physics, Melville, NY, 2000), pp. 377-380.

[34] T.C. Germann, J.-B. Maillet, B.L. Holian, N.G. Jensen, and P.S. Lomdahl, in Proceedings of the Twelfth Symposium (International) on Detonation (2002, in press); http://www.sainc.com/onr/detsymp/PaperSubmit/FinalManuscript/pdf/Germann-13.pdf.

[35] D.H. Robertson, D.W. Brenner, and C.T. White, in High Pressure Science and Technology -- 1993, S.C. Schmidt, J.W. Shaner, G.A. Samar, and M. Ross, eds. (AIP Press, New York, 1994), p. 1369.

[36] J.R. Asay and D.B. Hayes, J. Appl. Phys., 46 (1975) 4789.

[37] D.H. Robertson, D.W. Brenner, and C.T. White, Mat. Res. Soc. Symp. Proc., 296 (1993) 183.

[38] F. J. Owens and J. Sharma, J. Appl. Phys., 51 (1979) 1494.

[39] F.P. Bowden and A.D. Yoffe, Initiation and growth of explosion in liquids and solids (Cambridge University Press, 1952).

[40] C.L. Mader, Numerical modeling of detonations (University of California Press, Berkeley, 1979).

[41] B.L. Holian, T.C. Germann, J.-B. Maillet, and C.T. White, Phys. Rev. Lett., 89 (2002) 285501; 90 (2003) 069902(E).

[42] B.F. Henson, B.W. Asay, P.M. Dickson, C.S. Fugard, and D.J. Funk, in Proceedings of the 11th Symposium (International) on Detonation; http://www.sainc.com/onr/detsymp/financcmt.html.

Chemistry at Extreme Conditions
M. Riad Manaa (Editor)

Chapter 10

A Multi-Scale Approach to Molecular Dynamics Simulations of Shock Waves

Evan J. Reed[1], Laurence E. Fried[2], M. Riad Manaa[2], J. D. Joannopoulos[1]

[1]Massachusetts Institute of Technology, Cambridge, MA 02139

[2]Lawrence Livermore National Laboratory, Livermore, CA 94550

1. INTRODUCTION

Study of the propagation of shock waves in condensed matter has led to new discoveries ranging from new metastable states of carbon [1] to the metallic conductivity of hydrogen in Jupiter, [2] but progress in understanding the microscopic details of shocked materials has been extremely difficult. Complications can include the unexpected formation of metastable states of matter that determine the structure, instabilities, and time-evolution of the shock wave. [1,3] The formation of these metastable states can depend on the time-dependent thermodynamic pathway that the material follows behind the shock front. Furthermore, the states of matter observed in the shock wave can depend on the timescale on which observation is made. [4,1] Significant progress in understanding these microscopic details has been made through molecular dynamics simulations using the popular non-equilibrium molecular dynamics (NEMD) approach to atomistic simulation of shock compression. [5]

The NEMD method involves creating a shock at one edge of a large system by assigning some atoms at the edge a fixed velocity. The shock propagates across the computational cell to the opposite side. The computational work required by NEMD scales at least quadratically in the evolution time because larger systems are needed for longer simulations to prevent the shock wave from reflecting from the edge of the computational cell and propagating back into the cell. When quantum mechanical methods with poor scaling of computational effort with system size are employed, this approach to shock simulations rapidly becomes impossible. For example, the computational work required for the simulation of a shock is of quadratic order in the simulation duration $\left(O\left(t^4_{\text{simulation}}\right)\right)$ for a tight-binding method of force evaluation requiring computational work that scales with the number of atoms N like $O\left(N^3\right)$. While NEMD is well suited for the study of short-timescale phenomena around the shock front, chemistry and phase transitions well behind the shock front remain almost completely unexplored due to these scaling difficulties.

This chapter presents a method that circumvents these difficulties by treating some aspects of the shock wave within continuum theory. [6] This method requires molecular dynamics simulation only of a small part of the shock wave at a given instant in time. The effects of the shock wave passing over this small molecular dynamics system are simulated by regulating the applied stress and energy that are obtained from a continuum theory description of the shock wave structure. Because the size of the molecular dynamics system is independent of the simulation time in this approach, the computational work required to simulate the shock is linear in the simulation time (computational work is of order $O(t_{\text{simulation}})$), circumventing the scaling problems of NEMD. This multiscale approach attempts to constrain the molecular dynamics system to the same thermodynamic states that are found in the macroscopic shock wave, ensuring that thermodynamic path-dependent processes are captured correctly. As a benefit of following the correct thermodynamic pathway, it can be shown that this multiscale approach requires no *a priori* knowledge of the system phase diagram, metastable states, chemical reaction or phase transformation rates, or sound speeds. The method can also detect the presence of material instabilities that lead to the formation of double shock waves and simulate these double shock waves in an approximate fashion.

The first sections of this chapter are devoted to a description of the method and practical details for its implementation and utilization. Subsequent sections extend the method to the detection and simulation of double shock waves, which are ubiquitous in condensed matter. Example applications are presented for a Lennard-Jones atomic potential system (which can provide a description of solid Argon), an empirical potential model of crystalline silicon, and a tight-binding atomic potential for the chemically reactive explosive nitromethane (CH_3NO_2).

2. MULTI-SCALE MODEL DERIVATION

We model the propagation of the shock wave using the 1D Euler equations for compressible flow, which neglect thermal transport.

$$\frac{d\rho}{dt} + \rho\frac{\partial u}{\partial x} = 0 \tag{1}$$

$$\frac{du}{dt} + \tilde{v}\frac{\partial p}{\partial x} = 0 \tag{2}$$

$$\frac{d\tilde{e}}{dt} + p\frac{d\tilde{v}}{dt} = 0 \tag{3}$$

Here, ρ is the density, u is the local material velocity, $\tilde{v} = \frac{1}{\rho}$ is the specific volume, p is the pressure, $\tilde{e}$ is the energy per unit mass, and complete time derivatives are $\frac{df}{dt} \equiv \frac{\partial f}{\partial t} + u\frac{\partial f}{\partial x}$. We take the variables in these equations to be instantaneous (microscopic) variables that include thermodynamic fluctuations, i.e. thermodynamic fluctuations are not averaged over. These equations represent the conservation of mass, momentum, and energy respectively everywhere in the wave. No explicit terms account for thermal transport in these equations. The validity of this approximation is discussed in the section on thermal transport below.

While the Euler equations are not rigorously applicable at elastic shock fronts which can be atomistically sharp in NEMD simulations, it will be shown that the correct dynamics will be approximated in these special regions. It is expected that these continuum equations provide a reasonable description of the shocked material in the region behind the shock front, after any very short length-scale phenomena occur at the shock front.

We seek solutions of the Euler equations which are steady in the frame of the shock wave moving at speed v_s by making the substitution $(x,t) \rightarrow x - \mathrm{v}_s t$. This substitution, and integration over x yields a variation of the Hugoniot relations:

$$u - u_0 = (\mathrm{v}_s - u_0)\left(1 - \frac{\rho_0}{\rho}\right) \tag{4}$$

$$p - p_0 = (u_0 - \mathrm{v}_s)^2 \rho_0 \left(1 - \frac{\rho_0}{\rho}\right) \tag{5}$$

$$\tilde{e} - \tilde{e}_0 = p_0\left(\frac{1}{\rho_0} - \frac{1}{\rho}\right) + \frac{(\mathrm{u}_0 - \mathrm{v}_s)^2}{2}\left(1 - \frac{\rho_0}{\rho}\right)^2. \tag{6}$$

Variables with subscripts 0 are the values before the shock wave, and we take $u_0 = 0$, i.e. the material is initially at rest in the laboratory frame. In the terminology of shock physics, Eq. (5) for the pressure is the Rayleigh line and Eq. (6) for the internal energy is the Hugoniot at constant shock velocity. The Hugoniot condition Eq. (6) is more commonly written as

$$\tilde{e} - \tilde{e}_0 = \frac{1}{2}(p + p_0)(\tilde{v}_0 - \tilde{v}) \tag{7}$$

by combining Eq. (5) with Eq. (6) to eliminate v_s. These equations apply to a system that has a time-independent steady-state in the reference frame moving at the shock speed v_s.

If the stress and energy of a molecular dynamics simulation can be constrained to obey Eq. (5) and Eq. (6), then the simulation proceeds through the same thermodynamic states that would occur in a steady shock. The goal of this section is to present a Lagrangian for the molecular dynamics system that performs these constraints. The fashion in which this constraint is done is not unique, and there may be other Lagrangians that perform the same basic constraint task with some different fluctuation or other properties. We choose the Lagrangian for the molecular dynamics simulation to be

$$L = \frac{1}{2}\sum_i m_i \dot{\vec{r}}_i^{\,2} - \phi(\{\vec{r}_i\}) + \frac{1}{2}MQ\dot{\tilde{v}}^2 + \frac{1}{2}M\frac{\mathrm{v}_s^2}{\tilde{v}_0^2}(\tilde{v}_0 - \tilde{v})^2 + Mp_0(\tilde{v}_0 - \tilde{v}) \tag{8}$$

where ϕ is the potential energy, $M = \sum_i m_i$ is the total system mass, Q is a mass-like parameter for the simulation cell size. The pre-shock material is taken to be at rest, i.e. $u_0 = 0$. All variables in Eq. (8) are instantaneous variables, i.e. they are not thermally averaged. Note that Q has units of mass^2/length^4. To enable simulations in systems with periodic boundary conditions, consider the use of the scaled coordinate transformations as in Ref. [11] and [7],

$\vec{r} \equiv A\vec{s}$
$\dot{\vec{r}} \equiv A\dot{\vec{s}}$
where A is a matrix containing the computational cell lattice vectors in columns and $\vec{s}$ is a column vector containing scaled coordinates relative to the computational cell. The values of the scaled coordinates range between 0 and 1. Choosing velocities to be purely a function of $\dot{\vec{s}}$ without a term involving $\vec{s}$ provides a computational cell where the velocity of a particle is independent of its position. The real-space velocities within this coordinate system can be recovered through the relation,
$\dot{\vec{r}}_{\text{real-space}} = A\dot{\vec{s}} + \dot{A}\vec{s}$.

For simplicity, assume that the computational cell is orthorhombic with box lengths a_x, a_y, and a_z. We take the shock to be propagating in the x direction. Planar shocks are described by the 1D Euler equations so only a_x is allowed to vary, providing a uniaxial strain condition. Computational cell dimensions transverse to the shock direction are fixed, as in NEMD simulations.

The Hamiltonian form of Eq. (8) is,

$$H = \sum_{i,\alpha} \frac{P_{i,\alpha}^2}{2m_i a_\alpha^2} + \phi(\{A\vec{s}_i\}) + \frac{P_{a_x}^2}{2Q} \frac{M}{(a_y a_z)^2} - \frac{1}{2} M \mathrm{v}_s^2 \left(1 - \frac{a_x}{a_{x,0}}\right)^2 - p_0 a_y a_z (a_{x,0} - a_x) \tag{9}$$

where the momentum of particle i in the direction $\alpha = \{x, y, z\}$ is $P_{i,\alpha} = m_i a_\alpha^2 \dot{s}_{i,\alpha}$, momentum of the computational cell lattice vector a_x is $P_{a_x} = \frac{Q}{M} \dot{a}_x (a_y a_z)^2$, and the value of a_x in the pre-shock state is $a_{x,0}$. In addition to the usual kinetic and potential terms associated with the atomic degrees of freedom, this Hamiltonian contains a degree of freedom associated with changes in the volume, a_x. The third term in Eq. (9) is a kinetic term for changes in volume and the last two terms represent an external potential for the volume.
The equation of motion for $P_{i,\alpha}$ is,

$$\dot{P}_{i,\alpha} = -a_\alpha \frac{\partial \phi}{\partial r_{i,\alpha}} \tag{10}$$

or,

$$\ddot{s}_{i,\alpha} = -\frac{1}{m_i a_\alpha} \frac{\partial \phi}{\partial r_{i,\alpha}} - 2 \frac{\dot{a}_\alpha}{a_\alpha} \dot{s}_{i,\alpha}. \tag{11}$$

The equation of motion for P_{a_x} is,

$$\dot{P}_{a_x} = \sum_i \frac{P_{i,x}^2}{m_i a_x^3} - \frac{\partial \phi}{\partial a_x}(\{A\vec{s}_i\}) - \frac{M \mathrm{v}_s^2}{a_{x,0}} \left(1 - \frac{a_x}{a_{x,0}}\right) - p_0 a_y a_z. \tag{12}$$

Multiplication of Eq. (12) by $\frac{1}{2}(a_{x,0} - a_x)$ and substitution into the Hamiltonian yields,

$$H=\sum_{i,\alpha}\frac{P_{i,\alpha}^2}{2m_i a_\alpha^2}+\phi(\{A\vec{s}_i\})+\frac{P_{a_x}^2}{2Q}\frac{M}{\left(a_y a_z\right)^2}-\frac{1}{2}M\mathrm{v}_s^2\left(1-\frac{a_x}{a_{x,0}}\right)^2-\frac{1}{2}p_0 a_y a_z\left(a_{x,0}-a_x\right)$$
$$+\left[\frac{\partial\phi}{\partial a_x}(\{A\vec{s}_i\})-\sum_i\frac{P_{i,x}^2}{m_i a_x^3}+\dot{P}_{a_x}\right]\frac{\left(a_{x,0}-a_x\right)}{2}$$

or,

$$H=e+\frac{P_{a_x}^2}{2Q}\frac{M}{\left(a_y a_z\right)^2}-\frac{1}{2}p_0 a_y a_z\left(a_{x,0}-a_x\right)-\frac{1}{2}p a_y a_z\left(a_{x,0}-a_x\right)+\frac{1}{2}\dot{P}_{a_x}\left(a_{x,0}-a_x\right) \qquad (13)$$

where the instantaneous energy e is defined as,

$$e\equiv\sum_{i,\alpha}\frac{P_{i,\alpha}^2}{2m_i a_\alpha^2}+\phi(\{A\vec{s}_i\})$$

and the instantaneous uniaxial pressure ($-\sigma_{xx}$ stress tensor component) is,

$$p\equiv-\frac{1}{a_y a_z}\left[\frac{\partial\phi}{\partial a_x}(\{A\vec{s}_i\})-\sum_i\frac{P_{i,x}^2}{m_i a_x^3}\right]. \qquad (14)$$

At $t=0$ when the system is in the pre-shock state, $a_x=a_{x,0}$ and $P_{a_x}^2=0$ leading to, $H=e(t=0)\equiv e_0$. The Hamiltonian Eq. (13) can be rewritten,

$$e-e_0=\frac{1}{2}(p+p_0)(v_0-v)-\dot{P}_{a_x}\frac{\left(a_{x,0}-a_x\right)}{2}-\frac{P_{a_x}^2}{2Q}\frac{M}{\left(a_y a_z\right)^2} \qquad (15)$$

where $v=a_x a_y a_z$ and $v_0=a_{x,0}a_y a_z$ in this case. Time averaging of this equation yields the time average of the Hugoniot energy condition, Eq. (7), plus terms of order $1/N$, where N is the number of particles in the system, resulting from the average of the last two terms in Eq. (15), i.e.,

$$\langle e-e_0\rangle=\left\langle\frac{1}{2}(p+p_0)(v_0-v)\right\rangle+O\left(\frac{1}{N}\right).$$

The time averages of the last two terms in Eq. (15) can be shown equal using the virial theorem. [8] The degree of adherence of the simulation energy to the Hugoniot condition increases with an increasing number of particles in the system and increases with the duration of the simulation until equipartition of energy among all degrees of freedom is achieved.

Time averaging of Eq. (9) can also be shown to lead directly to the time-averaged version of Eq. (6) with one additional term of order $O(1/N)$. This time-averaged microscopic Hugoniot relation differs from the Hugoniot relation calculated using time-averaged thermodynamic quantities by,

$$\frac{\mathrm{v}_s^2}{2}\frac{\left(\langle\tilde{v}^2\rangle-\langle\tilde{v}\rangle^2\right)}{\tilde{v}_0^2}$$

which goes to zero in the limit of vanishing volume fluctuations.

Eq. (12) for the motion for the system volume can be expressed in terms of the instantaneous pressure, Eq. (14), as,

$$\dot{P}_{a_x} = \ddot{a}_x \frac{Q}{M}\left(a_y a_z\right)^2 = a_y a_z\left(p - p_0\right) - \frac{M\mathrm{v}_s^2}{a_{x,0}}\left(1 - \frac{a_x}{a_{x,0}}\right). \tag{16}$$

Upon time-averaging, $\left\langle \dot{P}_{a_x} \right\rangle \equiv \lim_{\tau\to\infty}\frac{1}{\tau}\int_0^\tau \dot{P}_{a_x} = \lim_{\tau\to\infty}\frac{1}{\tau}\left[P_{a_x}(\tau) - P_{a_x}(0)\right] = 0$ since $P_{a_x}(t)$ is bounded, and this equation of motion reduces to the Rayleigh line Eq. (5),

$$\left\langle p - p_0 \right\rangle = \left\langle \mathrm{v}_s^2 \rho_0\left(1 - \frac{\rho_0}{\rho}\right)\right\rangle.$$

By choosing a small representative molecular dynamics sample of the shocked material, application of the Euler equations requires that macroscopic stress, thermal, and density gradients in the actual shock wave are negligible on the length scale of the molecular dynamics computational cell size. While the thermal energy is assumed to be evenly spatially distributed throughout the sample by the shock, thermal equilibrium within the internal degrees of freedom computational cell is not required.

Some physical intuition for the function of this constraint scheme can be achieved. The last two terms in the Hamiltonian Eq. (9) are potential energy terms for the motion of the computational cell volume, or a_x in this case. The energy associated with the shock is initially contained entirely in the potential energy of a_x at $t = 0$ because $\dot{a}_x(t = 0) = 0$. When the simulation begins, a_x oscillates in a potential that is determined by the last two terms and the second term in Eq. (9). These oscillations are damped through coupling with the atomic degrees of freedom resulting from the second term in Eq. (9), i.e. energy flows from the volume degrees of freedom into the atomic degrees of freedom. Since there are many more atomic degrees of freedom (typically at least 100 in simulations we have performed) than volume degrees of freedom (there are two: a_x and P_{a_x}), this flow of energy is irreversible. This heat flow continues until the volume oscillations are damped and statistical equipartition of energy among all the degrees of freedom is achieved.

The time-dependent properties of the molecular dynamics simulation are characteristic of a material element flowing through the shock wave, within the approximations made in the derivation of the method. Therefore, the spatial profile of the simulated shock wave can be reconstructed by calculating the position of a material element x at time t,

$$x(t) = -\int_0^t \left(\mathrm{v}_s - u(t')\right)dt' \tag{17}$$

where $u(\rho(t))$ is given by Eq. (4). In this fashion, the spatial dependence of all quantities in Eqs. (4), (5), and (6) can be determined for the steady shock wave.

3. STABILITY OF SIMULATED WAVES

The molecular dynamics constraint technique presented in the previous section is designed to simulate steady solutions of the Euler equations but there is no guarantee that all of the simulated solutions are physical. Some steady solutions are characterized by unbounded volume expansion, and others may not be the particular shock wave solutions desired. This section defines mechanical stability conditions that characterize shock waves and then shows that the molecular dynamics constraint technique naturally takes the system through states that satisfy these stability conditions.

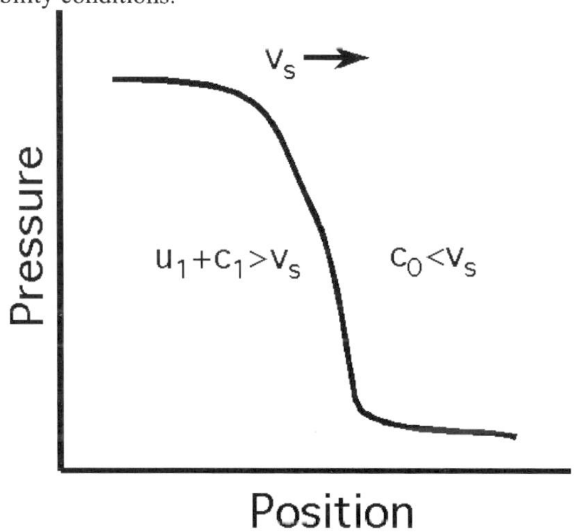

Figure 1. Schematic of a shock wave showing the conditions on sound speed, particle velocity, and shock speed required for mechanical stability in front of and behind the shock front.

There are two criteria for the mechanical stability of a shock wave. [9] The first criterion requires $v_s > c_0$, where c_0 is the speed of sound in the pre-shock material. The second criterion requires $u_1 + c_1 > v_s$, where the subscript 1 denotes the post-shock state.

The condition that $v_s > c_0$ can be motivated physically by considering the propagation of sound waves in front of the shock. If $v_s < c_0$, then pressure waves from behind the shock front can propagate out in front, resulting in an increase of the shock front width and eventual decay of the shock front. The stability condition behind the shock that $u_1 + c_1 > v_s$ can be motivated by considering the case where $u_1 + c_1 < v_s$. In this case, the shock front propagates faster than the speed of sound waves behind it. Compressive energy (in the form of a piston, etc.) behind the shock cannot reach the shock front, resulting in a decay of the shock pressure and eventual dissipation.

The equation of motion for the volume Eq. (12) can be shown to constrain the molecular dynamics system to thermodynamic states that satisfy the conditions for mechanical shock stability. As an example system, consider a shock from state A to state E of Figure 2. Figure 2 shows Rayleigh lines on a hypothetical shock Hugoniot. Eq. (12) indicates that volume increases or decreases depending on the relation between the stress of the molecular dynamics system (approximately given by the Hugoniot line in Figure 2) and that of the Rayleigh line stress (given by the straight lines in Figure 2). When the simulation begins at state A, shock compression will occur if the Rayleigh line is above the Hugoniot in pressure-volume space and the volume is initially slightly on the compressed side of the volume of state A. The slope of the Rayleigh line is $-v_s^2/\tilde{v}_0^2$. The Hugoniot and isentrope have a first-order tangent at point A, [9] providing a Hugoniot slope of $-c_0^2/\tilde{v}_0^2$ at state A. Therefore the stability condition $v_s > c_0$ must be satisfied at point A if compression proceeds up along the Rayleigh line since the slopes obey the condition $-v_s^2/\tilde{v}_0^2 < -c_0^2/\tilde{v}_0^2$ which implies $v_s > c_0$. If the shock speed is chosen such that $v_s < c_0$, then point A is a stable point of Eq. (12) and no compression will occur.

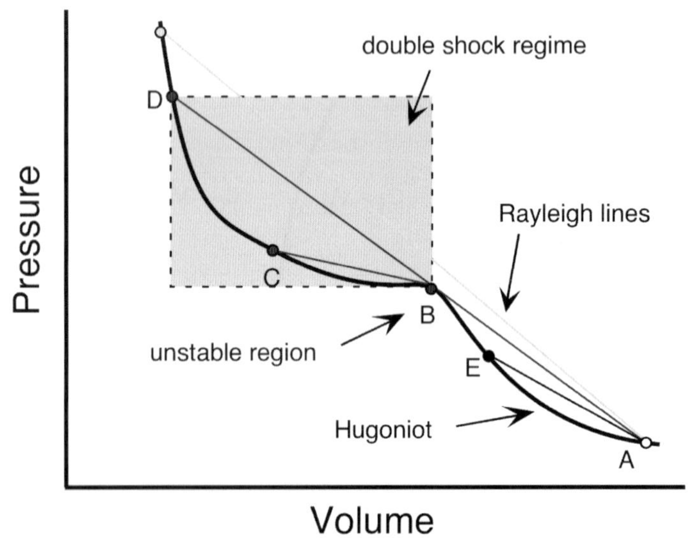

Figure 2. Rayleigh lines on a hypothetical Hugoniot. States on the Hugoniot outside of the gray box can be reached with a single Rayleigh line, or a single shock wave. States on the Hugoniot inside the gray box require two Rayleigh lines to be reached, or two shock waves.

The volume equation of motion Eq. (12) has stable points (where compression stops) at states where the Rayleigh line intersects the Hugoniot and the Rayleigh line slope magnitude is less than the Hugoniot slope magnitude. Point B in Figure 2 is an example of such a state. It can be shown that this condition on the Rayleigh line and Hugoniot slopes requires $u_1 + c_1 > v_s$ which is the stability condition behind the shock front. Ref. [9] contains an outline of this proof. Therefore the constraint Eq. (12) has stable points only where the shock wave mechanical stability conditions are met.

Prior knowledge of the local sound speeds is not required when beginning a simulation. If compression occurs at state A, then $v_s > c_0$. If compression stops at state B, then $u_1 + c_1 > v_s$. Note that as a consequence of the instability at point A of Eq. (12), runaway expansion on the tensile strain side of state A is also a valid solution of the steady state Euler equations. Such an expansion solution may have physical significance if there exists a larger volume where

Eq. (12) has a stable solution. Such solutions are expansion shocks and can be observed in materials where the Hugoniot has the property that $\left.\frac{\partial^2 p}{\partial v^2}\right|_{\text{Hugoniot}} < 0$ in some region. However, this chapter focuses on the compressive shock solutions. To allow only compressive shock solutions, a variety of techniques can be used to bias the simulation at point A to proceed along the compressive branch of the Rayleigh line rather than the expansive branch. These techniques are discussed within the section on Computational Details.

4. NEGLECT OF THERMAL TRANSPORT

Under some circumstances, the flow of heat within a shock can have an effect of the thermodynamic states the shocked material exhibits. Heat flow can be mediated by a variety of mechanisms including phonons, electrons, and photons. [10] Radiative heat flow is important only the most extreme shocks where temperatures are generally above a few thousand Kelvin. Electron mediated heat flow is also important above a few thousand Kelvin in insulating materials, but can be important at much lower temperatures in materials with a small electronic bandgap or in metals. For shocks in insulating materials where the temperatures are less than a few thousand Kelvin, phonons are the primary medium for heat flow.

The Euler equations do not contain any explicit terms related to the flow of heat. Therefore, care should be exercised when applying the method presented in this chapter to situations where significant thermal gradients exist. Photon and electron mediated heat flow propagates orders of magnitude faster than typical shock speeds. These heat flow mechanisms can cause pre-heating of the material in front of the shock among other effects. While all of the thermodynamic simulated states within the shock are not necessarily captured correctly in the molecular dynamics simulation correct in this case, the *final thermodynamic state* of the shock wave still obeys the Equations (4), (5), and (6) even with heat flow because these equations are based purely on the conservation of mass, momentum, and energy across the entire shock wave structure. Heat flow plays a role within the shock front structure, but does not affect these conservative properties across the shock wave. Therefore, the method presented here can still predict the correct final state of the shock if the assumption is made that the final state is not sensitive to the particular thermodynamic path through which it is reached.

While the treatment of phonon mediated heat flow is not strictly accounted for in this method, a rigorous statement can be made in this case. The mechanical shock stability conditions of the previous section can be used to show that phonon-mediated heat flow cannot occur in the forward direction through the shock front. Phonon propagation speeds in a particular direction are equal to or less than the sound speed in that direction. Therefore the stability condition $v_s > c_0$ in front of the shock front prevents heat behind the front from propagating out in front to pre-heat. A related statement can be made regarding the propagation of heat from a plastic wave to an elastic wave in a double shock scenario. (Double shock waves are discussed in subsequent sections.) In this scenario, continuum theory requires that there is a region in the shock where $v_s = c_1 + u_1$. This region marks the boundary between the first and second shock waves, and represents a region where heat is not allowed to flow forward from the second shock wave into the first shock wave. Therefore, the temperature of an elastic shock wave is unaffected by the temperature of the subsequent plastic shock wave within continuum theory.

The importance of the neglect of phonon-mediated heat flow can be estimated using a diffusion model,

$$\frac{\partial T}{\partial t} = D\frac{\partial^2 T}{\partial x^2} \tag{18}$$

where T is the temperature and D is the thermal diffusivity. Consider the case where the temperature has a steady profile in a reference frame moving at the shock speed, i.e. $T(x,t) = T(x - \mathrm{v}_s t)$. [10] Then Eq. (18) becomes,

$$-\mathrm{v}_s \frac{\partial T}{\partial x} = D\frac{\partial^2 T}{\partial x^2}$$

which has solutions of the form

$$T = (T_1 - T_0)\exp\left(-\frac{\mathrm{v}_s}{D}(x - x_0)\right) + T_0$$

where T_0 and T_1 are the temperatures in front of and behind the shock, respectively, and x_0 is the location of the point where the final post-shock temperature T_1 is reached. This system is equivalent to a constant temperature source with temperature T_1 (i.e., the post-shock material), moving at the shock speed. The characteristic length that the post-shock thermal energy diffuses forward in the propagation direction is given by,

$$\Delta x \sim \frac{D}{\mathrm{v}_s}. \tag{19}$$

This characteristic length can be compared to the characteristic length of temperature increase for a simulation using the method presented in this chapter. The latter can be determined using the time dependence of the temperature and Eq. (17). Heat conduction is not expected to be important if the length given by Eq. (19) is substantially less than the characteristic length scale for temperature increase in the simulation.

Typical values of Δx range from around tens of nm for metals like gold to a few Angstroms or less for insulating materials. Molecular solid energetic materials like nitromethane (to be discussed later) fall into the latter category where thermal transport is expected to play little role.

The addition of heat flow mechanisms to the continuum equations utilized here results in a breakdown of the locality of Equations (4), (5), and (6), i.e., the thermodynamic variables at a given point are not purely a function of other variables at that point. For example, the temperature of material in front of the shock will depend on the temperature behind the shock when radiative or electronic heat conduction mechanisms are at play. We speculate that it may be possible to extend the method presented in this chapter to solve for steady shock waves with heat flow by utilizing an iterative procedure. For example, the temperature profile determined during an initial simulation can be time-evolved using a relevant heat diffusion equation to obtain a temperature profile to be enforced for a subsequent simulation, and so on. If such an iterative procedure is carried out, a steady propagating wave with heat flow may be determined if the iterations converge on a time-dependence for the various thermodynamic quantities.

5. COMPUTATIONAL DETAILS

This section presents and discusses the practical issues associated with use of the constraint technique presented in this chapter. Some of these key issues include the need for energy conservation, techniques for ensuring the simulation initially proceeds along the compressive branch of the Rayleigh line, criteria for the choice of the empirical mass-like parameter Q in Eq. (8), and criteria for the choices of computational cell size and simulation duration.

5.1. Adherence to constraints

Unlike many popular molecular dynamics thermostating techniques, the technique presented in this chapter is *conservative* with respect to energy. The atomic degrees of freedom are coupled to the volume degree of freedom, which has a fixed amount of energy at the start of the simulation. In this respect, this technique is related to the Andersen technique for constant pressure simulations. [11] Therefore, it is important to ensure that the integration time-step is chosen to be sufficiently small to conserve the Hamiltonian value, Eq. (9). We utilize a Verlet-based integration algorithm to integrate the equations of motion for the volume (Eq. (16)) and the atoms (Eq. (11)). In the case of atomic equations of motion, the use of scaled coordinates results in a velocity dependent force which must be computed using velocities determined from the atomic trajectories within this Verlet scheme, leading to a sub-optimal integration algorithm. While this algorithm can be made to conserve energy sufficiently by choosing a sufficiently small time step, a different integration algorithm would enable use of a smaller time-step. An example of a higher-order accuracy integration algorithm can be found in Ref. [12]

Figure 3 shows the time-dependence of various temperatures for an example simulation of an elastic-plastic shock in the [110] direction in an approximately cubic perfect 25688 atom face centered cubic Lennard-Jones crystal. This computational cell size is large enough to prevent artificial influence of the periodic boundary conditions on the deformation mechanisms inside the computational cell. [13] In terms of the standard Lennard-Jones potential parameters, the initial volume per atom $\frac{v_0}{\sqrt{2}\sigma^3}=0.68$, initial stress $p_0=0$, and initial temperature $k_B T_0/\varepsilon=0.01$ with shock speed $\frac{v_s}{c_0}=1.87$ where the longitudinal sound speed in the [110] direction $c_0=9.5$. [13] To aid in physical intuition for some of the Lennard-Jones simulations presented in this chapter, we have utilized parameters for Argon: $k_B\varepsilon=119.8$ K, $\sigma=0.3405$ nm and mass $m=40$ atomic mass units. The Lennard-Jones simulations were performed using the spline potential of Ref. [14] to prevent numerical errors associated with a discontinuous potential at the cutoff. The spline parameters for this potential were chosen as in Ref. [14]. The Lennard-Jones simulations presented in this chapter utilized a timestep of 1.15×10^{-4} LJ time units, which resulted in conservation of energy to $10^{-3}\varepsilon$/atom for all of these simulations.

At the top of Figure 3 is the temperature of the atoms, showing the transition from elastic compression to plastic compression around 10 ps. The middle plot gives the temperature of the strain degree of freedom divided by the total number of degrees of freedom in the simulation. The peak strain degree of freedom temperature at the start of the simulation (about 17K) is approximately the amount of irreversible temperature increase the shock provides to the atomic degrees of freedom. The middle plot of Figure 3 shows the temperature of the strain degrees of freedom decrease with time as equipartition is approached. The bottom plot shows the temperature deviation from the initial Hamiltonian energy, Eq. (9), showing good energy conservation.

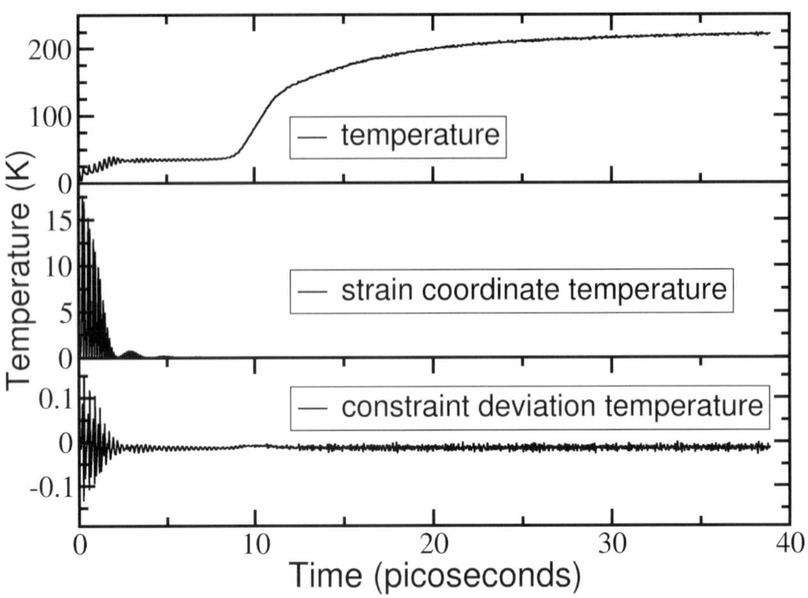

Figure 3: The time-dependence of various temperatures for an example simulation of an elastic-plastic shock in the [110] direction in a perfect 25688 atom face-centered cubic Lennard-Jones crystal. At top is the temperature of the atoms, showing the transition from elastic compression to plastic compression around 10 ps. The middle plot gives the temperature of the strain degrees of freedom divided by the total number of degrees of freedom in the simulation, showing the amount of irreversible energy the shock transfers to the atomic degrees of freedom. The bottom plot shows the temperature deviation from the initial Hamiltonian energy, Eq. (9), showing good energy conservation.

Figure 4 shows the time-dependence of the temperature, uniaxial stress in the shock propagation direction, and volume for a 2.8 km/sec shock in the [111] direction in a perfect 23400 atom Lennard-Jones crystal. The initial volume is 0.03851 nm^3/atom and initial temperature is 10K with zero initial pressure. The initial pressure for this and all other simulations in this chapter was obtained by averaging over the instantaneous pressure of a constant volume simulation for some duration. Figure 4 shows initial compression to the elastically strained state for the first 2 picoseconds. While the system is elastically compressed, slow changes can be seen in the volume and stress. After 2 picoseconds, plastic deformation and further compression occurs. This deformation is characterized by initial fast changes in temperature, stress, and volume followed by a slower relaxation period. While the characterization of compression as either elastic or plastic plays no role in the molecular dynamics Hamiltonian, it is possible to determine the nature of the compression by monitoring the radial distribution function, visual inspection of the computational cell, or other means. The *ab-initio* character of the multi-scale method requires no knowledge of the nature of any plastic deformation mechanism or chemical reactions that occur in the system.

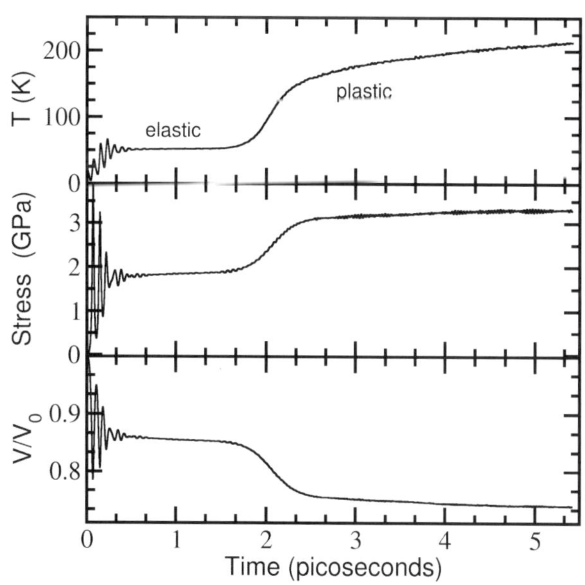

Figure 4: Time-dependence of temperature, uniaxial stress in the shock propagation direction, and volume calculated for an elastic-plastic shock in the [111] direction of a perfect Lennard-Jones crystal. After initial elastic compression, plastic deformation occurs around 2 picoseconds into the simulation. Lennard-Jones potential parameters have been chosen for Argon. See text for details.

Figure 4 shows that initial elastic compression is characterized by oscillations of the volume. These volume oscillations are damped within about 5 oscillations in this case. The damping of volume oscillations occurs by transfer of energy from the strain degrees of freedom to the atomic degrees of freedom, and the atomic temperature can be seen to increase while this process occurs.

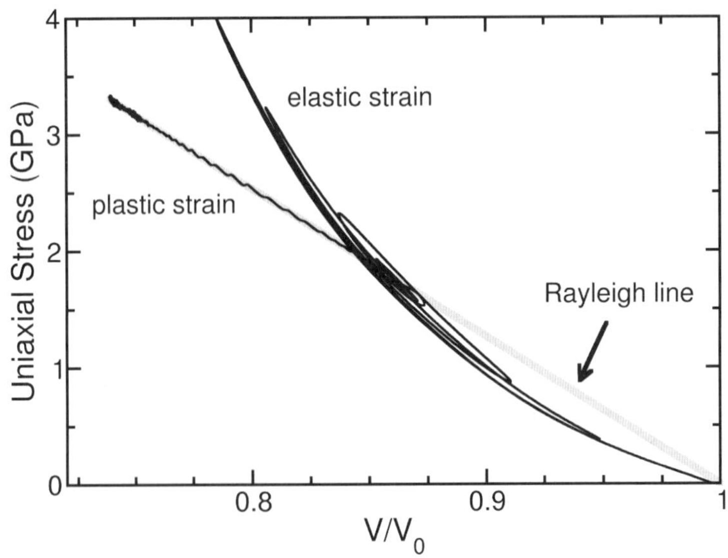

Figure 5: Uniaxial stress versus volume for an overdriven [111] direction shock simulation in a perfect Lennard Jones crystal. The gray line is the Rayleigh line, or constraint line provided by the volume equation of motion Eq. (16). The black line is the actual path of the simulation. The volume begins the simulation at $V/V_0 = 1$ and subsequently undergoes elastic oscillations around $V/V_0 \approx 0.85$. As the amplitude of these oscillations decays with time, the simulation trajectory approaches the Rayleigh line. After the oscillations have decayed away, plastic deformation and further compression occur. During this slower plastic wave, the simulation trajectory closely follows the Rayleigh line, ensuring the correct sequence thermodynamic states are sampled.

Figure 5 shows the uniaxial stress versus volume for an overdriven [111] direction shock simulation in a perfect Lennard Jones crystal. The gray line is the Rayleigh line, or constraint line provided by the volume equation of motion Eq. (16). The black line is the actual path of the simulation. The volume begins the simulation at $V/V_0 = 1$ and subsequently undergoes elastic oscillations around $V/V_0 \approx 0.85$. As the amplitude of these oscillations decays with time, the simulation trajectory approaches the Rayleigh line. After the oscillations have decayed away, plastic deformation and further compression occur. During this slower plastic wave, the simulation trajectory closely follows the Rayleigh line, ensuring the correct sequence thermodynamic states are sampled.

5.2. Choice of parameter Q

The observed initial elastic oscillations are of questionable physical significance. The damping rate of these oscillations is determined by the degree of coupling between the strain degrees of freedom and the atomic degrees of freedom. This coupling is determined by a variety of factors including the nature of the atomic potential in Eq. (9) and the magnitude of the mass-like parameter Q. The degree of coupling constitutes an effective viscosity. Oscillations are longer-lived for perfect crystalline systems at very low temperatures (1K for Argon), where 100 or more oscillations can occur. In this case, internal degrees of freedom are unavailable for transfer of energy from the strain degrees of freedom due to high symmetry conditions. Initial volume oscillations can also be only 1 or 2 oscillations in other systems like molecular solids at room temperature.

The magnitude of Q and the equation of state of the molecular dynamics system determine the frequency of the initial elastic oscillations in Figure 4. Figure 6 shows the time-dependence of the volume for three simulations of 2.2 km/sec shock waves in the [110] direction of a perfect 1400 atom nearly cubic Lennard-Jones crystal at about 1K. Each simulation was performed with a different mass-like parameter Q. If Q is chosen too large (top panel), long-lived oscillations can result. If Q is chosen too small (bottom panel) large amplitude oscillations that do not decay with time can result. An optimal value of Q results in fast decay of volume oscillations. We find that values of Q that provide fast volume oscillation damping result in volume oscillation frequencies that are resonant with internal vibrational degrees of freedom, i.e. the volume oscillation frequencies fall within the vibrational density of states of the atomic system. The number of oscillations required for equilibration in Figure 6 is significantly enhanced by the extremely low initial 1K temperature and perfect crystallinity.

Since the elastic oscillations are typically short-lived (representing a small fraction of the duration of the entire simulation because most of the time is spent in the plastic wave) they can generally be overlooked as long as no unphysical irreversible chemistry or plastic deformation occurs during deviations from the Rayleigh stress conditions in these oscillations. For example, plastic deformation may occur during overcompression periods if the timescale for volume oscillations is sufficiently slow. The choice of parameter Q should therefore be chosen to provide oscillations of sufficiently fast timescale to prevent any unphysical chemistry or plastic deformation from occurring during the oscillation damping process.

Figure 5 shows that the Rayleigh line is closely followed after the initial volume oscillations damp. Note that the timescale for plastic deformation in Figure 3 and Figure 4 is independent of the empirical parameter Q since the strain in this regime changes on a timescale much slower than the resonant volume oscillation frequency determined by Q.

The scaled coordinate scheme applies strain uniformly throughout the computational cell, which is typically several lattice units or more in each dimension. The volume degrees of freedom may therefore be poorly coupled into very short spatial wavelength phonons which may be important in transfer of energy. Shock fronts in NEMD simulations in perfect crystals can possess thickness as short as a few atomic spacings. However, shock wave experiments typically involve polycrystalline materials and non-planar shocks so that shock front thickness is likely to be far greater than the atomic length scale. The latter observation lends some validity to the approximation of uniform strain across the computational cell.

As an alternative to simulating elastic waves, the initial state of the simulation method described here can in principle be obtained directly from an NEMD simulation. NEMD simulations are well suited to simulating shock fronts with high spatial strain gradients for relatively short periods of time. The method presented in this chapter is well suited to reproducing long timescale dynamics behind the shock front where strain gradients are not appreciable. Therefore these two complementary methods could be combined by taking the input computational cell from some point in an NEMD simulation behind the shock front where strain and other gradients have sufficiently relaxed. A less ambitious method to damp initial elastic oscillations may be to utilize a modified Hamiltonian containing additional terms to provide enhanced coupling between volume and atomic degrees of freedom. The additional viscosity can be tuned to prevent elastic oscillations. We have not utilized these approaches to simulation initialization in this chapter.

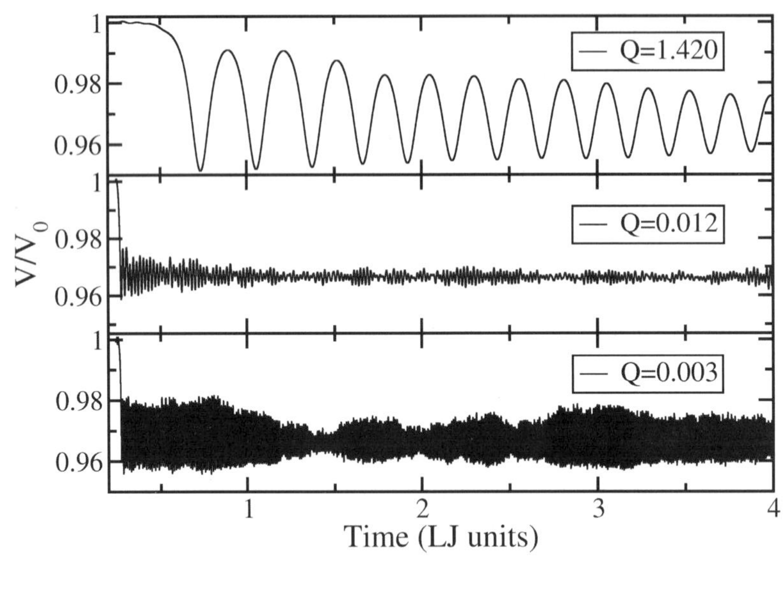

Figure 6. Depicted is the time-dependence of the volume for three simulations of 2.2 km/sec shock waves in the [110] direction of a perfect 1400 atom Lennard-Jones crystal. Each simulation was performed with a different mass-like parameter Q in Eq. (9), given here in reduced Lennard-Jones units. If Q is chosen too large (top panel), long-lived oscillations can result. If Q is chosen too small (bottom panel) large amplitude oscillations that do not decay with time can result. An optimal value of Q results in fast decay of volume oscillations.

5.3. Initialization bias for compressive shocks

Another practical issue associated with the use of this simulation technique is biasing the instability of the starting point. As discussed in the section on stability, as long as the shock speed exceeds the local sound speed, the volume equation of motion Eq. (16) can either force compression or expansion of the volume. While both of these steady solutions can potentially have physical significance, the solutions we focus on in this chapter are the compressive, shock-like solutions. Therefore some technique is required for biasing the initial instability so that only compression occurs. Note that this is simply a selection of the particular type of steady solution to be simulated (compressive shock versus expansion shock) and does not represent nor require an empirical parameter or extra degree of freedom.

There are a wide variety of techniques that can be utilized to bias the volume equation of motion to yield only compressive shock solutions. Here we present two techniques that we have implemented for various molecular dynamics systems. One of these techniques involves modifying the Hamiltonian to apply a constant external pressure at p_0 when $v > v_0$. In this case, the volume equation of motion (Eq. (16)) becomes,

$$\dot{P}_{a_x} = \ddot{a}_x \frac{Q}{M}\left(a_y a_z\right)^2 = a_y a_z\left(p - p_0\right) - \theta\left(a_{x,0} - a_x\right)\frac{M v_s^2}{a_{x,0}}\left(1 - \frac{a_x}{a_{x,0}}\right) \tag{20}$$

where θ is the Heaviside function. In this fashion, if a thermal fluctuation at the start of the simulation increases the volume, the system is forced back to a condition where $v < v_0$ where irreversible compression will occur. Another technique for ensuring compression occurs is to compressively strain the system a small amount at the start of the simulation. Some initial compressive strain provides an initial compressive force in Eq. (16) preventing expansion. Both of these techniques cause some small deviation from the conserved quantity given by Eq. (9), but we find that the magnitude of this deviation is negligible when compared with other errors like numerical integration errors. We find that the stress biasing technique of Eq. (20) works best for systems with small numbers of atoms. Thermal fluctuations can be large in such systems, and an appreciable strain can be required to utilize the strain bias technique. We have utilized the strain bias technique for all the Lennard-Jones simulations presented in this section, with an initial compressive strain of typically 10^{-3} to 10^{-4}.

5.4. Computational cell size

Of importance in using this molecular dynamics technique and other molecular dynamics techniques that utilize periodic boundary conditions are issues with artificially-induced correlations due to the finite size of the computational cell. Artificial phase transitions or other dynamics can be observed when the computational cell dimensions are sufficiently small to allow a particle to interact with its (correlated) periodic image. Artificial effects can be circumvented by making the computational cell sufficiently large that periodic atomic images are separated by a distance greater than the atomic correlation length in the material.

In addition to periodic image interaction effects, there is an additional factor in choosing the computational cell size that must be considered when using this shock molecular dynamics technique. The connection to continuum theory is based on the assumption that the simulated material element (molecular dynamics system) is sufficiently small that stress, density, and energy density in the shock wave do not vary appreciably across the length scale of computational cell. An alternative statement of this condition is that,

$$\dot{a}_x << c \tag{21}$$

where c is the sound speed of the material within the computational cell and $\dot{a}_x$ is the rate of change of the computational cell dimension in the shock propagation direction, using notation from the previous sections. If Eq. (21) holds, sound waves are able to equilibrate gradients in stress, density and energy density within a material element of the computational cell dimensions while the dimensions change. This condition is not unlike that required for adiabatic or reversible evolution of a material element. Eq. (21) provides a limit on the maximum computational cell size as a function of the strain rate.

Planar elastic waves in NEMD simulations in perfect crystals at low temperatures can exhibit considerable strain rates with spatial strain gradients that are exist across the atomic length scale. Within such waves, the condition provided by Eq. (21) breaks down for all but computational cell sizes with atomic scale dimensions. Breakdown of the condition Eq. (21) can result in thermodynamic conditions within the computational cell that may not exist in the shock. However, we utilize computational cells much larger than the atomic scale even for elastic waves because we find that the end state of elastic waves is insensitive to the particular thermodynamic pathway through which it was reached. Satisfaction of Eq. (21) is of greatest concern during plastic deformation or chemical reactions, where the thermodynamic pathway of the computational cell can have an effect of the states of matter formed. Eq. (21) is generally easier to satisfy for a large computational cell in these waves because the strain rates tend to be considerably smaller than those at elastic wave fronts.

Since gradients in stress, density, and energy density tend to decrease in magnitude with distance behind the shock wave, Eq. (21) is expected to become valid at some point behind the shock front and hold thereafter. Materials with relatively short atomic correlation lengths, like molecular solids, can be simulated with smaller computational cells making satisfaction of Eq. (21) possible with larger strain gradients. The *peak* strain rate during plastic deformation in Figure 4 (around 2 ps) has $\dot{a}_x \approx 1$ km/sec which is in marginal satisfaction of Eq. (21) because the stability condition $c + u > v_s$ implies $c > 2.3$ km/sec during this period. The degree of satisfaction of Eq. (21) improves monotonically as the deformation progresses. Better satisfaction of Eq. (21) during the peak strain rate could be achieved by utilizing a smaller computational cell.

5.5. Simulation duration

The molecular dynamics simulation duration is another factor that warrants some consideration. Ideally, the simulation duration can be made much longer than the timescales for all chemical reactions and phase transitions that occur to ensure that the true end state of the shock is achieved. However, it is not possible in general to determine when the absolute final thermodynamic state of the simulation has been achieved without knowing some details about the system, and this method requires no prior knowledge of these details. Furthermore, maximum simulation times for molecular dynamics are typically on the nanosecond timescale for classical interatomic potentials and much shorter timescales for quantum approaches. The timescales for chemical reactions and phase transitions is much longer for many materials of interest. For example, chemistry in a detonating explosive can occur for microseconds or longer behind the shock front. For these reasons, it is necessary to perform all the simulations on the same timescale when calculating points on a shock Hugoniot using the technique presented here.

It might be expected that the Hugoniots calculated with this method and experimental measurements made on the same timescale would be in agreement. This timescale correspondence is a very loose criterion and the quality of agreement between simulations and experiments on timescales before the final thermodynamic state is reached likely depends on details of the particular material system. Some qualitative agreement between simulations and experiments on intermediate timescales is demonstrated for silicon in a later section of this chapter. Simulation timescale issues are discussed further in the following sections on double shock waves.

6. TREATMENT OF MULTIPLE SHOCK WAVES

The sections above describe the simulation of a single stable shock wave. However, it is not always possible for a single shock to take the molecular dynamics system to some pressures or particle velocities. For example, Figure 2 shows how it may not be possible to connect a straight Rayleigh line to all final pressures when there is a region of negative curvature in the Hugoniot, $\left.\frac{\partial^2 p}{\partial v^2}\right|_{\text{Hugoniot}} < 0$. Such regions of negative curvature are common in condensed phase materials and may be a result of phase transformations or may be the shape of a single phase Hugoniot. In Figure 2, it is not possible to connect state A to any state between B and D with a single straight Rayleigh line. Therefore it is not possible for a single shock wave to compress the system to a pressure between the pressures of states B and D. However, state B is a special state where the Rayleigh line from A to B is tangent to the Hugoniot implying a condition of neutral shock stability there, i.e. $u_B + c_B = v_s$. Therefore the mechanical stability condition for the first shock wave breaks down at state B and a second shock wave with a different speed can form. In the case of Figure 2, two Rayleigh lines are sufficient to shock the material to a pressure between that of states B and D. The first Rayleigh line goes from A to B and a second forms from B to C. The mechanical stability criteria are satisfied at points A and C.

The presence of places in the shock Hugoniot where $u + c = v_s$ can be detected using the method presented in this chapter without any prior knowledge of the shock Hugoniot. Figure 7 illustrates this process. If a plot of the final pressure (or particle velocity or volume) as a function of shock speed for multiple single wave simulations at various shock speeds is discontinuous in some pressure region, a state exists on the shock Hugoniot where $u + c = v_s$ and a second shock wave can form. This result can be seen by a geometrical argument based on the schematic Hugoniot in Figure 2. In Figure 7, states B and D correspond to those shown in Figure 2. The first wave shock speed for double wave simulations is chosen to be the smallest shock speed that takes the material to state D. This choice ensures that the simulation will progress beyond state B. We take the thermodynamic state of transition between the first and second waves to be the state where the thermodynamic variables change most slowly. Such a region is illustrated in Figure 8.

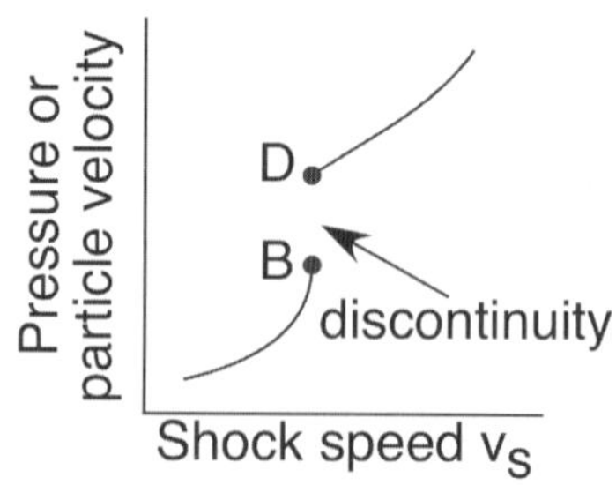

Figure 7. Schematic illustrating how regions on the Hugoniot where $u + c = v_s$ (which lead to the formation of a second shock wave) can be detected. If a plot of the final pressure (or particle velocity or volume) as a function of shock speed for multiple single wave simulations at various shock speeds is discontinuous in some pressure region, a state exists on the shock Hugoniot where $u + c = v_s$ and a second shock wave can form. States B and D in this plot correspond to those shown in Figure 2. The first wave shock speed for double wave simulations is chosen to the smallest shock speed that takes the material to state D.

Figure 8 shows the volume as a function of time for four overdriven single shock wave simulations in the [110] direction of a 25688 atom perfect Lennard-Jones face centered cubic crystal. Elastic compression is characterized by $V/V_0 \sim 0.9$ and plastic compression occurs for smaller volumes. As the shock speed decreases, the amount of time the molecular dynamics system spends in the elastically compressed state increases. This plot illustrates how the final thermodynamic state in the shock is a function of the simulation duration when slow chemical reactions or phase transitions occur. For example, on the 10-20 ps timescale, the 2.8 km/sec shock has an elastically compressed final state; on the 100 ps timescale, this simulation has a plastically compressed final state.

The choice of a particular simulation timescale enables determination of the velocity of the first shock wave and the thermodynamic state where the transition between the first and second waves occurs. For example, simulations performed for 60ps show plastic deformation for a 2.815 km/sec shock speed but no plastic deformation for 2.8 km/sec. Therefore the first (elastic) wave speed for a double wave simulation is 2.815 km/sec and the thermodynamic state at the transition between the two waves is the state where the slowest volume change occurs in the elastically compressed portion of the 2.815 km/sec simulation. (These choices were utilized to produce Figure 9, to be discussed later.) The dependence of the Hugoniot elastic limit on the simulation time is discussed in more detail in the next section.

Each of the single wave simulations performed to construct a plot like in Figure 7 has physical validity regardless of the presence or lack of regions on the Hugoniot where a double shock wave can form. For this reason, it is possible to perform a physically valid single shock wave simulation without any knowledge of the existence of double shock waves. This property is particularly useful when computationally expensive molecular dynamics methods like tight-binding are utilized where calculation of the entire shock Hugoniot can be prohibitively expensive.

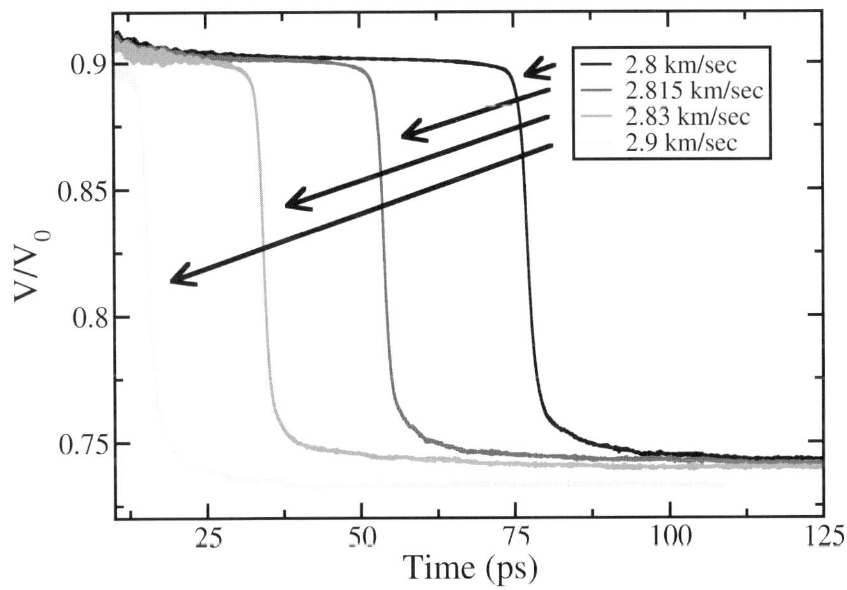

Figure 8. Volume as a function of time for four overdriven single shock wave simulations in the [110] direction of a 25688 atom perfect Lennard-Jones face centered cubic crystal. Elastic compression is characterized by $V/V_0 \sim 0.9$ and plastic compression occurs for smaller volumes. As the shock speed decreases, the amount of time the molecular dynamics system spends in the elastically compressed state increases. This plot illustrates how the final thermodynamic state in the shock is a function of the simulation duration when slow chemical reactions or phase transitions occur. For example, on the 10-20 ps timescale, the 2.8 km/sec has an elastically compressed final state; on the 100 ps timescale, this simulation has a plastically compressed final state.

The Hamiltonian Eq. (9) can be modified to constrain the molecular dynamics simulation to two or more Rayleigh lines. In the case of two lines, we utilize the form,

$$H=\sum_{i,\alpha}\frac{P_{i,\alpha}^2}{2m_i a_\alpha^2}+\phi(\{A\vec{s}_i\})+\frac{P_{a_x}^2}{2Q}\frac{M}{(a_y a_z)^2}-\theta(a_x-a_{x,1})\left[\frac{1}{2}M\mathrm{v}_{s,0}^2\left(1-\frac{a_x}{a_{x,0}}\right)^2+p_0 a_y a_z(a_{x,0}-a_x)\right]$$
$$-\theta(a_{x,1}-a_x)\left[\frac{1}{2}M(u_1-\mathrm{v}_{s,1})^2\left(1-\frac{a_x}{a_{x,1}}\right)^2-p_1 a_y a_z(a_{x,1}-a_x)+\frac{1}{2}M\mathrm{v}_{s,0}^2\left(1-\frac{a_{x,1}}{a_{x,0}}\right)^2+p_0 a_y a_z(a_{x,0}-a_{x,1})\right] \quad (22)$$

where $\theta(x)$ is the Heaviside function, $v_{s,0}$ and $v_{s,1}$ are the first and second wave shock speeds, respectively, and the quantities with subscript 1 are taken at the point of transition between the first and second waves. Specifically,

$$u_1 = v_{s,0}\left(1-\frac{\rho_0}{\rho_1}\right)$$

and

$$p_1 = p_0 + v_{s,0}^2\rho_0\left(1-\frac{\rho_0}{\rho_1}\right).$$

The equation of motion for the volume coordinate is,

$$\dot{P}_{a_x} = \ddot{a}_x\frac{Q}{M}(a_y a_z)^2 = \theta(a_x - a_{x,1})\left[a_y a_z(p-p_0) - \frac{Mv_{s,0}^2}{a_{x,0}}\left(1-\frac{a_x}{a_{x,0}}\right)\right]$$
$$+\theta(a_{x,1}-a_x)\left[a_y a_z(p-p_1) - \frac{M(v_{s,1}-u_1)^2}{a_{x,1}}\left(1-\frac{a_x}{a_{x,1}}\right)\right]$$

The extension to three or more waves can be accomplished in a similar fashion.

6.1. Time-dependence of the p-v space path

The formation and evolution of multiple waves becomes more complicated when chemical reactions or phase transitions occur. Volume decreasing phase transformations cause the pressure at point B in Figure 2 and Figure 7 to decrease with time. This common phenomenon is known as elastic precursor decay in elastic-plastic wave system. [9] The timescale for this pressure decay depends primarily on the timescale for the chemical reaction or phase transition that gives rise to the 2nd wave.

In a double shock wave with chemical reactions, unsteady behavior can lead to a *p-v* space path that is not necessarily well described by Rayleigh lines. However, we assume here that for a given period of time the *p-v* space path can be transiently approximated by a set of Rayleigh lines. This description is valid when the timescale of the pressure change at point B in Figure 2 is less than the time required for a material element to progress from the initial state to the final shocked state. A more quantitative version of this statement is formulated in the remainder of this section.

For the simulations performed using the method described in this chapter, the rate at which the pressure at point B (denote this pressure p_1) decreases can be determined using the so-called shock change equation. [15, 16] For purposes here, we assume the internal energy can now be expressed as $e = e(p(x,t), v(x,t), \lambda(x,t))$ where λ is a generalized reaction parameter for a reaction or phase transition, $0 \le \lambda \le 1$. The rate of pressure change in the moving frame of the shock wave at the metastable point B can be obtained from the so-called shock change equation,

$$\frac{dp_1}{dt} = \frac{\rho_1 c_1^2 \left[\sigma\dot{\lambda} - \frac{\partial p_1}{\partial x} \frac{\eta}{\rho_0 v_{s,0}} \right]}{1 + \rho_0 v_{s,0} (1-\eta)^{-1} \left. \frac{\partial u_1}{\partial p} \right|_{\text{Hugoniot}}} \tag{23}$$

where $\eta \equiv 1 - \frac{(u_1 - v_{s,0})^2}{c_1^2}$, c_1 is the local longitudinal sound speed, $\sigma \equiv \rho_1 \left. \frac{\partial \tilde{v}_1}{\partial \lambda} \right|_{S,p}$ where the derivative is taken at constant pressure along an isentrope, $v_{s,0}$ is the speed of the *first* shock wave of the pair, and all variables with subscript 1 are taken to be at the transition point between the first and second waves (state B in Figure 2). Equation (23) can be obtained by starting with Equations (1), (2), and (3) and calculating the pressure at a point moving at the shock speed, i.e. $\frac{dp_1}{dt} = \frac{\partial p}{\partial t} + v_s \frac{\partial p}{\partial x}$. A complete derivation can be found in Ref. [9].

Eq. (23) can be simplified considerably in the case of interest here. The stability condition at the transition state between the two waves is $u_1 + c_1 = v_s$ which leads to the result that $\eta = 0$ at that point. Furthermore the fact that the Rayleigh line for the first shock and the Hugoniot share a common tangent at the transition state of interest (giving rise to the condition $u_1 + c_1 - v_s$) leads the result,

$$\left. \frac{\partial u_1}{\partial p} \right|_{\text{Hugoniot}} = \left. \frac{\partial u}{\partial p} \right|_{\text{Rayleigh line}} = \frac{1}{\rho_0 v_{s,0}}$$

in this case, taking $u_0 = 0$. These simplifications lead the to the result,

$$\frac{dp_1}{dt} = \frac{\rho_1 (v_{s,0} - u_1)^2 \sigma\dot{\lambda}}{2} \tag{24}$$

Unfortunately, the parameter $\sigma\dot{\lambda} \equiv \rho_1 \dot{\lambda} \left. \frac{\partial \tilde{v}_1}{\partial \lambda} \right|_{S,p}$ cannot be determined directly from information obtained form a simulation using the method presented in this chapter since the stress condition of the material lies along a Rayleigh line, not constant stress. However, we estimate

$$\sigma\dot{\lambda} \approx \rho_1 \left. \frac{\Delta \tilde{v}}{\Delta t} \right|_{\min} \tag{25}$$

where $\left. \frac{\Delta \tilde{v}}{\Delta t} \right|_{\min}$ is the minimum rate of change in specific volume during the simulation. The minimum rate occurs just after the transition to the 2nd wave occurs, i.e. from 0.5 to 2 picoseconds in Figure 4, or state B in Figure 2. We expect that Eq. (25) provides an upper bound on the actual value of $\sigma\dot{\lambda}$ if $\dot{\lambda}$ has the form $\dot{\lambda} \sim (p_1 - p')$ since we calculate this parameter along a Rayleigh line rather than a constant stress. With these simplifications, Eq. (23) becomes,

$$\frac{dp_1}{dt} \approx \frac{1}{2} \left. \frac{\Delta \tilde{v}}{\Delta t} \right|_{\min} \rho_1^2 (v_{s,0} - u_1)^2 \tag{26}$$

This approximate form of the shock change equation enables the estimation of pressure decay of the first wave using information that can be obtained directly from the simulations. The approximation of the *p-v* space path by more than one Rayleigh line in the case of volume decreasing reactions is justified when the Rayleigh lines do not change appreciably during the simulation, i.e.

$$\frac{dp_1}{dt} \approx \frac{1}{2} \frac{\Delta \tilde{v}}{\Delta t}\bigg|_{\min} \rho_1^2 \left(\mathrm{v}_{s,0} - u_1\right)^2 << -\frac{\Delta p}{\Delta t} \tag{27}$$

where Δp and Δt are chosen to be the pressure change of the second shock wave and time duration of a given simulation respectively. Any overestimation of reaction rates through the use of $\frac{\Delta \tilde{v}}{\Delta t}\Big|_{\min}$ makes Eq. (27) more stringent. The rough and approximate criterion provided by Eq. (27) can be used to assess the validity of a two-wave simulation. All of the parameters in Eq. (27) can be determined from a two-wave simulation after it has completed. Alternatively, Eq. (27) provides a relation for the maximum duration of a simulation Δt can be performed without appreciable change in the *p-v* space through which the shock takes the material. This relation may be useful when multiple chemical reactions or phase transitions of disparate time scales exist, where a fast reaction gives rise to a large value of $\frac{dp_1}{dt}$ but slower reactions exist that prevent a final state from being reached before the Rayleigh line validity condition Eq. (27) breaks down. It is not necessary to have prior knowledge of the number, type, or any other details of chemical reactions or phase transitions to utilize the techniques presented in this chapter.

The Rayleigh line validity condition Eq. (27) can be shown to be valid for long wave propagation times. By considering a reaction rate of the form $\dot{\lambda} = \alpha(p_1 - p')$, the shock change equation, Eq. (24) gives,

$$p_1 = \left(p_1^0 - p'\right) \exp\left(\frac{\rho_1 \left(\mathrm{v}_{s,0} - u_1\right)^2 \sigma \alpha}{2} t\right) + p' \tag{28}$$

where p_1^0 is the initial pressure of the first shock wave. Note that $\alpha > 0$ and $\sigma < 0$ here. In a shock wave, the time between the arrival of the first shock and the final pressure of the second shock is attained (Δt in Eq. (27)) has an upper bound that is determined by the speed of the two shock waves and the particle velocity between them. It can be easily shown that Δt in this upper bound case scales linearly with the time that the shock waves have been propagating. The exponential time-dependence of p_1 and the linear time-dependence of Δt imply Eq. (27) is always satisfied after the shock has propagated for some period of time. During times when this condition is not satisfied, the *p-v* space path a material element follows is more complicated than straight Rayleigh lines, but such situations are transient. Therefore it is expected that the approximation of the *p-v* space path with a series of Rayleigh lines is valid for shock waves that have propagated for some period of time in most systems.

In practice, we find that the Rayleigh line validity condition Eq. (27) holds when the lifetime of the elastically strained state is appreciable, as it is for the slower shock speed simulations in Figure 8. It breaks down for simulations where relatively little time is spent in the elastically compressed state before plastic deformation occurs.

7. APPLICATION TO A LENNARD-JONES CRYSTAL

Figure 9 presents the calculated Hugoniot for shock waves propagating in the [110] direction of a Lennard-Jones face-centered cubic crystal of 25688 atoms. All results that follow are given in LJ units of σ, ε and m. The end states were taken around t=60. The integration time step was 1.15×10^{-4}, and volume mass-like parameter $Q=2.737\times10^{-3}$. $v_0=0.9617$, $T_0=0.01$. Longitudinal sound speed in the [110] direction $c_0=9.5$. For the double-wave simulations, the Hugoniot elastic limit (HEL) volume and shock speed were determined to be $v_1/v_0=0.9011$ and $v_S/c_0=1.818$. These are the volume of the transition between first and second waves and shock speed of the first wave, respectively. A choice of Q sufficient to ensure only elastic deformation occurs during the initial volume oscillations was verified by monitoring the radial distribution function during elastic oscillations during one of these simulations.

The black triangles in Figure 9 show results of single wave simulations. These simulations show that a gap exists between about $v/v_0=0.74$ and $v/v_0=0.9$. As in Figure 7, this gap indicates the existence of a shock instability leading to the formation of a second wave. Figure 9 shows good agreement with NEMD volume data in the double shock regime.

Figure 9 shows quantitative temperature agreement with NEMD for single wave simulations and double wave simulations with high plastic wave speeds ($v/v_0>0.76$), where we find qualitative agreement. Further study of both NEMD convergence (timescale issues), which is affected by slow plastic relaxation, and multiscale methods in this regime is desirable. In addition to timescale issues, a possible origin of the temperature difference is the difference in calculated HEL. NEMD simulations show a HEL volume of $v_1/v_0\approx0.91$, [17] which is slightly greater than the HEL for the multiscale simulations ($v_1/v_0=0.9011$). This difference is consistent with the observation of a higher temperature.

8. APPLICATION TO CRYSTALLINE SILICON

Crystalline silicon is another material that exhibits shock wave splitting due to phase transitions. Figure 10 shows shock speed as a function of particle velocity for shock waves propagating in the [110] direction in silicon described by the Stillinger-Weber potential. [18] This potential has been found to provide a qualitative representation of some condensed properties of silicon. Data calculated using the NEMD method are compared with results of the method presented in this chapter. NEMD simulations were done with a computational cell of size 920Åx12Åx11Å (5760 atoms) for a duration of about 10-20ps. Simulations with the multiscale method were done with a computational cell size of 19Åx12Åx11Å (120 atoms.) Both simulations were started at 300K and zero stress. Since the NEMD simulations were limited to the 10ps timescale, simulations with the multiscale method were performed to calculate the Hugoniot on this 10ps timescale for comparison. The final particle velocity in these simulations was taken to be a point of steady state after a few ps.

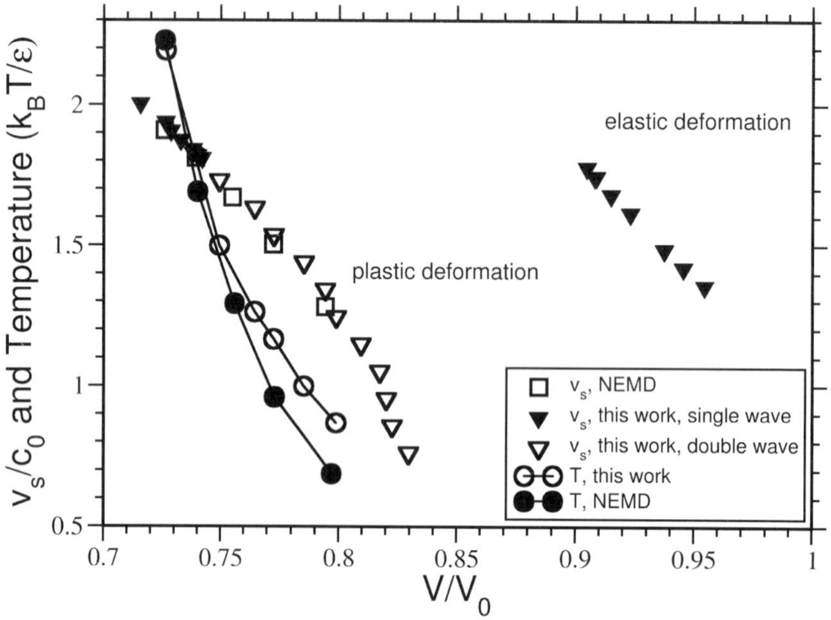

Figure 9: Calculated Hugoniot for shocks in the [110] direction of perfect 25688 atom Lennard-Jones face-centered cubic crystal. The NEMD shock speed and temperature data are from Ref.13. Here, $c_0 = 9.5$ in Lennard Jones units. See text for details.

Figure 10 indicates a single shock wave exists below 1.9 km/sec particle velocity. Above this particle velocity, the elastic shock wave precedes a slower moving shock characterized by plastic deformation. Agreement between the two methods is good for all regions except for the plastic wave speed for particle velocities less than 2.1 km/sec. The wide range of values for the plastic wave speeds in NEMD simulations in this regime is likely due to finite simulation cell size effects which are not present in the simulations shown in Figure 9 for Lennard-Jones.

The Rayleigh line validity condition Eq. (27) is satisfied for the simulations performed in the two-shock regime, giving a typical value for $-\frac{dp_1}{dt}$ of 0.1 GPa/ps, while $\frac{\Delta p}{\Delta t}$ is greater than 0.5 GPa/ps for all simulations in Figure 10.

One of the primary advantages of using the method outlined in this chapter is the ability to simulate for much longer times than is possible with NEMD. As an example, Figure 10 shows the result of a 5 ns simulation performed along a Rayleigh line corresponding to a shock speed of 10.3 km/sec. The uniaxially compressed elastic state required 5 ns to undergo plastic deformation. The difference in particle velocity between the 10 ps and 5 ns simulations at this shock speed is 0.8 km/sec, suggesting that the elastically compressed state is metastable with an anomalously large lifetime. While some caution should be taken when attributing physical significance to this result from the empirical potential of Stillinger and Weber, this result is qualitatively consistent with experimental observations of shocked silicon that indicate an anomalously high pressure elastic wave exists on the nanosecond timescale. [4]

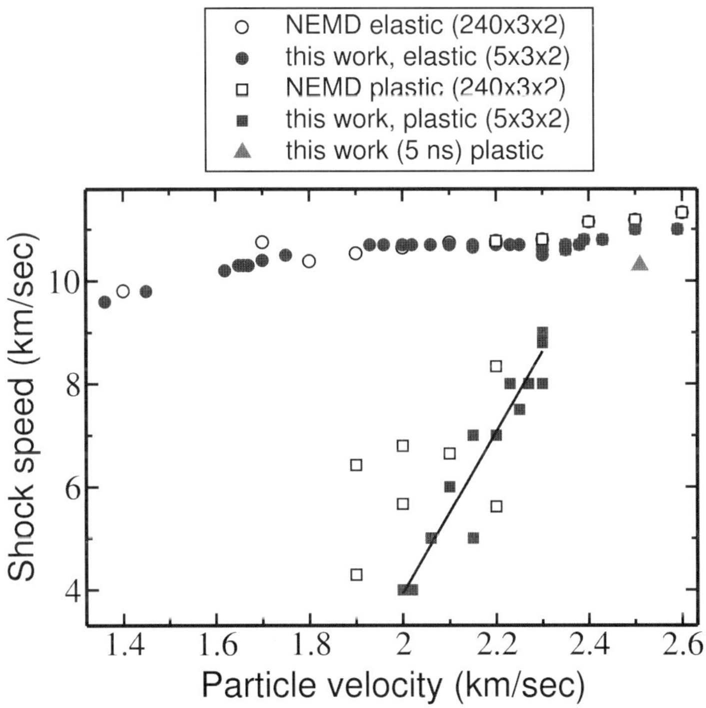

Figure 10. Hugoniot for shocks in the [110] direction of a 5760 atom Stillinger-Weber silicon perfect diamond structure crystal. The black line is an aid to the eye. The end state for all simulations was taken on the 10 picosecond timescale except for the red triangle data point which was taken after 5 nanoseconds.

The 5 ns simulation demonstrated a substantial computational savings over the NEMD method. For an $O(N)$ method of force evaluation, the computational cost of this simulation with the NEMD method would be at least 10^5 times greater than the multiscale method.

9. APPLICATION TO NITROMETHANE

Chemistry in detonating explosives can occur long after the shock front has passed and an accurate description of the chemical reactions in these materials generally requires use of a tight-binding or more accurate molecular dynamics approach. The multiscale method presented in this chapter has the biggest advantages over NEMD and other methods for simulations of long duration with computationally expensive molecular dynamics methods.

Furthermore, most practical energetic materials are molecular solids that have relatively short atomic correlation lengths under detonation conditions. This enables the use of small simulation cells and satisfaction of the strain rate condition Eq. (21) shortly behind the shock front.

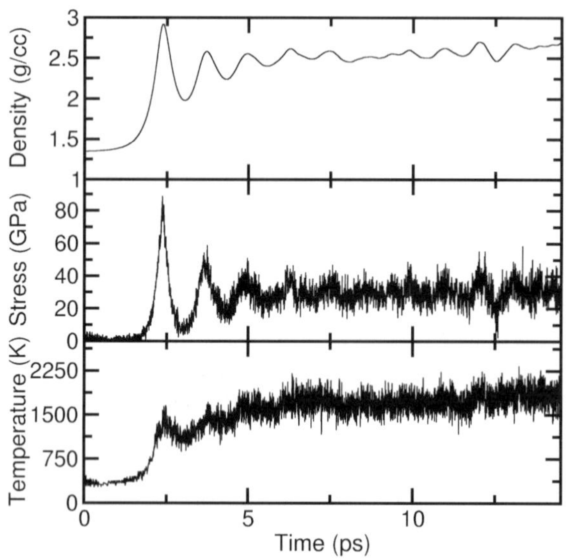

Figure 11. Time-dependence of density, uniaxial stress, and temperature for a 7 km/sec shock in nitromethane.

As an example study case, we have applied the multiscale method to nitromethane, experiencing shock compression along the c axis (longest axis of the primitive cell) with a shock speed of 7km/sec. The initial density is 1.34 g/cc, initial temperature is around 300K, and initial stress is around 0.5 GPa. The atomic energies and forces were computed using the SCC-DFTB method, [19] utilizing a supercell of solid crystalline nitromethane containing eight molecules (56 atoms). The supercell was obtained by doubling the primitive cell in the c lattice direction. The SCC-DFTB method is an extension of the standard tight binding approach [20] within the context of density-functional theory [21] and provides a self-consistent description of total energies, atomic forces, and charge transfer. The dynamics were followed up to 17.5 ps with an integration time step of 0.2 fs. Performing this simulation with the NEMD method would require 10^5 to 10^6 times more computational effort due to the roughly $O(N^3)$ scaling of the computational work with number of atoms.

Figure 11 shows the time profile of the density, stress, and temperature of the system throughout the simulation. Density oscillations are damped within a few oscillations. Examination of the simulation cell contents shows that at 4.9 ps, a proton transfer process occurs, which can be described as: $CH_3NO_2 + CH_3NO_2 \rightarrow CH_3NO_2H + CH_2NO_2 \rightarrow CH_3NO_2 + CH_2NO_2H$. This chemical event that leads to the formation of the so-called aci acid H_2CNO_2H moiety persists for over 4 ps of the simulation. There have been several experimental concurrences for the production of the aci ion in highly pressurized and detonating nitromethane. Shaw et al. [22] observed that the time to explosion for deuterated nitromethane is about ten times longer than that for the protonated materials, suggesting that a proton (or hydrogen atom) abstraction is the rate-determining step. Isotope-exchange experiments provided evidence that the aci ion concentration is increased upon increasing pressure, [23] and UV sensitization of nitromethane to detonation was shown to correlate with the aci ion presence. [24]

10. CONCLUSION

In this chapter we have presented a multi-scale method for molecular dynamics simulations of shock compression and characterized its behaviour. This method attempts to constrain the molecular dynamics system to the sequence of thermodynamic states that occur in a shock wave. While we have presented one particular approach, it is certainly not unique and there are likely a variety of related approaches to multi-scale simulations that have a variety of differing practical properties. These methods open the door to simulations of shock propagation on the longest timescales accessible by molecular dynamics and the use of accurate but computationally costly material descriptions like density functional theory. It is our belief that this method promises to be a valuable tool for elucidation of new science in shocked condensed matter.

REFERENCES

[1] C. S. Yoo, W. J. Nellis, M. L. Sattler, and R. G. Musket, Appl. Phys. Lett., 61 (1992) 273.

[2] W. J. Nellis, S. T. Weir, and A. C. Mitchell, Science, 273 (1996) 936; S. T. Weir, A. C. Mitchell, and W. J. Nellis, Phys. Rev. Lett., 76 (1996) 1860.

[3] M. D. Knudson and Y. M. Gupta, Phys. Rev. Lett., 81 (1998) 2938.

[4] Loveridge-Smith, A. Allen, J. Belak, T. Boehly, A. Hauer, B. Holian, D. Kalantar, G. Kyrala, R. W. Lee, P. Lohmdahl, M. A. Meyers, D. Paisley, S. Pollaine, B. Remington, D. C. Swift, S. Weber and J. S. Wark, Phys. Rev. Lett., 86 (2001) 2349.

[5] K. Kadau, T. C. Germann, P. S. Lomdahl, and B. L. Holian, Science 296, 1681 (2002); J. D. Kress, S. R. Bickham, L. A. Collins, B. L. Holian, and S. Goedecker, Phys. Rev. Lett., 83 (1999) 3896.

[6] E. J. Reed, L. E. Fried, and J. D. Joannopoulos, Phys. Rev. Lett., 90 (2003) 235503.

[7] M. P. Allen and D. J. Tildesley, Computer simulation of liquids (Oxford University Press, New York, 1989).

[8] H. Goldstein, Classical Mechanics (Addison-Wesley, Reading, MA, 1980).

[9] G. E. Duvall in Proceedings of the International School of Physics, Physics of High Energy Density (Academic Press, New York, 1971).

[10] Y. B. Zel'dovich and Y.P.Raizer, Physics of shock waves and high-temperature hydrodynamic phenomena (Academic Press, New York, NY, 1967).

[11] H. C. Anderson, J. Chem. Phys., 72 (1980) 2384.
[12] J. Fox and H. C. Andersen, J. Chem. Phys., 88 (1984) 4019.
[13] R. Ravelo, B. L. Holian, T. C. Germann, private communication.
[14] B. L. Holian, A. F. Voter, N. J. Wagner, R. J. Ravelo, S. P. Chen, W. G. Hoover, C. G. Hoover, J. E. Hammerberg and T. D. Dontje, Phys. Rev. A, 43 (1991) 2655.
[15] E. Jouget, Mecanique des explosifs (Octave Doin et Fils, Paris, 1917).
[16] W. Fickett and W. Davis, Detonation, (University of Califronia Press, Berkeley, CA, 1979).
[17] T. C. Germann, B. L. Holian, P. S. Lomdahl, and R. Ravelo, Phys. Rev. Lett., 84 (2000) 5351.
[18] F. H. Stillinger and T. A. Weber, Phys. Rev. B, 31 (1985) 5262.
[19] M. Elstner, D. Porezag, G. Jungnickel, J. Elsner, M. Hauk, T. Frauenheim, S. Suhai, and G. Seifert, Phys. Rev. B, 58 (1998) 7260.
[20] J. C. Slater and G. F. Koster, Phys. Rev., 94 (1954) 1498.
[21] P. Hohenberg and W. Kohn, Phys. Rev., 136 (1964) B864.
[22] R. Shaw, P. S. Decarli, D. S. Ross, E. L. Lee, and H. D. Stromberg, Combust. Flame, 50 (1983) 123; R. Shaw, P. S. Decarli, D. S. Ross, E. L. Lee, and H. D. Stromberg, Combust. Flame, 35 (1979) 237.
[23] R. Engelke, D. Schiferl, C. B. Storm, and W. L. Earl, J. Phys. Chem., 92 (1988) 6815.
[24] R. Engelke, W. L. Earl, and C. M. Rohlfing, J. Phys. Chem., 90 (1986) 545.

Chemistry at Extreme Conditions
M. Riad Manaa (Editor)

Chapter 11

Plastic Deformation in High Pressure, High Strain Rate Shocked Materials: Dislocation Dynamics Analyses

Mutasem Shehadeh and Hussein Zbib[1]

School of Mechanical and Materials Engineering,

Washington State University, Pullman, WA 99163-2920, USA

1. INTRODUCTION

Deformation of crystalline materials is determined to a large extent by underlying microscopic processes involving various defects such as dislocations, point defects and clusters. The interaction among these defects and the manner in which they interact with external agencies determine material strength and durability. In the case of high strain rate loading conditions, interaction between shock waves and dislocations play a major role in determining material behavior and the resulting microstructure [1]. Under shock loading conditions, a uniaxial strain state is produced in the material which results in a three-dimensional state of stress. This type of loading is achieved during plate impact, explosives or high intensity laser experiments. In general, shock waves are supersonic disturbances that lead to large changes in compression, particle velocity and internal energy in a nearly discontinuous manner [2]. The shock wave velocity depends on the elastic properties of the material that are pressure dependent [3], this indicates that under high pressure loading waves propagates faster than under ambient conditions.

Shock wave compression cannot only induce deformation in the form of high density of defects such as dislocations and twins but can also result in phase transition, structural changes and chemical reaction. These changes in the material are controlled by different components of stress, the mean stress and the deviatoric stress. The mean stress causes pressure-induced changes such as phase transformations while the deviators control the generation and motion of dislocations.

Plastic deformation in crystalline solids occurs mostly as a result of dislocation motion and multiplication. Dislocations move on specific planes in certain directions when the applied stress exceeds the critical value of shear stress. In general, dislocations move with velocities that increase as the rate of deformation increases. However, due to the relativistic effect, dislocations are limited to move at velocities less than the shear wave velocity. Under high strain rate loading (10^5/s-10^8/s), dislocations accommodate this velocity restriction by high

[1] Corresponding author

rates of dislocation multiplication. Weertman [4] and Lothe [5] derived expressions for total, kinetic and strain energies of dislocation lines as a function of velocity. Hirth et al. [6] derived an expression for the effective mass per dislocation length of a fast moving dislocation. In another work, Ross et al. [7] introduced a two dimensional computational methodology for high-speed dislocations. During impact loading, plastic deformation (occurs as a result of fast moving dislocations) is often localized in narrow band like structures. These bands generally lie on the slip planes and contain a high density of dislocations [8].

The advancements in experimental capabilities over the years have improved our understanding of the dynamic response of materials. Recently, laser based experiments have been used to study the plastic deformation in different materials. Short pulse duration (in the order of nanoseconds) has been used to generate strong pressure waves that propagate through the tested samples [9-13]. During plastic deformation, dislocations arrange themselves in certain entangled structures that depend on the loading condition and deformation history. Transmission Electron Microscopy (*TEM*) is utilized to study the recovered residual microstructures of the shocked samples. The observed microstructures consist of homogenously distributed dislocation cells and tangles, deformation micro bands and deformation twins.

The response of metal single crystal to high strain rate loading depends on many parameters such as crystal structure, peak pressure, pulse duration, crystallographic orientation, and stacking fault energy (*SFE*). Peak pressure and pulse duration were both observed to increase the dislocation density [14, 15]. Most metallic single crystals exhibit cubic symmetry that leads to anisotropy in their physical properties such as the elastic constants and wave propagation speed. The anisotropy leads to more complex wave propagation behavior as well as variations in shock-induced deformation. High intensity laser experiments on copper single crystals showed that the dislocation microstructures of the recovered samples are very sensitive to crystallographic orientation and level of pressure. It was observed that there is a threshold pressure at which the deformation mechanism changes from glide to twining and that this threshold pressure is orientation dependent [13, 16].

The deformation process under high strain rate is a multiscale dynamic problem. Therefore, experiments at each length and time scale need to be conducted in order to obtain more insight into the deformation process. However, the levels of pressure and temperature involved under extreme conditions make it difficult to address the deformation process using physical experiments. In fact, current experimental capabilities cannot address material response at pressure greater than 100 GPa. In addition, the cost of full-scale experiment in this area of research is high and escalating [17]. Consequently, theories and models applicable for each spatial and time scale need to be developed. Computer simulations can then be used to connect these regimes of different length and time scales.

In the atomistic scale, molecular dynamic (*MD*) simulations are used to investigate the response of single crystals to high strain rate loading. The effects of strain rate, domain size on the deformation of FCC single crystals are investigated in [19]. Kadau *et al.* [18] conducted a multimillion atoms *MD* simulation to study the shock induced phase transition in iron crystal. German *et al* [20] carried out *MD* study of shock waves in *FCC* single crystals in three different orientations. They found that the wave characteristics and patterns formation are orientation dependent. *MD* simulations have also been conducted to study the anisotropic effects on shock wave propagation and the induced plasticity patterns formation in Nickel single crystals [21].

At the macro scale, continuum models are used to describe the dynamic response of solids. These models can be (a) phenomenological obtained by fitting the experimental results, or (b) physically based obtained by incorporating the effects of the microstructures. Steinberg *et al.* [22, 23] introduced phenomenological models applicable to high strain rate loading. In these models, the shear modulus and yield stress are assumed to be functions of pressure, temperature and compression for elastic-perfectly plastic deformation. The Steinberg constitutive equations successfully regenerate the shock experiment data. One of the most notable constitutive equations for deformation in *FCC* metals under high strain rate is the Zerilli-Armstrong [24] constitutive equation, which is based on the introduction of viscous drag into a simple thermally activated dislocation model. Meyers *et al.* [25, 26] developed physically based models to describe the response of metals to high strain rate deformation. The effects of dislocation dynamics, twining, phase transformation, stacking fault, grain size and solution hardening are incorporated in the model. Nasser [27] suggested a physically based model to simulate the failure modes and the dynamic response of metals.

Computer simulations methodologies have been used to bridge the length scales from atomistic to macroscopic scales. Smirnova et al. [28] introduced a combined molecular dynamics and finite element approach to simulate the propagation of laser induced pressure in a solid. In the micorscale, discrete dislocation dynamics provide an efficient approach to investigate the collective behavior of many interacting dislocations. Recently, multiscale dislocation dynamics plasticity has emerged as an excellent numerical tool to simulate the collective behavior of dislocations in a bulk material. Dislocation dynamics can simulate sizes much larger than the current atomistic simulation capabilities. In our attempt to understand the response of *FCC* single crystal to high strain rates, we employ a multiscale model developed at Washington State University to study the interaction between stress waves and dislocations. In this study, the effects of high pressure, high strain rate, shock pulse duration, crystal anisotropy, and the dependence of elastic properties on pressure are investigated and presented in this chapter.

2. MULTISCALE DISLOCATION DYNAMIC PLASTICITY (MDDP)

This *MDDP* model is based on fundamental physical laws that govern dislocations motion and their interactions with various defects and interfaces. The multiscal model merges two length scales, the nano-microscale where plasticity is determined by explicit three dimensional dislocation dynamics analysis providing the material length scale, and the continuum scale where energy transport is based on basic continuum mechanics laws. The result is a hybrid elasto-viscoplastic simulation model coupling discrete dislocation dynamics (*DD*) with finite element analysis (*FE*)

In the macro level, it is assumed that the material obeys the basic laws of continuum mechanics, i.e. linear momentum balance and energy balance:

$$div\,\mathrm{S} = \rho\,\dot{\mathrm{v}} \tag{1}$$

$$\rho C_v \dot{T} = K\nabla^2 T + S\dot{\varepsilon}^p \qquad (2)$$

where $v = \dot{u}$ is the particle velocity, ρ, C_v and K are mass density, specific heat and thermal conductivity respectively. For elasto-visco-plastic behavior, the strain rate tensor $\dot{\varepsilon}$ is decomposed into an elastic part $\dot{\varepsilon}^e$ and plastic part $\dot{\varepsilon}^p$ such that:

$$\dot{\varepsilon} = \dot{\varepsilon}^e + \dot{\varepsilon}^p, \ \dot{\varepsilon} = \frac{1}{2}\left[\nabla v + \nabla v^T\right] \qquad (3)$$

For most metals, the elastic response is linear and can be expressed using the incremental form of Hooke's law such that:

$$\overset{\circ}{S} = \left[C^e\right] \dot{\varepsilon}^e, \ \overset{\circ}{S} = \overset{\bullet}{S} - \omega S + S\omega, \ \omega = W - W^p \qquad (4)$$

where C^e is, in general, the anisotropic elastic stiffness tensor for cubic symmetry, ω is the spin of the microstructure and it is given as the difference between the material spin W and plastic spin W^p. Combining (3) and (4) leads to:

$$\overset{\circ}{S} = \left[C^e\right]\left[\dot{\varepsilon} - \dot{\varepsilon}^p\right] \qquad (5)$$

In the nano-microscale, *DD* analyses are used to determine the plasticity of single crystals by explicit three-dimensional evaluations of dislocations motion and interaction among themselves and other defects that might be present in the crystal, such as point and cluster defects, microcracks, microvoids, etc. In *DD*, dislocations are discretized into segments of mixed character. Details of the model can be found in a series of papers by Zbib and co-workers. The dynamics of the dislocation is governed by a "*Newtonian*" equation of motion, consisting of an inertia term, damping term, and driving force arising from short-range and long-range interactions. Since the strain field of the dislocation varies as the inverse of distance from the dislocation core, dislocations interact among themselves over long distances. As the dislocation moves, it has to overcome internal drag, and local barriers such as the Peierls stresses. The dislocation may encounter local obstacles such as stacking fault tetrahedra, defect clusters and vacancies that interact with the dislocation at short ranges and affect its local dynamics. Furthermore, the internal strain field of randomly distributed local obstacles gives rise to stochastic perturbations to the encountered dislocations, as compared with deterministic forces such as the applied load. This stochastic stress field also contributes to the spatial dislocation patterning in the later deformation stages. Therefore, the strain field of local obstacles adds spatially irregular uncorrelated noise to the equation of motion. In addition to the random strain fields of dislocations or local obstacles, thermal fluctuations also provide a stochastic source in dislocation dynamics. Dislocations also interact with free surfaces, cracks, and interfaces, giving rise to what is termed as image forces. In summary, the dislocation may encounter the following set of forces:

- Drag force, $B\boldsymbol{v}$, where B is the drag coefficient and $\boldsymbol{v}$ is the dislocation velocity.
- Peierls stress $\boldsymbol{F}_{Peierls.}$
- Force due to externally applied loads, $\boldsymbol{F}_{external.}$
- Dislocation-dislocation interaction force $\boldsymbol{F}_{D.}$

- Dislocation self-force $\boldsymbol{F}_{self.}$
- Dislocation-obstacle interaction force $\boldsymbol{F}_{obstacel.}$
- Image force $\boldsymbol{F}_{image.}$
- Osmotic force $\boldsymbol{F}_{Osmotic}$ resulting from non-conservative motion of dislocation (climb) and results in the production of intrinsic point defects.
- Thermal force $\boldsymbol{F}_{thermal}$ arising from thermal fluctuations.

The DD approach attempts to incorporate all of the aforementioned kinematics and kinetics aspects into a computational traceable framework. In the numerical implementation, three-dimensional curved dislocations are treated as a set of connected segments. Then, it is possible to represent smooth dislocations with any desired degree of realism, provided that the discretization resolution is taken high enough for accuracy (limited by the size of the dislocation core radius r_0, typically the size of one Burgers vector). In such a representation, the dynamics of dislocation lines is reduced to the dynamics of discrete degrees of freedom of the dislocation nodes connecting the dislocation segments.

As mentioned above, the velocity $\boldsymbol{v}$ of a dislocation segment s is governed by a first order differential equation consisting of an inertia term, a drag term and a driving force vector [6] such that

$$m_s \dot{v} + \frac{1}{M_s(T,p)} v = \boldsymbol{F}_s \qquad with \quad m_s = \frac{1}{v}\left(\frac{dW}{dv}\right) \tag{6$_1$}$$

$$\boldsymbol{F}_s = \boldsymbol{F}_{Peirels} + \boldsymbol{F}_D + \boldsymbol{F}_{Self} + \boldsymbol{F}_{External} + \boldsymbol{F}_{Obstacle} + \boldsymbol{F}_{Image} + \boldsymbol{F}_{Osmotic} + \boldsymbol{F}_{Thermal} \tag{6$_2$}$$

In the above equation the subscript s stands for the segment, m_s is defined as the effective dislocation segment mass density, M_s is the dislocation mobility which could depend both on the temperature T and the pressure P, and W is the total energy per unit length of a moving dislocation (elastic energy plus kinetic energy). As implied by (6)$_2$, the glide force vector $\boldsymbol{F}_s$ per unit length arises from a variety of sources described above. The following relations for the mass per unit dislocation length have been suggested by Hirth et al [6] for screw (m_s)$_{screw}$ and edge (m_s)$_{edge}$ dislocations when moving at a high speed.

$$\begin{aligned}(m_s)_{screw} &= \frac{W_0}{v^2}(-\gamma^{-1} + \gamma^{-3}) \\ (m_s)_{edge} &= \frac{W_0 C^2}{v^4}(-16\gamma_l - 40\gamma_l^{-1} + 8\gamma_l^{-3} + 14\gamma + 50\gamma^{-1} - 22\gamma^{-3} + 6\gamma^{-5})\end{aligned} \tag{7}$$

where $\gamma_l = (1 - v^2/C_l^2)^{\frac{1}{2}}$, $\gamma = (1 - v^2/C^2)^{\frac{1}{2}}$, C_l is the longitudinal sound velocity, C is the transverse sound velocity, ν is Poisson's ratio, $W_0 = \frac{Gb^2}{4\partial}\ln(R/r_0)$ is the rest energy for the screw per unit length, G is the shear modulus. The value of R is typically equal to the size of the dislocation cell (about 1000 b, with b being the magnitude of the Burgers vector), or in the case of one dislocation is the shortest distance from the dislocation to the free surface. In the non-relativistic regime when the dislocation velocity is small compared to the speed of sound, the above equations reduce to the familiar expression $m = \beta \rho b^2 \ln(R / r_0)$, where β is a constant dependent on the type of the dislocation, and ρ is the mass density.

In *DD*, Equation (6) applies to every infinitesimal length along the dislocation line. In order to solve this equation for any arbitrary shape, the dislocation curve may be discretized into a set of dislocation segments as outlined by Zbib and co-workers. Then the velocity vector field over each segment may be assumed to be linear and, therefore, the problem is reduced to finding the velocity of the nodes connecting these segments. There are many numerical techniques to solve such a problem. Consider, for example, a straight dislocation segment s bounded by two nodes. Then within the finite element formulation, the velocity vector field is assumed to be linear over the dislocation segment length. This linear vector field $\boldsymbol{v}$ can be expressed in terms of the velocities of the nodes such that $\boldsymbol{v} = \left[\boldsymbol{N}^D\right]^T \boldsymbol{V}^D$ where $\boldsymbol{V}^D$ is the nodal velocity vector and $[\boldsymbol{N}^D]$ is the linear shape function vector. Upon using the Galerkin method, equation (6) for each segment can be reduced to a set of six equations for the two discrete nodes (each node has three degrees of freedom). The result can be written in the following matrix-vector form.

$$[\boldsymbol{M}^D]\dot{\boldsymbol{V}}^D + [\boldsymbol{C}^D]\boldsymbol{V}^D = \boldsymbol{F}^D \tag{8}$$

where $[M^D] = \mathrm{m}_s \int [N^D][N^D]^{\mathrm{T}} \mathrm{dl}$ is the dislocation segment 6x6 mass matrix, $[C^D] = (1/M_s) \int [\mathrm{N}^D][\mathrm{N}^D]^{\mathrm{T}} dl$ is the dislocation segment 6x6-damping matrix, and $\mathrm{F}^D = \int [\mathrm{N}^D] \mathrm{F}_s \mathrm{dl}$ is the 6x1 nodal force vector.

The integration is performed over the dislocation segment length *l*. Then, following the standard element assemblage procedure, one obtains a set of discrete system of equations, which can be cast in terms of a global dislocation mass matrix, global dislocation damping matrix, and global dislocation force vector. In the case of one dislocation loop and with ordered numbering of the nodes around the loop, it can be easily shown that the global matrices are banded with half-bandwidth equal to one. However, when the system contains many loops that interact among themselves and new nodes are generated and/or annihilated continuously, the numbering of the nodes becomes random and the matrices become un-banded. To simplify the computational effort, one can employ the lumped matrix method. In this method, the mass matrix $[\boldsymbol{M}^D]$ and damping matrix $[\boldsymbol{C}^D]$ become diagonal matrices (half-bandwidth equal to zero), and therefore the only coupling between the equations is through the nodal force vector $\boldsymbol{F}^D$. The computation of each component of the force vector is described below.

The motion of each dislocation segment contributes to the macroscopic plastic strain and spin via the relations:

$$\dot{\varepsilon}^P = \sum_{i=1}^{N} \frac{\mathrm{l}_i \mathrm{v}_{gi}}{2\mathrm{V}} (\mathrm{n}_i \otimes \mathrm{b}_i + \mathrm{b}_i \otimes \mathrm{n}_i) \tag{9}$$

$$W^p = \sum_{i=1}^{N} \frac{l_i v_{gi}}{2V} \left(n_i \otimes b_i - b_i \otimes n_i \right) \tag{10}$$

where l_i is the dislocation segment length, v_i is the dislocation glide velocity, n_i is a unit normal to the slip plane, V is the volume of the representative element, and N is the total number of dislocations segments within a given element. When the deformation process involves dislocations moving at speed larger than one-tenth the shear wave velocity, the effect of dislocation's effective mass per unit length becomes more and more pronounced. The expression for the dislocation effective mass derived by Hirth el al. [6] is implemented in this framework. The model development can be found in a number of papers by Zbib and co workers [29-33]. The *FE* part of the model is used to produce high stress waves that propagate in the material. The longitudinal wave velocity C_0 in slim bar considered here is given by:

$$C_0 = \sqrt{(E/\rho)} \tag{11}$$

where *E* is Young's modulus.

The time step in the analysis is dictated by the shortest flight distance for short-range interaction between dislocations in *DD* and the time step used in the dynamic *FE*. In this analysis, the critical time (t_c) and the time step (δt) for both *DD* and *FE* which yields a stable solution are given by $t_c = (l/C_0)$, $\delta t = (t_c/10)$ where l is the characteristic length which is the shortest dimension in the *FE* mesh. In our calculations, t_c is as small as 1×10^{-10} second and thus δt is one order of magnitude less. This time step is very suitable to simulate ultra high strain rate, short pulse duration conditions involved in laser based experiments.

3. DISLOCATION-SHOCK WAVES INTERACTION: SIMULATION-SETUP

MDDP simulations are performed to investigate the deformation process at high pressure and high strain rates in copper and aluminum single crystals. The simulations are designed to mimic uniaxial strain loading at extreme conditions of high strain rates > 10^6 /s, and short pulse durations of few nanoseconds (Loveridge *et al.* 2001). As illustrated in Fig. 1, the simulation setup consists of a block with dimensions 2.5 μm X 2.5 μm X25 μm. In order to achieve the uniaxial strain involved in shock loading, the four sides are confined so that they can only move in the loading direction. The bottom surface is rigidly fixed. To generate the stress wave a velocity-controlled boundary condition (v_p) with finite rise time (t_{rise}) is applied on the upper surface over a short period of time (t^*). In this case, v_p is related to the average strain rate and t^* is the pulse holding duration. The upper surface is then released and the simulation continues for the transmitted elastic wave to interact with existing dislocations sources.

The loading and the boundary conditions are summarized in the following equations:

$$u_z(t) = -C_1 t^2, \qquad 0 \le t \le t_{rise,} \qquad at\ z = L_z/2$$

$$u_z(t) = -C_2 t\ , \qquad t_{rise} \le t \le t^*, \qquad at\ z = L_z/2 \tag{12}$$

$$u_z(t) = 0, \qquad at\ z = -L_z/2$$

$$u_x(t) = 0, \qquad at\ x = -L_x/2, L_x/2$$

$$u_y(t) = 0, \qquad at\ y = -L_y/2, L_y/2$$

Where L_x, L_y, and L_z are the lengths of the computational cell in the x, y and z directions respectively and u_x, u_y, u_z are the corresponding displacement components, C_1 and C_2 are loading controlled constants related to the velocity of the upper surface. A typical loading history is shown in Fig. 2 by plotting the velocity history of the upper surface.

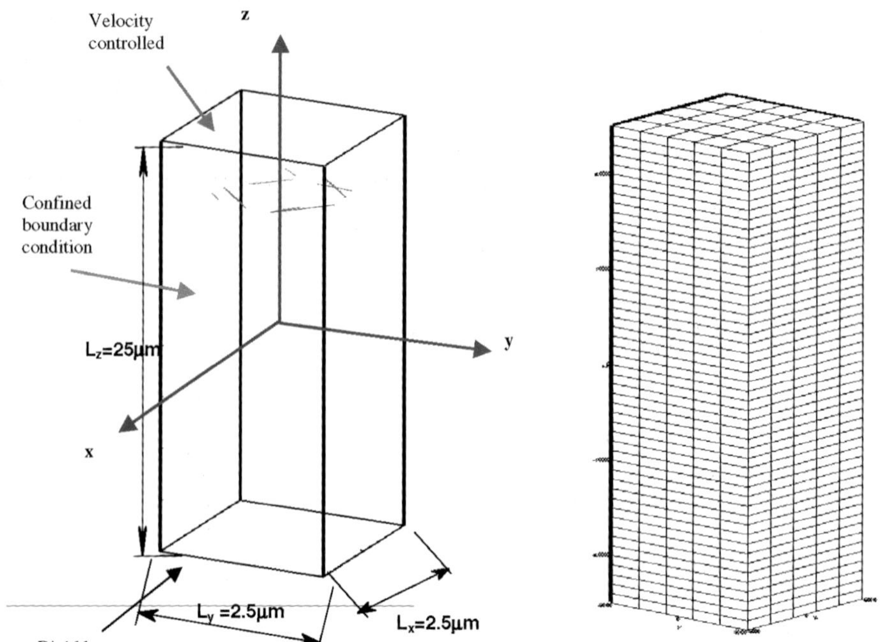

Fig. 1. Setup of the simulation cell and the finite element mesh.

When loading the crystalline materials under extreme uniaxial compression, the deviatoric stresses at the wave front may attain or even exceed the theoretical strength of the material in pure shear under normal conditions. The lattice responds to that by nucleating dislocation loops at the wave front. In the case of shock compression, the activation energy for loop generation is much lower than the thermal energy of the atoms and so the nucleation of dislocation loops is considered as thermally activated process [13, 41]. A homogenous nucleation mechanism was proposed by Meyers [42] in which he assumed that dislocations are homogenously generated at the front to accommodate the deviatoric stresses. The insertion of dislocations relaxes the deviatoric stresses and as the wave propagates, new layers of nucleated loops are generated. In real materials, the most favorable places for shear initiation are large radius substitutional atoms and micro particles [41]. *MD* simulations show that the dislocation densities around nucleated dislocation sources increase significantly when a shock wave with a finite rise time is applied compared to a steep wave front (Bringa, E. Private communication). Currently, a nucleation criterion is under implementation in our framework so that the simulations can be started from a defect free cell and end with high densities of dislocations. Alternatively, one can start from a random initial distribution of small dislocations loops that may act as dislocation sources.

In the present analyses, prismatic dislocation loops distributed on different slip planes are used as agents for dislocation generation. For copper, sources length of about 0.60 μm are used. It is worthy to mention that the boundary conditions of the computational cell sides are different in *FE* and *DD* parts of the code. In *DD*, periodic boundary condition for the representative volume element *RVE* is used to ensure both the continuity of the dislocation curves and the conservation of dislocation flux across the boundaries, by that we take into account the periodicity of single crystals in an infinite media. In *FE* analysis however, the sides are constrained to move only in the z direction so that a uniaxial strain consistent with the shock experiment is achieved. In order for the boundary conditions in *FE* and *DD* to be consistent, periodic *FE* boundary condition is implemented as well. The result of this implementation is discussed in the next section.

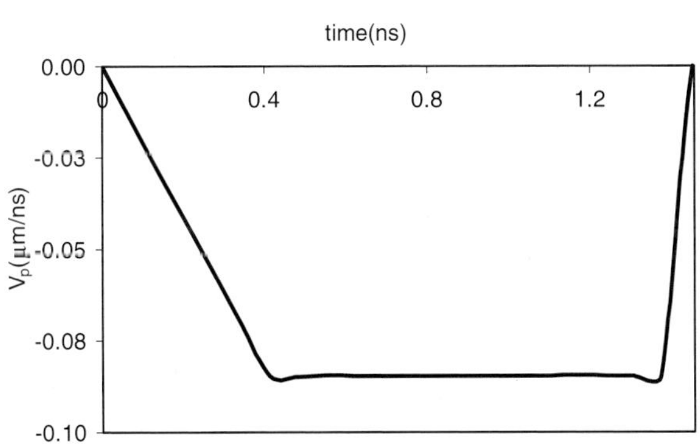

Fig. 2. The velocity loading history of the upper surface.

The main focus of this of this work is to investigate the interaction between the transmitted wave and dislocation sources. However, when the stress wave hits the rigid base, it reflects back to the material block and interacts with the dislocations again. In order to minimize the effects of the reflected waves, the length of the cell (25 μm) is chosen such that once the wave front reaches the bottom surface, the value of the stresses in the position where the dislocations are located is small so that dislocation relaxation process can take place well before the wave hits the bottom of the *RVE*. A better solution to isolate the effect of the reflected wave would be by implementing a suitable non-reflective *FE* boundary condition.

Under high strain rate loading, the elastic properties of metals depend on the applied pressure. In order to better simulate the physical problem, any constitutive model must account for this effect. In the *MDDP* code, the experimental results of Hayes et al, [3] for the shear modulus (G) and Poisson's ratio (ν) of isotropic copper are fitted in the constitutive equation to account for the dependence of elastic properties on pressure such that:

$$G = \begin{cases} G_0 + 0.89P & 0 < P < 60 \\ G_0 + 53.4 + 0.40P & 60 < P < 100 \end{cases} \tag{13}$$

$$\nu = \nu_0 + 1.70 \times \mathrm{P} \times 10^{-12} \tag{14}$$

where, P is in GPa, G_0 and ν_0 are the shear modulus and Poisson's ratio under normal static loading conditions. The implementation of the above nonlinear elastic model in the *FE* framework required the adjustment of the stiffness matrix at every time step. In passing we note that here the pressure is given by $\mathrm{P} = -\sigma_{33}$ as usually defined in the shockwave community, this definition is used through out this paper. In the continuum mechanics community, however, pressure is usually defined as $\mathrm{P} = -(\sigma_{33} + \sigma_{22} + \sigma_{11})/3$. It can be shown that these two definitions coincide only in the case of uniaxial strain of isotropic linear elastic material with Poisson's ratio equal to 1/2.

4. RESULTS AND DISCUSSION

4.1 Wave Propagation Characteristics

The application of uniaxial strain shocks results in propagating a three dimensional state of stress. The longitudinal shock wave is comprised of three distinct regimes: (a) a compressive loading regime of the wave front. In this regime, the pressure increases from its ambient value to the peak over a period of time called the shock rise time, (b) a plateau at which the peak pressure keeps a stable value for a period of a time equals to the pulse duration, and (c) a release part in which the sample returns to its ambient pressure. These features of the wave profile can be seen in Fig. 3 which shows the longitudinal shock wave propagating in copper sample shocked to a peak stress around 4.5 GPa for 1.50 nanoseconds pulse with 0.50 nanosecond rise time. The tail of the release wave shows fluctuations that are attributed to the *FE* mesh. The frequency of these fluctuations is proportional to the *FE* density [28, 34]. Under shock loading, the release wave moves at speed faster than that of the wave front, which leads to a decrease in the width of the plateau part as can be seen in Fig. 3(a). Contour plot of the propagating wave is shown in Fig 3(b).

As the wave propagates inside the *RVE*, it interacts with the dislocation sources and shock induced plasticity takes place leading to changes in the wave profile characteristics. In order to investigate the effects of dislocations activities on the wave profile, two simulations for defect free crystal and crystal with few initial dislocation sources were compared as illustrated in Fig. 4(a) by plotting the deviatoric stress in the loading direction for the two cases. Significant deviations in the plateau and the release parts are observed. This reveals that plastic deformation causes the material to soften, emphasizing that dislocations bring about stresses that have different values than those predicted by elasticity theory [35]. The energy used to generate plastic strains is converted into heat, which results in local temperature increase as shown in Fig. 4(b). Snapshots of shock wave induced plasticity are presented in Fig. 5 by plotting the effective plastic strain along the central line of the z-direction. The local values of plastic strain in the location where the dislocations are interacting with the stress wave keep changing till relaxation process takes place where the dislocation density saturates and so does the plastic strain. Moreover, several peaks appear in the plastic strain profile suggesting the existence of very high dislocation densities in these regions relative to other locations. These peaks show that patterns are formed as a result of the dislocation microstructures organization, which are affected by the periodic boundary condition imposed in our simulations.

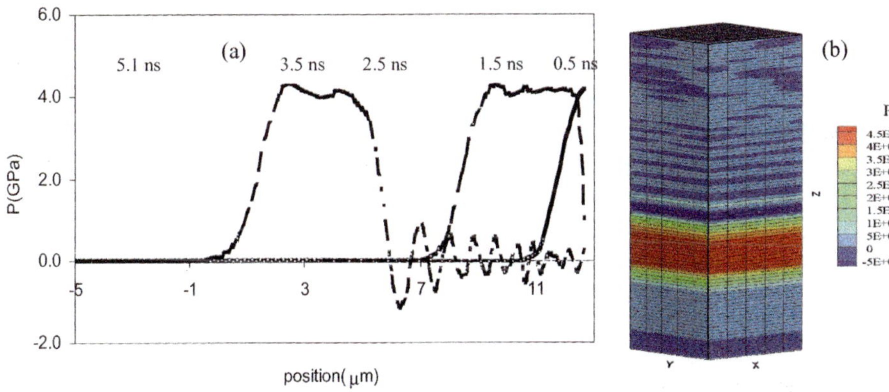

Fig 3. Shock wave propagating in copper single crystal shocked to 4.5 GPa peak pressure for 1.50 nanoseconds pulse duration.(a) snapshots (b) contour plot.

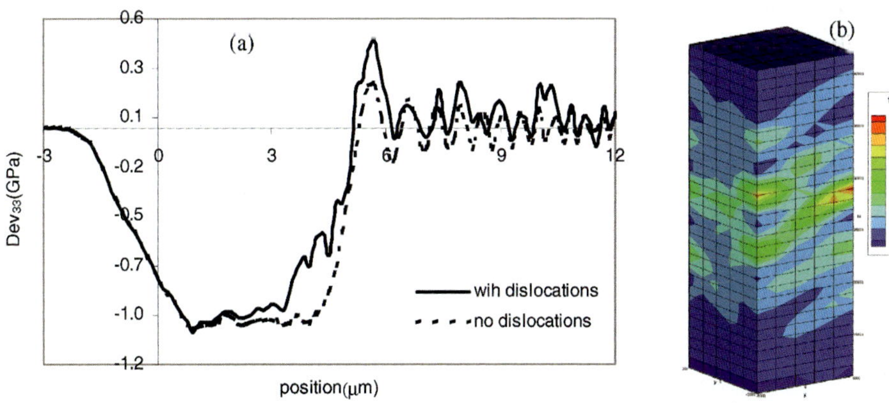

Fig 4. (a)The effect of dislocations activities on the deviatoric stress for copper shocked at 4.5 GPa peak pressure for 1.5 nanoseconds. (b) temperature rise from plastic deformation.

The deformed shape resulted from shock induced plastic strain is presented in Fig. 6 by plotting the deformed shape of a slice within the *RVE*. The deformed shape shows the formation of bands in the region where dislocation microbands are formed. This indicates that dislocation activities under high strain rate loading can be considered as sources for shear band formation.

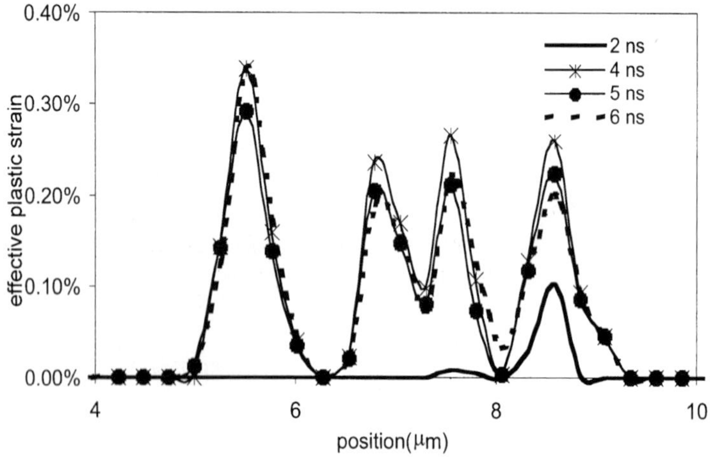

Fig. 5. Evolution of the effective plastic strain in a crystal oriented in the [011] direction shocked to a peak pressure of 4.5 GPa.

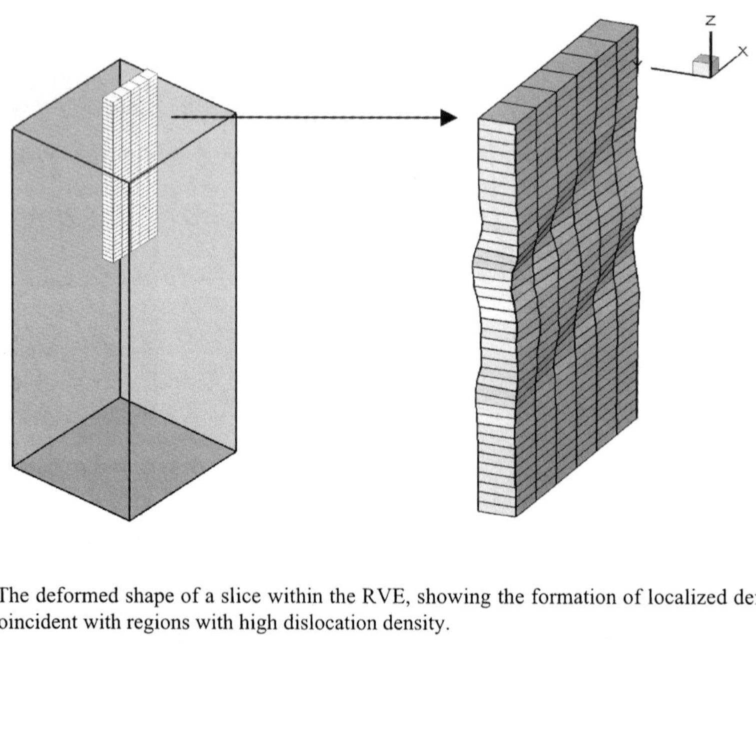

Fig. 6. The deformed shape of a slice within the RVE, showing the formation of localized deformation bands coincident with regions with high dislocation density.

The effect of the pressure dependent elastic properties on the wave profile (see equations 12 and 13) is shown in Fig. 7 by plotting the results of isotropic linear elastic constitutive equation and the nonlinear case. Clearly, the qualitative features of the two profiles are similar. However, as a result of the increase in the elastic properties, nonlinear elastic model leads to faster wave propagation and higher values of peak pressure.

For cubic symmetry materials, three independent elastic properties that are orientation dependent are required to describe the mechanical behavior of the material. This anisotropy effect increases significantly the number of the nonzero elements in the *FE* stiffness matrix leading to alteration in the calculated stress components and the wave speed. In order to test these anisotropy effects, we plot the wave profiles of three different orientations and compare it with the isotropic behavior with a loading axis in the [001] directions as shown in Fig 8. We observed that under the same loading condition, the peak stress of [111] and [011] orientations are slightly higher than those of the [001] which is lower that that of isotropic material. Furthermore, wave speed varies moderately with orientation with the fastest moving wave in the [111] followed by [011], isotropic medium and [001] respectively.

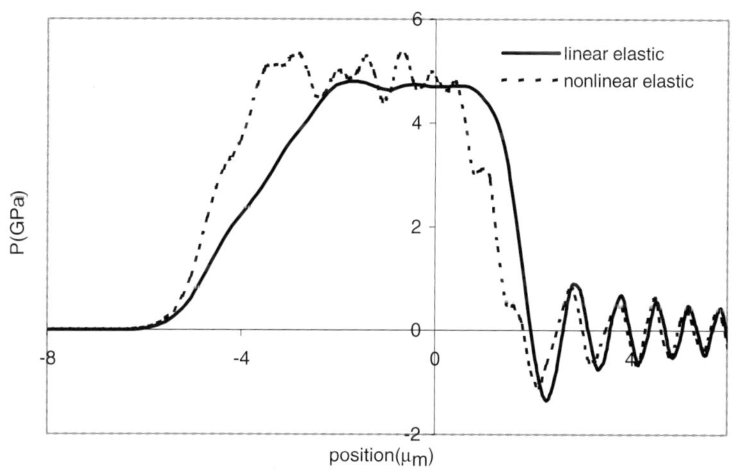

Fig. 7: The effect of pressure dependent elastic properties of the wave profile.

As mentioned previously, the boundary condition in *DD* and *FE* are different. Periodic boundary condition is used in *DD* analysis to take into account the periodicity of single crystals whereas confined boundary condition is used in the *FE* analysis to achieve the uniaxial state of strain. In order for the boundary conditions in *FE* and *DD* to be consistent, periodic *FE* boundary condition is implemented as well. This implementation of periodic *FE* boundary condition yields a relaxed state of stress with low peak pressure when compared to the experiment as illustrated in Fig. 9(a). Furthermore, both shear and longitudinal waves are generated which is discordant with plane wave characteristics as shown in Fig 9(b). Fig 10 shows the deformed shape when confined and periodic boundary conditions are used. In the confined case there is no distortion in the *RVE*. However, for the periodic case, considerable distortion in the *RVE* takes place because the nodes on one side of the cell are forced to follow the corresponding node on the opposite side leading to shear mode and shear wave propagation.

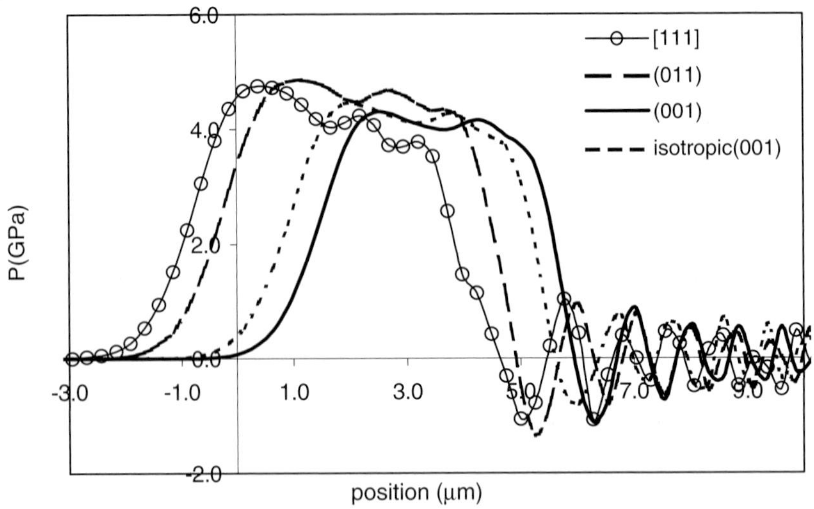

Fig 8. The effects of crystal anisotropy on the wave propagation.

4.2 Dislocation Histories

In materials with high dislocation mobility such as copper, dislocation patterns proceed through the rapid motion of dislocations in a very small volume of the specimen [36]. Under high strain rate deformation conditions, it is expected that the dislocations move at subsonic speed or even as fast as the shear wave velocity. The random motion of dislocations on their slip planes causes random changes not only in the local dislocation densities, but also in the dislocation velocities.

It is known that shock wave parameters namely peak pressure (strain rate) and pulse duration result in an increase in the mechanical properties of metals. Increasing the peak pressure results in increasing both the plastic strain and strain rate. Murr and Wilsdorf [37] observed that the dislocation density varies as a square root of the applied pressure. The result of the calculated dislocation density histories carried out at different strain rates reveals that the saturated dislocation density and the rate of dislocation multiplication increase with strain rate as illustrated in Fig 11.

Pulse duration is related to the time required for the dislocations to reorganize in certain patterns. During the shock time rise, dislocations are generated leading to permanent plastic deformation. However, some of the dislocations can possibly retrace their path during wave release reversibly [38]. Pulse duration may influence the amount dislocation reversibility and as a result the saturated dislocation density. Fig. 12 shows that the saturation density of dislocations increases with pulse duration in the nanosecond time scale. These results are consistent with the findings of Wright and Mikkola [39] on plate-impact experiments conducted on the microsecond range.

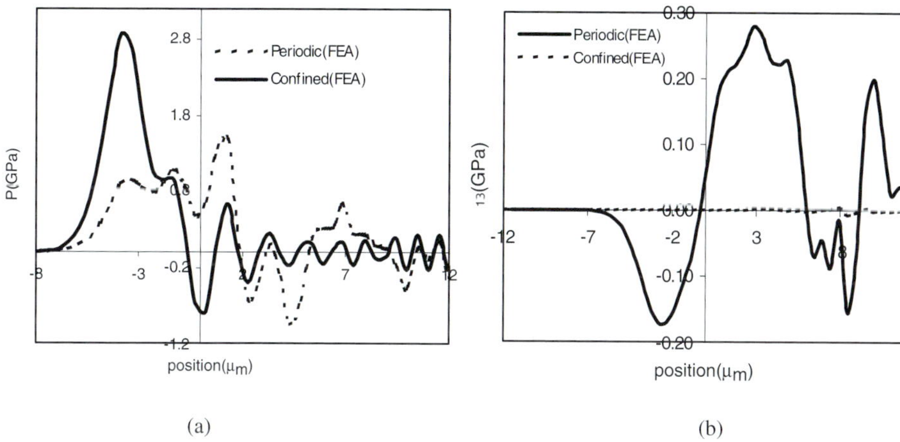

Fig. 9. The influence of *FE* boundary conditions on (a) the longitudinal wavel wave (b) Shear wave.

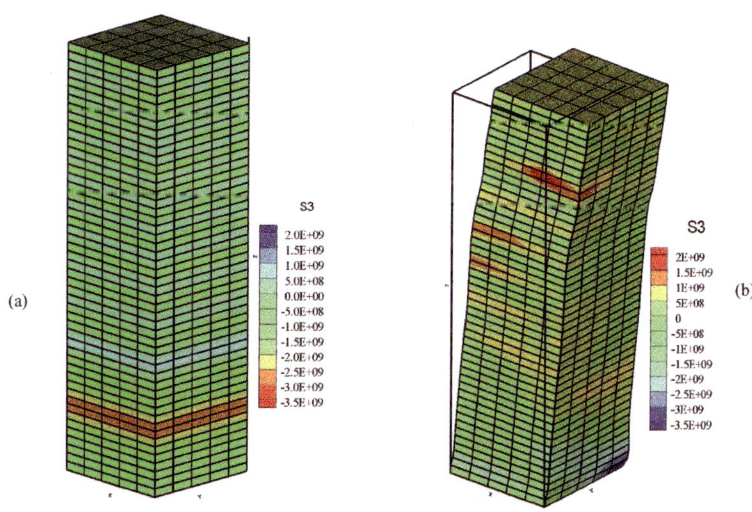

Fig. 10. The deformed shapes resulted from (a) confined boundary condition, (b) periodic boundary condition.

In FCC materials, there are 12 different slip systems, which can contribute to the deformation process. Dislocation density histories at a peak stress of 4.5 GPa for [001], [111] and [011] orientations and isotropic case with [001] orientation are calculated and plotted as shown in Fig. 13. It is clear that the dislocation density is very sensitive to crystal orientation with the highest density exhibited by [111] orientation followed by the isotropic media, [011] and [001] orientations respectively. This may be attributed to the number of slip systems activated and to the way in which these systems interact. The [001] orientation has the highest symmetry among all orientations with four possible slip planes {111} that have identical Schmid factor of 0.4082, which leads to immediate work hardening. The [011] orientation is also exhibits symmetry with 2 possible slip planes that have Schmid factor of 0.4082.

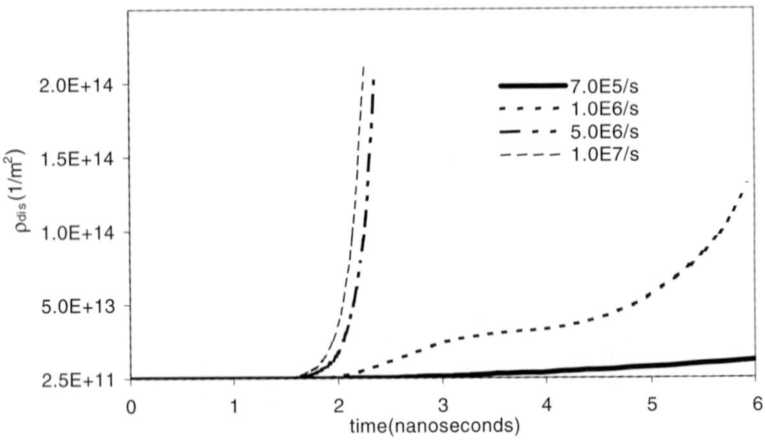

Fig 11. The influence of strain arte on the dislocation density histories.

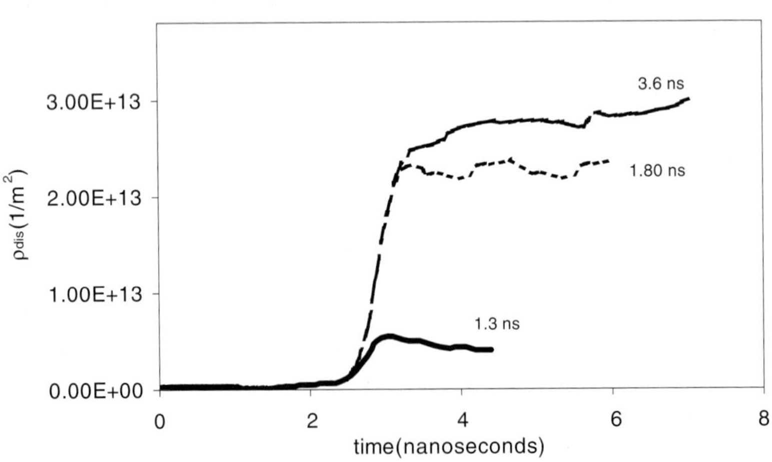

Fig 12. The influence of pulse on the dislocation density histories.

The dislocation density histories presented in Figs 11, 12 and 13 suggest the existence of three regimes during wave propagation. These are: a) no interaction regime, where the wave has not yet impacted the sources, b) the interaction regime characterized by avalanche of dislocations, and c) the relaxation regime. It is worthwhile to mention that the dislocation density histories presented here are the average values in the *RVE*. The local values of the dislocation density can be one order of magnitude higher than the average value. These features are discussed in the next subsection.

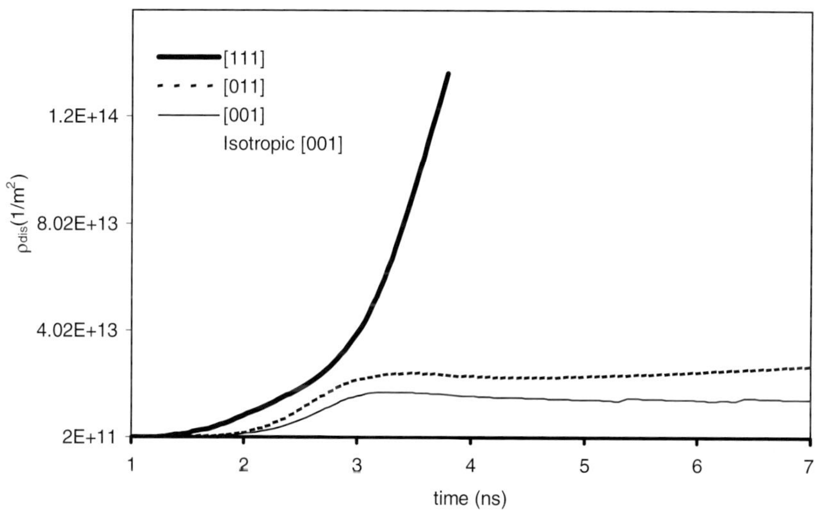

Fig. 13. The influence of crystal orientation on the dislocation density history in copper single crystal shocked for 1.5 nanoseconds.

4.3 Dislocation Microstructure

The dislocation microstructures generated by shocks depend strongly on the peak pressure and to a lesser extent on shock pulse duration. It is worthwhile to mention that for relatively low strain rates ($\sim 10^5$/s), the combination of the stress level and pulse durations (1-4 nanoseconds) is not sufficient for the dislocations to organize themselves in regular microstructure. However, as the strain rate increases, the state of stress renders so high that it allows the dislocations to form deformation band of submicron dimension coincident with the {1 1 1} planes. Morphologies of dislocations at different strain rates are illustrated in Fig. 14 within slices of the *RVE*. The lengths of these bands do not appear to be dependent on the strain rate. However, the thicknesses of the bands appear to correlate inversely with the applied strain rate.

The effect of pulse duration on the dislocation microstructure is mainly to give more time for dislocations to reorganize. In a previous study [34] we found that the microstructure at strain rates < 5×10^5/s consists of irregular dislocation entanglements. As the pulse duration increases, these entanglements become more distinguishable. However, at strain rates ≥ 5×10^5 /s, the microstructure consists of micro bands. These bands become more defined as the shock pulse duration increases. Within these bands, areas of high dislocation densities, surrounded by relatively lower dislocation density areas were observed.

In order to understand quantitatively the underlying microstructure of the material, it is very important to describe relevant features of the three dimensional form of the microstructure. The local values of dislocation density can give more insight into the local microstructure formation and improve our understanding of shock wave-dislocation interaction. In this study the local dislocation density distribution is investigated by extracting data from thin slices within the *RVE* as depicted in Fig. 16. The calculated local dislocation density distribution shows distinguished peaks of much higher dislocation densities compared to the average dislocation density. The average value of dislocation density in Fig. 16 is around 1.2×10^{14} /m^2, that is one order on magnitude lower than the maximum local dislocation density. The location of these peaks is where the dislocation microbands are formed.

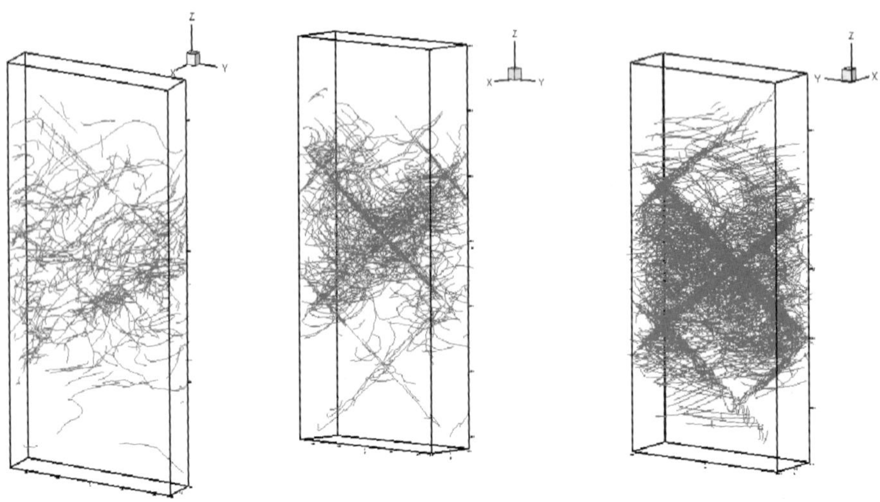

(a)

Fig 14. The effect of strain rate on dislocation microstructures of copper single crystals. The pulse duration of these simulations is 1.3 nanoseconds. The dislocation microstructures of each strain rate are shown in slices within that RVE. (a) $\dot{\varepsilon}=2\times10^5$/s, (b) $\dot{\varepsilon}=7\times10^5$/s, (c) $\dot{\varepsilon}=5\times10^6$/s

The local distributions of dislocation densities indicate that the dislocation microstructure is not homogenous. The local dislocation density plotted in Fig. 14 suggests the existence of two sub-areas with different dislocation density properties. In the first sub-area, the dislocation density is much higher than the average density; consequently distinguished peaks appear. In the second sub-area, low levels of dislocation densities that occupy very large portion of the computational cell were formed.

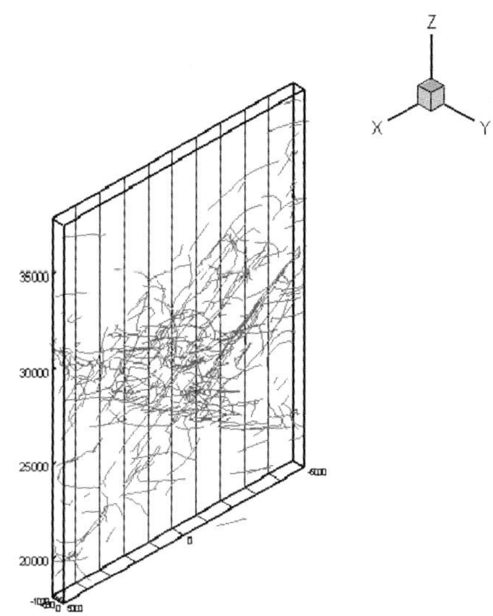

Fig. 15. A slice of the RVE showing the local dislocation distribution.

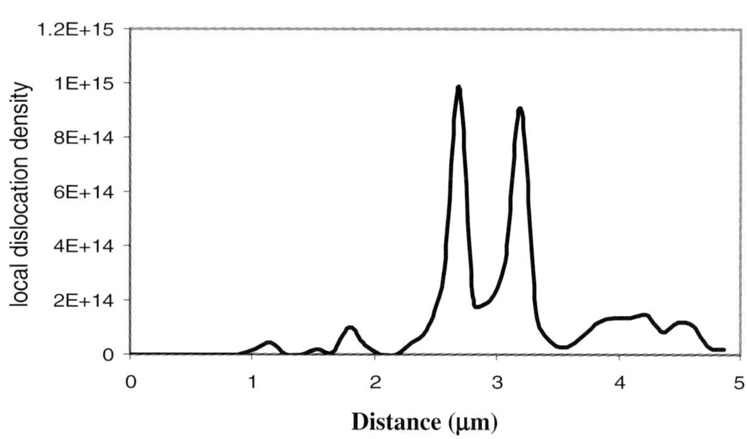

Fig 16. Local dislocation density in the slice section.

The dislocation density within the deformation bands is calculated at different peak pressure and pulse duration values. The calculated dislocation densities at different peak pressures are plotted in Fig. 17 and compared with the experimental observations of Murr [14] and the analytical predictions of Meyers et al [13]. Moreover, our calculations of the variation of dislocation density with pulse duration (constant peak pressure) is presented in Fig. 18 which illustrates that there is an increase in the dislocation density with pulse duration to a certain duration beyond which the dislocation density saturates. These results are in good qualitative agreement with the observations of Wright and Mikkola using plate impact experiment with pulses in the microsecond time scale [15, 39].

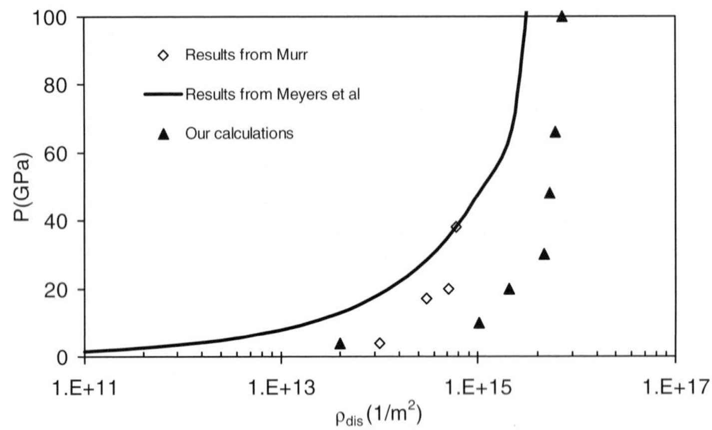

Fig. 17. The variation in the dislocation density with pressure.

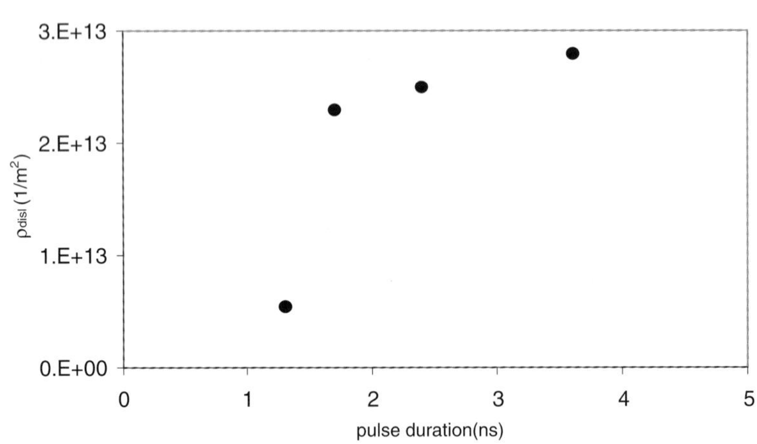

Fig. 18. The variation in the dislocation density with pulse duration. Peak pressure is 9 GPa.

4.4 Mesh Sensitivity Analysis

Different mesh sizes are used to perform mesh sensitivity analysis on the pressure profile. The number of elements in the x and y directions were kept constant (5×5) while the number of elements varies in the z direction such that 50, 100, 150 and 200 elements were used which results in a total number of 1250, 2500, 3750, and 5000 elements respectively. Fig 19 shows that as the number of element increases, the pressure profile converges to a unique value. The converged profile consists of a linearly increasing wave front, followed by a constant peak pressure plateau which is followed by the unloading. Further more, the mesh size affects wave propagation speed. Using coarse meshes result in underestimation of wave speed whereas when using fairly fine mesh, the wave speed reaches its theoretical value.

4.5 Calculations of Shock Wave Parameters

In this section we present a summary of the simulation results for copper and aluminum single crystals shocked with zero rise time pulses. As mentioned before, stress waves are generated by applying velocity controlled boundary condition (v_p) on the upper surface of the computational cell. The *FE* part of the code calculates the state of stress produced by the imposed particle velocity. The pressure produced is then used to find the corresponding particle velocity given in [40] which is denoted by U_p. For most metals, shock velocity (U_s) is directly proportional to the particle velocity via the relationship:

$$U_s = C_o + SU_p \tag{15}$$

where C_o is sound velocity at zero pressure, and S is an empirical parameter determined by experiments as given in Table 1. The theoretical longitudinal stress (σ_{33}) is then calculated using the momentum equation, which is given by:

$$\sigma_{33} = \rho_0 U_p U_s \tag{16}$$

In addition the longitudinal elastic wave velocities are calculated and then compared with the values given by the code. The properties of copper and aluminum used in these calculations are given in table 1.

Table 1. Material and shock properties used in the current computations.

Material	G(GPa)	v	ρ_o (kg/m^3)	Cp(J/kg.K)	K (W/m.K)	S
Cu	46.6	0.32	8900	385	398	1.49
Al	25.0	0.345	2700	900	274	1.34

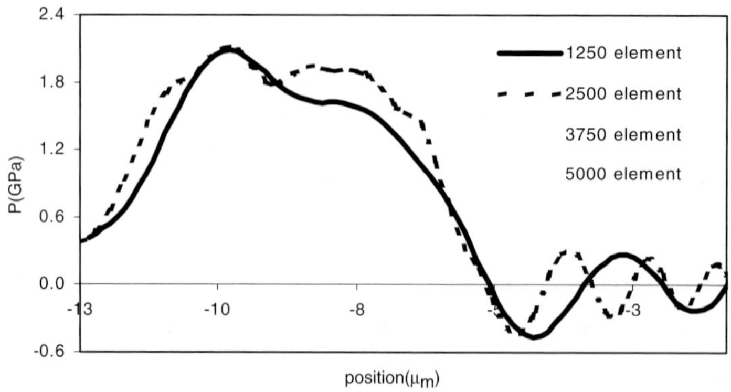

Fig. 19. The influence of FE mesh size on wave profile. Here the elements in x and y directions are constant(5×5), the number of elements increases in the z direction only.

Summary of the calculations is listed in table 2. This table displays the shock wave parameters calculated in the code and the corresponding theoretical values. The values of the axial stress (σ_{33}) obtained from our calculations are in a very good agreement with the corresponding theoretical values given by equation 16. However, the speeds of the longitudinal elastic wave are underestimated when compared to their theoretical values. This is attributed to the effect of *FE* mesh size. In fact when a fine mesh is used, the velocity approaches the theoretical wave speed value as shown in Fig. 19.

Table 2. Summary of the calculations for copper and aluminum.

Material	*Copper*				Aluminum			
Vp (m/s)	35.7	71.4	357.6	715.2	46.3	92.6	463.3	926.6
Ave Strain Rat (1/s)	5×10^5	1×10^6	5×10^6	1×10^7	5×10^5	1×10^6	5×10^6	1×10^7
Computed Axial Stress (σ_{33}) GPa	2.22	4.39	22.20	43.21	1.20	2.45	12.60	25.04
Longitudinal Elastic Wave Speed (m/s): Theoretical	3790	3970	4310	4820	5017	5170	5820	5875
Computes Elastic Wave Speed (m/s): From the code	3300	3346	3737	4155	4672	5144	5285	5651
Vp (m/s) Theoretical	57.5	113.3	524	908	70.44	143	649	1281
Theoretical axial stress (σ33) G.Pa	1.95	3.93	21.02	41.06	0.97	2.00	10.26	23.19
Dislocation Density (1/m^2)	4×10^{12}	7×10^{12}	1.6×10^{13}	2.5×10^{13}		2.5×10^{13}	2.0×10^{13}	2.1×10^{13}
Temp Rise	13	33	103	190	10	41	104	186

5. SUMMARY AND CONCLUDING REMARKS

Mutliscale simulations were carried out to study the pressure wave propagation and interaction with dislocations in *FCC* single crystals. These simulations were designed to mimic the loading conditions in recent laser based experiments, where the pulse duration is few nanoseconds. It is shown that avalanche of dislocations is a natural consequence of the interaction between dislocations and stress waves. The results of our calculations show that dislocation density is proportional to strain rate, pulse duration and crystal orientation; however, the dislocation microstructure is controlled mainly by strain rate. Dislocation micro bands coincident with the {1 1 1} planes were formed. The Inclusion of pressure-dependent elastic properties for isotropic media in our calculations leads to faster wave propagation speed. Incorporating the effect of crystal anisotropy in the elastic properties results in orientation dependent wave speed and peak pressure. Computer simulations of dislocation motion under impact loading hold a great promise for investigating deformation process of metals in regimes that cannot be probed by current experiments.

ACKNOWLEDGEMENT

The support of Lawrence Livermore National Laboratory to WSU is greatly acknowledged. This work was performed, in part under the auspices of the US Department of Energy by Lawrence Livermore National Laboratory (contract W-7405-Eng-48).

REFERENCES

[1] G.T. Gray, in High Pressure Shock Compression of Solids, J. R. Asay and M. Shahinpoor, Eds. (Springler-Verlag, NewYork, 1993) pp. 187-213.
[2] Y. M. Gupta, in Multiscale Modeling of Materials, V. Bulatov, T. D. de la Rubia, R. Philips, E. Kaxiras, and N. Ghoniem, Eds. (Material Research Society, 1998) pp. 139-149.
[3] D. Hayes, R. S. Hixson, R. G. McQueen, in Shock Compression of Condensed Matter, M. D. Furnish, L. C. Chhbildes and R. S., Hixson, Eds. (Melville, NewYork, 1999) pp. 483-488
[4] J. Weertman, Response of metals to high velocity deformation. P.G. Shewmon and V. F. Zackay, Eds., (New York: Interscience, 1961) p. 205.
[5] J. Lothe, Elastic Strain Fields and Dislocation Mobility, V. L. Indenbom, J. Lothe., Eds. (Amsterdam: north-Holland, 1992) p.447.
[6] J. P. Hirth, H. M. Zbib, and J. Lothe, Model. Simul. Mater. Sci., Eng., 6 (1998) 165-169.
[7] A. Ross, J. Th.M. De Hosson, E. Van der Giessen, Comp. Mater. Sci., 20 (2001) 1.
[8] C. S. Coffey, in Shock Waves and High Strain Rate Phenomena in Materials, M. A. Meyers, L. E. Murr, and K. P. Staudhammer, Eds, (Marcel Dekker, 1992) p.669-690.
[9] A. Loveridge-Smith et al. Phys. Rev.Lett, 86 (2001) 2349-2352.
[10] D. H. Kalantaret et al. in: 12th bienneial International Conference of the APS Topical Group on Shock Compression of Condensed Matter, Atlanta, Georgia, (2001) p.615.
[11] D. H. Kalantar et al. Phys. Plasmas, 10 (2003) 1569-1576.
[12] M. A. Meyers et al. in: Shock Compression of Condensed Matter, M.D. Furnish, N. N. Thadhani, and Y. Horie, Eds., Atlanta, Georgia, (2001) p.619.
[13] M. A. Meyers et al. Acta Mater., 51 (2003) 1211-1228.
[14] L. E. Murr, in: Shock Waves and High Strain Rate Phenomena in Metals, M. A. Meyers, L. E. Murr, Eds., (Plenum, NewYork) pp. 607-673.
[15] R. N. Wright and D. E. Mikkola, Metallurg. Trans. 16 A, 891-895.
[16] M. S. Schneider. Private communication.

[17] G. Mayer, in: Shock-Wave and High strain Rate Phenomena in Materials, M. Meyers, L. Murr, and K. Staudhammer, Eds., (Marcel Dekker, 1992) p.35.
[18] K. Kadau, T. C. Germann, P. S. Lomdhal, and B.L. Holian, Science, 296 (2002) 1681.
[19] M. F. Horstemeyer, M. I. Baskes, S. J. Plompton, Acta. Mater., 49 (2001) 4363.
[20] T. C. Germann, B. L. Holian, P.S. Lomdahl, Phys. Rev. Lett., 84 (2000) 5351-5354.
[21] O. Kum, J. App. Phys., 93 (2003) 3239-3247.
[22] D. J. Steinberg, S. G. Cochran, and M. W. Guinan, J. App. Phys., 51 (1980) 1498-1504.
[23] D. J. Steinberg, C. M. Lund, J. App. Phys., 65 (1989)1528-1533.
[24] F. G. Zerilli, R. W. Armstrong, Acta Mater., 40 (1992) 1803-1808.
[25] M. A. Meyers et al., Mater. Sci. Eng., A322 (2002) 194.
[26] M. A. Meyers, O. Vohringer, V. A. Lubarda, Acta Mater., 49 (2001) 4025-4039.
[27] S. Naser, in: Shock Wave and High strain Rate Phenomena in Materials, M. Meyers, L. Murr, and K. Staudhammer, Eds., (Marcel Dekker., 1992) pp.3-20.
[28] J. A. Smirnova, L.V. Zhigilei, B. J. Garrison, Comp. Sci. Commun., 118 (1999)11.
[29] H. M. Zbib, M. Rhee, J. P. Hirth, Int. J. Mech. Sci. 40 (1998)113-127.
[30] M. H. Zbib, T. D. de la Rubia, Int. J. Plasticity, 18 (2002)1133-1163.
[31] M. H. Zbib, M. Shehadeh, S. M. Khan, and G. Karami, Int. J. Mult. Comp. Eng., 1 (2003) 73-89.
[32] H. Yasin, H. M. Zbib, and M. A. Khaleel, Mater. Sci. Eng., A309 (2001) 294-299.
[33] M. Rhee, H. M. Zbib, A. Hansen, H. Huang, T.D. de La Rubia, Model. Simul. Mater. Sci. Eng., 6 (1998) 467-492.
[34] M. A. Shehadeh, H. M. Zbib, T. D. de la Rubia, submitted.
[35] J. Weertman, in: Shock Waves and high Strain Rate Phenomena in Metals, M. Meyers, L.E. Murr, Eds.(Plenum, NewYork, 1981) pp. 469-486.
[36] M. Zaiser, Mater. Sci. Eng., A309 (2001) 304.
[37] L. E. Murr, D. Kuhlmann-Wisldorf, Acta Met, 26 (1978) 847.
[38] J. N. Johnson, in High Pressure Shock Compression of Solids, J. R. Asay and M. Shahinpoor, Eds. (Springler-Verlag, NewYork, 1993) pp. 217-264.
[39] R. N. Wright, D. E. Mikkola, S. Larouchie, in: Shock Waves and High Strain Rate Phenomena in Metals, M. Meyers, and L. E., Murr (Plenum, NewYork, 1981) pp 703-716.
[40] M. A. Meyers, Dynamic Behavior of Materials, (New York: John Wiley & Sons, 1994).p. 2, 386, 394, 108.
[41] M. A. Mogilevskii, I. O. Mykin, Comb., Expl., & Shock Waves (English Translation of Fizika Goreniya i Vzryva), 21 (1985) p 376-382.
[42] M. A. Meyers, Scripta Metallurgica, 12 (1978) 21-26.

Chemistry at Extreme Conditions
M. Riad Manaa (Editor)

Chapter 12

Shock-Induced Chemistry in Hydrocarbon Molecular Solids

Mark L. Elert[a], Sergey V. Zybin[b], and Carter T. White[c]

[a]U. S. Naval Academy, Annapolis, MD 21402 USA

[b]The George Washington University, Washington, DC 20052 USA

[c]Naval Research Laboratory, Washington, DC 20375 USA

1. INTRODUCTION

1.1. Background

Shock waves produced by an impact cause a transient pulse of high pressure and high kinetic energy (or high "temperature" to use the term loosely) as they propagate through condensed phases. The products and reaction mechanisms of shock-induced chemical reactions may therefore be quite different than those produced under ambient conditions or static high pressure. Despite the potential for novel chemistry, however, little attention has been devoted to the subject of shock-induced chemical reactions in molecular solids. As a practical matter, shock waves are difficult and expensive to generate and control, and there is little likelihood that such a method could be feasible for large-scale synthesis. In addition, the challenges facing any experimental time-resolved study of shock-induced chemistry are significant due the extremely short time scales involved and the potentially destructive nature of the shock wave itself.

Although it is not practical as a general synthetic technique, there are some circumstances in which an understanding of shock-induced chemistry is of great importance. One obvious example is the reaction chemistry of high explosives. Another is the chemical response of common materials such as ceramics and metal alloys which may be subjected to shock waves due to proximity to an explosive device. It is also of significant interest to the astrophysical community to understand the chemical reactions, which might be induced by shock impact of comets or meteorites into planetary atmospheres. In this regard, the shock-induced chemistry of molecular hydrocarbons is of particular interest because of their central role in the development of life, and the ubiquity of small hydrocarbons such as methane in the atmospheres of many planetary bodies.

In this report we summarize some recent molecular dynamics studies of shock-induced chemistry in condensed-phase hydrocarbon systems. As mentioned above, experimental study of such systems can be difficult because of the short time scales and extremely small spatial extent of the reaction zones involved. These conditions, however, are well-suited to the technique of molecular dynamics simulations, which are currently limited by computational constraints to systems on the order of a few millions of atoms and a few tens of picoseconds.

1.2. Potential Energy Function

To study shock-induced chemistry via molecular dynamics, it is necessary to use a potential energy function which takes into account bond-breaking and bond-forming processes along with their associated energy changes. Early attempts at such a formulation were made by Tsai and Trevino [1-3] who used a compound Morse potential to describe the interaction between two atoms or molecular fragments labeled "A". In this model the A—A bond is metastable and will dissociate exothermically when stretched beyond a critical length. The potential function resembles a predissociation crossover between two adiabatic potential surfaces. (See figure 1.) This model provided a simple method for simulating energy release in a reactive molecular system. However, energy release is associated with bond breaking rather than bond forming, whereas the reverse is usually the case.

Another early attempt to incorporate chemical reactions into molecular dynamics of shock waves was the use of the LEPS (London, Eyring, Polanyi, Sato) potential [4], originally developed in the 1930's to model the H_3 potential energy surface. This method can be applied to systems in which each atom interacts with exactly two nearest neighbors, and is therefore suitable for modeling one-dimensional reactive chains [5-6]. It provides a more realistic treatment of energy release as a function of bond formation but is not readily extended to more complex systems.

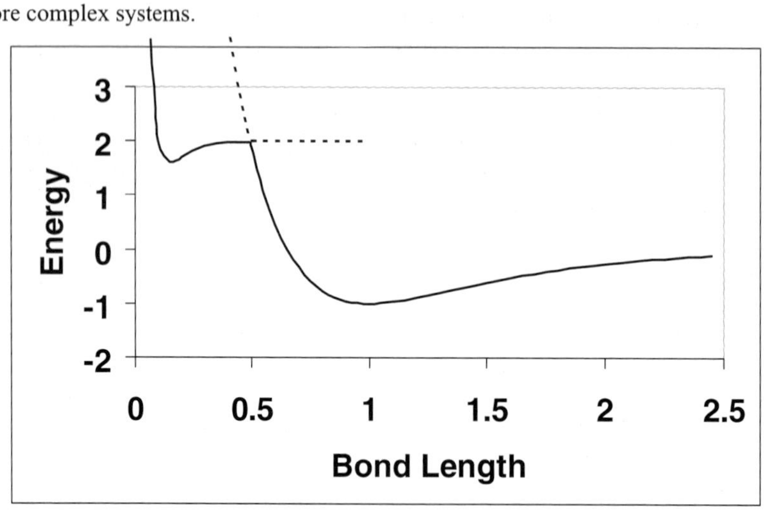

Figure 1. Compound Morse potentials of the form employed by Tsai and Trevino [1-3] to simulate energy release arising from chemical reaction. Energy and bond length are in arbitrary units.

In 1988, Tersoff [7] introduced an analytical expression for a many-body potential energy function based on bond order, which was able to accommodate reactive dynamics in a straightforward manner. In this formalism, the interaction energy E_{ij} between a pair of atoms i and j is given by

$$E_{ij} = V_R(r_{ij}) - B_{ij}V_A(r_{ij})$$

where V_R is a repulsive potential, V_A is an attractive bonding potential, and B_{ij} is a many-body term, which weakens the bonding energy as the local coordination of the atoms, i and j increases. This reactive empirical bond order (REBO) potential formalism was first applied to molecular dynamics simulations of shock phenomena using a simple diatomic model system in which the generic reaction

$$2AB \rightarrow A_2 + B_2$$

was chosen to be exothermic [6, 8-10]. The method was able to reproduce many of the features of a detonation.

The REBO formalism has since been extended to provide a realistic description of hydrocarbon systems [11, 12], including the hybridization states of the carbon atom and its corresponding geometric configurations, as well as accurate bond energies. The hydrocarbon REBO potential was first developed to study chemical vapour deposition on diamond [11] but has since been widely used for a variety of molecular dynamics simulations including the tribology [13] and fracture [14] of diamond and the dynamics of carbon nanotubes used as STM tips [15]. Even more recently, the hydrocarbon REBO potential has been extended [16] to include torsional effects and London dispersion forces. In this form, known as the adaptive intermolecular REBO (AIREBO) potential, it is suitable for use in condensed-phase molecular hydrocarbon systems. The AIREBO potential has been used [17-19] to study the tribology of diamond surfaces in the presence of anchored hydrocarbon chains. In addition, it is well-suited for the simulation of shock-induced chemistry in solid molecular hydrocarbons Some of those simulations will be described in this work.

2. SIMULATION RESULTS

We have carried out a number of molecular dynamics (MD) simulations on small hydrocarbon molecules in the solid phase using the AREBO potential. The studies were motivated, at least in part, by a desire to understand the possible chemical reactions of these materials under shock compression in planetary systems due to impact of comets or meteorites into planetary atmospheres. In addition, we were interested in learning more about the detailed kinetics of shock-induced chemistry in condensed phases, a subject in which little experimental information is available. For example, it is of some interest to examine the orientation dependence of shock impact, the spatial and temporal extent of the reaction zone, the time scale of energy thermalization, and the range of reaction products for each of the materials studied.

The results presented here include simulations of methane, acetylene, and anthracene. These systems span a range from one to fourteen carbon atoms, from isotropic to highly anisotropic molecules, and from saturated to unsaturated to aromatic organic structures. All three materials have some astrophysical significance. The results and discussion for each molecular system will be presented individually, and some general conclusions appear at the end of the article.

2.1. Methane

Methane is the most abundant hydrocarbon in the universe. In our solar system it is found in high abundance in the outer planets and their moons. For example, methane ice is thought to be a major constituent of the interiors of Neptune and Uranus. In that environment, methane may experience pressures up to 600 GPa (6 Mbar) and temperatures as high as 8000 K [20, 21]. As pressures and temperatures increase toward the center of those planets, methane will at some point begin to decompose. Experimental shock wave studies have been a major source of information on the fate of methane at high temperatures and pressures. Ree [22] proposed on the basis of extensive shock data on larger hydrocarbons that those molecules, and presumably hydrocarbons in general, decompose into carbon (primarily diamond) and molecular hydrogen under sufficiently extreme conditions. Shock studies on methane itself [23-25] confirmed this view and indicated that the decomposition threshold for methane occurs at about 23 GPa. More recent studies employing electrical conductivity measurements in two-stage light-gas gun shock experiments placed the decomposition threshold first above 42 GPa [26] and later at about 30 GPa [27]. Static high pressure experiments [28] using a laser-heated diamond anvil cell found a mixture of polymerized hydrocarbons, molecular hydrogen, and diamond in the pressure range between 10 and 50 GPa, with complete decomposition to carbon taking place only at the highest temperatures of the laser focus area ($\geq$ 1500 to 2000 K) at these pressures.

MD studies have also contributed to our understanding of the fate of methane at extreme temperature and pressure. First-principles simulations [29] which followed 16 methane molecules for two picoseconds at constant pressure found that pressures of 100 GPa were required for polymerization and that diamond formation occurred only above 300 GPa. Tight-binding simulations [30] of up to 1728 molecules for 1.2 ps found polymer formation at 43 GPa. The empirical AIREBO potential allows larger MD simulations for longer times, enabling better quantitative analysis of reaction products and contributing to our fundamental understanding of the initial reactions of methane decomposition.

AIREBO MD simulations of shock impact in solid methane were carried out [31] using a flyer plate configuration, in which a thin crystal of the material is allowed to strike a larger stationary crystal. Periodic boundary conditions were imposed in the directions transverse to the flyer plate velocity. Within the periodic boundaries, a cross section of 6x6 methane unit cells was included. The unit cell geometry matched the experimental data closely, but was optimized to give the most stable crystal at 0 K within the AIREBO potential model. The flyer plate impact velocity was varied to determine the threshold for reaction; above the threshold, product distribution could readily be analyzed in the rarefaction region behind the shock front. The reaction zone is quite narrow in these simulations since the impact energy of the flyer plate is dissipated as the shock wave moves through the target crystal.

For flyer plate impact speeds below 20 km/s, no appreciable reactivity is observed in methane. Near this threshold, the major reaction is hydrogen abstraction. At higher impact speeds some carbon-carbon bonds are formed. At 30 km/s impact speed, polymers up to eight carbon atoms long are observed. A snapshot of a simulation with a flyer plate impact speed of 25 km/s is shown in Figure 2. In this figure, the flyer plate has impacted the crystal from the left and the shock wave has propagated almost to the right-hand edge of the picture. C-C bonds are shown in red and H-H bonds in cyan, indicating the occurrence of chemical reactions. C-H bonds are not shown, so that atoms in unreacted methane molecules appear as dots (red for carbon, cyan for hydrogen).

Chemical reactions are essentially complete within 2 ps in these simulations, and product molecules retain their identity and integrity as they expand into a state of lower density following passage of the shock wave. Product distributions can be analyzed in the format of a mass spectrum, as shown in Figure 3. Here the relative abundances of products are shown for impact speeds of 20, 25, and 30 km/s, using a flyer plate with a thickness of six unit cells. The population of C_2 species (mostly C_2H_4) increases dramatically with increasing impact speed, and higher oligomers are apparent at 30 km/s. The population of free hydrogen atoms is higher than that of H_2 at all impact speeds, indicating that recombination of hydrogen atoms produced from separate abstraction reactions has not had time to occur on this timescale.

Profiles of the shock front can also be obtained at any time in the simulation by computing properties such as temperature and pressure as functions of longitudinal position along the propagation direction. Profiles of mass velocity, temperature, longitudinal stress (pressure) and density for the 25 km/s simulation, corresponding to the snapshot in Figure 2, are shown in Figure 4 for times up to 1 ps after flyer plate impact. It can be seen that this simulation produces reaction zone temperatures up to 9000 K, pressures up to 150 GPa, and a density of 1.0 g/cm^3, which is twice the initial density of solid methane at ambient pressure.

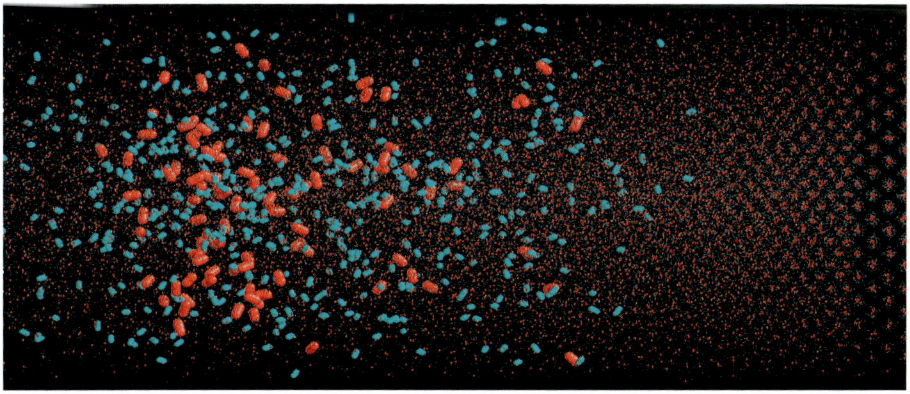

Figure 2. Snapshot of methane simulation with flyer plate impact from left at a speed of 25 km/s. Carbon-carbon bonds are shown in red, hydrogen-hydrogen bonds in cyan. Periodic boundary conditions are imposed along the two axes perpendicular to the shock direction. In this view the square periodic repeat cell has been rotated 45° around the propagation axis so that the material appears less dense along the top and bottom of the frame, making it easier to see the reaction products.

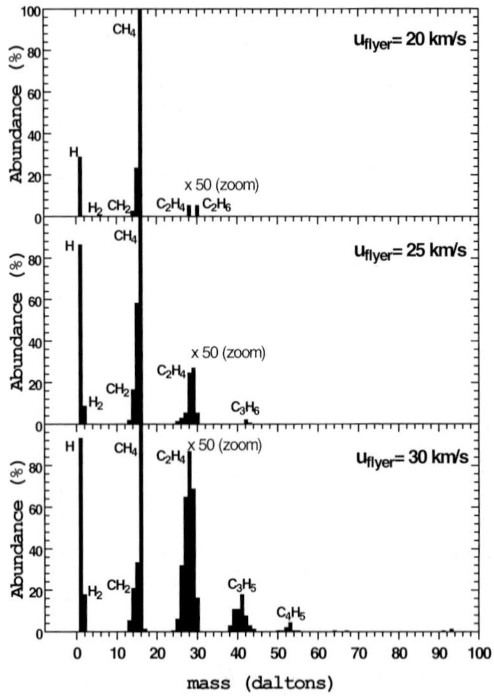

Figure 3. Product analysis for methane simulation at various flyer plate impact speeds. Abundances are relative to the parent CH_4 peak. Note that abundances for species with two or more carbon atoms have been scaled up by a factor of fifty.

The reaction threshold of 20 km/s impact speed found in these simulations corresponds fairly well with the threshold of 8.3 km/s piston speed reported by Kress et al. [30], since the flyer plate speed must be divided by two to obtain the equivalent piston velocity. Similarly, the reaction zone temperature of 6000 – 9000 K and pressure of 80 – 150 GPa seen in Figure 4 are in reasonable agreement with the values found in previous MD simulations of methane [29, 30] near the reaction threshold. All three MD simulations report polymerization thresholds somewhat higher than the experimental shock wave values, and this may be due in part to the difference in time scales of the two techniques. The MD studies follow molecular trajectories for times on the order of picoseconds, whereas experimental shock studies occur on time scales of nanoseconds to microseconds. Higher energies may be required to see significant reactivity on the shorter timescale of MD simulations. Two additional factors may contribute to the high threshold reaction pressure found in the AIREBO simulation [31]. First, in contrast to the two previous MD studies, this simulation uses a flyer plate configuration in which rapid rarefaction in the reaction zone limits the time available for product formation. Secondly, the AIREBO potential seems to overestimate intermolecular forces at high compression, leading to higher calculated pressure (along with lower density and higher shock speed).

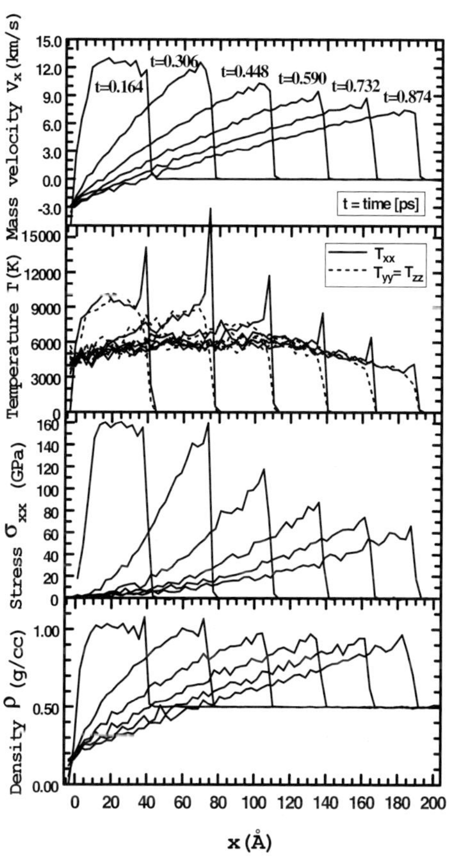

Figure 4. Profiles of mass velocity, temperature, longitudinal stress, and density as functions of longitudinal position at various times in methane simulation at 25 km/s flyer plate impact speed. The impact point of the flyer plate is at x=0, and the impact time is t=0.

Finally, the profiles of Figure 4 can also be analyzed in terms of the Rankine-Hugoniot relations

$$P = \rho_o \cdot v_p \cdot v_s$$

and

$$\rho_o / \rho = 1 - (v_p/v_s)$$

where P is the pressure or stress behind the shock front, ρ_o and ρ are the initial and shocked densities, v_p is the "piston" or mass velocity, and v_s is the shock wave speed. For the simulation depicted in Figure 4, P = 155 GPa, ρ_o = 0.50 g/cm^3, ρ = 1.00 g/cm^3, v_p = 12.5 km/s, and v_s = 24.5 km/s. These values can be seen to be consistent with the Rankine-Hugoniot relations above. In this context it should also be noted that linear extrapolation of the experimental Hugoniot curve for liquid methane [25] from the maximum measured mass velocity of 8.3 km/s up to v_p = 12.5 km/s would yield values of v_s and P significantly lower than those reported above. There is some uncertainty in this extrapolation, and of course it applies strictly to liquid rather than solid methane. To the extent that the difference is real, however, it may be an indication of the excessive "stiffness" of the inner repulsive wall of the AIREBO intermolecular potential mentioned earlier.

2.2. Acetylene

In contrast to methane, acetylene (C_2H_2) is anisotropic, highly unsaturated, and has a much lower 1:1 hydrogen: carbon ratio. The π bonding system allows for the possibility of facile addition polymerization reactions, and therefore acetylene is predicted to be much more reactive than methane upon shock compression. This is reflected in a significantly lower flyer plate velocity threshold for the onset of chemical reactions. Polymerization of acetylene is also exothermic; a simple bond energy calculation shows that the reaction

releases about 123 kJ per mole of acetylene. Although the energy release facilitates further reaction at the shock front, our simulations [31, 32] do not show evidence of sustained shock-induced detonation in solid acetylene.

Acetylene has been observed in the atmospheres of Jupiter and Titan [33, 34] and more recently has been identified in significant abundance in comet Hyakutake [35]. Following the discovery of acetylene in Hyakutake, photochemical experiments have demonstrated [36] that this molecule is a likely precursor of C_2, a widely observed component of comets. Acetylene itself may therefore be a ubiquitous constituent of comets. It has been proposed [37] that polymerization of acetylene in cometary impact on planetary atmospheres may be responsible for the formation of polycyclic aromatic hydrocarbons (PAHs) which may in turn be responsible for the colors of the atmospheres of Jupiter and Titan. Shock-induced polymerization of acetylene has been observed in the gas phase [38], and static high-pressure experiments have demonstrated polymerization of orthorhombic solid acetylene above 3 to 3.5 GPa at room temperature [39, 40], and above 12.5 GPa at 77 K [41].

MD simulations of shock-induced chemical reactions in solid acetylene using the AIREBO potential have been carried out [31, 32] using a flyer plate configuration similar to that described above for methane. Periodic repeat units up to 8x8 unit cells wide, and flyer plate thickness up to 24 unit cells, were employed in the simulations. Target crystals were made sufficiently long that all chemical reactions had ceased due to energy dissipation by the time the shock wave reached the edge of the crystal. Each simulation included up to 64512 atoms and dynamics were followed for up to 6 ps.

No chemical reactions were observed for flyer plate velocities below 10 km/s. For impacts at 10 and 12 km/s, only small oligomers of acetylene with even numbers of carbon atoms were found in the rarefaction region behind the shock front. This indicates that the dominant reaction mechanism was simple addition polymerization. At higher impact velocities, the population of reaction products showed a smooth monotonic decrease with increasing number of carbon atoms, with no preference for even-numbered carbon chain lengths. The maximum observed chain lengths increased with increasing flyer plate speed. Nominal chain lengths greater than about twenty carbons, however, which are observed at impact speeds at 25 km/s and higher, may be spurious because of "wrapping" across periodic boundaries. Simulated mass spectra of shocked acetylene in the reaction zone at flyer plate impact speeds of 12, 16, and 20 km/s are shown in Figure 5. Peak heights are percent abundances relative to the most abundant species. Peak heights for species with three or more carbon atoms are scaled up by a factor of 25. A flyer plate thickness of 6 unit cells was used for these simulations. Although some chemistry was still occurring in the rarefaction region, the chain length distributions were fairly constant by 2 ps after flyer plate impact. The abundances depicted in Figure 5 are these "final" product distributions.

Figure 6 shows the dependence of reactivity on the thickness of the incident flyer plate. The number of carbon atoms in chains of three or more carbons is chosen as a measure of "reacted" carbons, and the number of such carbon atoms per square Ångstrom of cross-sectional area is shown. A thicker flyer plate imparts more total kinetic energy and also lengthens the time over which the target crystal is maintained at maximum pressure before experiencing the release wave. Therefore it is not surprising that the fraction of carbon atoms

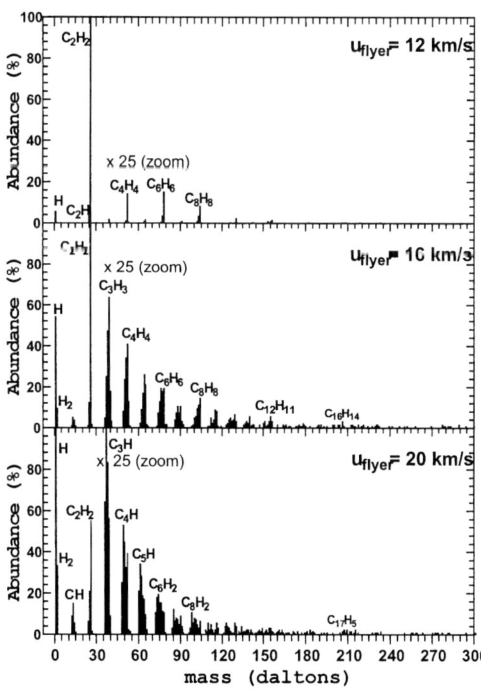

Figure 5. Simulated mass spectra (relative percent abundance vs. molecular mass) for shocked solid acetylene at flyer plate impact velocities of 12, 16, and 20 km/s.

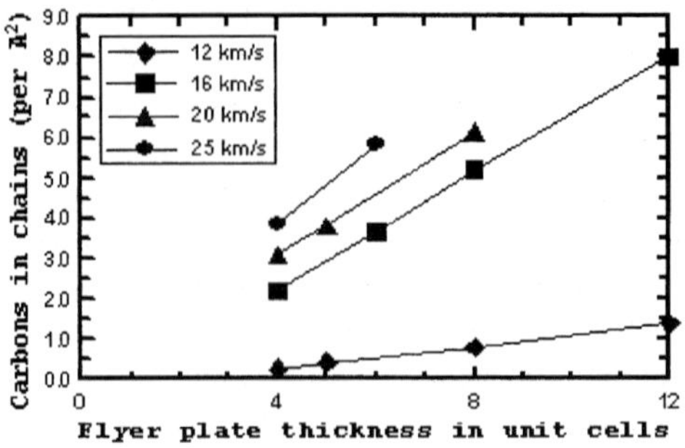

Figure 6. The number of carbon atoms found in chains of length three or more, per square Ångstrom of cross-sectional area, after shock impact in acetylene. The number of carbon atoms is plotted versus flyer plate thickness for various flyer plate impact speeds.

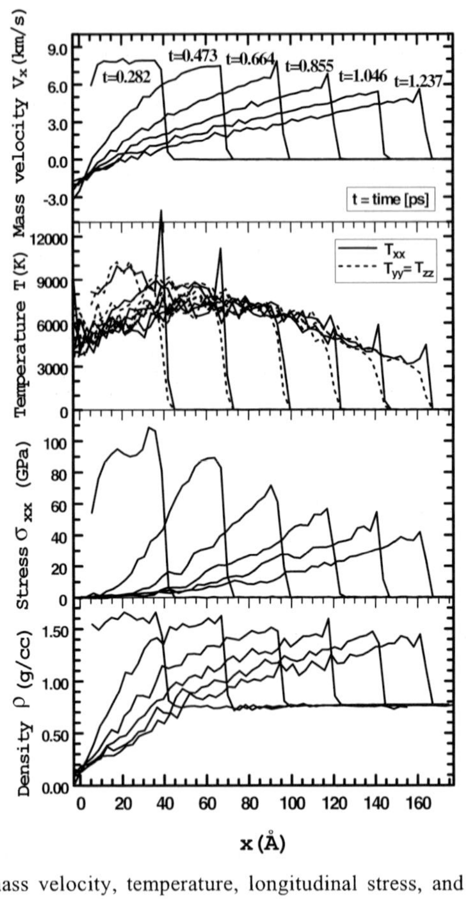

Figure 7. Profiles of mass velocity, temperature, longitudinal stress, and density as functions of longitudinal position at various times in acetylene simulation at 16 km/s flyer plate impact speed.

involved in reaction increases linearly with flyer plate thickness. The change in slope between 12 and 16 km/s impact speed is probably related to the change in reaction mechanism from simple addition to multistep fragmentation and recombination, which is evident in the product distributions of Figure 5.

Longitudinal profiles of shock wave properties in acetylene are shown in Figure 7 for a flyer plate impact speed of 16 km/s and a flyer plate thickness of six unit cells. Profiles for various times up to 1.2 ps after impact are depicted. At early times before the appearance of the release wave, when the mass velocity and density profiles are flat behind the shock front, it is possible to derive the parameters necessary for a Hugoniot analysis. As for methane, it is found that the Rankine-Hugoniot relations are satisfied. The Hugoniot parameters for several of the acetylene and methane simulations are collected in Table 1. Note that for a given flyer plate velocity, the temperature in the reaction zone is much higher for acetylene than for methane due to the exothermicity of the polymerization reactions.

Table 1. Hugoniot parameters for methane and acetylene from the AIREBO flyer plate impact simulations.

Molecular Crystal	v_{flyer} (km/s)	v_p (km/s)	v_s (km/s)	P (GPa)	ρ_o (g/cm^3)	ρ (g/cm^3)	T (K)
CH_4	16.0	8.0	17.0	70	0.50	0.90	4000
	20.0	10.0	20.4	100	0.50	0.95	6500
	25.0	12.5	24.5	155	0.50	1.00	9500
C_2H_2	12.0	6.0	12.3	56	0.75	1.50	4500
	16.0	8.0	15.2	90	0.76	1.60	9500
	20.0	10.0	18.5	140	0.76	1.70	16000
	25.0	12.5	21.8	210	0.76	1.85	24000

A snapshot of the same simulation at 1.5 ps is shown in Figure 8. As in the methane snapshot (Fig. 2) the shock wave is proceeding from left to right. In this figure, only carbon-carbon bonds for chains of three or more carbons are shown, so that C-C bonds of unreacted acetylene molecules do not obscure the view. It is evident that reactions have largely ceased in this simulation at 1.5 ps. (Carbon chains appearing just behind the shock front at the right of the figure are transitory "bonds" due to compression of the material but will not endure into the rarefaction zone.) The herringbone structure of the unreacted acetylene crystal at 0 K is visible ahead of the shock front.

The flyer plate velocity range employed in these simulations, from 8 to 25 km/s, is comparable to the speeds expected for cometary impact into planetary atmospheres [42]. It is clear from the simulations that significant shock-induced polymerization can occur in cometary acetylene under these conditions. Ring structures characteristic of polycyclic aromatic hydrocarbons (PAHs) were not found in the simulations, but unsaturated oligomer precursors would presumably react to form these more complex structures over a much longer time interval than could be followed in the shock impact simulations.

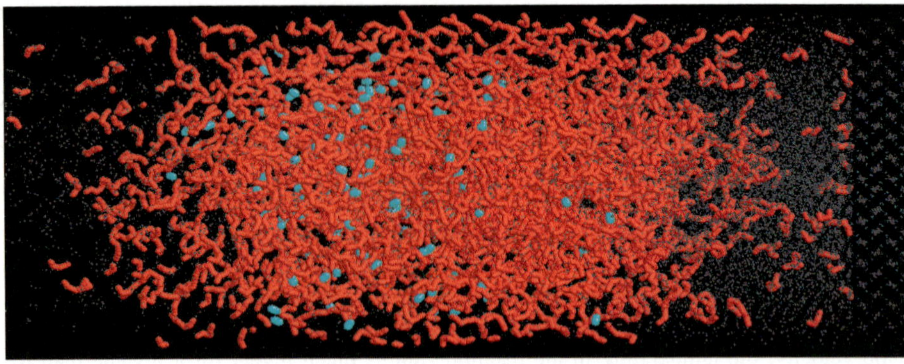

Figure 8. Snapshot of acetylene simulation at a flyer plate speed of 16 km/s, 1.5 ps after impact. As in Figure 2, carbon-carbon bonds are shown in red and hydrogen-hydrogen bonds in cyan, but in this figure only carbon-carbon bonds in chains of three or more carbons are depicted. The shock wave is proceeding from left to right and is visible near the right edge of the figure. Also as in Figure 2, the periodic repeat cell has been rotated 45° along the propagation axis so that the material appears less dense near the top and bottom of the snapshot.

2.3. Anthracene

Whether formed from acetylene or from some other sources, PAHs are widely distributed in the solar system. As mentioned earlier, PAHs are found in the atmospheres of Jupiter and Titan [37]. They have also been detected in meteorites, including the Martian meteorite Allan Hills 84001 [43], in interplanetary dust [44], and in circumstellar graphite grains [45]. The ubiquity of these complex organic structures and their stability under extreme conditions is a significant factor in discussions of the origin of life on earth and the possibility of its existence elsewhere.

To study the possible shock-induced chemistry of PAHs we have performed MD simulations [46] on the representative system anthracene, $C_{14}H_{10}$, composed of three fused benzene rings. Anthracene crystallizes in a monoclinic structure with the long axis of the molecules aligned close to the *c* crystallographic axis [47]. The structure therefore shows a high degree of anisotropy, leading to the possibility of orientation dependence of the shock-induced chemistry. As in the methane and acetylene simulations described above, the molecular positions and unit cell parameters were adjusted slightly to find the minimum-energy configuration within the AIREBO model potential. The resulting crystal structure is shown in Figure 9. For the anthracene simulations, two equal-sized crystal slabs were launched toward each other at a given relative velocity so that, upon impact, shock waves propagated through the material to both left and right, leaving the center of mass of the system unchanged. Simulations were performed with the shock propagation direction along the *a* and *c* crystallographic axes. The initial system configuration for shock impact along *a* is shown in Figure 10. Because of the monoclinic crystal structure, the shock front is not orthogonal to the shock propagation direction in these simulations.

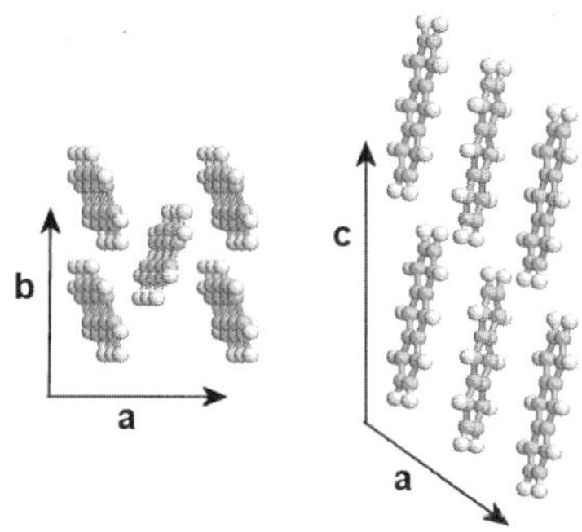

Figure 9. Crystal structure of anthracene. The unit cell parameters used for the AIREBO simulations were: $a = 0.848$ nm, $b = 0.604$ nm, $c = 1.157$ nm, _ = 123.0°.

The anthracene simulations included 40320 atoms or 1680 anthracene molecules and were continued for a time period of at least 5 ps. Periodic boundary conditions were employed as usual, but because the repeat directions were along the crystallographic axes, one of the periodic axes was not perpendicular to the shock direction.

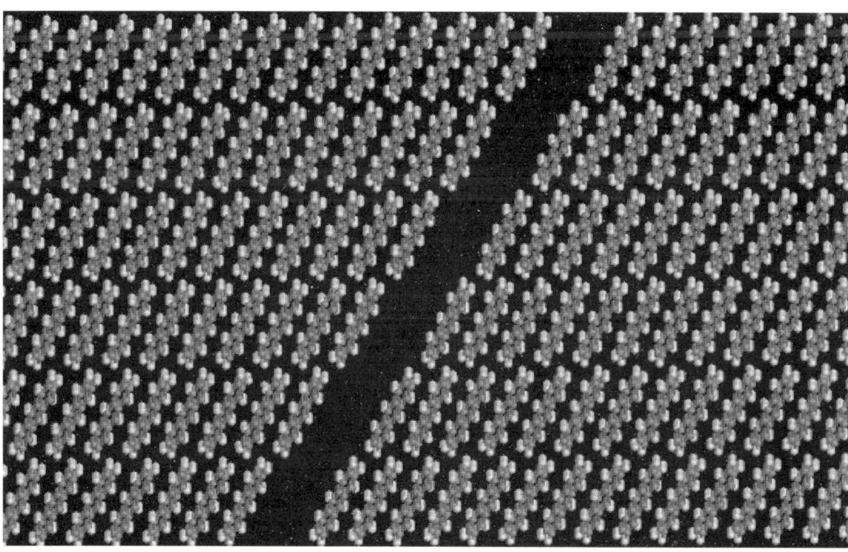

Figure 10. Initial anthracene crystal configuration for shock impact along *a* axis. Crystal segments are projected toward each other along the horizontal figure axis, closing the gap and initiating shock compression.

At a relative impact velocity below 8 km/s, no shock-induced chemical reactions were observed for anthracene in the simulations, although interesting orientation dependence of the shock wave structure and energy distribution was found in this regime [48]. Above the threshold, both fragmentation and dimerization reactions were observed. The peak density and pressure in the impact region were approximately 1.8 g/cm^3 and 3000 K, respectively, near the reaction threshold at 8 km/s. These values increased to 2.0 g/cm^3 and 7000 K at 12 km/s, as shown in Figure 11. Anthracene reactivity increased dramatically at higher impact velocities.

Warnes [49] observed a discontinuity in a plot of shock velocity vs. particle velocity for anthracene at a density of 1.8 g/cm^3, which he attributed to an intermolecular coupling reaction. More recently Engelke and Blais [50] have found direct evidence of dimerization in mass spectral analysis of shocked anthracene at 18 GPa. That pressure corresponds closely to a density of 1.8 g/cm^3 according to the pressure-volume Hugoniot data of Warnes [49]. This in turn matches the reaction threshold found in the AIREBO MD simulations.

To analyze the reaction products of shocked anthracene in more detail, the connectivity of carbon clusters was determined for each simulation 5 ps after impact, when temperature and pressure had subsided from peak values and most chemical reactions were complete. Hydrogens were ignored in this analysis, so that a properly connected cluster of 14 carbon atoms was regarded as unreacted anthracene whether or not it contained 10 hydrogen atoms. In fact, single hydrogen abstractions were fairly common, amounting to about 10% of "unreacted" anthracenes in the impact region at 10 km/s and 20% at 12 km/s, but the loss of multiple hydrogens was rare.

A snapshot of the impact region is shown in Figure 12 for a relative velocity of 12 km/s, with the shock direction along the crystallographic *a* axis as depicted in Figure 10. In this figure, unreacted anthracene molecules have been removed and only carbon atoms are shown. The diagonal orientation of the reaction zone, resulting from the crystal configuration shown in Figure 10, is apparent. A simulated mass spectrum (carbon atoms only) is presented in Figure 13 showing the distribution of reaction products for impact speeds of 10 km/s and 12 km/s, again with the shock direction along the *a* axis. Total reactivity is seen to increase markedly in going from the lower to the higher of these velocities. Dimerization (to 28 carbon

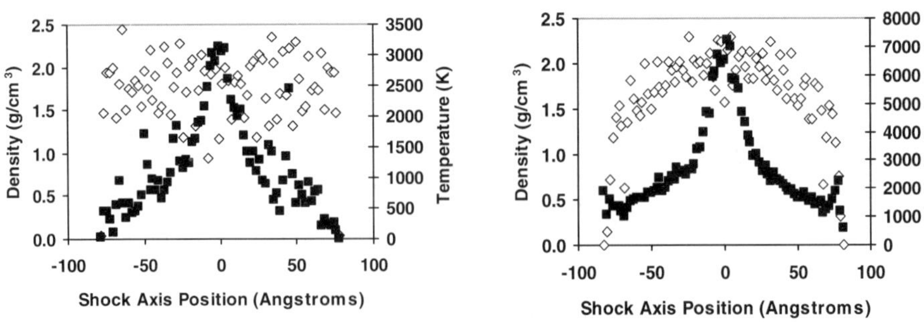

Figure 11. Density (open diamonds) and temperature (closed squares) profiles for the shock impact region in anthracene simulations at 8 km/s (left) and 12 km/s (right), 1.0 ps after impact of the two crystal plates.

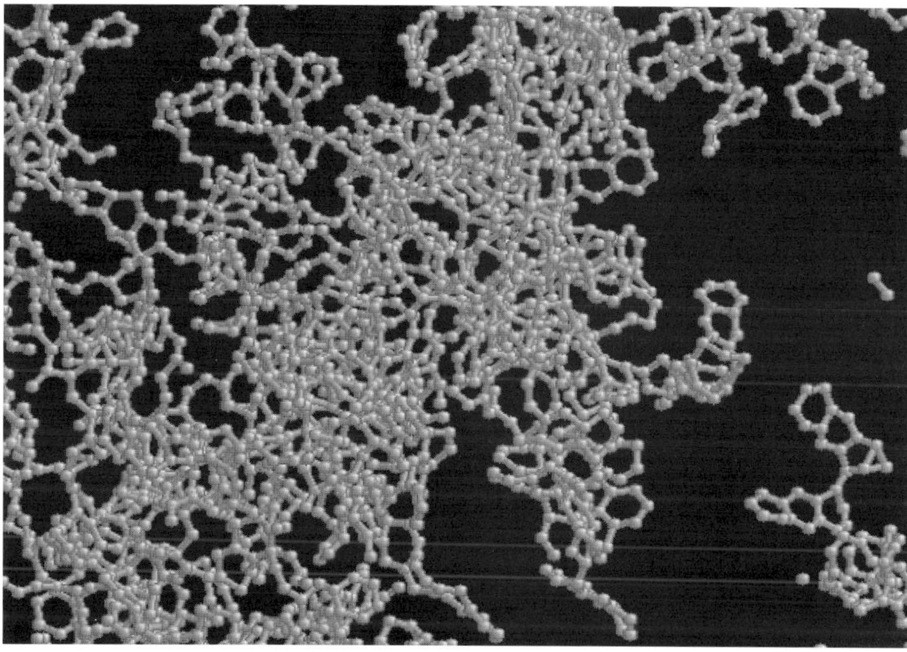

Figure 12. Snapshot of impact region after collision of anthracene crystals at a relative speed of 12 km/s, with shock direction along *a* axis. Unreacted anthracene molecules have been deleted, and hydrogen atoms are not shown.

atoms) and fragmentation are the primary reaction channels. At 12 km/s a few higher anthracene oligomers (not shown) are also formed.

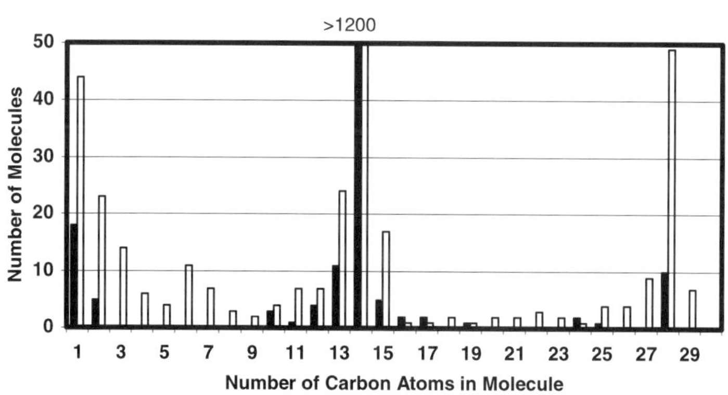

Figure 13. The number of molecules containing a given number of carbon atoms for anthracene shock simulations with shock direction along the *a* axis at impact velocities of ten (solid bars) and twelve (open bars) km/s. The peaks for unreacted anthracene molecules (14 carbons) are far off scale with more than 1200 molecules.

Orientation dependence of shock-induced chemistry in anthracene was investigated by comparing simulations in which the shock direction was oriented along the *a* crystallographic axis, as in Fig. 10, with simulations where the shock direction was along *c*. Figure 14 is a comparison of product distributions for anthracene shock simulations in the *a* and *c* directions at 12 km/s. For shocks along *c* the anthracene molecules strike each other nearly along the long molecular axis, or "end-on." Although dimerization and fragmentation continue to be the dominant reaction channels, end-on collisions produce substantially more small fragments and fewer simple dimers than do shock impacts along the *a* direction.

These anthracene simulations indicate that PAHs of moderate size may survive and even undergo polymerization reactions under shock impact conditions to be expected in cometary impacts on planetary atmospheres. Preliminary simulations of shock-induced chemistry in naphthalene [51] suggest a similar reaction threshold for the smallest PAH as well.

3. CONCLUSION

MD simulations using a robust empirical potential can provide valuable information on the behavior and reactivity of shocked materials at extreme conditions. The AIREBO potential used in the simulations described here is much faster to evaluate than first-principles methods, enabling the simulation of many thousands of molecules for chemically relevant time periods of several picoseconds. At the present time this potential is limited to hydrocarbons, primarily because the inclusion of other atoms types with substantially different electronegativities would require the addition of electrostatic terms. Such extensions are currently being pursued by several groups, however, and within a few years it is likely that

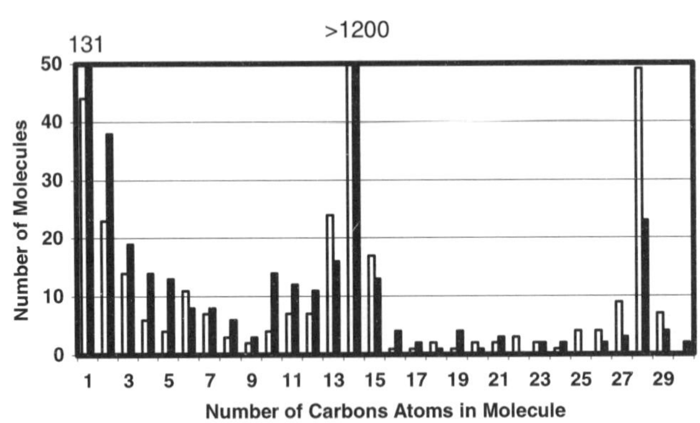

Figure 14. Product distributions for anthracene shocked along the crystallographic *a* axis (open bars) and *c* axis at a relative impact speed of 12 km/s. The *c* axis peak for single carbon atoms is off-scale at 131, and the number of unreacted anthracene molecules (14 carbons) is far off-scale for both impact directions.

MD simulations of shock-induced chemistry for condensed-phase explosives, and for biologically and astrophysically relevant molecules such as amino acids, will be attainable. Such simulations will provide a unique contribution to the understanding of complex chemical phenomena occurring at ultrafast timescales under experimentally challenging conditions of high temperature and pressure.

ACKNOWLEDGEMENT

The MD studies described here were made possible by the consistent support of the Office of Naval Research. MLE received additional funding from the USNA/NRL Cooperative Program for Scientific Interchange and the Naval Academy Research Council.

REFERENCES

[1] D. H.Tsai and S. F. Trevino, J. Chem. Phys., 79 (1983) 1684.
[2] S. F. Trevino and D. H. Tsai, J. Chem. Phys., 81 (1984) 248.
[3] D. H. Tsai and S. F. Trevino, J. Chem. Phys., 81 (1984) 5636.
[4] J. Tully, J. Chem. Phys., 73 (1980) 6333.
[5] M. L. Elert, D. M. Deaven, D. W. Brenner, and C. T. White, Phys. Rev. B, 39 (1989) 1453.
[6] M. L. Elert, D. W. Brenner, and C. T. White, Shock Compression of Condensed Matter – 1989, S. C. Schmidt, J. N. Johnson, L. W. Davison (eds.), Elsevier Science Publishers B. V. (1990) 275.
[7] J. Tersoff, Phys. Rev. B, 37 (1988) 6991.
[8] D. W. Brenner, M. L. Elert, and C. T. White, Shock Compression of Condensed Matter – 1989, S. C. Schmidt, J. N. Johnson, L. W. Davison (eds.), Elsevier Science Publishers B. V. (1990) 263.
[9] C. T. White, D. H. Robertson, M. L. Elert, and D. W. Brenner, Microscopic Simulations of Complex Hydrodynamic Phenomena, M. Mareschal and B. L. Holian (eds.), Plenum Press (1992) 111.
[10] D. W. Brenner, D. H. Robertson, M. L. Elert, and C. T. White, Phys. Rev. Lett., 70 (1993) 2174.
[11] D. W. Brenner, Phys. Rev. B, 42 (1990) 9458.
[12] D. W. Brenner, O. A. Shenderova, J. A. Harrison, S. J. Stuart, B. Ni, and S. B. Sinnott, J. Phys.: Condens. Matter, 14 (2002) 783.
[13] J. A. Harrison, C. T. White, R. J. Colton, and D. W. Brenner, Phys. Rev. B, 46 (1992) 9700.
[14] O. A. Shenderova, D. W. Brenner, A. Omeltchenko, X. Su, and L. H. Yang, Phys. Rev. B, 61 (2000) 3877.
[15] J. A. Harrison, S. J. Stuart, D. H. Robertson, and C. T. White, J. Phys. Chem. B, 101 (1997) 9682.
[16] S. J. Stuart, A. B. Tutein, and J. A. Harrison, J. Chem. Phys., 112 (2000) 6472.
[17] A. B. Tutein, S. J. Stuart, and J. A. Harrison, J. Phys. Chem. B, 103 (1999) 11357.
[18] A. B. Tutein, S. J. Stuart, and J. A. Harrison, Langmuir, 16 (2000) 291.
[19] P. T. Mikulski and J. A. Harrison, J. Am. Chem. Soc., 123 (2001) 6873.
[20] W. B. Hubbard, Science, 214 (1980) 145.
[21] D. J. Stevenson, Ann. Rev. Earth Planet Sci., 14 (1982) 257.
[22] F. H. Ree, J. Chem. Phys., 70 (1979) 974.
[23] M. Ross and F. H. Ree, J. Chem. Phys., 73 (1980) 6146.
[24] W. J. Nellis, A. C. Mitchell, M. Ross, and M. van Thiel, High Pressure Science and Technology, vol. 2, B. Vodar and Ph. Marteau (eds.), Pergamon, Oxford (1980) 1043.
[25] W. J. Nellis, F. H. Ree, M. van Thiel, and A. C. Mitchell, J. Chem. Phys., 75 (1981) 3055.
[26] H. B. Radousky, A. C. Mitchell, and W. J. Nellis, J. Chem. Phys. 93 (1990) 8235.
[27] W. J. Nellis, D. C. Hamilton, and A. C. Mitchell, J. Chem. Phys., 115 (2001) 1015.
[28] L. R. Benedetti, J. H. Nguyen, W. A. Caldwell, H. Liu, M. Kruger, and R. Jeanloz, Science, 286 (1999) 100.
[29] F. Ancilotto, G. L. Chiarotti, S. Scandolo, and E. Tosatti, Science, 275 (1997) 1288.
[30] J. D. Kress, S. R. Bickham, L. A. Collins, and B. L. Holian, Phys. Rev. Lett., 83 (1999) 3896.

[31] M. L. Elert, S. V. Zybin, and C. T. White, J. Chem. Phys., 118 (2003) 9795.
[32] M. L. Elert, S. V. Zybin, and C. T. White, Shock Compression of Condensed Matter – 2001, M. D. Furnish, N. N. Thadhani, and Y. Horie, (eds.), AIP Press (2002) 1406.
[33] K. S. Noll, R. F. Knacke, A. T. Tokunaga, J. H. Lacy, S. Beck, and E. Serabyn, Icarus, 65 (1986) 257.
[34] A. Coustenis, B. Bezard, and G. Gautier, Icarus, 80 (1989) 54.
[35] T. Y. Brooke, A. T. Tokunaga, H. A. Weaver, J. Crovisier, D. Bockelée-Morvan, and D. Crisp, Nature, 383 (1996) 606.
[36] O. Sorkhabi, V. M. Blunt, H. Lin, M. F. A'Hearn, H. A. Weaver, C. Arpigny, and W. M. Jackson, Planet. Space Sci., 45 (1997) 721.
[37] C. Sagan, B. N. Khare, W. R. Thompson, G. D. McDonald, M. R. Wing, J. L. Bada, T. Vo-Dinh, and E. T. Arakawa, Astrophys. J., 414 (1993) 399.
[38] J. N. Bradley and G. B. Kistiakowsky, J. Chem. Phys., 35 (1961) 264.
[39] K. Aoki, S. Usuba, M. Yoshida, Y. Kakudate, K. Tanaka, and S. Fujiwara, J. Chem. Phys., 89 (1988) 529.
[40] K. Aoki, Y. Kakudate, M. Yoshida, S. Usuba, K. Tanaka, and S. Fujiwara, Synth. Met., 28 (1989) D91.
[41] C. C. Trout and J. V. Badding, J. Phys. Chem. A, 104 (2000) 8142.
[42] C. F. Chyba, P. J. Thomas, L. Brookshaw, and C. Sagan, Science, 249 (1990) 366.
[43] T.Stephan, C. H. Heiss, D. Rost, and E. K. Jessberger, Lunar Planet. Sci., 30 (1999) 1569.
[44] S. J. Clemett, C. R. Maechling, R. N. Zare, P. D. Swan, and R. M. Walker, Science, 262 (1993) 721.
[45] S. J. Clemett, S. Messenger, X. D. F. Chillier, X. Gao, R. M. Walker, and R. N. Zare, Lunar Planet. Sci., 27 (1996) 229.
[46] M. L. Elert, S. V. Zybin, and C. T. White, Shock Compression of Condensed Matter – 2003, M. D. Furnish and Y. Gupta, (eds.), AIP Press (2004), to be published.
[47] A. I. Kitaigorodskii, Organic Chemical Crystallography, Consultants Bureau, New York (1961) 420.
[48] S. V. Zybin, M. L. Elert, and C. T. White, Shock Compression of Condensed Matter – 2003, M. D. Furnish and Y. Gupta, (eds.), AIP Press (2004), to be published.
[49] R. H. Warnes, J. Chem. Phys., 53 (1970) 1088.
[50] R. Engelke and N. C. Blais, J. Chem. Phys., 101 (1994) 10961.
[51] M. L. Elert, S. M. Revell, S. V. Zybin, and C. T. White, unpublished.

Chemistry at Extreme Conditions
M. Riad Manaa (Editor)

Chapter 13

At the Confluence of Experiment and Simulation: Ultrafast Laser Spectroscopic Studies of Shock Compressed Energetic Materials

D. S. Moore, D. J. Funk, and S. D. McGrane

Los Alamos National Laboratory, Los Alamos, NM 87545 USA

ABSTRACT

Large-scale molecular dynamics simulations are producing information on shock-induced reactions on picosecond (ps) to nanosecond (ns) time scales and approaching micron spatial scales. We describe experiments using ultrafast laser methods to produce experimental data on similar time and space scales to help benchmark the simulations as well as motivate their expansion to larger scales and more complicated materials.

1. INTRODUCTION

A molecular description of detonation, particularly initiation, has been pursued for decades, with little success. One difficulty has been obtaining high-quality data at the appropriate length and time scales and with molecular specificity. What are the appropriate time and space scales? Detonation waves have typical velocities of 6-8 km/s, or equivalently 6-8 nm/ps. Recent molecular dynamics studies suggest that reactions in shocked energetic materials can occur in times as short as a few ps [1-6]. Energy transfer studies on molecular systems also reveal similar fast time scales [7-11]. Therefore, appropriate spectroscopic probes should have ps or better time resolution. Also, shock rise time measurements with sub-ps resolution require samples with surface uniformity better than 6-8 nm over the probed area.

To date, traditional shock production methods have not provided the required short time resolution, because of the difficulty in achieving time synchronicity between the shock arrival and the spectroscopic probe. Still, certain types of experiments have inferred the time response in a shocked sample, such as was achieved using coherent Raman probes of shocked molecular liquids [12-18]. The immediate appearance of molecular hot bands in the shocked material was used to estimate an upper bound on vibrational relaxation times of a few ns. The bandwidth of the molecular vibrational features was used to measure ps-scale vibrational phase relaxation times, assuming homogeneous broadening. These experiments, however, did not provide sufficient time resolution to answer important questions regarding the mechanism of shock loading, or the mechanism of energy transfer from the shock into molecular vibrations.

One method to achieve the requisite short time resolution is to utilize laser pulses to both initiate the shock process and probe the shocked material. Lasers have been used for decades to drive shocks. [19-33] Direct laser drive is being pursued to achieve fusion, but has not yet succeeded. Nevertheless, such studies produced a wealth of important data on the mechanisms of interaction of high power laser pulses with solid targets. Other researchers have tried to use lasers to initiate high explosives, either directly [34-37] or via the launch of a flyer [38-39]. Again, these studies provided very important background information for the work described in this chapter, which will focus on shock wave studies utilizing table-top ultrafast laser systems.

There were several early lower pressure (few GPa) interferometric and optical studies using ps laser drive on thin samples (tens of μm sandwiched between thick windows), and also using confined laser ablation as the shock drive mechanism [19-26]. These researchers showed that a Michelson interferometer with the sample in one leg could be used to infer the time-dependent pressure in the shocked sample. The decaying nature of their shocks (triangular or saw-tooth shape in time) due to the short laser pulse used was problematic.

Dlott and his coworkers have extended this early work to the tens of ps time resolution range using both ablation methods and photo induced reaction of a decomposable material to drive a shock through a very thin layer of a sample (nano-gauge) sandwiched between an impedance matched material [27-30]. Both Dlott et al. [27-30] and Campillo et al. [21-26] used targets with large surface areas so that they could be rasterred perpendicular to the shock axis between events, exposing fresh material for the next shock.

There have also been a large number of spectroscopic studies of shocked molecular materials. Some of these have involved UV/visible emission and absorption, where the observables have been band edge shifts with pressure and broad, sometimes non-thermal, emission bands [40-48]. The molecular-specific information content in such UV/visible spectra is limited, so that a number of researchers have utilized instead vibrational spectroscopic probes. Both Campillo et al. and Dlott et al. have monitored vibronic absorption or fluorescence structure in laser dyes used as probes of local shock structure and shock strength [22-30]. Spontaneous Raman has been used to measure temperature (via Stokes/anti-Stokes intensity ratios) in shocked energetic materials, as well as in attempts to identify intermediates in shock-induced reactions [49-59]. Time-resolved infrared spectral photography (TRISP) was used in a similar way [60-61]. In these studies, the time resolution achieved was in the ns to μs regime, because typically 0.1 to 1 mm thicknesses of shocked materials were probed at some time after the shock had traversed the sample, or during and after shock ring up to a steady state. Dlott et al. have utilized coherent Raman techniques both in nano-gauges to measure shock rise times in molecular materials [27-29, 62-63], and to determine dynamics of molecules and polymers under shock loading [7-11, 27, 30].

We have utilized this wealth of background information in the design of our experiments. We use ultrafast laser techniques to drive sustained shocks into thin films of energetic materials, which are then interrogated using several different kinds of ultrafast spectroscopic and interferometric probes. The remainder of this chapter will describe these experiments in detail, especially the ultrafast laser shock production and characterization methods and the spectroscopic and interferometric anomalies caused by working with thin films, and present recent spectroscopic evidence for the initial chemical reactions that occur directly behind the shock wave.

2. SAMPLE PREPARATION AND EXPERIMENTAL DESIGN

Our desired sub-ps temporal resolution places very strict requirements on the target design and fabrication. A typical shock velocity of 5-8 km/s, or equivalently 5-8 nm/ps, implies a requirement of nm scale target surface uniformity to achieve sub-ps temporal resolution in our experiments. We settled on an experimental design that is illustrated in Fig. 1. The targets are based on thin glass substrates (microscope cover slips) 120-150 μm thick (Fisher Scientific). A metal layer used for shock production is vapor plated onto this substrate. We have tried several thicknesses and kinds of metal. We primarily utilize aluminum because it has a sufficiently short electron-phonon coupling time (chromium and nickel have shorter times), it has well-known shock properties, and it has a relatively small shock impedance mismatch (compared to nickel and chromium) to organic materials. The shock-driving laser is focused through the substrate and is partially absorbed in the skin depth of the Al. A variety of processes occur, including multiphoton and avalanche ionization, plasma formation, time-dependent plasma optical density, plasma expansion and ion heating, electron-electron relaxation, and electron-phonon coupling. [64-66].

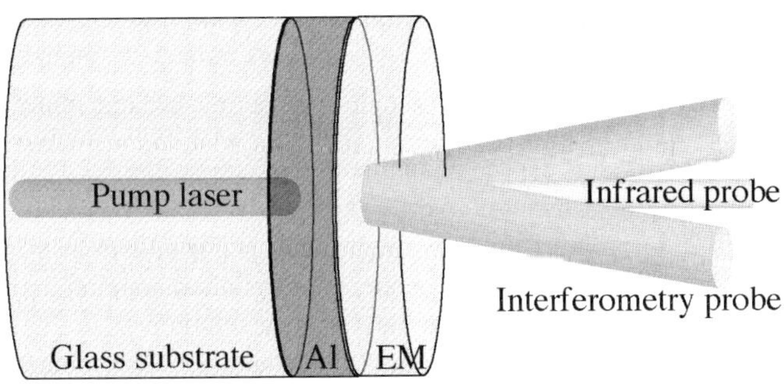

Figure 1: Cross sectional representation of a portion of the target, showing the aluminum layer on a thin glass substrate and an energetic material layer on top of the aluminum. The pump laser is focused through the glass substrate onto the Al interface, launching a shock that runs through the Al towards the probe laser beams. The probe beams pass through the energetic material, reflect from the Al, and pass again through the energetic material and on to the detector. Typical sizes are: Substrate 22 mm diameter 120-150 μm thick; Al 0.25-2 μm thick, EM 100-1000 nm thick; pump laser 0.2-6 mJ, shaped pulse (see text), ~100 μm focus diameter; Interferometry probe < 1 μJ, 120-170 fs, ~500 μm focus diameter; Infrared probe ~40 μm diameter (see text)

The shock diagnostic laser pulses typically examine the aluminum surface opposite the substrate. If we call the substrate side the front of the target, then the shock emerges from the back at some time after the drive pulse is absorbed at the front of the aluminum layer. The aluminum layer thickness and the shock velocity determine the shock transit time. Each shock driving laser pulse (ca. 100 μm in focal diameter at the target) destroys the sample (after times much longer than our experiments) at that particular location. A series of experiments can be performed by rastering the sample transverse to the shock direction to a fresh sample area. Our samples are typically 22 mm in diameter, and the extent of damage around the focal spot is usually limited to ca. 100-400 μm, so that rastering 300-600 μm between shots allows many hundreds to thousands of experiments to be performed on a single target.

This experimental design implies that each shock event produces a single snapshot in time of the surface position and/or infrared absorption spectrum (or other spectroscopic probe) at a given shock-driving laser / probe laser delay time. The complete time history can be built up stepwise by performing multiple such experiments at a sequence of delay times. This requirement – of repetitively shocking and measuring – implies that the target must be uniform, especially in aluminum layer thickness (and energetic material layer thickness) over the entire target, to the few nm level. Aluminum layer thickness variations contribute directly to the temporal uncertainty because of the variation of shock transit time and therefore appearance of the shock at the front of the target. Laser energy variations from shot to shot also contribute to temporal uncertainty because of the correlation between shock velocity and drive laser energy. This latter temporal uncertainty contribution could, in principle, be accounted for by measuring the drive laser energy for each shock event.

For experiments designed to study chemical processes, the energetic or other reactive materials are coated in a thin layer on the back (free) surface of the vapor-plated aluminum layer. The shock produced by laser absorption and confined expansion of the plasma at the substrate/aluminum surface at the back of the target transits the aluminum and then is transmitted into the organic material layer, sending a release wave back through the aluminum layer. These thin films have been produced, for the polymeric materials reported on here, using spin-casting methods. The films are spin cast at 2500 rpm for 5-25 s from various low volatility solvents at several concentrations depending on the film thickness desired. PMMA is cast from a toluene solution of 7.5-10% by mass PMMA (Acros, MW 93,300) with ~0.5% surfactant (BASF Pluronic L-62) to promote uniform smoothness. NC is cast from magic solvent (50% MEK, 20% 2-pentanone, 15% n-butyl acetate, 15% cyclohexanone) at 2-10% polymer concentration (by mass). PVN is cast from n-butyl acetate solutions of 4-12% by mass. The thickness of the films and surface uniformity are monitored using both null ellipsometry (Rudolph Research, AutoEL) at helium neon laser (632.8 nm) wavelength and 70^0 incidence, and white light reflectometry spectral interference fringe analysis (Filmetrics). Both methods of film thickness measurement are sensitive to changes as small as a nanometer, and the best films produced have surface thickness variations of ≤ 1% across most of a 2 cm diameter sample.

Spin casting methods are used extensively in the microelectronics industry. We have adopted the best practices available to achieve full density thin films. The full density (ρ_0 = 1.186 g/cm^3) is used for PMMA, justified by measurement of the full refractive index at 632.8 nm (n = 1.49) of our spin coated thin films to within Δn = 0.02. The density of PVN was measured by gas pychnometry to be ρ_0 = 1.34 g/cm^3, somewhat different from the literature [67]. Although there is no literature value for n, the measured n = 1.50 ± 0.03 compares well with the precursor polyvinylalcohol n = 1.52 [68]. An uncertainty in n of ± 0.03 translates into a density uncertainty of 5% via the Lorentz-Lorenz equation, essentially $\rho \propto (n^2\text{-}1)/(n^2\text{+}2)$. Although the samples are likely to be full density, up to 5% porosity may be present (note that pores would have to be submicron due both to the submicron film thickness and to avoid observation via the interferometric microscopy performed on the samples – see below).

3. CONTROLLED SHOCK PRODUCTION USING ULTRAFAST LASERS

While femtosecond lasers previously have been successfully employed to generate ultrafast shocks, the pressure typically rises sharply over a few picoseconds then decays quickly over tens of picoseconds [69-71]. This property makes such laser pulses extremely valuable for micromachining via photoablation [72-73], but the highly transient nature and time dependent pressure of femtosecond shocks makes their use in studying physics of shock compressed materials problematic.

In general, a shock driving pulse's time dependent intensity profile must be controlled to obtain a simple time dependent pressure profile in the shocked sample. The actual pulse shape required depends on the details of the optical and physical processes involved in transforming the light pulse into material motion. These processes include the time dependent absorption due to multiphoton and avalanche ionization processes, electron-electron coupling, plasma production, time-dependent plasma optical density, plasma expansion, energy transfer, and electron-phonon coupling [64-66]. While the laser pulse is typically a Gaussian or sech^2 function of time, the desired pressure profile for studies of shock compressed materials is often a step wave, with a very sharp leading edge and a relatively long time at uniform pressure. Therefore, a driving laser pulse shape is needed that can produce simultaneously <u>both</u> the fast pressure onset required to resolve picosecond molecular- or phonon-mediated dynamics <u>and</u> the sustained constant pressure desired for simplifying analysis and for pressurizing material thicknesses exceeding tens of nanometers.

We developed a method of generating shocks using temporally shaped pulses that can be simply implemented in common tabletop chirped pulse amplified lasers [74]. Driving a shock using a chirped, amplified pulse with the temporally leading (in our case, reddest) spectral range bluntly removed results in a pressure rise time of <10-20 ps (after transiting 0.5-2 μm of aluminum) and a sustained constant pressure for a few hundred picoseconds. Since these spectrally-modified pulses are generated by stretching a 100 fs seed pulse, a fraction of the amplified pulse can be recompressed to < 200 fs for use in probing the effects of shock loading.

In our embodiment of this idea, a Ti: sapphire femtosecond laser provides the seed pulse for a chirped pulse amplifier. The seed pulse is centered at 800 nm with a bandwidth full width at half max (FWHM) of ~9.5 nm and transform limit pulse length of ~90 fs (sech^2). The chirped pulse amplifier utilizes grating based pulse stretcher and compressor, with a regenerative amplifier and two stages of further amplification, to produce up to 50 mJ per pulse at 10 Hz repetition rate, and, usually, ~110 fs pulse length. Halfway through the stretcher, where the seed pulse is spectrally dispersed, amplitude and/or phase modulation of the spectrum can be used to produce a variety of pulse shapes [75-76]. At this point, we simply blocked a portion of the red end of the spectrum, which corresponds to the leading temporal edge of the chirped pulse.

Figure 2 shows a schematic diagram of the grating based stretcher, and indicates where the red end of the spectrum is blocked. Since the spectral block is prior to the amplification stages, the amplified pulse energy does not change as a function of the wavelength range blocked. However, the blocked spectrum is temporally shorter and care must be taken not to exceed the damage threshold of amplifier materials. We used a beam splitter placed after the amplification stages, but prior to the compressor stage, to remove 80% of the spectrally modified, chirped pulse to drive shocks, while allowing the remaining fraction to be recompressed and used for spectroscopy and shock diagnostics.

The effect of this spectral clipping on shock generation was studied by measuring the spectra of the modified pulses, their time dependent intensity profiles, and the time dependent surface motion they produce in shocked aluminum and polymer-coated aluminum thin films. Spectra were measured using diffuse scattering into a fiber coupled CCD spectrograph (resolution 0.3 nm). A temporal width (FWHM) of the compressed pulse was measured by autocorrelation (this is only approximate due to the artificial symmetry induced by the single shot autocorrelator). The temporal intensity profile of the chirped pulse was determined by cross correlation with the compressed pulse in a 1 mm BBO crystal, via measurement of the intensity of the second harmonic generated at the wavevector sum of the two individual beams as a function of delay between pulses. The spatial interferometry technique used to measure surface motion is described below.

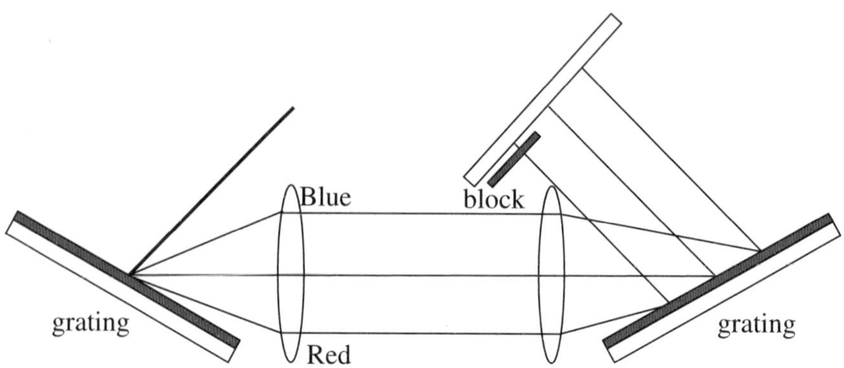

Figure 2: Schematic representation of a grating based pulse stretcher. The incident pulse enters at the top left, is dispersed by the left grating, collimated by the left lens, focused by the right lens, subtractively dispersed by the right grating (gratings are anti-parallel), and reflected back through the system by the mirror. In our application, a portion of the red end of the spectrum is sharply cut off in the spectrally-collimated region between the second grating and the mirror.

Figure 3 shows the spectra, pulse intensity versus time, and aluminum free surface position versus time using these spectrally modified chirped pulses to drive shocks in a 250 nm thick Al layer vapor plated onto a thin glass substrate.

The cross correlation signal shown in Fig. 3(b) is proportional to the intensity of the chirped pulse, $I_1(t)$, as the second harmonic depends on the product of the intensities $I_1(t)I_2(t)$, and the intensity of the recompressed pulse, $I_2(t)$, is essentially a delta function on these time scales. Comparison of Fig. 3(a) and (b) illustrates how spectral clipping removes the corresponding fraction of the temporal intensity profile from the chirped pulse; clipping about one third of the spectrum allows a sharply rising pulse intensity to be obtained.

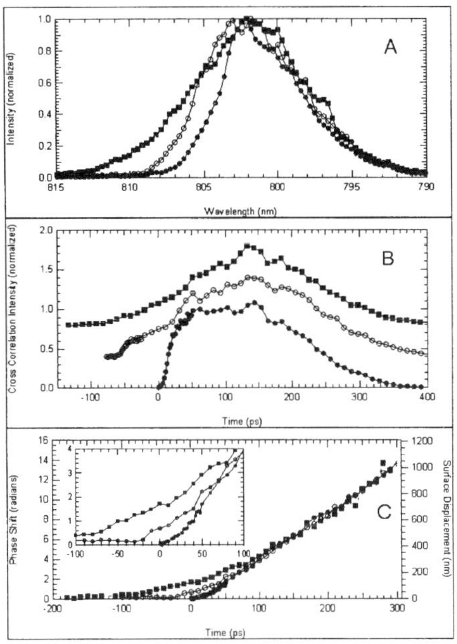

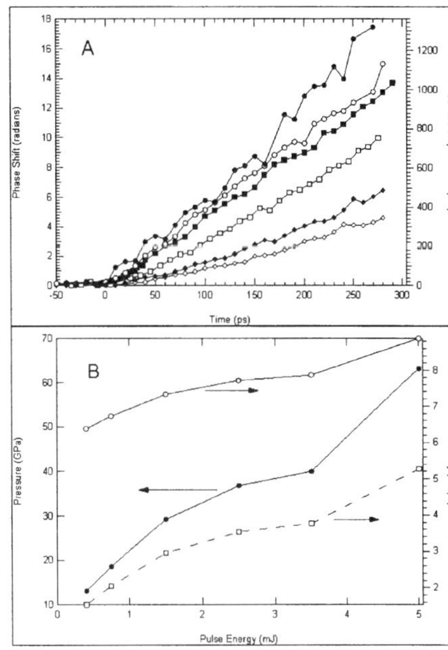

Figure 3 (left). (a) Spectra of chirped amplified pulses; ■, unaltered; ❍, spectrally clipped at 806 nm, ●, spectrally clipped at 804 nm. (b) Cross correlation (with sub-200 fs pulse) measurements of time dependent intensity for chirped pulse spectra of (a) (offset vertically for clarity). (c) Phase shifts and motion of aluminum free surface for shocks generated with 2.5 mJ pulse energy, using chirped pulse spectra of (a), inset magnifies short times.

Figure 4 (right). (a) Phase shifts and motion of aluminum free surface for chirped pulse spectrum clipped at 804 nm and pulse energies: ●, 5.0; ❍, 3.5; ■, 2.5; □, 1.5; ◆, 0.75; ◊, 0.4 mJ. (b) Calculated ●, pressure; ❍, shock velocity; □, bulk particle velocity as a function of pulse energy in aluminum films, determined with the aluminum Hugoniot [22] and the free surface velocities from the slopes in (a) at times >100 ps.

The shocks generated in 250 nm thick aluminum films by pulses of the three spectral contents shown in Fig. 3(a) were characterized using spatial interferometry. Fig. 3(c) shows the time dependent phase shift and corresponding surface displacement in the aluminum thin films. The one-dimensional surface displacement Δx is related to the phase shift $\Delta\phi$ geometrically by the equation $\Delta x=\Delta\phi\lambda(4\pi n\cos\theta)^{-1}$, where λ is the probe wavelength, n is the refractive index of the transparent medium (in this case air) and θ is the angle of incidence. The phase shift plotted is the value determined at the peak of the spatial intensity, but note that the measured phase shift has a Gaussian spatial distribution with a FWHM of ~150 μm. The inset to Fig. 3(c) illustrates how clipping the spectrum can remove the initial >100 ps of pressure ramping and achieve pressure rise times of 10-20 ps (determined by fitting free surface velocity to a tanh function as in Ref. [70]). We have also found that the measured rise times do not change appreciably with run distance (using Al samples of 1 and 2 μm thickness), indicating good constancy of the shock parameters, at least at these time and length scales.

The time dependent phase shifts in aluminum films were measured for the spectrum clipped at 804 nm at various pulse energies. The results are shown in Fig. 4(a). Figure 4(a) illustrates the final pressures achieved in the aluminum and the corresponding shock and particle velocities in the bulk film derived from the free surface velocity measured. The slopes of phase shift versus time, $\Delta\phi/\Delta t$, of Fig. 4(a) at times >100 ps determine the free surface velocity $u_{fs,}$ and the aluminum particle velocity u_p in the bulk material is $\approx u_{fs}/2$ (to within several percent at these pressures). The shock speed u_s and pressure P in the aluminum can be determined from the particle velocity, the aluminum Hugoniot (experimental u_s vs.u_p relation), and the Hugoniot-Rankine equations [77] that account for conservation of mass, energy, and momentum across a shock discontinuity. For aluminum, $u_s = c_s + 1.34\ u_p$, where c_s is the speed of sound, 5.35 nm/ps [78]. Pressures are calculated using the relation $P = u_p u_s \rho$, where ρ is the density of unshocked aluminum, 2.7 g/cm^3. Our particular methodology (substrate) has so far limited pulse energies to 5 mJ or less, but this is not a fundamental limit and higher pressures could be obtained with higher intensities.

One feature available via 100 fs laser pulse shock drive that we gave up by switching to longer temporally shaped pulses is the fortuitous production of planar shocks. Planar shocks can be produced using a particular combination of laser pulse energy, pulse length, substrate material, and substrate thickness [79]. Figure 4 shows the extent of the planarity that is achievable, equivalent to a few atomic layers across nearly 80 μm diameter! We found that the planarity was achieved via flattening of the drive pulse during transit through the substrate via non-linear absorption and self-focusing processes. These processes are much different for the longer temporally shaped pulses, and we have not yet found a combination of substrate material and thickness to achieve flattening, as the self-focusing term appears to dominate and causes formation of a sharp spike in intensity at the center of the drive spatial profile rather than flattening, using glass substrates.

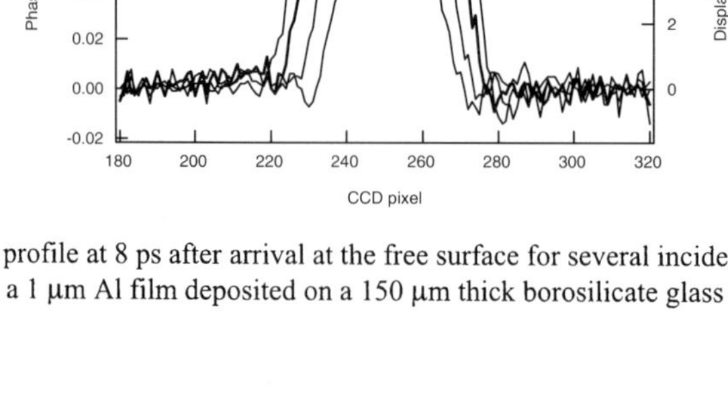

Figure 5: Shock profile at 8 ps after arrival at the free surface for several incident 110 fs pulse length laser energies in a 1 μm Al film deposited on a 150 μm thick borosilicate glass microscope cover slip.

4. ULTRAFAST INTERFEROMETRY

In the experiments described here, two separate techniques have been used for interferometric characterization of the shocked material's motion: frequency domain interferometry (FDI) [69, 80-81] and ultrafast 2-d spatial interferometric microscopy [82-83]. Frequency domain interferometry was used predominantly in our early experiments designed to measure free surface velocity rise times [70-71]. The present workhorse in the chemical reaction studies presented below is ultrafast interferometric microscopy [82]. This method can be schematically represented as in Figure 6. A portion of the 800 nm compressed spectrally-modified pulse from the seeded, chirped pulse amplified Ti: sapphire laser system (Spectra Physics) was used to perform interferometry. The remainder of this compressed pulse drives the optical parametric amplifier used to generate tunable fs infrared pulses (see below).

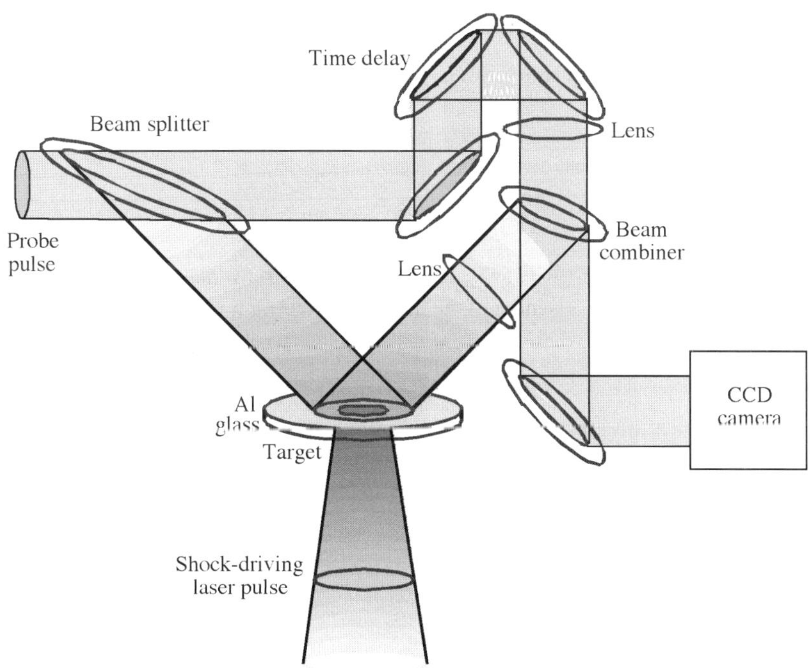

Figure 6: Schematic diagram of the ultrafast interferometric microscopy system.

The interferometric microscope is a modified Mach-Zehnder design with the sample in one arm (the sample arm) and a variable delay, to control temporal overlap, in the other (reference) arm. The probe pulse in the sample arm is focused onto the target at an incidence angle of either 32.6° or 76.0° to a spot size of ~500 μm to circumscribe the laser shocked region. The probe pulse can be s- or p-polarized relative to the plane of incidence using half-wave plates to rotate the polarization. A lens is used to image the sample surface (at ca. 2 pixels/μm) onto a CCD camera (Photometrics Sensys). A duplicate imaging lens is used in the reference arm. The sample and reference arms are recombined at a slight angle to produce an interference pattern on the CCD. This interferogram is transferred to and stored in a computer, and all interferograms from a time series, built up by adjusting the time delay in the shock driving laser arm, are post processed off line. In practice, three images are obtained at 1 Hz: (a) a "reference" interferogram, I_r, taken before the pump pulse arrives, (b) a "pump" interferogram, I_p, taken "during" the experiment, and (c) a post-shot interferogram to observe the damage to the material after the experiment. Analysis is conducted using the FFT method first developed by Takeda, and described in [84]. Two 2-D data maps are constructed, one containing the amplitude or reflectivity information, the other containing the phase information, which is composed of optical property and surface position data. Each time data point is obtained by averaging the 2-D map values in an area that is approximately 20 microns in diameter at the center of each laser experiment.

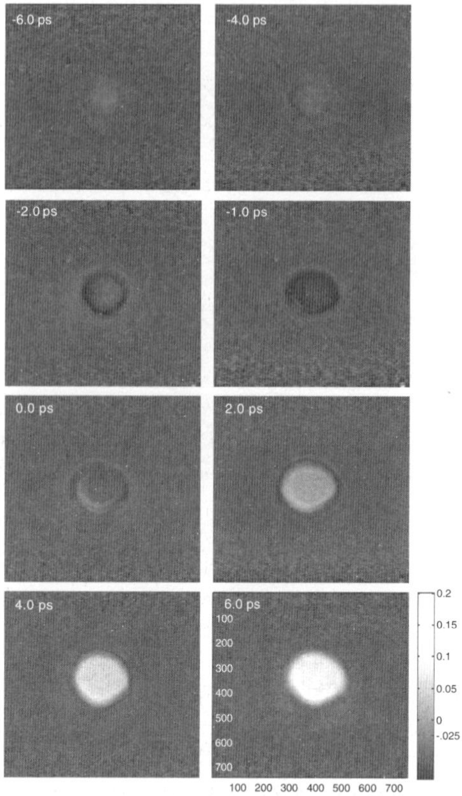

Figure 7: Phase images from the interferometric microscope (p-polarized, 32.6 incidence angle, 800 nm wavelength) during breakout of a 4.7 GPa shock from 250 nm thick Al, using 130 fs pulse length shock drive. The image z-axis scale is phase shift in radians.

To demonstrate the capabilities of the interferometer, Figure 7 presents phase images at several times during shock breakout from a 250 nm thick Al layer using 130 fs pulse length shock drive (hence the flatness). The noise level in the regions outside the shock determines the ultimate measurable surface displacement. For the experimental parameters used here (800 nm wavelength, 6 pixel fringes, 1024 pixel total frame area), the single pulse phase shift noise level is ~3 mrad, which is equivalent to 0.5 nm of surface displacement.

The *negative* phase shift that occurs during shock breakout in Al is very apparent and unexpected (we use the convention in our data analysis that material motion alone would yield a *positive* phase shift). We have also conducted experiments with an identical Al sample under similar conditions, but with a probe wavelength of 400 nm. No experiment done with the probe wavelength at 400 nm resulted in an observable negative phase shift during shock breakout. We also conducted experiments using Ni as the shocked metal, and observed only a positive phase shift with either 400 or 800 nm probe wavelength [71].

Reflection from an air-metallic interface is governed by Maxwell's equations and the appropriate boundary conditions leading to the Fresnel relations for the reflection amplitudes of s- and p-polarized light. Thus, upon reflection from a stationary metallic surface, the electric field undergoes a phase shift, ϕ_n, with magnitude r_n, that can be accurately calculated from knowledge of the complex index of refraction, polarization state, and the angle of incidence of the light striking the sample. Moreover, this phase shift will be influenced by any time-dependent changes in the complex index of refraction of the material. Thus, we hypothesized that the differences in the p-polarized 800 nm probe data and the 400 nm probe data result from the pressure induced shift of the *U(200)* interband transition in aluminum. The proof of this assertion is given in detail in Ref. [71], which also shows that the transient changes in the optical properties of the sample (the negative phase portion of the signal) are approximately linearly proportional to the acceleration of the surface. Therefore, the experimental phase plot obtained from Fig. 7 was modeled using a phase due to optical dynamics, $\phi_\eta(t)$, and a phase due to surface motion, $\phi_x(t)$. The 10%-90% rise time of the pressure profile from the fit to the experimental phase plot is estimated at 3.7 ps. See Ref. [70] for a complete discussion of the rise time measurements.

The main point to be made here is that the phase shift data obtained from spectral interferometry has two contributions: surface motion and optical effects. These two contributions to the phase versus time data can be separated by performing these experiments at two angles of incidence and two polarizations, at technique we term ultrafast dynamic ellipsometry. The optical effects during shock breakout in nickel films were "hidden" because they produce phase shifts of the same sign as that caused by surface motion. Ultrafast dynamic ellipsometry allowed that contribution to be measured [71]. In our experiments on bare metals, the observed optical effects are due to changes in the material's complex conductivity under shock loading. We will see below that this is only one of several kinds of optical effects that can be observed in these and other materials.

5. EFFECTS OF THIN FILM INTERFERENCE ON ULTRAFAST INTERFEROMETRY

Thin film interference arises when multiple reflections, from index of refraction variations, overlap in space and time and waves superpose. The connection with studies of shocked materials is that a shock wave moving through a transparent material represents an interface between an ambient density and a higher-density, and thus different refractive index, material. Ideally, there are two films produced, the shocked and the unshocked, with different refractive indices; and the thickness of each change with time during passage of the shock.

The general thin film interference problem considers reflection of a light ray from a reflecting surface covered by an arbitrary number of thin films. At each interface, there is a change in refractive index that leads to partial reflection. The interference of multiple reflections at each of the internal interfaces affects the net reflectivity and phase of light that exits the film. This interference depends on the wavelength, polarization, and angle of incidence of the light, as well as on the details of the film structure.

The angle of incidence, θ, changes at each refractive index, n, discontinuity as given by Snell's law $\theta_i = \arcsin(n_{i-1} \sin(\theta_{i-1})/n_i)$. Following the treatment found in references [85-89], a matrix M_i is formed for each layer as given by equation 1, where $g_i = 2\pi n_i d_i \cos(\theta_i)/\lambda$, and $q_i = n_i \cos(\theta_i)$ for s polarized light or $q_i = n_i/\cos(\theta_i)$ for p polarized light, d_i represents the thickness of layer i, and λ is the wavelength of light.

$$M_i = \begin{pmatrix} \cos(g_i) & i\sin(g_i)/q_i \\ i\,q_i \sin(g_i) & \cos(g_i) \end{pmatrix} \tag{1}$$

A net matrix is formed by multiplying all of the individual layer matrices $M = M_j M_{j-1} \ldots M_r$, where j is the outermost layer and r is the reflective layer. The reflection amplitudes, r, are formed from the elements of the matrix by equation 2, where $M_{1,2}$ is the matrix element of M at row 1 column 2, the q subscript j is for the outermost layer, and r is for the reflective layer.

$$r = \frac{M_{1,1}q_j - M_{2,2}q_r + M_{1,2}q_r q_j - M_{2,1}}{M_{1,1}q_j + M_{2,2}q_r + M_{1,2}q_r q_j + M_{2,1}} \tag{2}$$

The reflection amplitudes for s and p polarization are used to determine the reflectivity (amplitude modulus squared), phase (complex argument), and ellipsometric variables Δ and φ. If the reflection amplitude is expressed as $r = |r| \exp(i\phi)$, then ϕ is the phase change, $\Delta = \phi_p - \phi_s$ and $\varphi = \arctan(|r_p|/|r_s|)$ [86]. This treatment is used both for the static ellipsometric measurements of the thickness and refractive index and for modeling the dynamic problem.

In the shocked system, an additional time dependent phase change is observed due to the motion of the reflecting surface. The phase change $\Delta\phi_{sm}(t)$ due to surface motion at time, t, is given by the geometric relation:

$$\Delta\phi_{sm}(t) = \Delta z(t) 4\pi \cos(\theta)/\lambda \tag{3}$$

The surface motion, $\Delta z(t)$ is either that of the Al reflective surface freely moving into air, or the reflective surface at the Al/dielectric interface.

The time dependent phase shift observed experimentally $\Delta\phi_{exp}(t) = \Delta\phi_{sm}(t) + \Delta\phi_{int}(t)$, where $\Delta\phi_{int}(t)$ is the difference in phase for the sample before the shock and the sample during the shock at time t. The phase change $\Delta\phi_{int}(t)$ arises because the film structure changes from an initial thickness d_0 of PMMA, to a thickness d_0-$u_s t$ of unshocked PMMA and a thickness $(u_s$-$u_p)t$ of shocked PMMA, where u_s is the shock velocity and u_p is the particle velocity of the Al/PMMA interface. The sequence of films and time dependence is illustrated in Fig. 8. Note also that the spot size of the light is much greater than the film thickness, allowing multiple reflections to overlap spatially and temporally (600 nm light transit time ~ 3 fs in a material with $n \sim 1.5$).

The required parameters in the calculation are $n_{shocked}$, u_s, and u_p. The shocked refractive index is given by the Gladstone-Dale equation $n_{shocked}=1+(n-1)\rho_{shocked}/\rho$, where ρ is the initial density of the PMMA (1.186 g/cm^3) and n=1.487. Conservation of mass provides the shocked density $\rho_{shocked}=\rho/(1- u_p /u_s)$ for a one-dimensional shock compression. Previous studies have validated the Gladstone-Dale model for shocked PMMA up to 22 GPa. [90-91]

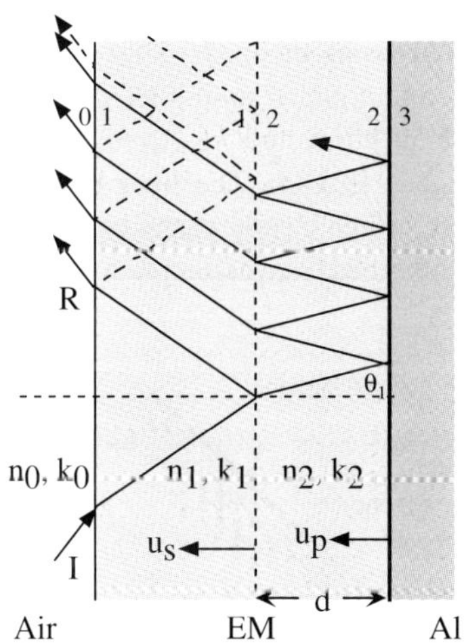

Figure 8: Diagram of thin film structure and time dependent thicknesses as shock transits sample from right to left. Shock velocity is u_s and Al interface velocity is u_p. Arrows indicate path of light partially reflected off interfaces, leading to thin film interference.

Measurements on the initial films were made to the extent possible, but the use of AFM or other sub-optical resolution surface techniques was limited to the Al layer. Atomic force microscopy on the Al films shows tightly packed grains on the scale of hundreds of nanometers laterally and a root mean squared surface roughness of 8 nm over a 100 μm by 100 μm region. These grains are too small to be resolved in the optical microscopy. The Al films show helium neon wavelength (632.8 nm) ellipsometric values typical for chemical vapor deposited films with a 4 nm aluminum oxide layer [92]. Agreement to within the experimental error of a few percent between the PMMA thin film refractive index determined ellipsometrically and bulk refractive index [93] indicates that the PMMA films are essentially full density. No voids are apparent in the microscopy. The magnitude of the possible error in pressure associated with small errors in refractive index and therefore density is discussed above in section 2.

6. USE OF DYNAMIC ELLIPSOMETRY TO MEASURE SHOCK STATES

PMMA was chosen as a test material due to its well-characterized properties as a common window material in shock studies. It has been reported to obey the Gladstone-Dale refractive index model at shock pressures up to 22 GPa, where a reversible transition to opacity has been reported, and a kink in the Hugoniot is present [78, 90-91]. All measurements reported are below this transition pressure. Also, there is a possibility that such reactions do not occur on these ultrashort time scales, which leads to the possibility of using these techniques to measure unreacted Hugoniots in energetic materials to higher pressures than other methods.

The time dependent phase shifts of a 625 nm PMMA film measured using *s* and *p* polarization at both 32.6^0 and 76.3^0 angles of incidence are shown in Fig. 9. The data in Fig. 9 were fit for $n_{shocked}$, u_s, and u_p directly. Times before 10 ps were excluded to prevent surface acceleration from affecting the outcome; the shocks were assumed steady. The sum of the residuals squared, $(\Delta\phi_{exp}(t) - \Delta\phi_{theory}(t))^2$, was mapped in the relevant three dimensional parameter space of $n_{shocked}$, u_s, and u_p and a global minimum was clearly identified. The fit parameters were: u_p=2.45 km/s, u_s=6.50 km/s, and $n_{shocked}$=1.77. The pressure is directly experimentally accessible as $P = u_s u_p \rho$ = 19 GPa. The lines in Fig. 9 are the theoretical predictions based on the fit parameters, which clearly characterize all four data sets. The fit parameters are very near the values determined from assumption of the bulk PMMA Hugoniot

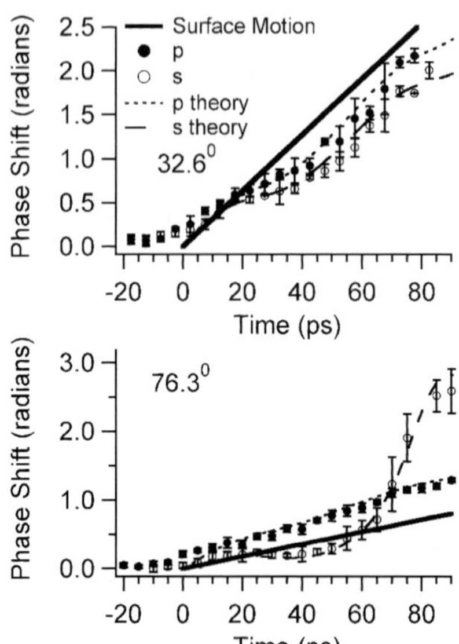

Figure 9. Phase shifts for 625 nm PMMA on Al during shock. The lines are theoretical predictions for u_p =2.45 km/s, u_s=6.5 km/s, $n_{shocked}$=1.77; the parameters determined by simultaneously fitting all four data sets. *P*=19GPa. The solid line is surface motion only, the dotted (*p* polarization) and dashed(*s* polarization) lines are calculated including thin film interference. Experimental points are ● *p* and ❍, *s* polarization at (a) 32.6^0 and (b) 76.3^0.

($u_s = c + s\,u_p$, where c=2.59 km/s and s=1.54 up to 22 GPa), the Gladstone-Dale shocked refractive index, and fitting only u_p: u_p=2.4 km/s, u_s=6.3 km/s, and $n_{shocked}$=1.79. For further details, see Ref. [94]. Additional tests of the agreement between the thin film and bulk PMMA Hugoniots were performed by interferometrically measuring PMMA and Al films (at a single incidence angle and a single polarization) as a function of shock strength. Data were taken at 32.6^0 with p-polarized light for bare Al films and for Al coated with PMMA thin films, each under identical shock driving conditions (laser energy, pulse shape, and spot size), which were chosen to encompass a range of shock strengths. The use of a low incident angle eases the difficulty of producing samples with large (3 cm^2) areas of thickness variation less than 10 nm. The impedance matching methods of Hugoniot determination applied to our data are commonly used, but are somewhat indirect and deserve some clarification [78, 95].

The first method assumes the thin film Al Hugoniot equals that of the bulk. The reflected Hugoniot is taken as the shock release isentrope. The reflected Hugoniot originates from the measured free surface velocity, u_{fs}, that is the velocity measured in expansion into air where $P \approx 0$. Any P- u_p point will lie on the intersection of this reflected Hugoniot unloading curve and the Hugoniot of the material it unloads into. The intersection point expresses graphically the mathematical requirement of conservation of pressure and particle velocity at the interface, the standard method of impedance matching in shock problems. The experimental data corresponding to this method are shown in Figure 10a, which plots the u_p fit from the measured PMMA interferometry data, against that determined by impedance matching from the measured bare Al u_{fs}. The caveat of this measurement is that both the Al and PMMA Hugoniots are assumed equal to that of the bulk, and this assumption is then tested by the data. Indeed, the data are very well described by the bulk Hugoniots. This is evidenced by the agreement between the experimental points in Fig. 6a and the line, which is not a fit, but represents the particle velocity given by the PMMA bulk Hugoniot [78].

A second method of Hugoniot determination with the same data does not require any *a priori* knowledge of the PMMA Hugoniot. The Al loading and unloading curves are determined by the Hugoniot, reflected Hugoniot, and Al u_{fs}. The P- u_p point in the thin film PMMA is determined by the PMMA u_p and the P at which this u_p crosses the Al unloading curve. Therefore, measurements of the bare Al u_{fs} and the PMMA u_p at multiple pressures allow the PMMA Hugoniot to be mapped out. The important feature of this method is that no PMMA Hugoniot is assumed. The circles show the experimental P- u_p results; the solid line is the Hugoniot of bulk PMMA. [78] Unfortunately, the pressure determined by impedance matching magnifies the relatively small noise in the PMMA u_p, limiting the quality of the independent confirmation of the agreement between thin film and bulk Hugoniot. However, along with the more precise agreement between the thin film and bulk Hugoniots established in Fig. 10a, it is strongly suggestive that the thin film and bulk material shock properties are essentially the same [94].

For comparison, the open circle of Fig. 10b is the single data point found by the most direct method of Hugoniot determination (as described above – ultrafast dynamic ellipsometry): the fit of the interferometric data from two incidence angles and two polarizations allows u_s, u_p, and therefore $P = u_s u_p \rho$, to be determined directly, and the P- u_p point plotted clearly agrees with bulk PMMA Hugoniot.

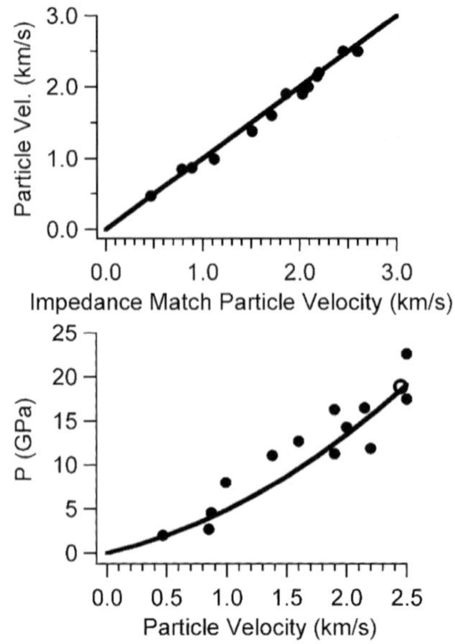

Figure 10. Thin film Hugoniot measurements. (a) The experimental PMMA u_p, ●, are plotted versus the u_p found by impedance matching from the Al Hugoniot and the experimental Al u_{fs} into the bulk PMMA Hugoniot. The solid line *is not a fit*, but is the particle velocity expected from the bulk PMMA Hugoniot. (b) The experimental PMMA u_p, ●, are plotted versus P found by impedance matching from the Al Hugoniot and the experimental Al u_{fs} into the experimental PMMA u_p. The solid line is the bulk PMMA Hugoniot. The open circle is the u_p and $P=\rho\, u_p u_s$=19 GPa directly fit from the data of Fig 9, as described in Section 5.

7. THIN FILM INTERFERENCE EFFECTS ON INFRARED REFLECTION SPECTRA

The same thin film interference effects, as discussed above for microscopic interferometry in the visible region, also occur in the infrared region. During passage of a shock through a transparent material with infrared absorptions, these interference effects cause time-dependent changes to the IR absorption spectra when these are obtained in reflection, as in our experiments. Therefore, such effects complicate the interpretation of infrared reflectance spectra obtained in shock-compressed thin film materials and must be carefully accounted for in any analysis attempting to unravel shock-induced energy transfer or reactivity.

In order to calculate the effects, the spectral complex refractive index, i.e., n and k at all the wavelengths of interest throughout the infrared, of the energetic material must be known. As most energetic materials, particularly the energetic polymers discussed below, do not have complex refractive index spectra available, we obtain them using angle dependent IR ellipsometry. The complex refractive index spectra are determined by simultaneously fitting all the angle and polarization FTIR spectra for each material, assuming that all of the data can be fit with a single complex refractive index spectrum. The use of a single refractive index is valid only if the material is isotropic and the density does not depend on the thickness. These conditions seem to be valid for the materials examined here, judging by the agreement between the measured and predicted reflectivity spectra.

Fitting for the complex refractive index versus wavenumber requires knowing the absolute reflectivity, film thickness, and angle. Film thicknesses are measured independently as described above. Angles are computer controlled to reproducible positions (< 0.5°). Reflectivity is measured relative to an identical aluminum thin film without the organic substrate. Correspondingly, reflectivity is calculated as the ratio of the reflectivity of the thin organic film on aluminum to the reflectivity of bare aluminum. Bare aluminum reflectivity is calculated using the literature complex refractive index spectrum [92], and including the effect of 4 nm aluminum oxide (thickness determined with 632.8 nm null ellipsometry) through its reported refractive index spectrum [96]. The system calculated is a planar multilayer of 1 μm aluminum, 4 nm aluminum oxide, and a uniform thickness of the organic film, in a medium of nitrogen gas (n=1.00).

A reflectivity function is defined, having angle, polarization, and thickness as independent variables, and fit for the 2 parameters – the real and imaginary parts of the refractive index (hereafter called n and k) versus wavenumber – using the Levenberg-Marquardt algorithm [97]. The large quantity of data and range of thicknesses employed allow a reasonably accurate fit simply using n = 1.5 and k = 0.01 as starting points for the search. However, this method sometimes introduces physically-unrealistic discontinuities in some regions of the n spectrum. To aid the convergence of the iterative fitting procedure, the Kramers-Kronig transform (KKT) of the fitted k spectrum is used to determine the n spectrum. A numerical implementation of the MacLaurin formula, Eq. 4, is utilized for its speed and accuracy advantages for KKT.

$$n(\omega_i) = n_{base} + \frac{2}{\pi}\sum_j \Delta\omega_i \frac{\omega_j k(\omega_j)}{\omega_j^2 - \omega_i^2} \tag{4}$$

The baseline value n_{base} is determined by comparison to the initial simultaneous n and k fit (no KKT), which matches most of the spectrum both at and away from the absorption peaks. The k spectrum and its n spectrum determined by KKT are then used as the initial values in the next iteration of the fitting procedure. The resultant k spectrum is again transformed to determine the n spectrum and the procedure iterated until the input and output spectra differ negligibly. Throughout the fitting procedure, no approximations are made regarding the lineshape of any absorptions.

Figure 11 shows the complex index spectra of the three polymers of interest: polymethylmethacrylate (PMMA), polyvinyl nitrate (PVN), and nitrocellulose (NC). These were obtained as described above from sets of IR reflection data obtained every 5° from 25 to 80° at both s and p polarization for each material. The spectra for PMMA agree substantially with those found in the literature [98].

These IR complex index component spectra were used to calculate the spectral effects that would be observed in a shock compression experiment. Figure 12 shows the time-dependent IR reflectance spectra calculated for normal incidence and p polarization in a 1 μm thick PMMA film during passage of the shockwave, assuming no pressure shift of the band frequencies. The uniaxial shock compression ratio $V_0/V = 1/(1-u_p/u_s)$ was 1.5, as expected for

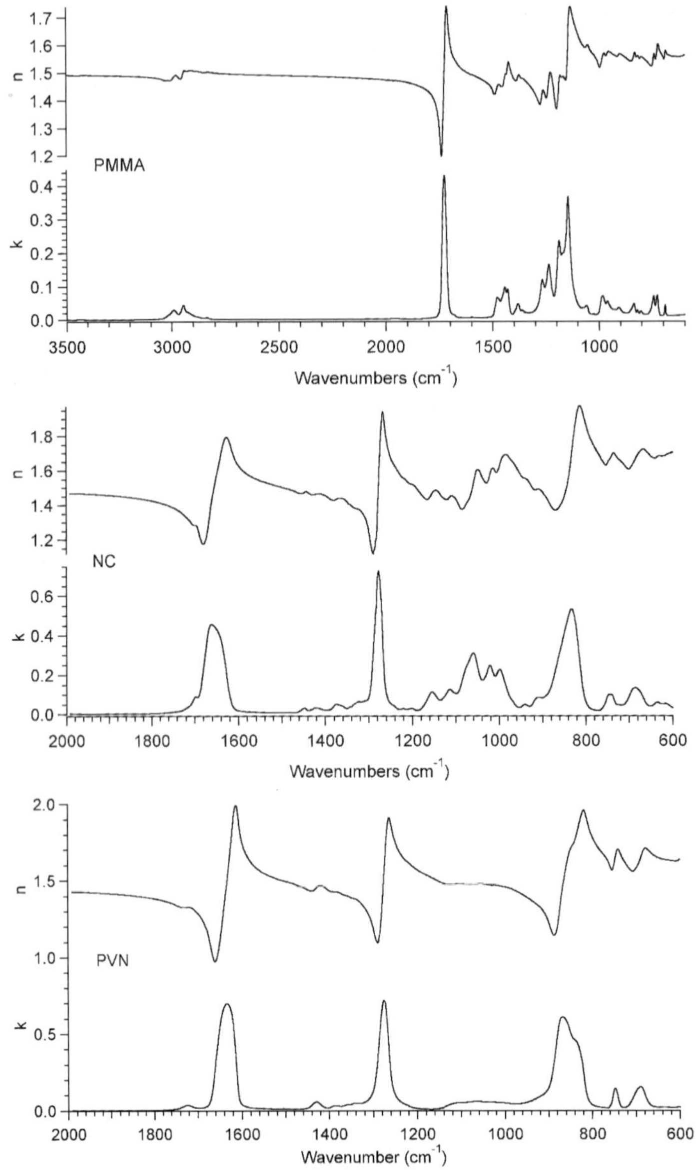

Figure 11: Complex index component spectra for PMMA, NC, and PVN.

a 6600 m/s shock velocity, 2600 m/s particle velocity, and 20 GPa pressure. The CO stretch band exhibits an absorption peak shift from 1723 to 1719 cm^{-1} as well as an increase in transmission from 0.02 to 0.25. Figure 13 shows the effect of including a +20 cm^{-1} shift on all peaks. This shift is a reasonable estimate for PMMA at 20 GPa, though different modes will be shifted differently; the Grüneisen parameters are available for only a few modes [99], which are taken as representative for the simple considerations here. The shifting and broadening of the bands due to shock induced temperature rise is neglected in these simulations, but will further complicate the analysis of real experimental data.

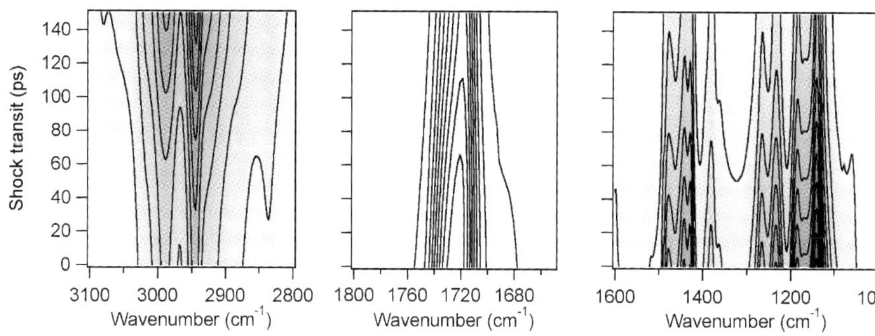

Figure 12. Simulation of the spectral effects in IR reflectance spectra caused by passage of a 20 GPa shock wave through 1000 nm of PMMA. In the end panels, the gray scale is the same as for Fig. 1 and contours are drawn every 0.05 transmission units to ease observing the spectral changes. In the middle panel the gray scale is not shown, and the contours are drawn every 0.1 transmission units starting at transmission=0.1.

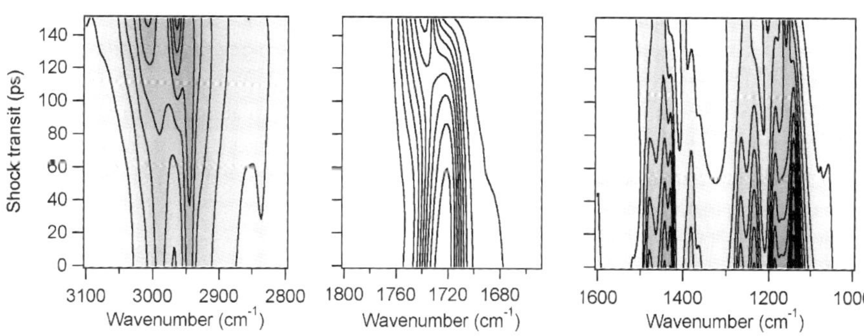

Figure 13: Same as Figure 12, except for a +20 cm^{-1} shift of the central vibrational frequencies of the shock-compressed PMMA

The general features to be gleaned from these simulations are that absorption bands shift, change shape, and change both absolute and relative peak intensities due entirely to thin film interference effects during passage of the shock wave. This conclusion is qualitatively confirmed in the comparison of simulation and shock data for NC in Figure 14. Even static experiments at various thicknesses show surprising spectral features (see Ref. [100] Figs. 4 and 5). These effects can all be accounted for with thin film equations and knowledge of the material optical constants. Unfortunately for our interest in shocked materials, the optical constants at the pressures and temperatures achieved by shock compression are typically unknown.

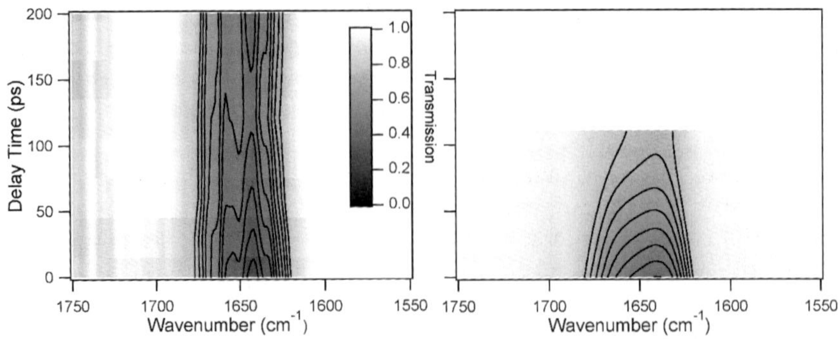

Figure 14: Comparison of simulated (right panel) and experimental (left panel) time-resolved IR reflectance spectra in shock-compressed nitrocellulose. Shown is the NO_2 ν_{as} mode spectral region. The simulation ends near 120 ps when the shock reaches the NC free surface, as the rarefaction had not yet been included.

The specific directions of the thin-film interference effect induced changes in frequency, bandwidth, and peak transmission upon shock compression depend upon specific conditions of n, k, and thickness at any given wavelength probed. Usually vibrational frequencies are modified by compression. Static high-pressure experiments using diamond anvil cells are capable of measuring the change in vibrational frequency and bandwidth in many materials. However, a shock wave also increases the temperature, which can additionally modify the vibrational spectra through broadening, softening of the frequencies, and the appearance of hot bands [101]. Furthermore, using static means to measure vibrational spectra at both high pressure and high temperature is particularly difficult for reactive materials such as energetics, which rapidly react under such conditions so that there is little chance to obtain corroborative data for shock experiments. Nevertheless, we have included simulations of spectral changes due to thin film effects combined with a particular vibrational frequency shift in the compressed material, to illustrate the kinds of data that may be expected. Since it is very difficult to obtain the complex index in the shock state, we have had to make very simple assumptions regarding the change in vibrational spectra upon shock loading. Further experimental work or detailed molecular dynamic calculations are necessary to predict realistic spectra under shock conditions.

8. ULTRAFAST INFRARED ABSORPTION

Optical parametric amplification methods (Spectra Physics OPA 800) were used to generate tunable signal and idler pulses that were then focused into a difference frequency generation (DFG) crystal (AgGaS) to generate tunable mid-infrared with a frequency bandwidth of >130 cm^{-1} FWHM and a pulse duration of ~170 fs. The DFG was performed close to the sample, ~0.5 m, to avoid absorption lines from atmospheric water broadening the pulse appreciably. The signal and idler were removed with a pair of long pass filters with cut on wavelengths at 2.5 and 5 microns. The interferometry probe (800 nm) and mid-IR probe (5-9 μm) were overlapped in time by monitoring the mid-IR transmission of a 250 micron thick Si wafer in the target position as a function of relative delay. The cross correlation derived in this manner was used to set the relative interferometry probe/mid-IR probe delays to within 1 ps. The size and spatial overlap of the mid-IR pulse with the pump pulse was monitored by measuring the change in reflection for various size holes generated by shocking Al films, and by transmission through pinholes.

The infrared pulse traversed a metallic beamsplitter (approximately 50/50) to form a reference pulse and a pulse that was focused on the sample (spot size ~45 μm diameter) with a 2" focal length BaF_2 lens. The mid-IR was transmitted through the thin polymer film, reflected off the Al film, and transmitted through the polymer film again before being collimated with a 3" focal length BaF_2 lens. Both the sample and reference mid-IR pulses were then imaged collinearly through a monochromator (Oriel 0.125 m, 75 g/mm grating blazed at 7 micron, 120 micron entrance slit) but displaced vertically to produce two stripes on a liquid nitrogen cooled 256x256 pixel HgCdTe focal plane array (SE-IR, Indigo chip). Differences in collection efficiency between sample and reference arms were corrected by placing an Al mirror (identical to the Al mirror on which the samples were deposited) at the target position. Ratioing the sample arm stripe to the reference arm stripe allowed single shot spectra to be taken with a typical accuracy of <5% across a 150-200 cm^{-1} range. Pulse to pulse variability of the mid-IR could not be improved with the 10 Hz Ti: Sapphire amplifier pump lasers (532 nm Quanta Ray GCR-170 and PRO-190) employed, and the infrared intensity and spectral fluctuations are the primary noise source in the experimental data.

Wavelength calibration of the infrared spectra was performed with weak water line absorptions (occurring primarily between the sample and the detector) in the reference and sample spectra in the range 1500-1750 cm^{-1}. Calibration in the range 1150-1350 cm^{-1} began by using the peak absorption from the FTIR spectrum to provide the absolute wavenumber at the peak. Determining the wavenumber width per pixel involved scanning the center wavelength to move the peak absorption to the edges of the image and reading the monochromator setting (backlash compensated), which agreed well with the water line calibration in the 1500-1750 cm^{-1} range. The pixel sizes at the magnifications employed in each series of experiments were 0.72 cm^{-1} and 1.24 cm^{-1} for the ranges 1150-1350, 1500-1750 cm^{-1}, respectively. Resolution with the 120 μm slit width is approximately 13 nm, or 2.0 and 3.4 cm^{-1} at the central wavenumbers 1270 and 1630 cm^{-1}, respectively.

Analysis of the infrared spectra requires accounting for thin film interference effects, which, upon shock compression, change the composite reflectivity. To analyze the thin film interference effects, the infrared complex refractive index spectra for ambient samples of PVN, and the inert polymethylmethacrylate (PMMA), were determined as described in the section above. The only modification to the description above is the inclusion here of the dispersive rarefaction wave that releases the pressure [102].

The effects of shock loading on the infrared absorption resonance (measured in reflectance) associated with the 1270 cm^{-1} symmetric nitro group stretch [103], $\nu_s(NO_2)$, of 940 nm thick PVN films are shown as a function of shock pressure in Fig. 15 [104]. The spectral data are distributed on the x-axis and the y-axis defines the time since the shock wave has entered the polymer film, determined interferometrically. The calculated spectra include only thin film effects [100, 103], and do not account for pressure and temperature shifts or chemical reaction. The areas of disagreement between the calculated and observed shocked spectra are thus the areas of most interest. In Fig. 15, spectra at 7 and 9 GPa exhibit loss of absorption followed by recovery, as expected for thin film interference. In contrast, data at 17 and 18 GPa show a permanent loss of absorption, as expected for chemical reaction. The 9 GPa peak reflectance increases from 0.14 to 0.73, then recovers to 0.20 as the rarefaction wave releases the sample pressure. In contrast, when the sample is shocked to 17 GPa, the reflectance changes to 0.79, and then recovers slightly to 0.71. The 7 and 9 GPa data show slightly greater reflectance than the fully shocked calculated spectra, which we attribute to temperature and pressure effects on the index of refraction that are unaccounted for in the calculation. Recovery of the reflectance upon rarefaction (at 7 and 9 GPa) indicates that the origin of the spectral changes is predominantly thin film interference, whereas lack of recovery upon rarefaction (at 17 and 18 GPa) indicates chemical reaction [104].

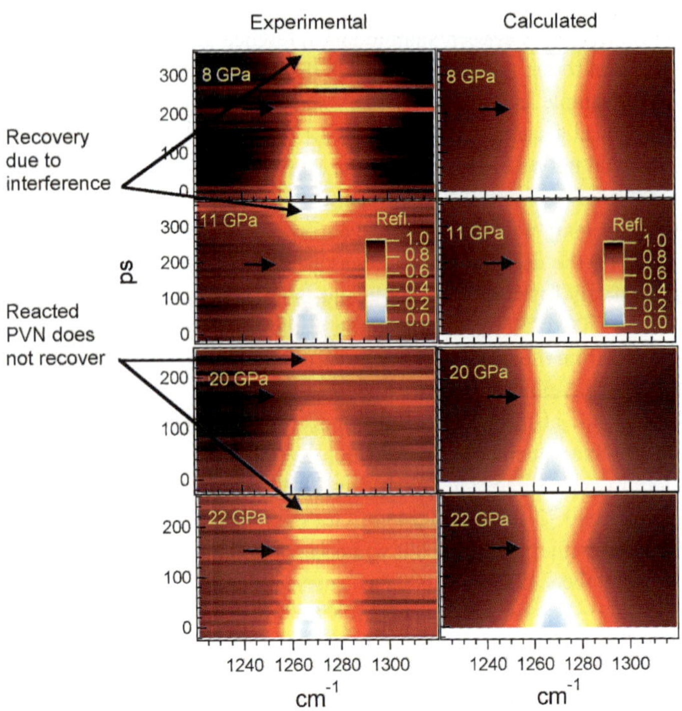

Figure 15. Time resolved infrared absorption spectra of $\nu_s(NO_2)$ in 940 nm thick PVN films during shock loading, in reflectance units. Calculated plots include only thin film interference effects, excluding pressure and temperature shifts and chemical reaction, making the differences between the experimental and calculated spectra the primary subjects of interest. Shock pressures were determined by interferometry. Arrows show the time at which the shock has fully traversed the film and rarefaction begins.

The effects of shock loading on the 1625 cm^{-1} antisymmetric nitro group stretch band, $\nu_{as}(NO_2)$, of 940 nm PVN films are shown in Fig. 16 (top). The increase in reflectance at the

$\nu_{as}(NO_2)$ resonance is less complete, when fully shocked, than observed in the symmetric stretch. There is no clear evidence of recovery; apparently the temperature broadening in the 7 GPa data permanently reduces the peak absorption, and the 9 GPa data was not acquired at times far enough into the rarefaction to show recovery. In all cases, the fully shocked PVN exhibits greater increases in reflectance than the thin film interference calculations predict, and the loss of absorption persists into the rarefaction. To more clearly understand the $\nu_{as}(NO_2)$ mode behavior, especially during the rarefaction, we repeated the 18 GPa experiment with a 700 nm PVN film; the infrared absorption spectra are shown in Fig. 16 (bottom). The loss of the 700 nm PVN $\nu_{as}(NO_2)$ absorption is far more obvious, and clearly does not recover during rarefaction. The differences seen in Figs. 16 (top) and 16 (bottom) illustrate that changing film thickness affects the thin film interference effects as well as the time available for reaction.

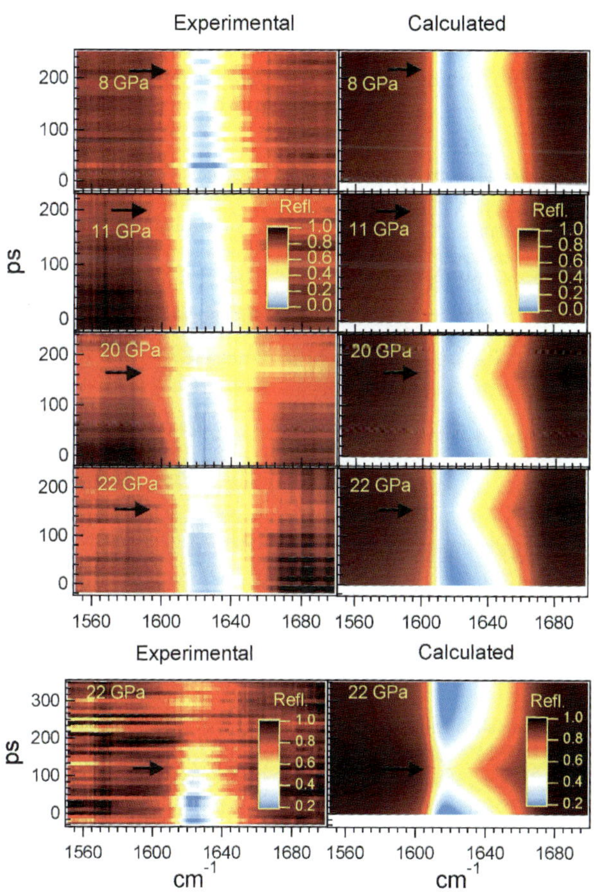

Figure 16. As Fig. 15, but the $\nu_{as}(NO_2)$ spectral region of (top) 940 nm PVN films shocked between 7 and 18 GPa, (bottom) 700 nm PVN film shocked to 18 GPa.

The above evidence for chemical reaction, i.e., loss of $\nu_{as}(NO_2)$ and $\nu_s(NO_2)$ mode absorption strength, occurred at times when the rarefaction is traversing back through the shocked sample. This fact raised the possibility that the rarefaction plays an essential role in the loss of absorption. To address this possibility, a multilayer film experiment was performed. A 440 nm window of inert PMMA was spin-coated onto 575 nm of PVN. The PMMA layer maintains the pressure in the PVN after the shock wave has passed.

Figure 17 shows the time dependent IR absorption spectra (measured in reflectance) in the $\nu_{as}(NO_2)$ mode region during shock and rarefaction. The caption provides the times that the shock or rarefaction reaches various interfaces (using shock and particle velocities from Ref. 94). Only half of the loss of PVN absorption occurs by 100 ps, when all the PVN is shocked. The rest of the absorption loss occurs while the shock is traversing the PMMA. At the same time, there is a larger increase in absorption spread over higher frequencies than was apparent in the single PVN layer data. Again, there is no recovery of the initial PVN absorption spectrum after rarefaction. The PMMA absorption has changed slightly, which is likely caused by both residual temperature broadening and by inaccurate accounting of thin film effects due to the significant refractive index changes in the reacted PVN.

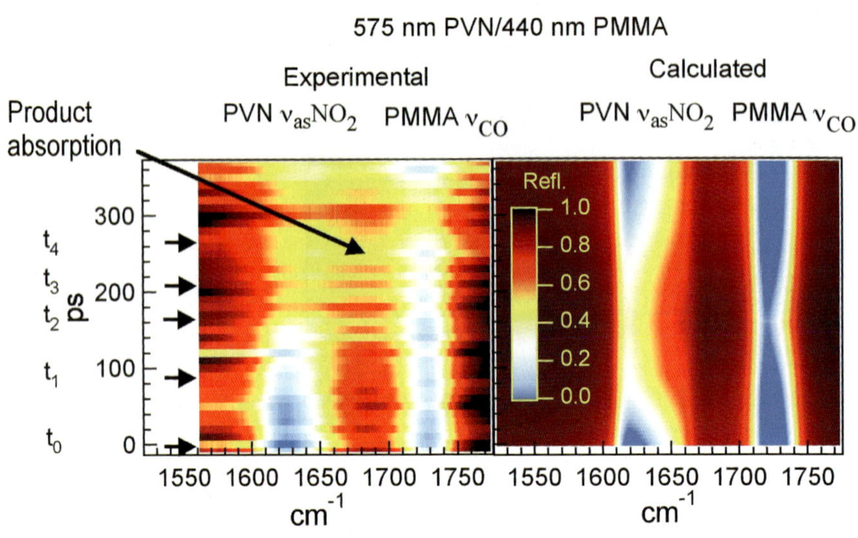

Figure 17. 575 nm PVN + 440 nm PMMA stacked film shocked to 18 GPa. The 1624 cm^{-1} peak is PVN $\nu_{as}(NO_2)$ and the 1728 cm^{-1} peak is PMMA's carbonyl stretch. Arrows denote timings: t_0, shock enters PVN; t_1, shock has fully transited PVN; t_2, shock has transited PMMA and rarefaction begins; t_3, head of the rarefaction fan reaches PVN/PMMA interface; t_4, head of the rarefaction fan reaches PVN/Al interface.

The similar permanent loss of absorption in the $\nu_{as}(NO_2)$ and $\nu_s(NO_2)$ modes on shock loading indicates chemical reaction affecting the nitro group. The fact that the absorptions do not completely disappear may be due either to partial reaction quenched by the rarefaction or to formation of products that have absorptions in the same spectral regions. Indeed, both nitrous acid [105], a possible product from bimolecular reaction, and nitrogen dioxide [106], a possible product from unimolecular decomposition, are expected to have absorptions in both the $\nu_{as}(NO_2)$ and $\nu_s(NO_2)$ mode spectral regions. The spectra of such likely products are not yet known at these pressures and temperatures, and further experiments and calculations are required to quantitatively explain the reaction kinetics and identify the contributing reactions.

The observation that the reaction requires an induction time of tens of picoseconds can be used to differentiate between proposed mechanisms of how shock wave energy localizes to cause chemical reaction. This induction time is expected for mechanisms that involve vibrational energy transfer, such as multiphonon up-pumping [107], where the shock wave excites low frequency phonons that multiply annihilate to excite the higher frequency modes involved in dissociation. It is also consistent with electronic excitation relaxing into highly excited vibrational states before dissociation, and experiments are underway to search for electronic excitations. On the other hand, prompt mechanisms, such as direct high frequency vibrational excitation by the shock wave, or direct electronic excitation and prompt excited state dissociation, should occur on sub-picosecond time scales, in contrast to the data presented here.

The infrared absorption results presented above demonstrate that it is possible to spectroscopically monitor shock induced chemical reactions on picosecond time scales at the beginning of the reaction zone. This demonstration opens the door to further probing of such events with the myriad of ultrafast laser based spectroscopic tools now available, promising to provide more insight into the effects of extreme pressure and temperature jumps at the molecular scale.

9. CONCLUSIONS

Just as the most accurate results available from conventional gas-gun investigations of bulk material properties and reactions require the utmost care in manufacture, assembly, and measurement of the targets, and in recording, calibrating, and evaluating pin and optical probe data, so too must the utmost care be given to the similar components and techniques used in experiments involving laser generated shocks. In order to obtain meaningful and reproducible data, we have spent very much effort in optimizing the laser pulse shape to produce supported shocks (at least for the length of time necessary for our experiments), in careful preparation and characterization of thin film targets, and in careful measurements of surface motion, which was greatly aided by the extension to two dimensions of a new interferometric tool. Such careful effort uncovered, quite serendipitously, the role of optical effects, such as shock induced changes in material complex conductivity and thin film interference, first in the unexpected negative phase shift during shock break out from a thin Al film, then also in the oscillatory nature of the phase shift recorded during shock transit through a thin dielectric material. The careful study and eventual understanding of these phenomena led eventually to a robust interferometric tool we term ultrafast dynamic ellipsometry, which is capable of measuring very small changes in complex conductivity that might result from melt or a phase transition, as well as the shock and particle velocities, and index of refraction, in shocked transparent dielectric materials. The number of uses of this ultrafast dynamic ellipsometry tool in studies of shock-compressed materials is just beginning to be investigated.

Our understanding of thin film interference helped us to apply the above methodology to our ultrafast infrared reflectometry studies on shocked energetic materials. In fact, the time-resolved IR reflection spectra obtained during shock transit and concomitant chemical reaction cannot be understood without accounting for thin film interference. The methodology requires, however, that the spectral complex index be known, or measured, for all the reactants, intermediates, and products, which are present, or might be present, in the shocked thin film. This latter requirement seems quite daunting, especially for short-lived intermediates, at the pressures and temperatures of the shock compressed material. How do you produce a static high pressure and high temperature sample of HONO, for example? We have great hope that the theoretical methods discussed elsewhere in this book will eventually help us to accurately calculate the necessary spectral constants so that accurate simulations will finally be able to extract quantitative reaction rates and pathways from our ultrafast time-resolved spectroscopic data. In the meantime, we will continue to obtain the highest quality spectral data possible, and will apply sensible models to achieve at least a qualitative understanding of the first steps in the shock induced chemistry of energetic materials.

ACKNOWLEDGMENT

The authors would like to thank Darren Naud for PVN synthesis, Michael Oldenburg for the refractive index measurement of bulk PVN, Jose Archuleta for the density measurement of bulk PVN, Joe Tiee and Roger Petrin for loan of the infrared camera, Kevin Gahagan for early interferometry results, Deanne Idar for support, and Tim Germann, Ed Kober, Brad Holian, and Alejandro Strachan for stimulating discussions. This work was performed at Los Alamos National Laboratory under Department of Energy contract W-7405-ENG.

REFERENCES

[1] M.R. Manaa, L.E. Fried, C.F. Melius, M. Elstner, and Th. Frauenheim, J. Phys. Chem. A, 106 (2002) 9024-9029.
[2] C. Cavazzoni, G.L. Chiarotti, S. Scandolo, E. Tosatti, M. Bernasconi, M. Parrinello, Science, 283 (1999) 44.
[3] F. Ancilotto, G.L. Chiarotti, S. Scandolo, E. Tosatti, Science, 275 (1997) 1288.
[4] S.R. Bickham, J.D. Kress, L.A. Collins, J. Chem. Phys., 112 (2000) 9695.
[5] J.D. Kress, S.R. Bickham, L.A. Collins, B.L. Holian, S. Goedecker, Phys. Rev. Lett., 83 (1999) 3896.
[6] A. Strachan, A. C. T. van Duin, D. Chakraborty, S. Dasgupta, W. A. Goddard, Phys. Rev. Lett., 91(2003) 98301/1-4.
[7] H. Kim, S.A. Hambir, D.D. Dlott, Shock Waves, 12 (2002) 79-88.
[8] I.-Y.S. Lee, J.R. Hill, H. Suzuki, D.D. Dlott, B.J. Baer, E.L. Chronister, J. Chem. Phys., 103 (1995) 8313-8321.
[9] S. Chen, X. Hong, J.R. Hill, D.D. Dlott, J. Phys. Chem., 99 (1995) 4525-4530.
[10] J.C. Deak, L.K. Iwaki, S.T. Rhea, D.D. Dlott, J. Raman. Spectrosc., 31 (2000) 263-274.
[11] J.C. Deak, L.K. Iwaki, D.D. Dlott, J. Phys. Chem. A, 103 (1999) 971-979.
[12] D.S. Moore, S.C. Schmidt, M.S. Shaw, J.D. Johnson, J. Chem. Phys., 90 (1989) 1368-1376.
[13] S.C. Schmidt, D.S. Moore, M.S. Shaw, J.D. Johnson, J. Chem. Phys., 91 (1989) 6765-6771.
[14] S.C. Schmidt, D.S. Moore, Acc. Chem. Res., 25 (1992) 427-432.
[15] D.S. Moore, S.C. Schmidt, M.S. Shaw, J.D. Johnson, J. Chem. Phys., 95 (1991) 5603-5608.

[16] S.C. Schmidt, D.S. Moore, M.S. Shaw, J.D. Johnson, J. Chem. Phys., 98 (1993) 9379-9388.
[17] D.S. Moore, S.C. Schmidt, M.S. Shaw, J. Chem. Phys., 101 (1994) 3488-3494.
[18] S.C. Schmidt, D.S. Moore, M.S. Shaw, J. Chem. Phys., 107 (1997) 325-336.
[19] P.S. Peercy, E.D. Jones, J.C. Bushnell, G.W. Gobeli, Appl. Phys. Lett., 16 (1970) 120-122.
[20] E.D. Jones, Appl. Phys. Lett., 18, (1971) 33-35.
[21] P.E. Schoen, A.J. Campillo, Appl. Phys. Lett., 45 (1984) 1049-1051.
[22] A.L. Huston, B.L. Justus, A.J. Campillo, Chem. Phys. Lett., 122 (1985) 617-621.
[23] A.L. Huston, B.L. Justus, A.J. Campillo, Chem. Phys. Lett., 118 (1985) 267-270.
[24] B.L. Justus, A.L. Huston, A.J. Campillo, Appl. Phys. Lett., 47 (1985) 1159-1161.
[25] R.D. Griffin, B.L. Justus, A.J. Campillo, L.S. Goldberg, J. Appl. Phys., 59 (1986) 1968-1971.
[26] A.J. Campillo, R.D. Griffin, P.E. Schoen, Opt. Commun., 57 (1986) 301-306.
[27] D.D. Dlott, Ann. Rev. Phys. Chem., 50 (1999) 251-278.
[28] J. Franken, S.A. Hambir, D.D. Dlott, J. Appl. Phys., 85 (1999) 2068-2074.
[29] D.D. Dlott, S. Hambir, J. Franken, J. Phys. Chem. B, 102 (1998) 2121-2130.
[30] D.D. Dlott, in Energetic Materials: Initiation, Decomposition, and Combustion, P. Politzer, J.S. Murray, Eds.,(Elsevier Science, Amsterdam, 2003).
[31] S.S. Mao, X.L. Mao, R. Greif, R.E. Russo, J. Appl. Phys., 89 (2001) 4096-4098.
[32] M.A. Shannon, R.E. Russo, Appl. Surf. Sci., 96-98 (1996) 149-153.
[33] A. Benuzzi-Mounaix, M. Koenig, J.M. Boudenne, T.A. Hall, D. Batani, F. Scianitti, A. Masini, D. Di Santo, Phys. Rev. E, 60 (1999) R2488-2491.
[34] S. Dufort, A. Delpeuch, in Proceedings Eighth Symposium (International) on Detonation, (NSWC MP 86-194, White Oak, MD 1986) pp. 847-854.
[35] A.E. Delpeuch, in Proceedings Ninth Symposium (International) on Detonation, (OCNR 113291-7, Arlington, VA, 1989), pp. 172-179.
[36] D.L. Paisley, in Proceedings Ninth Symposium (International) on Detonation, (OCNR 113291-7, Arlington, VA, 1989), pp. 1110-1117.
[37] A.M. Renlund, P.L. Stanton, W.M. Trott, in in Proceedings Ninth Symposium (International) on Detonation, (OCNR 113291-7, Arlington, VA, 1989), pp. 1118-1127.
[38] D.L. Paisley, D.C. Swift, R.P. Johnson, R.A. Kopp, G.A. Kyrala, in Shock Compression of Condensed Matter – 1991; M.D. Furnish, N.N. Thadhani, Y. Horie, Eds. (Elsevier Science BV: Amsterdam, 1990) pp. 1343-1346.
[39] D.B. Stahl, R.J. Gehr, R.W. Harper, T.D. Rupp, S.A. Sheffield, D.L. Robbins, in Shock Compression of Condensed Matter – 1989, Schmidt S.C.; Johnson J.N.; Davison L.W. Eds. (Elsevier Science BV: Amsterdam, 1990) pp. 1087-1090.
[40] G.E. Duvall, K.M Ogilvie, R. Wilson, P.M. Bellamy, P.S.P. Wei, Nature, 296 (1982) 846-847.
[41] C.S. Yoo, Y.M. Gupta, J. Phys. Chem., 94 (1990) 2857-2865.
[42] J.M. Winey, Y.M. Gupta, J. Phys. Chem., 101 (1997) 9333-9340.
[43] R. Engelke, D. R. Pettit, S. A. Sheffield, J. Phys. Chem. A, 101 (1997) 1696.
[44] M. Nicol, S.W. Johnson, N.C. Holmes in Shock Waves in Condensed Matter – 1987, S.C. Schmidt, et al., Eds (Elsevier, Amsterdam, 1988), pp. 471-476.
[45] X.A. Shen, Y.M. Gupta, J. Appl. Phys., 70 (1991) 7549-7553.
[46] C.S. Yoo, Y.M. Gupta, J. Phys. Chem., 96 (1992) 7555-7560.
[47] Y.A. Gruzdkov, Y.M. Gupta, J. Phys. Chem. A, 102 (1998) 8325-8332.
[48] Z. A. Dreger, Y. A. Gruzdkov, Y. M. Gupta, J. J. Dick, J. Phys. Chem. B, 106 (2002) 247.
[49] F. Boisard, C. Tombini, A. Menil, In Proc. 7th Symp. Intl. Detonation, J.M. Short Ed. (Naval Surf. Weapons Cent., White Oak, MD, 1981) pp 1010-1015.
[50] N.C. Holmes, W.J. Nellis, W.B. Graham, G.E. Walrafen, Phys. Rev. Lett., 55 (1985) 2433-2436.
[51] A.M. Renlund, W.M. Trott, in Shock Compression of Condensed Matter – 1989; Schmidt S.C.; Johnson J.N.; Davison L.W. Eds., (Elsevier Science BV: Amsterdam, 1990) p. 875.
[52] G.I. Pangilinan, Y.M. Gupta, J. Phys. Chem., 98 (1994) 4522-4529.
[53] J.M. Winey, Y.M. Gupta, J. Phys. Chem. B, 101 (1997) 10733-10743.
[54] G.I. Pangilinan, Y.M. Gupta, J. Appl. Phys., 81 (1997) 6662-6669.

[55] S.C. Schmidt et al., in Advances in Chemical Reaction Dynamics; Rentzepis P.M.; Capellos C. Eds., (Reidel: Dordrecht, 1985) pp. 425-454.
[56] D.S. Moore, S.C. Schmidt, in Proceedings Ninth Symposium (International) on Detonation, (OCNR 113291-7, 1989) p. 180.
[57] W.M. Trott, A.M. Renlund, J. Phys. Chem., 92 (1988) 5921-5925.
[58] T. Kobayashi, T. Sekine, H.-L. He, J. Chem. Phys., 115 (2001) 10753-10757.
[59] K.G. Nakamura, K. Wakabayashi, A. Matsuda, K.-I. Kondo, Appl. Surf. Sci., 197-198, (2002) 17-20.
[60] A. M. Renlund, S. A. Sheffield, W. M. Trott in Shock Waves in Condensed Matter, Y. M. Gupta Ed. (Plenum, New York, 1986) p. 237-42.
[61] W. M. Trott, A. M. Renlund, App. Opt., 24 (1985) 1520.
[62] D.E. Hare, J. Franken, D.D. Dlott, J. Appl. Phys., 77 (1995) 5950-5960.
[63] G. Tas, S.A. Hambir, J. Franken, D.E. Hare, D.D. Dlott, J. Appl. Phys., 82 (1997) 1080-1087.
[64] M.J. Schmitt, R.A. Kopp, D.S. Moore, and S.D. McGrane, "Analysis of laser-driven shocks in confined and unconfined geometries," in Shock Compression of Condensed Matter – 2003, M.D Furnish, et al. eds (AIP, Melville, NY, 2004) in press.
[65] D. von der Linde, K. Sokolowski-Tinten, Appl. Surf. Sci., 154/155, (2000) 1.
[66] B.C. Stuart, M.D. Freit, S. Herman, A.M. Rubenchik, B.W. Shore, M.D. Perry, Phys. Rev. B, 53 (1996) 1749.
[67] W. Diepold, Explosivstoffen, 1, (1970) 2.
[68] K. M. Abd El-Kader, A. S. Orabi, Polymer Testing, 21 (2002) 591.
[69] R. Evans, A.D. Badger, F. Failliés, M. Mahdieh, T.A. Hall, P. Audebert, J.-P. Giendre, J.-C. Gauthier, A. Mysyrowics, G. Grillon, A. Antonetti, Phys. Rev. Lett., 77 (1996) 3359.
[70] K. T. Gahagan, D. S. Moore, David J. Funk, R. L. Rabie, and S. J. Buelow, and J.W. Nicholson, Phys. Rev. Lett., 85 (2000) 3205-8.
[71] D.J. Funk, D.S. Moore, K.T. Gahagan, S.J. Buelow, J.H. Reho, G.L. Fisher, and R.L. Rabie, Phys. Rev. B, 64 (2001) 115114.
[72] C. Momma, S. Nolte, B.N. Chichkov, F. v. Alvenslebeen, A. Tünnermann, Appl. Surf. Sci., 109/110 (1997) 15.
[73] M. Lenzner F. Krauz, J. Krüger, W. Kautek, Appl. Surf. Sci., 154/155 (2000) 11.
[74] S.D. McGrane, D.S. Moore, D.J. Funk, R.L. Rabie, Appl. Phys. Lett., 80 (2002) 3919-3921.
[75] A.M. Weiner, Prog. Quant. Electr. 19 (1995) 161.
[76] L. Xu, L. Li, N. Nakagawa, R. Morita, M. Yamashita, IEEE Phot. Tech. Lett., 12 (2000) 1540.
[77] Y. B. Zel'dovich, Y. P. Raiser, Physics of Shock Waves and High Temperature Hydrodynamic Phenomena, (Academic Press, New York, 1966).
[78] S. P. Marsh, LASL Shock Hugoniot Data, (University of California, Berkeley, CA, 1980).
[79] D. S. Moore, K. T. Gahagan, J. H. Reho, David J. Funk, S. J. Buelow, R. L. Rabie, and T. Lippert, Appl. Phys. Lett., 78 (2001) 40-42.
[80] E. Tokunaga, A. Terasaki, and T. Kobayashi, Opt. Lett., 17 (1992) 1131.
[81] J.P. Geindre, P. Audebert, A. Rousse, F. Falliès, J.-C. Gauthier, A. Mysyrowicz, A. Dos Santos, G. Hamoniaux, and A. Antonetti, Opt. Lett., 19 (1994) 1997.
[82] K.T. Gahagan, D.S. Moore, D.J. Funk, J.H. Reho, and R.L. Rabie, J. Appl. Phys., 92 (2002) 3679.
[83] S.R. Greenfield, J.L. Casson, A.C. Koskelo, in High Power Laser Ablation III, Phipps, C.R. Ed. (SPIE Vol. 4065, 2000), pp. 557-566.
[84] M. Takeda, H. Ina, S. Kobayashi, J. Opt. Soc. Am., 72 (1982) 156-160.
[85] M. Born and E. Wolf, Principles of Optics 4th ed., (Pergamon Press, New York, 1970).
[86] R. M. A. Azzam, in Handbook of Optics, Vol. II, edited by M. Bass (McGraw-Hill, New York, 1995).
[87] J. Macdonald, Metal-Dielectric Multilayers, (Elsevier, New York, 1971).
[88] H. Anders, Thin Films in Optics, (The Focal Press, New York, 1965).
[89] A. Vasicek, Optics of Thin Films, (North-Holland, Amsterdam, 1960).

[90] J.L. Wise and L.C. Chhabildas, in Shock Waves in Condensed Matter 1985, edited by Y. M. Gupta, (Plenum Press, 1985) p. 441.
[91] L.C. Chhabildas and J.R. Asay, J. Appl. Phys., 50 (1979) 2749.
[92] D.Y. Smith, E. Shiles, and M. Inokuti, in Handbook of Optical Constants of Solids, E. D. Palik, ed., (Academic Press, San Diego, CA, 1998) p. 369.
[93] T. Ishigure, Y. Koike, J. W. Fleming, J. Lightwave Tech. 18, (2000) 178.
[94] S.D. McGrane, D.S. Moore, D.J. Funk, J. Appl. Phys., 93 (2003) 5063-5068.
[95] R.F. Trunin, Shock Compression of Condensed Materials, (Cambridge University Press, 1998).
[96] F. Gervais, in Handbook of Optical Constants of Solids II, edited by E. D. Palik, (Academic Press, San Diego, CA, 1998) p. 761.
[97] D.W. Marquardt, J. Soc. Ind. Appl. Math., 11 (1963) 431-441.
[98] A. Solera and E. Monterrat, Polymer, 43 (2002) 6027.
[99] J. J. Flores and E. L. Chronister, J. Raman Spec., 27 (1996) 149.
[100] D.S. Moore, S.C. McGrane, D.J. Funk, Appl. Spectrosc., (in press 2004)
[101] D.S. Moore, J. Phys. Chem. A, 105 (2001) 4660-4663.
[102] G.R. Gathers, Selected Topics in Shock Wave Physics and Equation of State Modeling, (World Scientific Publishing, River Edge, NJ, 1994).
[103] D.S. Moore, S.D. McGrane, J. Mol. Struct., 661 (2003) 561.
[104] S.D. McGrane, D.S. Moore, D.J. Funk, submitted (2004).
[105] M. Krajewska, A. Olbert-Majkut, Z. Mielke, Phys. Chem. Chem. Phys., 4 (2002) 4305.
[106] Sadtler Research Laboratories, The Sadtler Handbook of Infrared Spectra, (Sadtler Research Laboratory, Philadelphia, 1978).
[107] D. D. Dlott, M. D. Fayer, J. Chem. Phys., 92, (1990) 3798.

Chapter 14

The Equation of State and Chemistry at Extreme Conditions: Applications to Detonation Products

Joseph M. Zaug, W. Michael Howard, Laurence E. Fried, Alexander F. Goncharov, Wren B. Montgomery, and Jonathan C. Crowhurst

University of California
Lawrence Livermore National Laboratory
P. O. Box 808, Livermore, CA 94551, USA

1. INTRODUCTION

Laboratory experiments conducted on materials held in excess of several kbar provide insight into a realm of chemical and material properties that are significantly different from those encountered under ambient conditions. Such studies extend and test the theoretical framework which permits progress from properties at the atomic and molecular level to macroscopic behavior, constitute a potential source of novel materials and new tools for chemical transformation, and are important adjuncts to progress in other disciplines. There is no question that an improved general knowledge of electronic, physical, and chemical behavior of relevant constituent materials at high density is required, for example, for a less fragmented description of the processes that precede and govern dynamic exothermic chemical reactions. The same arguments are routinely made with regard to the structure and evolution of the major planets [1] and the deep interior of the Earth.

Dynamical simulation based on approximate Born-Oppenheimer potentials plays a large and increasingly important role in chemistry and in the biological and materials sciences. More generally, knowledge of an effective interatomic potential function underlies any effort to predict or rationalize the properties of solids and liquids. While there exists an extensive body of experimental techniques and experience on computational methods appropriate to ambient conditions, the regime of strong repulsive interactions at very high densities has not been as extensively investigated [2-8]. The experiments discussed here are aimed both at enlarging the family of properties conveniently measured at high pressure and, principally, at providing the data appropriate to a critical test of the theory of the interatomic potential in simple substances at high density.

2. HIGH PRESSURE EXPERIMENTAL METHODS

High pressure Raman and FTIR spectroscopy methodologies were employed to study chemical reactions at extreme pressure and temperature conditions. Angle dispersive synchrotron x-ray diffraction and impulsive stimulated light scattering (ISLS) were used to determine equations of state. ISLS is a nondestructive optical process for determining sound velocities [9-23], fluid flow velocities [24], compressibilities [22], acoustic damping and structural relaxation rates [13, 23, 25-33], elastic constants [21, 22], energy transfer processes [34-43], chemical reaction kinetics [29, 44-49], population density kinetics [50, 51], electron-hole transport and decay rates [52-59], binary mass diffusivity [60, 61], temporal and spatial coherence properties of laser beams [62, 63] and thermodynamic properties (*i.e.,* equations of state [15, 21-23] and thermal diffusivity [9, 40, 55, 64-67]) on a very broad class of samples. Temperature and/or pressure induced phase transitions can be accurately characterized with ISLS techniques [20, 21, 68, 69]. A partial list of samples that have been investigated includes proteins [27, 28, 31, 80], thin film polymers [17, 18, 29], semiconductors [52-58], superconducting ceramics [16], metals [12], fluids [15, 19, 23, 26, 30, 32 64, 67], fluid mixtures [27, 60, 65], solutions [19, 25, 36-40, 43-49], glassy state amorphous structures [46, 66, 67], liquid crystals and crystalline solids [10-11, 13-14, 20-22, 33-35, 50, 68, 69]. In addition, ISLS experiments may be conducted on high-energy plasmas [42], and flames [70]. Unlike traditional frequency-domain measurement techniques (such as, pulse-echo and including the latest in resonant ultrasound spectroscopy, and Raman, including CARS), ISLS can effectively resolve heavily damped, or overdamped, modes that predominate in liquids and amorphous structures (glasses) at high-pressure. Traditional optical frequency-domain techniques (Raman spectroscopy) tend to produce broad frequency linewidth spectra, when the sample medium is viscous, which yield indeterminate and/or ambiguous time-resolved measurements. Furthermore, the frequency range of ISLS experiments depend primarily on short pulse widths of the excitation source. The temporal resolution of the excited material modes is limited only by the pulse width. (In a more limited sense, the coherence time of the light dictates the time resolution of the experiment [84]). The competing qualities (resolution vs. intensity) found with even the largest spectrometers are rendered to *non-issues* when using fast and gated electronic instruments within the context of time resolved spectroscopy. Analyzing ~30 μm thick samples with diameters on the order of ~50 μm are considered routine using laser-induced ultrasonic methods, whereas transducer-induced ultrasonic methods are limited to samples that are >1000 μm thick and 200 μm in diameter. When such small samples are illuminated with an ~80 picosecond pulsed laser system acoustic phonons can be excited and probed within a frequency range of 10 MHz - 10 GHz. With today's femtosecond systems, this range extends to THz (optical phonons), where nearly any motion of atoms or molecules (vibrations in condensed materials, molecular rotations and vibrations, or simply structural relaxations) can be time-resolved although at high frequencies frequency-domain techniques may well be preferable. ISLS can help resolve/characterize structural and liquid-glass transitions. It can also be used to detect ppm (parts per million) concentration changes in chemical binary solutions or mixtures, with temporal resolutions exceeding reaction or recombination times [77]. Perhaps one of the most convincing utilities of ISLS experiments is the determination of anisotropic properties of molecular environments. For all of the above, the essential requirements (aside from a good optical sample) are very short excitation pulse widths, and beam-material coupling mechanisms by which the optical pulses can initiate and monitor the time dependent results.

There is extensive literature describing experimental methods and computational models designed to provide insight into chemical behavior under ambient conditions. Yet, in terms of the universe that we live in one may argue that ambient conditions are a physical anomaly. Most chemical reactions occur under conditions of extreme temperature and pressure, the largest exception being reactions in biological systems. In general, knowledge of an effective interatomic potential function underlies any effort to predict or rationalize the properties of solids and liquids. The regime of strong repulsive interactions in condensed phases at very high densities, where many body interactions play a primary role has not been so extensively investigated.

Percy W. Bridgman [85] (1882-1961) opened the field of high-pressure studies with his piston-cylinder apparatus during the early part of the 20th century. His ideas were followed up, beginning in the mid 1940's, by groups at Norwell and General Electric who included high-temperature to the foray of evolving high-pressure devices. Diamonds were first proposed as having potential for high-pressure studies in 1887 [86]. However, it wasn't until Alvin van Valkenburg* (1913-) and C. E. Weir *et al.*, introduced [87] the modern diamond anvil cell (DAC), during the same time Schalwlow and Townes presented optical masers, that this potential became reality. Weir's device§ opened a new era of high-pressure optical experimental characterizations of material properties. Smaller (palm size) diamond cells were designed to increase access in reciprocal lattice space illuminated by x-ray producing equipment [89]. In turn, these smaller DAC's were found useful in phonon scattering of laser light (Brillouin scattering) [90]. The first ISLS experiment involving a DAC was conducted in 1988 on room temperature liquid methanol and ethanol up to 6.82 GPa [15].

The combination of DAC technology with ISLS experiments offers a powerful methodology for comprehensive studies of effective interatomic potential surfaces in the region of high repulsion. Results from such experiments are providing a means to further refine molecular dynamics calculations and more accurate determinations of intermolecular potential functions and correlations of high-frequency molecular equilibrium relaxation times. The experiments described in this body of work demonstrate the versatility found with ISLS when hydrostatic pressure is used to increase density. The results show that extrapolation and/or computational modeling from ambient properties for deduction of high-pressure chemical behavior (*i.e.,* sound velocities, the equation of state, lattice constants, compressibilities) can be inaccurate. The logical progression will be the incorporation of computational methods with new high-pressure ISLS data to accurately characterize the interatomic potentials at high density.

Photoacoustic spectroscopy is often the most appropriate form of optical spectroscopy when material absorption is weak. ISLS spectroscopy offers the advantages of photoacoustic detection in a geometry that is compatible with the requirements of a DAC. The ISLS approach enables the experimenter to *tune in* material modes for observation, by adjustment of the excitation pulse width and wavelength. Control of the probe wavelength permits

* It should be noted that Prof. John Jamieson, University of Chicago, made a diamond piston-cylinder device around the same time as Valkenburg.

§ It should be noted that as early as 1956, [88] optical measurements were made using NaCl and synthetic sapphire windows.

observation of each contributing mechanism of the resulting diffraction grating. In this way material modulations can be optimized to increase the amount of diffracted probe light (signal). The excitation beams' intersection angle can also be configured to control the acoustic wavelength (grating spacing), or modulation of the induced longitudinal, quasilongitudinal and/or surface waves in the material. Picosecond pulse-widths are short enough to excite low-frequency material modes or acoustic phonons. Shorter femtosecond pulses may excite higher frequency material modes, or optical phonons, in materials. This enables observation of individual vibrational oscillations.

The optical generation of ultrasonic waves has been successfully employed to study a wide array of physical and chemical systems. ISLS represents one form of dynamic, or transient, grating experiments that have evolved from early optical endeavors. Since 1973, ISLS experiments have been successfully applied to a wide number of scientific problems including: determination of orientational relaxation times and singlet lifetimes for dye molecules in solution, thermal diffusion measurements of solutions (liquids and solids), phonon and excited-state phonon studies of crystals, characterization of acoustic behaviors in solids near structural phase transitions, characterization of various phases of liquid crystal thin films, energy transport in molecular solids, polariton scattering, observation of protein motions in hemoglobin and myoglobin, and nondestructive characterization of thermal and mechanical properties of thin films and thin film coatings. In addition, multiple-pulse ISLS experiments have been successfully applied to manipulate molecular motions along excited-state potential energy surfaces

In the remaining body of this chapter, a description of the ISLS experiment is provided along with a description of our high-pressure vibrational spectroscopy instrumentation. Subsequent to the experimental descriptions, theoretical formalisms are presented.

3. ISLS EXPERIMENTS

The beam configuration of an impulsive stimulated light scattering experiment is displayed in Figure 1. The form may be considered, with respect to nonlinear optics, as one version (partially degenerate) of the four-wave mixing experiment. Two 100-picosecond near-infrared (1.064 μm) parallel-polarized excitation beams, or pulses, converge spatially and temporally at an angle θ in a sample medium. The wave vectors of these pulses are given by $\mathbf{k}_1$ and $\mathbf{k}_2$ where $|\mathbf{k}_1| = |\mathbf{k}_2|$. (In this chapter, both vector and tensor quantities appear in bold typeface). The crossed light produces a spatially periodic electric field, which in turn produces a spatial modulation (grating) of material properties (population of excited electronic states, polarization, or vibration, of atoms, temperature, or molecular orientation). In an absorbing material, rapid radiationless relaxation may heat the sample at the interference maxima. Subsequently, in an anisotropic medium, thermal expansion impulsively launches one quasilongitudinal and two quasitransverse pairs of counterpropagating ultrasonic waves. Along symmetry directions pure longitudinal waves are generated. The material strain created by the acoustic waves causes a time-dependent and spatially periodic variation in the index of refraction of the sample (standing strain wave), which, in effect, modulates the optical properties of the sample (index of refraction and/or absorption).

Phonon excitations are dependent on the condition that the temporal width of the excitation pulses, τ_P, is short compared to the single acoustic oscillation periods $(1/\tau_P > 1/\tau_A)$. Hence the interaction term "impulsive" is incorporated into the name ISLS. Material modes where $\tau_P > \tau_A$ will not be efficiently excited. The acoustic wavelength and wave vectors describing the two coherently excited elastic waves are $(\lambda_A, \pm \mathbf{k}_A)$ where

$$d = \lambda_A = \lambda_E / 2\sin(\theta/2) \tag{1a}$$

and

$$\mathbf{k}_A = \mathbf{k}_1 - \mathbf{k}_2 . \tag{1b}$$

The diffraction grating only contains the fundamental wave vector $\mathbf{k}_A$, and so, with exception to a non-linear refraction response, higher order scattering vanishes. As mentioned above, density variations affect the index of refraction, n, (both real and imaginary parts) thus changing the optical properties of the material. Hence, density variations give rise to a transient, or dynamic, optical diffraction grating that remains long after the excitation pulses have departed from the sample. (A frequency offset between the two-excitation pulses would result in a traveling grating). In non-absorbing materials, density changes occur due to electrostriction. The broad frequency spectrum of the short excitation pulses provides a mix of optical Fourier components that couple with material modes to generate high-frequency phonons. In this way, the crossed electric field terms of the excitation pulses produce an electrostrictive force density. The elastic deformations resulting from this force are again impulsively launched counterpropagating acoustic waves, whose wavevector matches the optical pattern. If the polarization of the two-excitation pulses is not parallel then acoustic waves of particular polarizations along selected wave-vector directions can be produced. Within isotropic materials, pure transverse modes (shear waves) can be excited with perpendicular polarized pulses. (There is an inherent cost because transient diffraction gratings are significantly reduced in intensity for non-parallel beam polarizations). In other words, ISLS can generate acoustic modes (longitudinal, quasilongitudinal, quasishear, and shear) within the constraint of what the medium can support, in any orientation, in materials of any symmetry. However, the photosentivities vary widely, some to the point where the diffracted intensities are too weak to detect.

Since material properties govern the time-dependent behavior of the laser-induced grating, a third variably delayed *probe* beam, with wavelength and wave vector $(\lambda_P, \mathbf{k}_P)$, is monitored as it coherently diffracts off the grating $(\lambda_D, \mathbf{k}_D)$ at the phase matched Bragg angle. Measuring the scattered diffraction intensity, I, of the probe pulse as a function of delay time, given stable physical conditions of the experiment (*i.e.*, grating spacing, pressure and

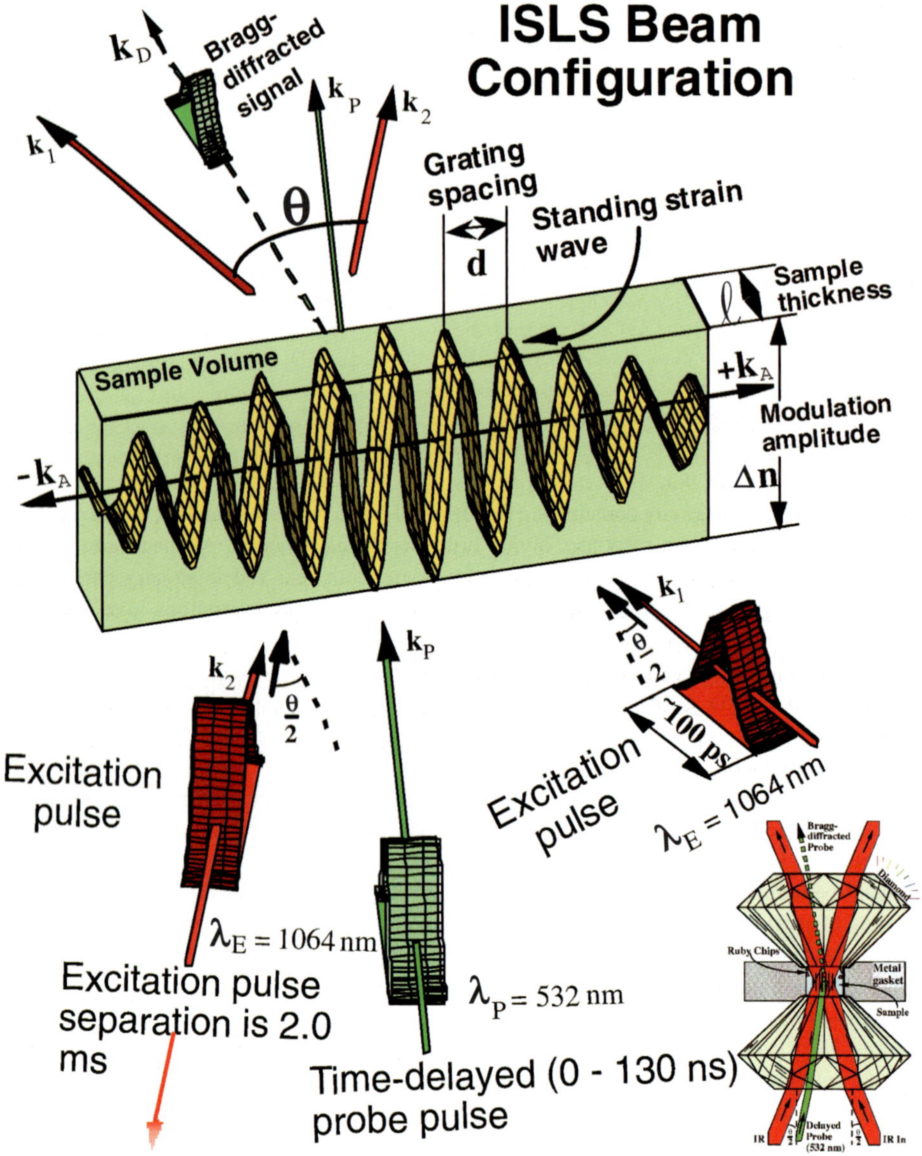

Figure 1. Laser-induced ultrasonic wave excitation and detection using ISLS. Two 100 picosecond infrared pulses converge spatially and temporally within a transparent sample medium. The time-dependant intensity or strength of the optical transient diffraction grating is monitored by a third frequency doubled pulse which, in the case for relatively low scattering strength, is systematically delayed in time. The time response of stronger gratings can be monitored using a CW laser.

temperature), serves to determine the frequency, f_A, and attenuation, γ, of the induced acoustic standing waves and any observable structural modes. Dispersion properties are monitored by measuring sound velocity, u, as a function of f_A (under constant physical conditions) by varying λ_A (*i.e.*, changing the intersection angle θ). The sound velocity is determined by

$$u = \lambda_A \cdot f_A \tag{2}$$

where again λ_A (in microns if $\lambda_E = 1.064\ \mu m$) is the grating spacing or period. The grating spacing is determined by measuring f_A of a temperature-independent standard. The confirmed literature value for the velocity of the standard u_S, and the measured frequency provides an accurate determination of $\lambda_A \pm 0.2\%$ in the sample, provided the angle θ remains constant and the standard has a low thermal expansion coefficient with no dispersion. (For this study glass was the secondary standard and was calibrated against water and fused silica) [90-93]. If the diffracted signal intensity is high enough, then a record of the grating modulation can be taken using a cw probing source in conjunction with very fast transient recording devices. At a typical pulse repetition frequency (PRF) of 0.5 KHz, one second of averaging provides a substantial s/n advantage over averaging each shot from a systematically delayed pulsed probing source. In this way, the data acquisition time can be reduced by factors of 1000 or more. The grating intensity, or diffractive strength, depends inherently on material modes coupling to the optical electric field and absorption of the excitation laser pulses.

The superposition of two intersecting optical pulses can generate grating-like modulations in the optical properties of the excited material. For example, the modulation amplitudes for n, the real part of the index of refraction, and *K*, the (spatial) absorption coefficient as functions of the probe wavelength are

$$\Delta n\ (\lambda_P) = (\partial n/\partial T)_\rho \cdot \Delta T \qquad \textit{Thermal grating} \tag{3a}$$

$$\Delta K\ (\lambda_P) = -\ (\sigma_o - \sigma_n) \cdot \Delta N \qquad \textit{Population density (excited-state) grating} \tag{3b}$$

where ΔT is the temperature amplitude, $(\partial n/\partial T)_\rho$ is the thermo-optic coefficient, σ_o and σ_n are the absorption cross-sections for the ground and nth excited electronic-energy states, and ΔN is the change in the population density where $(N_n + \Delta N) - (N_o - \Delta N) = \Delta N$. To generalize, any modulation within the sample volume of a material property with amplitude ΔA will be accompanied by an optical grating with amplitudes

$$\Delta n\ (\lambda_P) = (\partial n/\partial A)\ \Delta A \tag{4a}$$

$$\Delta K\ (\lambda_P) = (\partial K/\partial A)\ \Delta A\ . \tag{4b}$$

The complex refractive index includes both optical parameters

$$\tilde{n} = n + iK/2|\boldsymbol{k}_P| \tag{5a}$$

where $\mathbf{k}_P$ is the wave vector at which the optical properties are measured, and as above its amplitude can generally be described by the material response amplitude to optical excitation by

$$\Delta\tilde{n} = (\partial\tilde{n} / \partial A)\cdot\Delta A\,. \tag{5b}$$

Furthermore, in a non-magnetic (where the permeability $\mu = 1$) medium, ñ is related to the complex frequency-dependent dielectric constant $\tilde{\varepsilon}$ (ω_P), and the dielectric susceptibility χ_ε, by

$$\tilde{n}^2 = \tilde{\varepsilon} = 1 + 4\pi\chi_\varepsilon \tag{6a}$$

and

$$\Delta\tilde{n} \approx \Delta\tilde{\varepsilon} \,/\, 2\tilde{\varepsilon}^{1/2} = \Delta\chi_\varepsilon \,/\, 8\pi(1 + \chi_\varepsilon)^{1/2} \tag{6b}$$

where, $\tilde{\varepsilon} = \varepsilon + i\frac{4\pi\sigma_c}{\omega_P}$ and χ_ε couples the polarization to the electric field by $\mathbf{P} = \varepsilon_0\chi_\varepsilon\mathbf{E}$ [94]. The permittivity of a vacuum ε_0 has SI units of (C^2/J m). The specific conductivity σ_c ($1/(\Omega{\cdot}m)$) couples the electric field to the electric current density by $\mathbf{J} = \sigma_c\mathbf{E}$. From the relations described in (6b), it becomes evident that optically generated gratings correspond to spatial modulations of ñ, $\tilde{\varepsilon}$, or χ_ε. The parameters ΔA, $\tilde{\varepsilon}$, and χ_ε are tensorial. This means that the value of χ_ε depends on the material orientation to the electric field (anisotropic interactions). In general, **P** and **E** can be related by higher-rank susceptibility tensors, which describe anisotropic mediums. The refractive index n, and absorption coefficient *K*, can be joined to specify the complex susceptibility when $K(\lambda_P) \ll 4\pi/\lambda_P$ such that

$$\tilde{\chi}_\varepsilon = (n + i\,K / 2|\mathbf{k}_P|)^2 - 1 \approx n^2 - 1 + iKn / |\mathbf{k}_P|\,. \tag{7a}$$

If the absorption at λ_P is sufficiently small then,

$$\Delta\tilde{\chi}_\varepsilon = 2n\Delta n + i(n / |\mathbf{k}_P|)\cdot\Delta K \tag{7b}$$

Therefore, now we have one tensorial material parameter $\tilde{\chi}_\varepsilon$, composed of the primary optical properties that are driven by the macroscopic electric field $\mathbf{E}(\mathbf{r},t)$. A simple form of the constitutive equation is now available for the frequency-domain relating $\mathbf{P}(\mathbf{r},t)$ to $\mathbf{E}(\mathbf{k},\omega)$.

$$\mathbf{P} = \varepsilon_0\,(\chi^{(1)}\cdot\mathbf{E} + \chi^{(2)}:\mathbf{E}^2 + \chi^{(3)}\vdots\mathbf{E}^3 + \ldots) \tag{8}$$

where the n dots indicate a contraction of Fourier components, and $\chi^{(n)}(\omega_\sigma)$ is now the complex electric susceptibility tensor. The ω_σ term is defined as the sum of the optical frequencies in the susceptibility tensor. The definition holds for all orders of nonlinearity. If $\mathbf{E}(\mathbf{k},\omega)$ is well behaved, a Fourier transformation of (8) will yield a similar expansion for the time-domain where time dependent response functions replace the susceptibility. These constitutive equations are central to conventional nonlinear optic methodologies. The explicit

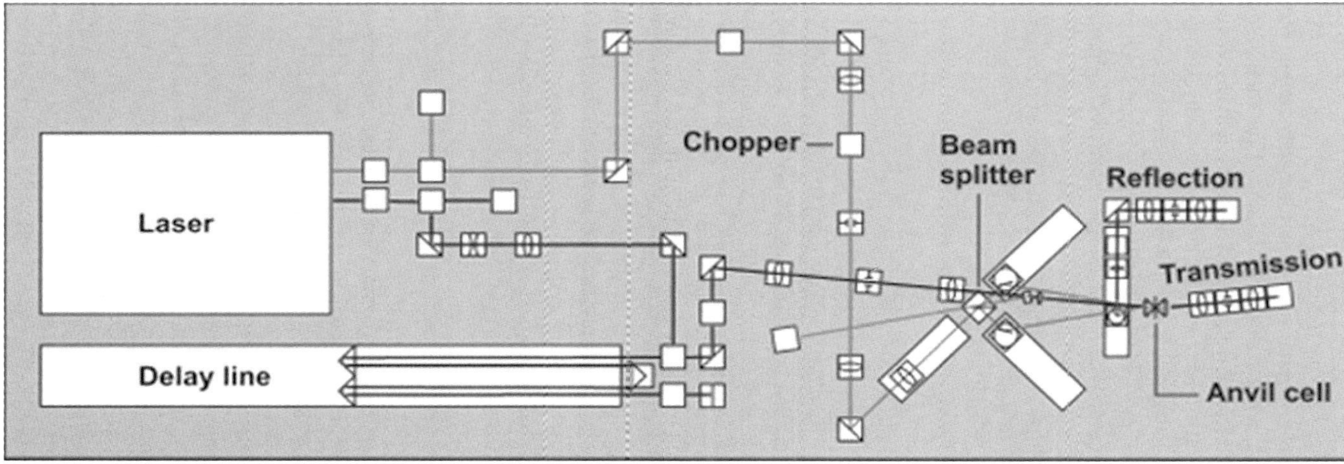

Figure 2. ISLS system. A Time-Bandwidth Jaguar Nd:YAG laser provides a continuous train of 100 ps 1064 nm excitation and 532 nm probe pulses at a fixed pulse repetition frequency. Every other near IR pulse is removed using a mechanical chopper wheel to provide a non-transient grating based parasitic scattering background that is then subtracted from grating diffracted signals. The 532 nm probe pulses are time-of-flight delayed by using a mechanical translator. Diffracted probe pulse signals are collected using a spatial filter and a digitally amplified photo multiplier tube. Use of a boxcar integrator increases the signal-to-noise ratio and a Labview® data acquisition board and software is employed to digitize and normalize time-domain ultrasonic data. The system can be used in either transmission or reflection geometries.

forms for the susceptibilities can only be obtained through a full quantum-mechanical calculation.

As a laser spectroscopic technique, ISLS has a spectral resolution ultimately limited only by the laser pulse width. The very high spatial and temporal resolutions of the method are limited only by the excitation wavelength and the pulse width. Because the output is coherent and directional, spatial, temporal, and spectral filtering can be employed to discriminate against background radiation due to possible luminescence or incoherent scattering related to the optical quality of the sample. What is even more convenient is that the signal becomes ever more spatially separated from the initial probe beam and spurious scattered light with further distance away from the sample. Hence, the signal can be obtained in a virtually dark background. This owes to the extraordinarily high sensitivity of the technique. Furthermore, the method lends itself to *in-situ* characterization of materials in physical situations very different from ambient conditions. This has allowed for interesting studies of high-energy plasmas [42] and flames [70].

4. RAMAN AND FTIR EXPERIMENTS

Our near-simultaneous Raman and Fourier transform infrared (FTIR) microscope was designed to accommodate long working distance and hot DACs (see Figure 3). A motorized rail permits rapid switching between probes. A Bruker Optics Vector-33 interferometer is also used as an IR glow bar source and data is collected via an external IR LN cooled CdTe detector. Raman fluorescence data is collected via a fiber optic channel and fed into a fast spectrograph and CCD detector. The nominal IR resolution is adjustable and typically, we choose between 1 cm^{-1} or 4 cm^{-1} resolution. Pressure is determined during melting and recrystallization studies by monitoring a calibrated optical pressure gauge such as the SrB_4O_7: $Sm2^+$ $\lambda_{0\text{-}0}$ line [95] and/or the temperature-dependant ruby (Al_2O_3: $Cr3^+$) fluorescence line [96]. Temperature was monitored using type-k thermocouples secured against the gasket and a diamond anvil with gold foil.

5. COMPUTATIONAL METHODS

We apply an accurate and numerically efficient equation of state for multiphase mixtures in the Cheetah thermochemical code to detonation, shocks, and static compression. The Cheetah thermochemical code is used to determine the properties of reacting energetic materials. We present a library of parameters for fluid and condensed high pressure molecules in Ref. [97]. We call this library "EXP6". Cheetah supports a wide range of elements and condensed detonation products. We have applied a Murnaghan [98] equation of state (EOS) form to a variety of metals, metal oxides and other solids. We have also matched phase transition data for many of these solids. We have applied the EXP6 equation of state to numerous formulations containing the elements C, H, N, and O. We find that the EXP6 equation of state library improves significantly on previous equation of state libraries, without fitting to detonation properties.

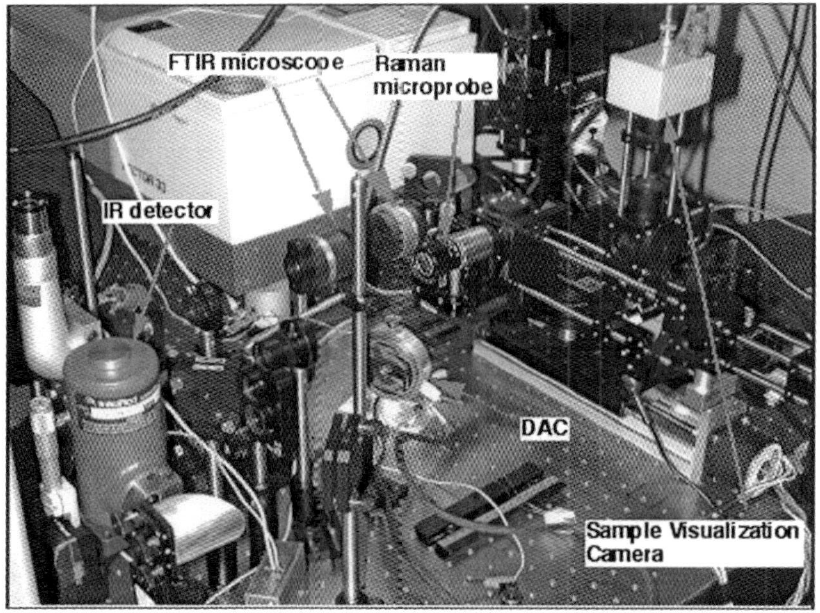

Figure 3. Raman/FTIR system. A custom designed vibrational spectroscopy instrument that accommodates a large variation of geometrically different diamond cells for both Raman and FTIR spectroscopy. For Raman, 488 nm or 514nm CW light is injected into the instrument stage through a fiber optic cable. Spectra are collected in reflection geometry and coupled to a second fiber optic line that is fed into a spectrometer with a CCD array. The entire system can be rapidly switched over to FTIR data collection by use of a motorized stage. A standard glow bar source housed in a Bruker® Vector 33 interferometer provides the near and mid IR source that is focused with a reflecting mirror into the DAC encapsulated sample. IR absorption data are collected in transmission with a second reflecting objective and focused into a CdTe detector. Care is taken to correctly aperture the IR light before and after the DAC. A minimum sample diameter of approximately 30 microns can be interrogated with this system. The microscope stage also includes a CCD camera to allow visual acuity of the DAC sample in either Raman or FTIR mode.

While there exists an extensive body of experimental techniques and experience on computational methods appropriate to ambient conditions, the regime of strong repulsive interactions at very high densities has not been as extensively investigated. The experiments discussed here are aimed both at enlarging the family of properties conveniently measured at high pressure and, principally, at providing the data appropriate to a critical test of the theory of the interatomic potential in simple substances at high density. We present new experimental data for the equation of state of CH_3OH and C_2H_5OH and CH_2O_2. We find that CH_2O_2 is present during the detonation of some common explosives. These developments will help to further improve the accuracy of the Cheetah code in the future.

5.1. Introduction to computations

The energy content of an energetic material often determines its practical utility. An accurate estimate of the energy content is essential in the design of new materials [99] and in the understanding of quantitative detonation tests [100]. The Cheetah thermochemical code is used to predict detonation performance for solid and liquid explosives. Cheetah solves thermodynamic equations between product species to find chemical equilibrium for a given pressure and temperature. The useful energy content is determined by the anticipated release mechanism. Since detonation events occur on a microsecond timeframe, any chemical reactions slower than this are not relevant when considering a detonation. Another way of looking at energy release mechanisms is through thermodynamic cycles. Detonation can be thought of as a cycle that transforms the unreacted explosive into stable product molecules (chemical equilibrium) at the Chapman-Jouget state [101]. This is simply described as the slowest steady shock state that conserves mass, momentum, and energy. Similarly, the deflagration of a propellant converts the unreacted material into product molecules at constant enthalpy and pressure. Understanding energy release in terms of thermodynamic cycles ignores the important question of the time scale of reaction. The kinetics of even simple molecules under high pressure conditions is not well understood. Diamond anvil cell and shock experiments promise to provide insight into chemical reactivity under extreme conditions. Despite the importance of chemical kinetic rates, chemical equilibrium is often nearly achieved when energetic materials react. This is a consequence of the high temperatures produced by such reactions (up to 6000K). We will begin our discussion by examining thermodynamic cycle theory as applied to high explosive detonation. This is a current research topic because high explosives produce detonation products at extreme pressures and temperatures: up to 40 GPa and 6000K. Relatively little is known about material equations of state under these conditions. Nonetheless, shock experimentation on a wide range of materials has generated sufficient information to allow reasonably reliable thermodynamic modeling to proceed.

One of the attractive features of thermodynamic modeling is that it requires very little information regarding the unreacted energetic material under elevated conditions. The elemental composition, density, and heat of formation of the material are the only information needed. Since elemental composition is known once the material is specified, only density and heat of formation needs to be predicted. The Cheetah thermochemical code offers a general-purpose, easy to use, thermodynamic model for a wide range of materials.

Chapman-Jouget (C-J) detonation theory [101] implies that the performance of an explosive is determined by thermodynamic states -the Chapman-Jouget state and the connected adiabat. Thermochemical codes use thermodynamics to calculate these states, and hence obtain a prediction of explosive performance. The allowed thermodynamic states behind a shock are intersections of the Rayleigh line (expressing conservation of mass and momentum), and the shock Hugoniot (expressing conservation of energy). The C-J theory states that a stable detonation occurs when the Rayleigh line is tangent to the shock Hugoniot. This point of tangency can be determined, assuming that the equation of state P = P(V,E) of the products is known. The chemical composition of the products changes with the thermodynamic state, so thermochemical codes must simultaneously solve for state variables and chemical concentrations. This problem is relatively straightforward, given that the equation of state of the fluid and solid products is known.

One of the most difficult parts of this problem is accurately describing the equation of state of the fluid components. Efforts to achieve better equations of state have largely been based on the concept of model potentials. With model potentials, molecules interact via idealized spherical pair potentials. Statistical mechanics is then employed to calculate the equation of state of the interacting mixture of effective spherical particles. Most often, the exponential-6 potential is used for the pair interactions:

$$V(r) = \frac{\varepsilon}{\alpha - 6}\left[6\exp(\alpha - \alpha r / r_m) - \alpha (r_m / r)^6\right] \tag{9}$$

where, r is the distance between particles. r_m is the minimum of the potential well. ε is the well depth, and α is the softness of the potential well.

The JCZ3 EOS was the first successful model based on a pair potential that was applied to detonation [102]. This EOS was based on fitting Monte Carlo simulation data to an analytic functional form. Hobbs and Baer [103] have recently reported a JCZ3 parameter set called JCZS

The exponential-6 model is not well suited to molecules with a large dipole moment. Ree [104] has used a temperature-dependent well depth $\varepsilon(T)$ in the exponential-6 potential to model polar fluids and fluid phase separations. Fried and Howard have developed an effective cluster model for HF [98]. The effective cluster model is valid to lower temperatures than the variable well-depth model, but it employs two more adjustable parameters.

Many materials produce large quantities of solid products upon detonation. The most common solid detonation product is carbon, although some explosives produce aluminum and aluminum oxide [105]. Uncertainties in the equation of state and phase diagram of carbon remain a major issue in the thermochemical modeling of detonation. van Thiel and Ree have proposed an accurate Mie-Gruneisen equation of state for carbon [106]. Fried and Howard [107] have developed a simple modified Murnaghan equation of state for carbon that matches recent experimental data on the melting line of graphite. There is considerable uncertainty regarding the melting line of diamond. Fried and Howard argue based on reanalysis of shock data that the melting line of diamond should have a greater slope. Shaw and Johnson have derived a model for carbon clustering in detonation [108]. Viecelli and Ree have derived a carbon-clustering model for use in hydrodynamic calculations [109, 110].

In the present approach, we apply an accurate and numerically efficient equation of state for the exp-6 fluid based on Zerah and Hansen's hypernetted-mean spherical approximation (HMSA) [111] equations and Monte Carlo calculations to detonation, shocks, and static compression. Thermal effects in the EOS are included through the dependence of the coefficient of thermal expansion on temperature, which can be directly compared to experiment. We find that we can replicate shock Hugoniot and isothermal compression data for a wide variety of solids with this simple form.

The exp-6 potential has also proved successful in modeling chemical equilibrium at the high pressures and temperatures characteristic of detonation. However, in order to calibrate the parameters for such models, it is necessary to have experimental data for molecules and mixtures of molecular species at high temperature and pressure. Static compression data, as well as sound speed measurements, provide important data for these models. We validate Cheetah through several independent means. We consider the shock Hugoniots of liquids and solids in the "decomposition regime" where thermochemical equilibrium is established. We argue that this regime is reached for most organic materials above 50 GPa shock pressures. We also validate the code against high explosive overdriven shock Hugoniots, and more traditional metrics such as the detonation velocity and pressure. Overall, we find that Cheetah offers a highly accurate representation of high-pressure equation of state properties with no empirical fitting to detonation data.

The nature of the Chapman-Jouget and other special thermodynamic states important to energetic materials is strongly influenced by the equation of state of stable detonation products. Cheetah can predict the properties of this state. From these properties and elementary detonation theory the detonation velocity and other performance indicators are computed. Thermodynamic equilibrium is found by balancing chemical potentials, where the chemical potentials of condensed species are just functions of pressure and temperature, while the potentials of gaseous species also depend on concentrations. In order to solve for the chemical potentials, it is necessary to know the pressure-volume relations for species that are important products in detonation

We now specify the equation of state used to model detonation products. For the ideal gas portion of the Helmholtz free energy, we use a polyatomic model including electronic, vibrational, and rotational states. Such a model can be conveniently expressed in terms of the heat of formation, standard entropy, and constant pressure heat capacity of each species. The heat capacities of many product species have been calculated by a direct sum over experimental electronic, vibrational, and rotational states. These calculations were performed to extend the heat capacity model beyond the 6000K upper limit used in the JANAF thermochemical tables (J. Phys. Chem. Ref. Data, Vol. 14, Suppl. 1, 1985). Chebyshev polynomials, which accurately reproduce heat capacities, were generated.

Experimental observables were placed into categories. We took the first category to be the volume along the shock Hugoniot and reshocked states. The second was the temperature along the shock Hugoniot and reshocked states. The third was the volume under static compression. The last category was the sound speed under static compression. For each category, we determined an average error.

The Figure of merit is a weighted average of the category errors. We nominally assign a weight of 40% to shock volumes, 25% to shock temperatures, 25% to static volumes, and 10% to the speed of sound. Depending on the degree of chemical reactivity the optimization procedure is weighted more to shocks than static measurements, although we find below that we reproduce both well. A stochastic optimization algorithm was employed to minimize the figure of merit function. Our final parameters are listed in Ref. [97]. In the following subsections we analyze the performance of the resulting equation of state in reproducing a wide range of experimental measurements. Results for nitrogen are fully discussed in [111]. Although the parameters in that work are slightly different than those used here, the comparison to experiment is similar.

Other workers [112, 113] have shown that a chemical equilibrium model of hydrocarbons based on an exponential-6 fluid model using Ross's soft-sphere perturbation theory is successful in reproducing the behavior of shocked hydrocarbons. Our model of the supercritical phase includes the species H_2, CH_4, C_2H_6, and C_2H_4. We have chosen model parameters to match both static compression isotherms and shock measurements wherever possible. The ability to match multiple types of experiments well increases confidence in the general applicability of our high-pressure equation of state model.

We now specify the sources of experimental data used in the calculations that follow. Shock Hugoniot data for oxygen comes from Nellis *et al.* [114] and Marsh [115]. Static equation of state data comes from Weber [116], while sound velocity data comes from Straty *et al.* [117], and Abramson *et al.* [118]. Nitrogen shock Hugoniot data comes from Zubarev *et al.* [119], and Nellis *et al.* [120, 121]. Static equation of state data comes from Malbrunt and Robertson *et al.* [122,123], and low-pressure static EOS and sound velocity data comes from Robertson *et al.* [123] and Kortbeck *et al.* [124]. Shock data for pure methanol and ethanol comes from Marsh [115], and static sound speed data comes from Zaug and Crowhurst *et al.*, [125]. Shock Hugoniot data of formic acid comes from Trunin *et al.* [126], adiabatic sound velocities come from Crowhurst and Zaug [127], room temperature equation of state data from angle dispersive x-ray powder diffraction experiments of Goncharov and Zaug [128], and high-pressure melt data from Montgomery *et al.* [129]. Published thermodynamic parameters (ambient pressure) come from Stout *et al.* [130] and Wilhoit *et al.* [131].

6. FLUID EQUATIONS OF STATE

Measurements of the speed of sound in supercritical oxygen have been made using ISLS along two isotherms of 30° and 200° C, and in a 1:1 molar mixture of $N_2 + O_2$ along a 250° C isotherm. (See Figure 4a.) Each oxygen isotherm was followed up to near the freezing points (5.9 and 12GPa). Starting with known values of density, ρ, and specific heat, C_P, the thermodynamic equation of state is calculated by recursive numerical integration of

$$\left(\frac{\partial \rho}{\partial P}\right)_T = (c_o)^{-2} + \left(\frac{\partial^2 T}{\partial C_P}\right) \quad and \quad \left(\frac{\partial C_P}{\partial P}\right)_T = \left(\frac{\partial^2 V}{\partial T^2}\right)_P \tag{10}$$

where P, c_o, α, T, and V are respectively the pressure, zero frequency sound speed, thermal expansion coefficient, absolute temperature, and specific volume. In this work, initial values of ρ and C_P were taken from the EOS of Wagner *et al.* [132].

An overview of previous work on oxygen is given by Wagner and Schmidt (W&S) [132]. These authors have generated a thermodynamic potential based on experimental densities up to 0.08 GPa and at 130°C up to 0.03GPa. In addition, they used combined density and heat capacities measured to 30°C and 0.03GPa. Other data, not used by Wagner and Schmidt, are those of Tsiklis and Kulikova [133] who measured densities to 1GPa and 400°C. The latter were used above 0.2 GPa by Belonoshko and Saxena (B&S) [134] to constrain a molecular dynamics simulation (based on an exponential-6 potential), which was in turn used to construct a P-V-T surface. A Shock Hugoniot for the 1:1 fluid mixture provides P-V-E data between 9.89 and 24.0GPa [135].

The data presented here are currently insufficient to make a "positive" determination of the equation of state of O_2 or the mixture. The high-pressure sound speed data, especially at higher temperatures, do not extend to the lower pressures at which values for C_p and ρ, are known. Further, the small variations in speed of sound within the experimentally useful range of temperatures used here are small enough to be confounded with the uncertainties in the measurements of pressure. Consequently, several approximations have been made to yield a reasonably accurate EOS. The results are then compared with other data.

The assumptions made are that the sound speeds are linear in T over the stated range, that the W&S EOS correctly predicts the speeds up to 0.5 GPa, and that the form of the interpolating function is suitable to the task. At pressures higher than 0.7 GPa the speeds are assumed to vary linearly between 30° and 200° C, and an artificial data set is calculated at six temperatures from 30° to 100℃, based on the previous fits at the two stated end temperatures. Each isotherm is then fit individually, with the fits forced to conform to the W&S EOS for pressures between 0.02 and 0.05 GPa. The result is a velocity field in P and T in which the velocities are linear interpolations in T above 0.7 GPa, fairing into the W&S EOS below that. The usual equations are then iteratively solved to obtain the densities, heat capacities, entropies, etc. The results are reasonable, the densities increasing monotonically while remaining below those of the β phase. The heat capacities, C_P, are fairly constant in pressure, varying by at most 5% for each isotherm. They undergo several oscillations with increasing pressure, which probably derive from the cross over of dc/dT from a positive to a negative value at 0.5 g/cm^3.

At 30° C the O_2 densities determined here are 8% higher than the B&S results up to 0.5 GPa, then cross at about 1.5 GPa and are then uniformly lower than B&S, by 10% at 6 GPa. B&S densities are, however, always less than that of the solid, β phase. Given reasonable values of C_P (at 0.5 GPa), either from W&S results or those determined here, the speeds of sound inferred from the B&S EOS are uniformly low by about 10% (see Figure 4a). In comparison, this discrepancy is due to their higher compressibilities below ~4 GPa and higher densities above 2 GPa. In order to make their speeds of sound agree (approximately) with results here at 30° C it is necessary to assume an initial C_P at 0.5 GPa of 9.2 J/K/mole at 30° C which is about 5 times lower than expected.

Speeds of sound were measured at 30° C and 1.5 GPa at frequencies of 1.3, 0.77 and 0.27 GHz. Velocities matched to within the uncertainties, *i.e.* ±0.2% for the higher frequency and ±0.5% for the two lowest. The ISLS velocities fair nicely with those of the W&S model and are lower than the extrapolation of W&S. More dispersion may exist at lower frequencies. Between 22° C and 122° C the fluid β-phase boundary is well fit by the straight line P(GPa) = 0.0270 T(° C) + 5.153 with a two σ uncertainty on the slope of 10^{-4} GPa/°C. Each point of equilibrium was established by a visual observation of the simultaneous presence of both phases. Among observations, the volume of solid varied from approximately 5 to 95% of the sample; no correlation was apparent between the deviations of the data from the fit and the fraction of solid. Since one expects that any impurities will be concentrated in the fluid, this fact suggests strongly that impurities had no significant effect on the measurements

The measured oxygen velocities fit well to the form $\Sigma A_i \ln P^i$ with i={0...4}. The 30° - 200° C fitparameters are A_0=2.0438-1.8665, A_1=0.7764-0.8462, A_2=0.1040-0.140, A_3=0.0078- -0.0020, and A_4=0.0010- -0.0016. In such fits the data were supplemented by points at lower pressures generated from the W&S EOS. Additionally, the curve at 30° C was constrained to lie along the 200° C isotherm above 7 GPa. N_2-O_2 fit parameters from 1.3 to 6.5 GPa at 250° C are A_0=2.0058, A_1=0.4490, A_2=0.8424, A_3= -0.2605, and A_4= -0.0015. A 1:1 molar ratio of N_2-O_2 at 25° C forms δ-N_2 at approximately 4.3GPa [136], which accounts for the significant increase in velocity observed at 7.1GPa. The calculated points in Figure 4a were derived from an accurate EOS for exp-6 type fluids [111] based on HSMA integral equation theory and Monte Carlo calculations.

According to simple theories, substances should behave the same when all variables are suitably scaled and the critical parameters are the most common scaling factors chosen. Figure 4b shows Mills et al. [137] –25.5° C data, which is equivalent to oxygen at 30° C when scaling by the critical temperatures. The N_2 sound speeds are reduced using critical pressures and densities.

$$c_{O_2}(equivlent) = c_{N_2}\frac{\sqrt{\dfrac{P_c(O_2)}{P_c(N_2)}}}{\sqrt{\dfrac{Mw(O_2)}{Mw(N_2)}} * \sqrt{\dfrac{\rho_c(O_2)}{\rho_c(N_2)}}} \tag{11}$$

Since O_2 and N_2 have the exact same compressibility factor ($P_cV_c / RT_c = 0.292$), and no dipole moment, it may not be too much of a surprise that the sound speeds correlate well with the empirical law of corresponding states. This result suggests that N_2 and O_2 molecules are approximately spherical up to 2.2GPa.

The sound velocity of pure methanol (CH_3OH) and pure ethanol (C_2H_5OH) was measured along a 250° C isotherm up to 3.9 GPa. After each data point was taken the sample was cooled and the velocity was again measured and compared to previous measurements of uncooked methanol. No appreciable velocity difference between data sets was observed.

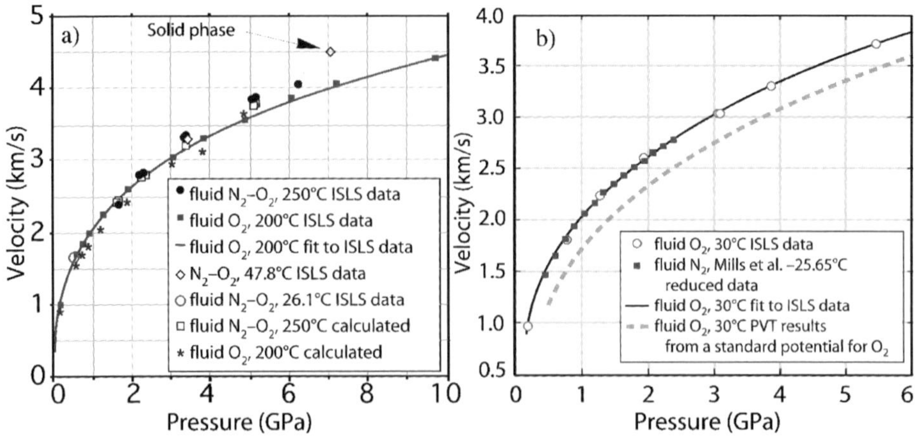

Figure 4. (a) ISLS sound speed data and corresponding calculations for oxygen and 1:1 molar ratio of fluid oxygen to nitrogen. (b) Example of the law of corresponding states for O_2 and N_2. The N_2 data [137] are reduced by the critical pressure, temperature and density and compared against ISLS O_2 data at 30° C [118]. The dashed line is a molecular dynamic result using a standard potential [134]. For O_2, a C_p at low pressure, where reasonably known, was used to start the integration necessary to generate the sound speeds.

A methanol model was previously implemented in the Cheetah code. The model is based on a combination of shock Hugoniot data and sound speeds determined via ISLS. High-pressure and temperature equation of state data on pure ethanol was not available, so Impulsive Stimulated Light Scattering measurements were made of the sound speed of ethanol at 250° C. Results are shown in Figure 5. A Cheetah exponential-6 potential model was fit to the ISLS measurements. The ~3% difference between data sets shows the utility of Cheetah and the consistency between static and dynamic equation of state measurements. High precision ISLS measurements easily resolve ethanol velocities from 2-3% lower methanol velocities.

The Cheetah thermochemical code uses assumptions about the interactions of unlike molecules to determine the equation of state of a mixture. The accuracy of these assumptions is a crucial issue in the further development of the Cheetah code. We have tested the equation of state of a mixture of methanol and ethanol in order to determine the accuracy of Cheetah's mixture model. Cheetah uses an extended Lorenz-Berthelot mixture approximation [138] to determine the interaction potential between unlike species from that of like molecules:

$$\varepsilon_{ij} = \sqrt{\varepsilon_{ii}\varepsilon_{jj}}$$

$$r_{m,ij} = K_{ij}(r_{m,ii} + r_{m,jj})/2 \qquad (12)$$

$$\alpha_{ij} = \sqrt{\alpha_{ii}\alpha_{jj}}$$

where, ε is the attractive well depth between two molecules and r_m is the distance of maximum attraction between two molecules. The parameter α controls the steepness of the repulsive interactions and K is a non-additive parameter, typically equal to unity.

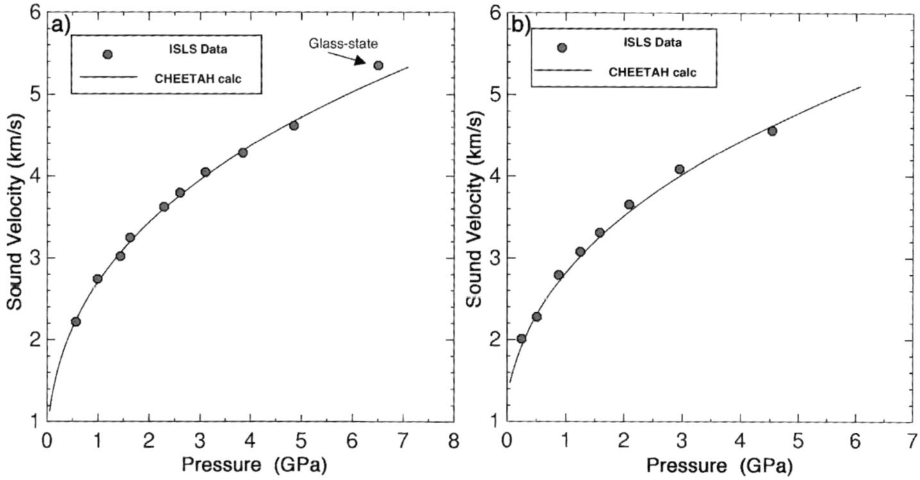

Figure 5. (a) ISLS sound speed data and corresponding calculations for supercritical methanol along a 250° C isotherm (b) Data and corresponding calculations for supercritical ethanol along a 250° C isotherm. The difference between MeOH and EtOH sound speeds is typically less than 3 % in this pressure/temperature regime. Raman spectra taken from MeOH at 6.51 GPa indicate that a liquid to glass transition occurred and accounts for the discontinuous increase in velocity compared to the fluid state.

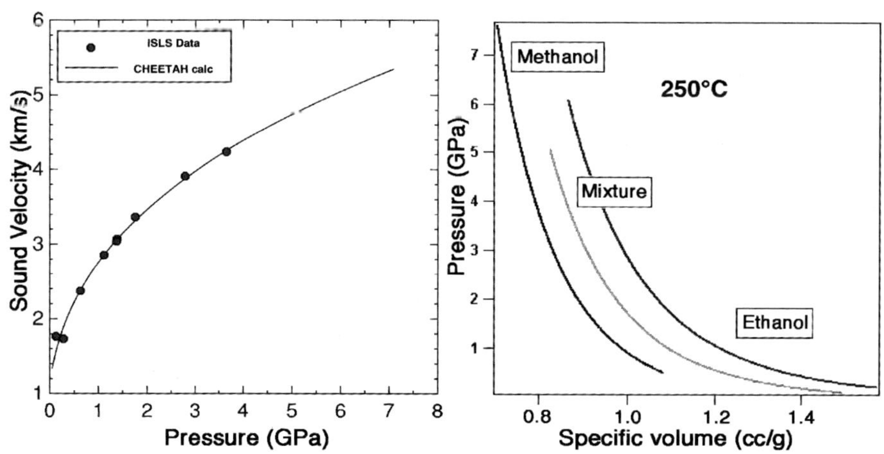

Figure 6. (a) ISLS sound speed data and corresponding prediction (line) for a 50:50 volumetric mixture of supercritical methanol-ethanol along a 250°C isotherm (b) Corresponding EOS calculated isotherms for supercritical methanol, ethanol and a 50:50 volumetric mixture [125].

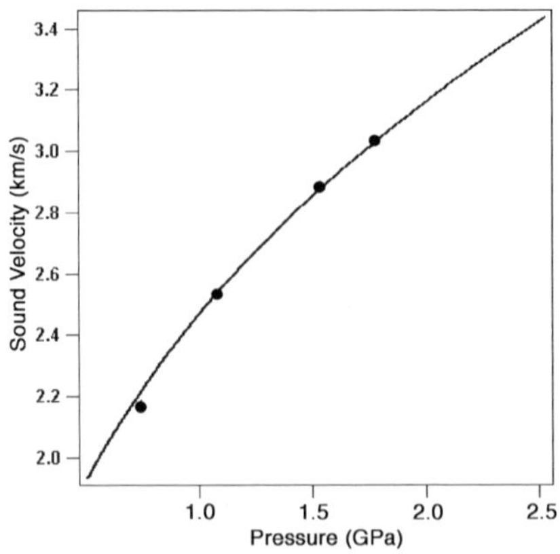

Figure 7. ISLS sound speed data and corresponding Cheetah calculation (line) for pure formic acid along a 140°C isotherm [127].

7. EXTREME CHEMISTRY

When measuring high-pressure and temperature sound velocities in supercritical organic fluids, one must verify that chemical reactions do not occur. Some of our preliminary measurements on formic acid gave indication through anomalous velocities and altered Raman spectra that reactions occurred above certain pressure-temperature conditions. Thus we began development of a phase stability diagram for formic acid using FTIR and Raman spectroscopic techniques to differentiate between liquid, solid, and reacted states.

Formic acid is a simple monocarboxylic acid. A study of solid formic acid provides insight into the nature of hydrogen bonding with pressure. Unlike other carboxylic acids, formic acid does not form dimers in the solid state, but instead forms an infinite length network of hydrogen-bonded chains, linked by the hydroxyl group. Formic acid has *cis*- and *trans*-conformations that form chains. A phase transition was previously reported by Shimizu to occur at 4.5 GPa [139]. A subsequent study proposed a high-pressure crystal structure consisting of a more complex phase, which combines *cis*- and *trans*- isomers of HCOOH in symmetrically flat layers [140]. Our x-ray powder diffraction data indicates the low-pressure phase is stable to well over 30 GPa. Rather than a cis/trans conformational change it is most probable that Shimizu observed mode coupling between the O-D stretch and C=O stretch Raman bands resulting in the observed frequency inflection at 4.5 GPa.

Pure (99.99%) and neat formic acid was loaded into a membrane DAC chamber consisting of two counter opposed 500 μm diamonds (synthetic type II anvils) and a pure Ir disk indented to ~30 microns thick and cut with a 220 micron EDM spark erode cutter. A Eurotherm® control system is used to power an external heating ring surrounding the DAC. The metal membrane capillary pressure was repeatedly adjusted to maintain a constant sample pressure. Sample temperature was monitored using type-K thermocouples lodged between diamond and a metal containment gasket using gold leaf foil. The temperature precision was approximately ±0.5 K and the absolute accuracy decreases with increasing temperature and was approximated to be +0 K and –4 K up to 575 K.

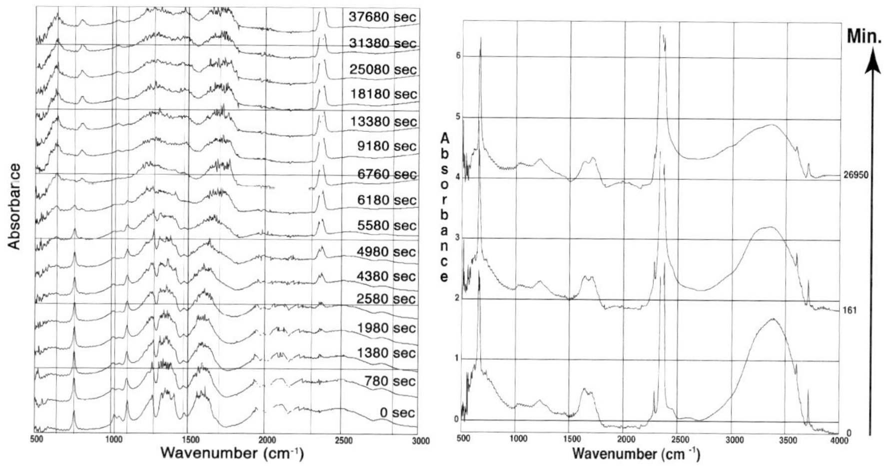

Figure 8. (a) Time-resolved FTIR spectra of pure formic acid at 5.9 GPa and 473K. Note the formation of CO_2 (662 cm^{-1}, 2364 cm^{-1}, and combination modes 3598 $cm^{-1}$and 3705cm^{-1} not shown in plot). (b) Time-resolved FTIR spectra of the products in (a) after rapid decompression to 0.2 GPa and temperature reduction to 298K.

Samples were heated at 1 K/min until melting was observed. Some samples were further heated at the same increasing rate until decomposition was observed. Changes in sample composition and structure were monitored by Raman and FTIR spectroscopy. Other samples were heated to achieve complete chemical decomposition. Temperature invariant FTIR spectral features indicated equilibrium was reached. A secondary indicator of a fully completed reaction was the evolution of a completely black and opaque sample. Some samples were cooled after melting had occurred, providing data on the solids of the system. There is not a smooth trend in pressure dependant crystallization temperatures due to inconsistent cooling rates. Constant pressure-temperature reactions were executed at 5.9 GPa and 4.2 GPa at 473K and 496 K respectively. In Figure 8a we display time-resolved FTIR spectra of formic acid held at 5.9 GPa and 473 K and in Figure 8b we show the evolution of the resultant products after temperature was reduced to 298 K followed by a rapid reduced to 0.2 GPa. The IR absorption background level decreased remarkably at this point. The reaction products are not apparently quenchable down to low pressure and room temperature conditions. The α-phase CO_2 band at 662 cm^{-1} provides evidence that the sample chamber remained sealed. There is a C-O bend mode at 1222 cm^{-1} and O-H and C=O and bending modes at 1638 cm^{-1} and 1710 cm^{-1} respectively. Over time, the I_{1638}/I_{1710} ratio decreases to less than unity. The broad background from 550-900 cm^{-1} provides evidence that H_2O is present. The number of O-H bonds (1638 cm^{-1}, 3345cm^{-1} and broad background centered around 720 cm^{-1}) decreased over the course of 18.7 days while the sp^2:sp^3 carbon bond ratio

(3226 cm^{-1}: 2950 cm^{-1} C-H bonds) also seemed to be decreasing. When solid polymer-like reaction products, intensely orange in color, were exposed to air, they appeared to be photosensitive: attempts to measure Raman vibrational spectra using low-intensity (< 2 mW over a 5 μm diameter area) visible light from an argon laser resulted, after prolonged exposure to the laser light, in photochemical oxidation of the solid product where the nature of carbon bonds become completely graphitic in nature. In some instances, relatively short laser light exposures (< 30 sec) yielded diamond-like carbon bonding spectra only to become graphitized with continued laser light exposure.

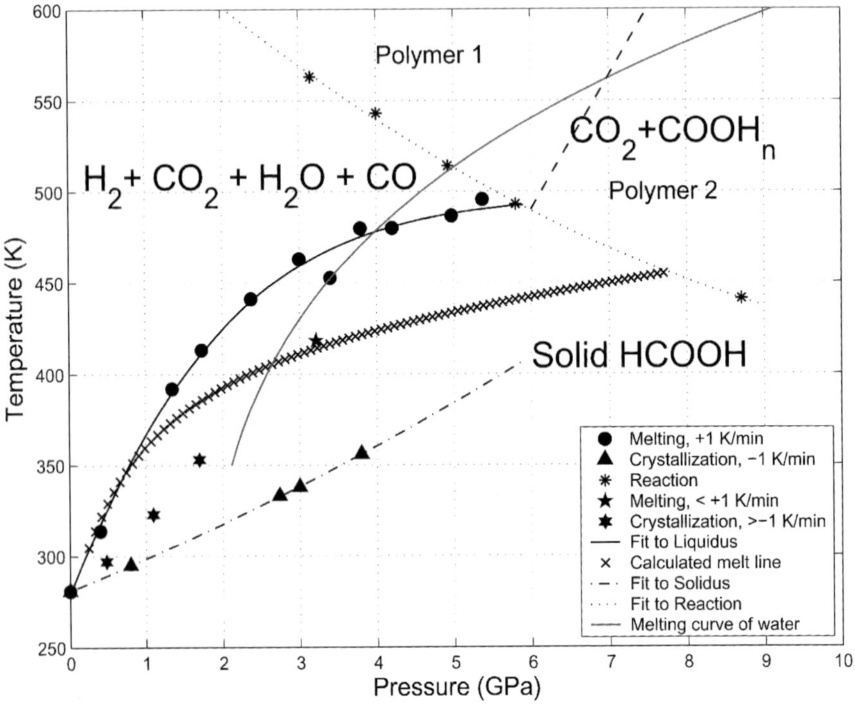

Figure 9. Reaction phase stability diagram of pure formic acid. Water melt curve is from F. Datchi *et al.*, Phys. Rev. B, 10 (2001) 6535.

The phase and chemical stability of formic acid is summarized in Figure 9. Below 5.5 GPa, we have observed that solid formic acid will melt and simultaneously begin to chemically react forming liquid CO_2, CO and H_2O. Due to experimental difficulty, we cannot provide direct evidence for the creation of molecular hydrogen though it would seem necessary in order to form CO_2 and CO. The molar concentration of these species is dependent on pressure, temperature and cooking time. As mentioned above with increased heating, a second decomposition reaction occurs producing an orange colored solid reaction product. Threshold temperatures required to produce polymer-like solid products are inversely proportional to pressure. Above 6 GPa, CO_2 production is accompanied with solid products resulting in a chemical triple point. At pressures under 5.5 GPa and 498 K, CO_2 and CO production occurs from the following reactions.

$$HCOOH_{solid} \longrightarrow HCOOH_{liquid} \longrightarrow CO_2 + H_2 \tag{13}$$

$$CO_2 + H_2 \longrightarrow CO + H_2O \tag{14}$$

Below 4.5 GPa we observe, in some cases, H_2O and CO. At room temperature where gas phase formic acid is a dimer, reaction (14) has a standard energy of 152 kJ/mol and a standard entropy change of 42.4 J/mol·K. From $\Delta G = \Delta H - T\Delta S$ we know the activation barrier for this gas phase reaction increases with temperature. If we observe CO and H_2O at elevated temperatures then this implies a reduced activation barrier in the high-pressure liquid state and/or a significant change in ΔH. Reactions 13 and 14 are catalyzed from metal substrates and Ir, our metal support gasket, is considered a particularly good catalyst for these reactions [141]. Evidence of CO from FTIR is experimentally more difficult where $I\nu_3(CO_2)/I(CO)=213$ and this may partially explain why we see no spectral evidence above 5 GPa where IR background absorption levels from polymer-like products are relatively high.

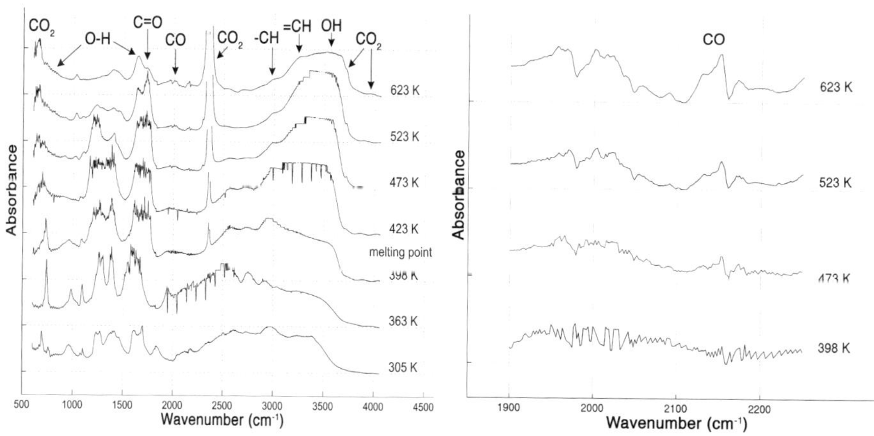

Figure 10. (a) Time-resolved FTIR spectra of pure formic acid at 3.0 GPa and a heating rate of 1K/min. At least three products from liquid formic acid, CO_2, H_2O and CO can be deduced from spectral assignments. (b) Expanded region in (a) centered around 2100 cm^{-1} showing the telltale IR absorption peaks of δ-CO.

The formation of hydrocarbons from thermal decomposition of formic acid at room pressure and high temperature (1696 K) has been reported by Muller *et al.* [142]. In our study we also find evidence of hydrocarbons and note how their spectral features depend on reaction conditions. At 3 GPa and room temperature, the nature of O-H bonds from formic acid become more covalent-like with increasing temperature. Once a reaction occurs the long-range order of the O-H network in crystalline formic acid becomes disrupted with bond distances increasing to a more hydrogen-like bond length centered at frequency of approximately 3500 cm^{-1}.

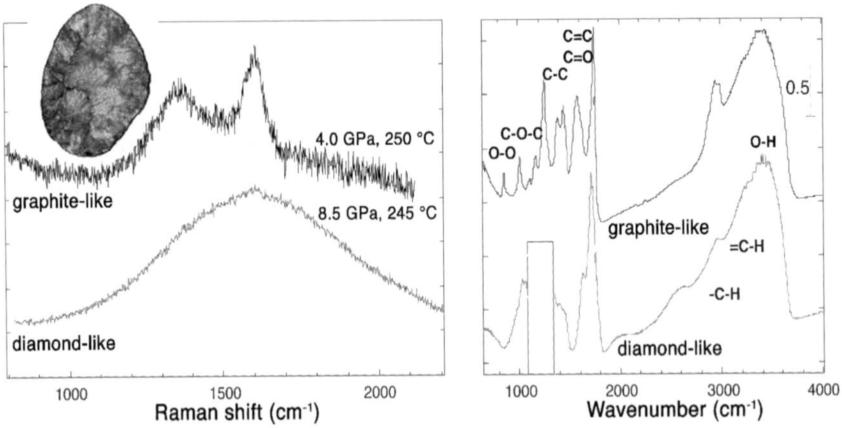

Figure 11. (a) Raman spectra (ambient conditions) of the C-H bend region of two different recovered products from formic acid. The inset is a photomicrograph of the 4.0 GPa products, intensely orange in color. (b) FTIR spectra of the samples described in (a).

We have yet to demonstrate the existence of the hydrocarbons produced in this pressure-temperature regime. In Figure 10a there is indication of H_2O products where a broad shoulder evolves at 600cm^{-1}- 800 cm^{-1}, and an O-H stretch mode forms at 1689 cm^{-1}. Figure 10b shows and expanded region of Figure 10a centered near 2100 cm^{-1} where a weak absorption doublet 2130cm^{-1} and 2148cm^{-1} intrinsic to δ-CO is observed. We also note CO spectral features appear at 4.2 GPa and T > 500 K. In this pressure regime, the infinite length hydrogen bond chains break following reaction (13) to form liquid HCOOH, where subsequently and CO_2 and presumably molecular hydrogen from. CO_2 combines with H_2 to produce CO, and some of the H_2 reduces the remaining HCOOH, producing amorphous hydrocarbons. A similar decomposition sequence occurs if the system is maintained at a fixed temperature and pressure.

At pressures above 5 GPa, for example at 8.3 GPa, there is no indication of CO formation. As the temperature is increased, CO_2 and hydrocarbon bands simultaneously appear, perhaps suggesting that formic acid is reduced by hydrogen created in reaction (13), and that reaction (14) does not occur at pressures over 5.5 GPa. Moreover hydrogen bond lengths remain invariant with temperature above 5.5 GPa and coincidently we see no evidence of water. As the reaction phase diagram shows, there seems to be two separate and identifiable reaction regimes delineated by the dotted curve in Figure 10. When thermally driven toward complete decomposition, each reaction region generates a different polymer-like product. Figures 11a and 11b show Raman and FTIR spectra of reacted samples recovered at STP respectively. The formation of C--C and C=C bonds at higher pressures, as indicated in Figure 11b where absorption occurs at 1027 cm^{-1} and 1585 cm^{-1} respectively, suggests that thermal decomposition of high-pressure formic acid may form what are perhaps complex organic compounds. Recovered samples appeared photochemically sensitive and their spectra may indicate how the nature of product carbon bonding depends on reaction conditions. The low-pressure (4 GPa) product (``Polymer 1") shows graphite-like sp$^2$ bonding, while the high-pressure (8.5 GPa) product (``Polymer 2") has a more diamond-like (sp^3) bonding nature. FTIR spectra indicates the presence of O-O, C-C, =C-H, and C=C bonds in Polymer 1 and Polymer 2 clearly contains –C-H bonds. Further analysis of recovered products from high-pressure DAC reactions will be conducted using conventional analytical chemistry techniques such as mass spectroscopy or perhaps nuclear magnetic resonance spectroscopy. It is an important challenge to systematically study chemical products prior to exposure of atmospheric oxygen and hydrogen. Chemical kinetic studies for the reactions discussed above are underway in our laboratory.

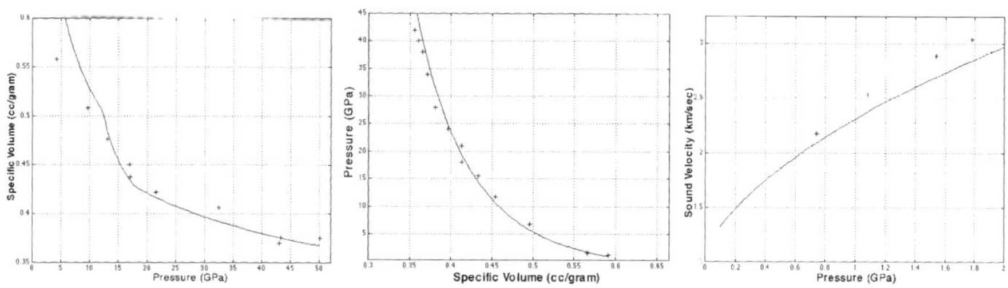

Figure 12. Comparison of (a) shock Hugoniot data (b) static cold compression, and (c) ISLS adiabatic sound velocities, and corresponding calculated curves. The respective summation of average errors between experiment and calculations are 4.0%, 0.5% and 6.5%.

8. GIBBS FREE ENERGY EQUATION OF STATE

In Figure 9 we present a calculated melt curve that compares favorably with our experimental results. This melt curve is the result of a minimized two-phase Gibbs free energy equation of state made to match accepted thermodynamic parameters and all available high-pressure experimental data including shock Hugoniot data [126], static cold compression volumes and compressibility from x-ray [128], and adiabatic ISLS sound velocity measurements [127]. Comparisons of these data are provided in Figure 12a,b, and c.

The pressure dependant term for our Gibbs free energy was derived using a Murnagham form for volume. Given the relatively low temperature of melting for formic acid to 6 GPa, we chose to set thermal expansion in our model to zero thus resulting in a simplified temperature independent expression for the high-pressure Gibbs free energy component. Our equation of state is based on an explicit functional form for $G(P)$. The Gibbs free energy expression derived here appears to be appropriate from $0 \leq P$ (GPa) ≤ 6 where formic acid was observed to melt.

We begin by breaking the Gibbs free energy into a reference component [$G_0(T)$] accounting for properties at 1-atmosphere and a second pressure dependant component:

$$G(P,T) = G_0(T) + \Delta G(P) \tag{15}$$

First we consider $G_o(T)$ where from $G = H - TS$ we have $G_0(T) = H_0(T) - TS_0(T)$. The enthalpy and entropy are conveniently expressed in terms of the constant pressure heat capacity at 1 atmosphere $C_{p,0}(T)$:

$$H_0(T) = \Delta H_0 + \int_{T_0}^{T} C_{p,0}(T)\,dT \tag{16}$$

$$S_0(T) = \Delta S_0 + \int_{T_0}^{T} \frac{C_{p,0}(T)}{T}\,dT. \tag{17}$$

In our present study $T_0 = 298.15$ K thus ΔH_0 and ΔS_0 are respectively, the standard heat of formation and entropy. In our computational model we consider the heat capacity for the liquid phase to be temperature independent and set it to 99.036 J/mole·K from literature data. For the solid phase, we employ a single Einstein oscillator to compute heat capacity:

$$C_{p,0}(T) = 7.5R * E(\Theta / T) \tag{18}$$

where the Debye temperature Θ is set to 281K and the Einstein expression is

$$E(x) \equiv \frac{x^2 e^x}{(e^x - 1)^2} \tag{19}$$

At the limit $T \to \infty$, $C_p = 62.358$ J/mole·K or 7.5R. Integration of equation 16 yields

$$H_0(T) = \Delta H_0 + \theta \left[\frac{1}{e^x - 1} \right]_{x_0} \tag{20}$$

where $x \equiv \Theta/T$ and $x_0 \equiv \Theta/T_0$.We evaluate $S_0(T)$ analytically to yield:

$$S_0(T) = \Delta S_0 \left[\frac{x}{e^x - 1} - \ln(1 - e^x) \right]_{x_0}^{x} \tag{21}$$

This completes our definition of $G_0(T)$. The pressure dependant component of equation 15 follows from $dG = VdP - SdT$ where we define $\Delta G(P)$ through postulation of a form for $V(P)$. Our result is

$$\Delta G(P) = \int_{P_0}^{P} V(P)\,dP. \tag{22}$$

The Murnagham form uses the relation

$$V(P) = V_0 \left[n\kappa_0 P + 1 \right]^{-1/n} \tag{23}$$

This form is derived by assuming a linear pressure dependence for the bulk modulus: $B = B_0 + nP$, where $B_0 = 1/\kappa_0$. The thermodynamic values that best minimize $G(P,T)$ to all available experimental data are provided in table 1.

The heat capacity for the liquid is taken from JANAF tables, while that for the solid is taken from ref [131]. The enthalpy and entropy for the liquid at STP are also taken from experimental data [130, 131], as is V_o [128]. The enthalpy and entropy of the solid are determined by replicating the melt temperature and enthalpy of melting at 1 ATM pressure. The V_o for the solid is determined by an extrapolation of the static compression data [128] to 1 ATM. The B_o and n for the solid are fit simultaneously to the static compression data and the measured slope of the melt curve. Likewise, the B_o and n values for the liquid-phase are fit to sound speed data [127] and the slope of the experimental melt curve [129]. The relatively high error in fitting sound speed data was due to compromising these parameters relative to fitting the measured slope of the melt curve.

Table 1. Two-phase parameters of formic acid used to calculate a high-pressure melt curve.

Parameter	Liquid-Phase	Solid-Phase
C_p (J/mole·K)	99.036	62.358
ΔH_0 (kJ/mole)	-425.100	-436.750
ΔS_0 (J/mole·K)	131.840	88.241
V_o (cc/g)	1.22	1.57
B_0 (GPa)	1.63	10.67
n	6.65	5.60

9. CONCLUSIONS

In the present chapter, we have reviewed our recent efforts to combine experimental and theoretical efforts to refine our knowledge of interatomic potentials and chemical processes at extreme conditions of pressure and temperature. We have demonstrated using selected molecular systems that our equation of state model can be used to accurately predict properties of non-polar and polar fluids including fluid mixtures. The accuracy of the equation of state of polar fluids is significantly enhanced by using a multi-species or cluster representation of the fluid.

We have measured sound velocities of various supercritical fluid systems. An attempt to carry forward such measurements on higher temperature isotherms of formic acid was frustrated by chemical reaction toward products that may include carbon dioxide, carbon monoxide, water, hydrogen and differentiated solid-like products at even higher temperatures and pressures. Nonetheless, the diamond anvil cell provides a unique opportunity to study the chemistry and kinetics of fluids under extreme conditions. We also find that CH_2O_2 is present during the detonation of some common explosives.

ACKNOWLEDGEMENTS

The authors thank Dr. P. Pagoria for preparation of pure CH_2O_2 and Don Hansen for technical assistance. We express our gratitude toward Dr. R. Simpson, Dr. L. Terminello, and Dr. P. Allen for their sustained support for our research. This work was performed under the auspices of the U.S. Department of Energy by the Lawrence Livermore National Laboratory under contract # W-7405-Eng-48.

REFERENCES

[1] D. J. Stevenson, Ann. Rev. Earth Planet. Sci., 10 (1982) 257.
[2] A. J. Cohen, and R. G. Gordon, Phys. Rev. B, 14 (1976) 4593.
[3] M. Waldman, and R. G. Gordon, J. Chem. Phys., 71 (1979) 1325.
[4] R. LeSar, and R. G. Gordon, J. Chem. Phys., 78 (1983) 4991.
[5] E. H. Abramson, J. M. Brown, and L. J. Slutsky, Ann. Rev. of Phys. Chem., 50 (1999) 279.
[6] M. Parinello and R. Carr, Phys. Rev. Lett., 55 (1985) 2471.
[7] J. A. Barker, in: Simple Molecular Systems at Very High Density, A. Polian, P. Loubeyre, and N. Boccara, Eds., NATO ASI Series B: Physics 186 (Plenum Press, NY, 1989) p. 331, ibid, p. 341.
[8] A. M. Stoneham, and J. H. Harding, Annu. Rev. Phys. Chem., 37 (1986) 53.
[9] H. Eichler, and H. Stahl, J. Appl. Phys., 44 (1973) 3429, ibid, 44 (1973) 5383.
[10] K. A. Nelson, and M. D. Fayer, J. Chem. Phys., 72 (1980) 5202.
[11] S. De. Silvestri, J. G. Fujimoto, E. P. Ippen, E. B. Gamble Jr., L. R. Williams, and K. A. Nelson, Chem. Phys. Lett., 116 (1985) 146.
[12] A. F. Goncharov, J. C. Crowhurst, and J. M. Zaug, Phys. Rev. Lett., 92 (2004) 155502.
[13] T. P. Dougherty, G. P. Wiederrecht, and K. A. Nelson, J. Opt. Soc. Am. B., 9 (1992) 2179.
[14] G. Eyring, M. D. Fayer, J. Chem. Phys., 81 (1984) 4314.
[15] J. M. Brown, L. J. Slutsky, K. A. Nelson, and L.-T. Cheng, Science, 241 (1988) 65.
[16] C. D. Marshall, I. M. Fishman, and M. D. Fayer, Phys. Rev. B, 43 (1991) 2696.
[17] A. R. Duggal, J. A. Roggers, and K. A. Nelson, J. Appl. Phys., 72 (992) 2823.
[18] J. A. Roggers, and K. A. Nelson, SPIE-INT. Soc. Opt. Eng., 1861 (1993) 314.
[19] K. A. Nelson, R. D. Miller, D. R. Lutz, and M. D. Fayer, J. Appl. Phys., 53 (1981) 1144.
[20] M. R. Farrar, L.-T. Cheng, Y.-X. Yan, and K. A. Nelson, IEEE J. QE., 22 (1986) 1453.
[21] E. H. Abramson, L. J. Slutsky, and J. M. Brown, J. Chem. Phys., 100 (1994) 4518.
[22] J. M. Zaug, E. H. Abramson, J. M. Brown, and L. J. Slutsky, Science, 260 (1993) 1487.
[23] J. M. Zaug, L. J. Slutsky, and J. M. Brown, J. Phys. Chem., 98 (1994) 6008.
[24] M. Fermigier, P. Jenffer, J. C. Charmet, and E. Guyon, J. Phys. Lett., 41 (1980) 519.
[25] D. W. Phillion, D. J. Kuizenga, A. E. Siegman, Appl. Phys. Lett., 27 (1975) 85.
[26] K. A. Nelson, R. Casalegno, R. D. Miller, and M. D. Fayer, J. Chem. Phys., 77 (1982) 1144.
[27] L. Genberg, Q. Bao, S. Gracewski, and R. D. Miller, Chem. Phys., 131 (1988) 81.
[28] L. Genberg, L. Richard, G. McLendon, and R. D. Miller, Science, 251 (1991) 1051.
[29] D. E. Hare, and D. D. Dlott., Appl. Phys. Lett., 64 (1994) 715.
[30] S. Palese, L. Schilling, P. R. Staver, W. T. Lotshaw, and R. J. Miller, J. Phys. Chem., 98 (1994) 6308.
[31] L. Genberg, Q. Bao, S. Gracewski, and R. J. Miller, Chem. Phys., 131 (1981) 81.
[32] R. J. Miller, R. Casalegno, and K. A. Nelson, Chem. Phys., 72 (1982) 371.
[33] X. Wen, S. Chen, and D. D. Dlott, J. Opt. Soc. Am. B, 8 (1990) 813.
[34] J. R. Salcedo, A. E. Siegman, D. D. Dlott, and M. D. Fayer, Phys. Rev. Lett., 41 (1978) 131.
[35] J. R. Salcedo, and A. E. Siegman, IEEE J. Quant. Electron., 15 (1979) 250.
[36] K. A. Nelson, D. R. Lutz, M. D. Fayer, and L. Madison, Phys. Rev.B, 24 (1981) 3261.
[37] R. D. Miller, M. Pierre, and M. D. Fayer, J. Chem. Phys., 78 (1983) 5138.
[38] R. S. Moog, M. D. Ediger, S. G. boxer, and M. D. Fayer, J. Phys. Chem., 86 (1982) 4694.
[39] D. D. Dlott, M. D. Fayer, J. Salcedo, and A. E. Siegman, Topics in Chemical Physics, vol. 4; Picosecond Phenomena (Springer-Verlag, New York, 1978) 230.
[40] J. R. Andrews, and R. M. Hochstrasser, Chem. Phys. Lett., 76 (1980) 207.
[41] P. Günter, Phys. Rept., 93 (1982) 199.
[42] C. Joshi, Y. Kitagawa, and A. Lal, Int. J. Nonlinear Opt. Phys., 1 (1992) 1.

[43] D. R. Lutz, K. A. Nelson, C. R. Gochanour, M. D. Fayer, Chem. Phys., 58 (1981) 325.
[44] F. Rondelez, H. Hervet, and W. Urbach, Chem. Phys. Lett., 53 (1980) 138.
[45] D. M. Burland, Acc. Chem. Res., 16 (1983) 218.
[46] F. Rondelez, H. Hervet, and W. Urbach, Chem. Phys. Lett., 53 (1978) 138.
[47] C. Bräuchle, and D. M. Burland, Agnew. Chem. Ed. Engl., 22 (1983) 582.
[48] D. M. Burland, IEEE J. Quant. Electron., 22 (1986) 1469.
[49] F.-W. Deeg, J. Pinsl, and C. Bräuchle, IEEE J. Quant. Electron., 22 (1986) 1476.
[50] R. J. Miller, M. Pierre, T. S. Rose, and M. D. Fayer, J. Phys. Chem., 88 (1984) 3021.
[51] A. Von Jena, and H. E. Lessing, Opt. Quant. Electron., 11 (1979) 419.
[52] C. A. Hoffman, K. Jarasiunas, and H. J. Gerritsen, Appl. Phys. Lett., 36 (1978) 536.
[53] R. K. Jain, and M. B. Klein, Appl. Phys. Lett., 35 (1979) 454.
[54] R. K. Jain, Opt. Eng., 21 (1982) 199.
[55] S. C. Moss, J. Ryan Lindle, H. J. Mackey, and A. L. Smirl, Appl. Phys. Lett., 39 (1981) 227.
[56] H. J. Eichler, and F. Massmann, J. Appl. Phys., 53 (1982) 3237.
[57] D. H. Auston, C. V. Shank, and P. LeFur, Phys. Rev. Lett., 35 (1975) 1022.
[58] C. A. Hoffman, H. J. Gerritsen, and A. V. Nurmikko, J. Appl. Phys., 51 (1979) 1603.
[59] M. G. Moharam, T. K. Gaylord, and R. Magnusson, J. Appl. Phys., 50 (1979) 5642.
[60] H. Hervet, W. Urbach, and F. Rondelez, J. Chem. Phys., 68 (1978) 2725.
[61] D. Pohl, Phys. Lett. A, 77 (1980) 53.
[62] H. J. Eichler, U. Klein, D. Langhans, Appl. Phys., 21 (1980) 215.
[63] H. J. Eichler, G. Enterlein, and D. Langhans, Appl. Phys., 23 (1980) 299.
[64] E. V. Ivakin, A. M. Lazaruk, I.P. Petrovich, and A. S. Rubanov, Sov. J. Quant. Electron., 7 (1977) 1382.
[65] K. Thyagarajan, and P. Lallemand, Opt. Commun., 26 (1978) 54.
[66] D. Pohl, Phys. Rev. Lett., 43 (1979) 143.
[67] H. Eichler, G. Enterlein. P. Glozbach, J. Munschau, and H. Stahl, Appl. Optics., 11 (1972) 372.
[68] M. M. Robinson, Y-X Yan E. B. Gamble Jr., L. R. Williams, J. S. Meth, and K. A. Nelson, Chem. Phys. Lett., 112 (1984) 491.
[69] M. R. Farrar, L.-T Cheng, Y-X. Yan, and K. A. Nelson, IEEE J. Quant. Electron., 22 (1986) 1453.
[70] J. T. Fourkas, T. R. Brewer, H. Kim, and M. D. Fayer, J. Chem. Phys., 95 (1991) 5775.
[71] F. V. Bunkin, M. I. Tribelskil, Sov. Phys. Usp., 23 (1980) 105.
[72] K. A. Nelson, J. Appl. Phys., 53 (1982) 6060.
[73] L. Cheng, Y. Yan, and K. A. Nelson, J. Chem. Phys., 91 (1989) 6052.
[74] A. M. Weiner, D. E. Learid, G. P. Weidderecht, and K. A. Nelson, Science, 247 (1990) 1317.
[75] G. P. Wiederrecht, T. P. Dougherty, L. Dhar, K. A. Nelson, Mat. Res. Symp. Proc., 293 (1993) 431.
[76] M. D. Fayer, Ann. Rev. Phys. Chem., 33 (1982) 63.
[77] H. J. Eichler, P. Günter, and D. W. Pohl, Optical Sciences; vol. 50, Laser-Induced Dynamic Gratings, (Springer-Verlag, Berlin, 1986).
[78] Y. R. Shen, IEEE J. Quant. Electron., 22 (1986) 1196.
[79] M. D. Fayer, IEEE J. Quant. Electron., 22 (1986) 1437.
[80] R. D. Miller, Annu. Rev. Phys. Chem., 42 (1991) 581.
[81] L. Dhar, J. A. Roggers, and K. A. Nelson, Chem. Rev., 94 (1994) 157.
[82] R. Y. Chiao, C. H. Townes, and B. P. Stoicheff, Phys. Rev. Lett., 12 (1964) 592.
[83] E. Garmire, and C. H. Townes, Appl. Phys. Lett., 5 (1964) 84.
[84] T-S Yang, R. Zhang, and A. B. Myers, J. Chem. Phys., 100 (1994) 8573.
[85] P. W. Bridgman, Amer. Acad. Proc., 49 (1914) 627; ibid. page 4.
[86] H. Carvill Lewis, Geol. Mag., 4, (1887).
[87] C. E. Weir, E. R. Lippincott, A. Van Valkenburg, and E. N. Bunting, J. Res. Natl. Bur. Stand. Sec A, 63 (1959) 55.
[88] E. Fishman, and H. G. Drickamer, Anal. Chem., 28 (1956) 804.

[89] L. Merrill, and W. A. Bassett, Rev. Sci. Instrum., 45 (1974) 290.
[90] C. H. Whitfield, E. M. Brody, and W. A. Bassett, Rev. Sci. Instrum., 47 (1976) 942.
[91] American Institute of Physics Handbook (McGraw-Hill, New York, 1957).
[92] W. D. Wilson, J. Acoust. Soc. Am., 31 (1959) 1069.
[93] M. Greenspan, and C. E. Tshiegg, J. Acoust. Soc. Am., 31 (1959) 75.
[94] M. Born, and E. Wolf, in: Principles of Optics, (Pergamon Press 6th ed., Oxford, 1980).
[95] F. Datchi, R. LeToullec, and P. Loubeyre, J. Appl. Phys., 81 (1987) 3333.
[96] H. K. Mao, J. Xu, and P. M. Bell, J. Geophys. Res., 91 (1986) 4673.
[97] L. E. Fried, W. M. Howard, P. C. Souers in 12th Symposium (International) on Detonation, J.L.M.a.J.M. Short, Eds., NSWC, Indian Head, San Diego, CA, (2002).
[98] L. E. Fried, and W.M. Howard: J. Chem. Phys., 110 (1999) 12023.
[99] L. E. Fried, M. R. Manaa, P. F. Pagoria, and R. L. Simpson, Ann. Rev. Mat. Sci., 31 (2001) 291.
[100] P. C. Souers, J. W. Forbes, L.E. Fried, S. Anderson, S. Dawson, P. Vitello, and R. Garza, Prop. Explos. Pyrotech., 26 (2001) 180.
[101] W. Fickett, W. C. Davis, Detonation, (University of California Press, Berkeley, 1979).
[102] M. Cowperthwaite and W. H. Zwisler, Sixth Detonation Symposium, (1976) p.162.
[103] M. L. Hobbs, M. R. Baer, and B. C. McGee, Prop. Explos. Pyrotech. 24 (1999) 269.
[104] F. H. Ree, J. Chem. Phys., 84 (1986) 5854.
[105] H. D. Jones, and F. J. Zerilli, J. Appl. Phys., 69 (1991) 3893.
[106] M. van Thiel, and F. H. Ree, Int. J. of Thermophysics, 10 (1989) 227.
[107] L. E. Fried, and W. M. Howard, Phys. Rev. B, 61 (2000) 8734.
[108] M. S. Shaw, and J. D. Johnson, J. Appl. Phys., 62 (1987) 2080.
[109] J. A. Viecelli, and F. H. Ree, J. Appl. Phys,. 86 (1999) 237.
[110] J. A. Viecelli, and F. H. Ree, J. Appl. Phys., 88 (2000) 683.
[111] L. E. Fried, and W.M. Howard, J. Chem. Phys., 109 (1998) 7338.
[112] F. H. Ree: J. Chem. Phys., 70 (1979) 974.
[113] W. J. Nellis, F.H. Ree, R. J. Trainor, A. C. Mitchell, and M. B. Boslough: J. Chem. Phys., 80 (1984) 2789.
[114] W. J. Nellis, and A. C. Mitchell, J. Chem. Phys., 73 (1980) 6137.
[115] S. P Marsh, LASL Shock Hugoniot Data, (University of California Press, Berkeley, 1980).
[116] L. A. Weber, NTIS Report n78, 29 (1977) 16119.
[117] G. C. Straty, and B. A. Younglove, J. Chem. Thermodyn., 5 (1973) 305.
[118] E. H. Abramson, L. J. Slutsky, and J. M. Brown, J. Chem. Phys., 110 (1999) 10493.
[119] V. N. Zubarev, and G. S. Telegin, Sov. Phys. Dokl, 7 (1962) 34.
[120] W. J. Nellis, and A. C. Mitchell, J. Chem. Phys., 73 (1980) 6137.
[121] W. J. Nellis et al., J. Chem. Phys., 94 (1991) 2244.
[122] P. Malbrunt and B. Vodar, Physica, 66 (1973) 351.
[123] S. L. Robertson and J. S. E. Babb, J. Chem. Phys., 50 (1960) 4560.
[124] P. J. Kortbeck, N. J. Trappeniers, and S. N. Biswas, Int. J. Thermophys., 9 (1991) 103.
[125] J. M. Zaug, J. C. Crowhurst, and D. W. Hansen et al., submitted, (2004).
[126] R. F. Trunin, M. V. Zhernokletoc, N. F. Kuznetsov, O. A. Radchenko, N. V. Sychevskaya, and V. V. Shutov, Khim. Fiz., 11 (1992) 424.
[127] J. C. Crowhurst, J. M. Zaug, L. E. Fried et al., submitted, (2004).
[128] A. F. Goncharov and J. M. Zaug, from xrd experiments conducted at HPCAT (2003, 2004).
[129] W. B. Montgomery, J. M. Zaug, W. M. Howard, L. E. Fried, J. C. Crowhurst, A. F. Goncharov, and R. Jeanloz, submitted, (2004).
[130] J. W. Stout and L. H. Fisher, J. Chem. Phys., 9 (1940) 163.
[131] R. C. Wihoit, J. Chao, and K. R. Hall, J. Phys. Chem. Ref. Data, 14 (1985) 123.
[132] W. Wagner, K. M. de Reuck, R. Schmidt, J. Ewers , R. B. Stewart, R. T. Jacobsen, in: Oxygen International Thermodynamic Tables of the Fluid State, vol. 9, (Blackwell Scientific Publications, Oxford, 1987).
[133] D. S. Tzikilis and A. I. Koulikova, Zh. Phiz. Khimii, 39 (1965) 1752.

[134] A. B. Belonoshko, and S. K. Saxena, Geoch. Cosmochim. Acta, 55 (1991) 3191.
[135] G. L. Schott, M. S. Shaw, and J. D. Johnson, J. Chem. Phys., 82 (1985) 4264.
[136] B. J. Baer and M. F. Nicol, J. Phys. Chem., 94 (1990) 1073.
[137] R. L. Mills, D. H. Liebenberg, and J. C. Bronson, J. Chem. Phys., 63 (1975) 1198.
[138] T. M. Reed, and K. E. Gubbins, Applied Statistical Mechanics (McGraw-Hill, New York, 1973) p 131.
[139] H. Shimizu, Physica, 139 (1986) 479.
[140] D. R. Allan and S. J. Clark, Phys. Rev. Lett., 82 (1999) 3464.
[141] P. G. Maiella and T. B. Brill, J. Phys. Chem., 102 (1998) 5886.
[142] J. A. Muller and E. Peytral, Memoires Presentes a la Societe Chemique, 34 (1920).

Chemistry at Extreme Conditions
M. Riad Manaa (Editor)

Chapter 15

Theoretical and Computational Studies of Energetic Salts

Dan C. Sorescu,[a] Saman Alavi,[b] and Donald L. Thompson[b]

[a] U. S. Department of Energy, National Energy Technology Laboratory, Pittsburgh, PA 15236

[b] Department of Chemistry, University of Missouri, Columbia, MO 65211

ABSTRACT

Descriptions of classical and quantum mechanical methods for simulating energetic salts are presented. An overview of recent applications of these methods for predictions of gas- and condensed-phase properties, chemical reactivities, and phase transitions is given. The limitations and some suggestions for further developments of the methods are also discussed.

1. INTRODUCTION

1.1. Solid-Phase Ionic Energetic Materials

1.1.1. Crystalline Phases and Structural Properties

Among the various types of potential energetic materials being considered, ionic systems represent a class of particular importance. The interest in these materials is mainly for use as strong oxidizers in propellant formulations and as monopropellants. Classical examples of such materials are nitrate salts such as potassium nitrate, which was used originally as the oxidizer component in gunpowder although more recently its practical applications are as a plant fertilizer and in automobile airbags. Other important systems are perchlorates, which are now used extensively as solid oxidizers for rocket propulsion;[1] the most important being ammonium perchlorate (AP), $NH_4^+ClO_4^-$. However, this use of AP presents significant environmental problems. For example, contamination of ground and surface water by the perchlorate anion can lead to significant toxicity risks for humans. Also, the combustion of AP propellants in rocket engines results in the formation of highly toxic HCl. The release of atomic chlorine as a result of the reaction between HCl and hydroxyl radicals can lead to damage to the stratospheric ozone layer with important environmental implications. Another problem resulting from the release of HCl is that it provides nucleation sites for atmospheric water vapor, causing an increase in the secondary smoke signature of AP-containing propellants. This is problematic for military applications since it makes launching sites more easily detected. The environmental concerns and the need for low-signature propellants provide strong motivation for the development of new ionic oxidizers that eliminate chlorine and produce less environmental hazardous combustion products.

An alternative salt that is more environmentally benign is ammonium nitrate (AN) (NH_4NO_3), [1] which has been used in many applications since its discovery in 1659. [2] The interest in AN is largely due to its low cost of production and because it is relatively safe to handle. Currently, AN is the most important commercial ammonium salt, with a reported production of 50-70 million metric tons per year in 1992.[3] Its main use is as an agricultural fertilizer, but it is also used in as an explosive in industries such as mining. Recently, interest has been expressed for rocket propellants in which AN serves as an important source of oxygen.[1] In propellant applications AN could provide a cleaner burning fuel without chlorine by-products and lower detectable smoke signature than an AP propellant.

Applications of AN as an oxidizer in both rocket-propulsion and high-performance propellants are hindered by some undesirable physical and chemical properties. First, AN has lower energetic performance per unit mass than perchlorate salts. Additionally, its high hygroscopicity can lead to particle degradation and caking. From a thermodynamic point of view AN is a very challenging system because at atmospheric pressure it exists in at least five phases below the melting point at 442 K. Brief lists of properties of the crystal phases of AN are given in Table 1[4-12] while atomic configurations of various crystalline phases are pictorially illustrated in Figure 1. Of these phases only the lower-temperature phases V and IV are translationally and rotationally ordered. In phases I and II the ammonium and nitrate ions are rotationally disordered while in phase III only the ammonium ions are disordered. The transitions among these phases are quite important for propellant applications. In particular, the most problematic is the phase transition between phases IV and III (see Table 1), which occurs close to the room temperature at about 305 K. This phase transition takes place with a significant volume change of about 3.84% (see Figure 2) and with an important crystallographic reorganization. As a result, during the daily thermal cycles or during transportation and handling the material can undergo several back and forth phase transitions that lead to changes in the grain sizes and ultimately to caking or powdering, which degrade the burning rate and mechanical properties. Defects formed in the propellant grains can cause undesirable burning behavior.[1] *If AN is to be used as a propellant component we need to understand the mechanisms for the phase transitions so that we can develop ways of controlling them.*

The results of more than 50 years of research show that some of these problems can be overcome.[13] Particularly, identification of various phase stabilizing agents which will eliminate the IV$\rightarrow$III phase transition, leading to a direct IV $\rightarrow$ II transition, and agents which will shift the IV $\rightarrow$ III phase transition from 323 K to below 273 K have received a lot of attention. For example, transition metals such as Ni,[10] Cu,[10, 14] and Zn;[15] Mg,[16] and Ca salts;[17] and ammonium phosphates,[18] ammonium sulfate,[19] and potassium nitrates [20] have been shown to function as stabilizing agents in solid solutions with AN and to extend the thermal stability ranges of crystal phases of AN. Despite this progress, detailed mechanisms of the phase transitions in AN and the role played by various doping agents on the phase-stability ranges have not been determined. *Moreover, an understanding of the dependence of the phase transition on various factors such as moisture content, thermal history, heating mode or the grain size is needed if AN is to be used for rocket propulsion, where precise control of the physical-chemical properties of the material is crucial.* Clearly, further research is needed. *An atomic-level understanding of the interplay between the structural, electronic, thermochemical, and dynamic properties of AN would be valuable in developing ways of improving the performances of these propellants.*

A new class of ionic compounds with high oxygen or nitrogen content and high densities were obtained after the first report in 1991 of the synthesis of dinitramic acid, $NH(NO_2)_2$, and its dinitramide salts.[21] Among these, one of the most important compounds to date is ammonium dinitramide (ADN), ($[NH_4]^+$ $[N(NO_2)_2]^-$).[21-24] The interest in this compound is due to several factors. First, it is a halogen-free compound, thus it is attractive from both the military and environmental points of view; it burns with a low plume signature and without HCl formation as in the case of AP. Secondly, ADN has a higher energy content than AN and consequently better performance in propellant applications. Additionally, there are no phase transitions at normal temperatures and pressures as in AN to affect the crystalline and volume properties, thus ADN burns more readily and predictably than AN, and without a residue.[25] These characteristics make ADN a strong candidate for use in solid-propellant formulations.

Table 1. Stability Range and Crystallographic Parameters of Ammonium Nitrate in Different Phases.

Phase	AN V	AN IV	AN III	AN II	AN I	Liquid
Temp. range (K)	0-255	255-305	305-357	357-398	398-442	> 442
Cryst. ordering	ordered	ordered	disordered	disordered	disordered	
Symmetry Type	Orthor.	Orthor.	Orthor.	Tetragonal	Cubic	
Space Group	Pccn[a]	Pmmn[c,d]	Pnma[f]	P$\bar{4}$21m[e]	Pm3m[h,i]	
a(Å)	7.8850	5.7507	7.7184	5.7193	4.3655	
b(Å)	7.9202	5.4356	5.8447	5.7193	4.3655	
c(Å)	9.7953	4.9265	7.1624	4.9326	4.3655	
Z	8	2	4	2	1	
Space Group	Pccn[b]	Pmmn[e]	Pnma[g]			
a(Å)	7.9804	5.7574	7.6772			
b(Å)	8.0027	5.4394	5.8208			
c(Å)	9.8099	4.9298	7.1396			
Z	8	2	4			

[a] Choi and Prask, Ref.[4]. [b] Ahtee *et al.* Ref. [5]. [c] Choi *et al.*, Ref. [6]. [d] Choi *et al.* Ref. [7]. [e] Lucas *et al.*, Ref. [8]. [f] Lucas *et al.* Ref. [9]. [g] Choi and Prask, Ref. [10]. [h] Ahtee *et al.* Ref. [11]. [i] Yamanoto *et al.*, Ref. [12].

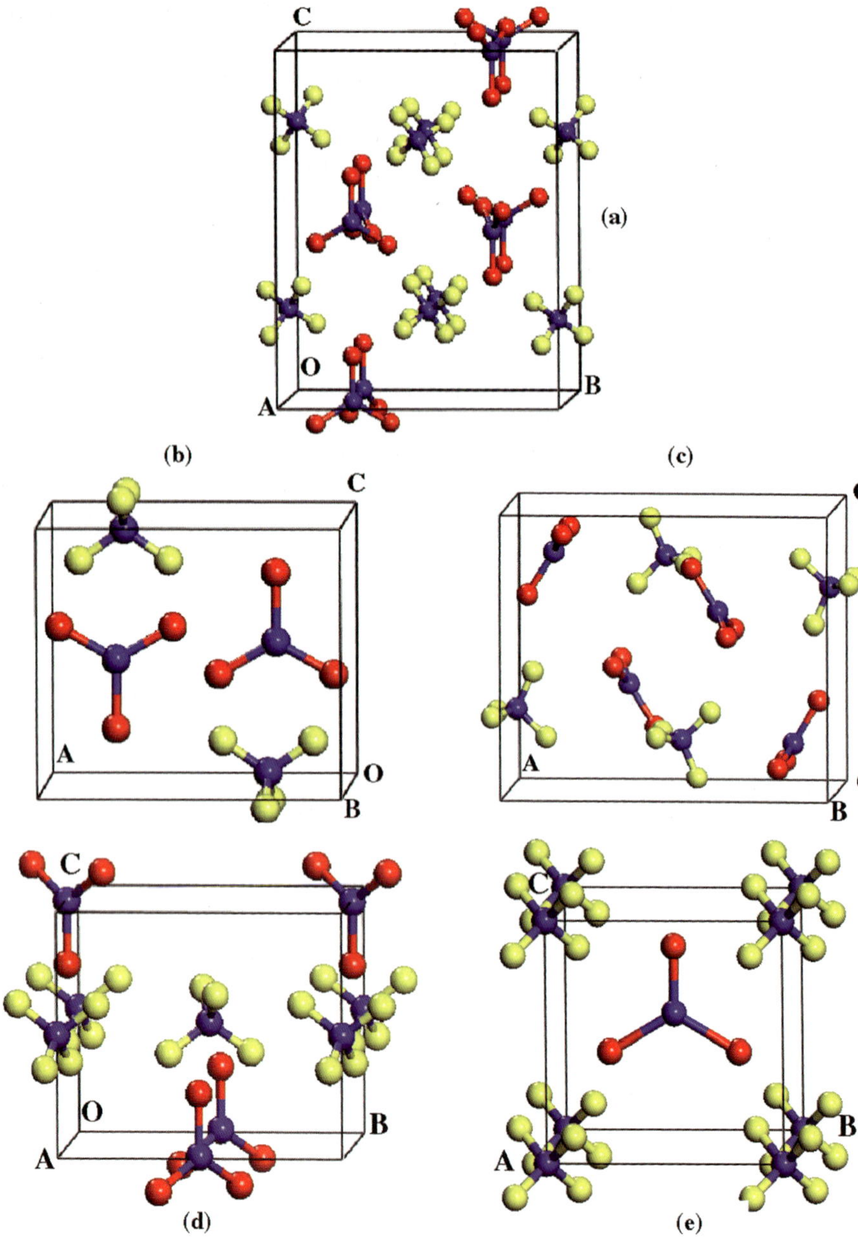

Figure 1. Crystal structure configurations of AN in phases: (a) V; (b) IV; (c) III; (d) II; (e) I; based on the data reported in Refs.[4-12].

A very important trait of the dinitramide anion (DN), $[N(NO_2)_2]^-$, is that it can form salts with a variety of inorganic and organic cations.[26] For example, salts obtained by combination of the dinitramide ion with metallic lithium, potassium, cesium, sodium, mercury, and iron cations have been reported. [27-29] Additionally, salts formed with organic cations such as cubane-1,4-diammonium and cubane-1,2,4,7-tetraammonium,[30] 3,3-dinitroazetidinium and 1-*i*-propyl-3,3-dinitroazetidinium,[31] hexaaquomagnesium, hexaaquomanganese, hexaaquozinc dehydrate,[32] hexammonium and the ethane-1,2-diammonium,[33] hydrazinium and the hydroxylammonium,[34] guanidinium and hydroxyguanidinium,[35] biguanidinium,[36] melaminium,[37] and N-guanylurea[38] have been prepared and their structures determined by X-ray diffraction. Among these systems the N-guanylurea-dinitramide salt, also known as FOX-12, has been identified as a particularly promising candidate for propellant applications. In contradistinction to dinitramide salts of Li, K, Cs, and biguanidinium, which are highly hygroscopic, FOX-12 is not soluble in water and is not hygroscopic. Additionally, FOX-12 is less shock sensitive than common energetic materials such as RDX and HMX and has a thermal stability superior to that of ADN.[38] Overall, the ability of the dinitramide anion to form salts with a great variety of cations giving materials with high oxygen content makes this a very promising candidate for the development of new energetic materials.[26, 27]

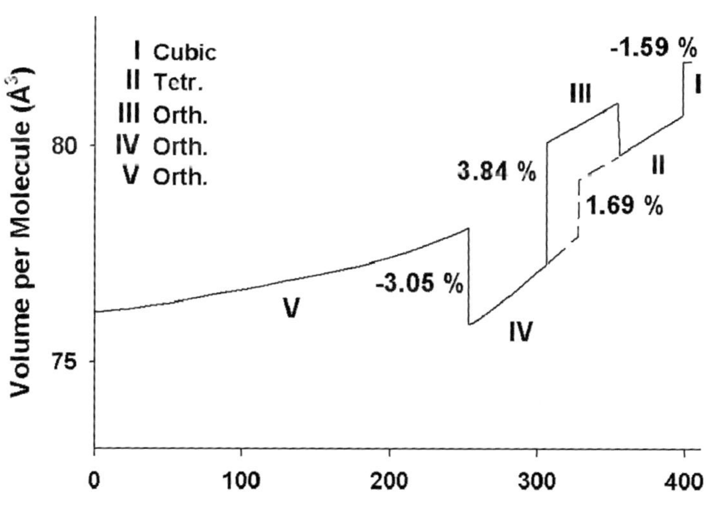

Figure 2. The volumetric thermal expansion of ND_4NO_4, given as specific cell volume versus temperature, over the temperature range from 10 to 393 K. The dashed line indicates the IV-II transition. The magnitude of the volumetric change at each transition point is also included. Reproduced with permission from Ref.[4].

The large variety of dinitramide salts are also quite interesting from a fundamental point of view and thus there have been several studies of their structural and thermal properties. Two polymorphs of ADN, denoted α and β (see Figure 3), have been found to exist.[24, 39] The α phase is stable from atmospheric pressure to 2.0 GPa over a large range of temperatures and melts above 95 °C. The high-pressure β phase is stable above 2.1 GPa, in the temperature interval -75 °C to 120 °C. It has been determined that this phase has a monoclinic symmetry,[24] but no detailed crystallographic information is available. Above 140 °C and in the pressure range 1 to 10 GPa, ADN decomposes to AN and N_2O apparently by molecular rearrangement, although no detailed mechanism has been determined.

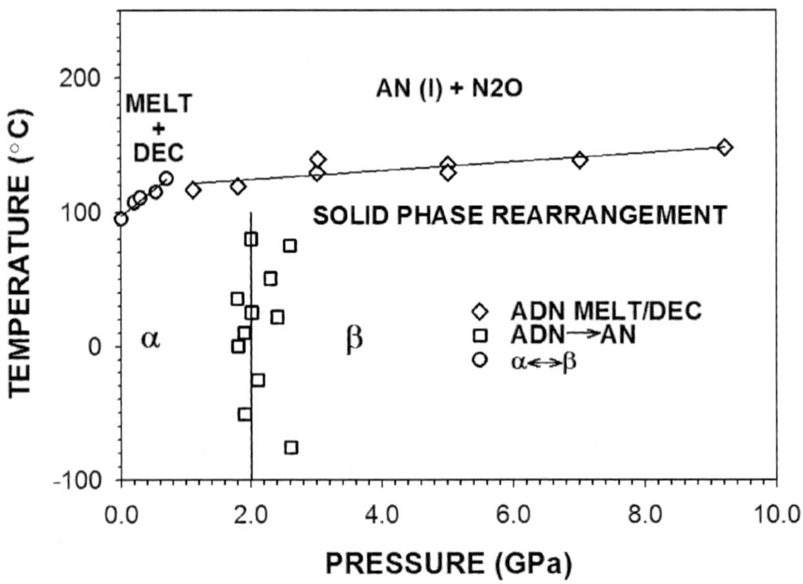

Figure 3. Pressure-temperature-reaction phase diagram for ammonium dinitramide (ADN) showing the estimated thermodynamic stability fields for the α and β polymorphs and the liquidus curve. The monoclinic α phase is stable up to about 2.0 GPa between -75 and 120 °C. The high-pressure β phase, which is also monoclinic, is stable above 2.1 GPa between -75 and 120 °C. Above 140 °C between 1.0 and 10.0 GPa, ADN undergoes a molecular rearrangement to form ammonium nitrate (AN) and N_2O. The α-β transition pressure is estimated to be 2.0±0.2 GPa and is the result of a least squares fit of the data points. Reproduced with permission from Ref. [39].

The structure of the low-pressure α-phase of ADN has been characterized by X-ray diffraction experiments.[27] The crystal has a monoclinic symmetry with space group $P2_1/c$ and has four formula units ($[NH_4]^+[N(NO_2)_2]^-$) per unit cell (see Figure 4). All hydrogen atoms of the ammonium ion are involved in extensive hydrogen bonding. Three of the protons participate in two-dimensional hydrogen bonding in the *ab* plane. The fourth ammonium proton is hydrogen bonded along the *c*-axis direction and links the *ab* sheets. The two halves of the dinitramide anion are asymmetric as reflected by the lengths of both the N-N and N-O bonds. Specifically, the lengths of the two N-N bonds are $r(N_1\text{-}N_2)=1.376$ Å and $r(N_1\text{-}N_3)=1.359$ Å. The internal N-O bond lengths are $r(N_2\text{-}O_{2b})=1.227$ Å and $r(N_3\text{-}O_{3b})=1.223$ Å, and they are significantly shorter than the outer N-O bonds: $r(N_2\text{-}O_{2a})=1.236$ Å and $r(N_3\text{-}O_{3a})=1.252$ Å (see Figures 4(b) and 4(c)).

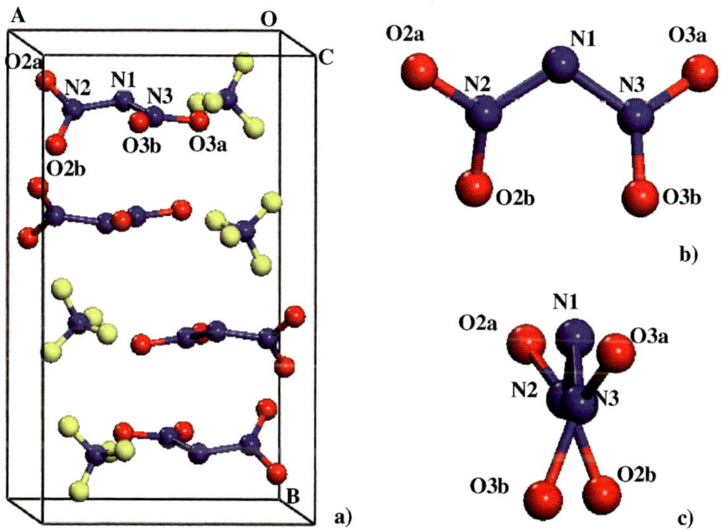

Figure 4. (a) Representation of the α-ADN crystal unit cell with monoclinic space group $P2_1/c$ and Z=4 molecules per unit cell. Insets (b) and (c) detail front and lateral views of the dinitramide ion.

An interesting finding of the experimental studies is the large conformational variations in the dinitramide ion in various salts. Gilardi *et al.*[30-35] found that among the entire series of 23 dinitramide salts investigated, only the lithium salt has C_2 symmetry and the dinitroazetidinium salt has mirror-image symmetry, while the remaining salts have C_1 symmetry. In the majority of those dinitramide salts both the N-N and N-O bond distances are not equivalent and the corresponding nitro groups are twisted from the central NNN plane by varying amounts. The difference in the N-N bond lengths varies from 0.002 Å for guanidinium dinitramide,[35] to 0.017 Å for AND[27] and is as large as 0.068 Å in the hydroxyguanidinium salt.[35] Additionally, the pseudotorsion angles, defined as the torsional angle between the closest N-O bonds in the dinitramide ion belonging to different nitro groups, such as $O_{2b}\text{-}N_2\cdots N_3\text{-}O_{3b}$ (see Figure 4 b), have wide variations ranging from 2.1° in hydroxyguanidinium,[35] to 37.9° in AND [27] and to 43.4° in 1-*i*-propyl-3,3-dinitroazetidinium.[31] *These data indicate that there is a strong correlation between the conformation adopted by the dinitramide ion and the local electronic and steric environment. However, a direct explanation of these observations as superposition of crystal packing effects and interatomic interactions is still unresolved and thus invites theoretical investigations.*

1.1.2. Thermal Stability and Dissociation Mechanisms

Due to their importance in propellant formulations the thermal properties and decomposition mechanisms of AP, AN, and ADN have attracted considerable interest.

Ammonium Perchlorate. The thermal decomposition of AP has been the subject of several studies, and the results have been summarized in several excellent articles. [40-43]At room temperature AP has a stable phase with orthorhombic crystal symmetry, space group Pnma, with four formula units per cell. [44] The ClO_4^- and NH_4^+ ions are essentially tetrahedral in structure and are linked by N-H···O hydrogen bonds. However, the NH_4^+ are not rigidly fixed in the lattice; starting from temperatures as low as 10 K they undergo rotational motion, which increases in complexity as the temperature increases.[44-46]

Ammonium perchlorate decomposes over the wide temperature range of 200 °C to 440 °C by two different mechanisms.[40, 43] Between 200 °C and 300 °C the decomposition takes place by an autocatalytic process which ceases after about 30% decomposition. The decomposition proceeds via a second mechanism in the high-temperature regime (300-430°C), where the reaction is not autocatalytic and the decomposition goes to completion. Bircumshaw and Newman [47, 48] were the first to report that simultaneous with decomposition, sublimation of AP takes place throughout both the low- and the high-temperature decomposition regions.

According to Singh *et al.*[43] the overall decomposition of AP takes place through three major pathways: (a) an electron transfer from the perchlorate anion to the ammonium cation; (b) a proton transfer to form perchloric acid and ammonia; and (c) thermal breakdown of the perchlorate anion. They suggested that in the low-temperature decomposition regime the proton transfer is the rate determining process.[43] In this regime decomposition proceeds at a faster rate than sublimation such that NH_3 and $HClO_4$ remain adsorbed on the surface, from which oxidation reactions of ammonia or bimolecular decomposition of $HClO_4$ can occur (see Scheme I).

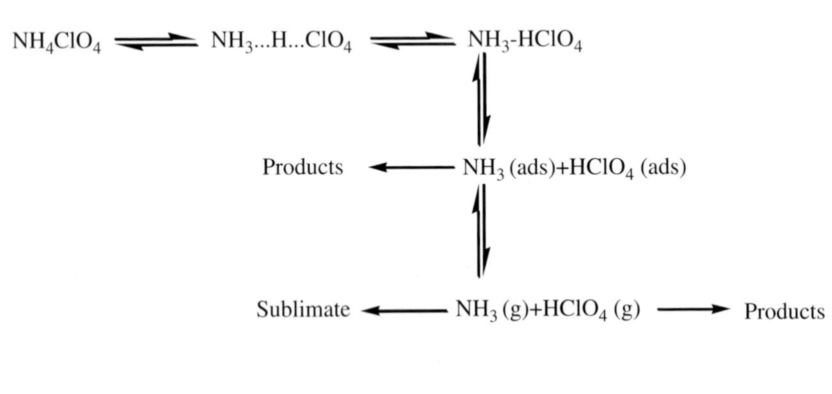

Scheme I (from Singh *et al.*, Ref. [43]).

The overall decomposition of AP leads to a large number of products such as Cl_2, ClO, $HClO_3$, O_2, N_2, NO, NO_2, N_2O, H_2O, and HCl, indicating a rather complex process. The reported values of the activation energy for the thermolysis of AP cover the range 9-44 kcal/mol, corresponding to various experimental conditions[43, 48-51] or various kinetic models. *The complexity of the decomposition mechanism raises several questions for further investigation. For example, the precise identification of various limiting reaction steps in different temperature regimes needs further study. Also, the energetics of various intermediate species involved in these processes need to be determined. Moreover, since the stability of AP is found to be extremely sensitive to various catalysts (such as copper carbonate, chromium carbonate, CuO, and Cr_2O_3) and inhibitors (such as Ca, Ba, Sr, Cd, and CaO),[43] the precise effects of these additives on both the thermochemical and kinetic properties need to be investigated in order to develop a rational design of new energetic materials.* Some of these can be readily addressed by theoretical methods.

Ammonium Nitrate. Like AP, the decomposition of AN occurs via complex decomposition mechanisms. Studies performed by Oxley and coworkers indicate two modes of decomposition.[52-54] In the temperature range 200-300 °C, decomposition starts through an endothermic dissociation to ammonia and nitric acid and the formation of the nitronium ion is the rate-determining step (See Scheme II).

$$NH_4NO_3 \rightleftharpoons NH_3 + HNO_3$$

$$HNO_3 + HA \xrightleftharpoons{-A} H_2ONO_2^+ \longrightarrow NO_2^+ + H_2O$$

where HA = NH_4^+, H_3O^+, HNO_3

$$NO_2^+ + NH_3 \xrightarrow{slow} [NH_3NO_2^+] \longrightarrow N_2O + H_3O^+$$

$$NH_4NO_3 \rightleftharpoons N_2O + 2\,H_2O$$

Scheme II (from Oxley *et al.*, Ref.[54]).

Above 290 °C, a free–radical decomposition mechanism has been shown to be dominant and homolysis of nitric acid forming NO_2 and HO- was proposed to be the rate-determining step.[52, 53]

Oxley *et al.*[53, 54] have shown that the thermal stability of AN is significantly influenced by the type of additives used in mixtures with AN. For example, basic additives such as carbonate, formate, oxalate, and mono-phosphate salts significantly raise the temperature of the AN exotherm and enhance AN stability. However, stability seems to be increased even in the absence of an increase in pH of the AN solution, as is the case for urea additives which upon decomposition form ammonia which only then increases the basisity of the medium.[54] As the most efficient stabilizers were found to contain carbon, it was speculated that formation of carbon dioxide may be an important factor in the ability of a compound to stabilize AN.[54] *These results indicate that a full understanding of the mechanism involved in AN decomposition and the influence of various basic or acidic additives on the thermal stability of AN is not yet available.* Oxley *et al.*[54] have also pointed out that the identified additives that stabilize AN do not render it totally non-detonatable, making its use in fertilizer formulations problematic. *Consequently, further studies to correlate the type of additives with the stability and detonation properties of AN and to determine the corresponding mechanistic steps involved are clearly necessary.*

Ammonium Dinitramide. The thermal decomposition of ADN has been investigated in a large number of experimental studies.[55-60] The mechanism appears to be quite complex and highly dependent on experimental temperature and pressure conditions. Here we summarize only the main models that have been proposed. Brill and coworkers [56] have proposed a mechanism for the gas-phase decomposition that involves sublimation to NH_3 and $HN(NO_2)_2$ via the reactions:

$$NH_4N(NO_2)_2 \rightarrow NH_3 + HNO_3 + N_2O$$

and

$$NH_4N(NO_2)_2 \rightarrow NH_3 + HN(NO_2)_2.$$

Vyazovkin and Wight [57, 58] have proposed that ADN decomposition involves two parallel channels. The first one is a molecular rearrangement mechanism with the formation of AN and N_2O instead of NH_3:

$$NH_4N(NO_2)_2 \rightarrow N_2O + NH_4^+ + NO_3^-,$$

while the second mechanism begins with N-N bond rupture and leads to the formation of NO_2 and the mononitramide ion

$$NH_4N(NO_2)_2 \rightarrow NO_2 + NH_4^+ + NNO_2^-.$$

This mononitramide ion can subsequently dissociate via [61]

$$NNO_2^- \rightarrow NO^- + NO.$$

An alternative mechanism was suggested by Oxley and coworkers.[59] They concluded that above 160°C the decomposition occurs by a free-radical mechanisms while an ionic mechanism is important below this temperature. The decomposition of ADN leads to nitrous oxide, nitric acid or nitrate and nitrogen gas. It was assumed that the first step in ADN decomposition is hydrogen transfer to form ammonia and dinitramic acid. Several proposed decomposition pathways for ADN decomposition that involve conversion of the dinitramide ion to N_2O have been proposed but they await confirmation.[59]

This brief review of various experimental studies indicates there is still much to be resolved about the decomposition mechanisms of ADN. *Particularly, there is significant need for evaluation of the energetics of the individual steps that have been proposed and intermediates involved in them. Significant clarification of the reaction mechanisms can be achieved by theoretical calculations as illustrated by the recent results described in subsequent sections of this chapter.*

1.2. Liquid-Phase Ionic Energetic Materials

Many ionic materials, particularly those composed of organic cations, are liquids at room temperature. While much of the interest in *room-temperature ionic liquids* (RTILs) is due to their potential use as solvents, because of their environmental and solvation characteristics, they are also of interest as propellants. For example, they are being considered as replacements for undesirable propellants such as the extremely toxic hydrazines. The challenge is to design the salts with sufficient oxygen (in the anion) to consume the fuel in the organic cation. For the most part, the RTILs that are currently being studied as solvents, lack the reactivity properties to serve as propellants; however, the nitrate and perchlorate salts of organic cations such as ethylammonium or 1 ethyl-3-methylimidazolium [62] may be of use. Ethylammonium nitrate (EAN) [63] has a melting point of 12°C. [64] This ionic liquid is used as a conductive solvent for electrochemical analysis [65] and protein crystallizing reagent. [66] More complicated nitrates have been developed as liquid propellants for artillery applications. An example of such a material is LGP 1846, which is a mixed nitrate salt of hydroxylammonium and *tris*-(2-hydroxyethyl)amine.[64] Imidazolium, triazolium, pyridinium, and other ring-based cations are candidates for forming energetic salts with low melting points.

1.3. General Remarks on Theoretical Simulations of Ionic Energetic Systems

The above discussion illustrates the progress made by experimentalists in studies of crystal structures, phase stability, and thermal decomposition reaction mechanisms of the energetic ionic materials. There has also been progress in making materials with improved properties for practical propellant applications, *e.g.*, phase-stabilizing agents have been identified, some methods for protecting against caking and hygroscopicity have been explored, and some preliminary efforts to improve sensitivity and burning properties have been made. These data provide the basis and motivation for theoretical studies to expand upon what is known and to eventually develop predictive models.

There are several areas that can benefit from insight provided by computational studies. We can arbitrarily break the overall theoretical problems into the following categories:

- Evaluation of the energetic, thermodynamic, and kinetic parameters for compounds, intermediates, and reactions.
- Determinations of structural and energetic properties of crystal phases and liquid properties, and the interplay between these properties and the chemical identity of cations and anions.
- Predictions of phase diagrams, *i.e.*, calculations of equilibria between solid-solid and solid-liquid phases as functions of temperature and pressure; and the development of microscopic (atomic-level) mechanisms for the transitions between phases.
- Evaluation of the effects of catalysts and inhibitors on phase stability and thermal decomposition mechanisms.
- Assessment of the sensitivity of ionic energetic materials in various phases in both neat form, mixtures, and with dopants.

In the following we review the methods that have been used to address some of these issues. For the most part, we focus on methods that provide an atomic-level view of the physical and chemical processes based on models derived from quantum chemistry computations. We also discuss how to establish relationships between atomistic properties at different time and length scales to macroscopic properties relevant to practical applications. Our goal here is to give a status report and some perspective of what needs to be done in the future.

2. COMPUTATIONAL METHODS AS APPLIED TO SIMULATIONS OF IONIC ENERGETIC MATERIALS

2.1. General Areas of Practical Impact for Atomistic Computational Studies

One of the major challenges for theory is to establish direct relationships between the atomistic properties and the corresponding physico-chemical properties of ionic materials. In this section we review some of the practical areas where computational methods can be used to gain insight in various types of properties of ionic systems. We start by observing that there are several types of properties that need to be considered. First, it is necessary to assess the structural and energetic properties of the compound of interest. If possible this analysis should be done for the various phases of the material, *i.e.*, solid, liquid, and gas, such that a comprehensive understanding of the role played by both intramolecular and intermolecular interactions in determining the equilibrium properties is available. Also, we need to establish the regions of thermodynamic stability for these compounds. For condensed phases this requires an evaluation of the state of the system as functions of pressure and temperature, *i.e.*, the phase diagram. All these properties can be computed given knowledge of the intra- and intermolecular interactions. Predicting and simulating the transitions between phases is a critical, cutting-edge problem, and we will describe basic methods for doing this. Earlier in this chapter we noted that the evolution of the material among different phases such as solid-solid for AN or solid-liquid for ADN significantly affects, even determines, their use in practical applications. Consequently, the ability to accurately describe such transitions on the atomic scale and develop an understanding of how they depend on the interatomic interactions, which can be affected, *e.g.*, by doping with a phase-stabilizing agent, are essential.

Besides the prediction of structural information of various phases, calculations of thermochemical quantities such as the heat of formation in gas, liquid, and solid phases, the heat of vaporization for liquids, the heat of sublimation for ionic solids or the lattice energy for ionic solids represent important quantitative measures related to stability and phase-transformation properties of the compounds of interest. Calculations of such properties are feasible and we will describe the methods for performing them.

Perhaps the most important aspect of energetic salts that needs to be understood for their energetic applications is the mechanisms of thermal decomposition. The immediate challenge is to use computations, since experimental measurements are in many cases not feasible, to determine the initial chemical reactions for various conditions, *i.e.*, phase, temperature, and pressure. This is critical for understanding both combustion and detonation. Quantum chemistry methods can be used to compute bond-dissociation energies and transition-state barriers for isolated molecules (and we will discuss below some results for energetic salts); however, a similar capability for condensed phases is less developed. In any case, the mapping of critical regions of potential energy surfaces for reactions facilitate the analysis and interpretation of experiments and various kinetic models. Methods for calculations of detonation properties of materials are developing along various directions, but clearly atomistic simulations will be an important approach because of the insights to be gained. Determining the basic aspects of the propagation of a shock wave through an ionic material and the manner in which it initiates the chemistry is one of the major, long-term objectives in the general scheme of the theory we describe here. More immediate goals include modeling chemical reactions in condensed phases, transport phenomena, and phase transitions; and we present a review of the status of this work. The roles played by specific external stimuli such as shock impact, heating, or electrical discharge on the initiation of detonation can be determined given accurate models by using atomistic simulations. Structural, electronic, and electrostatic properties can all be investigated to identify the most sensitive parameters responsible for detonation properties.

Given the complexity and diversity of the problems involved a wide range of computational methods is needed. We will limit our discussion to the case of molecular dynamics simulations and quantum chemistry methods. Specific examples will be used to illustrate the benefit of various approaches for particular problems.

2.2. Quantum Chemistry Calculations: Applications to Dinitramide, Ammonium Dinitramide, and Ammonium Perchlorate

Ab initio calculations have been primarily performed using the well-known Gaussian package,[67] but in a few instances CADPAC [68] and demon [69] programs have been used.[70, 71] The main applications have been in studies of gas-phase species relevant to the condensed-phase ionic materials. For example, through gradient optimization methods the geometries of various molecular and ionic species at different stationary points of the potential energy surface have been determined in the gas phase.[70-73] Harmonic vibrational frequencies were calculated to characterize these stationary points and determine the zero-point energy corrections. Furthermore, infrared and Raman intensities have been calculated and compared with the experimental data for crystals and solutions.[29]

Besides geometric and vibrational properties, identification of the relative energies of compounds, or the energy differences between points on the potential energy surface of a particular compound have also been undertaken.[70-73] Calculations of the bond dissociation energy, reaction energy, electron affinity, heat of formation, and enthalpy of deprotonization are practical examples of the type of properties that have been determined for salts by using quantum chemistry methods.

Michels and Montgomery,[70] for example, have investigated the structure and thermochemistry of hydrogen dinitramide and dinitramide anion using restricted Hartree-Fock and second order Möller-Plesset (MP2) levels [74] with 6-31G** and 6-311+G** basis sets.[75] In the most stable configuration of hydrogen dinitramide the proton is attached to the central nitrogen of the dinitramide anion. This structure has C_s symmetry with the plane of symmetry perpendicular to the plane of the three nitrogen atoms containing the N-H bond.

The optimized configuration of the dinitramide ion was determined to have C_2 symmetry,[70] with the nitro groups twisted such that the oxygen atoms are positioned below and above the plane of the nitrogen atoms. There are small barriers of less than 3 kcal/mol for rotation in either direction about the N-NO_2 bond. This characteristic has been used to explain the large variety of twist angles observed for the series of dinitramide salts.[30-36]

Using similar *ab initio* molecular orbital calculations Mebel *et al.* have investigated the gas-phase structure, thermochemistry, and decomposition mechanism of ammonium dinitramide.[73] Structure optimizations were done at the MP2 level with 6-31G*, 6-311G**, and 6-311+G** basis sets. Accurate energy levels and heats of reactions were evaluated by using MP4(SDQ), MP4(SDTQ), and QCISD(T) levels [76] as well as the G1 and G2 approaches.[77] The calculations predict that in the gas-phase the ion-pair structure $[NH_4^+][N(NO_2)_2]^-$ is not stable, rather the stable structure is the H-bonded acid-base pair $[NH_3]\cdot[HN(NO_2)_2]$, with the isomeric structure $[NH_3]\cdot[HON(O)NNO_2]$ only 2.3 kcal/mol less stable. Calculations by Mebel *et al.* also predict that these conformations may be involved in the early stage of ADN decomposition, based on reactions $HN(NO_2)_2 \rightarrow HNNO_2 + NO_2$ and $HON(O)NNO_2 \rightarrow HON(O)N \rightarrow NO_2$, respectively.[73] However, in the high-temperature regime, due to the difference in entropic contributions, they concluded that elimination of NO_2 is the dominant reaction.[73]

A major development in computational chemistry of the last decade was the emergence of density functional theory (DFT).[78-80] The main advantage of DFT is that electron correlation effects for atomic and molecular systems are considered explicitly in calculations but the computational requirements remain relatively similar to those needed by Hartree-Fock calculations. Consequently, the DFT method has attracted a lot of interest in recent years for applications to systems of increasing complexity and size.

One of the earliest applications of DFT to an ionic system was reported by Politzer and coworkers.[72, 81] Using non-local DFT calculations and a Gaussian DZVPP basis set (approximately equivalent to 6-31G**) they investigated the structure of the dinitramide anion and the energetics of some possible decomposition steps. The exchange and correlation functionals were included through the generalized gradient approximation (GGA).[82, 83] They obtained good agreement between the calculated structure of the dinitramide anion and the experimental crystallographic results despite the fact that calculations were done for the isolated anion and thus neglect the inter-ionic interactions present in the crystal phase. In the crystal structure the inner N-O bond lengths are shorter than the outer ones, and the NO_2 groups of the dinitramide ion are rotated out of the N-N-N plane by about 27° while the two N-N-O angles of each nitro group have unequivalent values.[72] The significant difference of about 10° between the two N-N-O angles was attributed to two main effects: steric interference between oxygen atoms and the increased conjugation between the NO_2 groups and the lone pairs on the central nitrogen atom.[72] Politzer and co-workers calculated 112.4° for the N-N-N angle in the gas-phase dinitramide ion and 27° for the out-of-plane rotation of NO_2 groups; the electron pairs are tetrahedrally distributed, thus allowing the twist of the NO_2 groups and optimizing the conjugation to the central nitrogen. This corresponds to sp^3 hybridization of the central nitrogen atom. If the central nitrogen were sp^2 hybridized, dinitramide would be planar with the lone pairs of the central nitrogen in different orbitals, one sp^2 orbital in the N-N-N plane and a p-orbital perpendicular to it. In the experimental ADN crystallographic structure,[27] the N-N-N angle is 113.2°, intermediate between the expected values for tetrahedral and trigonal planar geometries. Consequently, Gilardi *et al.* concluded that the hydridization of the central nitrogen atom lies between sp^2 and sp^3 in the crystal.[27]

Besides the structural properties, Politzer *et al.*[72] also investigated decomposition steps of the dinitramide ion involving the elimination of NO_2, NO_2^-, and NO_2^+ and an internal rearrangement of the dinitramide ion suggested by Doyle's [84] mass spectrometric studies. They found that the dissociation of $N(NO_2)_2^-$ to NNO_2^- and NO_2 has the lowest activation energy of 49.8 kcal/mol. These products can then react to give N_2O and NO_3^-.

Calculations of structural parameters of the dinitramide ion have also been performed by Pinkerton and coworkers.[85-87] Using DFT calculations at the B3LYP/6-311+G* level they investigated the flexibility of the dinitramide ion for a series of 27 dinitramide salts.[85] A major focus of these calculations was to determine the energy required to distort single molecular anions from the calculated minimum energy structure to that found in the experimental crystal structure. The results indicate that the local environment specific to each crystal produces significant geometrical perturbations. Consequently, the non-planar geometry of dinitramide ion was attributed to a superposition of resonance and steric repulsion, which have opposing energetic effects. A similar conclusion was reached by Shlyapochnikov *et al.*[29] Based on structural parameters for several dinitramide rotamers calculated by using B3LYP-6311+G* they concluded that changes in the valence angles cannot be explained solely by steric effects; instead, conjugation of the π-orbitals of the nitro groups with the p-orbitals of the central nitrogen atom must be considered, which causes delocalization of the negative charge and thus stabilizes the anion. Shlyapochnikov *et al.*[29] estimated the magnitude of this interaction by comparing the energy difference of configurations where the overlap of *p*-orbitals of the central N-atom with the π orbitals of the nitro groups is the most and the least efficient; 20 kcal/mol was estimated for the energy of conjugation, which is similar to that determined by Pinkerton *et al.*[85]

Further details about the nature of the bonding in the dinitramide ion have been obtained by Pinkerton and coworkers [86, 87] and by Shlyapochnikov *et al.*[29] based on topological theory of atoms-in-molecules.[88, 89] Within this theory the electron density, ρ(r), and its first (∇ρ) and second ($\nabla^2\rho$) derivatives can be used to define molecular structure at the equilibrium state of a system. Several features have emerged from these studies.[86, 88] First, the local asymmetry of the nitro groups was directly probed by the atomic charge distribution. As this asymmetry is more pronounced in crystal structures, it follows that a higher delocalization of charge takes place in this case, with a corresponding increase in the effective charges of O atoms. Additionally, a bond critical point and an atomic interaction line between the inner oxygen atoms belonging to different nitro group has been evidenced in topological analyses, indicating a bonding type of interaction between these atoms. This atomic interaction line was found to have similar properties for the isolated dinitramide ion and in the crystal.[86] Topology analysis confirms that all hydrogen atoms of the NH_4^+ ions are involved in hydrogen bonds in the ADN crystal.[86] Additionally, all but one of the O-atoms participate in hydrogen bonding, with one of outer O-atoms participating in two hydrogen bonds. The calculated topology of electron density confirms that the central N-atom is sp^2-like. Similar conclusions were reached for biguanidinium dinitramide and biguanidinium *bis*-dinitramide systems.[87]

We pointed out in Section 1.1 that the mechanism of thermal decomposition of ADN is very complex and the initial reaction depends on the temperature and pressure, thus various initial steps have been proposed by different groups of investigators.[56, 57, 59, 61] The main decomposition processes that have been proposed are (a) sublimation of ADN to NH_3 and $HN(NO_2)_2$, (b) an ionic mechanism for the formation of AN and N_2O, (c) dissociation of the dinitramide ion with the formation of NO_2 and the mononitramide ion which subsequently dissociates to NO^- and NO, and (d) conversion of the dinitramide ion to NO_3^- and N_2O.

The chemical decomposition of ADN has been significantly clarified by the work of Politzer, Seminario, and Concha,[90] who performed a very impressive theoretical study of the energetics of the decomposition. Using DFT Becke-3 (B3) exchange and Perdew 86 (P86) correlation functions [83, 91] and 6-31+G** basis sets, they calculated the structures, energies at 0 K, and enthalpies at 298 K for 37 molecules, ions, and transitions states that are involved in thermal decomposition of ADN. This level of theory has been shown to give accurate bond dissociation energies with average errors within 1.9 kcal/mol. [92] Politzer *et al.* developed the comprehensive map of the various pathways for the decomposition of ADN that is shown in Figure 5. The sublimation of ADN to NH_3 and $HN(NO_2)_2$ (reaction 1 in Figure 5) requires only 44 kcal/mol, while sublimation to gas-phase ions (reaction 2 in Figure 5) requires 144 kcal/mol. Consequently, sublimation is predicted to be the main initial decomposition pathway for ADN. Following reaction 1 (see Figure 5), the next most probable steps were found to be reactions 20 and 9. These correspond to the loss of NO_2 from $HN(NO_2)_2$ (with a dissociation energy of 40.7 kcal/mol) and from its tautomer O_2N-NN(O)OH (with a dissociation energy of 36.3 kcal/mol). From the $HNNO_2$ product there is a chain of reactions (23)+(25) leading to $N_2O + H_2O$ and the sequence of reactions (24)+(38) leading to NO. From NN(O)OH reaction can proceed through reactions (15)-(17) with the formation of $HONO_2$ as illustrated in Figure 5.

Lin and coworkers [73, 93] have also performed detailed theoretical studies of the decomposition of gas-phase ADN. They determined the optimized structures of possible ADN configurations at the MP2 level for the 6-31+G** basis set. They also studied the decomposition mechanism of dinitramic acid. The N-NO_2 bond-fission energy was determined with high-level modified Gaussian-2 (G2M)[94] calculations, along with the structure of the four-center transition state for the N_2O decomposition pathway. By determining points with maximum free energy on the N-NO_2 bond-fission pathway using RRKM theory, they were able to calculate the rate constant for the N-NO_2 bond-fission decomposition pathway.

Rate constants for the N_2O decomposition pathway have been calculated by Alavi and Thompson [95] using RRKM theory. High-level G2M calculations [94, 96] were used to determine the energies of the reactants and transition-state structures involved this pathway. The activation energies for this decomposition pathway are close to those of the N-NO_2 bond-fission pathway, however, the pre-exponential factor for the gas-phase N-NO_2 bond-fission pathway is at least two orders of magnitude larger than the N_2O elimination channel, making the N-NO_2 bond fission the dominant reaction for ADN decomposition.

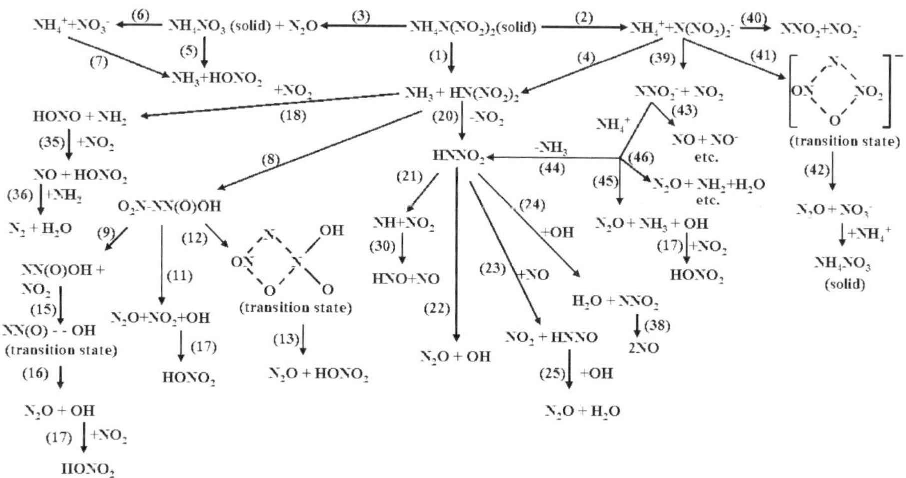

Figure 5. Possible steps involved in decomposition of ammonium dinitramide (after Politzer *et al.*, Ref. [90]).

A similar comprehensive theoretical investigation of the individual steps involved in the thermal decomposition of AP was done by Politzer and Lane.[96] They showed that for systems containing several oxygen and/or chlorine atoms in close proximity, accurate reaction energetics can be calculated by using DFT with Becke-3 (B3) exchange and Perdew-Wang 91 (PW91) correlation functionals [97] and larger 6-311+G(2df) basis sets. Based on these calculations, the energy minima and enthalpies at 298 and 300 K were computed for 37 atoms, molecules, radicals, and ions involved in the thermal decomposition of AP.[96] They predicted that the initial reaction step is the sublimation of AP to NH_3 and $HClO_4$, $NH_4ClO_4(s) \rightarrow NH_3(g) + HOClO_3(g)$, with a heat of reaction of ΔH(298 K) = 45 kcal/mol. Moreover, the intermediate $H_3N \cdots H \cdots OClO_3$ was predicted to be involved in sublimation. This intermediate is about 14.3 kcal/mol lower in enthalpy than the final gaseous products; the activation energy for sublimation is predicted to be 31 kcal/mol. Politzer and Lane [96] pointed out that this is in accord with the much earlier experimental observation by Jacobs and Whitehead [40] that the activation energy for sublimation is significantly less than the overall enthalpy change.

The selected results just presented demonstrate the kinds of information that can be obtained by using *ab initio* molecular orbital and DFT calculations. The studies to date have focused for the most part on structural and energetic properties of the various atomic, ionic, and molecular species that may be involved in the thermal decomposition of energetic salts. Also, theoretical calculations have been used to obtain quantitatively descriptions of the various elementary steps postulated in mechanisms of the dissociation processes of these salts and to predict the most probable initial steps. For both ADN and AP, quantum chemistry calculations have predicted that the initial reaction step is sublimation in which protons transfer from NH_4^+ to, respectively, $HN(NO_2)_2^-$ and ClO_4^-. Proton transfer appears to play an important role in the physical and chemical properties of these salts. In the next section, we briefly review some recent results of studies of proton transfer in energetic salts.

2.3. Quantum Chemical Calculations of Proton Transfer in AN, ADN, and HAN Clusters

Experimental studies, which we briefly summarized in Sec. 1.1.2, indicate that the first step of the decomposition mechanism for many energetic salts is proton transfer between the ion pair to give the acid H_A and base B, with possible sublimation of the resulting neutral acid-base pair. It is therefore important to understand the structures and energetics of the hydrogen-bonded acid-base complexes. For the energetic salts studied to date, the stable form of single formula-units in the gas phase are the neutral-pair A_H···B complexes. The ion-pair forms of the complexes A^-···H_B^+ are higher in energy and actually constitute transition states in the double proton-transfer between the acid and base molecules. For the cases studied the additional Coulombic interaction gained upon formation of the single ion-pair unit is not sufficient to compensate for the energy required to break the A_H bond. Calculations have been carried out for AN, ADN, and HAN; the predicted structures of the neutral- and ion-pair configurations are shown in Figures 6 and 7.

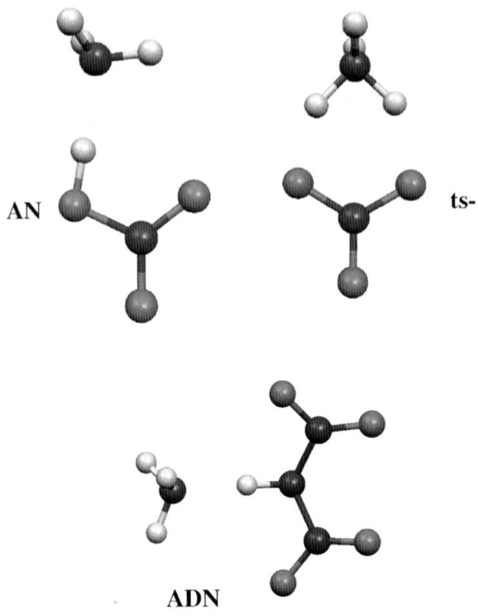

Figure 6. The structures of hydrogen-bonded gas-phase AN, and ADN molecules. The most stable complexes are the neutral-pair species. The structure of the ion-pair, which is labeled as ts-AN, is also shown.

Quantum chemistry studies of proton transfer in gas-phase AN predict that the stable molecular form is the hydrogen bonded $H_3N{\cdots}HONO_2$ neutral pair.[98-100] Calculations at the MP2 (B3LYP) level with the 6-311++G** basis set predict that the binding energy of the neutral-pair form is 11 kcal/mol (8.1 kcal/mol) greater than the $H_3NH^+{\cdots}ONO_2^-$ ion-pair.[100] Large basis sets with diffuse functions were used for these calculations to capture the long-range diffuse nature of the hydrogen bonding. In the dimer $(AN)_2$ proton transfer occurs from the nitric acid molecules to the ammonia molecules, and thus the dimer is composed of two pairs of NH_4^+ and NO_3^- ions arranged in a structure with C_{2h} symmetry. The additional electrostatic interactions gained from the presence of extra neighboring ions stabilizes the ionic structure relative to the neutral hydrogen-bonded form of the $(AN)_2$ complex.[100] The structure of $(AN){\cdot}NH_3$ and $(AN){\cdot}HNO_3$ complexes have also been studied and it has been determined that the former is composed of neutral-pair AN solvated by the NH_3 while the latter is composed of an ion-pair AN unit stabilized by HNO_3.

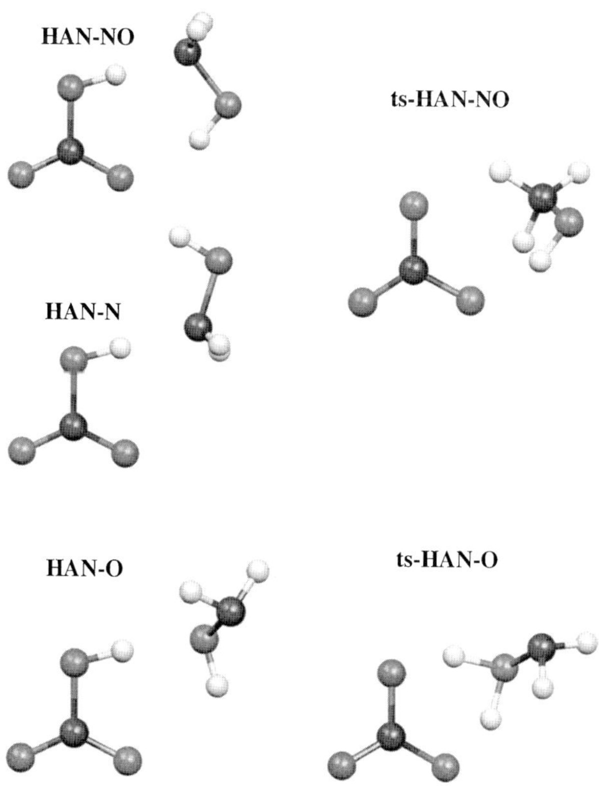

Figure 7. Atomic structures of the three neutral-pair and two ion-pair configurations of HAN molecule. The neutral-pair structures in order of stability are labelled as HAN-NO, HAN-N, and HAN-O. The ion-pair structures are labelled as ts-HAN-NO and ts-HAN-O.

The neutral-pair versus ion-pair nature of these hydrogen-bonded AN complexes can be characterized by calculating the strength of the Coulombic interactions in the acid and base moieties. Point charges from Mulliken population analysis[101] were assigned to the atoms of the separate acid and base units and the magnitude of electrostatic interactions between the species determined. The approximate values of the Coulombic interactions for AN, $(AN)_2$, $(AN)\cdot NH_3$, and $(AN)\cdot HNO_3$ calculated in this manner at the B3LYP level are +5, -167, -101, and –34 kcal/mol, respectively. The ionic $(AN)_2$ and $(AN)\cdot NH_3$ species have much larger electrostatic interactions than the neutral species. The binding energies for the same complexes are 12.4, 34.6, 20.3, and 18.5 kcal/mol, respectively. It is obvious that the binding energies do not greatly distinguish between the neutral-pair or ion-pair nature of the complexes. The main stabilizing factor of the ions is Coulombic interactions.

Theoretical studies at the B3LYP/6-311G** level of proton transfer in gas-phase ADN predict that the stable form is the acid-base pair. Like $(AN)_2$, the $(ADN)_2$ dimer is composed of ions.[102] Single ADN molecules "solvated" with ammonia or dinitramic acid molecules, *i.e.* $(ADN)\cdot NH_3$ and (ADN)·HDN, are composed of ion-pair ADN units.

Theoretical electronic structure calculations at the B3LYP/6-311++G** level of hydroxylammonium nitrate (HAN) predict that the hydrogen-bonded acid-base pair is the stable form of the monomer.[103] The HAN molecule has three possible neutral-pair and two ion-pair configurations; Figure 7 shows the predicted structures. The neutral-pair configurations are designated as HAN-NO, in which both the nitrogen and the oxygen atoms of hydroxylammonia participate in hydrogen bonding with the nitric acid, HAN-N, where the nitrogen atom of hydroxylammonia is both the proton donor and acceptor to nitric acid, and HAN-O, where the oxygen atom of hydroxylammonia is both the proton donor and acceptor. The most stable ion-pair structure is related to HAN-NO and has the nitric acid proton transferred to the NH_2 group of HAN. The less stable ion-pair is related to HAN-O and has the nitric acid proton transferred to the OH group. The most stable ion-pair structure is 13.6 kcal/mol higher in energy than the HAN-NO neutral-pair structure from which it is formed.

2.4. Quantum Chemistry Calculations Applied to Solid-Phase Ionic Energetic Materials

2.4.1. General Aspects

The studies discussed above have been concerned with ionic systems in the gas phase or in small clusters. This work has provided information about the structural and energetic properties of these systems as well as their reactions. However, the practical interest in these materials is in condensed phases. The inclusion of intermolecular interactions is essential for realistic descriptions of these materials. It is important to consider the electrostatic, long-range interactions. A practical way to consider these interactions is to perform simulations of solids or liquids with periodic boundary conditions.

An important step towards accurate descriptions of solid ionic energetic materials has been done by Sorescu and Thompson.[104-106] They used DFT and the pseudopotential method to investigate the structural and electronic properties of AND [104, 105] and AN [106] in solid phases. The advantage of using the pseudopotential approximation is that only the valence electrons are represented explicitly in the calculations, while the valence-core interactions are described by nonlocal pseudopotentials. The pseudopotentials used in these studies were norm-conserving of the form suggested by Kleinman and Bylander [107] and optimized using the scheme of Lin *et al.*[108] The occupied electronic orbitals were expanded in a plane-wave basis set subject to a cutoff energy. A gradient-corrected form of the exchange correlation functional (GGA) was used in the manner suggest by White and Bird.[109] The periodicity of the crystals studied is considered by using periodic boundary conditions in all three dimensions. Calculations were performed using the commercial version of the CASTEP code.[110]

There are several areas where this computational approach is useful. First, direct information about optimized crystallographic lattice parameters and the geometrical parameters of ionic systems can be determined. This can be done for uncompressed lattices or when external isotropic or anisotropic compression is applied to the crystal. Additionally, from the variation of the total energy of the lattice with respect to compression or expansion the corresponding elastic properties can be determined. Similarly, by analysis of the energy variations as functions of atomic displacements around equilibrium positions the phonon spectrum can be evaluated. Besides the structural, elastic, and phonon-modes parameters, other important energetic and electronic properties can be evaluated. Lattice energies and cohesive energies are examples of the former category. Among the list of electronic properties some representative examples are the band structure and the total or partial density of states. Furthermore, additional insight can be obtained from population analyses of charge distribution, bond order, and electron and spin density maps.

2.4.2. Structural and Electronic Properties

Using the plane-wave DFT computational approach Sorescu and Thompson [104, 106] obtained good agreement with experiment for the predicted lattice parameters of ADN and AN crystals. For α-ADN the errors of the calculated lattice parameters relative to experimental values are less than 1.62%,[104] while for AN phases V, IV, III, and II the errors range from 1.96-2.2%.[106] The larger differences observed for AN were attributed to dynamic contributions such as the orientational disorder of ammonium ions specific to phases III and II and to the neglect of temperature effects in the calculations. The calculations predicted for ADN, in agreement with the X-ray data,[27] the nonequivalence of the two halves of the dinitramide ion. Specifically, the nonequivalence of the N-N bond lengths and of the N-N-O angles of each nitro group as well as the out-of-plane twist of the nitro groups were found.

Calculations of the self-consistent band structures predict relatively large band gaps for the optimized lattices. For example, in the case of AN, the band gaps at the $\Gamma(0,0,0)$ point for phases V, IV, III, and II have values between 3.37-3.51 eV while for ADN the band gap is about 3 eV. These results indicate that both these two materials are electrical insulators at ambient conditions.

2.4.3. Pressure-Induced Effects

Another important set of results from these studies is the dependence of the structural and electronic properties of ADN and AN crystals on compression. The effects of compression on phase-IV AN were studied in the range 0-600 GPa.[106] Anisotropic effects were observed for the lattice dimensions *a, b,* and *c*. Particularly, compressibility effects were found to be similar for the *a* and *c* directions but different for the *b* direction which has the highest compressibility. These effects can be explained by the fact that there are relatively strong N-H···O hydrogen bonds in the *ac* plane and weaker H···O bonds along the *b* direction. The lattice compression also produces significant effects on the band structure and the density of states. Particularly, broadening of the occupied bands starting from the top of the valence band and shifts toward negative values were noticed with the increase of pressure. Additionally, the band gap is decreased from 3.25 eV at zero pressure to about 2.0 eV at 600 GPa. Finally, the compression of the lattice leads to strong charge redistribution among the atoms of the crystal.

In the case of ADN, computational studies revealed several regimes over the pressure range 0-150 GPa.[105] Between 0 to 10 GPa the $P2_1/c$ symmetry is maintained with small changes of lattice angles. Above 20 GPa significant deformations of lattice shape and sizes take place with increasing pressure. Correspondingly, the symmetry of the crystal was found to change from monoclinic to triclinic. Pictorial illustrations of the crystal structures at 0, 20, 50, and 150 GPa are shown in Figure 8. The lattice compression induces electronic changes that are similar to those observed for AN. There is also significant broadening of the electronic bands with shifts towards lower energies over the pressure range 0-150 GPa as reflected by evolution of the density of states plots at 0, 10, 40, and 150 GPa presented in Figure 9. Over this pressure range there is a corresponding drop in the band gap from about 3 eV to 2.3 eV. Lattice compression causes significant charge redistribution and delocalization. These changes indicate a decrease in the ionic character of the crystal with a concomitant increase in the covalent character with increasing pressure.

2.4.4. Transport Properties

The results discussed above illustrate how quantum chemistry methods can be successfully applied not only to gas-phase systems but also to compute structural and electronic properties of ionic energetic crystals at ambient pressure and for hydrostatic compression. While these calculations provide critically important fundamental information, it is necessary to include thermal effects to fully understand the nature of the materials. This can be done by performing molecular dynamics (MD) simulations. The critical component of a MD calculation is the force field. Traditionally, classical analytical force fields are developed, often with parameters determined from empirical data or simply estimated. However, recently it has become feasible to use *ab initio* potentials and forces. For example, in the Car-Parrinello extended Lagrangian approach to *ab initio* molecular dynamics (AIMD), the electronic configuration is described by using the Kohn-Sham formulation of the DFT with the Kohn-Sham orbitals expanded in plane-wave basis sets. In this case the forces exerted on atoms are generated "on the fly" by directly performing the electronic structure calculation at each integration step as the simulation proceeds. The major limitations of this approach are the large computational requirements and the accuracy of the quantum mechanical calculation. However, continuing development of computer hardware and further optimizations of the computational algorithms will lead to an increase in the use of this powerful tool.

A recent example of an application of AIMD to a salt is the study by Rosso and Tuckerman,[112] who investigated the charge transport mechanism in solid AP in an ammonia-rich atmosphere. The calculations were performed using B-LYP functional [113, 114] with the electronic structure represented within the generalized gradient approximation of DFT. Two temperatures regimes were considered, one at 300 K and the second at 530 K. For the first regime, after an initial equilibration for 1 ps under NVT conditions the dynamics was followed for 10 ps under NVE conditions. In the high-temperature regime, equilibration was done over 3 ps while subsequent evolution was followed for 15 ps. Exposure to ammonia resulted in it being absorbed in interstitial sites of the crystal. Further, proton transfer between NH_4^+ ions and neutral ammonia occurred to form a short-lived $N_2H_7^+$ complex, which was found to be responsible for a marked increase in the conductivity, in agreement with experimental observations. For the pure AP crystal no proton transfer between the NH_4^+ and ClO_4^- ions was observed in the simulations. This indicates that the dominant mechanism for charge transport is by diffusion of NH_4^+ and ClO_4^- ions. Moreover, there is significant coupling between translational diffusion and rotational diffusion for both these ions.

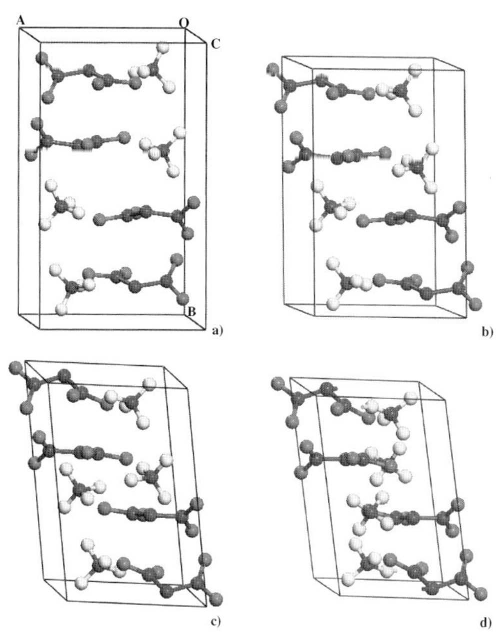

Figure 8. Pictorial views of the ADN crystal structures at (a) 0 GPa, (b) 20 GPa, (c) 50 GPa, (d) 150 GPa. Reproduced with permission from Ref. 105.

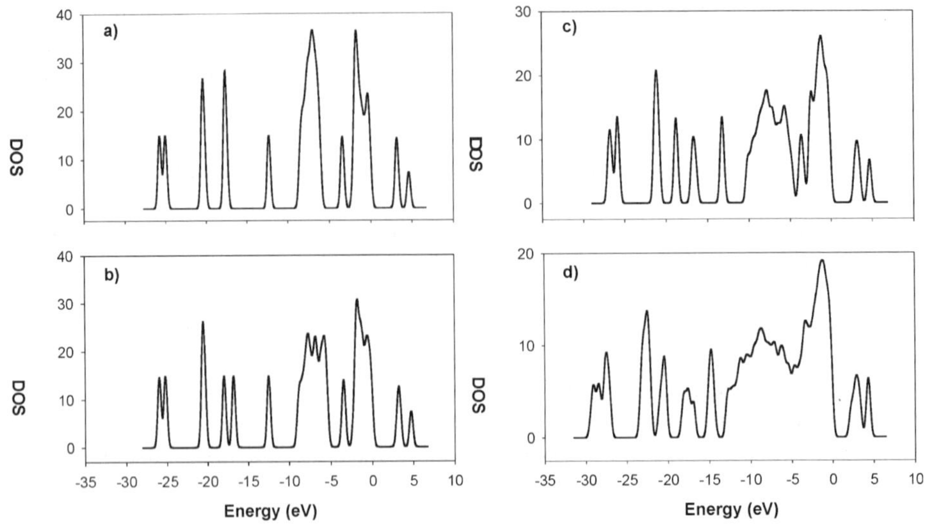

Figure 9. Calculated density of states for ADN at the pressures: (a) 0 GPa, (b) 10 GPa, (c) 40 GPa, (d) 150 GPa. Reproduced with permission from Ref.105.

3. CLASSICAL SIMULATIONS OF SALTS

3.1. General Aspects

As illustrated by the studies discussed in the previous sections, descriptions of the structural, energetic, and dynamic properties of energetic salts can be achieved by using quantum chemistry calculations by either simple energy optimizations to determine equilibrium or transition states configurations or by direct dynamics simulations. The first approach is currently performed routinely but applications of AIMD methods are still quite challenging due to the large computational resources required and the limitations in many applications to lower-levels of quantum theoretical method. Consequently, the approach that is commonly used consists of the development of a classical force field. The main assumption of this method is that the Born-Oppenheimer approximation for separation of the electronic and nuclear degrees of freedom is valid and the evolution of the system takes place without changes in the electronic state of the system. In many practical applications it is assumed that the system is in its ground electronic state. The properties of the system can then be evaluated by using either molecular dynamics (MD) or Monte Carlo (MC) methods. In the first case, Lagrangian or Hamiltonian formulations are used for integration of the equations of motion.[115] Furthermore, by coupling the system with different thermostats and barostats, the time evolution of systems in different statistical ensembles such as isothermal-isobaric, constant stress or constant temperature can be investigated.[116-119] In the MC approach, ensemble averages of observables are evaluated by performing random walks over the relevant phase space.[120, 121] Procedures such as importance sampling can be used to improve the efficiency of calculations. An advantage of the MC method is that it can be readily applied to a wide range of statistical ensembles such as isothermal-isobaric, grand-canonical, and constant-stress-isothermal.[120, 121]

An application of MD or MC methods requires a potential energy surface that accurately describes the most important regions of configuration space such as the equilibrium configurations and the transition states between them. For liquids, accurate descriptions of structural properties such as radial distribution functions, density, and energetic properties such as the heat of vaporization and heat capacity are prime examples of the data that must be reproduced by the force field. Similarly, for solids accurate predictions of the crystallographic parameters and crystal symmetries as functions of temperature and pressure and the energetic and elastic properties such as crystal lattice energy and the elastic coefficients are important quantities to be used to assess the accuracy of a given force field.

Simple classical intramolecular force fields are usually constructed in terms of the valence coordinates; *e.g.*:

$$V_{\text{intra}} = \sum_{bonds} k_r (r - r_0)^2 + \sum_{angles} k_\theta (\theta - \theta_0)^2 + \sum_{dihedrals} \frac{V_n}{2}[1 + \cos(m\Phi - \delta)] + \sum_{wags} k_\chi (\chi - \chi_0)^2 \tag{1}$$

in which the coordinates are the bond stretching (r), bond angle bending (θ), torsional motions described by the dihedral angle Φ formed by groups of four atoms and the out-of-plane motions of atoms or group of atoms (χ). In more elaborate force fields, anharmonic contributions and additional terms representing the couplings of various bond-bond and bond-angle interactions may be included. The intermolecular interactions, which in essence are many-body in nature, are often described by simple pairwise-additive functions. These interactions are usually of two types: van der Waals and electrostatic interactions. The intermolecular potential is often of the form

$$V_{\text{inter}} = \sum_{i<j}\left[\frac{A_{ij}}{r_{ij}^n} - \frac{B_{ij}}{r_{ij}^m}\right] + \sum_{i<j}\frac{q_i q_j}{r_{ij}} \tag{2}$$

Typical analytical representations for nonbonded interactions in Eq. (2) are Lennard-Jones (n=12, m=6) or 9-6 (where n=9 and m=6). Alternatively, the nonbonded interactions have been represented by the Buckingham potential:

$$V_{\text{Buck}} = \sum_{i<j} A_{ij} \exp(-B_{ij} r_{ij}) - C_{ij} r_{ij}^6 \tag{3}$$

The force constants in Equations (1)-(3) are fitted based on a database of structural, energetic, spectral, and elastic parameters. Traditionally, the database was obtained exclusively from experiments. However, due to significant improvements in the accuracy of quantum chemistry methods much of the necessary structural, energetic, and spectral data can be computed. The practical aspects of fitting force fields for gas- and condensed-phase systems have been described in many previous reviews,[122-126] thus, in the following we will detail only some recent developments specific to energetic salts.

3.2. Atomistic Models for Salts

3.2.1. Rigid-Ion Models

As discussed above useful analyses of energetic salts can be achieved by performing MD simulations using general force fields that include intra- and inter-molecular interactions. A first approximation of the dynamic properties can be obtained by using the rigid-ion approximation, which results in considerable savings in the cpu time needed for integration of the equations of motion. In the rigid-ion model, the geometries of the ions are held fixed at their experimental crystallographic values. Then, only the intermolecular interactions need to be included in the force field.

A very useful technique to test an intermolecular force field used for a particular crystal is a molecular packing calculation. The method has been well described in previous reviews and will not be repeated here.[127, 128] We will only outline its major characteristics. Given a particular crystal, the performances of an intermolecular potential can be assessed by performing lattice-energy minimization with respect to the structural degrees of freedom of the crystal. For a crystal with Z rigid-molecules per unit cell at arbitrary positions, the degrees of freedom are determined by the positions and orientations of the molecules inside the unit cell as well as the dimensions and angles of the unit cell. The number of degrees of freedom can be decreased (with a significant decrease in the required computing time) by symmetry-imposed constraints on either the lattice parameters or on subsets of atomic coordinates when the atoms occupy special positions. However, accurate force fields should lead to the same optimized lattice parameters when optimization is done with or without symmetry constaints.

Within the rigid-molecule approximation, Sorescu, Rice and Thompson [129-134] have developed a transferable intermolecular force field for simulations of a large database of energetic nitro compounds, that includes 30 monocyclic, polycyclic, and acyclic nitramines, among which are representative energetic compounds such as hexahydro-1,3,5-trinitro-1,3,5-*s*-triazine (RDX), 1,3,5,7-tetranitro-1,3,5,7-tetraaza-cyclooctane (HMX), and 2,4,6,8,10,12-hexanitrohexaazaisowurtzitane (HNIW). Additionally, these intermolecular potentials have proven to be transferable to a second database of 51 other compounds, which includes various types of nitroalkanes, nitroaromatic, nitrocubanes, polynitro-adamantanes, polynitropolycyclo-undecanes, polynitropoly-cyclododecanes, hydroxynitro derivatives, nitrobenzonitriles, nitrobenzotriazoles, and nitrate esters.[133] For all these systems the transferability of the potential was achieved for the repulsive-dispersive interactions represented by Buckingham exp-6 potentials; see Equation (3). The electrostatic Coulombic interactions were determined by using partial charges centered on the nuclei of the atoms. These have been determined for each individual isolated molecule by fitting to the quantum mechanical electrostatic potential surrounding the molecule.

Using molecular packing calculations Sorescu *et al.*[129-134] have shown that the set of potentials they developed reproduce the crystal structure parameters and the lattice energies, including the relative stabilities of different polymorphic phases. Equally good agreement with experimental data was obtained when thermal effects were considered. These effects have been studied using isothermal-isobaric MD (NPT-MD) simulations. For example, in the case of RDX,[129] HMX,[131] HNIW,[130] TNT (2,4,6-trinitrotoluene),[133] and nitromethane [135] it was found that at ambient pressure the calculated crystal structures at 300 K are in very good agreement with experimental data with almost no rotational and translational disorder of the molecules in the unit cell. It is important to note that these intermolecular potentials were also able to accurately reproduce the crystal structures of energetic nitramines under hydrostatic compression, particularly for pressures below 7.5 GPa.[132] In this low-pressure regime the main effect of compression was found to be a decrease in intermolecular distances without significant molecular deformations.[132]

The first extension of this approach to ionic energetic crystals was done by Sorescu and Thompson [104] for ADN. Using the rigid-ion approximation, the intermolecular potential used was composed by pairwise Lennard-Jones (LJ), hydrogen bonding (HB), and Coulombic (C) terms of the form

$$V_{inter} = \sum_{i<j} V_{ij}^{LJ} + V_{ij}^{HB} + V_{ij}^{C} \tag{4}$$

where

$$V_{ij}^{LJ}(r) = \mathring{a}_{ij}^{0}\left[\left(\frac{r_{ij}^{0}}{r}\right)^{12} - 2\left(\frac{r_{ij}^{0}}{r}\right)^{6}\right] \tag{5}$$

$$V_{ij}^{HB}(r) = \mathring{a}_{ij}^{0}\left[5\left(\frac{r_{ij}^{0}}{r}\right)^{12} - 6\left(\frac{r_{ij}^{0}}{r}\right)^{10}\right] \tag{6}$$

and

$$V_{ij}^{C}(r) = \frac{q_i q_j}{4\pi\varepsilon_0 r} \tag{7}$$

Here ε_{ij}^{0} is the energy minimum for the pair of atoms i and j, r_{ij}^{0} is the interatomic distance at the energy minimum, q_i and q_j are the electrostatic charges on the atoms, and ε_0 is the dielectric permittivity constant for vacuum. The assignment of atom-centered monopole charges was made by fitting a quantum mechanically derived electrostatic potential in the region surrounding the van der Waals surface of the dinitramide and ammonium ions using the CHELPG method proposed by Breneman and Wiberg.[136] These calculations were done by using Møller-Plesset perturbation theory at the MP2/6-311+G** level.

The LJ potential parameters used in this potential have been fitted such that the entire interionic potential reproduces the experimental structure of the α-ADN crystal and the calculated lattice energy reported by Politzer *et al.*[90] Using molecular packing calculations without symmetry constraints as implemented in the LMIN package [137] the lattice dimensions of the ADN crystal have been reproduced to within 1.86% of experiment and with only small translations of the ions inside the unit cell. Moreover, it was verified that no changes from the $P2_1/c$ space group crystal symmetry take place as a result of lattice minimization. Additional tests of this potential performed using NPT-MD simulations in the temperature range 4.2-350 K and ambient pressure indicate that the agreement with experimental values is maintained even when finite temperature effects were considered. For example, at 223 K agreement within 3.4% between the predicted lattice dimensions and the experimental values was obtained.[104] The results of MD simulations showed that little translational disorder occurs with an increase in temperature [see Figure 10(a)]. However, as reflected by radial distribution functions [see Figure 10(b)] of different O···H pairs there is a loss of spectral structure when the temperature is increased. This indicates that a large degree of rotational disorder of the ammonium ions take place with the increase in temperature. The thermal expansion coefficients for ADN were determined from the temperature dependence of the average lattice dimensions. It was concluded that thermal expansion is anisotropic with preferentially expansions along the *b* and *c* axes. As the dinitramide ions have their NNN plane lying in the *ac* plane, with opposite N atoms running along the *a* axis it follows that the largest expansions are along directions perpendicular to this axis, *i.e.*, along the *b* and *c* axes.

3.2.2. Flexible-Ion Models

Despite the success of the rigid-ion potentials described in the previous section there are also significant limitations of these models. Major among these, obviously, are the lack of intra-ion vibrational excitation and the impossibility to introducing interactions to allow chemical reactions. Thus, it is important to develop models in which both intra- and intermolecular interactions are included. An example of such a development was reported by Sorescu and Thompson [106] for phase V of AN, where the intramolecular potential was taken to be a superposition of harmonic bond stretches and bond angle bends. For nitrate ions these functions were augmented with a trigonometric torsion potential to describe the position of the nitrogen atom relative to the plane of the oxygen atoms. The force constants were parameterized to reproduce the *ab initio* calculated harmonic vibrational frequencies of NH_4^+ and NO_3^- ions. The intermolecular potential, containing 12-6 Lennard-Jones potentials and 12-10 hydrogen-bond terms, was taken directly from the ADN crystal model.[104] However, additional refitting of the N···O and H···O interaction potentials pairs were found to be necessary for accurate reproduction of the structure and energy of phase-V AN. Electrostatic charges of the ammonium and nitrate ions were determined by fitting *ab initio* electrostatic potentials calculated at the MP2/6-31G** level.

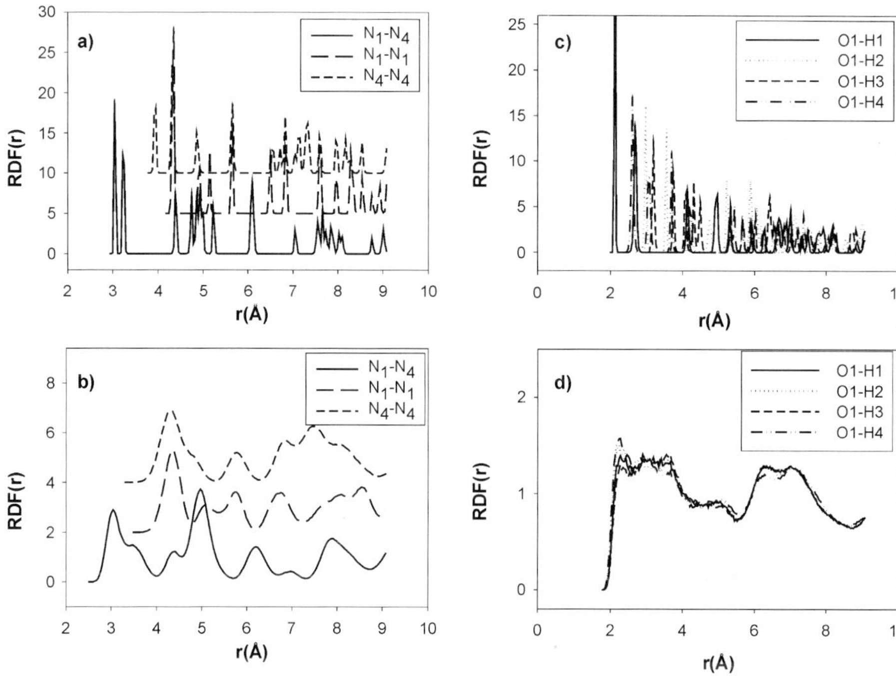

Figure 10. Radial distribution functions for nitrate-ammonium center-of-mass to center-of mass pairs (a) and the N-O···H pairs (b) as functions of temperature at (a, c) 4.2 K and (b, d) 350 K. Adapted with permission from Ref. 104. Labels N_1 and N_4 refer, respectively, to nitrate and ammonium nitrogen-atoms.

Molecular packing calculations using this set of potentials show that the lattice dimensions are reproduced to within 2.4%. The full matrix of the elastic coefficients was also reported. Additional tests performed using NPT-MD simulations over the temperature range 4.2-250 K showed that the model accurately reproduces the experimental unit cell dimensions with a maximum deviation of 3.4%. Additionally, it was found that in this temperature range there is little translational and rotational disorder of the ions in the crystal. As in the case of ADN, thermal expansion of phase-V AN was found to be highly anisotropic. The largest expansion coefficients correspond to the *a* and *b* axes in agreement with experimental neutron diffraction data reported by Choi and Prask.[4] These results can be understood based on the fact that the thermal properties of this phase are determined by the out-of-plane motion of the nitrate ions. As the planar nitrate ions are perpendicular on both the *a* and *b* axes it follows that the largest thermal expansions will take place along these axes.

The developments reported by Sorescu and Thompson [104, 106] for simulations of ionic energetic materials demonstrate that accurate classical force fields can be developed by using both experimental data and quantum chemistry results. The main use of these potentials was to predict nonreactive processes in ionic crystals. These potentials were developed to provide accurate descriptions of the structural, energetic, spectral, and elastic properties of these materials within a given phase as functions of temperature and pressure. However, the availability of these potentials has opened new areas of research; *e.g.*, MD simulations of transitions between solid-solid or solid-liquid phases have recently been carried out. In the next section, we describe the use of these potentials to predict of the melting temperatures of ionic crystals.

The main remaining limitation of these kinds of potentials is the lack of terms to represent chemical reactions. Some recent progress along these lines, particularly in the field of energetic materials, has been recently reviewed.[138] The most important developments have been achieved so far using reactive empirical bond order (REBO) potentials introduced by Brenner and coworkers.[139] The REBO potentials have been used mainly for simulations of shockwave propagation in simple diatomic or triatomic molecular crystals; however, it is likely that these kinds of potentials will soon be extended to more complex systems. A reactive potential based on somewhat different bond order concepts has been used to calculate the initial shock-wave induced chemical events in RDX.[140] To date, to our knowledge, REBO potentials have not been applied to ionic crystals.

4. SIMULATIONS OF PHASE TRANSITIONS IN ENERGETIC SALTS

4.1. Melting

A variety of methods have been proposed for calculating the melting points of molecular and ionic solids.[141-156] The thermodynamically rigorous method for calculating melting points involves determining the absolute values of the free energy of the solid and liquid phases at a specified pressure; the melting point is the temperature at which the free energies of the two phases are equal. This method requires separate MD calculations of the absolute values of the free energy for the separate solid and liquid phases.[143, 144] The potential energy function of the real system, U_{real} is coupled to the potential energy of an ideal system, U_{ideal}, through a coupling parameter λ. The potential energy of the system is written as,

$$U(\lambda) = (1 - \lambda)\, U_{ideal} + \lambda U_{real}. \tag{8}$$

The potential energy corresponds to that of the real system for $\lambda = 1$ and the ideal system for $\lambda = 0$. The partition function is

$$Q(N,V,T,\lambda) = \frac{1}{\Lambda^{3N} N!} \int d\mathbf{r}^N \exp[-\beta U(\lambda)] \tag{9}$$

The expression for the derivative of the Helmholtz free energy with respect to λ can be obtained from Eq. (9),[143]

$$\left(\frac{\partial A(\lambda)}{\partial \lambda}\right)_{N,V,T} = \left\langle \frac{\partial U(\lambda)}{\partial \lambda} \right\rangle \tag{10}$$

where the brackets represent an ensemble average. Integrating Eq. (10) gives the free energy for the real system $A(\lambda=1)$,

$$A(\lambda = 1) - A(\lambda = 0) = \int_0^1 d\lambda \left\langle \frac{\partial U(\lambda)}{\partial \lambda} \right\rangle. \tag{11}$$

The ideal system is such that it connects to the real system without going through a phase transition, and its Helmholtz free energy, $A(\lambda = 0)$, must be known at as a function of temperature and pressure. The ideal systems usually chosen for the real liquid and solid are, respectively, the ideal gas and ideal Einstein solid. In order to evaluate the integral in Eq. (11), a thermodynamically feasible pathway with no phase transitions must be designed to go from the real state to the ideal state. Recently, an idealized pathway for free energy calculations has been suggested that directly connects solid and liquid states.[157] Although rigorous in principle, the free energy calculations are often time consuming and not straightforward to implement for complex molecular and ionic solids; the method can be difficult and can be plagued by singularities.

A related thermodynamic method is to fit the temperature and density dependence of simulation data to equations of state for the solid and liquid.[158] The explicit forms of the pressure $P(T, \rho)$ and internal energy $U(T, \rho)$ equations for each phase are used to calculate the entropy and free energy. The condition $\Delta G = 0$ determines the solid-liquid equilibrium temperature as a function of pressure.

The equilibrium between solid and liquid phases at a given pressure can be studied by directly simulating the two-phase system in a *constant energy* NVE MD simulation [145, 153, 154] A fairly large simulation (between 1024 and 16000 particles have been used) is set up which includes separately equilibrated solid and liquid phases brought into contact at a boundary at a specified pressure and at a temperature above the melting point. As the constant energy simulation of these two phases progresses in time, some of the solid phase at the boundary will melt. This removes some of the kinetic energy from the system and lowers its temperature. The melting will continue until the temperature of the two-phase system converges to the true solid-liquid equilibrium value at the simulation pressure. This method has been used to determine the melting points for molecular solids,[151-154] including nitromethane,[154] for which 880 nitromethane molecules were required for the two-phase simulation. This method suffers from the drawback that long simulation times and large simulation boxes are needed to obtain converged values of the melting temperature. However, the contact between the two phases prevents superheating effects associated with the formation of the solid-liquid interfaces that are encountered in simulations of melting beginning with only the solid phase (see below).

The most straightforward method for calculating the melting point involves direct MD simulation of melting of a solid.[141, 142, 151-154] This can be achieved by a single variable-temperature constant-pressure MD simulation,[153, 154] or by a set of separate constant-temperature NPT simulations of the solid phase.[141, 142] In both experimental and theoretical melting studies, there is a free energy barrier associated with the formation of a solid-liquid interface which leads to hysteresis and superheating effects in the solid-liquid transition. Thermodynamic methods based on homogenous nucleation theory can be used to estimate the magnitude of superheating.[150] However, in simulations it is more convenient to avoid superheating effects by simulating melting of a solid phase in which molecular defects or voids have been introduced.[151-153] Introducing a void, i.e., a vacancy, is equivalent to putting Schottky defect in the solid-state structure.[159] As the simulation proceeds the structure of the solid collapses in the regions of the voids and local pockets of liquid-like structures form. The voids effectively lead to the formation of solid-liquid interfaces, which lower or eliminate the free energy barrier for the conversion of the solid to the liquid. The number of voids must be such that a sufficiently large effective solid-liquid interface is formed so as to eliminate the free energy barrier to melting, but at the same time a sufficient amount of the solid phase must be retained in the simulation box in order to maintain a true solid-liquid transition. If too many voids are introduced in the system, the solid-state structure will collapse and a discontinuous solid to liquid phase transition will not be observed. Upon introducing voids, the calculated melting point at first drops, but levels out in a plateau region as their concentration is increased. The number of molecules removed necessary to reach the plateau region is generally between 5 to 10% of the total number of molecules in the simulation cell. Upon removing more molecules for the solid lattice, the melting point will begin to drop, as the initial structure of the solid collapses.[148-151] This method allows the use of three-dimensional periodic boundary conditions and as such is more appropriate for constant pressure simulations than other methods which introduce free surfaces or grain boundaries.[152]

The direct simulation of ADN melting in a system with voids has been performed [141] using a slightly modified form of the force field described in Ref. [104]. A set of constant-temperature simulations were performed beginning with solid ADN with ~10% of the ion pairs removed from the simulation cell to create "interfacial area" to eliminate the free energy barrier to formation of the solid-liquid interface, thus preventing superheating. These simulations were done with a 3×2×4 supercell of ADN with 96 ion-pairs. The solid-liquid phase transition is characterized by a sharp drop in the density, qualitative changes in the structure of the radial distribution function, and a sudden jump in the value of the diffusion coefficient with temperature. The melting point of ADN was determined to be 366-368 K, which is in excellent agreement with the experimental value of 365-368 K. The important conclusion from this study is that the same force field used to predict solid-state properties of ADN can be used with little modification for predictions of the melting point. Simulations were done at different initial pressures to determine the solid-liquid boundary. This boundary is shown for ADN in Figure 11. After scaling the calculated melting points at different pressures with a constant 0.8 factor to account for the superheating effects, good agreement with the experimental phase boundary can be obtained.[24]

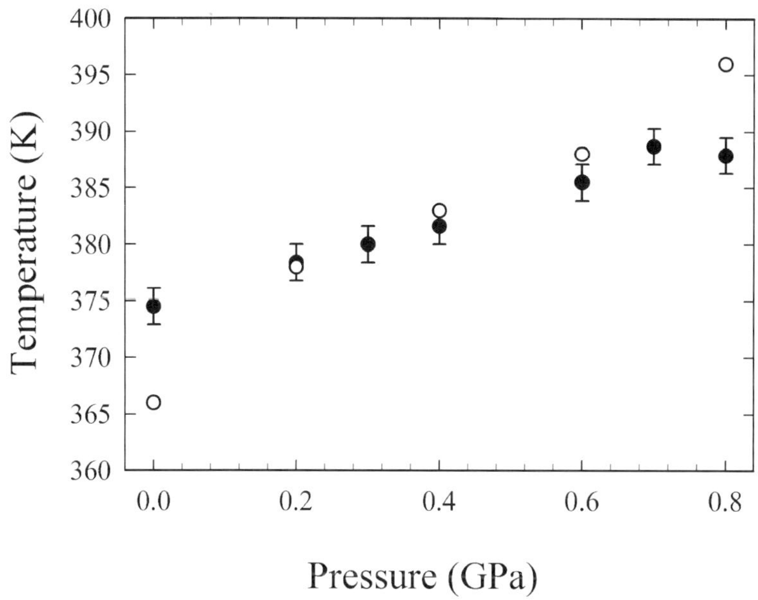

Figure 11. The experimental [24] (○) and calculated [141] (●) solid-liquid phase boundary of ADN. The calculated values of the melting point at each pressure were scaled by a factor of 0.80 to account for superheating. Adapted with permission from Ref. 141.

A theoretical study of the melting of AN was also performed.[142] This case is more complicated because of the numerous solid phases of AN at ambient pressure and also the fact that the force field for solid AN was determined for the low temperature phase V.[106] The rotationally disordered phase-I AN is in equilibrium with the liquid and a force field has not been determined specifically for this phase. Despite these difficulties, the melting simulation of AN starting from the phase-II structure was begun, but using the force field originally designed to reproduce structural and energetic features of phase V. Simulations were performed with 4×4×5 AN supercell with 160 ion-pairs. Calculations were performed with 0 to 16 voids in the supercell. As in the case of ADN, the solid-liquid phase transition was characterized by a sharp drop in the density and qualitative changes in the structure of the radial distribution function. The qualitative changes that occur in the nitrogen-nitrogen radial distribution function (for the nitrogen atom of the nitrate ions) of the supercell with no voids when going from 530 K to 550 K are shown in Figure 12(a). The sudden drop in density for the zero-void simulation (●) also occurs in between the range of 530 K and 550 K, see Figure 12(b). The melting point of the zero-void simulations illustrates the superheating effect discussed above. By introducing voids in the simulations, Figure 12(b) shows that the melting point shifts to lower values. The 16 void simulation predicts the melting point to be 445±10 K, which is in good agreement with the experimental melting point of 442 K. Again, in the case of AN we observe that a force field designed to reproduce solid-state properties can accurately predict the melting point of an ionic solid.

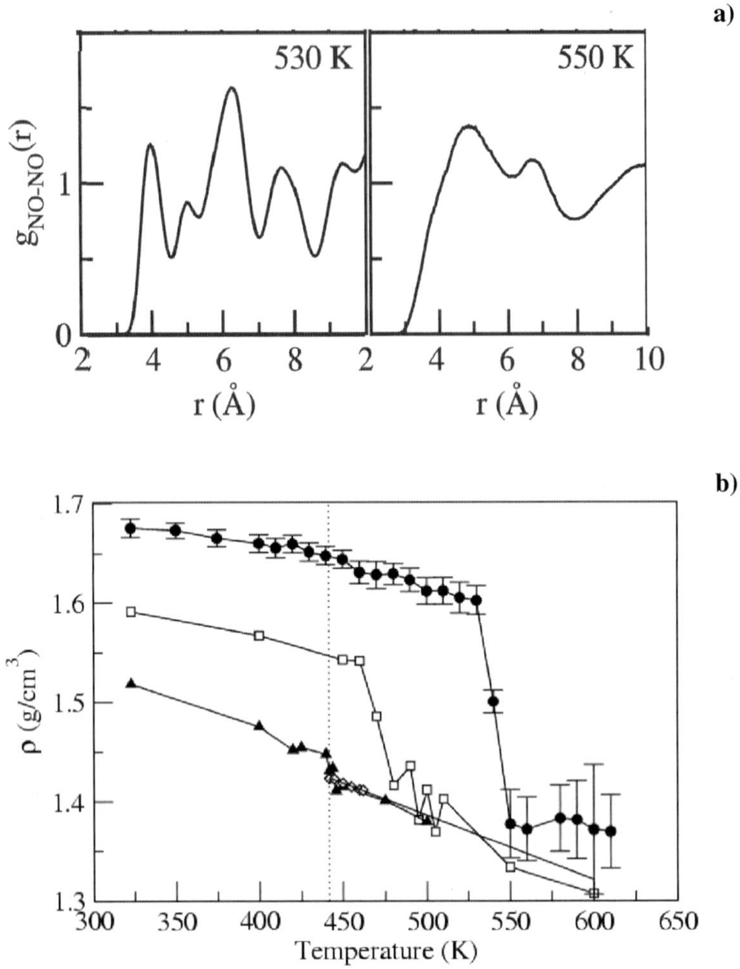

Figure 12. (a) The nitrogen-nitrogen (of the nitrate N-atoms) radial distribution functions at 530 K and 550 K. At 530 K the radial distribution function shows the long-range order characteristic of solid phases. This order is destroyed at 550 K. (b) The density of AN at different temperatures calculated for supercells with 0 (●), 8 (□), and 16 (▲) voids (defects). Introducing voids in the supercell simulation shifts the calculated melting point to lower values as it reduces the free energy barrier for forming the liquid-solid interface. Adapted with permission from Ref. [142].

In conclusion, for the molecular and ionic salts composed of small, relatively rigid molecules and ions, potential energy functions determined to reproduce solid-state properties (far from the phase transition) are capable of predicting melting points with good accuracy. Performing melting simulations with voids allows a straightforward use of periodic boundary conditions in a single NPT simulation and avoids problems associated with superheating. The simulations are straightforward to set up and require between 100 to 200 molecular or ion-pair units for a converged simulation. This is compared to the more than 800 molecules required for a two-phase NVE solid-liquid simulation. Performing a set of constant-temperature simulations over a range bracketing the melting point gives a sufficiently large set of data at each temperature from which converged averages of different thermodynamic quantities can be obtained.

4.2. Solid-State Phase Transitions

Solid-state phase transitions of salts have been studied by fitting the pressure and internal energy of each phase to an equation of state and determining the temperature for which $\Delta G = 0$ at each pressure. Simulations for each solid-state phase are performed separately in the NVE ensemble.[158] In general, the thermodynamic λ-integration method discussed in Sec. 4.1 can be used to study solid-state phase transitions as well.

Solid-state phase transitions often involve deformation of the unit cell along with rearrangements of the molecules inside the unit cell. Therefore a simple constant-pressure simulation which allows only isotropic expansion or contraction of the unit cell may not be able to reproduce all of the aspects of a solid-state phase transition. The technique of Parrinello and Rahman [160] introduces a time-dependent metric tensor in the Lagrangian of the system, which allows changes of both volume and shape of the unit cells. As such, simple solid-state phase transitions can be directly simulated with this technique. However, this method cannot be used for solid phases with very different unit cells.[145] The orientational order-disorder transitions in the solid state in some cases occur with little change in the unit cell parameters or molecular rearrangements. These orientational transitions are suitable for the Parrinello-Rahman technique.[161]

The complex phase behavior of solid-state AN has been described in Sec. 1.1.1. Understanding the nature of these phase changes is of paramount importance in the rational design of energetic materials based on AN.

The solid-state phase transitions in AN involve relatively small relocations of the positions of the centers of mass of the ions in the unit cell.[162] With the expansion of the lattice that occurs upon increasing temperature, the nature of the hydrogen-bonding network in AN phases is changed. In the high-density low-temperature phase V, the ammonium and nitrate ions form a three-dimensional hydrogen-bonded network. This network is reorganized into a set of loosely interacting hydrogen-bonded sheets in phase IV. As the density of the AN lattice is decreased upon further temperature rise phases II is formed, which has sets of weakly interacting hydrogen-bonding chains. The rotational disorder in phase I disrupts the hydrogen bonding network. Realistic modeling these complex changes will require an accurate description of the directional properties of the hydrogen bond and a careful consideration of restrictions on the volume of the unit cell in the simulation.

Phase II was used for the initial state in the AN melting studies described in Sec. 4.1. Based on NPT MD simulations and using a potential of the form given in Eq. (4), it was found that nitrate ions become rotationally unhindered prior to the melting of AN.[142] This provides some evidence that orientational aspects of the solid-state phase II $\rightarrow$ phase I transition are reproduced in the simulations. However, since only isotropic expansion of the simulation supercell was allowed in the NPT ensemble, the experimental changes that occur in the unit cell dimensions for the II→I phase transition were not observed in the simulations. Further careful studies will be required to quantify the solid-state phase transition.

5. SUMMARY AND SUGGESTIONS FOR FUTURE DEVELOPMENTS

The results presented in this chapter indicate that significant progress has been achieved in recent years in applying various computational methods for atomic-scale descriptions of the ionic energetic systems in either gas, liquid or solid phases. Using well developed computational chemistry methods it has been possible to determine accurate information about the structural properties of various ionic systems, the relative stabilities of different configurations in different phases, and the decomposition reaction mechanisms. These types of data allowed a direct correlation and interpretation of the corresponding experimental data. Moreover, in many instances the type of data obtained theoretically has substituted for the lack of information achievable through experimental means. This is particularly the case for example, for description of complicated reaction mechanisms where accurate identification of transition states for various pathways is experimentally very challenging.

Beside structural and energetic data, important steps have also been achieved computationally in description of the dynamic processes of ionic salts in condensed phases. Prediction of phase diagrams of ionic salts through description of equilibrium states at different values of temperature and pressure was one of the major objectives of the computational studies.

Among the computational methods used, first principles calculation methods are the first choice to accurately evaluate the structural, energetic, electronic, and even spectral data required to understand the behavior of ionic salts. On the other hand, dynamic problems which require large systems or long trajectories have been successfully analyzed using classical simulations methods. *Accelerated convergence of these methods is being improved through efficient use of quantum mechanical/molecular mechanics or ab initio molecular dynamics methods.* Recent results, related to prediction of the charge transport mechanism in ionic salts represent a significant forward step in the computational methodological development.

Despite this progress several areas require further development. For example, we have evidenced that only a limited number of force fields are presently available for treatment of ionic salts. It will be very beneficial that this gap will be filled and general, transferable sets of force fields for different classes of ionic systems will be available as is the case with other classes of energetic materials such as nitramines systems. We have also pointed out in this chapter that current classical force fields developed for ionic crystals are limited to description of nonreactive processes. Development of reactive force fields such as reactive empirical bond order potentials for the case of ionic systems will represent a major forward step for simulation of reactions and of combustion and denotation processes.

Alternatively, the progress made in computational hardware and in computational algorithms should allow further development of combined quantum mechanical/molecular mechanical methods for description of energetic and dynamical problems in ionic salts, for systems of increased size and complexity. The use of *ab initio* molecular dynamics methods for description of the properties of ionic systems is another area which is expecting to grow in importance and relevance. Predictions of the mechanisms of reactions not only in gas phase but also in liquid and solid phases is another area where significant progress can be made in the years to come. Finally, we have shown that a very large number of practical problems are related to interactions of ionic systems with various types of inorganic systems and in particularly with metallic systems. Such additive systems are currently used to stabilize different phases or to shift the temperature transition among different phases of ionic salts. Theoretical description of the heterogeneous interactions that appear at the interfaces of such systems with ionic salts will be essential in further development of these materials as oxidizers in rocket fuel systems or for propellant applications.

Based on the past and recent history of ionic energetic systems it can be expected that an even more important role will be played by these systems in the future. The particular steps of development might be uncertain but we can only foresee that the bright future of the ionic salts field will provide wonderful opportunities to theoretical and computational communities to answer some very difficult questions.

REFERENCES

[1] C. Oommen, S. R. Jain, J. Hazardous Mater., A67 (1999) 253.
[2] J. R. Glauber, De Natura Salium, Amsterdam, 1659, Pharmacopoeia Spagyrica, Amsterdam, 1667.
[3] R. Young, Ammonium Compounds (Ammonium Nitrate), in Kirk-Othmer Encyclopedia of Chemical Technology, 4th ed.; Kroschwitz, J. I., Exec. Ed.; (John Wiley: New York 1992) vol. 2, p. 698.
[4] C. S. Choi, H. J. Prask, Acta Cryst., B39 (1983) 414.
[5] M. Ahtee, K. J. Smolander, B. W. Lucas, A. W. Hewat, Acta Cryst., C39 (1983) 651.
[6] C. S. Choi, H. J. Prask, E. Prince, J. Appl. Cryst., 13 (1980) 403.
[7] C . S. Choi, J. E. Mapes, E. Prince, Acta Cryst., B28 (1972) 1357
[8] B. W. Lucas, M. Ahtee, A. W. Hewat, Acta Cryst., B35 (1979) 1038.
[9] B. W. Lucas, M. Ahtee, A. W. Hewat, Acta Cryst., B36 (1980) 2005.
[10] C. S. Choi, H. J. Prask, Acta Cryst., B38 (1982) 2324.
[11] M. Ahtee, K. K.Suonio, B. W. Lucas, A. W. Hewat, Acta Cryst., A35 (1979) 591.
[12] S. Yamamoto, Y. Shinnaka, J. Phys. Soc. Japan, 37 (1974) 724; Y. Shinnaka, J. Phys. Soc. Japan, 14 (1959) 1073.
[13] T. Urbanski, Chemistry and Technology of Explosives, (Pergamon, Oxford, 1983) pp. 450-475.
[14] W. Engle, Explosivstoffe, 1 (1973) 9.
[15] N. Eisenreich, W. Engel, J. Appl. Cryst., 16 (1983) 259.
[16] S. Varma, D. K. Sen, Technology, 2 (1965) 43.
[17] Y. I. Kilman, N. V. Antonova, V. V. Bogdanova, Khim. Tekhnol., 3 (1974) 27.
[18] F. V. Turchin, V. U. Sokolova, Khim. Prom., 68 (1955), 68.
[19] R. C. Saxena, S. Varma, Technology, 8 (1971) 277.
[20] H. H. Cady, Prop. Explos., 6 (1981) 49.
[21] J. C. Bottaro, R. J. Schmidt, P. E. Renwell, D. S. Ross, World Intellectual Property Organization, International Application Number PCT/US91/04268, December 26, 1991.
[22] T. B. Brill, P. J. Brush, D. G. Patil, Combust. Flame, 92 (1993) 178.
[23] M. J. Rossi, D. F. McMillen, M. Golden, Int. J. Chem. Kinet., 25 (1993) 549.
[24] T. P. Russell, G. J. Piermarini, S. Block, P. J. Miller, J. Phys. Chem., 100 (1996) 3248.

[25] I. B. Mishra, T. P. Rusell, Thermochim. Acta, 384 (2002) 47.
[26] J. C. Bottaro, P. E. Penwell, R. J. Schmitt, J. Am. Chem. Soc., 119 (1997) 9405.
[27] R. Gilardi, J. Flippen-Aderson, C. George, R. J. Butcher, J. Am. Chem. Soc., 119 (1997) 9411.
[28] V. A. Shlyapochnikov, G. I. Oleneva, N. O. Chereskaya, O. A. Luk'yanov, V. P. Gorelik, O. V. Anikin, V. A. Tartakovsky, J. Molec. Struct., 348 (1995) 103.
[29] V. A. Shlyapochnikov, M. A. Tafipolsky, I. V. Tokmakov, E. S. Baskir, O. V. Anikin, Yu. A. Strelenko, O. A. Luk'yanov, V. A. Tartakovsky, J. Molec. Struct., 559 (2001) 147.
[30] R. J. Butcher, R. D. Gilardi, J. Chem. Crystallogr., 28 (1998) 95.
[31] R. D. Gilardi, R. J. Butcher, J. Chem. Crystallogr., 28 (1998) 163.
[32] R. D. Gilardi, R. J. Butcher, J. Chem. Crystallogr., 28 (1998) 105.
[33] R. D. Gilardi, R. J. Butcher, J. Chem. Crystallogr., 28 (1998) 673.
[34] R. Gilardi, R. J. Butcher, J. Chem. Crystallogr., 30 (2000) 599.
[35] R. Gilardi, R. J. Butcher, J. Chem. Crystallogr., 32 (2002) 477.
[36] A. Martin, A. A. Pinkerton, R. D. Gilardi, J. C. Bottaro, Acta Cryst., B53 (1997) 504.
[37] R. Tanbug, K. Kirschbaum, A. A. Pinkerton, J. Chem. Crystallogr., 29 (1999) 45.
[38] H. Östmark, U. Bemm, H. Bergman, A. Langlet, Thermochim. Acta, 384 (2002) 253.
[39] See Russell et al., Ref. 24, for a detailed description of the two phases.
[40] P. W. M. Jacobs, H. M. Whitehead, Chem. Rev., 69 (1969) 551.
[41] H. Sahu, T. S. Sheshadri, V. K. Jain, J. Phys. Chem., 94 (1990) 294.
[42] G. Singh, I. P. S. Kapoor, J. Energ. Mater., 11 (1993) 293.
[43] G. Singh, I. P. S. Kapoor, S. M. Manna, J. Kaur, J. Hazard. Mat., A 79 (2000) 1.
[44] C. S. Choi, H. J. Prask, J. Chem. Phys., 61 (1974) 3523.
[45] R. M. Corn, H. L. Strauss, J. Chem. Phys., 79 (1983) 2641.
[46] D. J. J. Van Rensburg, C. J. H. Schutte, J. Molec. Struct., 229 (1972) 11.
[47] L. L. Bircumshaw, B. H. Newman, Proc. Roy. Soc. (London), A227 (1954) 115.
[48] L. L. Bircumshaw, B. H. Newman, Proc. Roy. Soc. (London), A227 (1955) 228.
[49] P. K. Gallagher, D. W. Johnson Kr., Thermochim. Acta, 6 (1973) 67.
[50] D. Dollimore, P. F. Rodgers, Thermochim. Acta, 30 (1979) 273.
[51] T. B. Brill, P. E. Gongwer, G. K. Williams, J. Phys. Chem., 98 (1994) 12242.
[52] K. R. Brower, J. C. Oxley, M. P. Tewari, J. Phys. Chem., 93 (1989) 4029.
[53] J. C. Oxley, S. M. Kaushik, N. S. Gilson, Thermochim. Acta, 153 (1989) 269.
[54] J. C. Oxley, J. L. Smith, E. Rogers, M. Yi, Thermochim. Acta, 384 (2002) 23.
[55] M. J. Rossi, J. C. Bottaro, D. F. McMillen, Int. J. Chem. Kinet., 25 (1993) 549.
[56] T. B. Brill, P. J. Brush, D. G. Patil, Combust. Flame, 92 (1993) 178.
[57] S. Vyazovkin, C. A. Wight, J. Phys. Chem. A, 101 (1997) 5653.
[58] S. Vyazovkin, C. A. Wight, J. Phys. Chem. A, 101 (1997) 7217.
[59] J. C. Oxley, J. L. Smith, W. Zheng, E. Rogers, M. D. Coburn, J. Phys. Chem. A, 101 (1997) 5646.
[60] A. S. Tompa, Thermochim. Acta, 357 (2000) 177.
[61] S. E. Barlow, V. M. Bierbaum, J. Phys. Chem., 92 (1990) 3442.
[62] T. Welton, Chem. Rev., 99 (1999) 2071.
[63] P. Walden, Bull. Acad. Imper. Sci. (St. Petersburg), (1914) 1800.
[64] J. S. Wilkes, Green Chem., 4 (2002) 73.
[65] J. B. Shotwell, R. A. Flowers II, Electroanalysis, 12 (2000) 223.
[66] J. A. Garlitz, C. A. Summers, R. A. Flowers II, G. E. O. Borgstahl, Acta Crystal, D 55 (1999) 2037.
[67] Gaussian 03, Revision A.1, M. J. Frisch, G. W. Trucks, H. B. Schlegel, G. E. Scuseria, M. A. Robb, J. R. Cheeseman, J. A. Montgomery, Jr., T. Vreven, K. N. Kudin, J. C. Burant, J. M. Millam, S. S. Iyengar, J.Tomasi, V. Barone, B. Mennucci, M. Cossi, G. Scalmani, N. Rega, G. A. Petersson, H. Nakatsuji, M. Hada, M. Ehara, K. Toyota, R. Fukuda, J. Hasegawa, M. Ishida, T. Nakajima, Y. Honda, O. Kitao, H. Nakai, M. Klene, X. Li, J. E. Knox, H. P. Hratchian, J. B. Cross, C. Adamo, J. Jaramillo, R. Gomperts, R. E. Stratmann, O. Yazyev, A. J. Austin, R.

Cammi, C. Pomelli, J. W. Ochterski, P. Y. Ayala, K. Morokuma, G. A. Voth, P. Salvador, J. J. Dannenberg, V. G. Zakrzewski, S. Dapprich, A. D. Daniels, M. C. Strain, O. Farkas, D. K. Malick, A. D. Rabuck, K. Raghavachari, J. B. Foresman, J. V. Ortiz, Q. Cui, A. G. Baboul, S. Clifford, J. Cioslowski, B. B. Stefanov, G. Liu, A. Liashenko, P. Piskorz, I. Komaromi, R. L. Martin, D. J. Fox, T. Keith, M. A. Al-Laham, C. Y. Peng, A. Nanayakkara, M. Challacombe, P. M. W. Gill, B. Johnson, W. Chen, M. W. Wong, C. Gonzalez, J. A. Pople, Gaussian, Inc., Pittsburgh PA, 2003.

[68] CADPAC: The Cambridge Analytic Derivatives Package Issue 6, Cambridge, 1995. A suite of quantum chemistry programs developed by R. D. Amos with contributions from I. L. Alberts, J. S. Andrews, S. M. Colwell, N. C. Handy, D. Jayatilaka, P. J. Knowles, R. Kobayashi, K. E. Laidig, G. Laming, A. M. Lee, P. E. Maslen, C. W. Murray, J. E. Rice, E. D. Simandiras, A. J. Stone, M.-D. Su, D. J. Tozer.

[69] D. R. Salahub, R. Fournier, P. Mlynarski, I. Papai, A. St. Amant, J. Ushio, in: Density Functional Methods in Chemistry, edited by J. K. Labanowski, J. W. Andzelm, (Springer, Berlin, 1991) Chap. 6.

[70] H. H. Michels, J. A. Montgomery, Jr. J. Phys. Chem., 97 (1993) 6602.

[71] P. Politzer, J. M. Seminario, Chem. Phys. Lett., 216 (1993) 348.

[72] P. Politzer, J. M. Seminario, M. C. Concha, P. C. Redfern, J. Molec. Struct. (Theochem), 287 (1993) 235.

[73] A. M. Mebel, M. C. Lin, K. Morokuma, C. F. Melius, J. Phys. Chem., 99 (1995) 6842.

[74] (a) C. M. S. Möller, Phys. Rev. 46 (1934) 618; (b) W. J. Hehre, R. Ditchfield, J. A. Pople, J. Chem. Phys., 56 (1972) 2257; (c) P. C. Hariharan, J. A. Pople, Theor. Chim. Acta., 28 (1973) 213.

[75] W. Hehre, L. Radom, P. v. R. Schleyer, J. A. Pople, Ab Initio Molecular Orbital Theory; (Wiley: New York, 1986).

[76] J. A. Pople, M. Head-Gordon, K. Raghgavachari, J. Chem. Phys., 87 (1989) 5768.

[77] (a) L. A. Curtiss, C. Jones, G. W. Trucks, K. Raghavachari, J. A. Pople, J. Chem. Phys., 93 (1990) 2537; (b) L. A. Curtiss, K. Raghavachari, G. W. Trucks, J. A. Pople, J. Chem. Phys., 94 (1991) 7221.

[78] W. Kohn, L. J. Sham, Phys. Rev. A, 140 (1965) 1133.

[79] R. G. Parr, W. Yang, Density-Functional Theory of Atoms and Molecules, (Oxford University Press, New York, 1989).

[80] J. M. Seminario, P. Politzer, Eds., Modern Density Functional Theory, (Elsevier, Amsterdam, 1995).

[81] J. M. Seminario, P. Politzer, Int. J. Quantum Chem., Symp., 26 (1992) 497.

[82] J. P. Perdew, Y. Yang, Phys. Rev. B, 33 (1986) 8800.

[83] J. P. Perdew, Phys. Rev. B, 33 (1986) 8822.

[84] R. J. Doyle, Org. Mass Spectrom., 28 (1993) 83.

[85] A. A. Pinkerton, J. P. Ritchie, J. Molec. Struct., 657 (2003) 57.

[86] J. P. Ritchie, E. A. Zhurova, A. Martin, A. A. Pinkerton, J. Phys. Chem. B, 107 (2003) 14576.

[87] (a) E. A. Zhurova, V. G. Tsirelson, A. I. Stash, A. A. Pinkerton, J. Am. Chem. Soc., 124 (2002) 4574. (b) E. A. Zhurova, A. Martin, A. A. Pinkerton, J. Am. Chem. Soc., 124 (2002) 8741.

[88] R. F. W. Bader, Atoms in Molecules: A Quantum Theory, International Series of Monographs on Chemistry, Vol. 22, (Oxford University Press, Oxford, 1990).

[89] F. W. Biegler-König, R. F. W. Bader, T. Tang, J. Comput. Chem., 3 (1982) 317.

[90] P. Politzer, J. M. Seminario, M. C. Concha, J. Molec. Struct. (Theochem), 427 (1998) 123.

[91] A. D. Becke, J. Chem. Phys., 98 (1993) 5648.

[92] J. J. M. Wiener, P. Politzer, J. Molec. Struct. (Theochem), 427 (1998) 171.

[93] (a) D. Chakraborty, C.-C. Hsu, M. C. Lin, J. Chem. Phys., 109 (1998) 8887. (b) J. Park, D. Chakraborty, M. C. Lin, Twenty-Seventh Symposium (International) on Combustion. (The Combustion Institute, Pittsburgh, 1998) p. 2351.

[94] A. M. Mebel, K. Morokuma, M. C. Lin, J. Phys. Chem., 103 (1995) 7414.

[95] S. Alavi, D. L. Thompson, J. Chem. Phys., 119 (2003) 232.
[96] P. Politzer, P. Lane, J. Molec. Struct., 454 (1998) 229.
[97] J. P. Perdew, Y. Wang, Phys. Rev. B, 45 (1992) 13244.
[98] M.-T. Nguyen, A. J. Jamka, R. A. Cazar, F.-M.Tao, J. Chem. Phys., 106 (1997) 8710.
[99] F.-M. Tao, J. Chem. Phys., 110 (1999) 11121.
[100] S. Alavi, D. L. Thompson, J. Chem. Phys., 117 (2002) 2599.
[101] R. S. Mulliken, J. Chem. Phys., 23 (1955) 2343.
[102] S. Alavi, D. L. Thompson, J. Chem. Phys., 118 (2003) 2599.
[103] S. Alavi, D. L. Thompson, J. Chem. Phys., 119 (2003) 4274.
[104] D. C. Sorescu, D. L. Thompson, J. Phys. Chem. B, 103 (1999) 6774.
[105] D. C. Sorescu, D. L. Thompson, J. Phys. Chem. A, 105 (2001) 7413.
[106] D. C. Sorescu, D. L. Thompson, J. Phys. Chem. A, 105 (2001) 720.
[107] L. Kleinman, D. M. Bylander, Phys. Rev. Lett., 45 (1980) 566.
[108] J. S. Lin, A. Qteish, M. C. Payne, V. Heine, Phys. Rev. B, 47 (1993) 4174.
[109] J. A. White, D. M. Bird, Phys. Rev. B, 50 (1994) 4954.
[110] M. C. Payne, D. C. Allan, T. A. Arias, J. D. Johannopoulus, Rev. Mod. Phys., 64 (1992) 1045.
[111] R. Car, M. Parrinello, Phys. Rev. Lett., 55 (1985) 2471.
[112] L. Rosso, M. E. Tuckerman, Solid State Ionics, 161 (2003) 219.
[113] A. D. Becke, Phys. Rev. A, 38 (1988) 3098.
[114] W. Yang, C. Lee, R. C. Parr, Phys. Rev. B, 37 (1988) 785.
[115] H. Goldstein, Classical Mechanics, 2nd Ed., (Addison-Wesley, Massachusetts,1989).
[116] H. C. Anderson, J. Chem. Phys., 72 (1980) 2384.
[117] S. Nosé, J. Chem. Phys., 81 (1984) 511.
[118] W. G. Hoover, Phys. Rev. A, 31 (1985) 1696.
[119] W. G. Hoover, Phys. Rev. A, 34 (1986) 2499.
[120] D. Frenkel, B. Smith, Understanding Molecular Simulation, (Academic Press, New York, 1996).
[121] D. P. Landau, K. Binder, A Guide to Monte Carlo Simulations in Statistical Physics, (Cambridge University Press, New York, 2000).
[122] D. L. Thompson, Ed., Modern Methods for Multidimensional Molecular Dynamics Computations in Chemistry, (World Scientific Pub., Singapore, 1998).
[123] T. D. Sewell, D. L. Thompson, Int. J. Mod. Phys. B,11 (1997) 1967.
[124] H. Sun, J. Phys. Chem. B, 102 (1998) 7338.
[125] W. D. Cornell, P. Cieplak, C. I. Bayly, I. R. Gould, K. M. Merz, Jr., D. M. Ferguson, D. C. Spellmeyer, T. Fox, J. W. Caldwell, P. A. Kollman, J. Am. Chem. Soc., 117 (1995) 5179.
[126] S. L. Mayo, B. D. Olafson, W. A. Goddard III, J. Phys. Chem., 94 (1990) 8897.
[127] A. J. Pertsin, A. I. Kitaigorodsky, A. I. The Atom-Atom Potential Method, Applications to Organic Molecular Solids; (Springer-Verlag: Berlin, 1987).
[128] D. E. Williams in Crystal Cohesion and Conformational Energies, Metzger, R. M., Ed.; (Springer-Verlag, Berlin, 1981) p. 3-40.
[129] D. C. Sorescu, B. M. Rice, D. L. Thompson, J. Phys. Chem. B, 101 (1997) 798.
[130] D. C. Sorescu, B. M. Rice, D. L. Thompson, J. Phys. Chem. B, 102 (1998) 948.
[131] D. C. Sorescu, B. M. Rice, D. L. Thompson, J. Phys. Chem. B, 102 (1998) 6692.
[132] D. C. Sorescu, B. M. Rice, D. L. Thompson, J. Phys. Chem. A, 102 (1998) 8386.
[133] D. C. Sorescu, B. M. Rice, D. L. Thompson, J. Phys. Chem. A, 103 (1999) 989.
[134] D. C. Sorescu, B. M. Rice, D. L. Thompson, J. Phys. Chem. B, 103 (1999) 6783.
[135] D. C. Sorescu, B. M. Rice, D. L. Thompson, J. Phys. Chem. B, 104 (2000) 8406.
[136] C. M. Breneman, K. B. Wiberg, J. Comput. Chem., 8 (1987) 894.
[137] K. D. Gibson, H. A. Scheraga, LMIN: A Program for Crystal Packing, QCPE, No. 664.
[138] D. C. Sorescu, B. M. Rice, D. L. Thompson in Energetic Materials, Part 1: Decomposition, Crystal and Molecular Properties, P. A. Politzer, J. S. Murray, Eds., (Elsevier, 2003) Chap. 6.
[139] (a) D. W. Brenner, D. H. Robertson, M. L. Elert, C. T. White, Phys. Rev. Lett., 70 (1993) 2174; (b) ibid, Phys. Rev. Lett., 76 (1996) 2202.

[140] (a) A. C. T. van Duin, S. Dasgupta, F. Lorant, W. A. Goddard III, J. Phys. Chem. A, 105 (2001) 9396; (b) A. C. T.van Duin, A. Strachan, S. Stewman, Q. Zhang, X. Xu, W. A Goddard III, J. Phys. Chem. A, 107 (2003) 3803; (c) A. Strachan, A. C. T. van Duin, D. Chakraborty, S. Dasgupta, W. A. Goddard III, Phys. Rev. Lett. 91, (2003) 098301; (d) Q. Zhang, T. Ça_in, A. van Duin, W. A. Goddard III, Y. Qi, L. G. Hector, Jr., Phys. Rev. B, 69 (2004) 045423.
[141] G. F. Velardez, S. Alavi, D. L. Thompson, J. Chem. Phys., 119 (2003) 6698.
[142] G. F. Velardez, S. Alavi, D. L. Thompson, J. Chem. Phys., 120 (2004) 9151.
[143] D. Frenkel, B. Smit, Understanding Molecular Simulation, (Academic Press, San Diego, 2000).
[144] (a) D. Frenkel, A. J. C. Ladd, J. Chem. Phys., 81 (1984) 3188; (b) E. J. Meijer, D. Frenkel, R. A. LeSar, A. J. C. Ladd, J. Chem. Phys., 92 (1990) 7570; (c) J. Anwar, D. Frenkel, M. Noro, J. Chem. Phys., 118 (2003) 728.
[145] (a) J. R. Morris, C. Z. Wang, K. M. Ho, C. T. Chan, Phys. Rev. B, 49 (1994) 3109; (b) J. R. Morris, X. Song, J. Chem. Phys., 116 (2002) 9352; (c) S. You, X. C. Zeng, J. R. Morris, J. Chem. Phys., 120 (2004) 1654 .
[146] A. B. Belonoshko, R. Ahuja, B. Johansson, Phys. Rev. Lett., 84 (2000) 3638.
[147] K. Lu, Y. Li, Phys. Rev. Lett., 80 (1998) 4474.
[148] Z. H. Jin, P. Gumbsch, K. Lu, E. Ma, Phys. Rev. Lett., 87 (2001) 0557031.
[149] L. Zhang, Z. H. Jin, L. H. Zhang, M. L. Sui, K. Lu, Phys. Rev. Lett., 85 (2000) 1484.
[150] (a) S.-N. Luo, T. J. Ahrens, Appl. Phys. Lett., 82 (2003) 1836; (b) S.-N. Luo, T. J. Ahrens, T. Çagin, A. Strachan, W. A. Goddard III, D. C. Swift, Phys. Rev. B, 68 (2003) 134206-1.
[151] (a) S. R. Phillpot, J. F. Lutsko, D. Wolf, S. Yip, Phys. Rev. B 40 (1989) 2831; (b) J. F. Lutsko, D. Wolf, S. R. Phillpot, S. Yip, Phys. Rev. B, 40 (1989) 2841.
[152] (a) J. Solca, A. J. Dyson, G. Steinebrunner, B. Kirchner, H. Huber, Chem. Phys., 224 (1997) 253; (b) J. Solca, A. J. Dyson, G. Steinebrunner, B. Kirchner, H. Huber, J. Chem. Phys., 108 (1998) 4107.
[153] P. M. Agrawal, B. M. Rice, D. L. Thompson, J. Chem. Phys., 118 (2003) 9680.
[154] P. M. Agrawal, B. M. Rice, D. L. Thompson, J. Chem. Phys., 119 (2003) 9617.
[155] J.-P. Hansen, L. Verlet, Phys. Rev., 184 (1969) 151.
[156] (a) A. J. C. Ladd, L. V. Woodcock, Chem. Phys. Lett., 51 (1977) 155; (b) A. J. C. Ladd, L. V Woodcock, Mol. Phys., 22 (1978) 649.
[157] G. Grochola, J. Chem. Phys., 120 (2004) 2122.
[158] B. Guillot, Y. Guissani, J. Chem. Phys., 116 (2002) 2047.
[159] C. Kittel, Introduction to Solid State Physics, 5th Ed., (Wiley, New York, 1976) p. 538.
[160] (a) M. Parrinello, A. Rahman, Phys. Rev. Lett. 45 (1980) 1196; (b) M. Parrinello, A. Rahman, J. Appl. Phys., 52 (1981) 7182.
[161] R. J. C. Brown, R. M. Lynden-Bell, J. Phys.: Condens. Matter, 6 (1994) 9903.
[162] N. G. Pasonage, L. A. K. Staveley, Disorder in Crystals, (Oxford University Press, Oxford, 1978).

Chemistry at Extreme Conditions
M. Riad Manaa (Editor)

Chapter 16

Computational Determination of the Energetics of Boron and Aluminum Combustion Reactions

Peter Politzer, Pat Lane and Monica C. Concha

Department of Chemistry, University of New Orleans, New Orleans, LA 70148, USA

1. BACKGROUND

It has long been recognized that the inclusion of aluminum particles as a fuel component of solid propellants enhances their performance [1-12]. One important reason for this is the large negative heat of formation of its ultimate combustion product, liquid Al_2O_3: −387.326 kcal/mole [13]. Thus the reaction,

$$2\ Al(s) + 1.5\ O_2(g) \rightarrow Al_2O_3(l) \quad (1)$$

is accompanied by a heat release of 7.2 kcal per gram of aluminum. While the actual formation of Al_2O_3 in the combustion chamber is a more complicated process, in which H_2O and CO_2 (produced from other ingredients of the propellant formulation) are believed to be the major oxidizing agents [9, 10], it is nonetheless a good source of energy. Another desirable feature is that particles of aluminum and droplets of its oxides help to dampen oscillatory combustion instabilities [6, 7].

From a thermochemical standpoint, boron is potentially an even better fuel component than aluminum; the heat of formation of liquid B_2O_3 is −299.560 kcal/mole [13], so that,

$$2\ B(s) + 1.5\ O_2(g) \rightarrow B_2O_3(l) \quad (2)$$

releases nearly 14 kcal per gram of boron. To put this in perspective, the complete combustion of liquid *n*-octane,

$$2\ C_8H_{18}\ (l) + 25\ O_2(g) \rightarrow 16\ CO_2(g) + 18\ H_2O(g) \quad (3)$$

yields 10.6 kcal per gram. There has accordingly been considerable interest in boron as a fuel additive to liquid or solid propellants [14-21], for example a slurry of boron particles in a liquid mixture of hydrocarbons. Several problems have been encountered, however. One of these (which also occurs with aluminum) is the formation of an oxide layer on the particle surfaces, which hinders their ignition. Another is that the production of boron oxyhydride intermediates, such as HOBO (g), impedes the combustion process and final condensation to B_2O_3(l).

It has been found that these issues can be at least partially addressed by the presence of fluorine in the propellant formulation [22-26]. For instance, it removes some of the oxide coating on the metal particles by conversion to gaseous metal fluorides or oxyfluorides. From the standpoint of producing energy, the inclusion of fluorine is likely to have a favorable effect, as can be seen from the comparisons given in Table 1. One means of introducing fluorine into the system is by substituting the difluoramino group, $-NF_2$, into the oxidizer and/or binder. Thus, the recently-synthesized HNFX (**1**) [28] could replace HMX (**2**) as the propellant oxidizing agent.

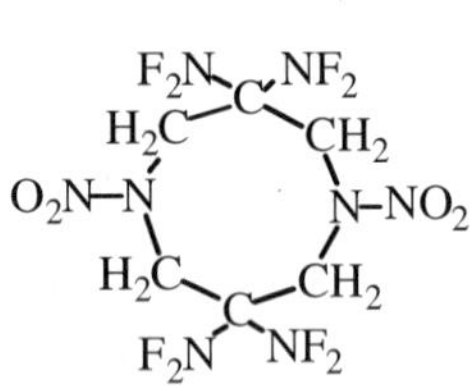

1

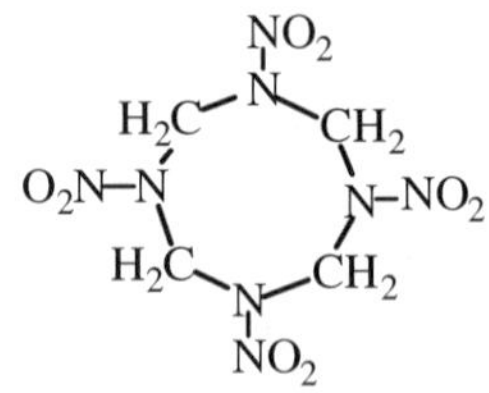

2

In order to better understand what is occurring in boron and aluminum ignition/combustion in oxygen and oxygen/fluorine environments, detailed mathematical modeling of these processes is being carried out [9-12, 19-21, 23, 26, 27, 29], in conjunction with experimental studies.

The modeling requires, as input, considerable amounts of thermodynamic and kinetic data (heats of formation, equilibrium constants, activation barriers, etc.), some of which are not known while others may be quite unreliable [19, 23, 27, 29-31]. For example, Yetter *et al* have cited two bond dissociation energies for which the uncertainty was ± 27 kcal/mole [19]. Belyung *et al* mention a reaction for which $\Delta H = 43 \pm 22$ kcal/mole [30]. The consequences for modeling efforts can be not only quantitative but even qualitative; the direction of exothermicity or equilibrium in a reaction step, or the relative stabilities of intermediates, may be predicted incorrectly.

We have sought to address these problems computationally. We have calculated the energies, enthalpies and free energies of about 120 atoms and molecules that have been implicated in the combustion of boron- and aluminum-containing propellant formulations. This large number of possible intermediates and products reflects the variety of oxidizers and binders that may be involved: difluoramines and nitramines such as **1** and **2**, ammonium perchlorate, ammonium dinitramide, polyazidooxetanes, etc. The results obtained were used to find the heats of formation of these species, and can be further applied to determining the heats of reaction, free energy changes and equilibrium constants for numerous possible steps in the combustion processes. The computed data are for both 298 K and 2000 K, the latter being more representative of the temperatures in the combustion chamber. For some of these reactions, we have also characterized the transition states and found the activation barriers.

This chapter will present a compilation and discussion of these thermodynamic and kinetic quantities. Thermodynamic properties of aluminum-containing molecules have also recently been the focus of two other extensive computational studies [31, 37], and there is some overlap with the ones included in this chapter.

Table 1. Some experimental heats of reaction at 298 K, in kcal/mole.[a]

Process	ΔH°	ΔH° per gram B or Al
2 B(s) + 1.5 O_2(g) → B_2O_3(l)	−299.56	−13.9
B(s) + 1.5 F_2(g) → BF_3(g)	−271.42	−25.1
B(s) + 0.5 O_2(g) + 0.5 F_2(g) → FBO(g)	−144.00	−13.3
2 Al(s) + 1.5 O_2(g) → Al_2O_3(l)	−387.33	−7.2
Al(s) + 1.5 F_2(g) → AlF_3(g)	−289.00	−10.7
Al(s) + 1.5 F_2(g) → AlF_3(g)	−105.10	−3.9
H_2(g) + 0.5 O_2(g) → H_2O(g)	−57.80	−28.7 (per g H_2)
0.5 H_2(g) + 0.5 F_2(g) → HF(g)	−65.14	−64.6 (per g H_2)

[a]Experimental heats of formation were taken from Tables 5 – 7 of this chapter, except for FAlO(g), for which a computed value from ref. 36 was used.

2. PROCEDURE

The data to be reported were computed using the CBS-QB3 method [38], which is one of the "composite" *ab initio* techniques that have been developed during the last two decades with the objective of achieving high levels of accuracy for energetic properties. The CBS-QB3 approach involves density functional B3LYP/6-311G(2d,d,p) geometry optimization and vibration frequency calculation, followed by several single-point higher-level energy corrections and including complete-basis-set (CBS) extrapolation. In tests comprising 125 dissociation energies, ionization potentials, electron affinities and proton affinities, the CBS-QB3 mean absolute error was 0.87 kcal/mole, with a root-mean-square error of 1.08 kcal/mole.

Using the Gaussian 98 code [39], the energy minimum at 0 K was determined for each atom and molecule, and then the enthalpy and free energy at 298 K and 2000 K. From the enthalpies at 298 K were found the gas phase heats of formation, as ΔH(298 K) for the reactions producing the molecules from their constituent elements. For the three relevant elements that are solids at 298 K (boron, carbon and aluminum), this requires knowing their experimental heats of sublimation at 298 K, which are 133.84, 171.29 and 78.8 kcal/mole, respectively [13].

The energy minima were confirmed by the absence of imaginary frequencies, the transition states by the presence of a single one [40]. That the latter lead to the desired products was verified by following the intrinsic reaction coordinate [41].

3. RESULTS

3.1. General

Our computed standard state enthalpies and free energies at 298 K and 2000 K are listed in Tables 2 - 4. The energy minima at 0 K upon which these are based can be found in our earlier papers, as can also the optimized geometries of the boron- [32, 35] and aluminum-containing molecules.[36] Some of the less familiar boron and aluminum molecules are shown in Figure 1.

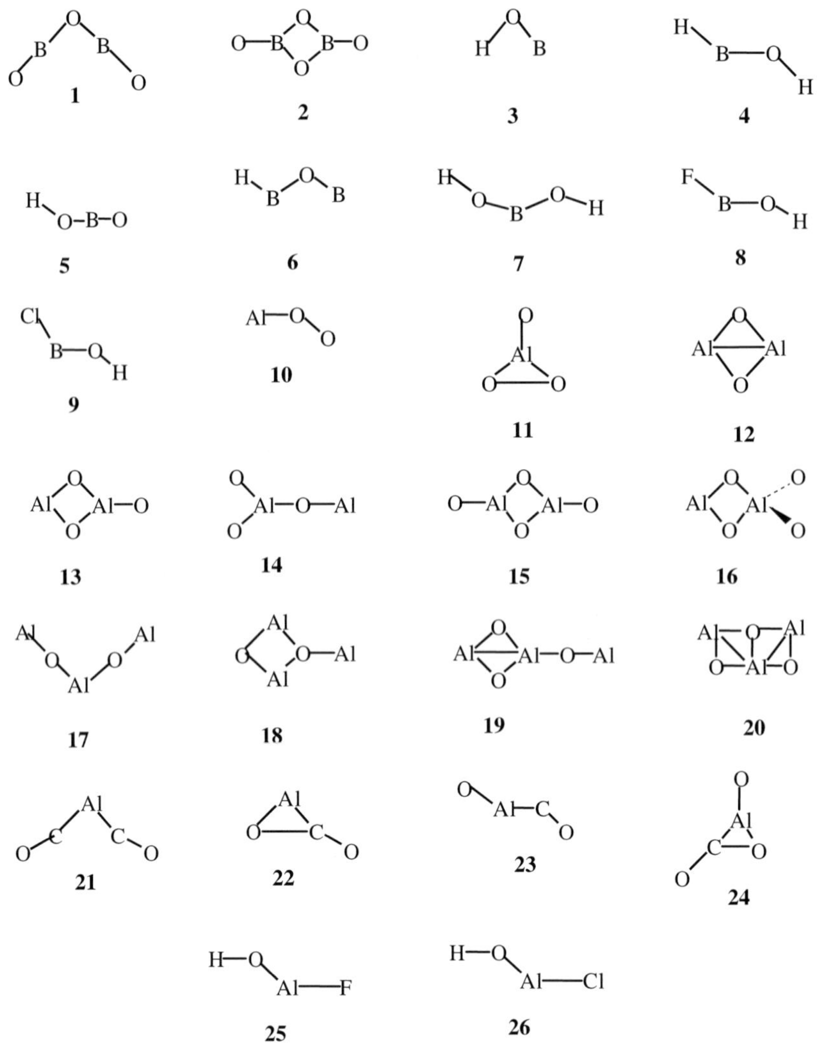

Figure 1. Some of the boron and aluminum molecules included in this work.

Table 2. Computed enthalpies and free energies at 298 K and 2000 K, in hartrees (1 hartree = 627.509 kcal/mole): Atoms and molecules that do not contain boron or aluminum.

Atom or Molecule	H°(298 K)	G°(298 K)	H°(2000 K)	G°(2000 K)
H	–0.49746	–0.51047	–0.48398	–0.60142
C	–37.78303	–37.79993	–37.76955	–37.91310
N	–54.51819	–54.53558	–54.50471	–54.65154
O	–74.98528	–75.00259	–74.97180	–75.11808
F	–99.64075	–99.65792	–99.62727	–99.77261
Cl	–459.68128	–459.69932	–459.66781	–459.81894
H_2	–1.16277	–1.17757	–1.14302	–1.28563
N_2	–109.39516	–109.41690	–109.37395	–109.56552
O_2	–150.16140	–150.18467	–150.13920	–150.34318
O_3	–225.18604	–225.21307	–225.15187	–225.40552
F_2	–199.34297	–199.36595	–199.31977	–199.52449
Cl_2	–919.45748	–919.48283	–919.43361	–919.65648
CH	–38.40930	–38.42941	–38.38845	–38.56845
NH	–55.14114	–55.16170	–55.12066	–55.30305
OH	–75.64635	–75.66658	–75.62620	–75.80586
H_2O	–76.33382	–76.35590	–76.30654	–76.51179
HO_2	–150.73627	–150.76225	–150.70567	–150.94401
H_2O_2	–151.37441	–151.40031	–151.33334	–151.59186
HF	–100.35661	–100.37632	–100.33670	–100.51248
HCl	–460.34477	–460.36595	–460.32403	–460.51106
CN	–92.58428	–92.60727	–92.56275	–92.76332
CO	–113.17895	–113.20138	–113.15750	–113.35417
CO_2	–188.36882	–188.39309	–188.33424	–188.56941
CF_4	–437.10070	–437.13177	–437.03811	–437.37757
NO	–129.74537	–129.76867	–129.72364	–129.92672
NO_2	–204.84844	–204.87633	–204.81529	–205.07208
N_2O	–184.44737	–184.47231	–184.41248	–184.65304
NF	–154.28158	–154.30575	–154.25866	–154.47054
NF_3	–353.76607	–353.79573	–353.71637	–354.02099
NCl	–514.30166	–514.32709	–514.27816	–514.50029
ClO	–534.76773	–534.79288	–534.74423	–534.96449
ClO_2	–609.85110	–609.88036	–609.81568	–610.08825
HCO	–113.70116	–113.72663	–113.67037	–113.90520
anti-OCOH	–188.86851	–188.89707	–188.82617	–189.13053
syn-OCOH	–188.86531	–188.89388	–188.82259	–189.12836
HNO	–130.31959	–130.34464	–130.28936	–130.51993
HONO	–205.47114	–205.49933	–205.42699	–205.70784
F_2CO	–312.71277	–312.74285	–312.66555	–312.96594

Table 3. Computed enthalpies and free energies at 298 K and 2000 K, in hartrees (1 hartree = 627.509 kcal/mole): Boron systems.

Atom or molecule	H°(298 K)	G°(298 K)	H°(2000 K)	G°(2000 K)
B	–24.59941	–24.61581	–24.58593	–24.72609
B_2	–49.29912	–49.32210	–49.27595	–49.48062
BH	–25.22892	–25.24843	–25.20759	–25.38446
BH_2	–25.85226	–25.87495	–25.82213	–26.03725
BN	–79.27894	–79.30244	–79.25664	–79.46241
BO	–99.88991	–99.91301	–99.86809	–100.07006
OBO[a]	–175.08305	–175.10855	–175.04626	–175.29556
BOB[a]	–124.64608	–124.67313	–124.60864	–124.87069
BBO[a]	–124.61987	–124.64654	–124.58264	–124.84204
BOB[b,c]	–124.57991	–124.60805	–124.54568	–124.80697
OBBO[a]	–199.96567	–199.99530	–199.91828	–200.21754
B_2O_3, **1**	–275.19064	–275.22303	–275.12946	–275.47544
B_2O_4, **2**	–350.24613	–350.28042	–350.16937	–350.55993
BF	–124.52862	–124.55139	–124.50604	–124.70760
BF_2	–224.34596	–224.37398	–224.31159	–224.57271
BF_3	–324.26170	–324.29234	–324.21346	–324.52052
BCl	–484.47692	–484.50115	–484.45345	–484.66741
BCl_2	–944.28557	–944.31629	–944.24957	–944.53391
BCl_3	–1404.16016	–1404.19308	–1404.10901	–1404.44060
HBO[a]	–100.56710	–100.59010	–100.53405	–100.75695
BOH, **3**	–100.49466	–100.51959	–100.46330	–100.69726
HBOH, **4**	–101.07814	–101.10442	–101.03608	–101.29852
HOBO, **5**	–175.76621	–175.79380	–175.72196	–175.99960
HBOB[b]	–125.24467	–125.27289	–125.19913	–125.48267
HBOB, **6**	–125.23051	–125.26036	–125.18392	–125.48224
$B(OH)_2$, **7**	–176.29493	–176.32444	–176.23947	–176.55190
$B(OH)_3$	–252.18089	–252.21264	–252.10116	–252.47743
FBO[a]	–199.79007	–199.81560	–199.75414	–200.00160
ClBO[a]	–559.76611	–559.79298	–559.72933	–559.98843
HB(O)F	–200.30822	–200.33743	–200.26240	–200.55341
HOBF, **8**	–200.32273	–200.35174	–200.27789	–200.56618
HB(O)Cl	–560.28292	–560.31374	–560.23608	–560.54120
HOBCl, **9**	–560.29264	–560.32303	–560.24694	–560.54720

[a]Linear. [b]B–O–B three-membered ring. [c]This is a triplet, in contrast to its two isomers which are singlets.

Table 4. Computed enthalpies and free energies at 298 K and 2000 K, in hartrees (1 hartree = 627.509 kcal/mole): Aluminum systems.

Atom or molecule	H°(298 K)	G°(298 K)	H°(2000 K)	G°(2000 K)
Al	–241.92645	–241.94412	–241.91298	–242.06165
AlH	–242.53999	–242.56131	–242.51779	–242.70865
AlH_2	–243.11372	–243.13798	–243.08138	–243.31182
AlN	–296.53813	–296.56342	–296.51457	–296.73592
AlO	–317.10263	–317.12743	–317.07938	–317.29647
OAlO[a]	–392.23988	–392.26793	–392.20110	–392.47379
AlO_2[b]	–392.19888	–392.22856	–392.16300	–392.44007
AlO_2, **10**	–392.17343	–392.20407	–392.13757	–392.42110
AlO_3, **11**	–467.34891	–467.38167	–467.29838	–467.62701
AlAlO[a]	–559.11419	–559.14553	–559.07505	–559.37127
AlOAl[a]	–559.24662	–559.27588	–559.20765	–559.48921
Al_2O[b]	–559.13963	–559.17069	–559.10324	–559.39115
AlOAlO[a]	–634.40336	–634.43682	–634.34961	–634.69076
OAlAlO[a]	–634.29189	–634.32543	–634.23798	–634.58029
Al_2O_2, **12**	–634.41599	–634.44675	–634.36495	–634.68140
OAlOAlO[a]	–709.55060	–709.58667	–709.48214	–709.87206
Al_2O_3, **13**	–709.55423	–709.58999	–709.48824	–709.87014
Al_2O_3, **14**	–709.50824	–709.54559	–709.44205	–709.83559
Al_2O_4, **15**	–784.70706	–784.74550	–784.62597	–785.05821
Al_2O_4, **16**	–784.67115	–784.70956	–784.59038	–785.02141
Al_3O_2, **17**	–876.48383	–876.52512	–876.41685	–876.83966
Al_3O_2, **18**	–876.46304	–876.50178	–876.39556	–876.80275
Al_3O_3, **19**	–951.71831	–951.75922	–951.63697	–952.08673
Al_3O_3, **20**	–951.69772	–951.73661	–951.61573	–952.05405
AlF	–341.82773	–341.85218	–341.80422	–342.01980
AlF_2	–441.63787	–441.66805	–441.60182	–441.88268
AlF_3	–541.52920	–541.56079	–541.47824	–541.80030
AlCl	–701.80532	–701.83125	–701.78135	–702.00846
$AlCl_2$	–1161.60101	–1161.63402	–1161.56399	–1161.86717
$AlCl_3$	–1621.47158	–1621.50727	–1621.41899	–1621.77412
AlOH[a]	–317.78532	–317.81227	–317.75008	–318.00748
HAlO[a]	–317.72260	–317.74778	–317.68641	–317.93235
AlCN[a]	–334.69594	–334.72474	–334.65869	–334.93292

AlNC[a]	–334.70587	–334.73497	–334.66854	–334.94482
AlCO[a]	–355.12128	–355.15076	–355.08361	–355.36341
AlOC[a]	–355.10666	–355.14116	–355.06902	–355.38286
AlCO[b]	–355.10686	–355.13788	–355.07145	–355.35663
$Al(CO)_2$, **21**	–468.32245	–468.35906	–468.25865	–468.64158
$AlCO_2$, **22**	–430.32264	–430.35563	–430.27353	–430.59982
OAlCO, **23**	–430.28596	–430.32409	–430.23591	–430.60042
$OAlCO_2$, **24**	–505.46883	–505.50511	–505.40492	–505.78469
FAlO[a]	–416.97176	–416.99978	–416.93348	–417.20441
ClAlO[a]	–776.95608	–776.98545	–776.91734	–777.19892
FAlCl	–801.61901	–801.65127	–801.58247	–801.87898
$HAlCl_2$	–1162.23331	–1162.26615	–1162.18362	–1162.51090
HAl(O)F	–417.53777	–417.56933	–417.48892	–417.80469
HOAlF, **25**	–417.59827	–417.62996	–417.55089	–417.86527
HAl(O)Cl	–777.51893	–777.55193	–777.46954	–777.79680
HOAlCl, **26**	–777.57988	–777.61304	–777.53199	–777.85801

[a]Linear, [b]Cyclic.

3.2. Heats of Formation, $\Delta H_f°(298\ K)$

Tables 5 – 7 contain the $\Delta H_f°(298\ K)$ calculated from the H°(298 K) in Tables 2 - 4. In Table 5 are atoms and molecules that contain neither boron nor aluminum. For most of these, quite reliable experimental $\Delta H_f°(298\ K)$ are available; however we include the computed values in order to (a) ensure consistency with the $\Delta H_f°(298\ K)$ of the boron and aluminum systems, and (b) to demonstrate the accuracy of the CBS-QB3 results. The average absolute deviation from experiment in Table 5 is 1.2 kcal/mole.

For the boron-containing molecules (Table 6), experimental $\Delta H_f°(298\ K)$ are often not known or else have large ranges of uncertainty. The latter is the case, for example, for B_2 (198.3 ± 8.0 kcal/mole), BOB (23 ± 25 kcal/mole) and $B(OH)_2$ (−114 ± 15 kcal/mole) [19]. Lias *et al* give 48 ± 15 kcal/mole for BH_2 and −20 ± 15 kcal/mole for BCl_2 [44]. For BN, reported [45] and calculated [46,47] dissociation energies indicate that $\Delta H_f°(298\ K)$ should be between 140 and 154 kcal/mole, in contrast to the NIST 114 kcal/mole [13]. Bauschlicher and Ricca have noted a discrepancy of more than 20 kcal/mole for BF_2 [48], and point out the unreliable nature of some of the measured BX_n data (X = F or Cl and n = 1 or 2). Two high-level computational techniques, multireference configuration interaction (MRCI) and multiconfiguration self-consistent-field (MCSCF), predict −59 and −61 kcal/mole for HBO [49], whereas the NIST listing has −47.4 kcal/mole [13]. When we are aware of such uncertainty in the NIST value, it is not included in Table 6.

The heat of sublimation of boron is also in dispute, as has been discussed by Bauschlicher and Ricca [48]; 133.84, 135 ± 1.2, 137.1 and 137.4 ± 0.2 kcal/mole have all been suggested.

Our results in Table 6 were obtained with the NIST 133.84 kcal/mole [13], which is the smallest of those mentioned. If one of the others were used, this would simply make each of our $\Delta H_f°(298\ K)$ more positive by that increment per boron atom. However, any error in the boron heat of sublimation will not affect heats of reaction calculated with our $\Delta H_f°(298\ K)$ in Table 6; it will cancel, provided that solid boron is not part of the reaction.

The average absolute deviation from the experimental $\Delta H_f°(298\ K)$ in Table 6, not including ClBO, is 2.2 kcal/mole. In view of the overall accuracy demonstrated by the CBS-QB3 procedure, plus the problems associated with some boron halide experimental data [48], we believe that the NIST $\Delta H_f°(298\ K)$ for ClBO, given in Table 6, should be viewed with skepticism.

The heats of formation of the aluminum systems (Table 7) show the same problem as do the boron; many of them are not known at all, or not known with satisfactory accuracy. This can be seen in the $\Delta H_f°(298\ K)$ data cited by Swihart and Catoire and by Allendorf *et al*, for instance: OAlO, −20.6 ± 8.0 [31]; HAlO, 8 ± 20 [31]; FAlOH, −137.2 ± 12 [37]; $HAlCl_2$, −84.0 ± 7.2 [37], etc. Another striking example is OAlO, for which have been reported −31 [44], −44.9 ± 3 [51] and −20.6 ± 7.6 kcal/mole [52].

Table 5. Calculated and experimental standard state gas phase heats of formation at 298 K, in kcal/mole: Atoms and molecules that do not contain boron or aluminum.

Atom or molecule	Calc.	Exp.[a]	Atom or molecule	Calc.	Exp.[a]
H	52.7	52.103	CO	−26.5	−26.417
C	-----	171.29	CO_2	−95.0	−94.054
N	112.6	112.97	CF_4	−225.1	−221. ± 6.
O	59.9	59.553	NO	20.7	21.58
F	19.3	18.97	NO_2	6.6	7.911
Cl	29.8	28.991	N_2O	17.9	19.61
O_3	35.2	34.099	NF	54.9	b
CH	143.1	142.	NF_3	−33.9	−31.570
NH	86.5	90.	NCl	78.2	---
OH	9.9	9.319	ClO	26.2	24.192
H_2O	−56.7	−57.799	ClO_2	24.5	25.
HO_2	4.1	0.5	HCO	10.6	10.40
H_2O_2	−31.5	−32.531	HNO	25.1	23.80
HF	−65.1	−65.141	HONO	−19.3	−18.34
HCl	−21.7	−22.06	F_2CO	−146.3	≥−149.1[c]
CN	106.2	104.			

[a]Experimental data are from reference 13 except where otherwise indicated. [b]There is substantial uncertainty associated with the literature values [42]. [c]Reference 43.

In Table 7, we list only those experimental values for which the quoted range of uncertainty is no greater than ± 4.0 kcal/mole. Our average absolute deviation from these is 3.0 kcal/mole. The largest discrepancy is for $AlCl_3$, 6.7 kcal/mole. Earlier generations of the CBS computational technique had a tendency to produce larger errors for molecules with three or more halogen atoms [53], and this may persist in CBS-QB3, at least for trichlorides; note that our $\Delta H_f°$(298 K) for BCl_3 is also significantly too negative (Table 6). However BF_3 and AlF_3 are treated quite accurately.

3.3. Heats of Reaction

The enthalpies in Tables 2 – 4 can be used to find ΔH°(298 K) and ΔH°(2000 K) for numerous reactions between these atoms and molecules. If solid or liquid boron or aluminum is involved, its heat of sublimation or vaporization at 298 K or 2000 K must be appropriately included. Alternatively, ΔH°(298 K) can be determined from the heats of formation in Tables 5 – 7. For B(s) and Al(s), $\Delta H_f°$(298 K) is zero by definition; for the liquids, it is 11.69 and 2.524 kcal/mole, respectively [13].

Table 6. Calculated and experimental standard state gas phase heats of formation at 298 K, in kcal/mole: Boron systems.

Atom or molecule	Calc.	Exp.[a]	Atom or molecule	Calc.	Exp.[a]
B	---	133.84	BCl_3	−101.2	−96.310
B_2	204.7	b	HBO[c]	−57.9	b
BH	103.6	105.80	BOH, **3**	−12.5	---
BH_2	77.3	b	HBOH, **4**	−13.8	---
BN	145.2	b	HOBO, **5**	−132.2	−134.
BO	2.2	0.	HBOB[d]	26.9	---
OBO[c]	−68.4	−68.	HBOB, **6**	35.8	---
BOB[c]	37.7	b	$B(OH)_2$, **7**	−99.2	b
BBO[c]	54.1	---	$B(OH)_3$	−239.7	−237.16
BOB[d]	79.2	---	FBO[c]	−141.3	−144.
OBBO[c]	−112.2	−109.	ClBO[c]	−90.3	−75.6
B_2O_3, **1**	−202.8	−199.80	HB(O)F	−101.6	---
B_2O_4, **2**	−187.0	---	HOBF, **8**	−110.7	---
BF	−27.9	−27.701	HB(O)Cl	−49.8	---
BF_2	−119.4	b	HOBCl, **9**	−55.9	---
BF_3	−272.7	−271.420			
BCl	40.5	b			
BCl_2	−9.7	b			

[a]Reference 13. [b]Value given in reference 13 has considerable uncertainty or has been questioned; see text. [c]Linear. [d]B-O-B three-membered ring.

We have used the first procedure to calculate ΔH°(298 K) and ΔH°(2000 K) for 27 gas phase reactions that are examples of possible steps in boron and aluminum combustion. Our results are compared, in Table 8, to values obtained with the experimentally-based temperature-dependent enthalpy relationships given in the NIST tables [13]. The average absolute deviations overall are 1.6 kcal/mole at 298 K and 1.5 kcal/mole at 2000 K; they are 2.4 and 2.2 kcal/mole when only the reactions in which boron or aluminum appears are considered.

Table 7. Calculated and experimental standard state gas phase heats of formation at 298 K, in kcal/mole: Aluminun systems.

Atom or molecule	Calc.	Exp.[a]	Atom or molecule	Calc.	Exp.[a]
Al	---	78.8[a]	AlF_2	−152.4	c
AlH	58.6	59.6 ± 0.8[b]	AlF_3	−290.4	−289.0 ±
AlH_2	63.4	c	AlCl	−15.4	−12.2 ±
AlN	132.7	d	$AlCl_2$	−57.4	c, d
AlO	18.9	16.1 ± 1.5[b]	$AlCl_3$	−146.4	−139.7 ±
OAlO[e]	−16.6	d	AlOH[c]	−44.7	−43.0 ±
AlO_2[f]	9.1	---	HAlO[e]	−5.3	c, d
AlO_2, **10**	25.1	---	AlCN[e]	68.8	71.92[h]
AlO_3, **11**	−34.4	---	AlNC[e]	62.6	66.49[h]
AlAlO[e]	44.3	---	AlCO[e]	42.3	---
AlOAl[e]	−38.8	−34.7 ± 4.0[g]	AlOC[e]	51.5	---
Al_2O[f]	28.3	---	AlCO[f]	51.4	---
AlOAlO[e]	−86.5	---	$Al(CO)_2$, **21**	1.9	
OAlAlO[e]	−16.6	---	$AlCO_2$, **22**	−33.4	---
Al_2O_2, **12**	−94.5	d	OAlCO, **23**	−10.4	---
OAlOAlO[e]	−128.3	---	$OAlCO_2$, **24**	−74.5	---
Al_2O_3, **13**	−130.6	---	FAlO[e]	−105.1	---
Al_2O_3, **14**	−101.7	---	ClAlO[e]	−59.4	d
Al_2O_4, **15**	−175.8	---	FAlCl	−104.6	c
Al_2O_4, **16**	−153.3	---	$HAlCl_2$	−89.4	c
Al_3O_2, **17**	−104.4	---	HAl(O)F	−95.5	---
Al_3O_2, **18**	−91.3	---	HOAlF, **25**	−133.5	c
Al_3O_3, **19**	−200.9	---	HAl(O)Cl	−47.7	---
Al_3O_3, **20**	−188.0	---	HOAlCl, **26**	−86.0	c
AlF	−65.4	−63.1 ± 0.7[b]			

[a]Reference 13. [b]Cited in reference 37. [c]Uncertainty greater than ± 4.0 (reference 37). [d]Uncertainty greater than ± 4.0 (reference 31). [e]Linear. [f]Cyclic. [g]Cited in reference 31. [h]Reference 50.

What is particularly interesting is how little the heat of reaction changes between 298 K and 2000 K [32,34-36]. Whereas the absolute enthalpies in Tables 2 – 4 are typically 15 – 40 kcal/mole more positive at 2000 K than at 298 K, the average absolute difference between ΔH°(298 K) and ΔH°(2000 K) is just 1.7 kcal/mole for the computed and 1.2 kcal/mole for the experimental. This can be understood in terms of the Kirchhoff equation [54], which is exact:

$$\Delta H(T_2) = \Delta H(T_1) + \int_{T_1}^{T_2} \Delta C_P(T)dT \tag{4}$$

$\Delta C_P(T)$ is the change in the sum of the heat capacities at constant pressure in going from the reactants to the products. It tends to be small in magnitude [32], leading to the similarity between ΔH°(298 K) and ΔH°(2000 K). We have used eq. (4) to evaluate the latter in a few instances [32], taking $\Delta C_P°(T)$ to be the average of $\Delta C_P°$(298 K) and $\Delta C_P°$(2000 K). For this temperature range, we observed no general improvement in accuracy over simply assuming that ΔH°(2000 K) ≈ ΔH°(298 K).

3.4. Free Energy Changes

In Table 9 are listed ΔG°(298 K) and ΔG°(2000 K) for the same reactions as in Table 8; they were calculated with the free energies in Tables 2 – 4. Also included are experimental values obtained using,

$$\Delta G° = \Delta H° - T\Delta S° \tag{5}$$

taking the ΔH° from Table 8 and the ΔS° from temperature-dependent entropy relationships provided in the NIST tables [13]. The accuracy of our computed ΔG°(298 K) is like that of ΔH°(298 K); the average absolute deviation is 1.7 kcal/mole overall, and 2.4 kcal/mole for reactions in which boron or aluminum appears.

In proceeding to examine the effect of temperature upon ΔG, it is useful to refer to the Gibbs-Helmholtz equation [54], which is also exact:

$$\frac{\Delta G(T_2)}{T_2} = \frac{\Delta G(T_1)}{T_1} - \int_{T_1}^{T_2} \left(\frac{\Delta H(T)}{T^2}\right)_P dT \tag{6}$$

It has already been demonstrated that ΔH changes very little between 298 K and 2000 K. If it is taken to be constant over that range, then eq. (6) becomes,

$$\Delta G(2000\ K) = 6.7080\ \Delta G(298\ K) - 5.7080\ \Delta H \tag{7}$$

ΔH could be set equal, for example, to the average of ΔH(298 K) and ΔH(2000 K). Eq. (7) indicates that ΔG(2000 K) ≈ ΔG(298 K) only if ΔG(298 K) ≈ ΔH, and Tables 8 and 9 show that this is often not the case; if ΔG(298 K) is just 1.0 kcal/mole greater in magnitude than ΔH, then ΔG(2000 K) and ΔG(298 K) will differ by 5.7 kcal/mole. It should be anticipated, therefore, that ΔG° (unlike ΔH°) will frequently change quite significantly in going from 298 K to 2000 K. This is confirmed by Table 9.

Table 8. Calculated and experimental heats of reaction at 298 K and 2000 K, in kcal/mole.[a,b]

Reaction	ΔH° (298 K)		ΔH° (2000 K)	
	Calc.	Exp.[c]	Calc.	Exp.[c]
$CO + H_2 \rightarrow HCO + H$	89.8	88.9	91.7	90.7
$CO + O_2 \rightarrow CO_2 + O$	–8.6	–8.1	–5.9	–5.4
$F + H_2 \rightarrow HF + H$	–31.7	–32.0	–31.6	–32.3
$Cl + H_2 \rightarrow HCl + H$	1.1	1.0	1.8	1.0
$HCO + O_2 \rightarrow CO_2 + OH$	–95.8	–95.1	–94.7	–94.0
$H + CO_2 \rightarrow CO + OH$	25.7	24.9	21.7	21.0
$HCO + Cl \rightarrow HCl + CO$	–88.7	–87.9	–90.0	–89.6
$O + H_2O \rightarrow 2OH$	16.6	16.9	16.3	16.7
$H + O_2 \rightarrow OH + O$	17.1	16.8	15.8	15.6
$HCO + F \rightarrow HF + CO$	–121.5	–120.9	–123.3	–122.9
$F + H_2O \rightarrow HF + OH$	–17.8	–17.0	–18.3	–17.6
$BF + H_2 \rightarrow BH + HF$	66.4	68.4	65.7	67.7
$BH + F_2 \rightarrow BF + HF$	–196.6	–198.6	–197.9	–200.4
$BO + F \rightarrow FBO$	–162.8	–163.	–162.4	–163.
$BO_2 + F \rightarrow FBO + O$	–32.3	–35.	–32.9	–36.
$BF + OH \rightarrow FBO + H$	–70.6	–74.	–66.4	–70.
$BO + OH \rightarrow BO_2 + H$	–27.8	–25.	–22.6	–20.
$BO + CO_2 \rightarrow BO_2 + CO$	–2.1	0.	–0.9	1.
$BH + OH \rightarrow BO + H_2$	–111.3	–115.	–111.3	–115.
$BF + F_2 \rightarrow BF_3$	–244.8	–243.7	–243.3	–242.9
$AlO + HF \rightarrow AlF + OH$	–9.3	–5.	–9.0	–6.
$AlH + HF \rightarrow AlF + H_2$	–58.9	–60.	–58.2	–59.
$AlO + H_2 \rightarrow AlOH + H$	–10.9	–7.	–7.3	–7.
$AlH + F_2 \rightarrow AlF + HF$	–189.1	–191.	–190.4	–192.
$AlOH + F \rightarrow AlO + HF$	–20.8	–25.	–24.3	–25.
$AlH + O_2 \rightarrow AlOH + O$	–43.4	–45.	–40.7	–45.
$AlF + H_2O \rightarrow AlOH + HF$	12.3	13.	15.0	13.

[a]Reference 35. [b]Reference 36. [c]Reference 13.

It is also evident that the agreement between the computed and the experimental $\Delta G°$ is not as good at 2000 K as at 298 K; the average absolute deviation is 4.5 kcal/mole (although it would decrease to 3.1 kcal/mole if the three worst cases, all involving aluminum, were omitted). This problem has been noted earlier [32,34-36]; it arises, presumably, because any error in the calculated and/or experimental $\Delta S°$(2000 K) is greatly magnified when multiplied by T = 2000 in eq. (5). We have explored other options for determining $\Delta G°$(2000 K), such as eq. (7) with $\Delta H°$ equal to the average of $\Delta H°$(298 K) and $\Delta H°$(2000 K) [32,35] and eq. (5) with T = 2000 K in conjunction with $\Delta S°$(298 K) and either $\Delta H°$(2000 K) or $\Delta H°$(298 K). The last of these gave the best results, with an average absolute deviation of 4.2 kcal/mole, somewhat better than was obtained with the G°(2000 K) in Tables 2 – 4. Thus, while the present approach does not give $\Delta G°$ at elevated temperatures with satisfactory reliability, it appears that at least a fair estimate may be achieved with,

$$\Delta G°(T) \approx \Delta H°(298\ K) - T\Delta S°(298\ K) \quad (8)$$

without actually doing any computations at the temperature T. However eq. (8) is likely to become a poorer approximation as T increases beyond 2000 K.

3.5. Equilibrium Constants

The equilibrium constant of a reaction is related to its standard free energy change by [54],

$$K_{eq} = exp(-\Delta G°/RT) \quad (9)$$

in which R is the gas constant. Thus the data in Tables 2 – 4 make it possible to find K_{eq} for numerous processes involving those atoms and molecules, at 298 K and at 2000 K. Fortunately, the uncertainty associated with the computed $\Delta G°$(2000 K) is not fully reflected in K_{eq}(2000 K). For example, an error of 4.0 kcal/mole in $\Delta G°$ changes K_{eq}(298 K) by a factor of 859, but K_{eq}(2000 K) only by 2.7.

Eq. (9) makes it possible to assess the relative amounts of the numerous boron- and aluminum-containing isomers in Tables 3 and 4 that are present at both 298 K and 2000 K. At 298 K, one in each pair or group usually greatly predominates, often overwhelmingly so. This can be seen by simply comparing their standard free energies. At 298 K, a difference of just 0.01 hartrees suffices to produce $K_{eq} \sim 10^4$; at 2000 K, on the other hand, it gives $K_{eq} \sim 5$. Thus, as was shown earlier [35,36], most of the isomers will be found in significant amounts at 2000 K.

An analogous situation prevails with respect to reactions, such as those in Tables 8 and 9; their equilibrium constants move in the direction of unity as the temperature increases. For example, the process,

$$F + H_2 \rightarrow HF + H \quad (10)$$

is virtually completely to the right at 298 K, with K_{eq}(298 K) $\sim 10^{23}$, whereas at 2000 K, appreciable amounts of the reactants are present, K_{eq}(2000 K) $\sim 10^3$. In contrast,

$$Al_2O + O_2 \rightarrow AlO + AlO_2 \quad (11)$$

essentially does not occur at 298 K, $K_{eq}(298\ K) \sim 10^{-30}$, but it is significant at 2000 K, where $K_{eq} \sim 10^{-4}$. The composition of the combustion system and the reactions that are occurring clearly depend very much upon the temperature.

Table 9. Calculated and experimental free energy changes at 298 K and 2000 K, in kcal/mole.[a,b]

Reaction	ΔG° (298 K)		ΔG° (2000 K)	
	Calc.	Exp.[c]	Calc.	Exp.[c]
$CO + H_2 \rightarrow HCO + H$	89.0	88.1	83.6	82.7
$CO + O_2 \rightarrow CO_2 + O$	−6.0	−6.1	6.2	2.7
$F + H_2 \rightarrow HF + H$	−32.2	−31.9	−34.9	−31.0
$Cl + H_2 \rightarrow HCl + H$	0.3	0.6	−5.0	−1.2
$HCO + O_2 \rightarrow CO_2 + OH$	−93.1	−92.8	−79.6	−81.4
$H + CO_2 \rightarrow CO + OH$	22.3	21.1	6.8	3.0
$HCO + Cl \rightarrow HCl + CO$	−88.7	−87.5	−88.5	−83.9
$O + H_2O \rightarrow 2OH$	15.9	15.6	11.4	7.7
$H + O_2 \rightarrow OH + O$	16.3	15.0	13.0	5.7
$HCO + F \rightarrow HF + CO$	−121.2	−120.1	−118.5	−113.7
$F + H_2O \rightarrow HF + OH$	−18.2	−17.7	−21.3	−21.9
$BF + H_2 \rightarrow BH + HF$	65.4	67.3	60.4	62.3
$BH + F_2 \rightarrow BF + HF$	−196.6	−198.6	−195.2	−196.7
$BO + F \rightarrow FBO$	−153.5	−153.	−99.7	−96.
$BO_2 + F \rightarrow FBO + O$	−32.5	−35.	−32.3	−33.
$BF + OH \rightarrow FBO + H$	−67.8	−71.	−56.2	−56.
$BO + OH \rightarrow BO_2 + H$	−24.7	−22.	−13.2	−11.
$BO + CO_2 \rightarrow BO_2 + CO$	−2.4	−1.0	−6.4	−8.
$BH + OH \rightarrow BO + H_2$	−110.2	−114.	−103.8	−105.
$BF + F_2 \rightarrow BF_3$	−235.3	−233.1	−181.0	−171.7
$AlO + HF \rightarrow AlF + OH$	−9.4	−6.	−10.5	−9.
$AlH + HF \rightarrow AlF + H_2$	−57.8	−59.	−52.9	−54.
$AlO + H_2 \rightarrow AlOH + H$	−11.1	−6.	−16.8	2.
$AlH + F_2 \rightarrow AlF + HF$	−189.0	−191.	−187.7	−188.
$AlOH + F \rightarrow AlO + HF$	−21.1	−26.	−18.1	−33.
$AlH + O_2 \rightarrow AlOH + O$	−43.2	−44.	−46.3	−37.
$AlF + H_2O \rightarrow AlOH + HF$	12.2	14.	7.3	20.

[a]Reference 35. [b]Reference 36. [c]Reference 13.

3.6. Reaction Mechanisms

We have determined the transition states and activation barriers at 298 K for a number of reactions that are likely to be involved in boron and/or aluminum combustion [23,30]. The structures of the transition states are shown qualitatively in Figure 2; the details of their geometries (bond lengths and angles, etc.) have been reported earlier [33,35,36]. The internal reaction coordinates were determined to verify that they lead to the desired products [41]. The CBS-QB3 computed activation barriers, ΔH_{act}(298 K), and heats of reaction, ΔH(298 K), are in Table 10. While we have made no attempt to survey the literature pertaining to each of these processes, we do have some comments about certain ones of them.

3.6.1. Reaction 1:

The doublet ($^2A''$) potential energy surface for H + O_2 has been studied in detail by configuration interaction computational procedures [55,56], which show the formation of a bent intermediate, HO_2, with almost no activation barrier; however subsequent dissociation to O + OH has $D_e > 60$ kcal/mole. (The hydroperoxyl radical, HO_2, is of considerable interest in several different areas [57].) We examined an alternative pathway, going through a linear quartet transition state that leads directly to O + OH.

3.6.2. Reaction 5:

This process, which has been studied extensively in the past [58], involves three steps but only two transition states, the *anti* intermediate being obtained without a barrier.

3.6.3. Reaction 6:

A combination of analyses, theoretical (MRCI and MCSCF) and experimental (rate constant determination at several temperatures), indicates an activation barrier at 0 K of 8.5 kcal/mole [59]. Our value at 0 K is 7.3 kcal/mole [33], which is very good agreement.

3.6.4. Reaction 8:

For complementary studies (MRCI//MCSCF) of the H–F–B–O potential energy surface, see Soto [60,61].

3.6.5. Reaction 11:

Another proposed mechanism, based in part upon the measured temperature dependence of the rate constant, involves the formation of a BO_3 intermediate [62,63].

3.6.7. Reaction 13:

The dissociaton energies of O_2 and AlO are nearly the same, both around 120 kcal/mole (Tables 5 and 7). Thus $\Delta H°$(298 K) ≈ 0. We found this to be true of ΔH_{act}(298 K) as well, which is consistent with the conclusion reached by Garland and Nelson from the observed variation of the rate constant with temperature [64].

3.6.8. Reaction 14:

It was mentioned earlier that the oxidation of aluminum by CO_2 is believed to be one of the major routes to Al_2O_3 in the combustion chamber [9,10]. Reaction 14 may be one of the initial steps in this process. For the same reason, there is considerable interest in the complexes of Al and AlO with CO and CO_2 [30,66-68], such as those in Tables 4 and 7.

3.6.9. Reaction 16:

Our finding this to proceed through an intermediate complex confirmed a suggestion by Belyung *et al* [30]. Even though $\Delta H°(298\ K)$ for the second step is predicted to be essentially zero, its calculated $K_{eq}(298\ K)$ is 3.8×10^3.

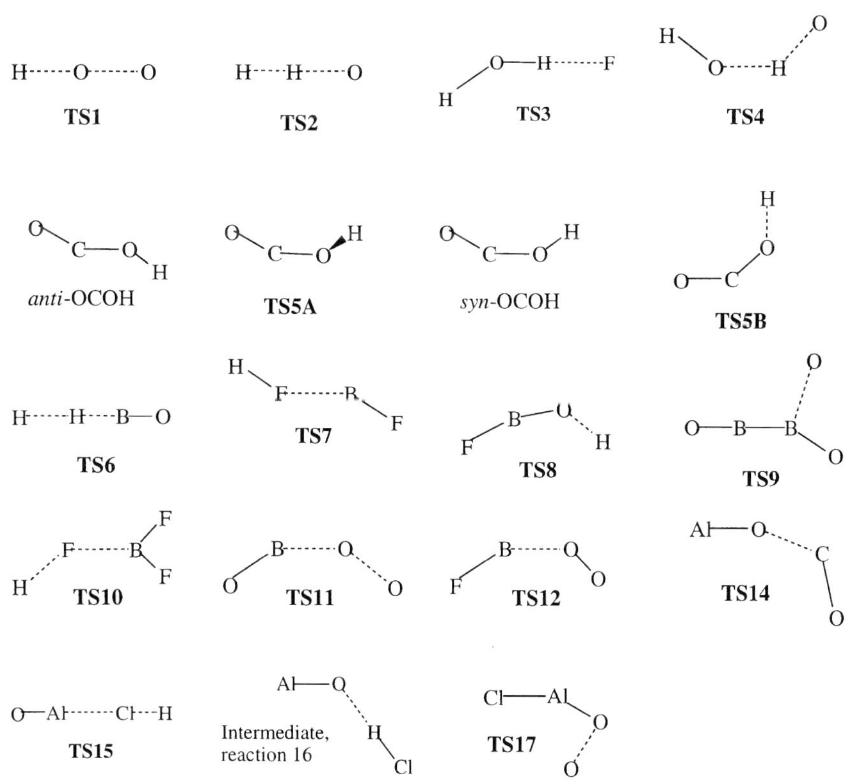

Figure 2. Structural drawings of some transition states and intermediates (Table 10).

Table 10. Calculated activation enthalpies and overall heats of reaction, at 298 K, in kcal/mole.

Reaction	ΔH_{act}[a]	$\Delta H°_{overall}$	
		Calc.	exp.[b]
1. $H + O_2 \rightarrow TS1 \rightarrow O + OH$	38.9	17.1	16.8
2. $O + H_2 \rightarrow TS2 \rightarrow H + OH$	10.2	2.7	1.9
3. $F + H_2O \rightarrow TS3 \rightarrow HF + OH$	13.4	–17.8	–17.0
4. $O + H_2O \rightarrow TS4 \rightarrow 2OH$	17.7	16.6	16.9
5. $CO + OH \rightarrow CO_2 + H$		–25.7	–24.9
(a) $CO + OH \rightarrow$ *anti*-OCOH	0.0	–27.1	---
(b) *anti*-OCOH $\rightarrow$ TS5A $\rightarrow$ *syn*-OCOH	8.1	2.0	---
(c) *syn*-OCOH $\rightarrow$ TS5B $\rightarrow CO_2 + H$	24.3	–0.6	---
6. $BO + H_2 \rightarrow TS6 \rightarrow H + HBO$	6.3	–7.5	---
7. $H + BF_2 \rightarrow TS7 \rightarrow HF + BF$	61.0	–26.2	---
8. $H + FBO \rightarrow TS8 \rightarrow FBOH$	16.3	–22.1	---
9. $O + B_2O_2 \rightarrow TS9 \rightarrow BO + BO_2$	2.0	–13.8	–19.
10. $HF + BF_2 \rightarrow TS10 \rightarrow H + BF_3$	22.6	–35.5	---
11. $BO + O_2 \rightarrow TS11 \rightarrow O + BO_2$	34.7	–10.7	–8.
12. $BF + O_2 \rightarrow TS12 \rightarrow O + FBO$	11.3	–53.5	–57.
13. $Al + O_2 \rightarrow AlO + O$	0.0	0.0	–3.
14. $Al + CO_2 \rightarrow TS14 \rightarrow AlO + CO$	14.4	8.6	5.
15. $AlO + HCl \rightarrow TS15 \rightarrow ClAlO + H$	12.3	–3.9	---
16. $AlO + HCl \rightarrow AlOH + Cl$		–12.0	–8.
(a) $AlO + HCl \rightarrow AlO{\bullet}HCl$	0.0	–12.6	---
(b) $AlO{\bullet}HCl \rightarrow AlOH + Cl$	0.0	0.5	---
17. $AlCl + O_2 \rightarrow TS17 \rightarrow ClAlO + O$	28.3	15.9	---

[a]The results for reactions 1 – 12 are from references 33 and 35; the remainder are from reference 36.
[b]Reference 13.

4. SUMMARY

Although the enthalpies in Tables 2 – 4 were computed specifically for 298 K and 2000 K, they are actually much more widely applicable in view of the fact that heats of reaction change so little with temperature. $\Delta H°(298\ K)$ and $\Delta H°(2000\ K)$ can certainly be used to estimate $\Delta H°(T)$ for $298\ K < T < 2000\ K$, and $\Delta H°(2000\ K)$ is likely to be a reasonable prediction for temperatures well beyond 2000 K.

There is much greater uncertainty with regard to ΔG° at high temperatures, although Tables 2 – 4 provide at least a rough approximation at 2000 K as does eq. (8) at other temperatures. Fortunately, as was pointed out, this is likely to be sufficient to obtain the correct order of magnitude of the equilibrium constant.

Thus, the data in Tables 2 – 4 can be used to calculate, with reasonable (or better) accuracy over a wide range of temperatures, thermodynamic properties of numerous reactions implicated in the combustion of boron- and aluminum-containing propellant formulations. The kinetics can be addressed as well, as was shown in a number of instances. The determination of transition states and activation barriers can be rather time-consuming. However computational methodology continues to improve (e.g. a new version of CBS-QB3 [69], and the IRCMax technique for transition state geometries and activation barriers [70]) as does processor technology. Computational analyses can be expected to become an increasingly effective and important tool for characterizing and elucidating propellant combustion processes.

ACKNOWLEDGEMENTS

We dedicate this chapter to the memory of Dr. E. Sheldon-Rahmel. We also gratefully acknowledge the financial support provided by the Ballistic Missile Defense Organization and the Office of Naval Research through contract # N00014-95-1-1339, program officers Dr. Leonard H. Caveny (BMDO) and Dr. Judah Goldwasser (ONR).

REFERENCES

[1] R. Friedman and A. Macek, *Ninth Symposium (International) on Combustion*, (The Combustion Institute, Pittsburgh, 1963) p. 703.
[2] A. F. Belyaev, Y. V. Frolov and A. I. Korotkov, Combust. Explos. Shock Waves, 4 (1968) 182.
[3] C. K. Law, Combust. Sci. Tech., 7 (1973) 197.
[4] V. M. Gremyachkin, A. G. Istratov and O. I. Leipunskii, Combust. Explos. Shock Waves, 11 (1975) 313.
[5] A. G. Merzhanov, Y. M. Grigorjev and Y. A. Gal'chenko, Combust. Flame, 29 (1977) 1.
[6] E. W. Price, K. J. Kraeutle, J. L. Prentice, T. L. Boggs, J. E. Crump and D. E. Zurn, *Behavior of Aluminum in Solid Propellant Combustion*, (NWC TP 6120, Naval Air Warfare Center, China Lake, CA, 1982).
[7] E. W. Price, Prog. Astronaut. Aeronaut., 6 (1984) 479.
[8] J. F. Driscoll, J. A. Nicholls, V. Patel, B. K. Shahidi and T. C. Liu, AIAA J., 24 (1986) 856.
[9] S. Yuasa, S. Sogo and H. Isoda, *Twenty-Fourth Symposium (International) on Combustion*, (The Combustion Institute, Pittsburgh, 1992) p. 1817.
[10] J. F. Widener and M. W. Beckstead, AIAA 98-3824, 34th AIAA/ASME/SAE/ ASEE Joint Propulsion Conference, (Cleveland, 1998).
[11] Y. Liang and M. W. Beckstead, AIAA 98-3825, *ibid.*
[12] P. Bucher, R. A. Yetter, F. L. Dryer, T. P. Parr and D. M. Hanson-Parr, *Twenty-Seventh Symposium (International) on Combustion*, (The Combustion Institute, Pittsburgh, 1998) Vol. 2, p. 2421.
[13] W. G. Mallard and P. J. Linstrom, eds., *NIST Chemistry Webbook*, NIST Standard Reference Database No. 69, (National Institute of Standards and Technology, Gaithersburg, MD, 1998)., (http://webbook.nist.gov).
[14] A. Macek and J. M. Semple, Combust. Sci. Tech., 1 (1969) 181.
[15] M. K. King, Combust. Sci. Tech., 5 (1972) 155; 8 (1974) 255.
[16] C. W. Burdette, H. R. Lander and J. R. McCoy, J. Energy, 2 (1978) 289.

[17] S. R. Turns, J. T. Holl, A. S. P. Solomon and G. M. Faeth, Combust. Sci. Tech., 43 (1985) 287.
[18] P. Antaki and F. A. Williams, Combust. Flame, 67 (1987) 1.
[19] R. A. Yetter, H. Rabitz, F. L. Dryer, R. C. Brown and C. E. Kolb, Combust. Flame, 83 (1991) 43.
[20] L. Pasternack, Combust. Flame, 90 (1992) 259.
[21] R. C. Brown, C. E. Kolb, S. Y. Cho, R. A. Yetter, F. L. Dryer and H. Rabitz, Int. J. Chem. Kinet., 26 (1994) 319.
[22] V. V. Golovko, E. N. Kondratyev and D. I. Polishshuk, in *Combustion of Boron-Based Solid Propellants and Solid Fuels*, K. K. Kuo and R. Pein, eds., (Begell House/CRC Press, Boca Raton, FL, 1993) p. 272.
[23] R. C. Brown, C. E. Kolb, R. A. Yetter, F. L. Dryer and H. Rabitz, Combust. Flame, 101 (1995) 221.
[24] C. L. Yeh and K. K. Kuo, Prog. Energy Combust. Sci., 22 (1996) 511.
[25] R. O. Foelsche, M. J. Spalding, R. L. Burton and H. Krier, Mat. Res. Soc. Symp. Proc., 418 (1996) 187.
[26] R. A. Yetter, F. L. Dryer, H. Rabitz, R. C. Brown and C. E. Kolb, Combust. Flame, 112 (1998) 387.
[27] W. Zhou, R. A. Yetter, F. L. Dryer, H. Rabitz, R. C. Brown and C. E. Kolb, Combust. Flame, 112 (1998) 507.
[28] R. S. Miller, Mat. Res. Soc. Symp. Proc., 418 (1996) 3.
[29] A. Fontijn, Combust. Sci. Tech., 50 (1986) 151.
[30] D. P. Belyung, G. T. Dalakos, J.-D. R. Rocha and A. Fontijn, *Twenty-Seventh Symposium (International) on Combustion*, (The Combustion Institute, Pittsburgh, 1998) Vol. 1, p. 227.
[31] M. T. Swihart and L. Catoire, Combust. Flame, 121 (2000) 210.
[32] P. Politzer, P. Lane and M. C. Concha, J. Phys. Chem. A ,103 (1999) 1419.
[33] P. Politzer, P. Lane and M. C. Concha, Proc. 36th JANNAF Combust. Subcomm. Mtg., (CPIA Publ. 691, II, 2000) p.331.
[34] P. Politzer, M. C. Concha and P. Lane, J. Mol. Struct. (Theochem), 529 (2000) 41.
[35] P. Politzer, P. Lane and M. Concha, Recent Res. Devel. Phys. Chem., 4 (2000) 319.
[36] P. Politzer, P. Lane and M. E. Grice, J. Phys. Chem. A, 105 (2001) 7473.
[37] M. D. Allendorf, C. F. Melius, B. Cosic and A. Fontijn, J. Phys. Chem. A, 106 (2002) 2629.
[38] J. A. Montgomery, Jr., M. J. Frisch, J. W. Ochterski and G. A. Petersson, J. Chem. Phys., 110 (1999) 2822.
[39] M. J. Frisch, G. W. Trucks, H. B. Schlegel, G. E. Scuseria, M. A. Robb, J. R. Cheeseman, V. G. Zakrezewski, J. A. Montgomery, R. E. Stratmann, J. C. Burant, S. Dappich, J. M. Millam, A. D. Daniels, K. N. Kudin, M. C. Strain, O. Farkas, J. Tomasi, V. Barone, M. Cossi, R. Cammi, B. Mennucci, C. Pomelli, C. Adamo, S. Clifford, J. Ochterski, G. Petersson, P. Y. Aayala, Q. Cui, K. Morokuma, D. K. Malick, A. D. Rubuck, K. Raghavachari, J. B. Foresman, J. Cioslowski, J. V. Ortiz, B. B. Stefanov, G. Liu, A. Liashenko, P. Piskorz, I. Komaromi, R. Gomperts, R. L. Martin, D. J. Fox, T. Keith, M. A. Al-Laham, C. Y. Peng, A. Nanayakkara, C. Gonzalez, M. Challacombe, P. M. W. Gill, B. G. Johnson, W. Chen, M. W. Wong, J. L. Andres, M. Head-Gordon, E. S. Replogle and J. A. Pople, Gaussian 98, Revision A.5, Gaussian, Inc.(Pittsburgh, PA, 1998).
[40] W. J. Hehre, L. Radom, P. v. R. Schleyer and J. A. Pople, *Ab Initio Molecular Orbital Theory*, (Wiley-Interscience, New York, 1986).
[41] C. Gonzalez and H. B. Schlegel, J. Phys. Chem., 94 (1990) 5523.
[42] S. S. Xantheas, T. H. Dunning, Jr. and A. Mavridis, J. Chem. Phys., 106 (1997) 3280.
[43] R. L. Asher, E. H. Appelman and B. Ruscic, J. Chem. Phys., 105 (1996) 9781.
[44] S. G. Lias, J. E. Bartmess, J. F. Liebman, J. L. Holmes, R. D. Levin and W. G. Mallard, J. Phys. Chem. Ref. Data, 17 (1988) Suppl. 1.
[45] D. R. Lide (ed.), Handbook of Chemistry and Physics, 78th ed., (CRC Press, New York, 1997).
[46] J. M. L. Martin, T. J. Lee, G. E. Scuseria and P. R. Taylor, J. Chem. Phys., 97 (1992) 6549.
[47] K. A. Peterson, J. Chem. Phys., 102 (1995) 262.

[48] C. W. Bauschlicher, Jr. and A. Ricca, J. Phys. Chem. A, 103 (1999) 4313.
[49] M. Page, J. Phys. Chem., 93 (1989) 3639.
[50] G. Meloni and K. A. Gingerich, J. Chem. Phys., 111 (1999) 969.
[51] R. D. Srivastava and M. Farber, Chem. Rev., 78 (1978) 627.
[52] M. W. Chase, Jr., J. Phys. Chem. Ref. Data, Monograph 9 (1998).
[53] G. A. Petersson, D. K. Malick, W. G. Wilson, J. W. Ochterski, J. A. Montgomery and M. J. Frisch, J. Chem. Phys., 109 (1998) 10570.
[54] S. Glasstone, Thermodynamics for Chemists, (D. Van Nostrand, Princeton, NJ, 1947).
[55] S. P. Walch, C. M. Rohlfing, C. F. Melius and C. W. Bauschlicher, Jr., J. Chem. Phys., 88 (1988) 6273.
[56] S. P. Walch and C. M. Rohlfing, J. Chem. Phys., 91 (1989) 2373.
[57] W. J. Lemon and W. L. Hase, J. Phys. Chem., 91 (1987) 1596.
[58] K. S. Bradley and G. C. Schatz, J. Chem. Phys., 106 (1997) 8464, and references cited.
[59] N. L. Garland, C. T. Stanton, H. H. Nelson and M. Page, J. Chem. Phys., 95 (1991) 2511.
[60] M. R. Soto, J. Phys. Chem., 99 (1995) 6540.
[61] M. R. Soto, Mat. Res. Soc. Symp. Proc., 418 (1996) 181.
[62] R. C. Oldenbourg and S. L. Baughcum, Advances in Laser Science I, AIP Conference Proceedings 146, W. C. Stwalley and M. Lapp, eds., (American Institute of Physics, New York, 1986).
[63] C. T. Stanton, N. L. Garland and H. H. Nelson, J. Phys. Chem., 95 (1991) 8741.
[64] N. L. Garland and H. H. Nelson, Chem. Phys. Lett., 191 (1992) 269.
[65] D. F. Rogowski, A. J. English and A. Fontijn, J. Phys. Chem., 90 (1986) 1688.
[66] J. M. Parnis, S. A. Mitchell, T. S. Kanigan and P. A. Hackett, J. Phys. Chem., 93 (1989) 8045.
[67] M. J. McQuaid and J. L. Gole, Chem. Phys., 234 (1998) 297.
[68] J. A. Montgomery, Jr., M. J. Frisch, J. W. Ochterski and G. A. Petersson, J. Chem. Phys., 112 (2000) 6532.
[69] D. K. Malick, G. A. Petersson and J. A. Montgomery, Jr., J. Chem. Phys., 108 (1998) 5704.

Chapter 17

Chemistry of Detonation Waves in Condensed Phase Explosives

Craig M. Tarver and M. Riad Manaa

Lawrence Livermore National Laboratory, P. O. Box 808, L-282, Livermore, CA 94551, U.S.A.

1. INTRODUCTION

Detonation of high density, high energy solid organic explosives produces self-sustaining waves traveling at speeds approaching 10,000 m/s that reach approximately 40 GPa pressures and 6000 K temperatures in nanoseconds. This pressure-temperature-time frame is unique and thus very difficult to study experimentally and theoretically. However, a great deal of progress has been made in understanding the extreme chemistry that occurs with the reaction zone of a condensed phase detonation wave. The Non-Equilibrium Zeldovich - von Neumann - Doring (NEZND) theory was developed to identify the non-equilibrium chemical processes that precede and follow exothermic chemical energy release within the reaction zones of self-sustaining detonation waves in gaseous, liquid and solid explosives [1-10]. Prior to the development of the NEZND model, the chemical energy released was merely treated as a heat of reaction in the conservation of energy equation in the Chapman-Jouguet (C-J) [11,12], Zeldovich - von Neumann - Doring (ZND) [13-15], and curved detonation wave front theories [16] and in hydrodynamic computer code reactive flow models [17]. NEZND theory has explained many experimentally observed detonation wave properties. These include: the induction time delays for the onset of chemical reaction; the rapid rates of the chain reactions that form the reaction product molecules; the de-excitation rates of the initially highly vibrationally excited products; the feedback mechanism that allows the chemical energy to sustain the leading shock wave front at an overall constant detonation velocity; and the establishment of the complex three-dimensional Mach stem structure of the leading shock wave fronts common to all detonation waves. When the leading shock front of a detonation wave compresses an explosive molecule, thermal energy must be transported into the vibrational modes of the explosive molecule before exothermic reactions can occur. The induction time for the onset of the initial endothermic reactions can be calculated using high pressure, high temperature transition state theory. First principle molecular dynamics studies of the primary chemical reactions are being done at the atomistic scale. These high-temperature, high-density calculations show the evolution of intermediate decomposition products and final stable detonation reaction products, such as H_2O, CO_2, N_2, CO and solid carbon. These reaction products are initially created in highly vibrationally excited states that must be de-excited as chemical and thermal equilibrium are attained at the Chapman-Jouguet

(C-J) state. Since the chemical energy is released well behind the leading shock front of a detonation wave, a physical mechanism is required for this chemical energy to reinforce the leading shock front and maintain its overall constant velocity. This mechanism is the amplification of pressure wavelets in the reaction zone by the process of de-excitation of the initially highly vibrationally excited reaction product molecules. The C-J state determines the energy delivery of the detonating explosive to its surroundings and thus must be accurately determined. Today's computers are still not large or fast enough to include all of these non-equilibrium processes in large scale two- and three-dimensional hydrodynamic calculations so phenomenological high explosive reactive flow models must still be developed. NEZND theory, molecular dynamics atomistic scale simulations, and high explosive reactive flow modeling studies are discussed in this chapter.

2. NEZND THEORY OF DETONATION

Figure 1 illustrates the various processes that occur in the NEZND model of detonation in condensed explosives. At the head of every detonation wave is a three-dimensional Mach stem shock wave front. There are several definitions of the width of a shock wave. Zeldovich and Raizer [18] defined shock wave width as the distance at which the viscosity and heat conduction become negligible. Behind the shock front in solid explosives, the phonon modes are first excited, followed by multi-phonon excitation of the lowest frequency vibrational (doorway) modes and then excitation of the higher frequency modes by multi-phonon up-pumping and internal vibrational energy redistribution (IVR) [19]. Internal energy equilibration is being studied in shocked liquid and solid explosives by Dlott et al. [20] and Fayer et al. [21]. After the explosive molecules become vibrationally excited, chemical reactions begin.

For gaseous explosives, the non-equilibrium processes that precede chemical reaction are easily measured, because they occur in nanosecond or longer time frames. Velocities, pressures and temperatures are calculated using the perfect gas law [2]. The high initial densities of liquids and solids make the measurement and calculation of the states attained behind a shock wave much more difficult, because the processes now take tens and hundreds of picoseconds and the perfect gas law does not apply. The distribution of the shock compression energy between the potential (cold compression) energy of the unreacted liquid or solid and its thermal energy is a complex function of shock strength.

The induction time for the initial endothermic bond breaking reaction can be calculated using the high pressure, high temperature transition state theory. Experimental unimolecular gas phase reaction rates under low temperature (<1000K) shock conditions obey the usual Arrhenius law:

$$K = Ae^{-E/RT} \qquad (1)$$

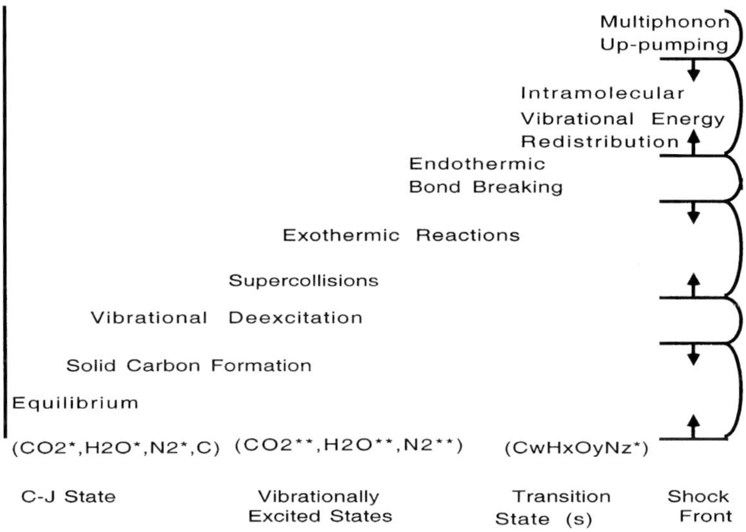

Figure 1. The Non-Equilibrium Zeldovich - von Neumann - Doring (NEZND) Theory of Detonation.

where K is the reaction rate constant, A is the frequency factor, E is the activation energy, R is the gas constant, and T is temperature, but "fall-off" to slower rates of increase at high temperatures [22]. Nanosecond reaction zone measurements for solid explosives overdriven to pressures and temperatures exceeding those attained in self-sustaining detonation waves have shown that the reaction rates increase very slowly with shock temperature [23]. Eyring [24] attributed this "falloff" in unimolecular rates at the extreme temperature and density states attained in shock and detonation waves to the close proximity of vibrational states, which causes the high frequency mode that becomes the transition state to rapidly equilibrate with the surrounding modes by IVR. These modes form a "pool" of vibrational energy in which the energy required for decomposition is shared. Any large quantity of vibrational energy that a specific mode receives from an excitation process is shared among the modes before reaction can occur. Conversely, sufficient vibrational energy from the entire pool of oscillators is statistically present in the transition state vibrational mode long enough to cause reaction. When the total energy in the vibrational modes equals the activation energy, the reaction rate constant K is:

$$K = (kT/h)e^{-s}\sum_{i=0}^{s-1}(E/RT)^i e^{-E/RT}/i! \qquad (2)$$

where k, h, and R are Boltzmann's, Planck's, and the gas constant, respectively, and s is the number of neighboring vibrational modes interacting with the transition state. The main effect of rapid IVR among s+1 modes at high densities and temperatures pressures is to decrease the rate constant dependence on temperature. Reaction rate constants have been calculated for detonating solids and liquids using Eq. (2) with realistic equations of state [4].

For the lower temperatures attained in shock initiation of homogeneous liquid and solid explosives, reaction rate constants calculated using Eq. (2) are larger than those predicted by Eq. (1). Reaction rate constants from Eqs. (1) and (2) are compared to induction time results for liquid nitromethane, and single crystal pentaerythritol tetranitrate (PETN) in Figs. 2 and 3, respectively [6]. Despite uncertainties in the calculated shock temperatures for various equations of state (EOS), it is clear that Eq. (2) agrees quite well with both sets of data using reasonable values of s. Thus high pressure, high temperature transition state theory accurately calculates induction times for shock initiation and detonation of homogeneous liquid and heterogeneous solid explosives. Then the exothermic chain reactions begin.

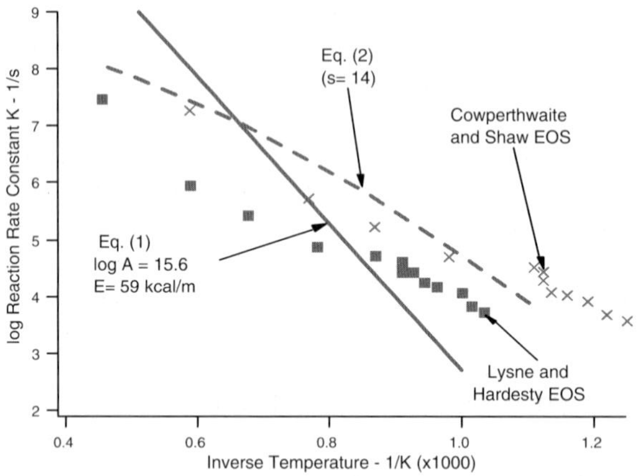

Figure 2. Reaction rate constants for nitromethane as functions of shock temperature.

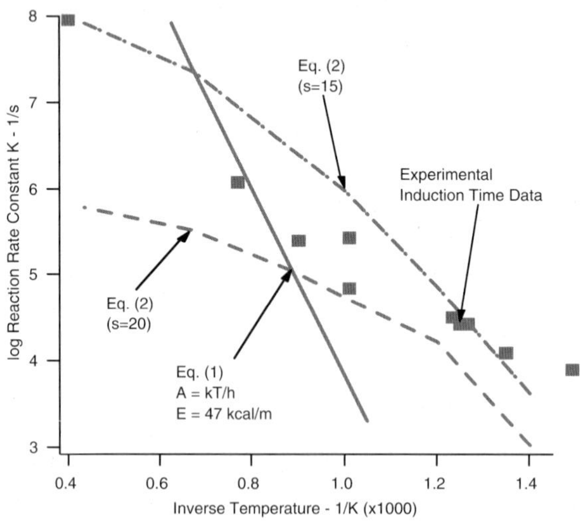

Figure 3. Reaction rate constants for single crystal PETN as functions of shock temperature.

3. ATOMISTIC STUDIES OF EARLY CHEMISTRY IN HOT DENSE MEDIA

Detailed knowledge of the reaction chemistry of energetic materials at high pressure and temperature is of considerable importance in understanding processes that these materials experience under detonation conditions. Answers to basic questions such as: (a) which bond in a given energetic molecule breaks first, and (b) what type of chemical reactions (unimolecular versus bimolecular, etc.) dominates early in the decomposition process, are still largely unknown. The most widely studied, and archetypical example of such materials is nitromethane (CH_3NO_2), a clear liquid with mass density 1.13 g/cm^3 at 298 K. Static high-pressure experiments [25, 26] showed that the time to explosion for deuterated nitromethane is approximately ten times longer than that for protonated nitromethane, suggesting that a proton or hydrogen atom abstraction is involved in the rate determining step. Isotope-exchange experiments, using diamond anvil cells, also gave evidence [27] that the aci ion concentration ($H_2CNO_2^-$) increases with increased pressure, and with addition of a minute amount of base. Other studies [28,29] also suggested that reactions occur more rapidly and are pressure enhanced when small concentrations of bases are present, giving further support to the aci ion production. Shock wave studies of the reaction chemistry are still inconclusive and at odds: mass spectroscopic studies suggesting condensation reactions [30]; time-resolved Raman spectroscopy suggesting a bimolecular mechanism [31]; and UV-visible absorption spectroscopy indicating no sign of chemical reaction [32] or the production of $H_3CNO_2^-$ intermediate for amine-sensitized nitromethane [33].

Molecular dynamics (MD) simulations provide a useful tool to unravel underlying mechanisms of condensed-phase processes at varying conditions. Several recent studies have been conducted on nitromethane using first principles or classical potentials for the molecular forces. Sorescu et al. have successfully investigated the dynamics of both the solid [34] and liquid [35] phases of nitromethane at ambient conditions, using a classical potential for the inter- and intra-molecular forces. Politzer's group simulated the structure of liquid nitromethane at ambient pressure and temperature [36], and the shocked liquid up to 143 kbar and 600 K [37] using density functional/MD approach. For the latter case, good agreement with experimental results was observed for the stretching frequencies of C-N and NO_2. Tuckerman and Klein [38] used Car-Parrinello ab initio/MD to obtain the frequency spectrum and barrier to methyl group rotation at T = 0 K and T = 285 K, which were in good accord with the experimental findings. A recent Car-Parrinello ab initio/MD study [39] of the pressurized liquid (by a factor of 3.0 at 150 K) demonstrated the occurrence of a proton transfer reaction between two closely spaced nitromethane molecules leading to CH_2NO_2 and $CH_3NO(OH)$. To date, however, there are no computational studies on nitromethane in regions of extreme conditions of pressure and temperature such as those encountered during detonation, despite the successful implementation of these methodologies on other systems. For example, ab initio/MD simulations have been successfully demonstrated recently for water, ammonia, and methane at temperatures up to 7000 K and pressures of up to 300 GPa [40,41]. Semi-empirical, tight-binding methods have also emerged as viable tools, as has been recently demonstrated in the studies of shocked hydrocarbons and the explosive HMX [42-44].

We have simulated the initial decomposition steps of hot (T = 3000 K), dense (ρ = 1.97g/cm^3) nitromethane at constant-volume and temperature conditions using ab initio density-functional, molecular dynamics calculations. This state is in the neighborhood of the C - J state, which is achieved behind a steady detonation front when the material has fully reacted. It could also be achieved through sudden heating of nitromethane in a diamond anvil cell under constant volume conditions. Our results emphatically show that the first chemical event is a proton extraction to form $CH_3NO_2H^+$, the aci ion $H_2CNO_2^-$, and the aci acid H_2CNO_2 H, establishing for the first time these decomposition species for neat nitromethane. The results are uniquely associated with the condensed-phase rather than the energetically favored C-N decomposition expected in the gas-phase.

We conducted three separate MD simulations on a single unit cell (simulation A), two (simulation B) and four unit cells (simulation C). Here, results from simulation C are discussed. To examine the early steps of the simulation, Figure 4 displays the variation of the N-O, C-H, and O-H bond distances with time for a single nitromethane molecule. As shown, the C-H bond clearly undergoes a significant stretch that eventually leads to a hydrogen ejection and subsequent capture by the oxygen of a nearby nitromethane molecule, leading to the formation of $CH_3NO_2H^+$ and $CH_2NO_2^-$ species. We note that time-resolved Raman measurements of shocked nitromethane up to 14 GPa have demonstrated an increase in the vibrational frequencies, the largest being exhibited in the CH_3 stretching mode [45]. The observed frequency hardening (and broadening) was attributed to strong intermolecular interactions at the shock pressure. This observation was corroborated by molecular dynamics simulations at 14.3 GPa and temperature of 600 K [37].

A snapshot of the MD simulation at 59 fs where the formation of $CH_3NO_2H^+$ and $CH_2NO_2^-$ takes place is shown in Figure 5. This process of proton transfer is initially facilitated by enhancement in the C-N double bond character, and an accelerated rotation of the methyl groups (CH_3), rotations that are omnipresent even at ambient temperatures [34,37]. Rotation of the methyl group serves to align a C-H bond with a vibrationally excited N-O, which undergoes a stretch from 1.24 to 1.45 Å. Interestingly, this process is not described by a bimolecular collision, rather it involves three nitromethane molecules as follows:

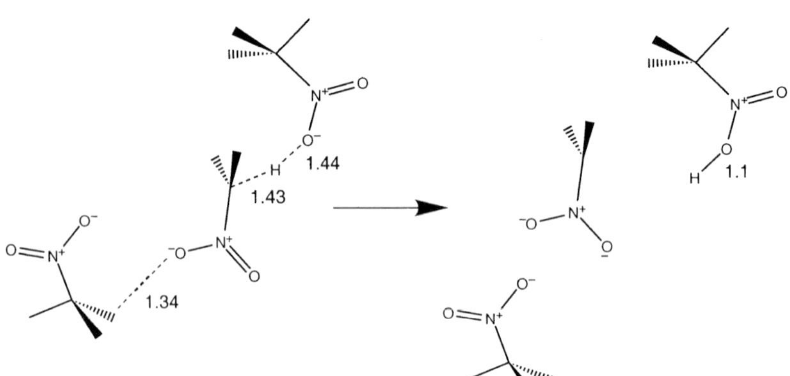

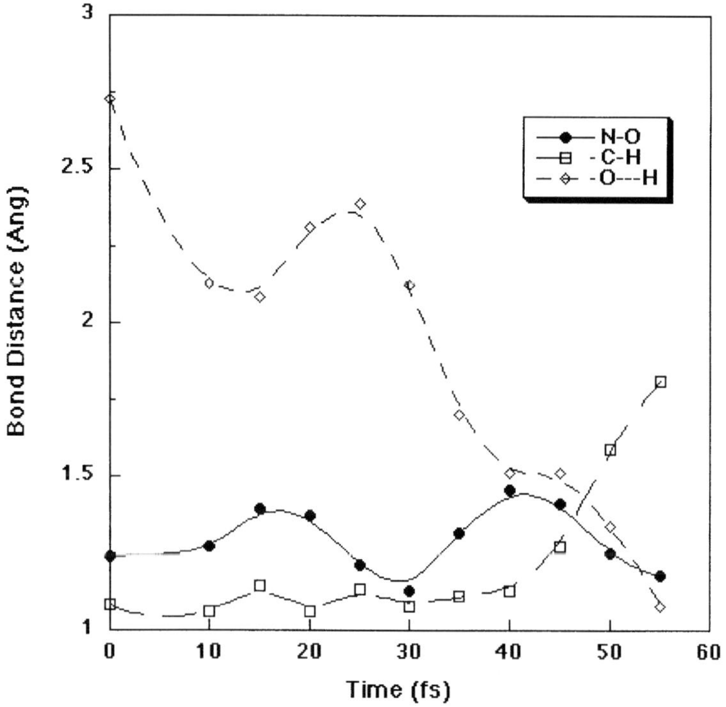

Figure 4. Time variation of intramolecular C-H and N-O, and the intermolecular O···H bonds for a single nitromethane molecule.

The proton transfer process described above is uniquely associated with the condensed fluid phase of nitromethane. This bond specificity is remarkable, since in the gas phase the C-N bond is the weakest in the molecule (D_0 = 60.1 kcal/mol) [46], and is therefore expected to be the dominant dissociation channel and the initial step in the decomposition of nitromethane, even at high temperature. In contrast, the C-H bond is the strongest in the nitromethane molecule.

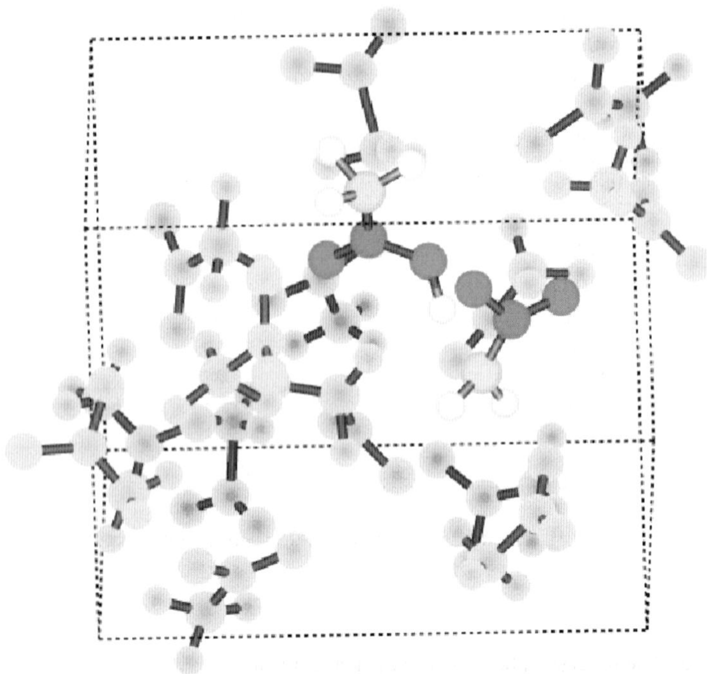

Figure 5. A snapshot of the MD simulation at 59 fs. The formation of CH_3NO_2H and CH_2NO_2 due to intermolecular hydrogen abstraction is clearly evident.

The formation of $CH_3NO_2H^+$ and $CH_2NO_2^-$ via proton extraction was observed in all three simulations of different supercell sizes A, B, and C. In simulation A, the event occurs at 785 fs of the simulation time, and in simulation B at 780 fs. This shows that the process itself is independent of the system size and the initial conditions of the atoms, since these simulations started with different initial velocities. We performed Mulliken charge analysis and listed the net charges on the atoms of the two moieties $CH_3NO_2H^+$ and $CH_2NO_2^-$ in Table 1, and for a nearby unreacted molecule. The total molecular charges for $CH_3NO_2H^+$ and $CH_2NO_2^-$ indicate that these species are only partially ionized in the dense fluid phase, while CH_3NO_2 remains neutral. We notice that the negative charge on the carbon atom of $CH_2NO_2^-$ is larger than in $CH_3NO_2H^+$, while the opposite trend is exhibited for the positive charge on nitrogen. This is a manifestation of electronic charge redistribution in the region between the C and N atoms. Upon contraction of the C-N single bond in nitromethane, electronic charge transfer from a hydrogen atom into this region serve to enhance the C-N double bond character while weakening the C-H bond. This crucial charge reordering is particularly significant in the case of intra-molecular hydrogen transfer discussed below. The net negative charges on oxygen atoms also increased in $CH_2NO_2^-$, rendering these atoms as even more pronounced proton acceptor sites. A significant increase in the net charge on the transferred hydrogen (H2) is consistent with its bonding with one of the oxygen atoms. The total net charge on $CH_2NO_2^-$ is –0.48 $|q_e|$, while $CH_3NO_2H^+$ has a net positive charge of 0.51 $|q_e|$.

Table 1. Calculated Mulliken charges of $CH_3NO_2H^+$ and $CH_2NO_2^-$.

Atom	Q (CH_3NO_2)	Q ($CH_3NO_2H^+$)	Q ($CH_2NO_2^-$)
C	-0.59	-0.54	-0.60
H1	+0.28	+0.35	+0.34
H2	+0.32	+0.28	
H3	+0.32	+0.36	+0.37
N	+0.42	+0.36	+0.29
O1	-0.35	-0.39	-0.46
O2	-0.38	-0.37	-0.42
H2		+0.46	
Total	+0.02	+0.51	-0.48

The second event in simulation C is a second proton transfer, observed at 60 fs, and is due to an intramolecular O-H attraction. The fast vibrations of the methyl group induce an umbrella inversion type motion that, along with a stretch in a N-O bond, leads to the hydrogen transfer from the methyl to the nitro group. This reaction yields the aci acid form of nitromethane. The process of bond rearrangement is illustrated as:

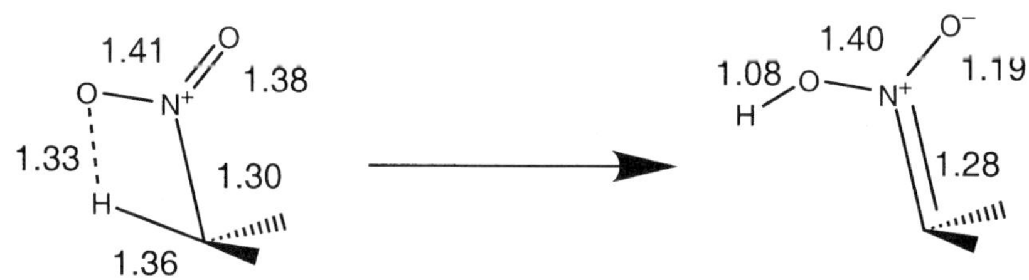

It is noteworthy that all three simulations (A, B, and C) have yielded the same results in the formation of $CH_3NO_2H^+$, $H_2CNO_2^-$, and CH_2NO_2H. Experimental concurrence for the production of the aci ion in highly pressurized and detonating nitromethane abound. Shaw et al. [25] observed that the time to explosion for deuterated nitromethane is about ten times longer than that for the protonated materials, suggesting that a proton (or hydrogen atom) abstraction is the rate-determining step. Isotope-exchange experiments provided evidence that the aci ion concentration increases with increasing pressure [27] and UV sensitization of nitromethane to detonation was shown to correlate with the aci ion presence [47]. Finally, we note that a recent electronic structure study of solid nitromethane determined a significant C-H stretch upon compression, which eventually lead to proton dissociation [48], and the Car-

Parrinello ab initio/MD study of Decker et al. [39] that showed a proton transfer reaction of liquid nitromethane when compressed by a factor of 3.0 at 150 K.

Having formed $CH_3NO_2H^+$, $CH_2NO_2^-$, and CH_2NO_2H, a series of steps then leads to the formation of water, initiated by a reaction between CH_2NO_2H and $CH_3NO_2H^+$ to form $CH_2NO_2H...HOH$ and CH_2NO^+ complexes in the period of 73 - 110 fs of the simulation:

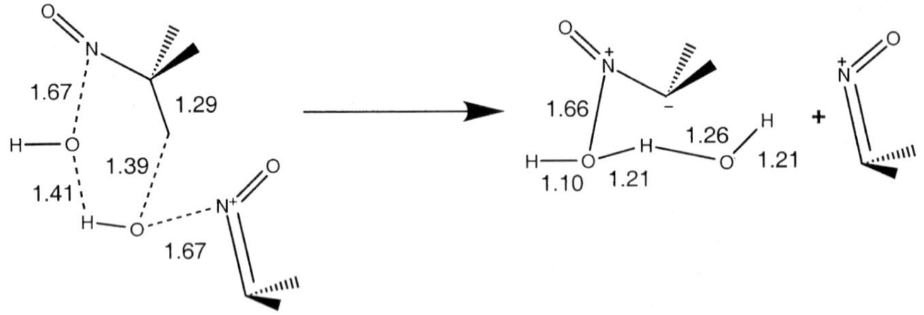

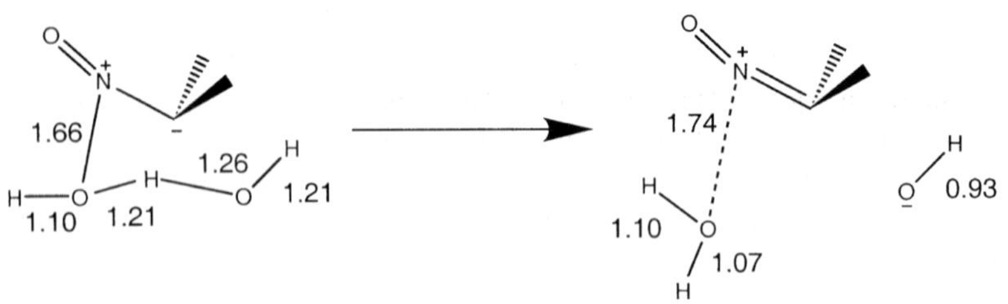

The hydroxy ion plays the role of an intermediate catalyst in the production of water. It later reacts with CH_2NO^+ to produce the aci acid form CH_2NO_2H.

The simulations discussed above are but of a single event in the vast pressure-temperature-volume phases of a reacting energetic materials. Nonetheless, they serve to provide insights on the early condensed-phase chemistry occurring at these extreme conditions. Being computationally tractable, similar simulations on nitromethane and other energetic materials at varying degree of density and temperature will help better understand early decomposition events, which, at least in the case of nitromethane, indicate a completely different mechanism than usually assumed in gas-phase studies. It is hoped that studies along the lines suggested above, in conjunction with previous work [49-51], will provide a better understanding of the electronic structure aspects and dynamics of the initiation and detonation process of nitromethane at the microscopic level.

4. RELAXATION TOWARD EQUILIBRIUM AFTER CHEMICAL REACTION

Following the induction and endothermic initial bond breaking processes, exothermic chain reaction processes follow in which reaction product gases (CO_2, N_2, H_2O CO, etc.) are formed in highly vibrationally excited states [2]. These excited products either undergo reactive collisions with unreacted explosive molecules or non-reactive collisions with other products in which one or more quanta of vibrational energy is transferred. Some collisions are "super-collisions" [52] in which several quanta of vibrational energy are transferred. Since reaction rates increase rapidly with each quanta of vibrational energy available, reactive collisions dominate and the main chemical reactions are extremely fast. Once the chain reactions are completed, the remainder of the reaction zone is dominated by vibrational de-excitation of the gaseous molecules and carbon formation.

This vibrational de-excitation partially process controls the length of the reaction zone and provides the chemical energy necessary for shock wave amplification during self-sustaining detonation [7]. As pressure wavelets pass through the subsonic reaction zone, they are amplified by discrete frequency vibrational de-excitation processes. The opposite effect, shock wave damping by a non-equilibrium gas that lacks vibrational energy after expansion through a nozzle, is a well-known phenomenon [18]. These pressure wavelets then interact with the main shock front and replace the energy lost during compression, acceleration and heating of the explosive molecules. The pressure wavelet amplification process provides the required chemical energy by developing a complex, three-dimensional Mach stem shock front structure, as shown in Fig. 1. This structure has been observed for gaseous, liquid and solid explosives [53] and is currently being replicated for gaseous explosives in two-dimensional and three-dimensional hydrodynamic computer simulations using multiple reaction chemical kinetic schemes [54]. Simulations of condensed phase detonation waves are also being attempted [55].

Since most condensed phase explosive formulations are under-oxidized, significant amounts of solid carbon particles form in the chemical reaction zone of self-sustaining detonation waves. These particles are diamond, graphite, or amorphous carbon depending on the temperatures and pressures attained in the reaction zone and have diameters of about 10 nanometers [56]. Since the solid carbon formation process is diffusion controlled as carbon atoms attempt to form chains and particles in the presence of several gaseous species, this process requires more time than gaseous product formation and equilibration. Thus the chemical energy release portion of a condensed phase detonation wave exhibits two energy release rates: a fast reaction taking tens of nanoseconds in which the main gaseous products are formed and a slower reaction for the solid carbon particle formation requiring several hundreds of nanoseconds. These rates have been measured by several nanosecond time resolved techniques including: embedded particle velocity and pressure gauges, electrical conductivity probes; and laser interferometry [57]. Chemical and thermal equilibrium at the C-J state is closely approached after the carbon particles form. A rarefaction wave in which the products expand and cool follows the detonation reaction zone. This expansion process does most of the useful work on surrounding materials. The C-J state and subsequent expansion can be accurately calculated using modern thermochemical equilibrium computer codes, such as the CHEETAH code [58].

Detonation reaction zones can be more complex for mixtures of explosive materials and for formulations containing metals that react with the product gases. Aluminum particles are added to organic explosives to provide later-time (microsecond to millisecond) energy release when the gaseous products penetrate the aluminum oxide outer layer and react with the aluminum to form Al_2O_3, which liberates a great deal of thermal energy [59].

Even with today's large, fast multiprocessor computers, all of the aforementioned chemical processes can not be included in practical one-, two-, and three-dimensional hydrodynamic code calculations of initiation and propagation of condensed phase detonations in large explosive charges. Therefore phenomenological reactive flow models of detonation, such as the Ignition and Growth model [60], have been developed to calculate the main features of shock initiation and detonation reaction zones and subsequent metal acceleration during reaction product expansion. The Ignition and Growth model is discussed in the next section.

5. THE IGNITION AND GROWTH MODEL OF DETONATION

All reactive flow models require as a minimum: two equations of state, one for the unreacted explosive and one for its reaction products; a reaction rate law for the conversion of explosive to products; and a mixture rule to calculate partially reacted states in which both explosive and products are present. The Ignition and Growth reactive flow model [60] uses two Jones-Wilkins-Lee (JWL) equations of state, one for the unreacted explosive and another one for the reaction products, in the temperature dependent form:

$$p = A\,e^{-R_1 V} + B\,e^{-R_2 V} + \omega C_V T/V \qquad (3)$$

where p is pressure in Megabars, V is relative volume, T is temperature, ω is the Gruneisen coefficient, C_V is the average heat capacity, and A, B, R_1 and R_2 are constants. The unreacted explosive equation of state is fitted to the available shock Hugoniot data, and the reaction product equation of state is fitted to cylinder test and other metal acceleration data. At the high pressures involved in shock initiation and detonation of solid and liquid explosives, the pressures of the two phases must be equilibrated, because interactions between the hot gases and the explosive molecules occur on nanosecond time scales depending on the sound velocities of the components. Various assumptions have been made about the temperatures in the explosive mixture, because heat transfer from the hot products to the cooler explosive is slower than the pressure equilibration process. In this version of the Ignition and Growth model, the temperatures of the unreacted explosive and its reaction products are equilibrated. Temperature equilibration is used, because heat transfer becomes increasingly efficient as the reacting "hot spots" grow and consume more explosive particles at the high pressures and temperatures associated with detonation. Fine enough zoning must be used in all reactive flow calculations so that the results have converged to answers that do not change with even finer zoning. Generally this requires a resolution of at least 10 zones for the detonation reaction zone. The insensitive solid explosive LX-17 (92.5% triaminotrinitrobenzene (TATB) and 7.5% Kel-F binder) has an experimentally measured reaction zone length of approximately three mm [61] so using 10 zones per mm spreads the reaction over 30 zones.

The Ignition and Growth reaction rate equation is given by:

$$dF/dt = I(1\text{-}F)^b(\rho/\rho_o\text{-}1\text{-}a)^x + G_1(1\text{-}F)^cF^dp^y + G_2(1\text{-}F)^eF^gp^z$$

$$0<F<F_{igmax} \qquad 0<F<F_{G1max} \qquad F_{G2min}<F<1 \qquad (4)$$

where F is the fraction reacted, t is time in μs, ρ is the current density in g/cm^3, ρ_o is the initial density, p is pressure in Mbars, and I, G_1, G_2, a, b, c, d, e, g, x, y, z, F_{igmax}, F_{G1max}, and F_{G2min} are constants. This three-term reaction rate law represents the three stages of reaction generally observed during shock initiation and detonation of pressed solid explosives [17]. The first stage of reaction is the formation and ignition of "hot spots" caused by various possible mechanisms (void collapse, viscous flow, friction, shear, etc.) as the initial shock or compression wave interacts with the unreacted explosive molecules. Generally the fraction of solid explosive heated during shock compression is approximately equal to the original void volume. For shock initiation modeling, the second term in Eq. (4) then describes the relatively slow process of the inward and/or outward growth of the isolated "hot spots" in a deflagration-type process. The third term represents the rapid completion of reaction as the "hot spots" coalesce at high pressures and temperatures, resulting in transition from shock induced reaction to detonation.

For detonation modeling, the first term again reacts a quantity of explosive less than or equal to the void volume after the explosive is compressed to the unreacted von Neumann spike state. The second term in Eq. (2) models the fast decomposition of the solid into stable reaction product gases (CO_2, H_2O, N_2, CO, etc.). The third term describes the relatively slow diffusion limited formation of solid carbon (amorphous, diamond, or graphite) as chemical and thermodynamic equilibrium at the C-J state is approached. These reaction zone stages have been observed experimentally using embedded particle velocity and pressure gauges and laser interferometry [57,61-63].

The Ignition and Growth reactive flow model has been applied to a great deal of experimental shock initiation and detonation data using several one-, two-, and three-dimensional hydrodynamic codes. In shock initiation applications, it has successfully calculated embedded gauge, run distance to detonation, short pulse duration, multiple shock, reflected shock, ramp wave compression, and divergent flow experiments on several high explosives at various initial temperatures (heating plus shock scenarios), densities, and degrees of damage (impact plus shock scenarios) [64-66]. Figure 6 shows ten experimental and calculated particle velocity histories for the shock initiation of an LX-17 target impacted by a Kel-F flying plate at 2.951 km/s [64]. Each gauge is about 0.85 mm deeper than the previous one. The initial shock pressure of 14.96 GPa increases at each gauge position, but the main growth of reaction is behind the shock wave front. When the growing pressure wave overtakes the initial shock, transition to detonation occurs, as shown on the last four gauges of Fig. 6. Figure 7 shows six experimental and calculated pressure histories for a PBX 9501 (95% HMX, 2.5 % BDNPA/F, and 2.5% Estane binder) impacted by an aluminium flyer plate at 0.697 km/s [65]. Since HMX-based plastic bonded explosives (PBX's) are more shock sensitive than TATB-based PBX's, the initial pressure is only 3.4 GPa, resulting in a transition to detonation at 12 mm. The growth of reaction occurs further behind the lead shock at these lower pressures.

Embedded gauges and the Ignition and Growth model are also used to study preheated, shocked explosives. Figure 8 contains four pressure histories in 190°C LX-04 (85% HMX and 15% Viton binder) impacted by an aluminium flyer plate at 0.92 km/s. This temperature is sufficient to cause the conversion of beta HMX to delta HMX before shock impact occurs. The LX-04 containing delta HMX is more sensitive than at temperatures of 25°C, 150°C, and 170°C at which the HMX remains in the beta phase [66].

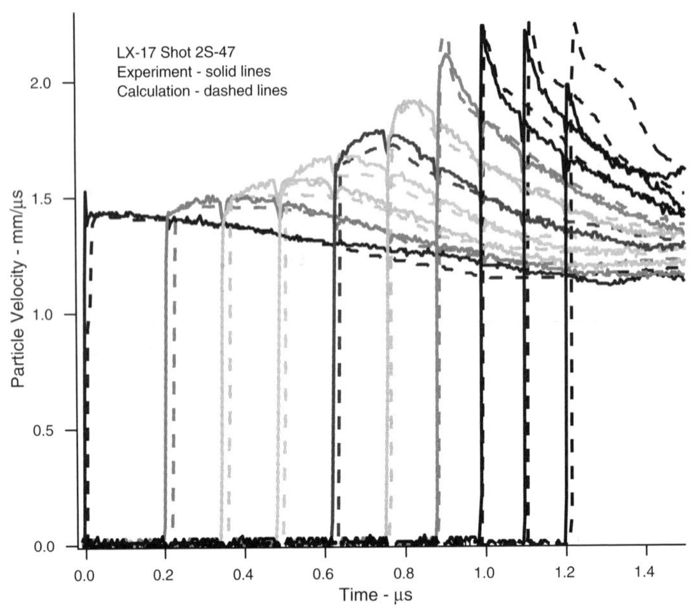

Figure 6. Particle velocity histories for LX-17 impacted by a Kel-F flyer at 2.951 km/s.

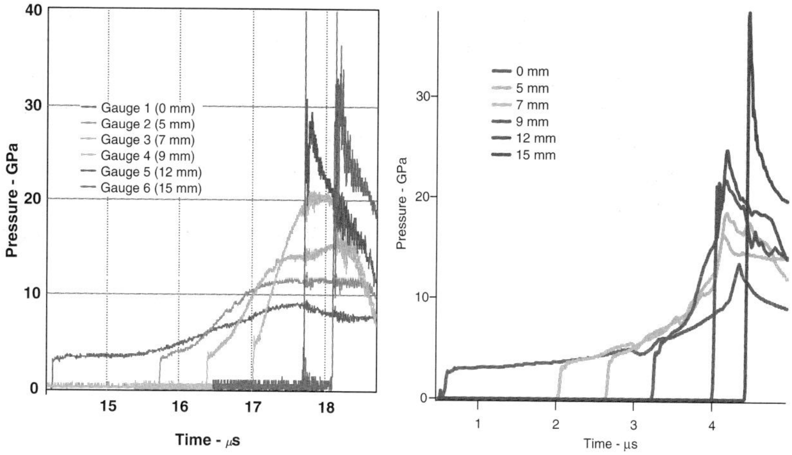

Figure 7. Experimental (left) and calculated (right) pressure histories in PBX 9501 impacted by an aluminium flyer at 0.697 km/s.

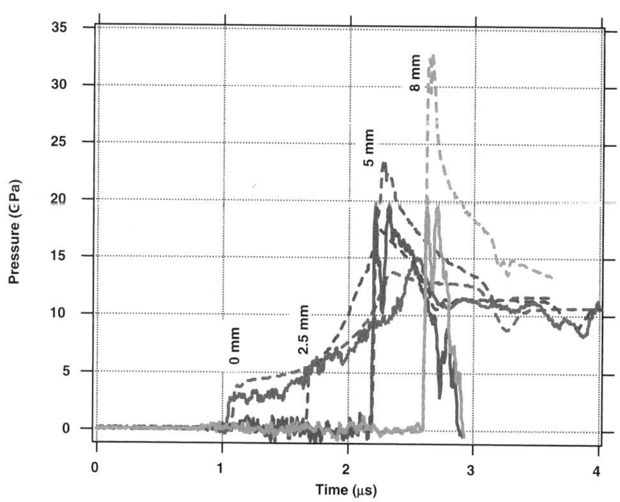

Figure 8. Experimental (solid) and calculated (dashed) pressure profiles in 190°C LX-04 impacted by an aluminium flyer at 0.92 km/s.

In detonation wave applications, the model has successfully calculated embedded gauge, laser interferometric metal acceleration, failure diameter, corner turning, converging, diverging, and overdriven experiments [57,61-63,67]. Figure 9 shows the measured and calculated interface velocity histories for detonating LX-17 impacting various salt crystals [61]. The von Neumann spike state, a relatively fast reaction, a slower reaction, and finally the initial expansion of the products are clearly evident in Fig. 9. Figure 10 illustrates the measured and calculated free surface velocities of 0.267 mm thick tantalum discs driven by 19.871 mm of detonating LX-17 [61]. The momentum associated with the LX-17 reaction zone, which is approximately 3 mm long, and early product expansion are accurately measured and calculated in these small-scale experiments. Figure 11 shows the radial velocity histories for a 2.54 cm radius LX-17 charge confined by 0.272 cm of copper in the LLNL cylinder performance test [67].

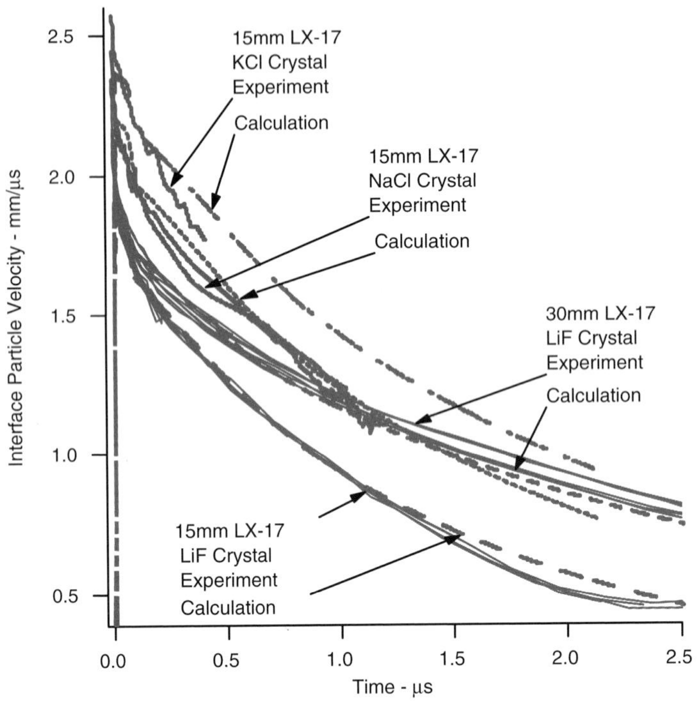

Figure 9. Interface particle velocity histories for detonating LX-17 and various salt crystals.

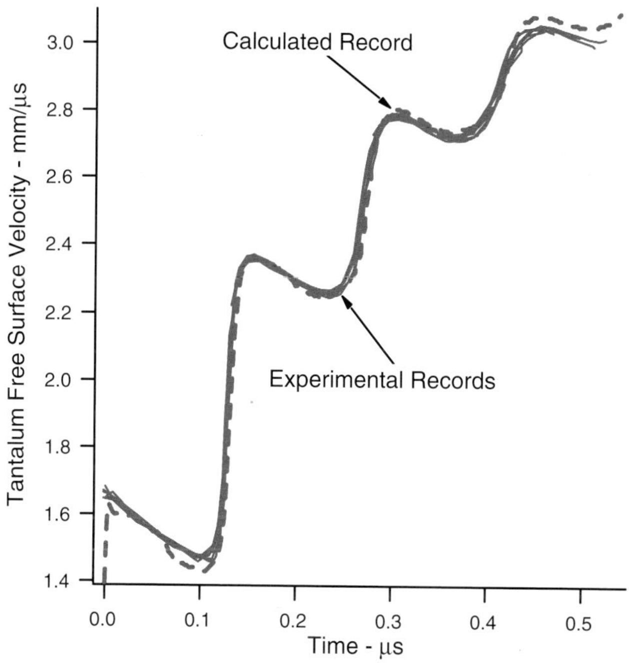

Figure 10. Free surface velocity of a 0.267 mm thick tantalum disk driven by 19.871 mm of LX-17.

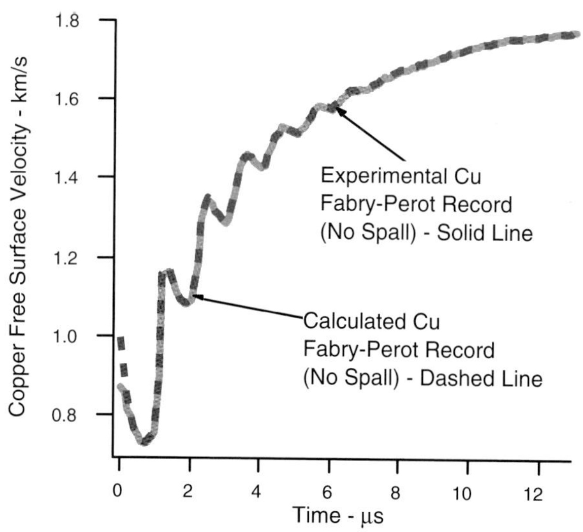

Figure 11. Experimental and calculated LX-17 copper cylinder test radial free surface velocities.

Another example of TATB detonation wave behavior is shown in Fig. 12 in which EDC35 (95% TATB and 5% Kel-F) is sandwiched between brass (left side) and beryllium (right side) [67]. Brass, like most metals, has a lower shock velocity than the detonation velocity of EDC35, so the brass shock front lags behind the detonation wave. Beryllium has a higher shock velocity than the EDC35 detonation velocity and thus pulls the detonation wave along at higher than velocity than normal. The resulting curved shape of the EDC35 detonation wave and the arrival times of the wave at both edges after various propagation lengths are very accurately calculated by the Ignition and Growth model.

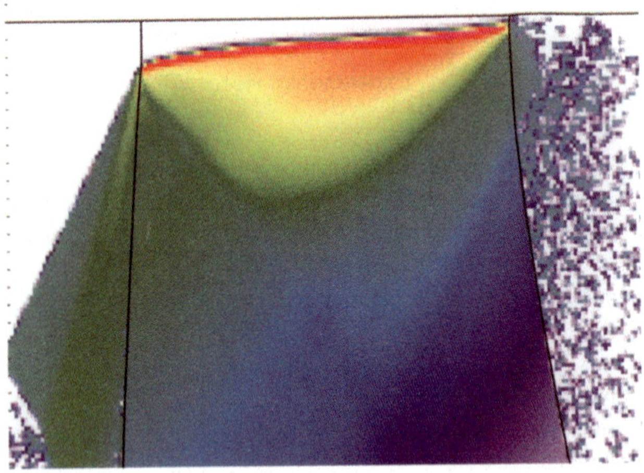

Figure 12. LX-17 detonation wave propagating between brass (left side) and beryllium (right side).

Since the main use of detonating solid explosives is to accelerate metals and other materials to high velocities, accurate measurements of the unreacted shock state (the "von Neumann spike"), the pressure profile of the reaction zone, and the subsequent expansion of the reaction products as they deliver their momentum are essential. Currently these properties are known to within a few percent with nanosecond resolution [67]. Improved accuracy and time resolution are future experimental and computational goals.

6. FUTURE RESEARCH

While a great deal has been learned in recent years about the extreme chemistry occurring in a detonation reaction zone, much more research is required to fully understand the non-equilibrium processes, the reaction pathways, and the equilibrium mixtures created within a detonation wave. A tightly coupled experimental and theoretical approach is required to produce such an understanding. Experimental efforts are underway to measure the rates of vibrational excitation by phonon up-pumping and IVR and to explain these rates using Rice-Ramsperger-Kassel-Marcus (RRKM) theory [68]. Molecular dynamics reaction pathway modeling is rapidly becoming more sophisticated, and larger scale systems can now be studied using parallel computers. More complete potentials are being developed to include the effects mentioned in the last section to better describe partial and complete equilibrium states.

Since chemical reaction rates and equilibrium concentrations are controlled by the local temperature in a region of molecules, the most urgent need in explosives research is for time resolved experimental measurements of temperature in all regions of reacting explosives: impact and shock induced hot spots; deflagration waves; reactive flows behind shock fronts; and detonation waves. Knowing the unreacted explosive temperature as a function of shock pressure will complete its EOS description and allow more accurate predictions of the induction time delay for the onset of bond breaking behind each individual shock front of a three-dimensional detonation wave. Accurate temperature measurements will enable molecular dynamics simulations to be done at the exact density and temperature conditions attained in various regions of a detonation wave. Temperature measurements in the vicinity of the C-J plane and in the subsequent reaction product expansion flow will eliminate the last remaining (and most important) unknown in the thermochemical equilibrium predictions. Improved potentials can be developed to predict the distribution of internal and potential energies under all of the conditions attained in the flows produced by detonation waves.

Since not all of the scenarios involving detonation waves can be tested experimentally, hydrodynamic computer models have to be improved to predict the safety and performance properties of the reactive flows produced by detonating explosives. Assuming that temperature data will soon become available, the next generation of hydrodynamic computer code reactive flow models for simulating detonation waves in one-, two-, and three-dimensions will be based entirely on temperature dependent Arrhenius rate laws, replacing current compression and pressure dependent rate laws [17]. A mesoscale model has been formulated in which individual particles of a solid explosive plus their binders and voids are meshed, shocked, and either react or fail to react using Arrhenius kinetics [69]. Using descriptions of individual particles is still impractical for larger scale simulations even with today's parallel computers, so a continuum Statistical Hot Spot reactive flow model is currently being developed in the ALE3D hydrodynamic computer code [70]. In this model, realistic numbers of hot spots of various sizes, shapes, and temperatures based on the original void volume, particle size distribution and temperature of the solid explosive are assumed to be created as the initiating shock front compresses the explosive particles. The hot spots then either react and grow into the surrounding explosive or fail to react and die out based on multistep Arrhenius kinetics rates [71]. The Statistical Hot Spot reactive flow model has accurately simulated for the first time the experimentally well-known phenomenon of "shock desensitization," in which a detonation wave fails to propagate in a precompressed solid explosive [70,72]. The coalescence of growing hot spots at high pressures and temperatures, the creation of additional surface area available to the reacting sites as the pressure rises, and the rapid transition to detonation are three of the most challenging current problems under investigation in hydrodynamic reactive flow modeling efforts.

ACKNOWLEDGMENTS

This work was performed under the auspices of the U.S. Department of Energy by the University of California, Lawrence Livermore National Laboratory under Contract No.W-7405-ENG-48.

REFERENCES

[1] C. M. Tarver, "On the Chemical Energy Release in Self-Sustaining Detonation Waves in Gaseous and Condensed Explosives," Ph. D. thesis, The Johns Hopkins University, Baltimore, MD, (1973).
[2] C. M. Tarver, Combust. Flame, 46 (1982) 111.
[3] C. M. Tarver, Combust. Flame, 46 (1982) 135.
[4] C. M. Tarver, Combust. Flame, 46 (1982) 157.
[5] C. M. Tarver, L. E. Fried, A. J. Ruggerio, and D. F. Calef, Tenth International Detonation Symposium, (Office of Naval Research ONR 33395-12, Boston, MA, 1993), p. 3.
[6] C. M. Tarver, Shock Compression of Condensed Matter-1997, S. C. Schmidt, D. P. Dandekar, and J. W. Forbes, eds., (AIP Press, 1998), p. 301.
[7] C. M. Tarver, J. Phys. Chem. A, 101 (1997) 4845.
[8] C. M.Tarver, Shock Compression of Condensed Matter - 1999, M. D. Furnish, L. C. Chhabildas, and R. S. Hixson, eds., (AIP Press, 2000), p. 873.
[9] C. M.Tarver, Shock Compression of Condensed Matter - 2001, N. Thadhani, Y. Horie, and M. Furnish, eds., (AIP Press, 2002), p. 42.
[10] C. M. Tarver, in High-Pressure Shock Compression of Solids VI, Y. Horie, L. Davidson, and N. N. Thadhani, eds. (Springer-Verlag, New York, 2003), p. 323.
[11] D. L. Chapman, Phil. Mag., 213 (1899) 5, 47, 90.
[12] E. Jouguet, Pure Appl. Math., 70 (1904) 6, 1, 347.
[13] Y. B. Zeldovich, J. Exper. Theor. Phys. (USSR), 10 (1940) 542.
[14] J. Von Neumann, Office of Science Research and Development, Report No. 549 (1942).
[15] W. Doring, Am. Physik, 43 (1943) 421.
[16] W. W. Wood and J. G. Kirkwood, J. Chem. Phys., 29 (1958) 957.
[17] E. L. Lee and C. M. Tarver, Phys. Fluids, 23 (1980) 2362.
[18] Y. B. Zel'dovich and Y. P.Raizer, Physics of Shock waves and High-Temperature Hydrodynamic Phenomena, (Academic Press, NY, 1966).
[19] R. E. Weston, Jr. and G. W. Flynn, Ann. Rev. Phys. Chem., 43 (1993) 559.
[20] X. Hong, S. Chen, and D. D. Dlott, J. Phys. Chem., 99 (1995) 9102.
[21] W. Holmes, W., R. S. Francis, and M. D. Fayer, J. Chem. Phys., 110 (1999) 3576.
[22] J. H. Kiefer and S. S. Kumaran, J. Chem. Phys., 99 (1993) 3531.
[23] L. G. Green, C. M. Tarver, and D. J. Erskine, Ninth Symposium (International) on Detonation, (Office of the Chief of Naval Research OCNR 113291-7, Portland, OR, 1989), pp. 670.
[24] H. Eyring Science, 199 (1978) 740.
[25] R. Shaw, P. S. Decarli, D. S. Ross, E. L. Lee, and H. D. Stromberg, Combust. Flame , 35 (1979) 237.
[26] R. Shaw, P. S. Decarli, D. S. Ross, E. L. Lee, and H. D. Stromberg, Combust. Flame, 50 (1983) 123.
[27] R. Engelke, D. Schiferl, C. B. Storm, and W. L. Earl, J. Phys. Chem., 92 (1988) 6815.
[28] J. W. Brasch, J. Phys. Chem., 84 (1980) 2084.
[29] D. L. Naud and K. R. Brower, High-Press. Res., 11 (1992) 65.
[30] N. C. Blais, R. Engelke, and S. A. Sheffield, J. Phys. Chem. A, 101 (1997) 8285.
[31] J. M. Winey and Y. M. Gupta, J. Phys. Chem. B. 101 (1997) 10733.
[32] J. M. Winey and Y. M. Gupta, J. Phys. Chem. A 101, 9333 (1997).
[33] Y. A. Gruzdkov and Y. M. Gupta, J. Phys. Chem. A, 102 (1998) 2322.
[34] D. C. Sorescu, B. M. Rice, and D. L. Thompson, J. Phys. Chem. B, 104 (2001) 8406.

[35] D. C. Sorescu, B. M. Rice, and D. L. Thompson, J. Phys. Chem. A, 105 (2001) 9336.
[36] J. M. Seminaro, M. C. Concha, and P. Politzer, J. Chem. Phys., 102 (1995) 8281.
[37] J. M. Seminaro, M. C. Concha, and P. Politzer, Int. J. Quant. Chem., 29 (1995) 621.
[38] M. E. Tuckerman and M. L. Klein, Chem. Phys. Lett., 283 (1998) 147.
[39] S. A. Decker, T. K. Woo, D. Wei, and F. Zhang, Twelfth International Detonation Symposium, San Diego, CA, (2002), in press.
[40] C. Cavazzoni, G. L. Chiarotti, S. Scandolo, E. Tosatti, M. Bernasconi, and M. Parrinello, Science, 283 (1999) 44.
[41] F. Ancilotto, G. L. Chiarotti, S. Scandolo, and E. Tosatti, Science, 275 (1997) 1288.
[42] S. R. Bickham, J. D. Kress, and L. A. Collins, J. Chem. Phys., 112 (2000) 9695.
[43] J. D. Kress, S. R. Bickham, L. A. Collins, B. L. Holian, and S. Goedecker, Phys. Rev. Lett., 83 (1999) 3896.
[44] M. R. Manaa, L. E. Fried, C. F. Melius, M. Elstner, and T. Frauenheim, J. Phys. Chem. A, 106 (2000) 9024.
[45] G. I. Pangilinan and Y. M. Gupta, J. Phys. Chem., 98 (1994) 4522.
[46] J. B. Pedley, R. D. Naylor, and S. P. Kirby, Thermochemical Data of Organic Compounds, 2nd ed. (Chapman, New York, 1986).
[47] R. Engelke, W. L. Earl, and C. M. Rohlfing, J. Phys. Chem., 90 (1986) 545.
[48] D. Margetis, E. Kaxiras, M. Elstner, T. Frauenheim, and M. R. Manaa, J. Chem. Phys., 117 (2002) 788.
[49] M. R. Manaa and L. E. Fried, J. Phys. Chem. A, 102 (1998) 9884.
[50] M. R. Manaa and L. E. Fried, J. Phys. Chem. A,103 (1999) 9349.
[51] E. J. Reed, J. D. Joannopoulos, and L. E. Fried, Phys. Rev. B, 62 (2000) 16500.
[52] V. Bernshtein and I. Oref, J. Phys. Chem., 100 (1996) 9738.
[53] J. H. L. Lee, in High-Pressure Shock Compression of Solids VI, Y. Horie, L. Davison, and N. N. Thadhani, eds., (Springer-Verlag, New York, 2003), p. 121.
[54] A. A. Vasil'ev and A. V. Trotsyuk, Combustion, Explosion and Shock Waves 39 (2003) 80.
[55] V. N. Gamezo, D. Desbordes, and E. S. Oran, Shock Waves, 9 (1999) 11.
[56] J. A. Viecelli and J. N. Goeski, J. Chem. Phys., 117 (2002) 11352.
[57] J. W. Kury, R. D. Breithaupt, and C. M. Tarver, Shock Waves, 9 (1999) 227.
[58] CHEETAH is discussed in detail in the J. M. Zaug et al. chapter of this book.
[59] W. C. Tao, C. M. Tarver, J. W. Kury, C. G. Lee, and D. L. Ornellas, Tenth International Detonation Symposium, (Office of Naval Research ONR 33395-12, Boston, MA, 1993), p. 628.
[60] C. M. Tarver, J. O. Hallquist, and L. M. Erickson, Eighth Symposium (International) on Detonation, (Naval Surface Weapons Center NSWC MP86-194, Albuquerque, NM, 1985), pp. 951.
[61] C. M. Tarver, J. W. Kury, and R. D. Breithaupt, J. Appl. Phys., 82 (1997) 3771.
[62] C. M. Tarver, R. D. Breithaupt, and J. W. Kury, J. Appl. Phys., 81 (1997) 7193.
[63] C. M. Tarver, W. C. Tao, and C. G. Lee, Propellants, Explosives, Pyrotechnics, 21 (1996) 238.
[64] C. M. Tarver, J. W. Forbes, F. Garcia, and P. A. Urtiew, Shock Compression of Condensed Matter-2001, M. D. Furnish, N. N. Thadhani, and Y. Horie, eds., (AIP Press, 2002), p. 1043.
[65] R. L. Gustavsen, S. A. Sheffield, R. R. Alcon, J. W. Forbes, C. M. Tarver, and F. Garcia, Shock Compression of Condensed Matter-2001, M. D. Furnish, N. N. Thadhani, and Y. Horie, eds., (AIP Press, 2002), p. 1019.
[66] P. A. Urtiew, J. W. Forbes, C. M. Tarver, K. S. Vandersall, F. Garcia, D. W. Greenwood, P. C. Hsu, and J. L. Maienschein, Shock Compression of Condensed Matter - 2003, M. D. Furnish, ed., (AIP Press, 2004), in press.
[67] C. M. Tarver and E. M. McGuire, Twelfth International Detonation Symposium, San Diego, CA, August (2002), in press.
[68] J. H. Kiefer, G. C. Sahukar, S. Santhanam, N. K. Srinivasan, and R. S. Tranter, J. Chem. Phys, 120 (2004) 918.
[69] J. E. Reaugh, "Grain-Scale Dynamics in Explosives," LLNL Report UCID-150388, (2001).

[70] A. L. Nichols and C. M. Tarver, Twelfth International Detonation Symposium, San Diego, CA, August (2002), in press.
[71] C. M. Tarver, Combust. Flame, 137 (2004) 50 (2004).
[72] A. L. Nichols, C. M. Tarver, and E. M. McGuire, Shock Compression of Condensed Matter - 2003, M. D. Furnish, ed., (AIP Press, 2004), in press.

Index